ALLUVIAL SEDIMENTATION

Alluvial Sedimentation

EDITED BY M. MARZO
AND C. PUIGDEFÁBREGAS

SPECIAL PUBLICATION NUMBER 17 OF THE
INTERNATIONAL ASSOCIATION OF SEDIMENTOLOGISTS
PUBLISHED BY BLACKWELL SCIENTIFIC PUBLICATIONS
OXFORD LONDON EDINBURGH BOSTON
MELBOURNE PARIS BERLIN VIENNA

© 1993 The International Association
of Sedimentologists
and published for them by
Blackwell Scientific Publications
Editorial Offices:
Osney Mead, Oxford OX2 0EL
25 John Street, London WC1N 2BL
23 Ainslie Place, Edinburgh EH3 6AJ
238 Main Street, Cambridge
  Massachusetts 02142, USA
54 University Street, Carlton
  Victoria 3053, Australia

Other Editorial Offices:
Librairie Arnette SA
2, rue Casimir-Delavigne
75006 Paris
France

Blackwell Wissenschafts-Verlag GmbH
Düsseldorfer Str. 38
D-10707 Berlin
Germany

Blackwell MZV
Feldgasse 13
A-1238 Wien
Austria

All rights reserved. No part of this
publication may be reproduced, stored
in a retrieval system, or transmitted,
in any form or by any means,
electronic, mechanical, photocopying,
recording or otherwise without the
prior permission of the copyright
owner.

First published 1993

Set by Semantic Graphics, Singapore
Printed and bound in Great Britain
at The Alden Press, Oxford

DISTRIBUTORS

Marston Book Services Ltd
PO Box 87
Oxford OX2 0DT
(*Orders*: Tel: 0865 791155
         Fax: 0865 791927
         Telex: 837515)

USA
Blackwell Scientific Publications, Inc.
238 Main Street
Cambridge, MA 02142
(*Orders*: Tel: 800 759-6102
         617 876-7000)

Canada
Oxford University Press
70 Wynford Drive
Don Mills
Ontario M3C 1J9
(*Orders*: Tel: 416 441-2941)

Australia
Blackwell Scientific Publications Pty Ltd
54 University Street
Carlton, Victoria 3053
(*Orders*: Tel: 03 347-5552)

A catalogue record for this title
is available from the British Library

ISBN 0-632-03545-5

Library of Congress
Cataloging-in-Publication Data

Alluvial sedimentation/
  edited by M. Marzo and C. Puigdefábregas.
       p.   cm.
  (Special publication number 17
  of the International Association of Sedimentologists)
    Includes bibliographical references
  and index.
    ISBN 0-632-03545-5
    1. Alluvium. 2. Sediment transport.
    I. Marzo, M. II. Puigdefábregas, C.
  III. Series: Special publication . . . of the
  International Association of Sedimentologists: no. 17.
  QE581.A435   1993
  551.3′53—dc20

Contents

ix   Preface

xi   Brian Rust (1936–1990). *In memoriam*
    M.R. Gibling

Sediment Transport

3   Entrainment of spheres: an experimental study of relative size and clustering effects
   C.S. James

11   A new bedform stability diagram, with emphasis on the transition of ripples to plane bed in flows over fine sand and silt
   J.H. van den Berg and A. van Gelder

23   In-transport modification of alluvial sediment: field evidence and laboratory experiments
   P.A. Brewer and J. Lewin

37   Bed material and bedload movement in two ephemeral streams
   M.A. Hassan

51   Bedform migration and related sediment transport in a meander bend
   J. Kisling-Møller

63   Sediment ice rafting and cold climate fluvial deposits: Albany River, Ontario, Canada
   I.P. Martini, J.K. Kwong and S. Sadura

77   Dynamics of bed load transport in the Parsęta River channel, Poland
   Zb. Zwoliński

Alluvial Facies

91   Morphology and facies models of channel confluences
   C.S. Bristow, J.L. Best and A.G. Roy

101 Interpretation of bedding geometry within ancient point-bar deposits
B.J. Willis

115 Geometry and lateral accretion patterns in meander loops: examples from the Upper Oligocene–Lower Miocene, Loranca Basin, Spain
M. Díaz-Molina

133 Alluvial ridge-and-swale topography: a case study from the Morien Group of Atlantic Canada
M.R. Gibling and B.R. Rust

151 Processes and products of large, Late Precambrian sandy rivers in northern Norway
S.-L. Røe and M. Hermansen

167 Crevasse splay sandstone geometries in the Middle Jurassic Ravenscar Group of Yorkshire, UK
R. Mjøs, O. Walderhaug and E. Prestholm

185 Grain-size distribution of overbank sediment and its use to locate channel positions
M.J. Guccione

195 Geometrical facies analysis of a mixed influence deltaic system: the Late Permian German Creek Formation, Bowen Basin, Australia
A.J. Falkner and C.R. Fielding

211 Computer modelling of flow lines over deformed surfaces: the implications for prediction of alluvial facies distribution
P.J. Weston and J. Alexander

Geomorphic and Structural Controls on Alluvial Systems

221 Geomorphic and structural controls on facies patterns and sediment composition in a modern foreland basin
J.F. Damanti

235 Quaternary alluvial fans in southwestern Crete: sedimentation processes and geomorphic evolution
W. Nemec and G. Postma

277 Palaeogeomorphological controls on the distribution and sedimentary styles of alluvial systems, Neogene of the NE of the Madrid Basin (central Spain)
A.M. Alonso Zarza, J.P. Calvo and M.A. García del Cura

293 Alluvial-fan sedimentation along an active strike-slip fault: Plio-Pleistocene Pre-Kaczawa fan, SW Poland
K. Mastalerz and J. Wojewoda

305 Present-day changes in the hydrologic regime of the Raba River (Carpathians, Poland) as inferred from facies pattern and channel geometry
B. Wyżga

Alluvial Stratigraphy

319 A revised alluvial stratigraphy model
J.S. Bridge and S.D. Mackey

337 Quantified fluvial architecture in ephemeral stream deposits of the Esplugafreda Formation (Palaeocene), Tremp-Graus Basin, northern Spain
T. Dreyer

363 Architecture of the Cañizar fluvial sheet sandstones, Early Triassic, Iberian Ranges, eastern Spain
J. López-Gómez and A. Arche

383 Effects of relative sea-level changes and local tectonics on a Lower Cretaceous fluvial to transitional marine sequence, Bighorn Basin, Wyoming, USA
E.P. Kvale and C.F. Vondra

401 Structural and climatic controls on fluvial depositional systems: Devonian, North-East Greenland
H. Olsen and P.-H. Larsen

425 Alternating fluvial and lacustrine sedimentation: tectonic and climatic controls (Provence Basin, S. France, Upper Cretaceous/Palaeocene)
I. Cojan

439 Control of basin symmetry on fluvial lithofacies, Camp Rice and Palomas Formations (Plio-Pleistocene), southern Rio Grande rift, USA
G.H. Mack and W.C. James

451 Simultaneous dispersal of volcaniclastic and non-volcanic sediment in fluvial basins: examples from the Lower Old Red Sandstone, east-central Scotland
P.D.W. Haughton

473 Siliciclastic braided-alluvial sediments intercalated within continental flood basalts in the Early to Middle Proterozoic Mount Isa Inlier, Australia
K.A. Eriksson and E.L. Simpson

489 Sedimentological response of an alluvial system to source area tectonism: the Seilao Member of the Late Cretaceous to Eocene Purilactis Formation of northern Chile
A.J. Hartley

501 Cyclicity in non-marine foreland-basin sedimentary fill: the Messinian conglomerate-bearing succession of the Venetian Alps (Italy)
F. Massari, D. Mellere and C. Doglioni

521 The impact of incipient uplift on patterns of fluvial deposition: an example from the Salt Range, Northwest Himalayan Foreland, Pakistan
T.J. Mulder and D.W. Burbank

Ores

543 Principles of a sediment sorting model and its application for predicting economic values in placer deposits
M. Nami and S.G.E. Ashworth

553 Alluvial basin-fill dynamics and gold-bearing aspect of Early Proterozoic strike-slip basins in French Guiana
E. Manier, D. Mercier and P. Ledru

569 Index

Preface

Most of the thirty-four papers contained in this Special Publication arise from the '4th International Conference on Fluvial Sedimentology' held in Sitges (Spain) in October, 1989. The Conference was hosted and organized by the Servei Geològic de Catalunya (Departament de Política Territorial i Obres Públiques, Generalitat de Catalunya) and by the Departament de Geologia Dinàmica, Geofísica i Paleontologia de la Universitat de Barcelona. The Conference was sponsored by these two institutions and by the Ministerio de Educación y Ciencia and the International Association of Sedimentologists, with the collaboration of the Institut Français du Petrole, Norsk Hydro, Société Nationale Elf Aquitaine and Shell España NV. The Ajuntament de Sitges and the Societat Recreativa El Retiro provided all the local facilities. The meeting was attended by 306 participants from academic and industrial research centres in twenty-three countries. The programme comprised ten field excursions as well as lecture and poster sessions. Fifty-five oral papers and 135 posters were presented, of which thirty are published here, together with four papers not presented at the meeting.

Originally six major themes were planned for the Conference: (i) sediment transport and bedforms; (ii) fluvial geomorphology and modern river systems; (iii) concepts and facies models; (iv) alluvial stratigraphy; (v) fluvial basins and tectonics; and (vi) applied fluvial sedimentology (gold, coal, hydrocarbons and reservoir modelling). The contents of this volume do not strictly correspond to the Conference programme and the contributions here have been arranged into five sections.

The first section contains seven papers dealing with various aspects of sediment transport and hydraulics in flume experiments and modern rivers, including: the effects of relative size and clustering on the entrainment of particles (James); the stability diagram for bedforms developed by unidirectional flows over fine sand and silt (van den Berg and van Gelder); the nature and origin of downstream changes in bed sediment texture (Brewer and Lewin); the characteristics of bed material and bedload transport in ephemeral (Hassan) and perennial streams (Zwoliński); the direction and degree of sediment transport and the dynamics of bedforms in a sandy meander bend (Kisling-Møller); and the effects of river ice rafting on sediment dispersal and deposition (Martini, Kwong and Sadura).

A second set of nine papers focuses on the analysis of alluvial facies. Three of these papers examine point-bar facies, including: a comparison of computer simulation with well-described outcrops in order to interpret the bedding geometry within ancient point-bar deposits (Willis); a study of the three-dimensional internal organization and accretion patterns in exhumed Miocene point-bar deposits (Diaz-Molina); and the description and interpretation of the different types of ridge-and-swale topography observed in an exhumed meanderbelt (Gibling and Rust). Two papers are dedicated to overbank facies: one paper deals with the characteristics and significance of some Jurassic crevasse splay sandstones (Mjøs, Walderhaug and Prestholm) and the other paper examines the usefulness of the grain-size distribution of overbank sediments to determine the position of their related feeding channel (Guccione). The remaining four papers in this section consider: the morphology and facies models of channel confluences (Bristow, Best and Roy); the characteristics and origin of some Late Precambrian sheet-like sandstone bodies (Røe and Hermansen); the fluvial facies associated with widespread coal seams in a deltaic setting (Falkner and Fielding); and the application of computer modelling to predict the distribution of alluvial facies (Weston and Alexander).

The third section in the book contains five papers dealing with geomorphic and structural controls on alluvial sedimentation. The first three papers consider: the effects of the interaction among structural style, bedrock lithology, climate and the drainage network on sediment composition and texture in a modern foreland basin (Damanti); the effects of sea-level variations and climatic changes on alluvial-fan sedimentology in a Quaternary piedmont system (Nemec and Postma); and the influence of structural lineations and the palaeomorphology of the basin margin on the distribution and sedimentology of several Miocene alluvial-fan systems (Alonso Zarza, Calvo and García del Cura). One paper illustrates the main syntectonic features developed in a Plio-Pleistocene alluvial fan located along an active strike-slip fault (Mastalerz and Wojewoda). The final paper examines the changes undergone by a modern river during the last 200 years in response to either climatic changes or human activities (Wyżga).

The fourth section contains twelve papers dealing with various aspects of alluvial stratigraphy and basin analysis. The first two papers in the section include a theoretical quantitative model of alluvial stratigraphy (Bridge and Mackey) and a quantified case study of the architecture of ephemeral stream deposits (Dreyer). The influence of tectonic uplift and subsidence as major controls on alluvial stratigraphy is considered in four papers dealing both with rifted basins (López and Arche; Mack and James) and with compressional settings (Hartley, Mulder and Burbank). Another set of three papers examines the complex simultaneous controls of tectonism and climate on the stratigraphic arrangement of fluvial and aeolian (Olsen and Larsen), fluvial and lacustrine (Cojan) and alluvial fan and fluvial (Massari, Mellere and Doglioni) deposits. The remaining three papers in this section consider: the effects of relative sea-level changes combined with local tectonism on the sequential arrangement of fluvial to transitional marine deposits (Kvale and Vondra); the synchronous dispersal of volcaniclastic, non-volcaniclastic and mixed sediments in fluvial basins and their sequential arrangement (Haughton); and the architecture of braided deposits intercalated within continental flood basalts (Eriksson and Simpson).

The final section in this volume contains two papers dealing with the exploration and exploitation of ores, including: a study on the Early Proterozoic gold-bearing alluvium in French Guyana (Manier, Mercier and Ledru); and a discussion on the applicability of a sediment sorting model for predicting economic values in placer deposits (Nami and Ashworth).

A large number of people put considerable efforts into the organization of the conference and in editing this volume we would like to thank especially the sixty-four reviewers of the manuscripts. The task was particularly difficult as most of the reviewers were also authors and some of them were asked to review two or even three papers. Their cooperative collaboration has been essential. Thanks are also given to Natàlia Gorga, Angela Lorenzo, Michèle Pereira, who kindly helped with the organization during the Conference, and Blanca Martínez, who helped in editing this volume.

M. MARZO
Barcelona, Spain

C. PUIGDEFÁBREGAS
Barcelona, Spain

Brian Rust (1936–1990)
In memoriam

Brian Rust, internationally known fluvial sedimentologist, died on 22 June 1990 of a virulent form of malaria contracted during field-work in Zambia.

Brian was born in Cork, Ireland, the son of a naval officer. After a period of military service in East Africa, he commenced his undergraduate education at the University of Oxford, participating as a geologist in an expedition to Guyana. His PhD at the University of Edinburgh was in the field of structural geology, with a study area in southern Scotland. During a postdoctoral fellowship in California, he studied the sedimentology of deformed rock units along the San Andreas Fault in an attempt to decipher their structural history. This led to an appointment in sedimentary geology at the University of Ottawa, Canada, where he taught from 1964 until his death.

Brian's research spanned a variety of subjects and field locations: modern rivers in the Yukon, Ontario, Iceland and Australia; ancient river deposits in Arctic and Atlantic Canada and Australia; coal-basin studies; and, in recent years, mineral deposits in sedimentary rocks. He was a very good field geologist, approaching outdoor work with enthusiasm, energy and a keen appreciation for the key elements of a problem, and he was particularly adept at comparing ancient and modern sediments based on his great store of knowledge of both. Something of an explorer at heart, he loved to visit new areas and find out what lay around the corner. He enjoyed working from boats, and was an impetuous navigator.

His scientific papers have been influential and widely quoted. Prominent studies include his 1969 paper with P.F. Williams on the Donjek River, his contributions to channel classification and braided-river models in the Proceedings of the 1st International Fluvial Conference at Calgary, and his recent work (with G.C. Nanson) on anastomosed rivers of Australia and their bedload mud aggregates. He considered the latter work his most important contribution to sedimentology.

Brian was a fine colleague and thesis supervisor, with a firm commitment to his students, whose writing skills he materially assisted through the example of his lucid style. We all benefited from his breadth of knowledge, generosity and keen sense of humour. He loved a good discussion but was always a fair critic. He was a man dedicated to his family and deeply concerned for social justice, in support of which he often spoke eloquently. He had received special permission to set up a Commonwealth student project on a Zambian, rather than a Canadian, topic because he felt that a project in Africa would better serve the student and the country. He was a gifted artist and a lover of music.

Contributors to the 4th Fluvial Conference will remember Brian's spirited discussions and enthusiastic participation in field excursions. He leaves a legacy of excellent field studies in fluvial sedimentology.

M.R. GIBLING
Halifax, Canada

Sediment Transport

Entrainment of spheres: an experimental study of relative size and clustering effects

C.S. JAMES

Department of Civil Engineering, University of the Witwatersrand, PO Wits 2050, South Africa

ABSTRACT

Entrainment of sediment particles is governed by fluid, flow and particle characteristics. A series of experiments was performed on spherical particles to study the effects of relative size under well-controlled conditions which standardized particle shape and bed geometry. The effect of particle size relative to bed roughness on the dimensionless critical shear stress required for entrainment was found to be consistent with incipient motion data for natural sediments. Experiments were also performed to investigate the effect on entrainment of the proximity of other particles in various clustering arrangements. In general, clusters were found to break up at lower shear stresses than required to move single particles but one arrangement was distinctly more stable than single particles.

INTRODUCTION

The hydraulic conditions under which sediment particles are just able to move depend on the size, shape and density of the particles and the microtopography of the bed. Under given hydraulic conditions, therefore, particles with different characteristics will be entrained with different relative ease. As entrainment is a continually repeated process during sediment transport it is therefore an important agent in the sorting of sediments by flowing water.

Quantitative description of entrainment as a sorting process requires understanding of the effects of the different sediment characteristics. Empirical information plays an important role in developing this understanding. Most incipient motion data currently available, however, were derived from natural sediments and are not entirely satisfactory for this purpose. Almost every natural sediment has a unique combination of particle characteristics and the effects of particular characteristics cannot easily be isolated. It is difficult, for example, to determine unambiguously the effects of particle size by studying the behaviour of natural sediments where particles with different sizes also have different shapes. A better understanding can be obtained from well-controlled experiments in which all characteristics except one are held constant, even if the situation is unnatural. This paper describes a series of experiments conducted to determine the hydraulic conditions necessary to entrain spherical particles in a known geometric packing configuration. A similar condition was studied by Coleman (1967) and Fenton & Abbott (1977), who studied effects of absolute size and relative protrusion for uniform spheres. This study extends their work by showing the effects of size relative to bed roughness.

It has been suggested (e.g. Reid & Frostick, 1987) that particle clustering inhibits entrainment and has a significant effect on bedload transport rates. This effect is attributed to mechanical particle interlocking as well as modification of the applied hydrodynamic forces. Experiments are described in this paper which were conducted with groups of spheres to study the effects on entrainment conditions of clustering *without* mechanical interlocking.

EXPERIMENTAL PROCEDURE

The experiments were performed in a 3.0 m long, 0.1 m wide flume with a hinged weir at the downstream end, shown in Fig. 1. A test panel was placed over the central 750 mm of the flume bed. This panel was covered with densely packed 5 mm glass beads which were glued down. The flume bed was sloped over 750 mm upstream and downstream of the test panel by smooth surfaces up to the level of the tops of the spheres.

For each experiment the test sphere (or cluster) was placed on the test panel about one-third of the panel length from its downstream end, and in the required position relative to the bed spheres (see Fig. 6). The flow in the flume was gradually increased until the test sphere moved from its original position. The discharge, flow depth and boundary shear stress (on the surface tangent to the tops of the spheres) were then measured. The experiment was performed five times for each test sphere or cluster. In each case the initial position of the sphere or cluster was slightly different and the weir was reset to give a different flow depth. For each experiment the boundary shear stress was measured at five different locations in the proximity of the initial position of the test sphere or cluster, and the average value calculated.

The discharge was measured using a propeller-type meter installed in the delivery pipe. The boundary shear stress was measured using a 1.0 mm internal diameter Pitot-static tube as a Preston tube, linked to a pressure transducer and analogue display unit.

Boundary shear stress was determined by measuring a pressure differential (Δp) with the Pitot-static tube against the boundary and relating this to shear stress (τ) according to an empirical relationship. Two such relationships have been proposed for rough beds. Hwang & Laursen (1963) proposed:

$$\frac{\Delta p}{\tau} = 16.531 \left\{ \left(\log \frac{30h}{k_s} \right)^2 \right.$$

$$- \ln \left(\log \frac{30h}{k_s} \right) \left[0.25 \left(\frac{a}{h} \right)^2 \right.$$

$$+ 0.0625 \left(\frac{a}{h} \right)^4 + 0.026 \left(\frac{a}{h} \right)^6 + \ldots \right]$$

$$+ \ln^2 \left[0.25 \left(\frac{a}{h} \right)^2 + 0.1146 \left(\frac{a}{h} \right)^4 \right.$$

$$\left. \left. + 0.0595 \left(\frac{a}{h} \right)^6 + \ldots \right] \right\} \quad (1)$$

in which k_s is the bed roughness (assumed to be the diameter of the bed spheres), h is the height of the centre of the tube above the theoretical bed (the reference level for the logarithmic velocity profile) and a is the inner radius of the tube. Ghosh & Jena (1971) proposed a similar relationship, namely

$$\frac{\Delta p}{\tau} = 16.531 \left\{ \left(\log 9u_* \frac{h}{\nu} \right)^2 - \ln \left(\log 9u_* \frac{h}{\nu} \right) \right.$$

$$\times \left[0.25 \left(\frac{a}{h} \right)^2 + 0.0625 \left(\frac{a}{h} \right)^4 \right.$$

$$\left. + 0.026 \left(\frac{a}{h} \right)^6 + \ldots \right] + \ln^2 \left[0.25 \left(\frac{a}{h} \right)^2 \right.$$

$$\left. \left. + 0.1146 \left(\frac{a}{h} \right)^4 + 0.0595 \left(\frac{a}{h} \right)^6 + \ldots \right] \right\} \quad (2)$$

in which ν is the kinematic viscosity of the water and u_* is the shear velocity.

Equations (1) and (2) were evaluated by comparing their predictions of shear velocities with shear velocities determined from measured velocity profiles. The velocity is defined by

$$u = u_* 5.75 \log(y/C) \quad (3)$$

in which u is the velocity at height y above the theoretical bed and C is a constant. The shear

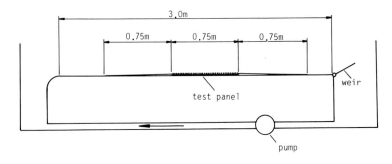

Fig. 1. Flume used for experiments.

velocity can therefore be calculated as

$$u_* = S/5.75 \quad (4)$$

in which S is the slope of the curve of the logarithm of y versus u.

The shear velocities were determined using equation (4) for four combinations of the flow depth and velocity representing the incipient motion condition for 5 mm single spheres. The velocity profiles are shown in Fig. 2 and the shear velocities are listed in Table 1.

The shear velocities were also computed for these four conditions using equations (1) and (2) together with the relationship

$$u_* = \sqrt{\tau/\rho} \quad (5)$$

in which ρ is the fluid density. The equations were applied with two assumptions regarding the distance of the theoretical bed below the surface tangent to the tops of the roughness elements (z_0), namely the recommended $0.2k_s$ and also $0.1k_s$, and assuming k_s to be represented by the bed sphere diameter. The computed shear velocities are listed in Table 1.

The results show that the equation of Hwang & Laursen (1963) with the assumption of $z_0 = 0.1k_s$ agrees well with the velocity profile results. Both the direct measurement relationships give more consistent results than the velocity profile approach, implying greater precision. The range of values obtained for each relationship is less than 3% of the mean value, compared with 16% for the velocity

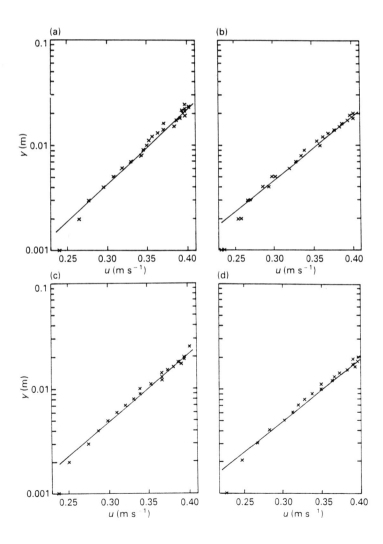

Fig. 2. Velocity profiles used for Preston tube calibration: (a) flow condition 1; (b) flow condition 2; (c) flow condition 3; (d) flow condition 4.

Table 1. Shear velocities (m s^{-1}) for calibration of Preston tube

Flow condition	Velocity profile	Hwang & Laursen		Ghosh & Jena	
		$z_0 = 0.2k_s$	$z_0 = 0.1k_s$	$z_0 = 0.2k_s$	$z_0 = 0.1k_s$
1	0.0245	0.0250	0.0278	0.0218	0.0235
2	0.0287	0.0246	0.0274	0.0215	0.0232
3	0.0270	0.0243	0.0270	0.0212	0.0229
4	0.0289	0.0247	0.0274	0.0212	0.0232
Average	0.0273	0.0247	0.0274	0.0214	0.0232

profile results. The relatively large variation associated with the velocity profile results could be caused by the subjectivity associated with estimating the positions of the straight lines in Fig. 2. The accuracy of the direct measurement techniques is limited by the accuracy of estimating a representative reading on the instrument within the range of turbulent fluctuations. The Hwang & Laursen relationship is explicit and easier to use than that of Ghosh & Jena, which requires iteration. Because of its overall advantages in accuracy, precision and ease of use, the Hwang & Laursen approach with $z_0 = 0.1k_s$ was used to determine τ for the remaining experiments.

RESULTS

Entrainment conditions were determined for thirteen single spheres with different sizes. In all cases the position of the test sphere relative to the bed spheres was as shown by arrangement 1 in Fig. 5. Sphere characteristics and entrainment conditions are summarized in Table 2. Entrainment conditions are given in terms of critical shear velocity and critical dimensionless shear stress (Shields parameter), $u_*^2/gD(S_s - 1)$, in which D is the sphere diameter and S_s its density relative to water. Tabulated values represent averages of the five independent experiments with each sphere.

The results for single sphere number 1, which is the same size as the bed spheres, are shown graphically in Fig. 3 on the Shields diagram used by Fenton & Abbott (1977). This shows that the results are consistent with those inferred from Coleman's (1967) data for the same flow conditions and geometric arrangement. The data of Fenton & Abbott included in Fig. 3 represent the results for their series C experiments, in which the geometric arrangement of spheres was identical to that in the present experiments and Coleman's experiments.

Relative size effects are shown in Fig. 4, in which critical dimensionless shear stress is plotted against relative size, D/K, where K is the diameter of the bed spheres. A very consistent trend with relative

Table 2. Entrainment conditions

Sphere no.	Diameter (mm)	Density (kg m^{-3})	u_{*c} (m s^{-1})	Shields
1	5.00	2750	0.0274	0.0087
2	6.30	2292	0.0245	0.0075
3	3.45	3256	0.0299	0.0117
4	9.85	2558	0.0269	0.0048
5	12.20	2482	0.0250	0.0035
6	16.00	2411	0.0242	0.0027
7	20.02	2411	0.0242	0.0021
8	13.80	2449	0.0255	0.0031
9	22.05	2476	0.0259	0.0021
10	2.58	2669	0.0239	0.0135
11	2.40	2763	0.0243	0.0142
12	1.90	2727	0.0271	0.0228
13	1.80	2727	0.0275	0.0252

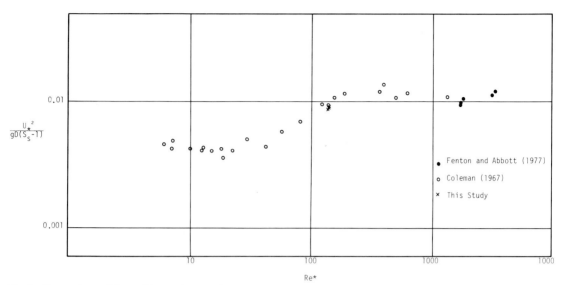

Fig. 3. Comparison of data with previous work.

size is apparent. This trend is similar to that of the data for natural sediments reported by Egiazaroff (1965), Day (1980) and Andrews (1983), although the position of the curve is lower because of the relatively greater exposure of the moving particles.

A 45° straight line of Fig. 4 would imply entrainment of all sphere sizes at the same shear stress. The data follow this trend for $D/K > 1.0$ but deviate from it for $D/K < 1.0$, suggesting that spheres with D/K approximately equal to 0.5 are more mobile than both smaller and larger sizes. This is consistent with the overpassing data of Everts (1973) and with the predictions of an analytical entrainment model (James, 1990).

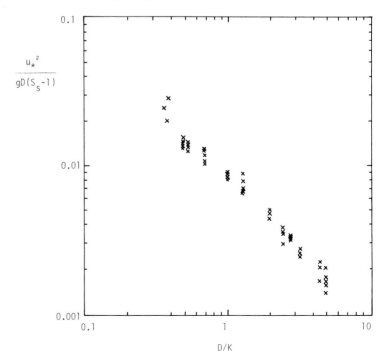

Fig. 4. Relative size effects.

To determine the effects of clustering, entrainment conditions for thirteen different geometric groupings of two, three and four spheres were measured, following a procedure similar to that for the single spheres. In all cases glass beads with the same diameter as the bed spheres were used. The geometric groupings are illustrated in Fig. 5, which includes the two possible positions for single spheres. When conducting these experiments it was assumed in all cases that movement of any one particle from the cluster implied a break-up of the cluster and defined the entrainment condition. Results are shown in Fig. 6.

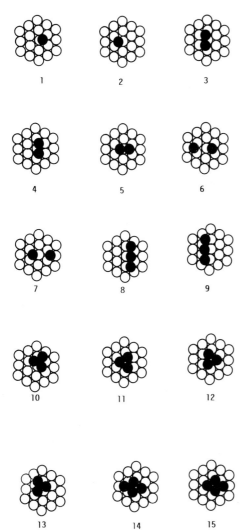

Fig. 5. Cluster geometries shown in plan. Unshaded circles represent bed spheres, shaded circles represent superimposed test spheres. Flow is from left to right.

Although there is considerable scatter in the data, which is more for the clusters than the single spheres, some general observations can be made. Clusters tend to break up at lower boundary shear stresses than are required to move isolated spheres. In most cases, apart from cluster 14 and, to a lesser extent, cluster 11, very little movement occurred once the cluster had broken up; the remaining particles required greater shear stresses for further movement. For cluster 14, as soon as the cluster was broken all the test spheres were very mobile and rolled away rapidly, implying that mutual stability had been imparted by the cluster formation.

The stability of individual spheres is affected significantly by their positions relative to the supporting bed spheres, and also the proximity and relative positions of other spheres in a cluster. These effects can be appreciated by pairwise comparison of the critical dimensionless shear stresses required for the break-up of different clusters.

The effect of sphere position relative to the bed spheres can be clearly seen by comparing the entrainment conditions of the single sphere in positions 1 and 2. The sphere in position 2 is more stable than that in position 1 because it has to move either over the top of the bed sphere directly downstream, or obliquely to move through a 'saddle' between adjacent bed spheres. The sphere in position 1 can move directly downstream through a saddle. It therefore has a smaller pivoting angle to overcome and movement is in the direction of the applied drag force. This effect is also apparent in comparisons of clusters 3 with 4, 8 with 9, 10 with 11, and possibly 6 with 7.

The effect of other nearby spheres on stability can also be seen by comparing the break-up conditions for different clusters. The most significant effect is the reduction in stability induced by placing spheres in line, transverse to the flow direciton. Spheres in clusters 3 and 4, and 8 and 9, for example, are much less stable than the single spheres in position 1 and 2. This must be due to a modification of the flow pattern around the individual spheres with consequent changes to the drag and lift forces. The stability is apparently not decreased further by extension of the line of spheres; the end spheres in clusters 3 and 8 moved at the same shear stresses as did those in clusters 4 and 9.

Stability is also influenced by arranging spheres in line in the direction of flow. For clusters 5, 6 and 7 the upstream sphere invariably moved first, which is consistent with the observations of Brayshaw

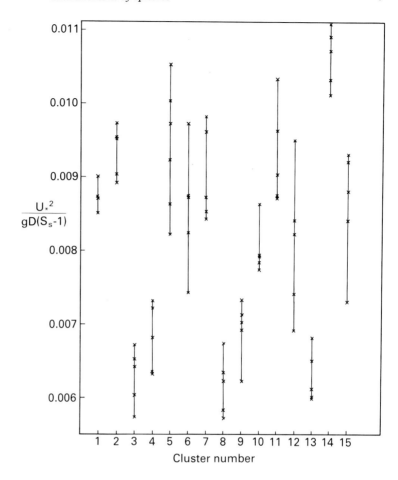

Fig. 6. Entrainment conditions for clusters ($D/K = 1$).

et al. (1983) on hemispheres in line. They found the lift and drag coefficients to be reduced for the downstream object and the lift increased on a close upstream one. The 'shielding' afforded by an upstream sphere is also effective for spheres positioned obliquely downstream. The transversely adjacent spheres in clusters 10 and 11 are much more stable than in clusters 3 and 4 because of the presence of the upstream sphere. A downstream sphere can also stabilize upstream spheres, but this is by physically obstructing movement rather than modifying the flow patterns. Cluster 12 is more stable than cluster 3 because the preferred directions of movement of the upstream spheres (determined by their positions relative to the bed spheres) are blocked by the downstream sphere, whose stability is also enhanced by shielding from upstream. The downstream sphere in cluster 13 on the other hand has no stabilizing effect because the preferred

directions of movement of the upstream spheres are outwards. The same argument applies when comparing clusters 15 with 11, and 14 with 10. In clusters 15 and 13 the downstream spheres actually appear to reduce the stability when compared with clusters 11 and 4 respectively.

Cluster 14 is the only one which is clearly more stable than single spheres. This can be attributed to the shielding afforded by the upstream sphere, and blocking of the preferred directions of movement of the transversely adjacent spheres by the downstream one.

SUMMARY AND CONCLUSIONS

A set of data has been produced which defines the entrainment conditions for spheres under known and well-controlled conditions.

The effect of relative size has been clearly shown and is consistent with observations of natural sediments.

The results indicate that spheres in groups of two or three are generally more susceptible to entrainment than single spheres. One cluster of four spheres did, however, have greater stability than single spheres. Stability is least for spheres arranged in line transverse to the flow direction, and greatest when these are shielded from upstream and obstructed from movement downstream.

ACKNOWLEDGEMENTS

The work described in this paper forms part of a research programme of COMRO, the research organization of The Chamber of Mines of South Africa. Their permission for its publication is gratefully acknowledged. The constructive review comments by Ian Reid and Peter Wilcock are greatly appreciated.

REFERENCES

Andrews, E.D. (1983) Entrainment of gravel from naturally sorted riverbed material. *Bull. Geol. Soc. Am.* **94**, 1225–1231.

Brayshaw, A.C., Frostick, L.E. & Reid, I. (1983) The hydrodynamics of particle clusters and sediment entrainment in coarse alluvial channels. *Sedimentology* **30**, 137–143.

Coleman, N.L. (1967) A theoretical and experimental study of drag and lift forces acting on a sphere resting on a hypothetical stream bed. In: *Proc. 12th Congr. IAHR*, Vol. 3, pp. 185–192. International Association for Hydraulic Research.

Day, T.J. (1980) A study of initial motion characteristics of particles in graded bed material. In: *Current Research, Part A* (Eds Blackadar, R.G., Griffin, P.J., Dumych, H. & Neale, E.R.W.) pp. 281–286. Geological Survey of Canada Paper 80-1A. Geological Survey of Canada, Ottawa.

Egiazaroff, I. V. (1965) Calculation of non-uniform sediment concentrations. *J. Hydr. Div., ASCE* **91**(HY4), 225–247.

Everts, C.H. (1973) Particle overpassing on flat granular boundaries. *J. Waterways, Harbors and Coastal Engineering Div., ASCE* **99** (WW4), 425–438.

Fenton, J.D. & Abbott, J.E. (1977) Initial movement of grains on a stream bed: the effect of relative protrusion. *Proc. R. Soc. Lond. A* **352**, 523–537.

Ghosh, S. & Jena, S.B. (1971) Boundary shear distribution in open channel compound. *Proc. Instn Civ. Engrs* **49**, 417–430.

Hwang, L. & Laursen, E.M. (1963) Shear measurement technique for a rough surface. *J. Hydr. Div., ASCE* **89**(HY2), 19–38.

James, C.S. (1990) Prediction of entrainment conditions for nonuniform, noncohesive sediments. *J. Hydraul. Res.* **28**(1), 25–41.

Reid, I & Frostick, L.E. (1987) Toward a better understanding of bedload transport. In: *Recent Developments in Fluvial Sedimentology* (Eds Ethridge, F.G., Flores, M. & Harvey, M.D.) pp. 13–19. Soc. Econ. Paleont. Miner., Tulsa, Spec. Publ. 39.

A new bedform stability diagram, with emphasis on the transition of ripples to plane bed in flows over fine sand and silt

J.H. VAN DEN BERG and A. VAN GELDER

Institute of Physical Geography, Utrecht University,
PO Box 80.115, Utrecht, The Netherlands

ABSTRACT

Channel deposits of the Yellow River (China), composed of silt and very fine sands, show in cross-section ripple lamination, even lamination and small scour-and-fill structures.

Flume experiments with silt conducted to simulate Yellow River flow showed that the scour-and-fill structures represent cycles of erosion due to a process of irregular scouring followed by ripple drift sedimentation. Cohesion plays an important role in the generation of these cycles, which only occurred at conditions of overall equilibrium sediment transport. At high flow velocities the experiments showed the simultaneous occurrence of ripples, scours and plane bed. Similar transitional bed stages were found in results of previous flume studies, but have never been incorporated in bedform stability diagrams. In this paper a new bedform existence diagram for unidirectional flow is proposed using our data and a large number of previous flume and field data with siliciclastic sediments in the size range of 10–5100 µm. This diagram shows the stability fields of various bedform states, including the ripple and scour to plane bed transition as a function of a grain-roughness related mobility parameter and a dimensionless grain size parameter. The use of these parameters gives the diagram the advantage that it can be used without having any information on the alluvial roughness. Therefore this presentation is especially suitable in palaeoflow analysis of preserved sedimentary structures.

INTRODUCTION

The paper presented here originated in an attempt to explain sedimentary structures and sequences in sediments of delta lobes of the Yellow River deposited after 1855. The Yellow River sediments range in size from clay to very fine sand. Most of them have silt grades (4–62 µm). When investigating shallow exposures in the streambelt areas, it appeared that specific structures like climbing-ripple lamination and small scour-and-fills were ubiquitous, whereas megaripple cross-stratification was absent. Even lamination occurred, but was restricted to channels.

Small (decimetre size) scour-and-fill sequences were common in river channel deposits. An example is shown in Fig. 1. The scour surfaces are quite irregular on a small (centimetre to decimetre) scale, sometimes with their steepest sides upstream. Apparently they developed from flute-like erosion of inhomogeneities on a slightly cohesive bed.

We tried to interpret the structures in terms of hydraulic conditions during deposition, but found problems because of the relative scarceness of flume experiments with natural sediments smaller than 100 µm. Therefore a series of test runs was conducted in the flume of the Earth Sciences Institute, Utrecht University. We used silt of 33 µm median diameter and simulated conditions of equilibrium transport as well as deposition from oversaturated suspension. Results of the equilibrium runs are plotted in a new type of bedform stability diagram, which gives less ambiguity when interpreting ancient sediments.

When plotting our data on existing diagrams problems arose when interpreting transitional bedforms. Furthermore we felt that a drawback of all published diagrams is that parameters used on the ordinate are always strongly related to bedform roughness or energy slope. This makes them of

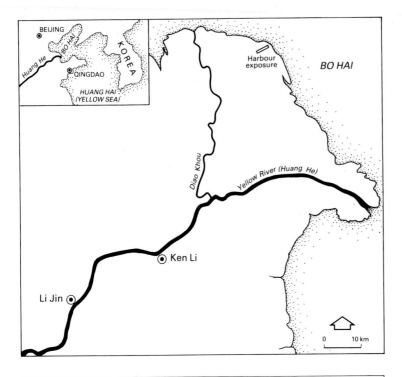

Fig. 1. Scour-and-fill structures as exposed in a harbour excavation of the Yellow River lower deltaic plain. The section is almost transverse to the flow direction. Bar size on rule is 0.1 m.

doubtful use for geologists, because in fossil sediments these factors cannot be estimated with sufficient accuracy.

As an alternative, we propose a new type of diagram in which a dimensionless 'mobility parameter' on the vertical axis is form-roughness independent. The horizontal axis of our diagram shows a dimensionless grain-roughness parameter, which can easily be reduced to the median grain size of the bed material. Besides our own experiments, the published data of 372 experiments with sizes ranging from 11 to 4080 µm were recalculated and plotted on our diagram.

FLUME EXPERIMENT

Previous work

Several series of flume experiments (e.g. Allen, 1984) indicate that in very fine sand- and silt-sized sediment, small-scale ripples seem to be the only type of constructive depositional bedform under unidirectional flow. With increasing flow velocities the ripple phase passes into the upper flat-bed phase and no dunes are formed.

Postma (1967) and Sundborg (1956) demonstrated that beds composed of natural sediments with a median grain size (D_{50}) smaller than 100 µm may develop a cohesiveness that depends on sediment composition, size and pore water content. The process of ripple movement under equilibrium conditions of flow and sediment transport involves ripple lee-side sedimentation and stoss-side erosion. In the case of a significant initial consolidation and cohesion of the bed the regular stoss-side erosion of migrating ripples is replaced by irregular scours.

The possible occurrence of such conditions might explain the generation of steep stoss-side slopes and deep so-called 'potholes' as described by Jopling & Forbes (1979) from flume experiments using a coarse silt (D_{50} = 45 µm) with a small proportion of clay. The depth of some of the scours below the nominal level of the bed exceeded the mean flow depth of 0.03–0.10 m. Possibly the 'rugged and irregular' bedform appearance as denoted by Kalinske & Hsia (1945) for some of their experimental runs with fine silt (D_{50} = 11 µm) refers to similar conditions.

The admixture of some clay-grade material to the bed material seems to be prerequisite to the formation of the irregular erosional features. In experiments using silt and very fine sand from which the original clay admixture was removed by Mantz (1978; D_{50} = 15 and 66 µm) and by Rees (1966; D_{50} = 10 µm) no irregular bedform topography was produced. The aim of our experiments was to reproduce the hydraulic and bed conditions which produce the scour-and-fill sequences found in the Yellow River sediments.

Experimental apparatus

Our experiments were made in a recirculating flume 7.96 m long, 0.45 m wide and 0.60 m deep (Fig. 2). The water circulation in the flume is generated by paddles. Guiding vanes are present in the corners of the flume to compensate for the increase in width of the flume and centrifugal forces exerted on the fluid in the bends. As a result the streamlines are parallel to the flume sides in bends. The test section of the

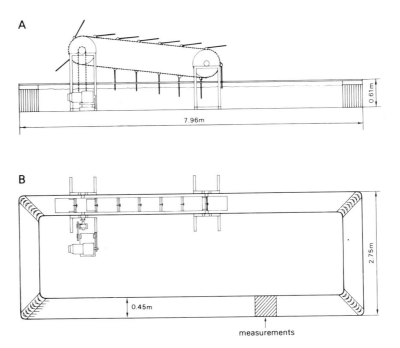

Fig. 2. Flume-plan (B) and cross-section (A).

flume is opposite the paddle section. The outer wall is made of transparent PVC to allow observations. A detailed description of the design of the flume and its flow characteristics for different discharge rates is given by Winkelmolen (1976).

Description of experiments

Several experiments have been conducted to simulate the overloaded suspended material conditions of the Yellow River overbank flow as well as the approximate equilibrium conditions of flow and sediment transport in the river channel. In this paper only some of the results of equilibrium runs will be presented. The sediment used was a siliciclastic clayey silt with a median grain size (D_{50}) for the particles larger than 2 μm (90.7% by weight) of 33 μm and a geometric standard deviation $\sigma_s = \frac{1}{2}(D_{84}/D_{50} + D_{50}/D_{16}) = 1.45$. The density of the sediment was 2650 kg m^{-3}.

The original aim of the tests was to study the bedform shape and bedform migration in a newly deposited and not yet consolidated sediment. Therefore, shortly before every test the sediment layer in the flume was thoroughly mixed using a fire-hose, while the discharge was maximum (depth-averaged flow velocity, $\bar{u}, \approx 0.75$ m s^{-1}). Subsequently the flow velocity was set to the value aimed for the test. When adjusting to a lower velocity, the flow became temporarily oversaturated with suspended bed material and a process of deposition in a ripple morphology took place. The measurements began when stoss-side erosion of ripples started in the test section. It was assumed that at this moment overall equilibrium conditions of sediment transport had set in (see also Fig. 3). Twelve runs were made at different discharges (Table 1). The length of the channel allowed full development of the bedforms. In all runs, ripple shape and migration showed no significant change beyond about 3 m from the head of the flume. This is in accordance with the findings of Winkelmolen (1976). The disturbance of the flow by the paddles seems to vanish as a result of the turbulence diffusing upwards from the channel bed before reaching the test section. All measurements of bedforms were made in the reach 4.0–6.5 m downstream from the entrance of the channel (see also Fig. 2). Current velocity measurements were conducted in the middle of the channel using Ott-type meters with 0.05 m diameter propellers attached in a frame. The vertical distances between the measuring points were 0.02, 0.05, 0.06 and 0.12 m. The lowest Ott-meter was placed at 0.03 m above the bed with an accuracy of 0.005 m. The water depth at the measuring section was determined from the level of the bed surface and the water surface as visible on the transparent outer wall of the flume.

Results and analysis

In all tests the lower half of the velocity profile approached the logarithmic distribution. Above a height of about 0.25 m from the bed the friction of

Table 1. Data flume experiments

Test	$\bar{u}$ (m s^{-1})	h (m)	Temperature (°C)	D_{50} (μm)	D_{90} (μm)	n	H ($\times 10^{-3}$ m)	U_c ($\times 10^{-6}$ m s^{-1})	$\bar{H}$	$\bar{U}_c$	Bedform state
1	0.639	0.43	16.0	33	52	8	7–18	92–154	13	106	Transition
2	0.525	0.42	16.8	33	52	10	11–25	24–83	19	55	Transition
3	0.426	0.41	17.0	33	52	8	15–29	8.3–47	20	25	Ripples
4	0.300	0.41	16.3	33	52	4	12–27	5.0–8.3	18	6.5	Ripples
5	0.761	0.45	16.7	33	52	3	22–37	189–239	28	217	Ripples
6	0.743	0.43	17.0	33	52	2	14–29	167–317	22	242	Ripples
7	0.665	0.39	17.8	33	52	3	13–40	73–156	23	122	Ripples
8	0.608	0.38	17.2	33	52	2	19–33	63–73	25	68	Ripples
9	0.393	0.37	17.5	33	52	6	25–37	19–36	26	28	Ripples
10	0.242	0.34	17.5	33	52	3	25–47	4.7–8.3	34	6.6	Ripples
11	0.198	0.33	19.0	33	52	3	11–30	0.5–1.0	19	0.75	Ripples
12	0.102	0.32	19.0	33	52	—	—	—	—	—	Flat bed (no motion)

$\bar{u}$, Depth-averaged flow velocity; h, water depth; D_{50}, median grain size; D_{90}, 90th percentile of grain size distribution; n, number of ripple measurements; H, bedform height; U_c, bedform celerity; $\bar{H}$, average bedform height; $\bar{U}_c$, average bedform celerity.

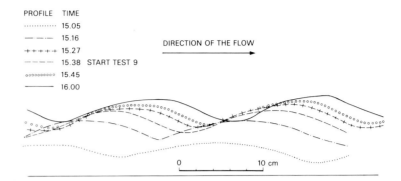

Fig. 3. Ripple migration just before and during test no. 9.

the side-walls resulted in a slight reduction of the current velocity near the water surface. In order to correct for this undesired influence the depth-averaged flow velocity was computed from a best-fit logarithmic profile using the measured values of the current velocity at a distance of 0.03, 0.05, 0.10, 0.16 and 0.28 m above the bed. The depth-averaged flow velocities computed according to the latter method were about 10% higher than actually measured.

Figure 3 shows the migration of bedforms in test 9. Ripple migration was accompanied by a general aggradation of the bed, resulting in a pattern of climbing ripples. Erosion occurred on some ripple stoss sides but was of minor importance. Similar processes occurred in the other tests. When equilibrium flow was maintained long enough it appeared that periods of gradually climbing ripples always alternated with intervals of irregular erosional scouring. During periods of accumulation the ripple migration was measured. The results of ripple measurements and the relevant hydraulic conditions are presented in Table 1.

A special run with maximum flow velocity ($\bar{u} = 0.75$ m s^{-1}) was conducted in order to observe a large number of succeeding cycles of depositional climbing ripples followed by erosional scouring. The sequence of bed morphologies resulting from the experiment is shown in Fig. 4. Due to the high flow velocity the ripple amplitude was low. Sometimes ripples were completely flattened and a transition to the upper plane bed phase occurred (Fig. 4, 42 min after start). Erosion and sedimentation occurred separately in patches of a few metres length. Taking the technical limitations into account, a good resemblance was found between the scour-and-fill structures seen in the field and those produced under controlled conditions.

Possibly the scour-and-fill cycles are related to local differences in the hydraulic roughness and initial consolidation of the bed. In patches of accumulation the bed is relatively smooth and little turbulence is created so that the sediment transport

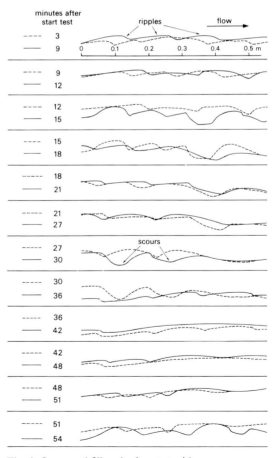

Fig. 4. Scour-and-fill cycles in a test with a depth-averaged flow velocity $\bar{u} = 0.75$ m s^{-1}.

of the flow remains slightly above the equilibrium capacity. In places with erosional scour holes much turbulence is produced. This stimulates erosion and a higher capacity of the flow to transport suspended sediment. Thus, small deviations in the overall equilibrium conditions lead to accumulation in rippled areas whereas at the same time erosion continues in patches with erosional scours. The presence of the depositional humps created by this process results in a local increase of the vertical flow velocity gradient or shear stress and a reduction of the deposition rate. Ultimately erosional processes will start and, because of the initial consolidation, scours will be formed, which enhance further erosion. An opposite reasoning holds for the moment that erosional patches change into smooth areas of deposition.

STABILITY FIELDS OF BEDFORMS UNDER UNIDIRECTIONAL CURRENTS

Disadvantages of existing diagrams

The hydraulic background of preserved sedimentary structures formed by unidirectional currents may be visualized in a graph of a sediment size versus some flow parameter. Studies of this kind have resulted in a number of bedform stability diagrams. Unfortunately most of these diagrams refer to grain sizes larger than 100 μm. Exceptions are formed by two graphs by Allen (1984, 1985), who revised the successful stability diagrams originally proposed by Simons & Richardson (1965) and Chabert & Chauvin (1963). In addition Allen included grain sizes as small as 10 μm. The diagrams show the stability fields for flume equilibrium conditions in a stream power–grain size plane and in a mobility–grain size plane. The stream power, ω, and mobility (= Shields entrainment) parameter, θ, are defined:

$$\omega = \tau \bar{u} = \frac{\rho g \bar{u}^3}{C^2} \quad (1)$$

$$\theta = \frac{\tau}{g(\rho_s - \rho) D_{50}} = \frac{\rho \bar{u}^2}{(\rho_s - \rho) C^2 D_{50}} \quad (2)$$

in which $\tau = \rho g h i = \rho g \bar{u}^2 C^{-2}$ = bed shear stress; g = acceleration of gravity; h = flow depth; i = energy gradient; C = Chézy coefficient; D_{50} = median grain size of the bed material; $\bar{u}$ = depth-averaged flow velocity; ρ_s, ρ = sediment, fluid density.

Allen's diagrams show that for $D_{50} < 10^{-4}$ m the bed state passes from a lower flat-bed state (no particle motion) into a rippled configuration, with increasing stream power or mobility. At high flow velocities this transforms into the upper flat-bed state. Allen's diagrams are most useful in predicting the bed state for given steady flow and sediment conditions, but are of limited value in palaeohydraulic analysis, as preserved sediments do not provide any information about the energy gradient ($\approx$ water surface slope), i, or alluvial roughness, C, during deposition.

At the highest values of the flow velocity reached in our experiments, ripples, scour forms and flat regions were present simultaneously. We have denoted this configuration as a transition state following proposals made by Vanoni (1975). In the bedform stability diagrams of Allen this bed state is not distinguished, but it might be incorporated, by inspecting the bedform description of the flume data used. Instead of doing so we present a new bedform stability diagram, which can be applied without any knowledge of bedform roughness or energy slope, and which therefore better serves the interests of geologists.

Bed configuration as related to particle mobility and grain size

The present analysis consists of a revision and extension of a graph proposed by Van Rijn (1984b). A basic element of this approach is the assumption that skin friction is of dominating influence on the existence and dimensions of bedforms. This is supported by the fact that the bed load transport and bedform migration can be predicted reasonably well without considering the influence of form drag (Van Rijn, 1984a; Van den Berg, 1986, 1987).

In the new graph (Fig. 5) a modified mobility parameter, related to grain roughness, θ', as proposed by Van Rijn (1984a), appears on the ordinate

$$\theta' = \frac{\rho \bar{u}^2}{(\rho_s - \rho)(C')^2 D_{50}} \quad (3)$$

with

$$C' = 18 \log \frac{4h}{D_{90}} \quad (4)$$

and a non-dimensional particle parameter, D_*, as introduced by Bonnefille (1963), on the abscissa

$$D_* = D_{50} \left(\frac{(\rho_s - \rho) g}{\rho \nu^2} \right)^{1/3} \quad (5)$$

in which C' = Chézy coefficient related to grain roughness; D_{90} = ninetieth percentile of the grain size distribution of the bed material; v = kinematic viscosity of the ambient fluid, which in the present data analysis was assumed to be that of clear fresh water value given by:

$$v = [1.14 - 0.031(T-15) + 0.00068(T-15)^2] 10^{-6} \quad (6)$$

with T = temperature in degrees Celsius.

It may be noted that at extremely high concentrations of suspended particles the apparent viscosity of the fluid increases. For instance the effect on viscosity of a sediment concentration of 10% by volume in a heavily silt laden flow would be a reduction of the clear water value of D_* by 16% as predicted by a function of Lee (1969). Such high concentrations, however, are far beyond the experimental conditions of the flume experiments considered, as the near-bed concentrations never exceeded 5% by volume.

In all, 372 flume experiments with median particle diameters ranging from 11 to 4080 μm were used. Table 2 gives the sources used, together with a summary of some basic data. In Fig. 5 some data points that duplicate and overlap others are omitted for clarity of presentation. The data used conformed to the following selection criteria.

1 For a particle parameter (D_*) larger than 2 only data with a flow depth (h) larger than 0.1 m and a flume width (b) larger than 0.3 m were considered. This was because Williams (1970) suggests that at smaller values than these dune development is obstructed. For values of the particle parameter $D_* < 2$, where dunes do not exist, these criteria were arbitrarily set at $h > 0.02$ m and $b > 0.2$ m.
2 A water temperature less than 40°C.
3 Natural siliciclastic river bed sediments, which in the case of very fine sands and silts implies the admixture of some clay.
4 Standing wave and antidune bedforms were not considered, as they are related to the flow depth and the behaviour of the water surface, and not to the

Table 2. Summary of experimental data used in Fig. 5

Source	Total runs	D_{50} (μm)	D_{90} (μm)	Flow depth (m)	Water temperature (°C)	Flume width (m)
Costello & Southard (1981)	7	510	600	0.15–0.16	29–31	0.92
Costello & Southard (1981)	8	660	700	0.15–0.16	30–31	0.92
Costello & Southard (1981)	9	660	800	0.14–0.16	28–29	0.92
Costello & Southard (1981)	9	790	950	0.14–0.16	27–28	0.92
Gee (1975)	3	310	440	0.10–0.11	21–24	0.61
Gee (1975)	3	1050	1260	0.10–0.15	24–27	0.61
Gilbert (1914)	8	1710	2500	0.11–0.17	–	0.31
Gilbert (1914)	2	3170	3720	0.10	–	0.31
Guy et al. (1966)	30	190	280	0.12–0.33	12–20	2.44
Guy et al. (1966)	15	270	560	0.15–0.33	14–19	2.44
Guy et al. (1966)	29	280	700	0.13–0.33	11–18	2.44
Guy et al. (1966)	20	320	600	0.16–0.23	10–34	0.61
Guy et al. (1966)	30	450	800	0.10–0.30	9–19	2.44
Guy et al. (1966)	39	930	1600	0.12–0.32	17–22	2.44
Hill et al. (1969)	7	88	130	0.23–0.25	22–38	0.61
Hill et al. (1969)	4	150	200	0.23–0.25	22–32	0.61
Hill et al. (1969)	5	310	380	0.16–0.20	24–32	0.61
Jopling & Forbes (1979)	16	45	105	0.02–0.10	28–31	0.20
Kalinske & Hsia (1945)	9	11	24	0.11–0.20	–	0.69
Kennedy & Brooks (1965)	4	142	220	0.11–0.17	18–25	0.85
Laursen (1958)	14	110	135	0.12–0.30	21–27	0.92
Nordin (1976)	40	250	450	0.30–0.86	10–25	2.38
Nordin (1976)	16	1140	1890	0.26–0.63	17–26	2.38
Taylor (1971)	12	228	395	0.10–0.18	22–38	0.85
USWES (1935)	5	4080	6100	0.10–0.12	20	0.71
Van den Berg, this paper	11	33	52	0.32–0.45	16–19	0.45
Vanoni & Brooks (1957)	7	137	200	0.16–0.17	17–25	0.61
Williams (1970)	10	1350	1800	0.15–0.22	22–28	0.61

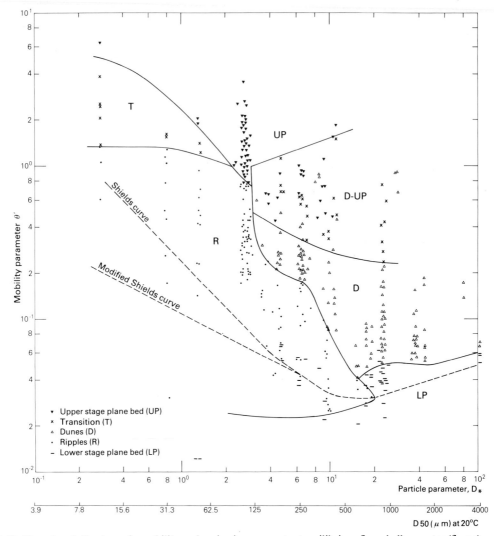

Fig. 5. Bedform in relation to grain mobility and grain size parameter: equilibrium flow shallow water (flume) conditions.

parameters in the graph.

The influence of the side-wall roughness of the flume on the flow characteristics was eliminated by using the method of Vanoni & Brooks (1957). Where the water temperature was not reported, a value of 20°C was assumed. In the case that the available grain size data were restricted to the median and geometric standard deviation σ_s, a lognormal grain size distribution was assumed and the value of the D_{90}-percentile was approximated by

$$D_{90} = \sigma_s^{1.3} D_{50} \qquad (7)$$

The initiation of general motion of particles along the bed is indicated in Fig. 5 by two curves. As Miller *et al.* (1977) found for silt or fine sand the original Shields curve overestimates the threshold values of the mobility parameter. A modified curve is therefore also presented (proposed by Zanke, 1987).

SUMMARY AND CONCLUSIONS

In the new stability field diagram presented here, some overlap of ripple and lower plane bed data points occurs near the Shields criterion. Presumably most points in this region refer to the rippled bed state. In many cases running times of the reported experiments under the low velocity conditions were too short for the development of ripples (Southard & Boguchwal, 1990).

The present knowledge of the occurrence of the transition bedform state in silt and fine sand is limited. Its existence is related to small admixtures of clay in the bed material. In clean silt or sand the transition state is missing and instead an abrupt boundary to upper plane bed is found (Southard & Boguchwal, 1990). Therefore the stability boundaries for the transition state as drawn in the graph should be considered as a rather crude and preliminary result. Also, the position of the boundary of dunes and transitional bedforms to upper plane bed for fine sands is questionable. This might still be a consequence of the shallow water depth of flume experiments that inhibits full development of dune bedforms. In order to investigate the reliability of the latter boundary in greater water depth, several series of river data with a flow depth over 1 m were analysed. In all 262 field observations with bed material particle sizes ranging from 80 to 5100 µm were used (Table 3).

Most of the field data do not allow a detailed division into bedform configurations similar to that achieved for flume tests. Therefore only two clear bedform states were considered: a flat (or rippled) bed, including all configurations with a bedform amplitude less than 0.06 m, and dune bedforms with an amplitude of more than 0.15 m. Transition bedforms or bedforms of intermediate height were discarded. The results of the analysis are shown in Fig. 6. Also shown are some of the bedform stability boundaries as drawn for flume conditions in Fig. 5. It appears that the position of the upper boundary of dune existence in fine sand as revealed from the flume data ($\theta' = 0.69\ D_*^{0.34}$) is equally applicable to greater water depths of natural river flows. It may be noted that dune bedforms in field situations were found at relatively small values of the mobility parameter; some data points even plotted below the threshold curve of initial particle motion (not indicated in Fig. 6). These dunes must be considered as relict bedforms, produced in periods of higher flow velocity, preceding the measurements.

It should be realized that the presented bedform diagrams, like those previously published, refer to steady and uniform flow and equilibrium suspended sediment concentration. Non-equilibrium conditions might result in some shift of the boundaries between the bedform stability fields. The transition state in fine sand and silt consisting of scours and areas of flat bed and ripples may even completely disappear when deposition from a flow overloaded with suspended bed material impedes any erosion of the bed. With regard to this it was noticed that the scour-and-fill structures were found in the Yellow River delta sediments only in deposits of the river channel (Fig. 1), and not in overbank sequences where flow decelerated and hence became overloaded.

ACKNOWLEDGEMENTS

The fieldwork in China, which formed the starting

Table 3. Summary of field data used in Fig. 6

Source	River or canal	No.	D_{50} (µm)	Depth-averaged flow velocity (m s)$^{-1}$	Flow depth (m)	Water temperature (°C)
Culbertson et al. (1972)	Rio Grande Conveyance Channel	15	180–280	0.90–1.69	1.0–1.5	1–2.2
Dinehart (1989)	North Folk Toutle	2	1900–5100	1.65–3.38	1.0–1.7	4.5–8.8
Hansen (1966)	Skive Karup	1	460	0.60	1.0	10
Mahmood et al. (1984)	Pakistan Link Canals	143	77–370	0.47–1.66	1.0–4.3	13.3–36.0
Shen et al. (1978)	Missouri	22	193–266	1.37–1.72	2.8–4.9	2.8–22.8
Shinohara & Tsubaki (1959)	Hii	1	1110	0.90	1.0	—
Simons (1957)	Americal Canal	6	96–715	0.42–0.77	1.0–2.5	21.7–26.1
Termes (1986)	Meuse	4	303–425	1.43–1.79	7.1–8.9	5
Termes (1986)	Rhine	36	300–1750	1.22–1.89	4.8–9.5	5
Termes (1986)	Zaire	7	345	1.37–1.55	14.0–15.0	27–29
Van den Berg (1987)	Rhine	25	690–3800	0.80–1.78	5.0–10.4	1–11

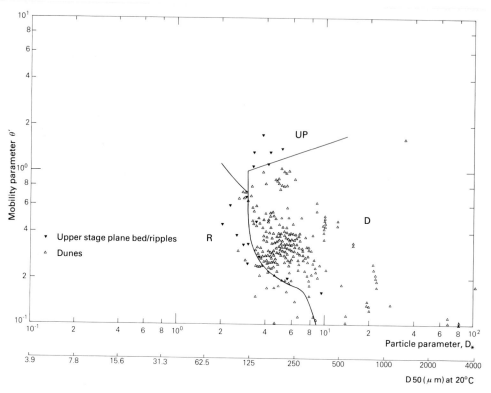

Fig. 6. Bedform in relation to grain mobility and grain size parameter: river flow conditions with water depths larger than 1 m.

point of this paper, was part of a research project on the Holocene sedimentary history of the Yellow River Delta. It was financed by the Royal Dutch Academy of Sciences, and the Ministry of Geology and Mineral Resources of China. The project was initiated by the Comparative Sedimentology Division of Utrecht, and we thank Djin Nio for his efforts. We also like to thank our colleagues of the Institute of Marine Geology, Qingdao, for their support in the field; and particularly Wei Helong, who participated in the flume experiments. The technical support of Hans Wijland is acknowledged with pleasure. We thank Joost Terwindt and Lynne Frostick for constructive criticism on the manuscript.

REFERENCES

ALLEN, J.R.L. (1984) *Sedimentary Structures, their Character and Physical Basis*, Vol. 1. *Developments in Sedimentology* 30. Elsevier, Amsterdam, 593 pp.

ALLEN, J.R.L. (1985) *Principles of Physical Sedimentology*. Allen and Unwin, London, 272 pp.

BONNEFILLE, R. (1963) Essais de synthese des lois du début d'entrainement des sédiments sous l'action d'un courant en régime continu. *Bull. du Centre de Recherches et d'essais de Chatou* **5**, 17–22.

CHABERT, J. & CHAUVIN, J.L. (1963) Formation des dunes et des rides dans les modéles fluviaux. *Bull. du Centre de Recherche et d'Essais de Chatou* **4**, 31–51.

COSTELLO, W.R. & SOUTHARD, J.B. (1981) Flume experiments on lower-flow-regime bed forms in coarse sand. *J. Sedim. Petrol.* **51**, 849–864.

CULBERTSON, J.K., SCOTT, C.H. & BENNETT, J.P. (1972) Summary of alluvial-channel data from Rio Grande conveyance channel, New Mexico, 1965–1969. US Geol. Surv. Prof. Paper 562-J.

DINEHART, R.L. (1989) Dune migration in a steep, coarse-bedded stream. *Water Resources Res.* **25**, 911–923.

GEE, D.M. (1975) Bed form response to nonsteady flows. *J. Hydr. Div., ASCE* **101**(HY3), 437–449.

GILBERT, G.K. (1914) The transportation of debris by running water. US Geol. Surv. Prof. Paper 86.

GUY, H.P., SIMONS, D.B. & RICHARDSON, E.V. (1966) Summary of alluvial channel data from flume experiments 1956–1961. US Geol. Surv. Prof. Paper 461-I.

HANSEN, E. (1966) Bed-load investigation in Skive-Karup river. *Technical Univ. Denmark Bull.* **12**, 1–8.

HILL, H.M., SRINIVASAN, V.S. & UNNY, T.E. (1969) Instability of flat bed in alluvial channels. *J. Hydr. Div., ASCE* **95**(HY5), 1545–1558.

JOPLING, A. & FORBES, D.L. (1979) Flume study of silt transportation and deposition. *Geogr. Ann.* **61A**, 67–85.

KALINSKE, A.A. & HSIA, C.H. (1945) Study of transportation of fine sediments by flowing water. Bull. Iowa Univ. Studies in Eng. 29.

KENNEDY, J.F. & BROOKS, N.H. (1965) Laboratory study of an alluvial stream of constant discharge. In: *Federal Inter-agency Sediment Conf. 1963 Proc., Misc. Pub. US Dept Agric.* **970**, Washington, pp. 320–330.

LAURSEN, E.M. (1958) The total sediment load of streams. *J. Hydr. Div., ASCE* **84**(HY1), Proc. Paper 1530.

LEE, D.I. (1969) The viscosity of concentrated suspensions. *Trans. Soc. Rheology* **13**, 272–288.

MAHMOOD, K., HAQUE, M.I., CHOUDRI, A.M., MASOOD, T. & MALIK, M.A. (1984) ACOP canals equilibrium data, vol. X: summary of 1974–1980 data. The George Washington Univ., Mech. and Environmental Eng. Dept, Rep. EWR-84-2.

MANTZ, P.A. (1978) Bed forms produced by fine, cohesionless granular and flakey sediment under subcritical water flows. *Sedimentology* **25**, 83–103.

MILLER, M.C., MCCAVE, I.N. & KOMAR, P.D. (1977) Threshold of sediment motion under unidirectional currents. *Sedimentology* **24**, 507–528.

NORDIN, C.F. (1976) Flume studies with fine and coarse sands. US Geol. Surv. Open File Report 76-762.

POSTMA, H. (1967) Sediment transport and sedimentation in the estuarine environment. In: *Estuaries* (Ed. Lauff, G.H.) pp. 158–180. Am. Ass. for the Advancement of Science, Washington.

REES, A.I. (1966) Some flume experiments with a fine silt. *Sedimentology* **6**, 209–240.

SHEN, H.W., MELLEMA, W.J. & HARRISON, A.S. (1978) Temperature and Missouri river stages near Omaha. *J. Hydr. Div., ASCE* **104**(HY1), 1–20.

SHINOHARA, K. & TSUBAKI, T. (1959) On the characteristics of sand waves formed upon the beds of the open channels and rivers. Report of Res. Inst. for Applied Mechanics, Kyushu Univ. pp. 15–45.

SIMONS D.B. (1957) *Theory of design of stable channels in alluvial material: Fort Collins, Colorado*. Thesis, Colorado State University.

SIMONS, D.B. & RICHARDSON, E.V. (1965) A study of variables affecting flow characteristics and sediment transport in alluvial channels. In: *Federal Inter-agency Sediment Conf. 1963 Proc., Misc. Pub. US Dept Agric.* **970**, Washington, pp. 193–207.

SOUTHARD, J.B. & BOGUCHWAL, A. (1990) Bed configurations in steady unidirectional water flows. Part 2. Synthesis of flume data. *J. Sedim. Petrol.* **60**, 658–679.

SUNDBORG, A. (1956) The river Klarälven: a study of fluvial processes. *Geogr. Ann.* **38**, 127–316.

TAYLOR, B.D. (1971) Temperature effects in alluvial streams. W.M. Keck Lab. of Hydraulics and Water Resources, California, Rep. KH-R-27.

TERMES, A.P.P. (1986) Dimensies van beddingvormen onder permanente stromingsomstandigheden. Delft Hydraulics Lab. Rep. M2130/Q232.

USWES (US WATERWAYS EXPERIMENT STATION) (1935) Studies of river bed materials and their movement with special reference to the Lower Mississippi river. US Army Corps of Engineers, Waterways Experiment Station, Paper 17.

VAN DEN BERG, J.H. (1986) *Aspects of sediment- and morphodynamics of subtidal deposits of the Oosterschelde, The Netherlands*. Thesis, University of Utrecht.

VAN DEN BERG, J.H. (1987) Bedform migration and bedload transport in some rivers and tidal environments. *Sedimentology* **34**, 681–698.

VANONI, V.A. (1975) *Sedimentation Engineering*. ASCE Manuals and Reports on Engineering Practice, Vol. 54. ASCE, New York, 745 pp.

VANONI, V.A. & BROOKS, N.H. (1957) Laboratory studies of the roughness and suspended load of alluvial streams. Sedimentation Lab., California Inst. Technol., Report E-68. ASCE, New York.

VAN RIJN, L.C. (1984a) Sediment transport, part 1: bed load transport. *J. Hydraul. Eng.* **110**, 1431–1456.

VAN RIJN, L.C. (1984b) Sediment transport, part 3: bed forms and alluvial roughness. *J. Hydraul. Eng.* **110**, 1733–1754.

WILLIAMS, G.P. (1970) Flume width and water depth effects in sediment transport experiments. US Geol. Surv. Prof. Paper 562-H.

WINKELMOLEN, A.M. (1976) An inexpensive infinite flume. *Geol. Mijnbouw* **55**, 51–60.

ZANKE, U. (1987) Sedimenttransportformeln für Bed-load im Vergleich. *Mitt. Franzius Inst. Wasserbau und Küsteningenieurwesen Univ. Hannover* **64**, 327–411.

In-transport modification of alluvial sediment: field evidence and laboratory experiments

P.A. BREWER *and* J. LEWIN

Institute of Earth Studies, University College of Wales, Aberystwyth SY23 3DB, UK

ABSTRACT

Downstream trends in bed sediment characteristics are analysed for two mid-Wales rivers, the Severn and Dyfi. Mean grain size and size distributions are compared for different sampling procedures on the Severn, and then for surface samples on both rivers. Size and skewness both decrease, but less clearly on the Dyfi, where there are pronounced 'jumps' below tributary inputs. Roundness values show little systematic change, and this is attributed especially to continuous input of more angular material along river reaches. Shape changes show some trends, more for the Dyfi than the Severn, and these are attributed to hydraulic sorting. Clast volume changes reflect those of shape.

The fact that field results indicate greater importance of hydraulic sorting than abrasion (though continued angular supply by bank erosion and tributary input are also important) is confirmed by laboratory experiments using a Kuenen-type abrasion tank and tumbling barrel. These do give different results, but neither reproduces the nature or extent of downstream changes in the field.

INTRODUCTION

It has been widely recognized that the physical characteristics of fluvial bed sediments in river channels alter systematically in a downstream direction. Sternberg (1875) was the first to quantify such a trend, fitting an exponential curve to the downstream decrease in size of Rhine bed sediments. Subsequent studies have established other trends for particle roundness (Wentworth, 1922; Krumbein, 1942; Goede, 1975), shape (Krumbein, 1941; Bradley, 1970), volume and various other derived indices (Sneed & Folk, 1958; McPherson, 1971).

The development of these trends has been attributed to two sets of processes acting within the channel, mechanical abrasion and hydraulic sorting. 'Abrasion' is a summary term covering a range of mechanical processes such as chipping, crushing, grinding and breakage, the combination of these processes leading to physical reduction of clast size through time and distance. Abrasion can operate in two forms: 'abrasion in place', where a clast is either vibrating within the bed or is stationary and being abraded by material impacting upon it (Schumm & Stevens, 1973); and 'progressive abrasion', where a clast is being abraded by virtue of its own downstream movement. The proportion of abrasion in place to progressive abrasion, and the degree to which each process is acting, is believed to be dependent on the characteristics of the clast, river bed, channel and fluid properties. Weathering of clasts whilst in temporary storage is also believed to be important in increasing their susceptibility to abrasion processes (Bradley, 1970).

'Hydraulic sorting' can be explained in terms of three groups of processes: selective entrainment, selective transport and selective deposition. A clast of a particular size, shape or specific gravity may be selectively entrained by a certain critical flow, and be incorporated into the mobile load of the channel. Other clasts may not be entrained by the same magnitude flow due to their greater size (Shields, 1936), their degree of exposure (Fenton & Abott, 1977), the nature of bed relief (Laronne & Carson, 1976) and bed packing (Church, 1978). These clasts will remain stationary until selectively entrained by a higher magnitude flow.

Once a clast is entrained, it may be selectively transported according to its size, density and shape (Russel, 1939). These clast characteristics and the channel hydraulic conditions combine to dictate whether particles are transported in suspension or as bed load, with sediment sorting possible within either transport mode (Rana et al., 1973). Selective deposition occurs as and where the competence of the flow reduces (Knighton, 1984). Larger clasts will generally be deposited first and thus create flow disturbances around them; deposition of finer sediments may then occur upstream or downstream of such clasts. Recent studies of mixed grain sediments and of pebble clustering have revealed how complex selective hydraulic factors are in natural environments (Laronne & Carson, 1976; Brayshaw, 1985; Komar, 1987).

The aims of this paper are twofold in nature: first, to examine whether any downstream sediment trends can be determined for relatively short (< 50 km) reaches on two Welsh gravel-bed rivers, the Severn and the Dyfi; and second, to evaluate to what extent hydraulic and mechanical processes in channels are responsible for producing such sediment trends. This evaluation is aided by isolating the abrasion process in the laboratory by use of a circular flume and tumbling barrel. The laboratory abrasion patterns and rates are then compared with the field data.

FIELD STUDY

Location

Table 1 gives comparative basin characteristics for the two study reaches. The 23 km Severn study reach comprises a small portion of the entire river, and lies between the relatively steep headwaters and the less steep alluvial reaches downstream. It is an unstable reach with braided and meandering river sections flowing within an alluvial valley ranging in width from 500 m at Llanidloes to some 1.5 km at Caersws. The potential sediment supply sources to this reach are from lateral bank erosion of thick Quaternary alluvial deposits and from two major tributaries which enter the reach at Caersws, the Trannon and Garno.

The Dyfi study reach, although only 8.5 km longer than the Severn reach, comprises over 80% of the length of the 'non-tidal' Afon Dyfi. The upper reaches, above Mallwyd, are dominated by degradation. The bedrock/boulder bed has a very narrow floodplain (< 10 m) with small terraces in places, and there is a shallow (10 m), 2 km long incised rock channel upstream of Dinas Mawddwy. In the lower reaches, the valley broadens to over 500 m at Machynlleth and the channel meanders over alluvial deposits. The potential sediment supply sources for this reach are the numerous tributaries flowing from the steep valley sides, and additionally, in the lower reaches below Mallwyd, from lateral bank erosion of alluvial deposits.

The underlying geology for both study reaches is Silurian turbiditic mudstones and grits of the Llandovery. Significant grain size and fabric variations can occur between different depositional units, and laterally within the same unit. However, since both rivers cut across many units and are reworking old floodplain deposits, it is assumed their sediment source material characteristics are statistically similar.

Sampling design

Standardization of sample location for such a study is vital, as significant variations can occur in clast parameters across channels at a section (Dawson, 1982) and laterally, vertically and longitudinally within a bar deposit (Ballantyne, 1978). The sampling of point bar features was undertaken in this study owing to their relative abundance on the two rivers and their ease of sampling compared to 'submerged channel' deposits. Sampled bars were spaced 2–3 km apart giving ten sites for the Severn and fifteen sites for the Dyfi. Sample bars were specifically chosen in the field upstream and downstream of major tributary inputs, to determine their influences on sediment trends for the mainstream channel.

The sample location on the bar was standardized by placing a 1 m^2 quadrat in a longitudinal position one-third of the way down the length of the bar from the bar head, and in a lateral position, one-third of the bar width from the water's edge at low water. The surface clasts were removed from this quadrat, this sample being termed the 'surface sample'. A 20 kg sample of the underlying deposits exposed by the removal of the surface sample was also taken, this sample being the 'bulk sample'.

Both samples were dry sieved at half phi intervals, -7 to -2.5 phi for the surface sample and -7 to 4 phi for the bulk sample. Previous investigations

Table 1. Basin characteristics for the Severn and Dyfi; data derived in part from NERC (1975)

Characteristic	River Severn (Abermule)	Afon Dyfi (Dovey Bridge)
Catchment area (km)	580	471
Mean stream length (km)	54.58	37.47
10–85% stream slope (m km^{-1})	3.60	5.22
Mean annual precipitation (mm)	1257	1863
Mean annual discharge (m^3 s^{-1})	11.69	20.31
Mean annual flood (m^3 s^{-1})	341.75	279.33
Max. flood (m^3 s^{-1})	616.90 (1960)	461.93 (1964)
Bedrock	Silurian grits and mudstones of the Llandovery	
Study reach range (from source) (km)	17.90–41.03	5.00–36.75
Study reach length (km)	23.13	31.75
Number of sample sites	10	15
Major tributaries (confluence distance from source, km)	Trannon (32.7)	Cywarch (12.8)
	Garno (33.0)	Cerist (13.8)
		Cleifion (16.9)
		Angell (20.9)
		Twymyn (27.4)
		Dulais (35.5)
		Dulais (36.0)

on downstream sediment trends have often selected a single size range for analysis, frequently the 16–32 mm fraction (Plumley, 1948; Hadley, 1960; Ouma, 1967). Therefore this size range was sampled for the surface and bulk samples to determine whether sediment trends could be determined for a single 'phi' size range. The following indices were calculated from measurements of the 'a', 'b' and 'c' axes, and from the roundness and weight of fifty individual clasts from each sample.
1 A measure of mean grain size ('b' axis).

2 Cailleux Roundness Index: $((R = 2r/b) * 1000)$ (Cailleux, 1947).
3 Zingg Shape Classification (b/a, c/b) (Zingg, 1935).
4 Volume $((\pi/6) * a.b.c)$ (Sneed & Folk, 1958).

For the River Severn both surface and bulk samples were taken; however, for the Dyfi only the surface sample was used, but this was additionally subsampled for analysis using the above indices for the fifty largest clasts per site.

SEDIMENT TREND RESULTS

Mean grain size

Figure 1 compares downstream mean particle size variations between four types of sample taken from the Severn. The whole surface sample shows a significant decline in mean grain size from -6.36 phi at site 1 to -4.64 phi at site 10 (Table 2: $R^2 = 79.8\%$). This decline in mean grain size is partly attributable to the downstream elimination of the coarser size fraction in the samples. Especially at the upper three sites, the size distribution is heavily weighted towards the coarse fraction. This is emphasized by a skewness value of 1.15 at site 3. However, as distance increases down the reach, the size distribution becomes more 'normal', skewness falling to 0.46 at site 10. However, there is also an increase in the percentage of fine material in the downstream bars. The < -4 phi fraction increases from 1% at site 1 to 35% at site 10.

This downstream decline in mean particle size and the change in form of the particle distribution can be explained in a number of ways. First, the gradual loss of the coarser fractions downstream may be due to a hydraulic sorting process, where the larger clasts are either not entrained, or if entrained, do not travel as far as the finer material. Second, it may be the result of severe abrasion physically reducing the clasts during transport downstream. The gradual emergence of the finer fractions also has several explanations. The lack of fine material in the upstream bars may be due to the sediment supply being largely of a coarser grade. The finer

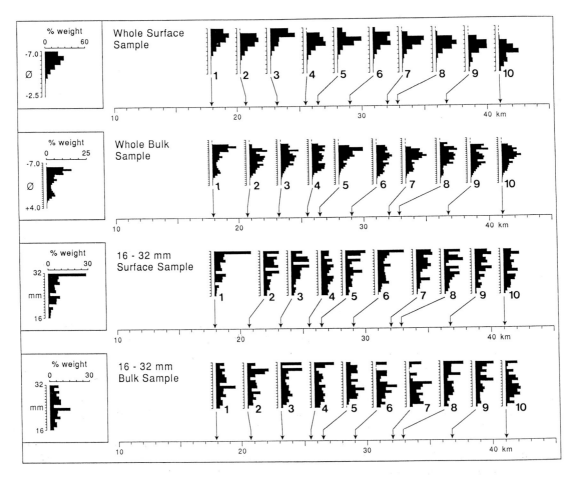

Fig. 1. Downstream mean sediment size changes ('b' axis) on the Severn: a comparison of sampling methods.

Table 2. Regression equations and correlations with distance downstream for mean particle size

Sample	Regression equation	r	R^2
Severn: whole surface	$y = -7.73 + 0.0678x$	0.894	79.8
Severn: whole bulk	$y = -3.44 + 0.0225x$	0.433	18.7
Severn: 16–32 mm surface	$y = 23.1 + 0.0089x$	0.174	3.0
Severn: 16–32 mm bulk	$y = 23.4 - 0.0185x$	-0.250	6.3
Dyfi: whole surface	$y = 6.92 + 0.0352x$	0.709	50.3
Dyfi: 16–32 mm surface	$y = 27.0 - 0.119x$	-0.730	53.4

gravel only begins to appear downstream either after the abrasion processes have produced finer gravels from larger clasts, or tributaries have supplied sediment of this grade. However, the upstream bars may have a deficit of finer gravels due to selective entrainment and transport processes being dominant in this reach, removing all the fine grade material. The gradual increase of finer gravels downstream is then attributable to the selective deposition of these size fractions in the lower reaches.

The subsurface bulk sample (Fig. 1) and the 16–32 mm fractions do not show significant decreases in mean grain size (Table 2: R^2 = 18.7, 3.0 and 6.3% respectively). However, the form of the bulk sample particle size distribution does change downstream, moving from a distinct bimodal distribution at site 1 to a nearly unimodal distribution at site 10. The only distinction to be made between the two 16–32 mm subsamples is that the 16–32 mm bulk sample has a greater percentage of finer clasts than the 16–32 mm surface sample. This apparent lack of finer gravels in the active surface layer may be due to the selective entrainment and transport of these sizes downstream, beyond the limits of the study reach.

Figure 2 compares mean particle size variations between the two study reaches. The whole surface sample for the Dyfi (R^2 = 50.3%) does not show the same significant downstream decrease in grain size as the Severn (R^2 = 79.8%). However, as in the Severn, the size distribution is initially highly skewed in the upstream sites (site 1 skewness = 1.30), but tends towards a more normal distribution in the lower sites (site 14 skewness = 0.44). Associated with this fall in skewness is a similar increase in the proportion of finer gravels in the downstream bars, as found on the Severn. The < -4 phi fraction increases from 0.06% at site 1 to 13.49% at site 15. The possible explanations for this change in form of sediment size distributions have already been outlined when discussing the Severn surface data.

The particle size distributions for the Dyfi surface sediments frequently show marked changes in form between two adjacent sites, upstream and downstream of a tributary input, as previously reported, for example, by Knighton (1982). Although the tributary sediments were not directly sampled, changes in sediment characteristics immediately below confluences have been used as an indicator of the nature of tributary sediment input (Church & Kellerhals, 1978; Nordin et al., 1980; Knighton, 1982). A coarsening of mean grain size of bar sediments downstream of tributaries is most likely the result of these channels providing sediment of a coarser grade than the mainstream. Such 'sediment size jumps' are clearly seen between sites 5 and 6 (Afon Cleifion), sites 8 and 9 (Afon Angell) and sites 10 and 11 (Afon Twymyn). Coarser sediment may be supplied by tributaries for a number of reasons. First, the supply sources for these tributary channels may be of a coarser grade than the main channel. Second, the tributary length may be shorter than the mainstream length at the point of confluence, to the extent that abrasion processes in the tributary have had less channel length than the mainstream in which to degrade the sediment.

The most significant 'sediment size jump' occurs between sites 6 and 7 (Afon Cleifion); the > -7 phi fraction rises from 13% at site 6 to 69% at site 7. However, by site 8, a further 4 km downstream, the > -7 phi fraction has fallen to 5%. Such a reduction in the percentage of coarse grades over such a short distance is unlikely to be solely the result of abrasion processes; a process of hydraulic sorting by size is a more likely explanation.

The mean stream gradient of the main Dyfi tributaries is 45 m km^{-1}, a high value in comparison to the relatively shallow mean tributary

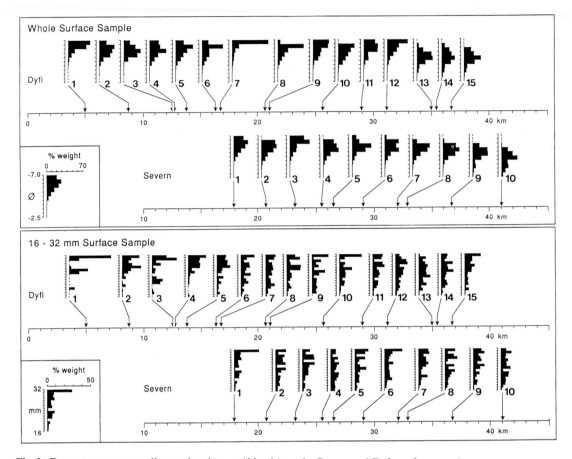

Fig. 2. Downstream mean sediment size changes ('b' axis) on the Severn and Dyfi: surface samples.

gradients of the Severn (15 m km^{-1}). The steep gradients and narrow valleys of these tributaries increase their competence to move coarse sediment during flood. These sediments do not appear to move a great distance downstream in the main channel, but instead go into storage within the channel and on gravel bars. Under flood conditions, the competence of the mainstream, however, appears generally to be lower than the tributaries, so that only the rarer floods in the mainstream entrain and transport the coarse tributary sediments.

Several tributaries of the Dyfi do not exhibit significant sediment size jumps, for example, the Cywarch, Cerist and Dulais. For the Severn also, no effect is seen as a result of the Trannon or Garno entering at Caersws. These tributaries must either have much lower sediment supply rates or are supplying sediment of a similar grade to that in the mainstream.

Roundness

The roundness of a clast at any location within a river channel is a result of the abrasion processes that have acted upon it in time and space. The roundness of a clast is increased by sand blasting, microchipping and granular removal by crushing or grinding, but reduced by larger scale chipping, cracking and subsequent fracturing. Lithology is known to be a major factor in determining abrasion rates in the laboratory (Kuenen, 1956) and field (Bradley, 1970). Features such as rock hardness, fabric and bedding structures influence initial roundness, the nature of granular removal and fracture planes.

Figures 3a–d illustrate the roundness trends for the Severn and Dyfi. It is clear from the histograms and Table 3 that there are neither any marked qualitative downchannel tendencies nor significant

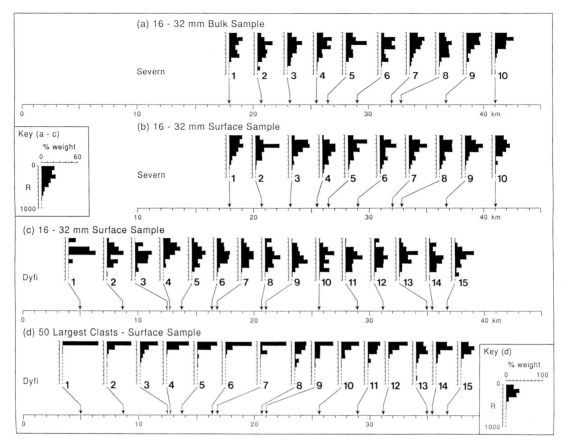

Fig. 3. Downstream mean roundness changes (Cailleux Index) on the Severn and Dyfi.

mean roundness trends for either study reach. The Severn samples show a negative correlation, indicating a decrease in roundness downstream, whereas the Dyfi samples conform to the commonly recognized increase in roundness downstream. The fact that no significant relationships exist for either study reach might suggest that the reaches are either high energy environments resulting in high breakage rates, or that there are simply no effective downstream modifications achieved over these distances, or that there are angular sediment inputs to the mainstream. It is believed that the last explanation is the more likely for the study reaches, whilst breakage may also be important.

Shape

Downstream changes in the form of particles may

Table 3. Regression equations and correlations with distance downstream for Cailleux Roundness Index

Sample	Regression equation	r	R^2
Severn: 16–32 mm bulk	$y = 441 - 4.17x$	−0.710	50.4
Severn: 16–32 mm surface	$y = 396 - 1.94x$	−0.325	10.6
Dyfi: 16–32 mm surface	$y = 340 + 3.69x$	0.609	37.1
Dyfi: 50 largest	$y = 88.1 + 5.80x$	0.769	59.1

be produced through hydraulic sorting by shape or mechanical abrasion. Sorting by shape operates under similar conditions to sorting by size, that is, clasts are selectively entrained, transported and deposited according to their shape. Clast abrasion by granular removal and chipping will remove sharp edges and slowly make the clast more spherical, that is, reduce the differences in length between the 'a', 'b' and 'c' axes. Clast abrasion by breakage, however, will rapidly change the clast shape, usually increasing the differences in length between axes, making them more bladed.

Figure 4 illustrates downstream shape changes according to Zingg's Shape Classification. The 16–32 mm Severn surface sample shows a downstream change in mean shape from 'bladed' clasts in the

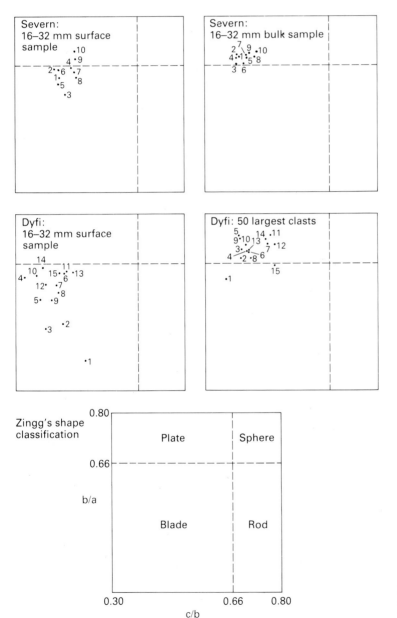

Fig. 4. Downstream mean shape changes (Zingg's Classification) on the Severn and Dyfi.

upper sites (1–8) to 'plates' in the lower sites (9 and 10). The 16–32 mm Severn bulk sample has a mean 'plate' shape for all ten sites, but the lower sites are slightly more spherical than the upstream sites. These small shape changes on the Severn reflect the relatively low energy environment of the reach.

The 16–32 mm surface sample for the Dyfi shows that the mean shape on all fifteen bars is bladed. However, there is a shape change trend similar to the Severn, that is, in the lower reaches of the Dyfi, the clasts become more 'platey' even though their mean shape remains bladed. The fifty largest clasts for the Dyfi surface sample are almost entirely 'plates'; however, the first and last sites are bladed. Setting these two sites aside, there is a tendency for the mean shape of these clasts to become more 'spherical plates', as with the 16–32 mm Severn bulk sample.

The magnitude of the shape changes on the Dyfi are greater than on the Severn. This may be a result of study reach length, the Dyfi being slightly longer and encompassing the majority of the river. However, if the reaches are made a similar length by removing the lowest four Dyfi sites, shape changes on the Dyfi still remain greater. This supports the notion that the dominant sediment supply to the Dyfi in the upper reaches is from high energy tributaries, supplying freshly broken, highly bladed gravels to the upper reaches, but that subsequent hydraulic sorting and further breakage processes produce the observed downstream shape changes.

Volume

Clast volume, as derived by Sneed & Folk (1958), gives an indication of the change in form of clasts downstream. If mean volume changes are additionally plotted with standard deviation data, an indication is also given as to downstream changes in the sorting of sediments. Downstream trends for volume with their standard deviations for the two reaches are shown in Fig. 5.

Within the 16–32 mm size range for the Severn surface and bulk sediments, there is no significant change in volume downstream, and there is only a slight fall in the standard deviation (sorting) of the surface sample. However, the bulk sample at every site has lower mean volume values than the surface sample, and this difference is related to shape; a disc (cf. bulk sample) of a given diameter will have a lower volume than a blade (cf. surface sample) of the same intermediate diameter.

Figure 5 also shows a rapid fall in volume for the 16–32 mm Dyfi surface sample. This again can be related to shape changes, from 'highly bladed' upstream to more 'platey blades' downstream. There is also a significant fall in the standard deviation for this fraction, that is, the sorting of the sediments improves downstream. The majority of this fall occurs over the first four sites; this rapid improvement in sorting is most likely the result of a hydraulic 'sorting-by-shape' process. Over the remainder of the study reach, there is little change in volume or sorting. Tributary inputs do not appear to influence these parameters for this size fraction.

The effect of tributary inputs on the volume and sorting of the fifty largest clasts is shown in Fig. 5. The three 'peaks' in the upper standard deviation curve correlate with the tributary inputs of the Cleifion, Angell and Twymyn. It is these three tributaries again that appear to have the greatest influence on the mainstream sediments. Because they are supplying sediment of a coarser grade, the size range of sediments in the mainstream is increased and the result is a fall in the sorting of these sediments (a standard deviation peak).

LABORATORY RESULTS

The field results obtained from the two study reaches suggest that abrasion processes are of limited importance in the two study reaches, though the picture is considerably obscured by sediment supply factors, and by the relative complexity of all the processes involved. However, in order to assess the potential relative importance of such processes, it is helpful to isolate and quantify them in the laboratory, as far as possible. In this study, the abrasion process has been simulated using a pump driven Kuenen-type abrasion tank (Kuenen, 1956), and a more conventional 34 cm diameter tumbling barrel, rotating at 54 r.p.m. Careful measurements were made of the weight losses and shape changes of twenty natural clasts, in runs performed using both pieces of equipment.

Figure 6a,b shows Zingg shape changes (Zingg, 1935) over 15 km (10 h) in the flume and 20 h (~ 70 km) in the tumbling barrel. Patterns of shape change are difficult to discern from these plots. There are however significant differences in the pattern of shape change produced by the two pieces of equipment. In the flume, nine clasts were initially 'platey': over the experimental period, four became

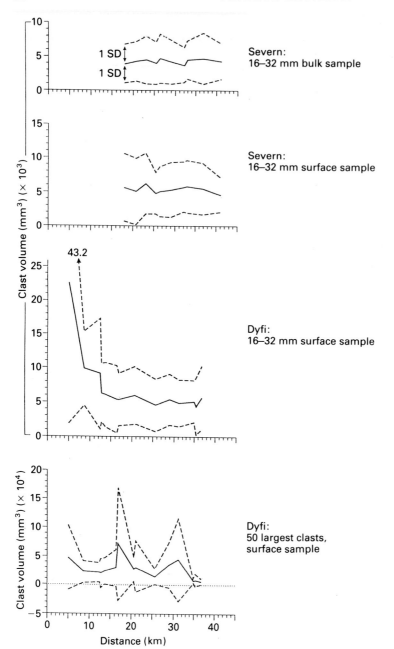

Fig. 5. Downstream volume (Sneed & Folk, 1958) with standard deviation on the Severn and Dyfi.

more spherical, two more rod-like, one more bladed and two remained relatively unchanged. In the barrel, however, out of the nine originally 'platey' clasts, six became more bladed, two more 'platey' and one remained relatively unchanged.

In general, abrasion in the flume leads to clasts becoming more 'platey' if originally bladed, and more spherical if originally 'platey'. In the barrel, however, abrasion tends to make originally 'platey' clasts more bladed, and bladed clasts remain relatively unchanged. The nature of the abrasion process is clearly different in the two pieces of equipment. Scanning electron micrograph analysis of microparticle wear has demonstrated that

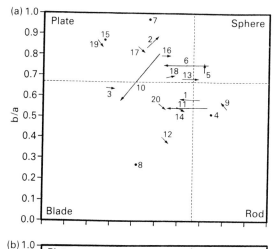

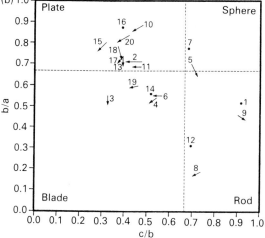

Fig. 6. Changes in shape (Zingg) of 20 natural clasts (a) after 15 km travel in the abrasion tank and (b) 20 h in the tumbling barrel. (●) No significant shape change; (→) magnitude and direction of shape change.

abrasion in the flume is the result of particle–bed impacts, the impact being either at a corner, edge or face. Therefore, granular removal will occur most rapidly in those areas where the number and force of impacts is highest. These areas tend to be corners primarily, then edges and faces, so through abrasion in the flume, clasts lose their corners and edges and become more 'platey' or spherical. Abrasion in the barrel is the result of particle–particle impacts. The abrasion of clast faces is maximized by such impacts, noticeable by the development of concave faces on cuboid experimental test clasts.

The large shape changes in the flume of clasts 1, 6, 10 and 11 are due to breakage during the run. Through breakage, all four clasts either became more 'platey' (6) or more bladed (1, 10 and 11). Breakage is clearly a most important factor in determining shape changes in both field and laboratory. We conclude that flume rather than barrel experiments result in more realistic simulation of natural trends in the field, but that overall, the rates of shape change through laboratory abrasion are considerably less than observed downstream field changes.

The natural clast weight losses, observed over these relatively short distances under laboratory conditions, are of a linear nature and not an exponential one (Sternberg, 1875). It may be that we are looking at only a relatively small (<40 km) downstream portion of a larger (>200 km) exponential trend, and therefore the weight losses we observe appear linear. However, considerable experimental work with both the tumbling barrel and abrasion tank indicate that weight loss can still be of a linear nature, even when varying clast characteristics and hydraulic conditions.

From a comparison of weight changes of these laboratory runs with the field results, it appears that changes in the field can be of a much greater magnitude. From the laboratory tests, rates of weight loss for the 16–32 mm size fraction range from 0.009 (barrel) to 0.201 g km^{-1} (tank). Weight losses from the similar size fraction in the field range from 0.133 g km^{-1} on the Severn to 0.775 g km^{-1} on the Dyfi.

Shape and weight analysis suggest consistently that laboratory simulation is unable to replicate the change rates and patterns observed in the field. We have already seen that such change rates are commonly quite low in any case. Large and angular material input and breakage, together with hydraulic sorting, appear to be more important in the field.

SUMMARY AND CONCLUSIONS

Field results obtained from the Severn and Dyfi have shown that their sediment trends are highly complex. Indications are that sediment supply is a significant controlling factor influencing sediment patterns for both reaches, but with sediment supply sources differing for each reach.

Sediment supply to the Severn reach is primarily from lateral bank erosion of extensive Quaternary floodplain and terrace deposits. The relatively low gradient, low energy tributaries do not appear to influence the mainstream sediment trends. In the Dyfi basin, by contrast, relatively high gradient, high energy tributaries, especially in the upper reaches, are a significant sediment supply source of coarse angular material. Clear 'sediment character jumps' occur downstream of several tributaries, the Cleifion, Angell and Twymyn. In the lower reaches of the Dyfi, additional sediment is supplied to the mainstream from the lateral erosion of Quaternary terrace and floodplain deposits.

Even though the sediment trends on both reaches have been affected and possibly obscured by the complexity of sediment supply processes, certain downstream patterns of mean particle size, roundness, shape and volume have been determined. The Severn surface sample shows the only statistically significant fall in mean particle size from both reaches. There is a fall in roundness downstream for both Severn samples, and this is attributed to the breakage of weathered clasts or supply of angular material from the eroded floodplain and terrace deposits. The surface clasts are more bladed than the bulk material, and downstream shape, volume and sorting variations are slight for both samples.

The sediment trends for mean particle size, roundness and volume are significantly influenced by anomalous sediment supplies from the upstream tributaries. These high energy channels input angular, coarse sediment which appears to go rapidly into storage within the channel and on bars. Shape changes on the Dyfi are greater than on the Severn, and this is believed to be partly a result of hydraulic sorting processes and breakage.

The mean particle size and shape trend results illustrate the differences in sediment characteristics that exist not only between the two study reaches, but also between different sampling methods employed on the same reach. It is clearly important when comparing results from different studies that care is taken to ensure that like is compared with like, that is, the same sampling methods and size fractions are compared.

The complexity of the sediment trends found on the Severn and Dyfi study reaches are to be expected in other mid-latitude gravel-bed rivers. We would expect to find sediment supply factors determining sediment trends in those channels which flow over and rework Quaternary deposits, and where high energy tributaries are significant sediment suppliers.

The results obtained from the field study have also suggested the important role of hydraulic sorting processes in producing downstream trends. From the simulated abrasion tests in the laboratory, it is clear that abrasion processes are unlikely to account for the majority of the decline in mean particle size and shape changes downstream. Hydraulic sorting by size and shape together with breakage, especially in the headwaters, seem to be responsible for the changes in sediment character downstream.

ACKNOWLEDGEMENT

We are grateful to the Natural Environment Research Council (UK) for provision of a research grant. This is Publication No. 150 of the Institute of Earth Studies, Aberystwyth.

REFERENCES

BALLANTYNE, C.K. (1978) Variations in the size of coarse clastic particles over the surface of a small sandur, Ellesmere Island, NWT, Canada. *Sedimentology* **25**, 141–147.

BRADLEY, W.C. (1970) Effect of weathering on abrasion of granitic gravel, Colorado River (Texas). *Bull. Geol. Soc. Am.* **81**, 61–80.

BRAYSHAW, A.C. (1985) Bed microtopography and entrainment thresholds in gravel-bed rivers. *Bull. Geol. Soc. Am.* **96**, 218–223.

CAILLEUX, A. (1947) L'indice d'emousse: definition et premiere application. *Comptes Rendus Sommaires de la Societe Geologique de France* **13**, 250–252.

CHURCH, M. (1978) Palaeohydrological reconstructions from a Holocene valley fill. In: *Fluvial Sedimentology* (Ed. Miall, A.D.) pp. 743–772. Can. Soc. Petrol. Geol. Memoir 5. Calgary.

CHURCH, M.A. & KELLERHALS, R. (1978) On the statistics of grain size variation along a gravel river. *Can. J. Earth Sci.* **15**, 1151–1160.

DAWSON, M.R. (1982) *Sediment variation in a braided reach of the Sunwapta River, Alberta.* MSc thesis (unpublished), University of Alberta, Alberta.

FENTON, J.D. & ABBOTT, J.E. (1977) Initial movement of grains on a stream bed: the effect of relative protrusion. *Proc. R. Soc. London* **352A**, 523–537.

GOEDE, A. (1975) Downstream changes in the pebble morphometry of the Tambo River, Eastern Victoria. *J. Sedim. Petrol.* **45**, 704–718.

HADLEY, R.F. (1960) Recent sedimentation and erosional history of Fivemile Creek, Fremont County, Wyoming. US Geol. Surv. Prof. Paper 352-A.

KNIGHTON, A.D. (1982) Longitudinal changes in the size and shape of stream bed material: evidence of variable transport conditions. *Catena* **9**, 25–34.

KNIGHTON, A.D. (1984) *Fluvial Forms and Processes.* Edward Arnold, London, 218 pp.

KOMAR, P.D. (1987) Selective grain entrainment by a current from a bed of mixed sizes: a reanalysis. *J. Sedim. Petrol.* **57**, 203–211.

KRUMBEIN, W.C. (1941) The effects of abrasion on the size, shape and roundness of rock fragments. *J. Geol.* **49**, 482–520.

KRUMBEIN, W.C. (1942) Flood deposits of Arroyo Seco, Los Angeles County, California. *Bull. Geol. Soc. Am.* **53**, 1355–1402.

KUENEN, P.H. (1956) Experimental abrasion of pebbles: 2. Rolling by current. *J. Geol.* **64**, 336–368.

LARONNE, J.B. & CARSON, M.A. (1976) Interrelationships between bed morphology and bed material transport for a small gravel bed channel. *Sedimentology* **23**, 67–85.

MCPHERSON, H.C. (1971) Downstream changes in sediment character in a high energy mountain stream channel. *Arctic Alpine Res.* **3**, 65–79.

NERC (1975) *Flood Studies Report. Vol. 4. Hydrological Data.* NERC, Whitefriars Press, London.

NORDIN, C.F., MEADE, R.H., CURTIS, W.F., BOSIO, N.J. & LANDIM, P.M.B. (1980) Size distribution of Amazon River bed sediment. *Nature* **286**, 52–53.

OUMA, J.P.B.M. (1967) Fluviatile morphogenesis of roundness: the Hacking River, New South Wales, Australia. *Int. Assoc. Sci. Hydrol.* **75**, 319–344.

PLUMLEY, W.J. (1948) Black Hills terrace gravels: a study in sediment transport. *J. Geol.* **56**, 526–577.

RANA, S.A., SIMONS, D.B. & MAHMOOD, K. (1973) Analysis of sediment sorting in alluvial channels. *J. Hydr. Div. ASCE* **99** (HY11), 1967–1980.

RUSSEL, R.D. (1939) Effects of transportation on sedimentary particles. In: *Recent Marine Sediments* (Ed. Trask, P.D.) pp. 32–47. Am. Ass. Petrol. Geol., Dover Publications, New York.

SCHUMM, S.A. & STEVENS, M.A. (1973) Abrasion in place: a mechanism for rounding and size reduction of coarse sediments in rivers. *Geology* **1**, 37–40.

SHIELDS, A. (1936) Anwendung der Ahnlichkeitsmechanik und der Turbulenenzforschung auf die Geschiebebewegung. *Mitteilung der Preussishen Versuchsanstalt fur Wasserbau und Schiffbau* **26**, 26 pp.

SNEED, E.D. & FOLK, R.L. (1958) Pebbles in the Lower Colorado River, Texas: a study in particle morphogenesis. *J. Geol.* **66**, 114–150.

STERNBERG, H. (1875) Untersuchungen uber Langen- und Querprofil geschiebefuhrender Flusse. *Zeitschrift fur Bauwesen* **25**, 483–506.

WENTWORTH, C.K. (1922) A field study of the shapes of river pebbles. *US Geol. Surv. Bull.* **730**, 103–114.

ZINGG, T. (1935) Beitrag zur Schotteranalyse. *Minerlogische und Petrographische Mitteilungen* **15**, 39–140.

Bed material and bedload movement in two ephemeral streams

M.A. HASSAN*

Department of Physical Geography, Institute of Earth Sciences, The Hebrew University of Jerusalem, Givat Ram, Jerusalem 91904, Israel

ABSTRACT

The characteristics of the stream bed material and the transport of individually tagged particles were examined in two ephemeral streams in the Negev and Judean deserts. The bed material shows a high spatial variability in texture between bedforms, between reaches and between the surface and the subsurface. Distance of movement and burial depth are gamma distributed. The study confirmed the lack of relation between the distance of movement and particle size. The movement of the tracer particles and the changes in the armouring ratio over time indicate that the armour was destroyed by large events and re-formed by small and medium events. Most of the tagged particles, after the first flow event, were found in bars. The vertical exchange rate shows that a high percentage of the particles become buried by events.

INTRODUCTION

Little is known about the characteristics and the behaviour of desert ephemeral streams. In contrast to perennial streams, ephemeral streams have dry beds between events and are hydrologically active for only a very short time. Thus, ephemeral streams offer a great opportunity to study the sedimentological and geomorphological characteristics of their beds. This is not the case in perennial streams where a large portion of the bed is permanently covered by flowing water.

In spite of the short duration and low frequency of desert stream events, volumes and sizes of the transported sediment are large. During a sediment mobilizing event, a portion of the exposed particles (stationary and mobile) becomes buried and a portion of the buried material becomes exposed on the surface (Schick *et al.*, 1987a,b). The bed structure, texture and morphology are determined by the interaction between the entrainment, transportation and depositional processes, which are modulated by bed material and channel characteristics.

Despite the relative simplicity of tracing individual particles in ephemeral streams, only a few research results have been reported. A study in a gravel bed stream started by Schick (1970) in the late 1960s is now the longest continuous tracing programme (Schick & Sharon, 1974; Schick *et al.*, 1987c). From the mid 1950s to the early 1960s Leopold *et al.* (1966) conducted a tracer programme in gravelly sand bed streams in New Mexico. These studies, as with many others conducted in perennial streams (Keller, 1971; Church, 1972; Laronne & Carson, 1976; Butler, 1977; Leopold & Emmett, 1981; Brayshaw *et al.*, 1983; Ashworth & Ferguson, 1986), confirmed the apparent lack of relation between distance of movement and particle size and highlighted instead the influence of the local sedimentological and geomorphological environment on the movement of individual particles.

In recent years armouring has become a major topic in fluvial sedimentology. 'Armour' is defined here as the relatively coarse surface layer, one particle diameter thick. There is disagreement between investigators about the way armouring develops and its influence on sediment movement.

*Present address: Department of Geography, University of British Columbia, Vancouver, British Columbia V6T 1Z2, Canada.

There are two approaches: the first is based on selective movement and winnowing of the fine sediment (Lane & Carlson, 1953; Gessler, 1970; Gomez, 1983, 1984; Sutherland, 1987) during which the surface becomes well packed and the armouring ratio (D_{50} of the surface/D_{50} of the subsurface) increases (Day, 1981); the second is based on the supposed equal mobility of bed material. Equal mobility is attained by sheltering the small particles and increasing the exposure of the large ones (Parker *et al.*, 1982a,b; Parker & Klingeman, 1982; Andrews & Erman, 1986; Andrews & Parker, 1987), hence the development of a coarse surface.

The objectives of this study are: first, to describe the characteristics of the bed material, including the existence of an armour layer, of two ephemeral streams; second, to examine the vertical exchange of tracer particles within the scoured layer; and third, to describe the erosion and deposition processes in an extreme event.

THE SITES

Studies were conducted at two ephemeral streams in the Negev and Judean deserts — Nahal Hebron and Nahal Og.

In Nahal Hebron, a site was located near Tel Shoqet, about 13 km north of Beer Sheva (Fig. 1A). The stream drains an elongated catchment of 250 km², predominantly underlain by limestone, within the southern subhumid parts of the Hebron Mountains. Based on a short record (1982–87), the average number of events is four and the mean annual flood is 11.5 m³ s⁻¹.

The study reach is 900 m long and 3–4 m wide and

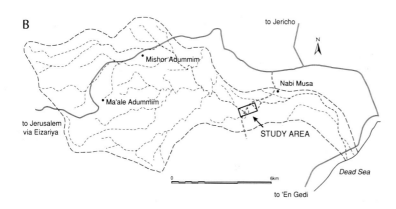

Fig. 1. Location and drainage network. (A) Nahal Hebron. (B) Nahal Og.

consists of two segments. The upper one is a 150 m long straight channel with an attenuated, barely discernible alternation of gravel bars and granular–sandy pools. The lower, a 750 m long meandering reach, has point and alternate bars with pools and riffles. The average slope of the two segments is 0.016.

In Nahal Og, a site was located near Nebi Musa, about 30 km east of Jerusalem (Fig. 1B). At this point the 5–12 m wide channel drains a 90 km² catchment in the eastern part of the Jerusalem Mountains, before crossing the Judean desert to the Dead Sea.

The 1000 m study reach consists of point and alternate bars, intrabar channels and a thalweg with small rocky waterfalls. The average slope is 0.014. Flow measurements were not made prior to 1986. Indirect estimates, based on inference from observations of flows and general models developed to calculate flows in the Negev desert (Ben-Zvi & Cohen, 1975), yield 10–15 m³ s⁻¹ for the mean annual flood and five for the mean number of events per year.

Twenty-one flow events occurred in Nahal Hebron over the 5 year study period (1982–87). The data for six of these, with a return period of 1 year or more, are summarized in Table 1. In Nahal Og, six flow events occurred during 1 year of the study period (1986–87), two of which were large enough to transport the marked particles.

METHODS

Methods used in the study include the following.
1 Tagged particles: 979 magnetically tagged particles (Hassan *et al.*, 1984) were used in Nahal Hebron and 252 in Nahal Og. The tagged particles were initially placed in lines on the surface across the channel.
2 Scour chains: six chain sections were installed along the study reach of Nahal Hebron. Each section consisted of four chains located across the channel. Three sections were installed in Nahal Og and each one consisted of six chains.
3 Erosion pins: ten sections of iron pins, 1 m long, were installed along the study reach of Nahal Hebron. Each section included four pins located across the channel. They served to determine the net change in the bed elevation.
4 Bed elevation cross-sections between events:

Table 1. Summary of the flow events, sediment movement and burial depth in the study sites

Event	Q_{max} (m³ s⁻¹)	Mean distance of movement (m)	Mean burial depth (cm)
Nahal Hebron			
8 Nov. 1982[a]	93.0	No data	No data
19 Jan. 1983	9.2	11.7	No data
23 Jan. 1983	33.0	62.6	19.7
17 Oct. 1984	18.0	13.9	16.8
1 April 1986	10.0	0.0[b]	0.0
8 Nov. 1986	49.8	65.2[c]	15.2
Nahal Og			
8 Nov. 1986	36.7	145.8	22.3
13 Dec. 1986	0.4	0.0[d]	0.0
20 Dec. 1986	0.8	0.0[d]	0.0
6 Jan. 1987	5.7	14.20[c]	15.2
9 Feb. 1987	1.0	[d,e]	0.0
14 Mar. 1987	0.7	0.0[d]	0.0

[a] Tracer particles were located after the event.
[b] No movement of any size fraction.
[c] The mean includes a high percentage of stationary particles.
[d] Fine sediment was transported by the event.
[e] Only three tracer particles were moved about 3 m each.

eight sections were measured recurrently in Nahal Hebron and four in Nahal Og.
5 Bed material samples, surface and subsurface: 1 m² of bed surface was painted and removed and then the immediately subjacent material (to a depth of one particle diameter) was removed. Sixty-three samples were taken from Nahal Hebron and twenty from Nahal Og.

BED MATERIAL

Nahal Hebron

Bed material varies between reaches along the stream channel and between bedforms. In the study reach size distributions were determined by bulk samples. Samples were taken from eleven separate bedforms and indicated a wide range of bed material sizes. Two typical samples are illustrated in Fig. 2A: the first (site 1) was taken from the straight upper reach and the second (site 4) from a pool located in the lower reach. In the first sample, the median size of the surface and subsurface layers is 68 mm and 38 mm, respectively. The ratio between the D_{50} of the surface and D_{50} of the

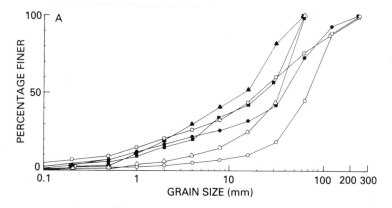

Fig. 2. Nahal Hebron. (A) Bed material size distribution. Site 4: (△) surface; (▲) subsurface; (■) 10–50 cm deep. Site 1: (○) surface; (●) subsurface; (□) 10–50 cm deep. (B) Bed surface.

subsurface is 1.79. In the layer 10–50 cm deep the median size is 20 mm (Fig. 2A). In the pool, the median size of the surface and the subsurface layers is 34 mm and 14 mm, respectively, but in the layer 10–50 cm deep D_{50} increased to 22 mm.

The eleven samples show a wide range of sediment size. These results led to the examination of the spatial changes in the bed along the stream channel. Such data are important to characterize the ephemeral stream bed, especially as our knowledge of this is limited. Areal samples were taken from a 5 km long reach located upstream of the study site. The reach that was chosen has no direct significant sediment contribution from tributaries. Forty-nine sites representing different bedforms were sampled and revealed similar results to those shown in Fig. 2A.

The median sizes of the surface and the subsurface layers ranged between 5 and 95 mm and between 5 and 60 mm, respectively (Fig. 3A). The armour ratio ranged from 0.78 to 5.3 while the mean and standard deviation were 2.9 and 1.8. Four samples, taken from a pool and bar tail located 3 km upstream of the study site, have finer surface material than subsurface material (Fig. 3A). Most of the samples taken from the riffles have ratios higher than 2 (Fig. 3A). The bed surface material of Nahal Hebron is well packed and the closed framework gravel (Church *et al.* 1987, their fig. 3.1a) is filled with granules, sand, small pebbles and silt (Fig. 2B). The packing and the coarse surface layer indicate that, generally, Nahal Hebron has a well armoured surface.

Three samples taken from the same site indicate

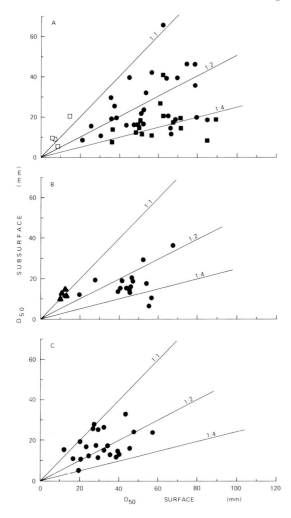

Fig. 3. Comparison of the subsurface D_{50} to D_{50} of the surface. (A) Nahal Hebron: (●) bar; (■) riffle; (□) pool. (B) Nahal Og: (●) bar; (▲) thalweg. (C) Fraser River: (●) bar (modified from Church *et al.*, 1987).

Nahal Og

Samples, similar to those of Nahal Hebron, were taken from various bedforms along the study reach. Figure 5A shows three of twenty samples taken, two of them (sites 10 and 3) from bars and one (site 17) from the thalweg. The median size of the surface layer at site 10 (Fig. 5A) is 56 mm, approximately seven times larger than that of the subsurface layer. Site 3 shows a similar trend, with an armour ratio of 2.36. In the thalweg, the median clast size ranged between 5.5 and 6.5 mm (site 17, Fig. 5A) and did not change with depth. In the bars, the median size of the surface and the subsurface clasts ranged between 5 and 70 mm and 5 and 27 mm, respectively (Fig. 3B). The armour ratio ranged between 1 and 7. Four samples all taken from the thalweg had a ratio of about 1. However, the mean ratio of the bar sites is 2.74 with a standard deviation of 1.93.

The armour ratio is similar to that of Nahal Hebron. Despite this similarity, the packing of the surface material of Nahal Og is poor (Fig. 5B), and large particles are fully supported by loose sandy matrix.

The bed material data from the two streams show a high degree of spatial variability in sediment size both between bedforms and between reaches. Similar variability has been measured in perennial streams (Bluck, 1982; Mosley & Tindale, 1985). The range of the armour ratios is similar to that reported by Church *et al.* (1987) from the Fraser River (Fig. 3C). It is important to emphasize here that the Fraser River is very much larger than the streams considered in this study. In addition, all of the Fraser River sites are barhead, which is the site most likely to be differentiated in the river. These facts may constrain the comparison between the desert streams and the Fraser River.

FLOW EVENTS AND DISTANCE OF MOVEMENT

The 8 November 1982 event of Nahal Hebron was the largest during the study period. Boulders of 0.5 m in diameter (painted as bench marks) were transported a few metres and a concrete ford ('Irish bridge'), 10 km downstream of the study site, was destroyed. The first group of tagged particles was emplaced after the 8 November 1982 event.

that the armouring ratio changed over the study period. The ratio before the 8 November 1982 event was 5.5 (Fig. 4). Two years later, after the 17 October 1984 event, it was only 2.6. The ratio subsequently increased to 3.3 (as measured before the 8 November 1986 event; Fig. 4). Though limited in scope, the data show a pattern that has already been described by Gomez (1983).

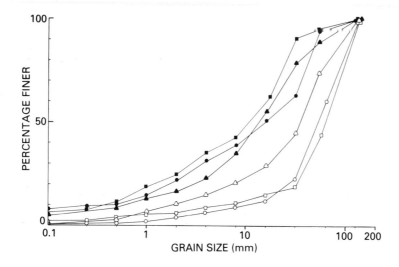

Fig. 4. Nahal Hebron. Size distribution of bed material over the study period. September 1982: (□) surface; (■) subsurface. November 1984: (△) surface; (▲) subsurface. September 1986: (○) surface; (●) subsurface.

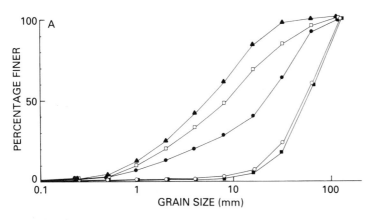

Fig. 5. Nahal Og. (A) Bed material size distribution. Site 17: (▲) surface. Site 3: (●) surface, (○) subsurface. Site 10: (■) surface; (□) subsurface. (B) Bed surface.

The 19 January 1983 event was small and resulted in the movement of most of the highly exposed tagged particles for short distances (Table 1). Sixty-three per cent of the particles were buried by the subsequent event of 23 January 1983.

Despite the relatively high magnitude of the 17 October 1984 event, the mean distance of clast movement was close to that of the 19 January 1983 event. However, mean burial depth was only 84% of that of the former event.

The 1 April 1986 event was similar in magnitude to the 19 January 1983 event but no movement was recorded in the study site.

The distance of movement per flow event varied significantly between events, without any apparent relation between particle size, roundness, shape or sedimentological environment of the particle before or after the event (Hassan, in press). These results concur with Einstein's (1937) laboratory experiments and with the field studies of Leopold et al. (1966) and others. Gamma functions accurately describe the distribution of the distance of movement. Two types of gamma curve were obtained: exponential for small events and full gamma curves for large events (Hassan et al., 1991).

Location maps (Hassan, 1990) of the tracer particles after each event illustrate the effect of bedforms (e.g. point and alternate bars, riffles) on the sediment movement. A high percentage of the moving sediment was trapped on bars and in pools, and a low percentage in riffles.

VERTICAL EXCHANGE

The burial depth of the tracer particles was found to vary with the peak discharge (Hassan, 1990) and with other attributes of the flow event, without any evident relation to the physical characteristics of each particle (size, shape, etc.). The gamma function was also found to fit the frequency distribution of burial depth (Hassan, 1988). Accordingly, the bedforms combine with sedimentological and hydraulic conditions to control the distance of clast movement and burial depth with, at best, a poor relation to the particle parameters (Hassan, 1988).

Of particles located on the surface before an event, afterwards, some were found buried by sediment within the scour layer and others remained exposed on the surface (Hassan, 1983, 1988; Schick et al., 1987a,b). The next event would expose some of the buried particles and bury a portion of the exposed ones.

Of 282 tagged particles (black generation: for details see Hassan, 1990), after the 19 January 1983 event, 66% remained exposed on the surface and 34% were buried (Fig. 6). The proportion of buried particles after the 23 January 1983 event was twice that of the 19 January 1983 event. The 17 October 1984 event slightly increased the percentage of the buried particles, from 63% to 65%.

Despite the magnitude of the 8 November 1986 event (Table 1), only a slight increase in the proportion of the buried particles was recorded, to 72% (Fig. 6). In this event 9.9% of the buried particles were exposed and 17.2% of the exposed ones buried.

The rate of burial in Nahal Og was higher than that of Nahal Hebron. The first event (8 November 1986) in Nahal Og after the injection of the marked particles was large, and this influenced the results of the next events: 87% of the particles became buried after the 8 November 1986 event.

Based on four events from Nahal Hebron and two events from Nahal Og, Fig. 7 shows that both the rate of exposure and burial generally increased with peak discharge. However, in spite of the high discharge of the 8 November 1986 event of Nahal Hebron, the burial and exposure rates were lower than those of the 23 January 1983 event and similar to the 17 October 1984 event values. Assuming a linear relation based on previous events (Fig. 7), the rates of exposure and burial, by the 8 November 1986 event, should be 40% and 80%, respectively. More data are needed to clarify this relation.

A comparison of the vertical exchange of the two streams shows that Nahal Og burial rates are higher than those of Nahal Hebron, probably for three reasons. First, based on field measurements, sediment transport rates in Nahal Og are higher than in Nahal Hebron, which allows high mixing rates; second, the first event in Nahal Og was large and a high percentage of the material became buried by the event; and third, the poorly developed armour layer in Nahal Og relative to Nahal Hebron promotes rapid scour and vertical exchange.

Schick et al. (1987a) developed a numerical model describing the rates of vertical exchange under uniform flows, and the proportion of pebbles buried by an event in relation to the total number of surficial pebbles was assumed to be constant. The model provides the possibility of calculating the number of events required to achieve equilibrium in vertical exchange. Seven events similar to the 23 January 1983 event of Nahal Hebron are needed to

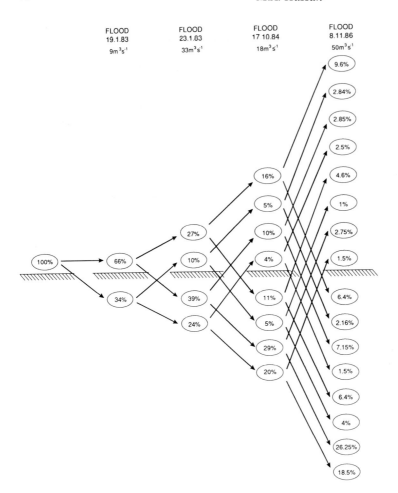

Fig. 6. Vertical exchange of tracer particles within the scour layer as a result of flow events in Nahal Hebron (282 total of injected particles).

achieve this equilibrium (Table 2). The return period of such an event is once in 2 years. Thus, 14 years are required to arrive at equilibrium or to natural conditions in the mixing of the tracer particles. Events of lesser magnitude require longer periods, up to 90 years (Table 2). The rates of vertical exchange depend on the sequence of events and on whether the first event is large or small.

THE 8 NOVEMBER 1986 EVENT IN NAHAL HEBRON

The 8 November 1986 event of Nahal Hebron was exceptional. Heavy, continuous rain in the Negev desert caused large floods in the area. Nahal Hebron flowed for several days and four peaks were recorded. The rising limb of each peak was very steep and within 10–15 min the flow reached a depth of 2–2.5 m.

The event deposited sediment along the entire reach. Data from the scour chains at the study site indicated that the event was mostly depositional with limited and very minor scour. A layer ranging between 0.0 and 50 cm in thickness, 17.5 cm on average, was deposited. Most of the sediment was deposited in pools and on bars. No sediment was found on riffles. Based on chains, erosion pins and stationary tagged particles (not moved by the event), a map showing the changes in the thickness of the deposited sediment was prepared (Fig. 8). Maximum depositional thickness was recorded near stations 40–45 m and 720–780 m (Fig. 8). Inspection of the channel upstream and downstream of the study site revealed that the deposited sediment in the study reach was the end of a long

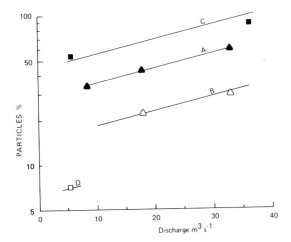

Fig. 7. Rate of vertical exchange of coarse particles within the scour layer as a function of peak discharge. Hebron, 8 November 1986. Nahal Hebron: A, percentage of burial; B, percentage of exposure. Nahal Og: C, percentage of burial; D, percentage of exposure.

lobe that started about 5 km upstream. Before and during the event tributaries deposited small alluvial fans in the main stream channel.

The median size of the deposited material changed along the study reach. Near stations 3 m and 100 m the median sizes were 74 mm and 24 mm, respectively. At station 180 m the median size was about 6 mm, after which there was no further change downstream.

Most of the tagged particles that moved were picked up from exposed positions. A small percentage of buried particles, most of them located before the event in the bend near station 180 m, were moved. Scouring was recorded in only a limited area (about 20 m^2) in this bend. The exposure rate was low as a result of the depositional phenomena of the event. On the other hand, the rate of burial was low mainly for two reasons: first, most of the exposed particles were concentrated in the riffles where no scouring or filling was recorded over the study period; and second, about two-thirds of the marked particles were buried by former events.

INTERPRETATION

Despite the magnitude of the 8 November 1986 event on Nahal Hebron, the mean burial depth and the vertical exchange rates were similar to those of the 17 October 1984 event. Likewise, the mean distance of movement was similar to that of the 23 January 1983 event. In addition, the event deposited material along the study reach with no scouring. This behaviour raises a number of questions: first, why was no sediment deposited in the riffles; second, why did the event only deposit material without antecedent scour; and third, how do these results influence the armour layer and its development on the stream bed?

In Nahal Hebron the riffle reaches are formed by the concentration of boulder and cobble size material, located between pools. Figure 8 illustrates that almost no sediment was deposited on or scoured from these riffles. Further, most of the tagged particles in these bedforms were found exposed on the surface. Erosion pins, painted boulders and large cobbles indicated that the activity in the riffles is restricted to throughput of material coming from upstream, while the underlying sediment remains essentially static. Accordingly, during such an event the riffles, which in Nahal Hebron account for about 30% of the study reach area (Fig. 8), remain almost unchanged. In contrast to the riffles, scouring and burial processes in the pools were detected even during small events (e.g. the 19 October 1983 event). A similar phenomenon was described by Meade (1985) in the East Fork River, Wyoming. In this river the sediment is stored in the pool areas separated by riffles. He reported that when the discharge increases, the sand size material scoured from pools moves across the gravel riffles into the next pool downstream.

The bed surface layer of Nahal Hebron, before the 8 November 1986 event, was well armoured (Fig. 4). High flows are needed to 'destroy' the

Table 2. Summary of number of events and number of years needed for equilibrium in vertical exchange

Event	Return period (years)	No. of events needed for equilibrium	No. of years needed for equilibrium
Nahal Hebron			
19 Jan. 1983	1	22[a]	22
23 Jan. 1983	2	7	14
17 Oct. 1984	1.5	15	23
8 Nov. 1986	6	15	90
Nahal Og			
6 Jan. 1987	0.5[b]	18	9

[a] Calculated based on Fig. 7, assuming linear relation.
[b] Estimated.

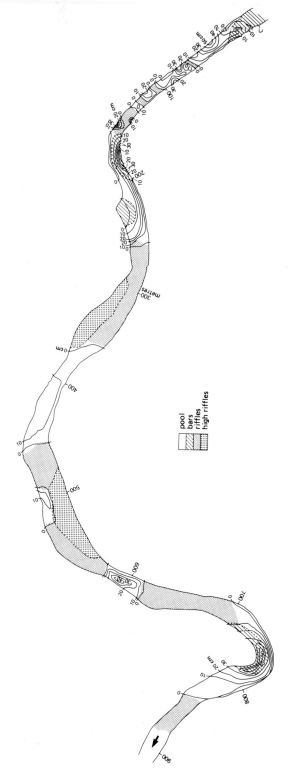

Fig. 8. Nahal Hebron. Bed map showing contours of deposition depths after the 8 November 1986 event. Please note that the width distances across the stream are exaggerated by four times in comparison with the scale along the channel (distances along the thalweg are marked).

armouring and to transport sediment from the subsurface as well as from the surface layer. The 8 November 1986 event was large enough to scour and transport material from the bed. The tracer particle data show that this process started near station 180 m. It appears that, before destroying the surface layer over a large area, a very large supply of material arrived from upstream. Deposition in response to this supply inhibited the scouring process. It seems that the incoming sediment consumed a high percentage of the power available for sediment movement. Therefore, the excess power was lower than the critical value needed to continue destroying the armouring. There is no direct evidence to prove this hypothesis, but inspection of the tributaries upstream of the study site shows that, in contrast to previous flows, in this event most of these tributaries transported sediment to the main stream channel. During the event several alluvial fans were deposited in the main stream channel; these fans blocked it when the flow was shallow, and then were remobilized with the increasing discharge in the main stream channel. Such activity covering all of the catchment area is rare in desert streams (Schick & Lekach, 1987). This contribution of sediment caused a net positive change in the sediment budget in the main stream channel. Therefore, the model of sediment transport process that was based mainly on local activity of the main stream channel had to be adjusted, in this event, to accommodate a contribution from the tributaries. In humid areas the fluvial system is generally adjusted to frequent contribution of sediment and water from the tributaries.

Bed samples taken over the study period may help to explain the development of armouring in this stream. Figure 4 shows that the armour ratio changed over the study period. Comparing two similar events, those of 19 January 1983 and 1 April 1986 (Table 1), indicates that the former, which occurred immediately after the 8 November 1982 event, transported sediment but the latter did not. Figure 4 shows that the bed surface before the 8 November 1982 event was well armoured. In this event, boulders 0.5 m in diameter were transported for a few metres. The magnitude of the event (Table 1) and its competence indicate that the armouring was destroyed. Thus, even small subsequent events were able to transport sediment as a result of the absence of an armour layer. This was the case in the 19 January 1983 and 23 January 1983 events. Over time, with selective movement (Gessler, 1970; Gomez 1983; Church, 1985) and the arrangement of the surface layer (Day, 1981; Raudkivi & Ettema, 1982) the armouring ratio increases. Unfortunately, we did not taken samples from the bed material after the 8 November 1982 event. However, the low ratio of armouring after the 17 October 1984 event (relative to the ratio before the 8 November 1982 event) supports the statements that the ratio after the 8 November 1982 event was low. On the other hand, in spite of the fact that the magnitude of the 17 October 1984 event was twice that of the 19 October 1983 event, the mean distance of movement of the two events was similar (Table 1). This is consistent with the 1 April 1986 event which did not move sediment. The armour ratio after that event increased without any real movement (Table 1 and Fig. 4). Accordingly, the combination of high armour ratio with unusual activity in the upstream tributaries caused the depositional phenomena that were observed in the 8 November 1986 event. It is reasonable to anticipate that activity in the bed in future competence events will be concentrated on new, poorly armoured deposits, at least during the initial events of the forthcoming series.

SUMMARY AND CONCLUSIONS

The characteristics of the stream bed material and the movement of individual tagged particles were examined in Nahal Hebron and Nahal Og, two ephemeral streams in the Negev and Judean deserts. In the two streams the bed material indicates a high variability in size between reaches, between bedforms and between bed surface and subsurface. In Nahal Hebron, the ratio between the median size of the surface layer and that of the subsurface ranged from 0.78 to 5.3 with a mean of 2.9. The samples with finer surface material were taken from pools or bar tail.

In Nahal Og, the ratio of armouring in the thalweg was about 1. In contrast, the mean ratio in the bars was 2.74.

The distance of particle movement varied between events and particles with very weak relation to particle size, shape, roundness or sedimentological situation of the particle before or after the event. The burial depth of particles also varied without any evident relation to the physical parameters of the particle. In both streams the distance of movement and burial depth are gamma distributed.

Flow events change the particle location relative to the bed surface. Some of the exposed particles become buried at the end of a competence event and a portion of the buried ones become exposed. Data from both streams show that a high percentage of the material changes position (i.e. becomes buried or exposed) from one event to another. Application of a numerical model that was developed by Schick et al. (1987a) to describe the rates of vertical exchange indicates that a tracer programme must continue for many years to arrive at equilibrium or to natural conditions in the mixing of marked particles.

The 8 November 1986 event of Nahal Hebron was exceptional due to input of sediment from tributary streams. Despite its magnitude it was a depositional event; 230 m^3 of material was deposited along the study reach. In this event, in contrast to previous small events, most of the tributary streams transported sediment to the main stream channel. During the event several alluvial fans were deposited in the main stream channel; these fans blocked it when the flow was shallow, and then were remobilized with the increasing discharge in the main stream channel.

Sampling bed material from Nahal Hebron indicates that the bed has become progressively armoured over the study period. The armouring was destroyed by large events and re-formed by small and medium events.

ACKNOWLEDGEMENTS

This work is based on a PhD thesis carried out by the author under the guidance of Professor Asher P. Schick at the Hebrew University of Jerusalem. Thanks are due to Professor A.P. Schick of the Hebrew University of Jerusalem and Professor M. Church, Professor O. Slaymaker and P. Jordon of the University of British Columbia for reviewing the first draft and Ian Reid and John Pitlick for their constructive comments; T. Grodek, S. Berkowicz, E. Hann and S. Sharoni assisted with the field work.

REFERENCES

Andrews, E.D. & Erman, D.C. (1986) Persistence in the size distribution of surficial bed material during an extreme snowmelt flood. *Water Resources Res.* **22**, 191–197.

Andrews, E.D. & Parker, G. (1987) The coarse surface layer as a response to gravel mobility. In: *Sediment Transport in Gravel-bed Rivers* (Eds Thorne, C.R., Bathurst, J.C. & Hey, R.D.) pp. 269–325. Proc. Pingree Park Workshop, 12–17 August 1985. J. Wiley & Sons, Chichester.

Ashworth, P.J. & Ferguson, R.I. (1986) Interrelationships of channel processes, changes and sediment in a proglacial braided river. *Geogr. Ann.* **68A**, 361–371.

Ben-Zvi, A. & Cohen, O. (1975) Frequency and magnitude of flows in the Negev. *Catena* **2**, 193–199.

Bluck, B.J. (1982) Texture of gravel bars in braided streams. In: *Gravel-bed Rivers* (Eds Hey, R.D., Bathurst, J.C. & Thorne, C.R.) pp. 339–355. John Wiley & Sons, New York.

Brayshaw, A.C. Frostick L.E. & Reid, I. (1983) The hydrodynamics of particle clusters and sediment entrainment in coarse alluvial channels. *Sedimentology* **30**, 137–143.

Butler, R. (1977) Movement of cobbles in a gravel bed stream during a flood season. *Bull. Geol. Soc. Am.* **88**, 1072–1084.

Church, M. (1972) Baffin Island sandurs: a study of the Arctic fluvial processes. Geol. Surv. Can. Bull. 216.

Church, M. (1985) Bedload in gravel-bed rivers: observed phenomena and implications for computation. In: *Ann. Con. Can. Soc. Civil Eng.*, pp. 17–37. SK, Saskatoon.

Church, M., McLean, D.G. & Wolcott, J.F. (1987) River bed gravels: sampling and analysis. In: *Sediment Transport in Gravel-bed Rivers* (Eds Thorne, C.R. Bathurst, J.C. & Hey, R.D.) pp. 43–88. Proc. Pingree Park Workshop, 12–17 August 1985. John Wiley & Sons, Chichester.

Day, T.J. (1981) An experimental study of armouring and hydraulic properties of coarse bed material channels. In: *Erosion and Sediment Transport in Pacific Rim Steeplands*, pp. 236–251. Int. Ass. Hydrol. Sci. Publ. 132, Christchurch.

Einstein, H.A. (1937) Bedload transport as a probability problem. In: *Sedimentation* (Ed. Shen, H.W.) App. C, Colorado State University, Colorado.

Gessler, J. (1970) Self stabilizing tendencies of alluvial channels. *J. Waterways and Harbors Div., ASCE* **96** (WW2), 235–249.

Gomez, B. (1983) Temporal variations in the particle size distribution of surficial bed material: the effect of progressive bed armouring. *Geogr. Ann.* **65A**, 183–192.

Gomez, B. (1984) Typology and segregated (armoured/paved) surfaces: some comments. *Earth Surf. Proc. Landf.* **9**, 19–24.

Hassan, M.A. (1983) *Transport and dispersion of coarse bed material, Nahal Hebron*. MSc thesis, Dept of Physical Geography, The Hebrew University of Jerusalem (in Hebrew).

Hassan, M.A. (1988) *The movement of bedload particles in a gravel stream and its relation to the transport mechanism of the scour layer*. PhD thesis, The Hebrew University of Jerusalem (in Hebrew).

Hassan, M.A. (1990) Scour, fill, and burial depth of coarse material in gravel bed streams. *Earth Surf. Proc. Landf.* **15**, 341–356.

Hassan, M.A. (in press) Structural controls of the mobility of coarse material in gravel bed rivers. *Israel J. of Earth Science*.

Hassan, M.A., Church, M. & Schick, A.P. (1991) Distance of movement of coarse particles in gravel bed streams. *Water Resources Res.* **27**, 503–511.

HASSAN, M.A., SCHICK, A.P. & LARONNE, J.B. (1984) The recovery of flood-dispersed coarse sediment particles, a three dimensional magnetic tracing method. In: *Channel Processes — Water, Sediment and Catchment Controls* (Ed. Schick, A.P.) pp. 153–162. *Catena Supplement* **5**. Catena Verlag, Germany.

KELLER, E.A. (1971) Bed movement experiments, Dry Creek, California. *J. Sedim. Petrol.* **40**, 1339–1344.

LANE, E.W. & CARLSON, E.J. (1953) *Some Factors Affecting the Stability of Canals Constructed in Coarse Granular Materials*. Int. Ass. Hydrol. Sci., Minneapolis

LARONNE, J.B. & CARSON, M.A. (1976) Interrelationship between bed morphology and bed material transport for a small gravel bed channel. *Sedimentology* **23**, 67–85.

LEOPOLD, L.B. & EMMETT, W.W. (1981) Some observations on the movement of cobbles on stream bed. In: *Erosion and Sediment Transport Measurements, Proc. Florence Symposium*. Int. Ass. Hydrol. Sci., pp. 45–59.

LEOPOLD, L.B., EMMETT, W.W. & MYRICK, R.M. (1966) Channel and hillslope processes in a semi-arid area, New Mexico. US Geol. Surv. Prof. Paper 352G, pp. 193–253.

MEADE, R.H. (1985) Wavelike movement of bedload sediment East Fork River, Wyoming. *Environ. Geol. Water Sci.* **7**, 215–225.

MOSLEY, P.M. & TINDALE, D.S. (1985) Sediment varability and bed material sampling in gravel bed rivers. *Earth Surf. Proc. Landf.* **10**, 465–482.

PARKER, G. & KLINGEMAN, P.C. (1982) On why gravel bed streams are paved. *Water Resources Res.* **18**, 1409–1423.

PARKER, G., DHAMOTHARAN, S. & STEFAN, H. (1982a) Model experiments on mobile paved gravel bed stream. *Water Resources Res.* **18**, 1395–1408.

PARKER, G., KLINGEMAN, P.C. & MCLEAN, D.G. (1982b) Bedload and size distribution in paved gravel-bed streams. *J. Hydr. Div., ASCE.* **108**, 544–571.

RAUDKIVI, A.H. & ETTEMA, R. (1982) Stability of armour layer in rivers. *J. Hydr. Div., ASCE* **108**, 1047–1057.

SCHICK, A.P. (1970) Desert floods – interim results of observations in Nahal Yael research watershed, 1965–1970. *Int. Assoc. Sci. Hydrol.* **96**, 478–493.

SCHICK, A.P. & LEKACH, J. (1987) A high magnitude flood in the Sinal Desert. In: *Catastrophic Flooding* (Eds Mayer, L. & Nash, D.) pp. 381–410. Binghamton Symposia in Geomorphology, Int. Series 18. Allen and Unwin, Boston.

SCHICK, A.P., HASSAN, M.A. & LEKACH, J. (1987a) A vertical exchange model for coarse bedload movement: numerical consideration. In: *Geomorphological Models — Theoretical and Empirical Aspects* (Ed. Ahnert, F.) pp. 73–83. *Catena Supplement* **10**. Catena Verlag, Germany.

SCHICK, A.P. LEKACH, J. & HASSAN, M.A. (1987b) Vertical exchange of coarse bedload in desert streams. In: *Desert Sediments: Ancient and Modern* (Eds Frostick, L.E. & Reid, I.) pp. 7–16. Geol. Soc. Lond. Spec. Publ. 35, Blackwell Scientific Publications, London.

SCHICK, A.P., LEKACH, J. & HASSAN, M.A. (1987c) Bedload transport in desert floods: observations in the Negave. In: *Sediment Transport in Gravel-bed Rivers* (Eds Thorne, C.R., Bathurst, J.C. & Hey, R.D.), pp. 617–641. Proc. Pingree Park Workshop, 12–17 August 1985. John Wiley & Sons, Chichester.

SCHICK, A.P. & SHARON, D. (1974) *Geomorphology and Climatology of an Arid Watershed*. Department of Geography, The Hebrew University of Jerusalem, Jerusalem, 161 pp.

SUTHERLAND, A.J. (1987) Static armour layers by selective erosion. In: *Sediment Transport in Gravel-bed Rivers* (Eds Thorne, C.R., Bayhurst, J.C. & Hey, R.D.) pp. 243–267. Proc. Pingree Park Workshop, 12–17 August 1985. John Wiley & Sons, Chichester.

Bedform migration and related sediment transport in a meander bend

J. KISLING-MØLLER

Institute of Geography, University of Copenhagen, 1350, Copenhagen K, Denmark

ABSTRACT

This case study deals with bedform transport and the dynamics of bedforms in an alluvial, sand bedded meander bend in the river Gels Å, western Jutland, Denmark.

The sediment transport in meander bends is found to be greatly influenced by the existence of bedforms. If the dune crests are oblique to the near-bed flow direction, a net transverse current is created in the separation cell in front of the dune. This process is of special importance to the direction and degree of material transport through meander bends. The use of dune orthogonals is presented to describe and quantify the transverse sediment transport in the leeside troughs in front of migrating dunes, and the method is tested with field data. On the basis of repeated surveys of the bed configuration the migration pattern, the rate of the transverse sediment transport and its variation through the meander bend are investigated.

INTRODUCTION

The secondary circulation is important for flow in meanders. The flow pattern is a result of the combined effect of the main downstream direction and the secondary circulation. The water will pass the bend in a corkscrew-like movement, and the near-bed flow direction will be up the point bar toward the convex bank. This has been treated by several authors (Rozowskii, 1961; Engelund, 1974; Hooke, 1974; Zimmerman & Kennedy, 1978; Bathurst et al., 1979; Bridge & Jarvis, 1982; de Vriend & Geldof, 1983; Dietrich & Smith, 1983; Alphen et al., 1984; Smith & McLean, 1984; Parker & Andrews, 1985; Thompson, 1986; Kisling-Møller, 1988)

The bedload transport is greatly influenced by this flow pattern because the sediment is swept up the point bar slope from the thalweg. It explains the processes of lateral transport in meander bends by which the locus of maximum longitudinal sediment transport shifts through the bend from one side to the other (Allen, 1970; Bridge & Jarvis, 1976; Bridge, 1977; Dietrich et al., 1979; Knudsen, 1981; Odgaard, 1981; Dietrich, 1982; Dietrich & Smith, 1984).

To obtain a more complete understanding of the transport processes in meanders it is, however, necessary to deal with bedforms and the sediment transport associated with them.

Bedforms in meanders have been the subject of many studies (e.g. McGowen & Garner, 1970; Jackson, 1975; Bridge & Jarvis, 1982; Bridge & Diemer, 1983; Bridge et al., 1986). But actual detailed observations of bedforms and bedform migration in meander bends have only been carried out by Dietrich and his co-workers (op. cit.). They point out how important bedforms are for understanding the processes of sediment transport in meanders. They give excellent sketches of bedforms in a bend, but no actual mapping of bedforms is carried out.

Where the dune crest is oblique to the mean flow direction the trough-wise current may produce sediment transport. In the trough a separation cell is created which sets the water particles in continuous circulation (e.g. Raudkivi, 1963; Allen, 1969, 1982; Paola et al.; 1986) (see Fig. 1). The reattachment point does not have a stationary position, but shifts backward and forward in a broad 'reattachment zone' in the trough. The intense turbulence in this

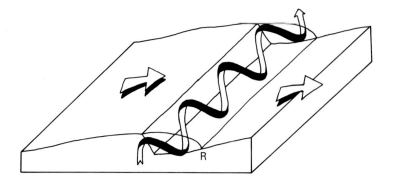

Fig. 1. Development of transverse current in the trough in front of a migrating dune. If the near-bottom current is oblique to the dune crest, the rotational flow pattern in the separation cell results in a net transverse current in the trough. 'R' denotes the location of the reattachment point.

zone allows a substantial sediment transport, and even migrating secondary bedforms, in the trough even if only a weak, resulting current is present.

In meander bends a transverse flow in the trough is very strong, both because of the secondary circulation pattern and because the dunes are not perpendicular to the near-bed flow direction. This produces a very intense lateral sediment transport.

The object of this paper is to investigate this lateral transport and to determine the rate and longitudinal distribution of the lateral transport in the meander bend.

This includes: (i) presentation of the use of dune orthogonals for describing the bedform-induced bedload transport in rivers in general and in meander bends in particular; (ii) mapping dune forms in a meander bend — their size and migration velocity and direction — in order to confirm the capability of dune orthogonals to quantify transverse bedload transport in dune troughs.

DUNE ORTHOGONALS

A migrating dune with cross-sectional area (A) and migration velocity (c) represents a sediment transport rate per metre width (q_b) equal to (after Simons et al., 1965):

$$q_b = \sigma \cdot A \cdot c \qquad (1)$$

where σ is the bulk density of the bed sediment, which is usually about 1700 kg m^{-3} for Danish alluvial rivers.

This relationship applies only under the assumption that the individual sediment grains are confined in a single dune, and no grains jump beyond this dune or are brought into suspension. This means that a single dune must maintain its size during migration.

The classical concept of bedform transport assumes that the direction of transport is perpendicular to the dune front. Any imaginary line perpendicular to the dune front will therefore act as a confining wall or 'orthogonal' through which no sediment can pass. Two such dune orthogonals are seen in the upper part of Fig. 2. At any position the sediment transport in the 'orthogonal-channel' is thus the transport rate of the migrating dune (equation (1)) multiplied by the local width of the orthogonal-channel. Under the assumption of no sediment transport across orthogonals, this transport will be constant even if the width of the orthogonal-channel is changing. The principle is

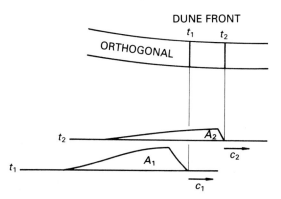

Fig. 2. Change of dune height during migration. Shown are dune positions at times t_1 and t_2. The dune is migrating along the dune orthogonals. According to the classical concept of bedform migration, the sediment transport is confined between the dune orthogonals. In this case the dune cross-sectional area is reduced from position t_1 to t_2 and it has clearly lost sediment. This loss is ascribed to net transverse transport in the trough.

analogous to the concept of wave orthogonals known from the construction of ocean wave refraction diagrams (e.g. Johnson *et al.*, 1948; Dunham, 1951.).

In the upper part of Fig. 2 the position of the dune front is shown at times t_1 and t_2. If we look at the area circumscribed by the two orthogonals and the dune positions at t_1 and t_2, the transport into this area at t_1 will, if dune size and migration celerity are kept constant, equal the transport out of the area at time t_2. If, however, sediment leaving the area is less than the amount brought into it, sediment transport in the orthogonal-channel must have been altered in some way. We can think of three reasons for this alteration.

1 Sediment is lifted into suspension and thus escapes bedform transport.
2 Sediment is deposited. This implies that the mean bed level is raised while the bedform transport is reduced.

In the present study both these possibilities are left out of consideration.
3 The third possibility is sediment transport in the leeside trough transverse to main dune migration direction.

In Fig. 2 we can see that the dune area has actually decreased from A_1 to A_2 during the migration from t_1-position to t_2-position. If we assume that both the migration velocity and the orthogonal-channel width are constant, the transport into the area is larger than the transport out of it. The deficit is ascribed to net transverse transport through the walls of the orthogonal-channel crossing the orthogonal. In this manner the magnitude of the transverse sediment transport in dune troughs can be determined.

METHODS

Sediment transport was measured at the bend entrance and exit with a bedload sampler of the pressure-difference type (e.g. Hansen, 1965). The measurements consisted of five to eight individual samplings in the cross-section.

The survey of bed configuration was carried out from a scaffold placed over the point bar (Fig. 3). The surface measured 2.4 by 3.6 m with 247 measuring points in a 0.2 by 0.2 m mesh. From the scaffold the field could be surveyed without disturbing the bedforms.

The size of the measurement mesh only allows recording of bedforms of a certain size, say longer than 0.6 m. Therefore ripples and the migration of them will not be recorded.

Water depth was measured with a measuring rod at 40 or 60 min intervals. Surveying was repeated six to nine times with the scaffold in four different positions (see Fig. 8).

Surveying was carried out in a 3-day period (23–25 August 1988) at Fields, I, II and III (see Fig. 8) and another period (18–20 April 1989) at Field IV when the scaffold was stationary for 3 days. Within a few centimetres the water level was iden-

Fig. 3. The survey scaffold in position on the point bar platform. By means of a measuring rod the bed configuration is surveyed repeatedly in a 0.2 by 0.2 m mesh without disturbing the bed configuration. Flow is from right to left.

tical for the two periods, and hydraulics and the sediment transport are assumed to be identical too.

On the basis of the 247 data points per survey a map of the bed configurations is constructed. On these maps the dune front positions at different times are located, and dune orthogonals drawn at right angles to them.

The orthogonals are viewed in sets of three; the one in the middle forms a line of section and the two neighbouring ones form the borders of the orthogonal-channel in which the sediment transport is supposed to take place (cf. Fig. 5). The section line is perpendicular to the dunes and curves according to the orthogonals.

The dune size is measured on the dune profiles. The migration rate is determined from the positions of the dune front in two successive surveys. The width of the orthogonal-channel is measured on a map of the orthogonals.

The magnitude of the transverse transport in the troughs is found from profiles of the secondary dunes in the trough in front of the main dunes. In this case the section lines are not stationary but migrate between surveys according to the migration of the main dune. The dunes in the troughs do not occupy the total width of the troughs between the main dunes, but only a fraction thereof. The sediment transport rates (g s^{-1} m^{-1}) of the transverse dunes are converted into total trough-wise transport (g s^{-1}) by multiplying by an assumed dune width of 0.3 m.

PRESENTATION OF STUDY AREA

The river Gels Å is a part of the Ribe Å system which has its outlet into the North Sea near the city of Ribe in western Jutland, Denmark (Fig. 4).

At the study site the catchment area is 325 km^2 consisting of glacial outwash plains and moraine landscapes. Mean annual precipitation is 800–900 mm, and the mean runoff is approximately 14 l s^{-1} km^{-2}. This corresponds to 4.5 m^3 s^{-1} at the study site varying from 1.5–2.5 m^3 s^{-1} during summer to 8–10 m^3 s^{-1} during winter (Det Danske Hedeselskab, 1985).

The bankfull width and mean depth are 13 m and 1.5 m respectively, and the river slope is 0.00045. The mean cross-sectional velocity is about 0.8 m s^{-1}.

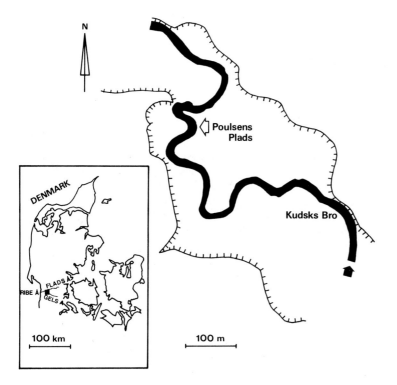

Fig. 4. Location of the study site in western Jutland, Denmark. The alluvial river Gels Å meanders freely in a 300–500 m wide valley filled with sandy glacial outwash material.

The yearly amount of bedload is approximately 6000 t with a mean grain size of about 250–500 µm (Bartholdy *et al.*, 1987).

The river meanders through a 300–500 m broad valley filled with sandy glacial outwash materials from the latest glaciation. Occasionally, the river encounters the valley wall where the meandering motion is restrained (Fig. 4). This is the case immediately downstream of the bend, and is probably the reason for the unusually small radius of curvature at the study site (radius/width = 1.4).

In the bend there is a clearly developed point bar which has a steep point bar slope in the apex and a shallow point bar platform with depths about 40–80 cm (see Fig. 8.). The point bar platform narrows toward the end of the bend and terminates at the exit. The thalweg shifts from the left side bank before the crossover to the right side in the bend. It is deepest and narrowest at the bend apex. Erosion is active at the concave bank, and most of the area between the thalweg and the cutside is occupied by slumps and slides from bank erosion.

DUNE ORTHOGONALS IN A SURVEY FIELD

The use of dune orthogonals is illustrated by the migration of a single dune front through Field I (for location of the fields in the bend, see Fig. 8).

In Fig. 5 the dune front is shown at five successive surveys ('0 min' through to '170 min'). The front is continuous and can be traced from the left side almost to the right side of the field. The migration celerity is approximately 0.9 m h^{-1}.

Dune migration direction is almost parallel to the mean flow direction at this location. The orthogonals are mainly parallel with few exceptions where orthogonal-channels are periodically expanded and narrowed. The migration celerity varies a little, but the dune front largely maintains its orientation. The dune is succeeded by a similar dune at a distance of about 2.4 m.

The dune orthogonals are constructed perpendicular to the migrating dune front at the specific surveying moments. The bold orthogonals in Fig. 5 represent the orthogonal-channel boundaries and the fainter orthogonals represent section lines.

Profiles of the dune along section 'A' are exemplified in Fig. 6. While the migration rate is almost constant, the size of the dune is clearly increasing. The increase in area is accomplished by an increase

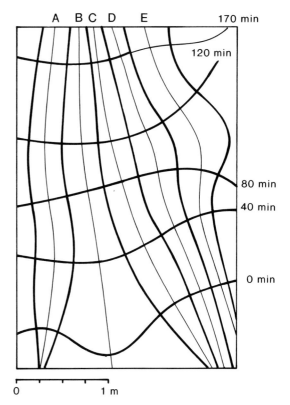

Fig. 5. Position of the dune front in Field I at five different surveying times with corresponding dune-orthogonals. The dune front is continuous throughout the total width of the field. Thick orthogonals indicate the borders between orthogonal-channels ('A'–'E'). Thin orthogonals indicate the section lines for the dune profiles representative of the orthogonal-channel in question. Profiles of the dune in orthogonal-channel 'A' are shown in Fig. 6.

in dune height as the dune length is actually decreasing. The average dune height is 0.12 m, and this is a typical value for all the main dunes in the bend. As the migration rates for individual dunes are almost constant as well, this means that an increase/decrease in transport rate is chiefly accomplished by changes in dune height.

Lateral transport in the field is provided by secondary dunes migrating in the troughs. Sections of the secondary dunes are located 0.4 m in front of the main dune and parallel to it. The individual secondary dunes cannot be followed for more than two successive surveys (Fig. 6). It is therefore necessary to use average values of dune size and celerity to obtain an average value for the lateral

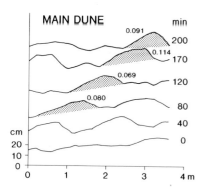 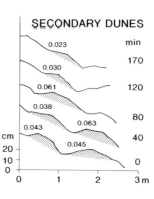

Fig. 6. Profiles of the migrating dune depicted in Fig. 5 in orthogonal-channel 'A' (left), and of the secondary dune migrating in the leeside trough in front of it (right). The secondary dunes are migrating toward the convex bank (from right to left on the figure). Individual secondary dunes can be followed through only two successive surveys. The times of surveys are indicated to the right of the respective profiles. The cross-sectional areas (m²) of the dune (stippled) are indicated.

transport in the trough at the specific dune position.

The migration direction of the dunes in the trough is toward the inner bank, and the average migration rate in the trough is 1.5 m h^{-1}. This is rather fast compared to the migration of the main dunes of 0.9 m h^{-1}. The average dune area, by contrast, is somewhat smaller, especially at the upstream end of the field. The average dune height for secondary dunes in Field I is 0.07 m, or a little more than half the height of the main dunes.

A sediment transport budget (Fig. 7) is worked out for each of the cells in the total area of Field I. The cells are contained between the orthogonal-channels and two successive positions of the dune front.

For each cell the longitudinal sediment budget is calculated from the input at the upstream side and the output at the downstream side. The difference between input and output represents the net contribution to the lateral transport in the trough at the given position. The average lateral transport for the different positions of the main dunes is calculated from the cross-sectional area and the migration of the secondary dunes (Fig. 6).

The average transport in the five orthogonal-channels 'A' to 'E' in Field I is 89 g s^{-1} and the average lateral transport in the troughs is 7.5 g s^{-1} or 8% of the longitudinal transport at this particular time and position on the point bar platform.

There are considerable differences in longitudinal sediment transport from one dune position to the next in the same orthogonal-channel. The largest relative change is found in orthogonal-channel 'C' where the transport decreases from 13 g s^{-1} at the upstream end to 0.3 g s^{-1} (zero on the map) at the downstream end. As mentioned earlier, this reduction is due to a reduction in dune area rather than migration rate, as can be judged from the map. It shows that the overall migration celerity remains almost constant.

A part of the dune front at the 0-min-survey is lagging behind, but this is compensated partly at the 40-min-survey and completely at the 80-min-survey, where the dune front has recovered its straight course. Apparently this is associated with a sudden increase in longitudinal transport when moving in the trough-transport direction (from right to left) in the migrating trough.

It is interesting to note that the lateral transport often is of the same order as, and sometimes even larger than, the longitudinal transport.

This detailed example of the use of dune orthogonals shows that this approach is able to give an excellent description of bedform migration and sediment transport in dune troughs.

BEDFORM MIGRATION ON THE POINT BAR PLATFORM

A similar procedure was carried out for each of the four survey fields. To give a description of the conditions over a longer period of time, the dune fronts in each field are averaged to make up 'artificial dunes' with straight, continuous crests, even though this is practically never the case. The dune orthogonals will consequently become parallel and the orthogonal-channels of constant width. It also means that the previous findings concerning Field I are altered accordingly.

In all cases the survey of one field includes the passage of more than a single dune front, sometimes two complete passages and sometimes one complete, and two half passages — one in the down-

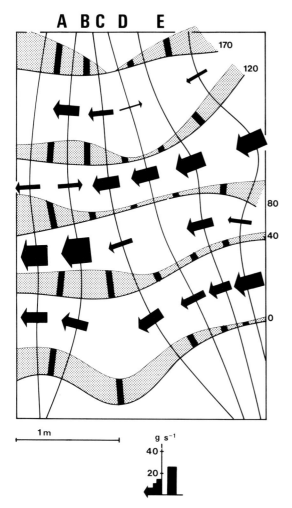

Fig. 7. Sediment transport rates of a single dune in Field I. The figure shows the successive positions of the dune front of the same dune, as it migrates through the survey field. (For location of fields, see Fig. 8.) The longitudinal sediment transport in the orthogonal-channels is indicated by the lengths of the bars while the transverse transport in the dune troughs is proportional to the thickness of the arrows. Flow is from bottom to top.

stream part of the field, starting in the middle, and one entering the field at the downstream end, migrating to the middle of the field by the end of the survey. Field III in the central area of the point bar was surveyed during 3 days giving many dune passages.

The orthogonal orientations in Fig. 8 show a very distinct picture. In Field I the orthogonals are parallel to the river banks, and the dunes follow rectilinear paths. They are probably formed in this area where the sediment is entering the point bar platform. At the entry of the bend the secondary circulation is apparently too weak to influence the bedform orientation.

Rotation of the bedforms is observed in the other fields, where the bedforms are heading towards the convex bank. The tendency toward rotation increases toward the convex bank and down the bend. At the downstream end of the point bar platform, the rotation is at a maximum, and the dune migration direction deviates from the mean flow direction by about 60°. Apparently the direction of dune migration again becomes current-parallel as the downstream end of the point bar is approached.

LATERAL SEDIMENT TRANSPORT IN THE MEANDER BEND

The lateral transport is closely connected with the orientation of the bedforms. In Field I, lateral transport exclusively takes place as migration of secondary dunes in leeside troughs, as the main dune migration is parallel to the river banks. The lateral transport here is $2 \, \text{g s}^{-1} \, \text{m}^{-1}$ while the longitudinal transport is $25 \, \text{g s}^{-1} \, \text{m}^{-1}$.

In the downstream part of Field II the lateral transport is $7 \, \text{g s}^{-1} \, \text{m}^{-1}$ rising gradually to $17 \, \text{g s}^{-1} \, \text{m}^{-1}$ at the upstream end of Field III. The large values at Field III arise because the migration direction has turned toward the convex bank, and adds to the lateral transport in the troughs. At Field IV the lateral transport has decreased to $3 \, \text{g s}^{-1} \, \text{m}^{-1}$ and this value is exclusively the effect of migration of the main dunes. There is no significant transport in the troughs in this field.

The bend area can be divided into a network of small bed areas (Fig. 9). For each area a sediment budget is worked out taking into account the sediment input from the upstream end, the lateral transport toward the inner bank (positive) or the outer bank (negative) and the output of sediment at the downstream end.

The values for the lateral transport achieved by the scaffold measurements comprise fourteen of these network areas (marked with bold edges in Fig. 9). These values are extrapolated to the total area of the point bar platform. Besides linear interpolation, three other assumptions have been made.

1 The lateral transport toward the inner bank is decreasing outward from a maximum at 2–3 m

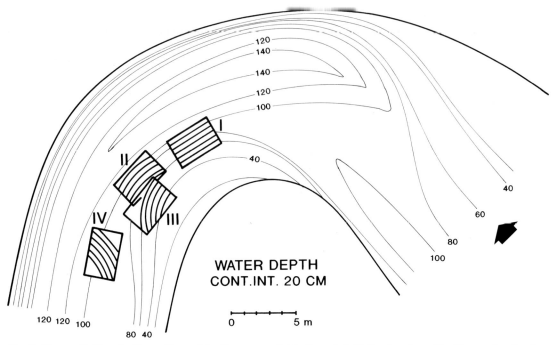

Fig. 8. The studied bend with positions of the four survey-fields. In each field the direction of bedform migration is indicated by dune orthogonals. The flow is from right to left.

from the inner bank to zero in the area around the thalweg.

2 The transport toward the inner bank ceases 2 m from the inner bank.

3 At the upstream end of the point bar some lateral transport exists toward the concave bank. This is because of the transverse bedslope gradient toward the concave bank in this area.

The resultant characterization of the sediment transport through the bend is compared to the measured gross material input and output of the bend. This is seen in Fig. 10, where the predicted

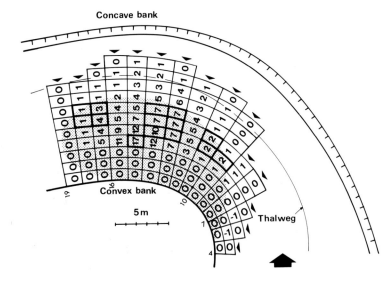

Fig. 9. Lateral sediment transport rates (g s^{-1} m^{-1}) distributed in network areas on the point bar platform. Negative values indicate sediment transport toward the concave bank and positive values transport toward the convex bank. Areas with actual measurements are indicated with thick edges, while values in the remaining areas are extrapolated. The stippling indicates the approximate extent of the point bar platform.

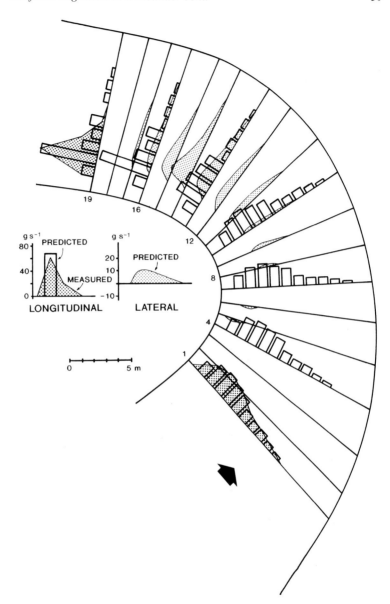

Fig. 10. Distribution of the longitudinal and lateral sediment transport in the bend. Predicted longitudinal transport is indicated as bars in selected cross-sections. Predicted lateral transport is indicated in each cross-section with light stippling (in some cross-sections the lateral sediment transport is zero). Measured transport at the bend entrance and the bend exit is indicated with dense stippling.

longitudinal and lateral transport in the bend are illustrated for nineteen cross-sections. The point bar comprises sections no. 6–17. The measured input to the bend is 209 g s^{-1} and the output 195 g s^{-1}, i.e. a difference much smaller than the error of measurement.

Considering the variety of uncertainties and approximations involved, the measured and predicted sediment transports at the bend exit are found to compare very well. The predicted and measured loci of maximum transport in the cross-section are in excellent agreement, and the cross-stream distribution shows a satisfying agreement with the measurements. The excellent comparison at the bend exit indicates that the results also apply to the bend and the point bar platform in general.

Figure 10 shows that the main part of the lateral transport is accomplished in a rather short interval around sections 13–16. The cross-stream distribution of the longitudinal sediment transport is not affected by lateral transport until below section 13.

This is confirmed in Fig. 11 which shows the

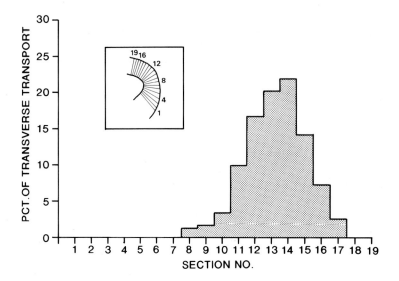

Fig. 11. Distribution of the lateral sediment transport in successive downstream cross-sections across the point bar platform. Numbering of cross-sections is indicated on the inset. The point bar comprises cross-sections 6–19. The majority (73%) of the total lateral sediment transport is taking place in cross-sections 12–15, comprising only 40% of the point bar length.

downstream distribution of the local lateral sediment transport expressed as a percentage of the total lateral transport in the bend. In the upstream end of the point bar platform (sections 6–11) the lateral transport is small. Only 16.3% of the total lateral transport in the bend is found in this area, comprising 40% of the length of the point bar platform. In a narrow zone a little more than halfway through the point bar platform (sections 12–15), the lateral transport is most intense. Actually 73.0% of the total lateral transport in the bend is taking place in these four sections that comprise only 40% of the platform length. The values are rapidly declining, and beyond section 16 only about 10% of the lateral transport is still to be completed.

SUMMARY AND CONCLUSIONS

A detailed description of the migration of a single dune front through a 2.4 by 3.6 m survey field shows that the use of dune orthogonals is able to give a detailed quantitative description of bedform migration and related sediment transport. Especially, the magnitude of the transverse transport in the leeside troughs of the migrating dunes seems to be characterized very well.

With the aid of dune orthogonals, a description of the bedform dynamics and the bedform transport in a meander bend is obtained. The bedforms (dunes) are gradually rotating toward the convex bank shortly after they have entered the point bar platform. The rotation is gradually increasing as the bedforms are moving downstream on the platform and approaching the convex bank. The maximum deviation from the mean flow direction of 60° is found shortly before the end of the platform.

Nearly all the sediment transport in the bend is transported across the point bar platform and hardly any sediment transport is found in the thalweg at the bend exit.

The lateral inward transport is not evenly distributed along the length of the point bar, but is concentrated in a zone around the bend apex.

ACKNOWLEDGEMENTS

The author thanks J.R. Boersma who reviewed the manuscript and had several helpful suggestions for improvements. He also wishes to thank J. Bartholdy for valuable discussion during the preparation of the paper.

REFERENCES

Allen, J.R.L. (1969) On the geometry of current ripples in relation to stability of fluid flow. *Geogr. Ann.* **51A**, 61–96.

Allen, J.R.L. (1970) A quantitative model of climbing ripples and their cross laminated deposits. *Sedimentology* **14**, 5–26.

Allen, J.R.L. (1982) *Sedimentary Structures. Their Character and Physical Basis*, Vol. II. Elsevier, Amsterdam, 633 pp.

ALPHEN, J.S.L.J., BLOKS, P.M. & HOEKSTRA, P. (1984) Flow and grainsize pattern in a sharply curved river bend. *Earth Surf. Proc. Landf.* 9, 513–522.

BARTHOLDY, J., HASHOLT, B. & PEJRUP, M. (1987) *Mineraletransport I Ribe Å Systemet (Sediment Transport in the Ribe Å System)*. University of Copenhagen, Institute of Geography, Copenhagen, 59 pp.

BATHURST, J.C., THORNE, C.R. & HEY, R.D. (1979) Secondary flow and shear stress at river bends. *J. Hydr. Div., ASCE* 105, 1277–1295.

BRIDGE, J.S. (1977) Flow, bed topography, grain size and sedimentary structure in open channel bends: a three-dimensional model. *Earth Surf. Proc.* 2, 401–416.

BRIDGE, J.S. & DIEMER, A. (1983) Quantitative interpretation of an evolving ancient river system. *Sedimentology* 30, 599–623.

BRIDGE, J.S. & JARVIS, J. (1976) Flow and sedimentary processes in the meandering river South Esk, Glen Clova, Scotland. *Earth Surf. Proc.* 1, 303–336.

BRIDGE, J.S. & JARVIS, J. (1982) The dynamics of a river bend: a study in flow and sedimentary process. *Sedimentology* 4, 498–541.

BRIDGE, J.S., SMITH, N.D., TRENT, F., GABEL, S.L. & BERNSTEIN, P. (1986) Sedimentology and morphology of a low-sinuosity river: Calamus River, Nebraska Sand Hills. *Sedimentology* 33, 851–870.

DET DANSKE HEDESELSKAB (1985) *Afstrømningsmålinger i Sønderjyllands Amtskommune 1984.* (The Danish Heat Society: *Run-off Measurements in Sønderjyllands Amtskommune 1984*) (in Danish). Det Danske Hedeselskab, Hydrometriske Undersøgelser, Slagelse, Denmark.

DE VRIEND, H.J. & GELDOF, H.J. (1983) Main flow velocity in short and sharply curved river bends. Communications on Hydraulic Report 83-6. Department of Civil Engineering, Delft Univ. Technology, The Netherlands.

DIETRICH, W.E. (1982) *Flow, boundary shear stress, and sediment transport in a river meander.* PhD dissertation, Univ. Washington, Seattle.

DIETRICH, W.E. & SMITH, J.D. (1983) Influence of the point bar on flow trough curved channels. *Water Resources Res.* 19, 1173–1192.

DIETRICH, W.E. & SMITH, J.D. (1984) Bed load transport in a river meander. *Water Resources Res.* 20, 1355–1380.

DIETRICH, W.E., SMITH, J.D. & DUNNE, T. (1979) Flow and sediment transport in a sand bedded meander. *J. Geol.* 87, 305–315.

DUNHAM, J.W. (1951) Refraction and diffraction diagrams. In: *Proc. 1st Conf. Coast. Eng.* (Ed. Johnson, J.W.) pp. 33–49. Council on Wave Research, The Engineering Foundation, Berkeley, California.

ENGELUND, F. (1974) Flow and bed topography in channel bends. *J. Hydr. Div., ASCE* 100, 1631–1648.

HANSEN, E. (1965) Bed load investigation in Skive-Karup River. Hydraulic Laboratory, Tech. Univ. Denmark.

HOOKE, R.L. (1974) Shear-stress and sediment distribution in a meander bend. UNGI Rapport 30. Department of Physical Geography, University of Uppsala, Sweden, 59 pp.

JACKSON, R.G. (1975) Velocity–bed-form–texture patterns of meander bends in the lower Wabash River of Illinois and Indiana. *Bull. Geol. Soc. Am.* 86, 1511–1522.

JOHNSON, J.W., O'BRIEN, M.P. & ISAACS, J.D. (1948) Graphical construction of wave refraction diagrams. US Hydrog. Off. Publ. 605.

KISLING-MØLLER, J. (1988) Sediment transport in a meander bend (abst.). In: *9th IAS Regional Meeting of Sedimentology, Leuven, Belgium, 1988* (Ed. Svennen, R.) pp. 117–118. KV, Leuven.

KNUDSEN, M. (1981) On the direction of the bed shear stress in channel bends. Inst. Hydrodyn. and Hydraulic Eng. Tech. Univ. Denmark, Prog. Rep. 53, pp. 3–7.

MCGOWEN, J.H. & GARNER, L.E. (1970) Physiographic features and stratification types of coarse-grained pointbars: modern and ancient examples. *Sedimentology* 14, 77–111.

ODGAARD, A.J. (1981) Transverse bed slopes in alluvial channel bends. *J. Hydr. Div., ASCE* 107, 1677–1694.

PAOLA, C., GUST, G. & SOUTHARD, J.B. (1986) Skin friction behind isolated hemispheres and the formation of obstacle marks. *Sedimentology* 33, 279–293.

PARKER, G. & ANDREWS, E.D. (1985) Sorting of bed load sediment by flow in meander bends. *Water Resources Res.* 21, 1361–1373.

RAUDKIVI, A.J. (1963) Study of sediment ripple formation. *J. Hydr. Div., ASCE* 89, 15–33.

ROZOWSKII, I.L. (1961) *Flow of Water in Bends of Open Channels.* Israel Program for Scientific Translations, Jerusalem. 233 pp.

SIMONS, D.B., RICHARDSON, E.V. & NORDIN, C.F., JR (1965) Bedload equation for ripples and dunes. US Geol. Surv. Prof. Paper 462-H.

SMITH, J.D. & MCLEAN, S.R. (1984) A model for flow in meandering streams. *Water Resources Res.* 20, 1301–1315.

THOMPSON, A. (1986) Secondary flows and the pool–riffle unit: a case study of the processes of meander development. *Earth Surf. Proc. Landf.* 11, 631–641.

ZIMMERMANN, C. & KENNEDY, J.F. (1978) Transverse bed slopes in curved alluvial streams. *J. Hydr. Div., ASCE* 104, 33–48.

# Sediment ice rafting and cold climate fluvial deposits: Albany River, Ontario, Canada

I.P. MARTINI,* J.K. KWONG† and S. SADURA*

*Department of Land Resource Science, University of Guelph, Guelph, Ontario N1G 2W1, Canada; and
†c/o Dames and Moore Ltd, 1144-10th Ave, Suite 200, Honolulu HI96816, USA

ABSTRACT

The Albany River is a typical meandering to anastomosing, non-glacial stream, which crosses vast, unconfined peatlands and is affected annually by ice drives, jams and ice rafting. The amount of sediment rafted by ice is large during spring breakup (approximately 3.5×10^4 t), but it is a single annual event, and the rafted load is less than 1% of the total yearly suspended load of the river. However, this material is deposited preferentially in specific parts of the channels and on non-forested overbank areas, where characteristic deposits develop. The rafted sediment consists of material placed in the ice in a variety of ways: (i) suspended load material is emplaced partly during freezeup, but mostly during breakup when flooding occurs; (ii) pebbles are lifted from the river bed by anchor ice; and (iii) bank material slumped onto the ice during spring thawing. Released rafted sediments have diagnostic features which are retained where not reworked by water flow. These features consist of patchy silt drapes over grassy banks, lenses of unsorted coarse sand and pebbles in fine overbank deposits, lenses of poorly sorted gravel and isolated pebbles and boulders strewn on levees and the treeless deltaic floodplain. Furthermore, the gravel deposits have a polymictic composition and usually contain angular pebbles of local carbonate and reworked, rounded cobbles and boulders of crystalline glacial erratics. All these sedimentary features constitute environmentally diagnostic 'textural inversions' and should be recognizable in ancient geological sequences. They and the presence of thick, widespread organic layers (peat, coals) should allow ready distinction between cold, non-glacial streams and the more typically braided glacio-fluvial deposits which may also show well defined rhythmic sedimentation where a suitable mixture of gravel and sand exists.

INTRODUCTION

Objectives

The objectives of this paper are to analyse the effect of river ice on sediment dispersal, and to present sedimentary features developed in cold climate, non-glacial subarctic rivers.

Much is known about fluvial sediments and sedimentary rocks (Miall, 1977, 1985; Collinson & Lewin, 1983; Ashley et al., 1985) and about ice-covers of rivers (Williams, 1965; Ashton, 1978, 1986; Tsang, 1982; Beltaos, 1984). However, river ice research has been generally aimed at avoiding damage either due to floods during breakup or caused by ice on structures and intakes to hydro-electric power stations. Knowledge about sediments and ice is seldom combined to evaluate the effect ice has on sediment distribution and properties (Collinson, 1971; Smith, 1979). This paper aims to fill this gap, examines the ice–sediment relationship and tries to determine whether it is of any geologic significance. It is, in fact, the general consensus among people working in arctic conditions and mountainous areas that the ice rafted load is volumetrically insignificant relative to the total sediment load of rivers (Smith, 1979). In addition, conflicting evidence exists on whether much of the rafted fluvial sediment is deposited in nearshore areas or whether it is transported offshore (Reimnitz & Bruder, 1972). The contention of this paper is that however small the ice rafted load is, in some

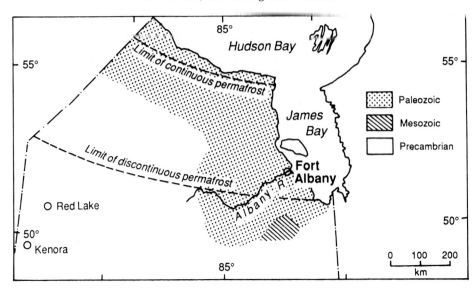

Fig. 1. Location map of the Albany River. The Hudson Bay Lowland is underlain by Palaeozoic and Mesozoic rocks.

environments it generates diagnostic deposits of geomorphologic and geologic importance.

Location and geomorphology

Pleistocene glacio-fluvial deposits and several modern non-glacial rivers have been studied over decades in Ontario, Canada, both in cold-temperate southern Ontario, and subarctic northern Ontario along the western coasts of James Bay (Fig. 1; Chapman & Thomas, 1968; Karrow et al., 1982; King & Martini, 1983; Lowden, 1983). This paper focuses on the results obtained from the lower reaches of the modern subarctic Albany River in northern Ontario.

The Albany River is the second largest river of the Hudson Bay Lowland. It has a drainage basin of 137 270 km^2 and an average discharge of 1500 m^3 s^{-1} with a nival regime (Fig. 2; Cumming, 1968; Hutton & Black, 1975). It carries an average suspended load of 120 mg l^{-1} during the spring floods (Water Survey of Canada, unpubl. data). For the remainder of the year, a small suspended load (average 30 mg l^{-1}) and a relatively large solution load (instantaneous measurements in July up to 110 mg l^{-1}) are transported (Water Survey of Canada, unpubl. data). The total suspended load carried annually by the river is estimated to be approximately 5–6 × 10^6 t, about half of which is carried during the spring runoff (1 April to 30 June).

The Albany River crosses one of the largest, low gradient boreal to subarctic peatlands of the world (Martini, 1989). Non-peat producing marshes occur at the river mouth, but fens and bogs develop up to 2.5–3 m of peat farther inland. The river crosses the peatland like a meandering ribbon and drains it, but does not significantly affect its development except for fostering formation of narrow, discontinuous swamps along its major channels (Martini & Glooschenko, 1985; Martini, 1989). In the lower reaches, the river is characterized by two dominant waterways (North Channel and South Channel), both featuring anastomosing (Smith & Smith,

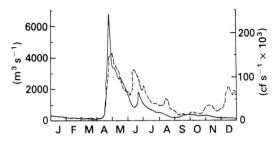

Fig. 2. Hydrograph of the Albany River at Hat Island (about 180 km upstream from Fort Albany); (——) 1976; (– –) 1984.

1980), shallow (some dry in summer) channels (Fig. 3).

The upper course of the river is on the Precambrian shield; the lower course (480 km) crosses over Palaeozic carbonate rocks which dip gently (less than 1 m km^{-1}) toward the centre of the Hudson Bay Basin, over Quaternary tills, and over glaciofluvial and glacio-marine deposits (Fig. 1). The area has been glaciated and is undergoing a mild postglacial (since about 7500 years BP) isostatic rebound (0.7 m century^{-1}; Webber et al., 1970) which leads to a slight entrenching of the channels. The river carries a mixed bedload of metamorphic and igneous clasts derived from glacial terrains mixed with locally derived Palaeozoic carbonate particles.

Methods of study

The breakup and ice jams of the Yellow Creek channel at Fort Albany were closely monitored over a period of 5 years (1983–88), with observations and measurements made over other parts of the lower reaches of the Albany River (Fig. 3). Helicopter reconnaissance flights and hand-held photographs were utilized for mapping the ice-cover and the distribution of ice floes. Autumn and spring ice conditions were determined through satellite images and information obtained from local people.

The study of local reaches in the spring consisted of determining the types and volume of ice and their sediment load. Ice samples (order of 1 kg each)

Fig. 3. Lower anastomosing reaches with bathymetric cross-sections of South Channel, and indication of major ice scours of banks.

were collected at pre-established equal distances along transects on remnant parts of ice jams and on major ice floe deposits, from the middle of the channel to the river bank. Some additional samples were collected from selected zones, such as overturned slabs of anchor ice. At each site, samples were taken both from the surface and from inside the ice floes, because their sediment concentration differed. Approximately twenty-five to thirty samples per year were treated. The concentration of fine sediments in the ice was determined at a field laboratory by measuring volume and weight of each block, then filtering the melt through ashless filter paper, to obtain the sediment weight. The sediments were subsequently analysed for grain size though sieving and Coulter Counter at the laboratories of the Canada Centre for Inland Waters (Burlington, Canada). The measured sediment concentrations in ice type and the visually estimated volume of the various types of ice allowed for an estimate of the amount of ice rafted sediment load.

The spring breakup studies were followed by field surveys of the ice rafted deposits during subsequent summers (low water stage), any by measuring and sampling bank sediment sequences which span several hundred years of history of the river.

ICE AND SEDIMENT DISPERSAL

The Albany River is a northward flowing river covered by ice for 6 months of the year from November to May. Breakup occurs every year in late April to early May, well before the ice is broken in James Bay. As a consequence, ice jams invariably develop along certain reaches of the river (at bends, shallowing zones, along secondary channels) and against the solid ice-cover of James Bay (Fig. 4). The dispersal of ice floes and their rafted sedimentary load during the ensuing ice drives varies from year to year depending on the type of ice breakup that develops and whether the northern channel or the southern channel breaks first (Fig. 3). During the 6 years of our monitoring, every breakup was different: in some years the breakup occurred later in the season after gradual warming and the ice-cover was weakened by melting. These 'thermal' breakups (Ferrick & Mulherin, 1989) generate weak ice drives and ice jams, do not usually lead to damaging large floods, and much of the ice-cover attached to shores and shoals rots in place. In other years the breakup occurred early in the season when abundant snow was melted by a rapid increase in temperature while the ice-cover was still strong. In this case the ice-cover is lifted and broken by flood waters. Strong 'dynamic' ice jams (Ferrick & Mulherin, 1989) and damaging floods occur with a widespread distribution of ice floes over the banks. In one case (winter 1985–86) a 'winter' breakup and ice jam occurred when water from a distant (more than 200 km) upstream reservoir was released after development of the ice-cover had started in early winter. This ice jam filled and blocked the secondary channels with ice and was incremented by the

Fig. 4. Ice jams in the Yellow Creek channel of the Albany River: in background the village of Fort Albany; in foreground an inactive channel with ice floes introduced from main channel.

following spring breakup and flood. It has been our experience and that of other workers in subarctic environments that, whereas the approximate time of breakup, location of ice jams and preferred areas of grounding of ice floes (and thus of release of ice rafted material) can be estimated with some confidence, the type of breakup and ice jams that develop and the intensity of the ice run and amount of material transported cannot be predicted from year to year (Zachrisson, 1988). This is further complicated in the Albany River because it has two major channel systems. In some years the North Channel breaks first, the flood waters are funnelled through it, and the ice-cover and ice jams of the South Channel melt in place (Fig. 5). In other years the roles of the two channels are reversed.

During ice runs, that is, periods when the ice floes (blocks) are moved downstream, part of the rafted load is removed from the ice floes by melting of the underside or by overturning. Much of the rafted

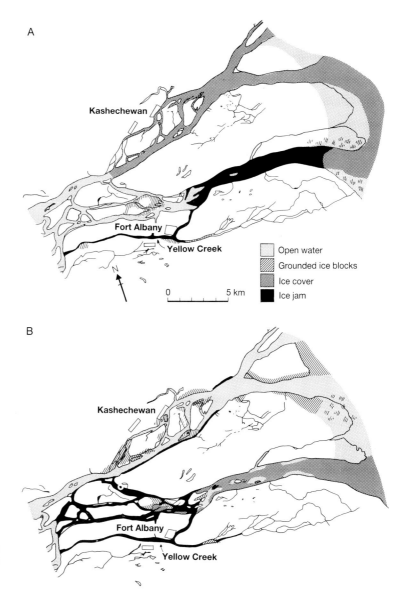

Fig. 5. Ice distribution map of the lower reaches of the Albany River. (A) 26 April 1984; (B) 1 May 1985.

load, however, is transported and released where ice floes are grounded, most of them within the channels themselves: even at high floods the floating ice blocks are trapped by trees growing on levees. Preferred ice floe accumulation zones occur not only in channels where ice jams form, but also high on the banks where floes become stranded during rapidly receding floods in slump scours and terraces, and in secondary intermittently active or dead-end abandoned channels. These channels receive flood waters and ice floes only during spring high floods (Fig. 6). Only in areas where arboreal vegetation is not present either because of fires, or because it has not yet developed such as in the coastal plains, can the ice flocs acquire a wider distribution in the overbank zone (Fig. 5).

Fig. 6. Remnant of an ice jam at Fort Albany showing an agglomeration of various types of ice and ice rafted material.

ICE TYPES AND RAFTED SEDIMENTS

Different types of ice carry different sedimentary loads (Fig. 7). Four main types of ice are found in the Albany River: black ice, snow slush ice, frazil ice and anchor ice. *Black ice* forms primarily by congelation (freezing) of the river water (Michel, 1978). If freezing is slow, fine material (derived from suspended load or wind blown) is extruded along the crystalline freezing front. This type of ice is generally clear (black) and carries little fine sediment distributed into laminae. *Snow slush ice* forms the upper part of the ice-cover and develops from a mixture of snow and flood water (Michel & Ramseier, 1971; Gerard, 1983). Usually it acquires flood and wind blown materials. *Frazil ice* forms as flocs of ice in supercooled water (Michel, 1971; Williams & Mackay, 1973; Ashton, 1980, 1986; Tsang, 1982). A small amount of suspended sediment may be trapped in developing pans of ice flocs. *Anchor ice* is formed by frazil ice flocs which get attached to the colder parts of the stream bottom (Michel, 1971; Smith, 1980; Tsang, 1982; Gerard, 1983; Ashton, 1986). In gravelly beds, pebbles are the colder objects which project into the supercooled water. If the anchor ice thickens enough to acquire buoyancy it floats, selectively lifting the coarser particles from the river floor (Fig. 8A; Österkamp, 1975).

Other important ways in which ice can obtain large quantities of fine and coarse sediments are, for instance, by flood water lifting the ice-cover frozen to the bottom near banks and in shallower reaches, by addition of turbid waters on top of the ice-cover during early spring floods, and by slumps and solifluction from thawing banks.

ICE RAFTED DEPOSITS

Surface occurrences

Rafted deposits acquire characteristic features during release from ice floes depending on the type and amount of rafted sediment, the attitude of the grounded ice floes, and the rate of melting, thus the amount of meltwater generated.

When a relatively small amount of fine material is present, it does not protect the surface of the grounded ice blocks from melting, rather it enhances melting by absorbing radiant energy (Shumskii, 1964). As the ice melts, the long vertical black

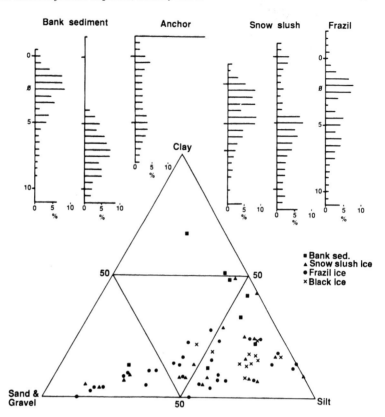

Fig. 7. Grain size of sediments carried by different types of ice, compared with samples from the river bank. The histograms are those of representative samples. ϕ, Phi.

ice crystals become loose, giving a characteristic 'candling' character to the ice block. The fine surficial sediment separates into small pellets which can be trapped in the spaces between the crystals and can be released on the ground as thin drapes of pelletized silt and sand (Fig. 8B,C). These drapes are widespread on banks in the spring. The continuity of the drapes is later broken by vegetation growth, but they are remarkably persistent and recognizable during the subsequent summer.

When the amount of fine sediment increases, but does not completely bury the grounded ice blocks, it is released slowly through solifluction as it is mixed with meltwater. In part it is sorted in small rills which form microalluvial fans at the base of the grounded ice blocks, and in part it is pelletized as the material is trapped temporarily along the edge of the melting ice by surface cohesion. As these cohesion blobs of sediment are released to the ground they maintain their pelletized form.

Pebbles of anchor ice generate two main types of deposits depending on whether the block lies horizontally or is imbricated. In the first instance disseminated pebbles are heated more than the surrounding ice, thermally drill through the ice block, and are released as a scatter on the ground. In the case of imbricated ice blocks, disseminated pebbles are freed by melting of ice around them and roll to the base forming elongated mounds on the ground. These mounds are generally poorly sorted and have a polymictic composition due to the mixed source of crystalline glacial erratics and locally derived carbonates.

When the amount of ice rafted material is large and buries the ice blocks, release occurs by melting from below. In secondary channels where large ice jams have occurred, some ice may still be present at the end of July, slowly melting and releasing sediments. This type of release generates relatively thick (20–35 cm) lensing layers of poorly sorted material, containing internal, better sorted, irregularly distributed lenses (Fig. 8D). The overall sorting of these deposits may be worse than that of the source area, because sediments derived from different types of piled-up ice blocks mix upon release.

Ice rafted material deposited in the main chan-

Fig. 8. Ice rafted sediments: (A) pebbles rafted by anchor ice; (B) pellets of argillaceous silt in the interstices of candle ice; (C) deposited fine sediment drapes and polymictic pebble mound; (D) unsorted ice rafted sediment released from a large floes accumulation in a secondary channel.

nels of the river may be reworked by water flow into bar deposits and may lose its diagnostic characteristics (Fig. 9A). However, the material released in secondary, dead-end or semi-abandoned channels, on levees, and over the deltaic plain, are not modified and persist in the stratigraphic record. Diagnostic features in such environments are the occurrence of anomalously large size clasts, isolated or in clusters, immersed in fine deposits derived from suspended flood loads, and the presence of thin lenses of sand and pelletized silt drapes (Fig. 9B). In many instances where pebble deposits occur, both within the channels and on overbank areas, a characteristic composition–shape inversion develops which is typical of cold subarctic to arctic climates, that is, the harder and less frost-susceptible igneous and metamorphic clasts maintain the roundness they have acquired through their glacial ordeal, while softer, thinly bedded and more frost-susceptible carbonates generate angular clasts.

Vertical bank sections

As previously indicated, the amount of ice rafted material is not large, its distribution is not uniform, and material which can be subsequently reworked by water flow loses most diagnostic ice rafting features. It is therefore not surprising that it is difficult to recognize ice rafted material in many bank exposures of the river. It is also true, however, that rafted materials are recurringly deposited in preferred environments, and when sections through such deposits are found, they record the rafting events very well. Furthermore, the present day cold climate of the study area forms a continuum since Pleistocene times, and therefore the glacial–postglacial transition is recorded in the stratigraphic sections.

Typical bank sections of the Albany River are characterized by a lower unit composed of frost-shattered carbonate bedrock, overlain by till, which is in turn overlain by glacio-marine deposits generally fossiliferous, massive to laminated with some thin sandy interlayers and scattered lonestones. This is invariably sharply overlain by a sand or gravelly layer which represents the channel deposits in areas directly influenced by the river, or by tidal flat sediments in adjacent areas (Figs 9C & 10). These coarse deposits have been reworked by water flows and ice rafting contributions are generally masked. The cold climate setting is still recognizable, however, because of the polymictic composition of the deposits and the presence of angular, frost-shattered carbonate clasts. The channel deposits are overlain sharply by sand and silt indicating that the abandonment of the channels is by avulsion, without traces of lateral accretion surfaces. A few stringers of coarse sand and some lonestones indicate occasional reactivation of the channels and injection of ice floes. The sequence is capped by a levee-like deposit where organic rich layers increase in number and thickness upward. The organic matter is composed of subarctic and boreal plant species, primarily spruce and aspen. In the fine deposits of levees and abandoned channels, ice rafting is recorded by occasional lonestones and thin lenses of coarser sand. The levee deposits are generally heavily bioturbated, primarily by roots, worms and insects. Calcareous and iron oxide rhyzoconcretions are well developed in dryer parts of the levees. Ground ice affects the banks of this river by fostering solifluction and slumping at spring melting. However, only sporadic permafrost exists under the protection of some thicker moss hummocks, and no patterned ground features have been observed except for occasional convolutions and planar microvoids related to ice lenses.

GEOLOGICAL SIGNIFICANCE: AN ANALYSIS

Ice related processes (Newbury, 1968; Michel, 1971; Tsang, 1982; Beltaos, 1984) and some associated sedimentary features have been observed in several cold climate, non-glacial recent rivers (Collinson, 1971; Smith, 1976; Scott, 1978, 1979; Burrows & Harrold, 1983; King & Martini, 1983). However, the cold climate, non-glacial, fluvial facies have not been well documented in the pre-Pleistocene geologic record. The various papers dealing with early postglacial fluvial deposits do not report specific features ascribable to ice effects, and only brief reference is made to them in Lower Permian Coal Measures of Australia and Brazil (Church & Ryder, 1972; Martini & Johnson, 1987; Martini & Banks, 1989; Martini & Rocha Campos, 1991). Even for the glacio-fluvial channel deposits of Pleistocene and older times, the cold environment is inferred primarily by the stratigraphic context of the units studied, their association with diamictites, and by the abundance of braided gravelly deposits. Increasing circumstantial evidence is being gathered, however, from channel

Fig. 9. Ice rafted sediments: (A) channel sediments in part ice rafted but reworked by water flow into bar deposits; (B) boulder and thin sandy lens in organic rich levee deposits; (C) vertical section of a bank of the Albany River.

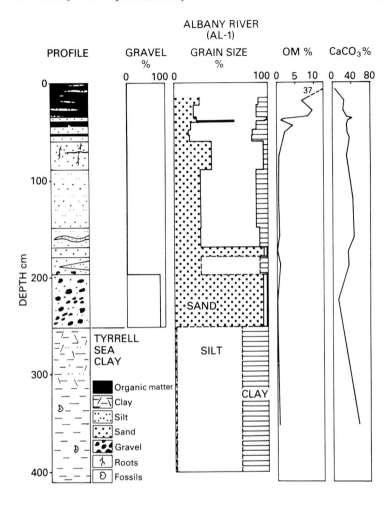

Fig. 10. Sedimentary characteristics of the bank section illustrated in Fig. 9C.

and overbank deposits which can readily discriminate between cold and warm settings of fluvial deposits and perhaps between glacial and non-glacial cold streams (Table 1).

In Pleistocene *channel deposits* the periglacial setting is occasionally indicated by diamict balls in gravels, and folds and faults ascribable to melting buried ice blocks, but these are common only in areas very close to the glacier (Price, 1983). They have not been observed in the subarctic, modern, non-glacial rivers we have studied in Ontario. The polymictic composition of the deposits and the presence of erratics (coarse clasts derived from outside the drainage basin) like those of Pleistocene glacial streams and of the Albany River may indicate the occurrence of glaciation and perhaps ice rafting, but cannot discriminate between glacio-fluvial settings and postglacial rivers reworking glacial deposits. Prudence is also required in using such lithic evidence for palaeoclimatic reconstructions because of possible tectonic and mass flow effects.

Another piece of circumstantial evidence for recognizing cold, non-glacial gravelly braided channel deposits has been suggested by Miall (1977), discussed by Smith (1985) and Miall (1984, 1985) and re-proposed by Morison & Hein (1987) in their work on Pleistocene, cold, non-glacial deposits of Alaska. They suggest that deposits not directly affected by glaciers develop more massive units and show little cyclicity in the gravel deposits of the type developed in the Donjek glacio-fluvial model (Miall, 1977). Such cyclicity has been documented with Markov analyses in Pleistocene outwash directly affected by thawing cycles of glaciers (Fraser, 1982; Bryant, 1983a,b). No such cyclicity is present in the deposits of the Albany River, but these are

Table 1. Comparison between glacio-fluvial and non-glacial cold, subarctic streams

Feature	Subarctic	Glacial
Braided streams	Occasional	Abundant
Anastomosing streams	Occasional	Rare to absent
Meandering streams	Common	Occasional
Ribbon streams	Common	Rare
Cyclicity of gravel deposits (Donjek type)	Rare	Local, well developed
Polymictic composition of coarse material	Common	Common
Selected shape modification of pebbles by permafrost	Common	Rare
Diamict balls in gravel deposit	Rare to absent	Occasional
Levee deposits	Abundant	Rare (distal)
Vegetation on levees and interchannel areas	Abundant	Rare (distal)
Peat deposits	Abundant	Rare to absent
Bioturbation	Abundant	Rare (distal)
Rhyzoconcretions	Abundant	Rare (distal)
Horizontal voids due to ice lensing	Common	Rare
Patterned ground features	Rare	Occasional
Synsedimentary, internal folds and faults	Rare to absent	Occasional

thin and whereas they can provide a good understanding of the ice related sediment dispersal and deposition, the vertical sequences that may develop must be inferred. Smith (1985) points out that the cyclicity presented by the Donjeck model is not recognizable in the proximal and distal areas of glacial outwash where more homogeneous sediments are deposited. Nevertheless, the more intense variations in floods and abundance of sediments in the glacial settings should and do foster development of repetitive vertical sequences in fluvial as well as other associated environments such as lakes (Smith & Ashley, 1985). Further testing of such hypotheses needs to be done in thicker sequences, not just in a few isolated outcrops or cores, but also analysing lateral regional variations.

Cold climate conditions are more clearly recorded and discriminated in *levees and overbank deposits*. Except in their most distal parts, glacio-fluvial settings usually have little vegetation, and, in unconfined areas, braided streams shift laterally and carry considerable amounts of coarse (gravel and sand) sediment. The Albany River instead shows many of the features of postglacial subarctic systems. These include the development of anastomosing to meandering, stable, ribbon-like channels, occasional avulsions (King & Martini, 1983), and coarse deposits restricted to multiple bars and pools within the channel themselves. The stability of the channel is due to fine overbank deposits, frozen ground conditions and plant roots. Narrow continuous shallow levees develop which locally contain abundant, highly diagnostic ice rafted materials, much organic matter, carbonate and iron oxide rhyzoconcretions, and some indication of horizontal voids associated with ice lensing. Furthermore, these ribbon-like, subarctic river systems cross vast, unconfined, boreal to subarctic peatlands characterized by plant associations which include graminaceous vegetation, mosses (*Sphagnum* sp.), tamarack (*Larix laricina*) and spruce (*Picea mariana* and *Picea glauca*).

There is no doubt that on the basis of typical organic, levee and channel deposits, and the presence of specific pedogenic features, cold climate streams may be readily distinguished from temperate to warm climate ones. Distinction between non-glacial and glacial systems is also possible if overbank deposits are preserved. Although the amount of sedimentation is less in the former than in the latter, deposits of the non-glacial systems should also be preserved in the geological record. Such deposits are important because, together with other evidence, they can contribute significantly to a better definition of cold palaeoclimatic zones which have existed even in geologic periods until now considered some of the hottest (Jurassic) in the planetary history (Frakes & Francis, 1987).

SUMMARY AND CONCLUSIONS

About 3.5×10^4 t y^{-1} of fine and coarse material is reworked and rafted by a variety of ice types in the Albany River. This is a relatively small amount (less than 1%) in respect of the total suspended sediment load of the river, but it is important on those reaches of the rivers where ice floes are preferentially accumulated. Ice rafted deposits contain highly diagnostic features such as coarse clasts and thin lenses of pelletized silt and sand which are preserved almost unmodified in organically rich, fine grained levee and abandoned channel fills.

It is also becoming increasingly clear that deposits of similar types of streams, for instance coarse grained, bedload dominated ones, which form under different climatic conditions, do not all look the same. Ribbon rivers like the Albany River have coarse deposits restricted to the active channels, and they are surrounded by and partially drain vast and rapidly aggrading unconfined peatlands. The organic deposits (peats, coals) containing specialized plant remains are reliable subarctic to boreal climatic and palaeoclimatic indicators. The coarse channel deposits themselves contain subtle differences which, although not discriminant when taken individually, certainly can help in distinguishing cold climate from temperate and warm conditions and perhaps cold non-glacial from periglacial settings. These include polymictic composition of clasts, textural–compositional relationships (such as rounded crystalline boulders associated with more frost-susceptible angular carbonate pebbles), and types and degree of cyclical gravelly sedimentation better developed in reaches directly affected by glacial melting cycles.

ACKNOWLEDGEMENTS

This project has been financially supported by the National Science and Engineering Research Council (Grant No. A7371) and the Ministry of Indian and Northern Affairs (Northern Training Program). Significant help in the field has been provided by R. Kelly and G. Wilson. Father G. St Onge of Fort Albany has provided information on the river stages during pre-breakup times. The external review by J. Lewin helped in improving the final version of the manuscript.

REFERENCES

ASHLEY, G.M., SHAW, J. & SMITH, N.D. (1985) *Glacial Sedimentary Environments*. SEPM Short Course No. 16. Soc. Econ. Paleont. Miner., Tulsa, 246 pp.

ASHTON, G.D. (1978) River ice. *Ann. Rev. Fluid Mech.* **10**, 369–392.

ASHTON, G.D. (1980) Freshwater ice growth, motion, and decay. In: *Dynamics of Snow and Ice Masses* (Ed. Colbeck, S.C.) pp. 261–289. Academic Press, New York.

ASHTON, G.D. (1986) *River and Lake Ice Engineering*. Water Resources Publications, Littleton, 485 pp.

BELTAOS, S. (1984) A conceptual model of river ice breakup. *Can. J. Earth Sci.* **11**, 516–529.

BRYANT, I.D. (1983a) The utilization of arctic river analogue studies in the interpretation of periglacial river sediments from southern Britain. In: *Background to Palaeohydrology* (Ed. Gregory, K.J.) pp. 413–431. John Wiley & Sons, Chichester.

BRYANT, I.D. (1983b) Facies sequences associated with some braided river deposits of late Pleistocene age from southern Britain. In: *Modern and Ancient Fluvial Systems* (Eds Collinson, J.D. & Lewin J.) pp. 267–275. Spec. Publs Int. Ass. Sediment. 6. Blackwell Scientific Publications, Oxford.

BURROWS, R.L. & HARROLD, P.E. (1983) *Sediment Transport in the Tanana River near Fairbanks, Alaska. 1980–1981* US Geol. Surv. Water Resources Investigations Report pp. 83–4064, Washington, DC.

CHAPMAN, L.J. & THOMAS, J.K. (1968) *The Climate of Northern Ontario*. Climatological Studies No. 6. Can. Dept Transport, Met. Branch, Toronto.

CHURCH, M. & RYDER, J.M. (1972) Paraglacial sedimentation: a consideration of fluvial processes conditioned by glaciation. *Bull. Geol. Soc. Am.* **83**, 3059–3072.

COLLINSON, J.D. (1971) Some effects of ice on a river bed. *J. Sedim. Petrol.* **41**, 557–564.

COLLINSON, J.D. & LEWIN, J. (1983) *Modern and Ancient Fluvial Systems*. Spec. Publs Int. Ass. Sediment. 6. Blackwell Scientific Publications, Oxford.

CUMMING, L.M. (1968) Rivers of the Hudson Bay Lowlands. In: *Earth Science Symposium on Hudson Bay* (Ed. Hood, P.J.) pp. 144–168. Geol. Surv. Canada, Paper 68-53, Ottawa.

FERRICK, M.G. & MULHERIN, D. (1989) *Framework for Control of Dynamic Ice Breakup by River Regulation*. CRREL Report 89-2. US Army Cold Regions and Engineering Laboratory, Springfield.

FRAKES, L.A. & FRANCIS, J.E. (1987) A guide to Phanerozoic cold polar climates from high-latitude ice-rafting in the Cretaceous. *Nature* **333**, 547–549.

FRASER, J.Z. (1982) Derivation of a summary facies sequence based on Markov chain analysis of the Caledon outwash: Pleistocene braided glacial fluvial deposits. In: *Research in Glacial, Glacio-fluvial and Glacio-lacustrine Systems* (Eds Davidson-Arnott, R., Nickling, W. & Fahey, B.D.) pp. 175–199. Proc. 6th Guelph Symp. on Geomorph., 1980. Geobooks, Norwich.

GERARD, R. (1983) River and lake ice processes relevant to ice loads. In: *Design for Ice Forces* (Eds Caldwell, S.R. & Crissman, R.D.) pp. 121–138. Am. Soc. Civil Engineers. New York.

HUTTON, C.L.A. & BLACK, W.A. (1975) *Ontario Arctic Watershed*. Map Folio No. 2. Env. Canada, Lands Directorate, Ottawa.

KARROW, P.F., JOPLING, A.V. & MARTINI, I.P. (1982) *Excursion 11A: Late Quaternary Sedimentary Environments of a Glaciated Area: Southern Ontario. Field Excursion*

Guidebook. Int. Ass. Sediment. 11th Int. Congress, Hamilton.

KING, W.A. & MARTINI, I.P. (1983) Morphology and recent sediments of the lower anastomosing reaches of the Attawapiskat River, James Bay, Ontario, Canada. *Sedim. Geol.* **37**, 295–320.

LOWDEN, D.J. (1983) *The effect of river ice on a point bar on the Grand River, Ontario.* MSc thesis, University of Guelph.

MARTINI, I.P. (1989) Hudson Bay Lowland: major geologic features and assets. *Geol. Mijnbow* **68**, 25–34.

MARTINI, I.P. & BANKS, M.R. (1989) Sedimentology of the cold-climate, coal-bearing, Lower Permian 'Lower Freshwater Sequence' of Tasmania. *Sedim. Geol.* **64**, 25–41.

MARTINI, I.P. & GLOOSCHENKO, W.A. (1985) Cold climate peat formation in Canada, and its relevance to Lower Permian coal measures of Australia. *Earth Sci. Rev.* **22**, 107–140.

MARTINI, I.P. & JOHNSON, D.P. (1987) Cold-climate, fluvial to paralic coal-forming environments in the Permian Collinsville Coal Measures, Bowen Basin, Australia, *Int. J. Coal Geol.* **7**, 365–388.

MARTINI, I.P. & ROCHA CAMPOS, A.C. (1991) Interglacial and early postglacial Lower Gondwana coal sequence in the Parana' Basin, Brazil. In: *Proceedings Gondwanaland International Symposium, Sao Paulo* (Eds Ulbrich, H. & Rocha Campos, A.C.) pp. 317–336. Universidade de Sao Paulo, Sao Paulo.

MIALL, A.D. (1977) A review of the braided-river depositional environment. *Earth Sci. Rev.* **13**, 1–62.

MIALL, A.D. (Ed.) (1978) *Fluvial Sedimentology*, pp. 597–604. Can. Soc. Petrol. Geol., Calgary, Memoir 5.

MIALL, A.D. (1984) Glaciofluvial transport and deposition. In: *Glacial Geology* (Ed. Eyles, N.) pp. 168–183. Pergamon Press, New York.

MIALL, A.D. (1985) Multiple-channel bedload rivers. In: *Recognition of Fluvial Depositional Systems and their Resource Potential* (Eds Flores, R.M., Ethridge, F.G., Miall, A.D., Galloway, W.E. & Fouch, T.D.) pp. 83–100. SEPM Short Course No. 19. Soc. Econ. Paleont. Miner., Tulsa.

MICHEL, B. (1971) *Winter Regime of Rivers and Lakes.* USA Cold Regions Res. Eng. Lab., CRREL Monographs III-Bla. US Army Cold Regions Res. Eng. Lab., Springfield.

MICHEL, B. (1978) *Ice Mechanics.* Les Presses de l'Université Laval, Quebec City. 499 pp.

MICHEL, B. & RAMSEIER, R.O. (1971) Classification of river and lake ice. *Can. Geotech. J.* **8**, 36–45.

MORISON, S.R. & HEIN, F.J. (1987) Sedimentology of the White Channel gravels, Klondike area, Yukon Territory: fluvial deposits of confined valley. In: *Recent Developments in Fluvial Sedimentology* (Eds Ethridge, F.G., Flores, R.M. & Harvey, M.D.) pp. 205–216. SEPM Spec. Publ. 39. Soc. Econ. Paleont. Miner., Tulsa.

NEWBURY, R.W. (1968) *The Nelson River: a study of subarctic river processes.* PhD thesis, Johns Hopkins University.

OSTERKAMP, T.E. (1975) Observations on Tanana River ice. In: *IAHR, Proceedings, Third International Symposium on Ice Problems, August 18–21, Hanover, New Hampshire*, pp. 201–208.

PRICE, R.J. (1983) *Glacial and Fluvioglacial Landforms.* Oliver and Boyd, Edinburgh, 242 pp.

REIMNITZ, E. & BRUDER, K.F. (1972) River discharge into an ice-covered ocean and related sediment dispersal, Beaufort Sea, coast of Alaska. *Bull. Geol. Soc. Am.* **83**, 861–866.

SCOTT, K.M. (1978) Effects of permafrost on stream channel behaviour in arctic Alaska. US Geol. Surv. Prof. Paper 1068.

SCOTT, K.M. (1979) Arctic stream processes — an annotated bibliography. US Geol. Surv. Water Supply Paper 2065.

SHUMSKII, P.A. (1964) *Principles of Structural Glaciology.* Dover Publications, New York, 497 pp.

SMITH, D.G. (1976) Effect of vegetation on lateral migration of anastomosed channels of a glacier meltwater river. *Bull Geol. Soc. Am.* **87**, 857–860.

SMITH, D.G. (1979) Effects of channel enlargement by river ice processes on bank-full discharge in Alberta, Canada. *Water Resources Res.* **15**, 469–475.

SMITH, D.G. (1980) River ice processes: thresholds and geomorphic effects in northern and mountain rivers. In: *Thresholds in Geomorphology* (Eds Coates, D.R. & Vitek, J.D.) pp. 323–343. Allen and Unwin, Boston.

SMITH, D.G. & SMITH, N.D. (1980) Sedimentation in anastomosed systems: examples from alluvial valleys near Banff, Alberta. *J. Sedim. Petrol.* **50**, 157–164.

SMITH, N.D. (1985) Proglacial fluvial environment. In: *Glacial Sedimentary Environments* (Eds Ashley, G.M., Shaw, J. & Smith N.D.) pp. 85–134. SEPM Short Course No. 16. Soc. Econ. Paleont. Miner., Tulsa.

SMITH, N.D. & ASHLEY, G. (1985) Proglacial lacustrine environments. In: *Glacial Sedimentary Environments* (Eds Ashley, G.M., Shaw, J. & Smith, N.D.) p. 246. SEPM Short Course No. 16. Soc. Econ. Paleont. Miner., Tulsa.

TSANG, G. (1982) *Frazil and Anchor Ice: A Monograph.* National Research Council of Canada, Subcommittee on Hydraulics of Ice-covered Rivers, Ottawa, 89 pp.

WEBBER, P.J., RICHARDSON, J.W. & ANDREWS, J.T. (1970) Post glacial uplift and substrate age at Cape Henrietta Maria, southeastern Hudson Bay, Canada. *Can. J. Earth Sci.* **7**, 317–325.

WILLIAMS, G.P. (1965) Correlating freeze-up and breakup with weather conditions. *Can. Geotech. J.* **11**, 313–326.

WILLIAMS, G.P. & MACKAY, D.K. (1973) The characteristics of ice jams. In: *Seminar on Ice Jams in Canada* (Ed. Williams, G.P.) pp. 17–35. Technical Memorandum 107. National Research Council of Canada, Associate Committee on Geotechnical Research, Ottawa.

ZACHRISSON, G. (1988) Reduction of damages from ice jam flooding in River Tornealven. In: *IAHR, Proceedings of the Nordic Expert Meeting on River Ice, Helsinki* (Eds Laasanen, O. & Forsius, J.) pp. 217–225. Nordic Hydrological Programme Report No. 21.

Dynamics of bed load transport in the Parsęta River channel, Poland

Zb. ZWOLIŃSKI

*Quaternary Research Institute, Adam Mickiewicz University,
Fredry 10, 61-701 Poznań, Poland*

ABSTRACT

Three channel cross-section types have been selected for three Parsęta River reaches. They represent a variety of channel patterns with straight sections and bends with a high and low curvature ratio. Mean annual discharges are 4.6, 13.6 and 35.7 m³ s⁻¹ for selected reaches. Median particle size is largely medium-grained sand ($0.24 \leq d_{50} \leq 0.81$ mm), occasionally more coarse-grained. These deposits display moderate and good sorting. The Parsęta bed load is most often transported in the form of sinuous and linguoid-shaped ripple marks, rarely straight-crested ones, and dunes with occasional streaming lineation. Bed material is transported as large bedforms (side, transverse, point and counterpoint bars). The most dynamic changes in the channel bed position take place within low curvature ratio bends ($CW < 1.5$), and the least changes along straight reaches. The magnitude of these changes is found to reduce downstream. In all cross-sections, the bed material was mobile across 39–90% (average 70%) of the channel width, irrespective of water stage and discharge phase. The bed load amounts calculated using a modified Meyer-Peter & Müller formula range from 0.15 to 0.78 kg s⁻¹. The analysis of the bed load rates reveals their twofold increase in the longitudinal profile of the river. Moreover, a twofold increase in bed load occurred along straight reaches compared with low and high curvature ratio bends. Finally, function models are offered for hydrotechnical applications.

INTRODUCTION

The sources of bed load delivery are connected with the processes of mechanical denudation in the catchment area and the processes of erosion in the river channel. Thus, the bed load consists of channel alluvial deposits, bank material, the bed load from tributaries, scoured bedrock deposits and, to a lesser extent, overbank surface deposits (Froehlich, 1982; Richards, 1982).

A general model of the contribution of particular types of fluvial transport (L_d, dissolved load; L_s, suspended load; L_b, bed load) for the Parsęta River takes the form:

$$L_d : L_s : L_b = 86.7 : 8.0 : 5.3 \quad (\%),$$

and assuming $L_s = 1$:

$$L_d : L_s : L_b = 10.75 : 1 : 0.66.$$

Studies of the internal structure of the total material discharge show the share of the bed load to vary between 0.73 and 14.72% (Zwoliński, 1989). An analysis of eighty-four transport ratios has shown that at high water stages, the share of suspended transport increases and bed load transport decreases, whereas at low water levels the tendency is reversed. This behaviour of the bed load indicates that it is sensitive to changes taking place in the river channel. This paper presents the characteristics and dynamics of the bed load in the Parsęta River channel. Although bed load is the smallest in comparison with the other types of load, it responds best to the hydraulic and morphological conditions of water flow in the river channel.

STUDY AREA AND METHODS

In order to examine the above mentioned problem, three reaches of the Parsęta River (Fig. 1A) were

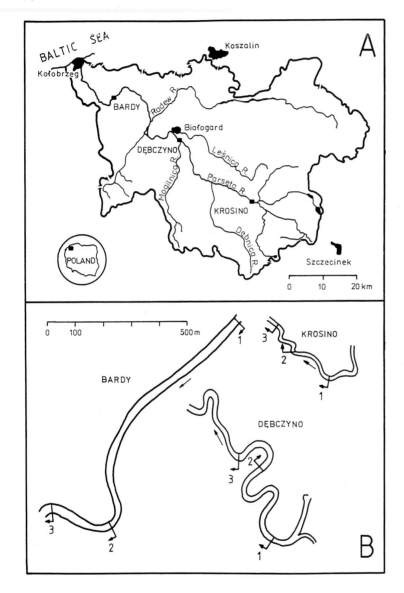

Fig. 1. The study area: (A) the Parsęta River drainage basin; (B) location and numbers of the Parsęta channel cross-sections.

selected for observation: near Krosino (98 km from the river mouth), Dębczyno (63 km) and Bardy (25 km). In each reach three channel cross-sections (Fig. 1B) were chosen representing various types of planar channel patterns, namely a straight reach and bends with a large and small curvature radius (Table 1). Apart from morphometric and hydrometric measurements, the cross-sections were used as sampling sites for bed deposits. The samples were taken using a cylindrical aluminium collector, at each full metre of the effective width of the channel (Zwoliński, 1985). Weight of samples varied from 40 to 230 g, depending on location of the sample site along the cross-section and distribution of hydrodynamical conditions, especially velocity, within the cross-section. Field studies were repeated between seven and ten times at particular cross-sections, thus obtaining a broad range of water discharges, from 18 to 98% of their frequency, including zones of low, medium and high water levels. For selected reaches discharges vary from 2.9 to 10.9 $m^3 s^{-1}$ at Krosino, from 7.8 to 25.2 $m^3 s^{-1}$

Table 1. Basic morphometric characteristics of investigated channel cross-sections on the study reaches of the Parsęta River at bankfull stage

Cross-section	R_c (m)	CW	W (m)	D (m)	F	S	Channel type
Krosino						0.0006	
1	25.0	1.67	14.95	1.86	8.04		High curvature bend
2	10.0	0.48	20.70	2.12	9.76		Low curvature bend
3	—	—	13.60	1.57	8.66		Straight reach
Dębczyno						0.0004	
1	65.0	3.33	19.50	1.67	11.68		High curvature bend
2	—	—	27.50	2.00	13.75		Straight reach
3	30.0	1.24	24.15	1.68	14.37		Low curvature bend
Bardy						0.0002	
1	—	—	30.75	1.66	18.52		Straight reach
2	70.0	2.69	25.95	1.98	13.10		High curvature bend
3	35.0	1.12	31.20	1.46	21.37		Low curvature bend

at Dębczyno and from 21.3 to 47.8 m³ s⁻¹ at Bardy. Mean annual water discharges are 4.6, 13.6 and 35.7 m³ s⁻¹, respectively.

The grain-size distribution of 1215 samples was determined by dry sieving at a 0.5 phi interval. For some samples with very fine fractions (< 0.1 mm), particle size was also determined by the pipette method using a Sartorius sedimentograph. The volume of bed load L_b passing the gauging profiles was calculated using the formula of Meyer-Peter & Müller (1948) with modifications of Zwoliński (1985, 1989):

$$L_b = W_E X (YV^{1.5}(d_{90}S)^{0.25} - AZd_{35})^{1.5} \quad \text{(kg s}^{-1}\text{)},$$

where: A is the constant in the Meyer-Peter & Müller formula, d_{35}, d_{90} the grain diameter for which 35 or 90% of the bed load is finer, S the channel slope, V the mean velocity, W_E the effective width of bed load transport, $X = 2080/(20397.832 - \gamma_w)$, $Y = 0.296\rho_w^{2/3}$, $Z = 81591.328\rho_w^{-1/3} - 39.227\rho_w^{2/3}$, and γ_w, ρ_w are the specific weight and density of water characteristic of a given water temperature.

TEXTURAL FEATURES OF BED LOAD

One of the most popular measures of the average grain diameter is the median, which supplies information about the basic character of the grain population of a sediment sample. The variation in the median grain diameter of all the measurements in the Parsęta River channel cross-sections is presented in Table 2. An analysis of the mean values of the grain diameter median shows that they vary only slightly. The mean values of the bed material median for cross-sections vary between 0.24 and 0.81 mm, and are largely included in the fraction of medium-grained sand, and only in fewer than twenty cases in the coarse-grained sand fraction. The variation of the median is much higher within particular cross-sections over time, with a maximum of almost 5 phi (from 0.125 to 4.0 mm) and an average of 1.5–2.0 phi.

No grain sizes characteristic of different morphological types of cross-sections were found. However, for cross-sections situated on straight reaches of the river and on bends with low curvature ratios, median values have a smaller range than on bends with high curvature ratios. Such median distributions along these cross-sections may be associated with less variable hydraulic conditions of water flow.

Table 3 gives a general presentation of mean textural parameters of the bed material in successive cross-sections calculated from all the measurements. The extreme mean cross-section values of mean grain diameters vary in a small range only over the period of study: from a range of 0.31 phi (cross-section 1 at Bardy) to 0.99 phi (cross-section 2 at Dębczyno).

The sorting parameters are poorly diversified, as standard deviation δ_I varies from 0.13 phi (cross-section 1 at Bardy) to 0.75 phi (cross-section 1 at Dębczyno), while the sorting degree γ varies from 0.28 phi (cross-section 1 at Bardy) to 1.04 phi (cross-section 3 at Dębczyno). The largest range of

Table 2. Variation of average value of grain diameter median d_{50} (mm) for channel cross-sections of the Parsęta River

Number of measurement	Cross-section location								
	Krosino			Dębczyno			Bardy		
	1	2	3	1	2	3	1	2	3
1	0.32	0.40	0.30	0.44	0.24	0.31	0.47	0.56	0.40
2	–	0.44	0.52	–	0.36	0.47	–	–	0.53
3	0.38	0.33	0.32	0.45	0.31	0.43	0.48	0.50	0.53
4	0.41	0.29	0.28	0.51	0.29	0.75	0.50	0.81	0.62
5	0.47	0.31	0.44	0.45	0.26	0.37	0.56	0.63	0.44
6	0.42	0.52	0.37	–	0.33	0.47	–	0.51	0.52
7	0.39	0.37	0.44	–	0.40	0.68	–	0.57	–
8	0.37	0.27	0.33	0.30	0.60	0.47	0.49	0.47	0.45
9	0.40	0.39	0.43	0.43	0.31	0.40	0.55	0.54	0.40
10	0.45	0.41	0.45	0.35	0.26	0.35	0.55	0.58	0.49
Min.	0.32	0.27	0.28	0.30	0.24	0.31	0.47	0.47	0.40
Max.	0.47	0.52	0.52	0.51	0.60	0.75	0.56	0.81	0.62

sorting parameters at the Dębczyno reach may be a result of the periodical scouring of bedrock. Generally, however, the Parsęta River bed material is characterized by moderate to good sorting ($\bar{\delta}_I = 0.67$ phi and $\bar{\gamma} = 1.45$ phi), and the extreme values of parameters are indicative of a poor diversification of the energetic conditions of water flow into the channel with tranquil discharges predominating.

The parameter of graphic skewness shows that the Parsęta River bed material is characterized by

Table 3. Extreme and mean values (phi) of mean channel cross-section values of textural parameters for bed-material deposits of the Parsęta River channel for all measurements

Textural parameter	Cross-section location								
	Krosino			Dębczyno			Bardy		
	1	2	3	1	2	3	1	2	3
Mz									
Min.	1.04	0.96	0.92	0.77	1.15	0.58	0.77	0.60	0.86
Mean	1.31	1.48	1.38	1.29	1.65	1.14	0.91	0.90	1.07
Max.	1.60	1.92	1.87	1.73	2.04	1.57	1.08	1.15	1.32
δ_I									
Min.	0.50	0.55	0.48	0.67	0.46	0.57	0.64	0.53	0.53
Mean	0.59	0.67	0.57	0.86	0.64	0.78	0.71	0.63	0.60
Max.	0.69	0.86	0.70	1.42	0.91	1.11	0.77	0.75	0.78
γ									
Min.	1.07	1.21	1.05	1.42	1.01	1.26	1.41	1.15	1.17
Mean	1.30	1.46	1.25	1.66	1.39	1.70	1.57	1.36	1.34
Max.	1.53	1.83	1.49	1.96	1.86	2.30	1.69	1.62	1.75
Sk_I									
Min.	–0.06	–0.12	–0.10	–0.22	–0.14	–0.18	–0.20	–0.15	–0.17
Mean	–0.01	–0.05	–0.05	–0.10	–0.04	–0.08	–0.13	–0.05	–0.11
Max.	0.03	0.06	0.01	0.01	0.08	0.13	–0.06	0.01	–0.05
K_G									
Min.	0.95	0.98	0.99	1.00	1.01	0.95	0.99	1.01	1.04
Mean	1.09	1.05	1.06	1.10	1.12	1.08	1.08	1.06	1.10
Max.	1.21	1.12	1.13	1.20	1.22	1.21	1.15	1.11	1.16

Mz, δ_I, Sk_I and K_G after Folk & Ward (1957), γ after Zwoliński (1984).

symmetric grain-size distributions ($\overline{Sk}_I = -0.07$ phi) with a tendency towards negatively skewed distributions. A greater contribution of coarser material (from bedrock) is most readily visible in the Dębczyno and Bardy reaches, which can be associated with the considerably greater stream power Ω in this part of the river course (Zwoliński, 1985).

NATURE OF THE MORPHODYNAMIC TRANSPORT ZONE

In a previous study (Zwoliński, 1985) it was observed that the median grain diameter changed together with five hydraulic parameters: the stream power Ω, the shear stress τ, the undimensional shear stress Θ, the Manning resistance coefficient n and the Darcy–Weisbach friction factor $f\!f$. A survey of formulae for the calculation of these parameters reveals that the only component parameter they have in common is the hydraulic radius of a channel cross-section (apart from the slope, which can be regarded as a constant quantity within one reach). Therefore, it seemed useful to trace relationships between the medians of single bed load samples (M) and the depths from which they had been taken (D_x). These relationships are presented graphically for the Bardy reach (Fig. 2), with the samples segregated for various types of cross-sections.

The diagrams are constructed on the basis of data for collected bed load samples. Thus, it may be presumed that samples occurring between the straight lines described by the equation $M/D_x = \pm 1$ indicate a morphodynamic transport zone. Therefore, samples above the straight line $M/D_x = 1$ and below that of $M/D_x = -1$ may be taken to define the morphodynamic zones of accumulation and erosion, respectively. However, this issue requires verification on other rivers, because with coarser or finer alluvial deposits the analysed clusters of sample points will move up or down. At the present stage of research the MD_x diagram can be recommended for the study of sandy channels.

It follows from the MD_x diagrams that for a cross-section on a bend with a higher curvature ratio (Fig. 2A) the median grain-size diameter increases proportionally with channel depth. For a cross-section located on a straight reach (Fig. 2B) the median grain size assumes a slightly variable value within a specific interval, irrespective of the depth. In turn, for a cross-section on a bend with a lower curvature ratio (Fig. 2C) the grain median is inversely proportional to the depth. The reverse relationships in bend cross-sections are closely correlated with velocity distributions in these cross-sections:

1 a bend with a higher curvature ratio: depth increases → velocity increases → grain median increases;

2 a bend with a lower curvature ratio: depth increases → velocity decreases → grain median decreases;

3 a straight reach: all three parameters have their specific values not differing significantly along the whole width of the cross-section.

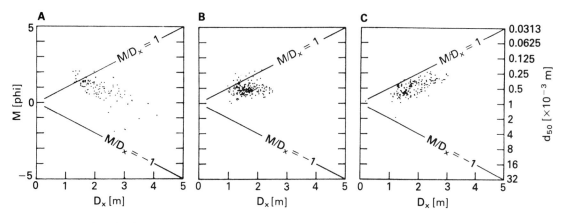

Fig. 2. M/D_x relationships for bed load samples on study reach of the Parsęta River at Bardy: (A) bend with higher curvature ratio; (B) straight river reach; (C) bend with lower curvature ratio.

While the regularities expressed in points 1 and 3 seem obvious and consistent with prevalent opinions, that of point 2 demands an additional explanation. It follows from the existence of a separation zone at the maximum channel depth and from the hydraulic implications of this separation, i.e. the shift of the maximum flow velocities towards the inside bank (Hickin, 1978; Bridge & Jarvis, 1982; De Vriend & Geldof, 1983) and back flows along the outside bank (Zwoliński, 1985, 1989). It is true for most cases of water discharges in bends with lower curvature ratios, including the highest discharges.

TRANSPORT FORMS OF BED LOAD

Morphometric measurements of the cross-sections under study allow the estimation of the extent of reworking of alluvial deposits. The most dynamic changes take place in bends with a low curvature ratio, as they affect a depth of reworked sediment of over 2 m. The smallest changes occur in cross-sections located on straight reaches, with the extent of reworked sediment depth up to 0.5 m (Zwoliński, 1989). Moreover, changes in the channel bed position have been observed to diminish downstream. An implication of these morphological changes of the channel bed is larger bed load along straight reaches of the river and in its lower course.

The Parsęta River bed load is usually transported in the form of sinuous and linguoid-shaped ripple marks, and sometimes in the form of straight-crested ones and dunes. At times streaming lineation on the flat, sandy bed can be observed. Among larger bed forms, it is noteworthy that bed materials were transported and/or accumulated within side, transverse, point and counterpoint bars.

The geometric parameters of dunes supply additional information on the morphological implications of channel cross-sections. Figure 3 presents parameters of dunes in the Parsęta River channel at Bardy for all three channel types with reference to the depth of their occurrence relative to nomogram lines. The length of the dunes can be stated to increase with channel depth more than their height. Allen (1970, 1982) seeks an explanation of this situation in the differences in the grain-size distribution of the bed load, since they reflect the properties of water flow in the channel, especially the thickness of the critical laminar sublayer and the flow velocity. Yalin (1977) has theorized that the dune length to depth ratio is $\lambda_d/D_x = 2\pi$ (i.e. about 6.28), while Jackson's (1975) findings prove that this ratio can vary between 4 and 9 given a wide range of depths. For the Bardy dunes the λ_d/D_x ratio varied from 0.16 to 0.83 (Fig. 3). Relatively low figures have also been reported by Ikeda & Iseya (1980) for the lower course of the Toshio River on Hokkaido Island, and by Bridge & Jarvis (1982) for a bend of the South Esk River in Scotland. Such low values of the λ_d/D_x ratio are indicative of complications in the flow of water within the cross-section caused by, for example, a low curvature ratio, because, as Jackson (1975) claims, the best correlation between dune length and depth is obtained in straight or nearly straight channels ($R_c/W > 10$). The points on the λ_d/D_x diagram correspond to the straight reach and the bend with a low curvature ratio at Bardy. This observation indicates a geomorphic convergence of these cross-section types (Zwoliński, 1989).

Interesting interpretations can be derived from the analysis of the channel depth to dune height

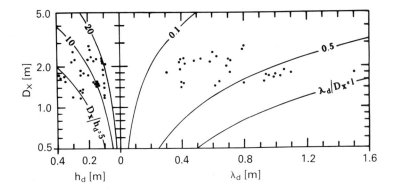

Fig. 3. Geometric parameters of dunes in reference to depth of their occurrence in the Parsęta River channel at Bardy: λ_d, wavelength of dune; h_d, height of dune; D_x, depth.

ratio D_x/h_d. The forms examined in the Parsęta River channel can be categorized, following Jackson's (1975) findings, as transverse bars ($D_x/h_d < 5$), a transition zone ($5 < D_x/h_d < 10$) and dunes ($D_x/h_d > 10$). On the basis of this division and the identification of the successive points in the D_x/h_d diagram (Fig. 3), it has been established that cross-section 1 at Bardy (a straight-channel reach) is represented by dunes, cross-section 2 (a bend with a higher curvature ratio) by a transition zone and dunes, and cross-section 3 (a bend with a lower curvature ratio) by dunes, a transition zone and a transverse bar. The occurrence of a transverse rather than point bar near the convex bank is one of the consequences of water flow in bends with low curvature ratios (Zwoliński, 1989).

RATE OF BED LOAD TRANSPORT

The morphological mapping of the Parsęta channel bed has shown that the bed load has never been transported in the whole channel width in the cross-sections under study, but only in a fragment, i.e. the effective width of bed load transport. The differences between channel width (W) and effective width (W_E) are described by an undimensional index of bed load transport efficiency $E_W = W_E/W$ (Table 4). For the Parsęta River the efficiency index averages 0.704, which means that bed load transport takes place, on average, in 70% of the channel width, although it ranged from 39 to 90% in individual measurements.

It should be emphasized that hardly any dependence can be detected between the efficiency index and the water level, the magnitude and stages of the discharge, the location and shape of a channel cross-section, or the course of the river. This statement and the high variability of the efficiency index bring the realization of two facts:
1 a complex character of bed load transport, determined by a number of local factors of various magnitudes and impacts; and
2 the necessity of accommodating the effective width of bed load transport, rather than channel width, in some palaeohydraulic calculations and the palaeohydrological retrodiction of fluvial processes.

The second fact is also corroborated by the lack of significant functional relations between channel width and the bed load, which is expressed by low correlation coefficients ($\bar{r} = 0.4$). The deficit in bed load transport was mostly recorded near the banks, especially on the concave ones, and in the zone of bed erosion during falling discharge, when lag deposits become exposed. Hence, it is the effective width of bed load transport that determines the efficiency of a channel cross-section.

The movement of bed material was recorded by use of cylindrical collectors in all the cross-sections under study irrespective of the water level and discharge. The bed loads (L_b) were found to vary between 15.7 and 778.6 g s^{-1} (Table 5), with bed load transport increasing roughly twofold in the

Table 4. Undimensional index of bed-material transport effectiveness E_W for channel cross-sections of the Parsęta River

Number of measurement	Cross-section location								
	Krosino			Dębczyno			Bardy		
	1	2	3	1	2	3	1	2	3
1	0.86	0.79	0.89	0.59	0.65	0.73	0.88	0.60	0.74
2	—	0.68	0.66	—	0.50	0.54	—	—	0.70
3	0.86	0.85	0.90	0.62	0.65	0.71	0.87	0.61	0.77
4	0.82	0.66	0.89	0.80	0.71	0.72	0.86	0.58	0.68
5	0.78	0.76	0.80	0.69	0.75	0.75	0.83	0.64	0.78
6	0.77	0.39	0.70	—	0.56	0.68	—	0.64	0.68
7	0.56	0.52	0.81	—	0.64	0.53	—	0.53	—
8	0.77	0.74	0.86	0.70	0.54	0.62	0.89	0.66	0.75
9	0.82	0.65	0.71	0.54	0.56	0.54	0.85	0.60	0.76
10	0.80	0.83	0.78	0.62	0.60	0.69	0.85	0.64	0.74
Min.	0.56	0.39	0.66	0.54	0.50	0.53	0.83	0.53	0.68
Max.	0.86	0.85	0.90	0.80	0.75	0.75	0.89	0.66	0.78

Table 5. Variation of bed load L_b (g s^{-1}) for channel cross-sections of the Parsęta River

Number of measurement	Cross-section location								
	Krosino			Dębczyno			Bardy		
	1	2	3	1	2	3	1	2	3
1	69	105	102	178	224	126	592	303	225
2	–	71	138	–	211	140	–	–	469
3	94	49	101	119	436	186	501	251	387
4	71	16	137	202	434	143	473	243	152
5	146	80	169	204	403	158	616	322	359
6	53	23	143	–	280	197	–	397	194
7	101	19	187	–	370	234	–	395	–
8	67	22	98	103	318	115	622	291	296
9	198	104	162	155	209	109	779	457	442
10	85	85	77	99	207	134	457	262	241
Min.	53	16	77	99	207	109	457	243	152
Max.	198	105	187	204	436	234	779	457	469

longitudinal profile of the river (Krosino $\overline{L}_b$ = 0.10, Dębczyno $\overline{L}_b$ = 0.21, Bardy $\overline{L}_b$ = 0.39 kg s^{-1}). The rise in the bed load transport rate results from appropriate decrease or increase in almost all the geometrical and hydraulic parameters of the channel.

Moreover, there is another double increase in the bed load in the cross-sections on the straight reaches ($\overline{L}_b$ = 0.31 kg s^{-1}) in comparison with those on bends ($\overline{L}_b$ = 0.18 kg s^{-1}). One of the factors accounting for the increased L_b values along the straight reaches is the near overlap of the effective width of bed load transport with the channel width ($\overline{E}_w$ = 0.75). Of the other two kinds of cross-sections, slightly smaller L_b values occur in those on bends with lower curvature ratios.

HYDROTECHNICAL APPLICATION: DISCUSSION

Previous analysis of the multiple correlation of the parameters of hydraulic geometry has shown that the bed load is principally correlated with only one, namely the mean velocity of water discharge (Zwoliński, 1989). Therefore, relationships were traced between the magnitude of the bed load and the mean velocity for the reaches under observation (Fig. 4) and for the different shapes of channel cross-sections (Fig. 5). The high significance and values of the coefficients of correlation between these magnitudes prove that changes in the mean velocity account well for changes in the bed load.

Colby (1964) has found the exponent of the power function of these dependences to be about 3. Similar results have been obtained for the Parsęta River. The relatively high calculated exponents indicate a higher variability of the bed load than that of the mean velocity. The higher exponent (4.35, Fig. 5) in the function for the cross-sections located on the straight reaches results from bed load transport extending over almost the whole width of the channel.

The derived $L_b = f(V)$ functions are important not only for geomorphological or hydraulic studies but also for hydrotechnical studies. Therefore, they can be considered valid within particular study reaches and types of channel cross-sections, with the reservation that they hold for the measured range of mean velocities. However, the relatively narrow interval of mean velocity variation ($0.22 \leq V \leq 0.66$ m s^{-1}) and the limitation of the function to specific river reaches and types of cross-section make their use rather restricted in practical application. A number of authors (e.g. Colby, 1964; Leopold & Emmett, 1977; Andrews, 1979; Przedwojski & Tschuschke, 1982) report a relatively slight diversification of velocity in relation to the bed load transport rate. In order to find a possibly universal formula for the bed load transport rate in the Parsęta River channel, the dependence of the bed load on some parameters of hydraulic geometry was examined. The function verified was:

$$L_b \approx f(Q, W, D, V, P, d_{50}, L_d, L_s, Q_L),$$

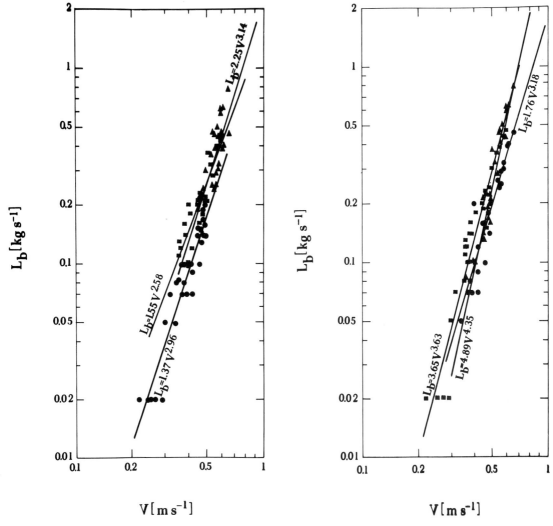

Fig. 4. Bed load L_b as function of mean water velocity V for the Parsęta River study reaches: (●) Krosino; (■) Dębczyno; (▲) Bardy.

Fig. 5. Bed load L_b as function of mean water velocity V for different shapes of the Parsęta River channel cross-sections: (●) bends with higher curvature ratio; (■) straight river reaches; (▲) bends with lower curvature ratio.

where Q is the water discharge, W the channel width, D the mean channel depth, V the mean velocity, P the wetted perimeter, d_{50} the median particle size, L_d the dissolved load, L_s the suspended load, and Q_L the total load discharge. This dependence was subjected to multiple regression analysis with the stepwise variable selection for all the eighty-four channel cross-sections checked. The result was that an optimal fit model can be obtained from the function:

$$\ln L_b = 0.085 + 3.047 \ln Q - 3.013 \ln D - 2.746 \ln P,$$

for $r = 0.968$ and $R = 93.6\%$. The bed loads obtained using this formula were then compared with the observed ones, i.e. those calculated by the modified Meyer-Peter & Müller formula (Fig. 6). The diagram indicates that the estimated loads match the observed ones to a great extent ($r = 0.969$). This means that the derived function can be applied to any given cross-section of the Parsęta River channel satisfying the following conditions:

$$2.89 \leq Q \leq 47.82 \quad (m^3 s^{-1}),$$
$$0.62 \leq D \leq 2.36 \quad (m),$$
$$10.52 \leq P \leq 37.62 \quad (m).$$

Thus, the $L_b = f(Q, D, P)$ function has a broader hydrotechnical application.

It should be noted that in this function the bed load depends on the discharge, mean depth and wetted perimeter. The information conveyed by discharge, or rather its mean velocity, relative to the bed load has been discussed earlier. The morphometric parameters D and P should be interpreted in hydraulic terms, namely mean channel depth decides the magnitude of the shear velocity and shear stress for a given channel cross-section. These two hydraulic parameters define the conditions of the removal and entrainment of grains in the bed load transport. The wetted perimeter, in turn, is closely connected with channel width ($\bar{r} = 0.96$), and hence with the effective width of bed load transport, and indirectly determines the resistance of the channel bed and banks to the flowing water and transported sediments. Thus, the discussed function is not only statistically significant, but also hydraulically justified.

SUMMARY AND CONCLUSIONS

The studies of the dynamics of the Parsęta River bed load prove it to be rather poorly diversified in both the sedimentological and hydraulic aspects. It may be presumed that this is a specific property of sandy alluvial channels in lowland areas.
1 The mean values of the bed material median for cross-sections of the Parsęta River are largely included in the fraction of medium-grained sand. The variation of the median is much higher within particular cross-sections over time. The Parsęta River bed material is characterized by moderate to good sorting.
2 No grain sizes characteristic of different morphological types of cross-sections were found. However, for cross-sections situated on straight reaches of the river and on bends with low curvature ratios, median values have a smaller range than on bends with high curvature ratios.
3 It follows from the MD_x diagrams that for a cross-section on a bend with a higher curvature

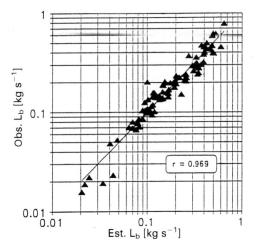

Fig. 6. Observed and estimated bed loads in the Parsęta River channel: obs. L_b, modified Meyer-Peter & Müller formula; est. L_b, $L_b \approx f(Q, D, P)$.

ratio the median grain-size diameter increases proportionally with channel depth. For a cross-section located on a straight reach the median grain size assumes a slightly variable value within a specific interval, irrespective of the depth. For a cross-section on a bend with a lower curvature ratio the grain median is inversely proportional to the depth.
4 The most dynamic changes of reworking of alluvial deposits take place in bends with a low curvature ratio. The smallest changes occur in cross-sections located on straight reaches. Moreover, changes in the channel bed position have been observed to diminish downstream.
5 The bed load has never been transported in the whole channel width in the cross-sections under study, but only in a fragment of the width. Measurements have shown that bed load transport takes place on average in 70% of the channel width.
6 The transport rates of bed loads were found to vary between 15.7 and 778.6 g s^{-1}, with bed load transport increasing roughly twice in the longitudinal profile of the river. At a very simplified count, 3014 t of bed load is transported annually in the Parsęta channel at Krosino, 6651 t at Dębczyno and 12 269 t at Bardy. Moreover, there is another doubling of the bed load in the cross-sections on the straight reaches in comparison with those on bends.

The continuity of traction at its specified effective width, as well as considerable reworking of alluvial deposits of up to 2 m in thickness, indicate the possibility of extracting the bed load deposits as mineral resources. The proposed dependence $L_b \approx f(Q, D, P)$ will come in useful as an easy criterion of selection of the most productive reaches and/or cross-sections of a river channel. A thorough study of all the controls of bed load transport is indispensable before any exploitation of transported sandy river deposits at present is attempted, because exploitation can be accompanied by the undesirable process of river channel incision. The lowering of the channel bed could eventually lead to a too rapid exhaustion of the sources supplying the bed load. The exploitation of this sand would not only be unfavourable from the economic point of view, but it would also induce drastic changes in the natural functioning of a river's channel system, which would in turn affect the whole catchment area.

ACKNOWLEDGEMENTS

This research has been supported by the Adam Mickiewicz University. The author wishes to express his appreciation to G.M. Ashley and P.A. Brewer for critical reading of and their comments on an early draft of the manuscript. Mistakes and omissions are his own.

REFERENCES

ALLEN, J.R.L. (1970) *Physical Processes of Sedimentation*. George Allen and Unwin, London.

ALLEN, J.R.L. (1982) *Sedimentary Structures: their Character and Physical Basis*, Vol. I, II. Developments in Sedimentology, 30A, B. Elsevier, Amsterdam.

ANDREWS, E.D. (1979) Hydraulic adjustment of the East Fork River, Wyoming to the supply of sediment. In: *Adjustments of the Fluvial System* (Eds Rhodes, D.D. & Williams, G.P.) pp. 69–94. Kendall/Hunt, Dubuque.

BRIDGE, J.S. & JARVIS, J. (1982) The dynamics of a river bend: a study in flow and sedimentary processes. *Sedimentology* 29, 499–541.

COLBY, B.R. (1964) Discharge of sands and mean-velocity relationships in sand-bed streams. US Geol. Surv. Prof. Paper 462-A.

DE VRIEND, H.J. & GELDOF, H.J. (1983) Main flow velocity in short river bends. *J. Hydr. Eng., ASCE* 109, 991–1011.

FOLK, R.L. & WARD, W.C. (1957) Brazos river bar: a study in the significance of grain size parameters. *J. Sedim. Petrol.* 27, 3–26.

FROEHLICH, W. (1982) Mechanizm transportu fluwialnego i dostawy zwietrzelin do koryta w górskiej zlewni fliszowej [Summary: The mechanism of fluvial transport and waste supply into the stream channel in a mountainous flysch catchment]. *Prace Geogr.*, IGPZ, PAN, 143.

HICKIN, E.J. (1978) Mean flow structure in meanders of the Squamish River, British Columbia. *Can. J. Earth Sci.* 15, 1833–1849.

IKEDA, H. & ISEYA, F. (1980) On the length of dunes in the lower Teshio River. *Trans. Japan. Geomorph. Union* 2, 231–238.

JACKSON, R.G. (1975) *A depositional model of point bars in the lower Wabash River*. Unpublished PhD thesis, University of Illinois.

LEOPOLD, L.B. & EMMETT, E.W. (1977) 1976 bedload measurements, East Fork River, Wyoming. *Proc. Natl Acad. Sci. USA* 74, 2644–2648.

MEYER-PETER, E. & MÜLLER, R. (1948) Formulas for bedload transport. *Proc. 2nd Meeting, Int. Ass. Hydraul. Struct. Res.*, Stockholm, pp. 39–64.

PRZEDWOJSKI, B. & TSCHUSCHKE, W. (1982) Analiza przydatności wzorów empirycznych do oceny transportu rumowiska wleczonego w warunkach cieków naturalnych w zlewni rzeki Warty [Summary: Usefulness analysis of empirical formulas for estimate of sediment transport in watercourse condition of Warta River Basin]. *Roczn. Akad. Roln. Poznań* 133, Melioracje 4, 69–80.

RICHARDS, K. (1982) *Rivers: Form and Process in Alluvial Channels*. Methuen, London.

YALIN, M.S. (1977) *Mechanics of Sediment Transport*. Pergamon Press, Oxford.

ZWOLIŃSKI, ZB. (1984) Zastosowanie stopnia wysortowania dla zróżnicowania osadów o zbliżonych wartościach miar dyspersji [Summary: Application of the sorting degree index to differentiation of deposits revealing similar dispersion parameters]. *Ann. Soc. Geol. Pol.* 54, 227–239.

ZWOLIŃSKI, ZB. (1985) *Geomorficzne dostosowywanie się koryta Parsęty do aktualnego reżimu rzecznego*. Unpublished PhD thesis, Adam Mickiewicz University.

ZWOLIŃSKI, ZB. (1989) Geomorficzne dostosowywanie się koryta Parsęty do aktualnego reżimu rzecznego [Summary: Geomorphic adjustment of the Parsęta channel to the present-day river regime]. *Dokument. Geogr.* 3/4.

Alluvial Facies

Morphology and facies models of channel confluences

C.S. BRISTOW*, J.L. BEST† and A.G. ROY‡

*Department of Geology, Birkbeck College, University of London, Malet Street, London WC1E 7HX, UK;
†Department of Earth Sciences, University of Leeds, Leeds LS2 9JT, UK; and
‡Département de Geographié, Université de Montréal, Quebec H3C 3J7, Canada

ABSTRACT

Channel confluences represent points of significant change within river networks that are of importance to geomorphologists, sedimentologists and engineers. At scales varying over four orders of magnitude confluences are characterized by distinct areas of scour and deposition: (i) tributary mouth-bars, (ii) a deep confluence scour, (iii) bars within areas of flow separation and (iv) post-confluence mid-channel bars. The morphology of these elements and their depositional facies are controlled predominantly by confluence angle, the discharge ratio between the two channels and modifications which occur at low stage. A review of confluence morphology from flume and field studies is presented alongside examples from the Brahmaputra River; these are then used to present tentative facies models which suggest that junctions may be represented by a unique assemblage of bedforms and sedimentary structures. Sedimentation at channel junctions is particularly important in multichannel braided or anastomosed rivers where channel confluences are most abundant and their likelihood of preservation is at its highest. Complete preservation of confluence sediments may be achieved by abandonment of one channel and domination of the confluence by the other channel or, more rarely, complete abandonment of the entire junction. Partial preservation of the deeper portions of confluence scour and fill at the base of channel sandstones is more likely.

INTRODUCTION

River channel confluences occur within every type of drainage network, but assume greater numerical importance in anastomosed and braided rivers. Confluence studies have considered the joining of individual channels within drainage networks (Morisawa, 1964; Howard, 1971; Mosley, 1976; Mosley & Schumm, 1977; Best, 1986; Roy & Woldenberg, 1987) and their influence on downstream hydraulic geometry (Richards, 1980; Roy & Roy, 1988). However, more recent attention has begun to document the morphology of the abundant channel confluences which occur within braided rivers (Ashmore, 1982; Ashmore & Parker, 1983; Klaasen & Vermeer, 1988) over a range of scales from the junction of two braided rivers (first order) through the joining of major second order channels around mid-channel bars to the confluence of small third order channels which dissect the tops of braid bars at low flow stage (terminology of Bristow, 1987a). It is a paradox that although the number of confluences per kilometre is greatest within braided rivers and it is within braided rivers that confluences may be expected to exert a significant impact upon sedimentation, they are ignored in current models of braided river deposition. Additionally, the temporal fluctuation of discharge within many braided rivers imparts an actively changing morphology which may change radically within a short period of time. In this paper we present morphological data from channel junctions several kilometres in width which significantly complement the existing data base on confluence morphology which is based upon channels several orders of magnitude smaller. Facies models of channel confluences are then presented which are based upon this combined experience of modern examples from natural rivers and flumes.

BED MORPHOLOGY AT CHANNEL CONFLUENCES

Within channel confluences four important depositional elements can commonly be identified (Fig. 1).

1 Bars which form across the mouth of each or both confluent channels (tributary mouth-bars).
2 A zone of scour where the confluent flows combine (confluence scour).
3 Bank-attached bars which owe their presence to the formation of a region of separated flow downstream from the lower junction corner (confluence flow separation bars).
4 Mid-channel bars deposited downstream from the confluence in the zone of flow recovery (post-confluence mid-channel bars).

These distinctive morphological elements have been linked to the junction flow dynamics by Best (1987) both in symmetrical 'Y' shaped junctions and in asymmetrical confluences in which the downstream post-confluence channel forms a linear extension of the upstream main channel (Fig. 1) (Best, 1986). Within asymmetrical confluences, morphology is commonly dominated by avalanche slopes on the mouth-bars, a central scour and a bank-attached bar within the flow separation zone. Often there is no post-confluence mid-channel bar. Within symmetrical 'Y' planform confluences the mouth-bars and scour are still present but the flow separation zone bars are often smaller and post-confluence mid-channel bars are more likely to be developed (Fig. 1).

Confluence mouth-bars

Significant bars develop across the mouth of the confluent channels (Fig. 1). The upstream stoss slope of the bar is usually very low angle but leeside dips can steepen sufficiently to become true avalanche faces. At low confluence angles and low discharge ratios (where discharge ratio is the ratio of tributary : mainstream flow) within asymmetrical junctions avalanche-faced mouth-bars from the main channel can build out well into the confluence (Best, 1986). However, as both junction angle and discharge ratio increase the deflection of the mainstream flow and its sediment load by the tributary becomes greater and the main channel mouth-bar face recedes upstream. At discharge ratios above unity the tributary mouth-bar migrates into the confluence (Fig. 2) and if there is only limited flow

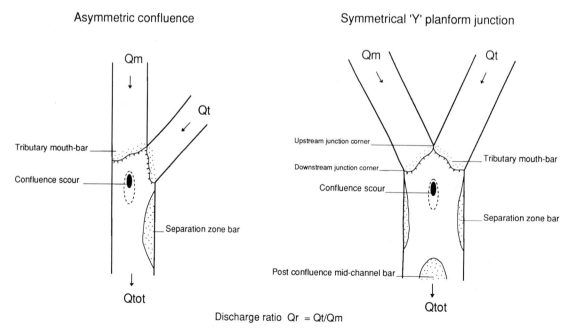

Fig. 1. Bed morphology developed at channel confluences indicating the major depositional and erosional elements within symmetrical and asymmetric confluences.

Fig. 2. A confluence mouth-bar with avalanche faces in the Brahmaputra River. The bar was formed by a third order channel which incised into the top of a second order bar at low flow stage and flowed into a chute channel. Flow within the chute channel is very slight and the mouth-bar has built out well into the chute channel. Scour depth is approximately 3 m.

within the main channel, extensive avalanche-bounded tributary mouth-bars can be formed (Fig. 2). The internal structure of this bar is likely to consist of very large sets of cross-stratification, the size of which is closely related to the depth of scour into which they dip. However, some basal infill of the scour at the bottom of the avalanche faces may show smaller sedimentary structures generated by flow outwards from the scour base (see below). Scour depth, and hence the cross-set size, may be expected to become more pronounced at higher junction angles (Best, 1986, 1988). Evidence of these mouth-bar faces in large channels comes from echo-sounder cross-sections of the Brahmaputra River (Fig. 3). On the echo-sounder trace large dune bedforms, up to 1 m in amplitude, occur on the top of the confluence mouth-bar upstream from the avalanche face edge, but terminate at the bar crest. The steep leeside dip of the mouth-bar appears to be 'smooth' and without any superimposed bedforms; this suggests true avalanching on these 13–14 m high bedforms. The 'true scale' depositional dip on the tributary mouth-bar slip face is around 18° (Fig. 3).

The formation of flow separation zones in the lee of avalanche faces on confluence mouth-bars, and the presence of counter-rotating helical flow cells which have been observed in flume studies by Mosley (1975, 1976) and Ashmore (1982), are likely to modify the avalanche foresets. Sediment tracing experiments by Best (1985, 1987) show that sediment transport is concentrated in two zones on the flanks of the confluence scour and along the face of the mouth-bars. Where the mouth-bar front is not so steep as to form a true avalanche face, sedimentary structures upon the lee face should reflect sediment transport along the faces. We have observed ripples migrating subparallel to the tributary mouth-bar avalanche face within the junction scour of a small confluence. Progradation and movement of the mouth-bar front during flood events may therefore give rise to a sequence of structures which include very large scale cross-stratification generated by migration of the mouth-bar face with sets of smaller scale planar/trough cross-stratification or ripple lamination generated by bedforms migrating subparallel to the avalanche face, especially near the foot of the foresets. The upper portions of mouth-bars are likely to contain frequent reactivation surfaces produced by successive flood events. In extreme cases where the momentum of one stream is many times that of the other, and where sediment loads are very high, mouth-bars can extend well into the channel confluence (Reid et al., 1989). These have been termed expansion bars (Baker, 1984) and may contain similar features to those described by Alam et al. (1985) with local reverse flows occasionally generating herring-bone cross-stratification.

Confluence scour

In both asymmetrical and symmetrical planform junctions the confluence scour originates near the

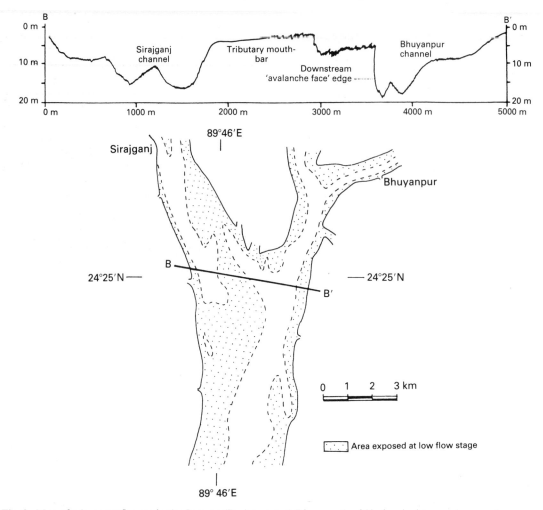

Fig. 3. Map of a large confluence in the Jamuna (Brahmaputra) River south of Sirajganj with an echo-sounder cross-section across the tributary bar top showing dune bedforms 1–2 m high. These dunes terminate at the sharply defined bar top edge where the tributary bar has a downstream avalanche slipface with a dip of approximately 18°. Migration of this slipface could produce cross-stratification up to 10 m high by migration into the 18 m deep confluence scour.

upstream junction corner and stretches downstream with an orientation which approximately bisects the junction angle. However, the scour orientation also varies with the channel discharge ratio and as the contribution of flow from the tributary increases for any given confluence angle, so the scour aligns itself more with this channel. The scour is formed and maintained by the interaction of three factors.

1 Erosion of the bed beneath the turbulent free shear layer generated between the confluent flows (Best, 1987) may initiate and form the bed scour.

2 Generation of flow separation and possible helical flow cells in the lee of the mouth-bar avalanche faces may also contribute to scour at the downstream reattachment line (Best, 1988).

3 The scour may be maintained by the diversion of sediment around the centre of the confluence, a pattern which becomes more pronounced at higher junction angles and discharge ratios (Best, 1988).

As a result the confluence scour is a zone of relatively little sediment transport with transport paths outwards from the centre of the scour and along the flanks of the mouth-bar faces.

Confluence scours are particularly important features as their scour depth often exceeds that generated at meander bends or in the lee of bedforms (Klaassen & Vermeer, 1988). Consequently, confluence scours are likely to produce the points of deepest erosion in the bed of river channels and hence the thickest part of any single channel sandbody. The confluence scour is also a potential site for the concentration of heavy minerals into placer deposits (Mosley & Schumm, 1977; Best & Brayshaw, 1985). Besides the influence of junction angle and channel discharge ratio, the absolute scour depth is related to the size of the channels concerned, total discharge being an important factor (Mosley, 1982; Sutherland, 1986). In the Brahmaputra at bankfull stage the maximum depth

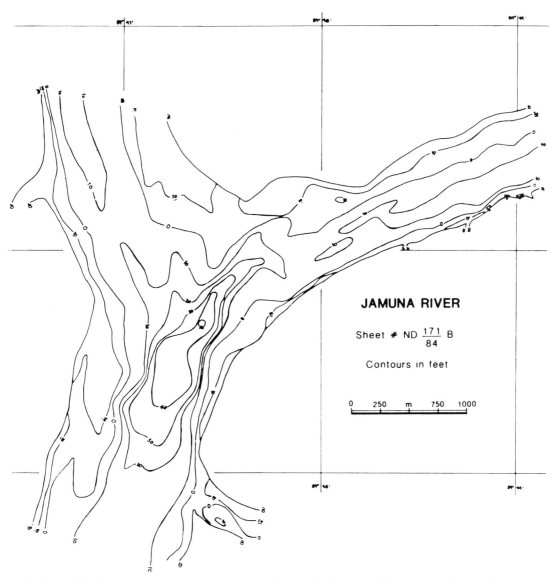

Fig. 4. Chart of the Bhuyanpur channel confluence in the Jamuna (Brahmaputra) River, just north of the Sirajganj confluence shown in Fig. 3. This chart shows the characteristic mouth-bar development across the Bhuyanpur channel (river left, entering from east), confluence scour and a possible confluence separation zone bar, river right, below the downstream junction corner.

recorded in a confluence scour is greater than 30 m. Figures 3 and 4 show large confluence scours within the Brahmaputra which have an orientation that approximately bisects the junction angle, with the scour commencing at the upstream junction corner. In most cases the scour is bounded by the depositional slopes of the tributary mouth-bars similar to those documented in laboratory flumes (Mosley, 1976; Best, 1987, 1988). Scour also increases with confluence angle whilst at junction angles less than 15° appreciable scour is absent (Mosley, 1976; Best, 1988). Under disequilibrium conditions in natural channels the scour is usually maintained but the absolute scour depth varies with changes in total discharge and/or discharge ratio. The eventual fill of the confluence scour is most likely to result from the migration of mouth-bars into the scour. Because abandonment of one of the upstream channels can occur frequently within braided rivers, the preservation potential of these scours and their fill by mouth-bar progradation is predicted to be high.

Confluence separation zone bars

The flow separation zone below the downstream junction corner has both lower fluid velocities and pressures than the surrounding flow (Best & Reid, 1984). Consequently sediment is entrained into this area and accumulates to form a bar. Because of the segregation of main channel and tributary sediment transport paths on either side of the confluence scour, the sediment within the flow separation zone bar is predominantly derived from the adjacent channel. Flume tank experiments (Best, 1985, 1987) demonstrate that the bar may reach an equilibrium state where flow depth is adjusted to provide the bed shear stress necessary to transport the sediment supplied to any one point. In both the flume and natural channels reverse flows have been observed in the separation zone which may generate sedimentary structures, such as cross-stratification within the bar top or pebble imbrication, on top of the bar with palaeocurrents reversed with respect to the main flow direction. The sediment entrained onto confluence separation zone bars is often relatively fine-grained, indicating the relatively low flow velocities in the flow separation zone. Separation zone bars are bank-attached and dip into the confluence scour but grade into the general bed elevation downstream where the effects of flow separation diminish. Although these bars have been documented at small junctions (Mosley, 1976; Best, 1988) their presence is far from clear at larger confluences. Figure 4 suggests the presence of a side-attached bar below the downstream junction corner and although its exact origin is unclear, it could be generated by large scale flow separation which has been observed on the downstream margin of some bars within the Brahmaputra (Bristow, 1987b).

Post-confluence mid-channel bars

Mid-channel bars are often formed downstream from the confluence scour. Deposition occurs in this area as flow expands and velocities decrease downstream from the confluence scour, a process which may be important in generating medial bars within braided rivers (Ashworth et al., 1992). Development of the mid-channel bar may also be assisted by the accumulation of material eroded from the scour and the convergence of sediment transport pathways after their segregation along either side of the confluence scour (Best, 1986). Post-confluence mid-channel bars accrete vertically as well as laterally (Bristow, 1987b), and may sometimes develop avalanche faces on their downstream margins. Post-confluence mid-channel bars may become exposed at low flow stage when their morphology can become modified and fine-grained sediment accumulates in the lee of the bar.

FACIES MODELS

From our knowledge of the morphology of river channel confluences across a range of scales in past work and the giant confluences described here, it becomes possible to suggest tentative models of confluence facies. Figure 5 illustrates a schematic symmetrical confluence at bankfull stage with well-developed mouth-bars. These bars prograde into the confluence and are bounded by large avalanche faces which dip into the confluence scour. Dune bedforms migrate parallel to the confluent channels and terminate at the mouth-bar crest, where sediment avalanches over the bar face. Bedform orientation on top of the mouth-bar is dependent upon the junction angle, discharge ratio, flow stage and position within the channel. The mouth-bar crest and avalanche faces which dip into the confluence scour are usually oblique to the tributary channel with obliquity increasing with junction angle. Occasional reverse flows occur near the upstream

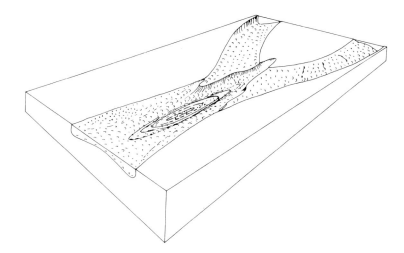

Fig. 5. Facies model for a symmetrical channel confluence at high flow stage.

junction corner (Baker, 1984; Alam *et al.*, 1985) due to flow stagnation and upstream flow from a tributary channel with exceptionally high discharge. On either side of the confluence scour, two bars are formed within the flow separation zones where recirculating flow produces upstream migrating bedforms on top of the separation zone bar. Further downstream, flow expansion and a reduction in velocity may result in sediment deposition and the initiation of a post-confluence mid-channel bar.

Effects of changes in stage

The schematic confluence in Fig. 5 is shown at low flow stage in Fig. 6. Water level within the tributaries has fallen and flow is now diverging across the top of the confluence mouth-bars. These low flow channels are incising into the bar top and building out their own smaller mouth-bars. At low flow stage the confluence scour is partially filled and the mouth-bars may build out well into the confluence, reworking and reactivating the avalanche face of one or both mouth-bars. Reverse currents have been observed on top of tributary mouth-bars in the Brahmaputra River during falling stage when large flow separation zones formed on the bar top. At low flow stage the tops of the flow separation zone bars and the tops of the post-confluence mid-channel bars become exposed; these bars may also be modified as flow is diverted around them.

A hypothetical cross-section through the confluence mouth-bars in the symmetrical confluence

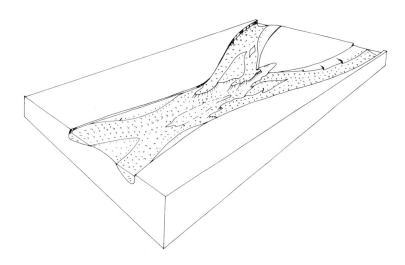

Fig. 6. Facies model for a symmetrical channel confluence at low flow stage.

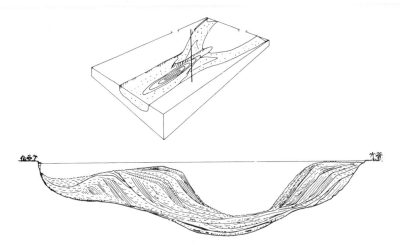

Fig. 7. Hypothetical cross-section of the symmetrical confluence shown in Figs 5 and 6.

depicted in Figs 5 and 6 is shown in Fig. 7. The sedimentary structures within the tributary mouth bar include: (i) giant sets of cross-stratification produced by progradation of the mouth-bar avalanche face, (ii) large scale scour-and-fill structures, and (iii) sets of cross-stratification normal to the dip of the avalanche faces produced by bedforms which migrate subparallel to the avalanche face. These are formed as a result of either flow deflection within the junction or flow along the avalanche faces which is generated by large scale flow separation over their leesides. All of these sedimentary structures are enclosed by bounding surfaces which are essentially surfaces of erosion or non-deposition between phases of mouth-bar build out and correlate with major changes in discharge. Mouth-bar progradation can occur at any time within the flood hydrograph, although reworking of the mouth-bar by smaller channels which incise into the bar top is most likely to occur on the falling stage of a hydrograph.

Non-coincident flood waves

Morphological changes which are caused by varying discharge ratio can be very pronounced at confluences. Two confluent rivers may have very different drainage basins and catchment areas which can cause the relative discharges of the confluent channels to fluctuate (Reid *et al.*, 1989). Non-coincident flood waves generated through different storm/interception conditions within the catchment area of each of the confluent channels may significantly alter the discharge ratio and lead to rapid and pronounced mouth-bar progradation (Reid *et al.*, 1989). This rapid change in discharge ratio may occur in perennial streams but is more common in ephemeral streams where the likelihood of coincident flood waves is much reduced. As a result, mouth-bar progradation from one channel and consequent asymmetric fill and preservation of the scour may be enhanced in ephemeral rivers. Furthermore, the migration of these mouth-bars and the possibility of the movement of the entire junction may result in a relatively high preservation potential for these fills. At channel junctions within individual multichannel rivers flood events are synchronous but discharge ratio between second and third order channels may change with channel migration and sedimentation.

Effects of changes in planform

An account has been given of the influence of confluence planform upon the distribution of characteristic morphological elements. In general, asymmetrical junctions possess larger flow separation zones than equivalent sized symmetrical junctions but commonly do not develop distinct mid-channel bars downstream from the confluence. This imparts a different facies distribution to the confluence (Fig. 8). Best (1986) has shown that an increase in confluence angle up to 90° results in increasingly pronounced avalanche faces and more intense scour at the channel confluence; increase in junction angle up to 90° also results in an increase in flow deflection and the width of the flow separation zone (Best, 1987). The increased width of the flow separation

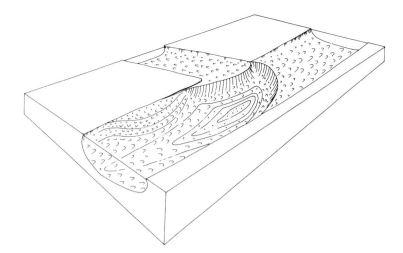

Fig. 8. Schematic facies model for an asymmetrical channel confluence at high flow stage.

zone in a strongly asymmetric junction produces a well-developed confluence flow separation zone bar (Fig. 8). Reverse circulation in the flow separation zone may lead to ripples and dunes migrating upstream on the bar top.

SUMMARY AND CONCLUSIONS

Four morphological elements can commonly be identified within channel junctions: (i) confluence mouth-bars, (ii) large central scours, (iii) confluence separation zone bars and (iv) post-confluence mid-channel bars. Within asymmetrical planform confluences morphology is dominated by avalanche slopes, a central scour and a bar formed below the downstream junction corner within the separation zone. Within symmetrical, 'Y' planform confluences all these elements may be present but post-confluence flow separation bars may not be so well developed and a mid-channel bar can be formed downstream of the post-confluence channel. Confluence scour often exceeds scour depths present in channel bends, in the lee of bedforms or around obstacles and therefore may form the deepest part of a river. Because of this deep erosion level at confluences these sites are likely to produce one of the greatest thicknesses of sediment in alluvial channel deposits. The scour is therefore a potentially important site for increased sandbody thickness in ancient fluvial deposits. Scour-fill is most likely to result from the migration of mouth-bars into the confluence scour. The sedimentary structures predicted to occur within tributary mouth-bar deposits include very large sets of cross-stratification relative to others within the channel, down-current dipping bed sets and large scale scours. It is possible that these facies suites may be of especial importance in braided river deposits where the density of confluences and the potential for mouth-bar progradation and confluence abandonment are both high.

ACKNOWLEDGEMENTS

C.S.B. would like to thank Texaco for funding the Brahmaputra River research and the Royal Society for a grant to attend the conference in Sitges. J.L.B. would like to thank the Royal Society for a grant to read this paper at Sitges whilst the FCAR and NSERC (Canada) provided continuing funds to support this research. The thoughtful and critical comments of Gail Ashley and Signe-Line Røe have much improved the final form of this paper.

REFERENCES

Alam, M.M., Crook, K.A.W. & Taylor, G. (1985) Fluvial herring-bone cross-stratification in a modern tributary mouth bar, Coonamble, New South Wales, Australia. *Sedimentology* **32**, 235–244.

Ashmore, P.E. (1982) Laboratory modelling of gravel braided stream morphology. *Earth Surf. Proc.* **7**, 201–225.

Ashmore, P.E. & Parker, G. (1983) Confluence scour in coarse braided streams. *Water Resources Res.* **19**, 392–402.

Ashworth, P.J., Ferguson, R. & Powell, M. (1992) Bedload

transport and sorting in braided channels. In: *Dynamics of Gravel Bed Rivers* (Eds Billi, P., Hey, R.D., Thorne, C.R. & Taconni, P.) pp. 497–513. John Wiley & Sons, London.

BAKER, V.R. (1984) Flood sedimentation in bedrock fluvial systems. In: *Sedimentology of Gravels and Conglomerates* (Eds Koster, E.H. & Steel, R.J.) pp. 87–96. Can. Soc. Petrol. Geol., Calgary, Memoir 10.

BEST, J.L. (1985) *Flow dynamics and sediment transport at river channel confluences.* Unpublished PhD thesis, Birkbeck College, University of London.

BEST, J.L. (1986) The morphology of river channel confluences. *Prog. Phys. Geogr.* **10**, 157–174.

BEST, J.L. (1987) Flow dynamics at river channel confluences: implications for sediment transport and bed morphology. In: *Recent Developments in Fluvial Sedimentology* (Eds Ethridge, F.G., Flores, R.M. & Harvey, M.D.) pp. 27–35. Soc. Econ. Paleont. Miner., Tulsa, Spec. Publ. 39.

BEST, J.L. (1988) Sediment transport and bed morphology at river channel confluences. *Sedimentology* **35**, 481–498.

BEST, J.L. & BRAYSHAW, A.C. (1985) Flow separation — a physical process for the concentration of heavy minerals within alluvial channels. *J. Geol. Soc. Lond.* **142**, 747–755.

BEST, J.L. & REID, I. (1984) Separation zone at open channel junctions. *J. Hydr. Eng., ASCE* **110**, 1588–1594.

BRISTOW, C.S. (1987a) Brahmaputra River: channel migration and deposition. In: *Recent Developments in Fluvial Sedimentology* (Eds Ethridge, F.G., Flores, R.M. & Harvey, M.D.) pp. 63–74. Soc. Econ. Paleont. Miner., Tulsa, Spec. Publ. 39.

BRISTOW, C.S. (1987b) *Sedimentology of large braided rivers ancient and modern.* Unpublished PhD thesis, University of Leeds.

HOWARD, A.D. (1971) Optimal angles of stream junctions; geometric, stability to capture and minimum power criteria. *Water Resources Res.* **7**, 863–873.

KLAASSEN, G.J. & VERMEER, K. (1988) Confluence scour in large braided rivers with fine bed material. *Int. Conf on Fluvial Hydraulics, Budapest.*

MORISAWA, M. (1964) Development of drainage systems on an upraised lake floor. *Am. J. Sci.* **262**, 340–354.

MOSLEY, M.P. (1975) *An experimental study of channel confluences.* Unpublished PhD thesis, Colorado State University.

MOSLEY, M.P. (1976) An experimental study of channel confluences. *J. Geol.* **84**, 535–562.

MOSLEY, M.P. (1982) Scour depths in branch channel confluences, Oahu River, Otago, New Zealand. *Proc. New Zealand Soc. Civil Engrs* **9**, 17–24.

MOSLEY, M.P. & SCHUMM, S.A. (1977) Stream junctions — a probable location for bedrock placers. *Econ. Geol.* **72**, 691–694.

REID, I., BEST, J.L. & FROSTICK, L.E. (1989) Floods and flood sediments at river confluences. In: *Floods: Hydrological, Sedimentological and Geomorphological Implications* (Eds Bevin, K. & Carling, P.) pp. 135–150. John Wiley & Sons, London.

RICHARDS, K.S. (1980) A note on the changes in channel geometry at tributary junctions. *Water Resources Res.* **16**, 241–244.

ROY, A.G. & ROY, R. (1988) Changes in channel size at river confluences with coarse bed material. *Earth Surf. Proc. Landf.* **13**, 77–84.

ROY, A.G. & WOLDENBERG, M.J. (1987) A model for changes in channel form at a river confluence. *J. Geol.* **94**, 402–411.

SUTHERLAND, A.J. (1986) Scouring at channel confluences. In: *Proc. 9th Australian Fluid Mechanics Conf. Auckland*, pp. 260–263. University of Auckland, Auckland.

Interpretation of bedding geometry within ancient point-bar deposits

B.J. WILLIS*

Department of Geological Sciences, State University of New York at Binghamton, Binghamton, NY 13901, USA

ABSTRACT

Few ancient river-channel deposits have been described in enough detail to allow interpretation of channel-bar geometry and migration. Such interpretations require an understanding of the interaction between the style of channel migration, temporal and spatial variation in channel-bar geometry and facies, and outcrop orientation. This interaction is modelled with the aid of a computer model which predicts thickness, bedding geometry, grain size and palaeocurrent orientations of deposits formed by a migrating, curved alluvial channel segment. Computer simulations are compared with several well-described ancient point-bar deposits. These comparisons suggest that bedding geometry and sedimentary characteristics observed in outcrops can be interpreted in terms of specific cross-sections through migrating channel bends of prescribed geometry and hydraulics.

INTRODUCTION

Lateral-accretion beds and their bounding surfaces observed in ancient fluvial channel deposits have long been recognized as the record of channel bar deposition. Channel bars are complex, three-dimensional bodies which change shape and position with time as the channel migrates. These changes in bar shape and position should be reflected in the facies and geometry of the bar deposit. Interpretation of bar deposits requires an understanding of the interaction between the style of channel migration, temporal and spatial variations in channel-bar surface geometry and facies, and outcrop orientation. Unfortunately, bar geometry, surface texture, bedform distribution and channel migration in modern rivers have not generally been studied in enough detail to allow a reliable estimate of internal bedding geometry and sedimentological characteristics within specific cross-sections (however see Bridges & Leeder, 1976; Jackson, 1976; Bridge *et al.*, 1986). Similarly, few studies of ancient deposits integrate observations of lateral variation in deposit thickness, bedding geometry, texture, sedimentary structures and palaeocurrent orientation in a way that allows for accurate interpretation of bar geometry and migration.

In this paper a theoretical model is used to predict variation in three-dimensional geometry and grain size distribution of point-bar surfaces in channels with varying geometry and hydraulics. Synthetic deposits are then produced by migrating and superimposing several of these bar surfaces. Lateral trends in bedding geometry, grain size and palaeocurrent orientation within specific cross-sections are examined and compared with palaeochannel outcrops from the Devonian Catskill Magnafacies (NY, USA) and the Miocene Siwalik Group (Potwar Plateau, Pakistan). These comparisons suggest that a three-dimensional perspective is needed to accurately interpret bedding geometry observed in ancient fluvial channel deposits.

SIMULATION OF POINT-BAR DEPOSITS

Three-dimensional bar surface geometry and grain size variations are controlled by the interaction of flow and sediment transport. Despite many recent

* Present address: Bureau of Economic Geology, University of Texas at Austin, University Station, Box X, Austin, Texas 78713-7508, USA.

advances in modelling this interaction (Engelund, 1974a,b; Bridge, 1977, 1982; Dietrich & Smith, 1983; Dietrich, 1987; Nelson & Smith, 1989a,b), a general model for bar development, geometry and migration has yet to be developed. The approximate mathematical model of Bridge (1976, 1977, 1982) predicts flow, bed topography and sediment size distribution for steady, non-uniform flow through non-circular channel bends. This model closely simulates point-bar geometry and textural characteristics of many modern river bends (Bridge & Jarvis, 1982; Bridge, 1984). Bridge (1978) demonstrated use of the model in predicting vertical sequences of grain size and sedimentary structures within channel deposits and in reconstructing palaeochannel hydraulics. Willis (1989) used this model to examine lateral variations in point-bar deposits resulting from different styles of channel migration and varying outcrop orientation; it will also be used in this paper.

Bed topography and surface grain size distributions for two simulated point-bars are shown in Fig. 1. The two channels differ only in bend sinuosity. In cross-sections transverse to the channel centreline, bed topography varies little at the bend entrance and exit. Bed topography shows maximum cross-channel variation at the bend apex. The less sinuous bend shows less variation in bed topography than the bend with greater sinuosity. Grain size coarsens up the bar surface in upstream bend segments, and fines up in the downstream segments. Grain size sorting over the bar is more pronounced for the higher sinuosity bend. The geometry of bar surfaces, the distribution of grain size and the palaeocurrent orientations within a given cross-section depend on both bar geometry and cross-section orientation.

For channels which migrate downstream and/or increase in sinuosity, bar migration will tend to produce deposits which thicken away from the channel belt axis (Fig. 2A,B). Successive lateral-accretion beds will also tend to steepen in dip and display wider lateral variation in mean grain size. Migration by translation removes upstream (coar-

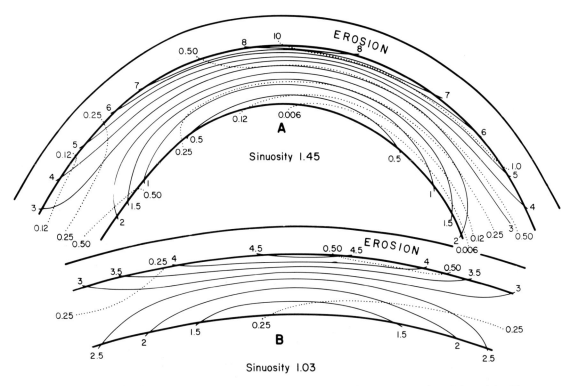

Fig. 1. Two simulated point-bars with different bend sinuosities. Channel flow progresses from left to right. The channels have the same discharge, mean centreline depth/width, and down valley slope. Solid contours display depth in metres; dotted contours (in phi increments) display grain size in millimetres.

sening upward) deposits in favour of downstream (fining upwards) deposits, thus producing the characteristic fining upward point-bar sequence. Palaeocurrent variation within the deposit depends on sinuosity variation and outcrop orientation relative to that of channel migration and the channel-belt axis. Willis (1989) presents a more detailed examination of predicted lateral trends in deposit characteristics.

Exceptions to the general trends described above are observed within a few specific modelled cross-section orientations and with variation of other parameters which affect bar geometry. Temporal variations in the ratio of channel width to mean channel depth can greatly affect lateral variations in deposit thickness and bedding inclination. Variation of this ratio along the bend and with bend migration has not been well documented. Similarly, variations in channel-bend wavelength can influ-

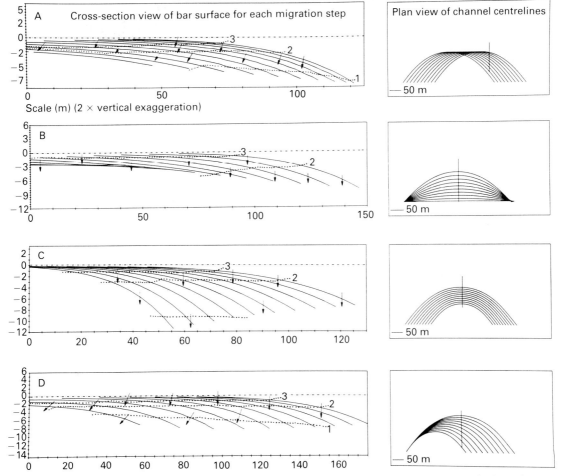

Fig. 2. Profiles displaying progressive changes in simulated bar geometry as viewed in a plane perpendicular to the channel belt axis. Solid lines show apparent bar geometry after each migration step. Grain size contours (in phi increments) are displayed by dotted lines. Arrows show downstream channel orientation relative to the cross-sections (up is into the cross-section plane). Cross-sections are displayed as viewed in the upstream direction. Note 2 × vertical exaggeration. Cross-section orientation relative to migrating channel centreline is displayed in box to the right of profiles. (A) Variation in apparent bar geometry due to down valley bend translation; cross-sections progress from bend crossover to bend apex positions. (B) Progression of apparent bar geometry due to sinuosity increase. All sections pass through the bend apex. (C) Variation in apparent bar geometry with increasing bend wavelength. All profiles pass through the bend apex. (D) Apparent bar geometry produced by increasing bend wavelength and progression from bend crossover to bend apex cross-sections.

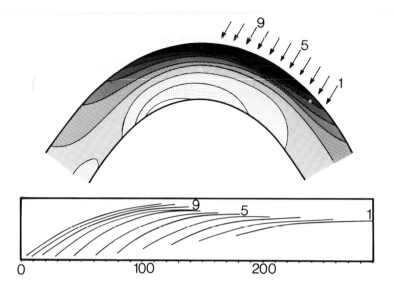

Fig. 3. Schematic diagram of a channel bend with maximum depth downstream of the bend apex (bend contoured for topography, no specific scale implied). Arrows indicate position and orientation of cross-section profiles displayed in the box below.

ence bar geometry (Fig. 2C). Progressive increase in wavelength during bar migration will tend to decrease lateral variations in deposit thickness and bed inclination (Fig. 2D). However, channel-bend wavelength is expected to remain relatively constant for a channel with constant discharge (Carlton, 1965; Dury, 1965; Schumn, 1968; Ackers & Charlton, 1970). Temporal variation in hydraulic conditions (e.g. discharge) can also greatly affect lateral trends in bedding geometry and grain size observed within deposits. Finally, the model used here predicts only symmetric point-bars with maximum channel depth located at the bend apex (Fig. 1). Many studies have reported asymmetry within channel bends associated with a shift in maximum depth downstream of the bend apex and/or asymmetry of the bend planform (see references in Hooke, 1984 and Dietrich, 1987). Downstream translation of bends with maximum channel depth downstream of the bend apex will tend to produce deposits that initially thicken and display increased bed dips, but display decreasing bed dips directly adjacent to the channel fill (Fig. 3). Channel-bend planform asymmetry can also significantly affect lateral variations in bedding geometry and palaeocurrent orientations. Because the complex array of controls on bar geometry and migration cannot be considered by the model used in this study, the synthetic deposits examined here provide a restricted view of possible deposit geometries and characteristics. Despite these limitations the model usefully summarizes expected variations within point-bar deposits.

INTERPRETATION OF BEDDING GEOMETRY

Four palaeochannel deposits (storeys) are described below in terms of deposit thickness, bedding geometry, grain size variation and palaeocurrent orientations. Bedding diagrams were prepared from photo mosaics, then digitized and replotted with two times vertical exaggeration to emphasize lateral variations in bedding geometry. Palaeocurrent orientations are plotted on bedding diagrams relative to the outcrop orientation. Sedimentological logs display variations in grain size, sedimentary structures and palaeocurrent orientations relative to north. Observed variations are compared with selected simulated deposits. Initial hydraulic parameters for simulated deposits were obtained using the methods of Bridge (1978) and Bridge & Diemer (1983).

Outcrop 1

This outcrop is located within the Upper Devonian Catskill Magnafacies near Otego, New York, USA. The outcrop contains a single-storey sandstone body which passes laterally into a fine-grained channel fill (Fig. 4). Description of the entire

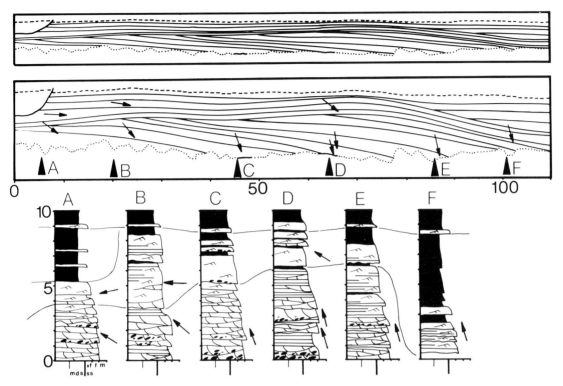

Fig. 4. Bedding geometry, palaeocurrent orientation and sedimentological logs of outcrop 1. Upper box shows bedding surfaces without vertical exaggeration; in the lower box bedding surfaces are displayed with 2 × vertical exaggeration to emphasize lateral deposit variations (scale in metres). Solid lines mark bedding surfaces, dashed line marks the top of the storey, thick black line marks the base of the storey (dotted line where the base is not exposed). Position of sedimentological logs is marked by triangles (A–F). Palaeocurrent orientations (arrows) are displayed relative to the outcrop in bedding diagrams (up is into the outcrop) but relative to north in the logs (up is to the north). The azimuth of the outcrop is N 95° E.

exposure and its regional setting has been presented by Bridge & Gordon (1985); here only the right half of the exposure is examined. Beds within most of the storey display a progressive increase in inclination towards the channel fill. An erosion surface within the inclined beds adjacent to the channel fill truncates the top of previously deposited beds. Above this erosion surface the beds have a convex-upward geometry. The story shows an overall fining upward trend and passes gradually into overlying mudstones. Large-scale cross-stratification and planar-stratification dominate lower in the storey but small-scale cross-stratification becomes common in the upper part of the storey. The convex-upwards beds are dominated by small-scale cross-stratification. Beds within the channel fill are mainly fine grained (mudstone) and horizontal, and disconformably overlap the adjacent, inclined, sandstone beds. Palaeocurrents are nearly parallel with the outcrop to the left of the bedding diagram and progressively change to nearly outcrop normal to the right. The relatively small channel fill (far left) is interpreted as a chute channel.

The progressive increase in bedding inclination observed in most of the storey is the expected sequence for deposition on a migrating channel point-bar (Fig. 2A,B). Bedding geometry and palaeocurrent variation indicate an outcrop orientation that is at a low angle relative to the palaeochannel belt axis and that the bar migrated by both downstream translation and expansion (sinuosity increase) (Fig. 5A). The convex-upward geometry of the upper beds of the storey suggests a cross-sectional view through both the upstream and downstream parts of the bar. The asymmetrical inclinations of these beds and their palaeocurrent

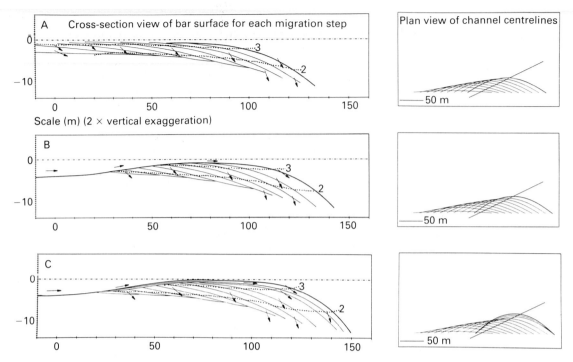

Fig. 5. Computer simulations used to interpret outcrop 1. In cross-section views, solid lines show bar topography after each migration step, dotted lines show variation of mean grain size in phi increments. Arrows show downstream channel orientation relative to the cross-section plane (up is into the outcrop). Deposit cross-sections presented as viewed in the upstream direction. Orientation of the cross-sectional plane relative to the channel centreline after each migration step is shown in plan view to the right of each deposit cross-section. Horizontal dashed-dotted line displays water surface. In simulations channel width, depth and discharge and bend wavelength were held constant. Channel bend sinuosity and position were varied in equal steps as shown in the plan views of the channel centreline for each migration step.

variations indicate a section almost parallel to the channel upstream of the bar but nearly normal to the channel downstream of the bar. The erosion surface between these beds and the underlying inclined beds at the upstream (left) end of the bar but conformity of these beds at the downstream (right) end of the bar suggest continued downstream bar migration (Fig. 5B). The conformity of the downstream beds (adjacent to the channel fill) also indicates that the bend apex did not traverse the outcrop plane. If downstream translation had continued until channel abandonment the deposit would have resembled Fig. 5B. However, preservation of a few convex-upward beds indicates that migration only by expansion occurred during the final stages of bar deposition (Fig. 5C). The finer grain size and dominance of small-scale cross-stratification within convex-upwards beds may also indicate slower channel flows during the final stages of bar migration. Finally the fine-grained, horizontal beds within the channel fill (which are disconformable with the bar deposit) suggest rapid abandonment of the channel.

The point-bar model describes this ancient channel deposit well; however it slightly underpredicts lateral palaeocurrent variations or slightly overpredicts bed inclinations, particularly on the downstream end of the deposit. This may reflect a slight increase in channel width (relative to channel depth) or a slight increase in channel bend wavelength that occurred during bend migration. Reconstructions indicate that the channel width and mean depths stayed relatively constant (75 m and 3.5 m respectively). Mean downstream velocities were about 0.45 m s^{-1} and channel discharge was 120 $\text{m}^3 \text{ s}^{-1}$. The bend wavelength was about 330 m. Bend sinuosity increased to 1.12 as the channel migrated about 120 m in the downstream direction.

Outcrop 2

This roadcut is also located within the Upper Devonian Catskill Magnafacies near East Gilford, New York, USA and has also been described by Bridge & Gordon (1985). The exposure is capped by a thick multistorey sandstone body (Fig. 6). Storey boundaries are defined by major erosion surfaces, disconformities in bedding geometry and/or abrupt shifts in palaeocurrent direction. Here only the uppermost storey will be discussed. This storey is underlain by a major, concave-upward erosion surface and grades upward into overlying mudstones. Palaeocurrents are generally

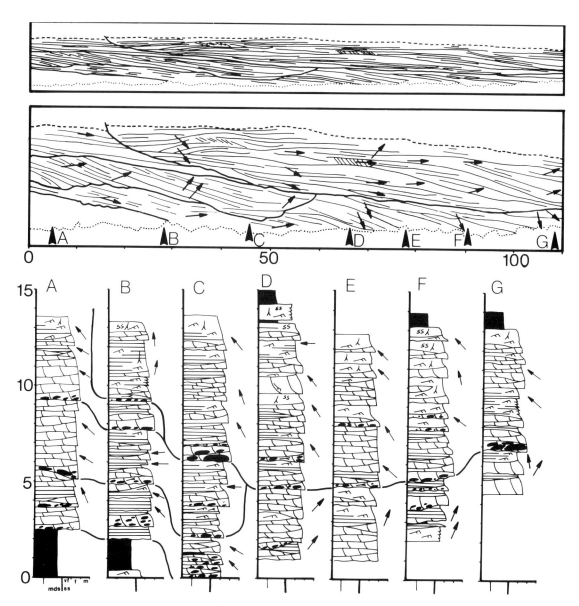

Fig. 6. Bedding geometry, palaeocurrent orientation and sedimentological logs of outcrop 2. Upper box shows bedding surfaces without vertical exaggeration; in the lower box bedding surfaces are displayed with 2× vertical exaggeration. See caption of Fig. 4 for explanation of symbols. Scale in metres. The azimuth of the outcrop is N 135° E.

from left to right, but progress from obliquely out of the outcrop to the left to obliquely into the outcrop to the right. Bedding surfaces dip generally upstream in the left side of the outcrop, become horizontal (parallel to the basal erosion surface) in the centre of the outcrop, and dip downstream to the right of the outcrop. A greater proportion of beds dip in the downstream direction. Beds are dominated by large-scale trough cross-stratification lower in the storey but fine in grain size and contain more planar-stratification and small-scale cross-stratification upward. Vertical fining becomes more pronounced in the downstream (right hand) side of the outcrop (Fig. 6, logs D–F), except to the far right (Fig. 6, log G) where coarser beds cap the channel fill deposit. In the upstream (left) part of the storey, beds do not show a pronounced fining upward trend (Fig. 6, logs C and D) except within beds directly adjacent to the storey margin (Fig. 6, log B) where beds also decrease in dip, display more planar-stratification and small-scale cross-stratification, and are channel filling. Near the top of the storey in the centre of the outcrop a few sets of large-scale cross-stratification dip at an oblique angle into the outcrop.

Bedding geometry and palaeocurrent variations within this outcrop indicate a bar cross-section oriented approximately parallel to the channel belt axis. Palaeocurrent variations provide only a minimum estimate of sinuosity, because in outcrops oriented parallel to the channel belt axis, palaeocurrent variation is dependent on both channel-bend sinuosity and the outcrop's distance from the channel belt axis. The channel deposit presented here migrated by both downstream translation and expansion (Fig. 7A). Had this channel migrated by expansion alone, an equal proportion of upstream dipping and downstream dipping beds would have been preserved (Fig. 7B). Conversely, had migration been dominated by downstream translation, all

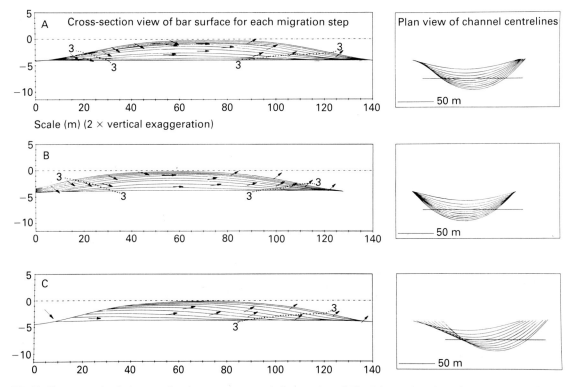

Fig. 7. Computer simulations used to interpret outcrop 2. See caption of Fig. 5 for explanation of symbols. Deposit cross-section presented as viewed toward the channel belt axis. In simulations channel width, depth and discharge and bend wavelength were held constant. Channel bend sinuosity and position were varied in equal steps as shown in the plan views of the channel centreline for each migration step.

bedding surfaces would dip in the downstream direction, and they would be truncated in the upstream direction by the upstream channel fill (Fig. 7C). Vertical fining of beds within the storey records decreased flows higher on the bar surface. Lack of coarsening-upwards deposits in the upstream end of the bar suggests the outcrop plane is some distance from the channel belt axis. The occurrence of a few sets of large-scale cross-stratification which dip into the outcrop near the storey top record inward rotation of dune crestlines.

Simulated deposits have a relatively horizontal base but the storey examined here has a broad concave-upward base. The model used here assumes a direct relation between bar topography and radius of bend curvature. Thus straight channels will have no bed topography and deposits from a migrating curved channel segment will increase in thickness progressively with distance from the channel belt axis. The concave-upward base of this storey suggests the initial, low sinuosity channel had greater bed topography than predicted by the bar model. Palaeochannel reconstructions indicate that the channel width, mean depth and mean velocity were 45 m, 2.6 m and 0.45 m s^{-1} respectively. Palaeochannel discharge was 60 m^3 s^{-1}. The bend wavelength is well constrained by the length of the storey (about 300 m) and bend sinuosity increased to nearly 1.2 as the channel migrated. The bar migrated only about 20 m downstream.

Outcrop 3

This outcrop is located within the Miocene Nagri Formation near Chinji Village, Potwar Plateau, northern Pakistan. The storey documented here occurs within a sandstone body which displays multiple storeys stacked laterally and to a lesser extent vertically. An asymmetrical, concave-upward erosion surface marks the base of the storey and its top is truncated upwards by the nearly horizontal major erosion surface at the base of the storey above (Fig. 8). Beds are generally concordant with the left storey margin but are discordant with the steeper right margin. A few beds directly adjacent to the concordant (left) margin display very low angle dips. These beds are truncated by more steeply dipping beds. The first few steeply inclined beds progressively increase in dip but beds throughout most of the storey generally decrease in dip away from the concordant margin. Beds become nearly horizontal directly adjacent to the discordant margin. In a few locations low angle discordance is observed between adjacent beds.

Palaeocurrent directions are generally out of the outcrop with orientations varying from 30° oblique to outcrop normal near the concordant margin to nearly outcrop normal near the discordant margin. The storey displays an overall fining upward trend. Large-scale trough cross-stratification is common in beds within most of the storey. Lower in the storey planar-stratification occurs in the top of beds. Higher in the storey planar-stratification and small-scale cross-stratification become common. The storey also fines laterally as the beds decrease in dip near the discordant margin. This lateral fining is associated with increased abundance of planar-stratification and small-scale cross-stratification.

The relationship of palaeocurrent variation to bedding dip and the fining-upward grain size trends suggests the channel migrated largely by expansion. The concordant margin and overlying, nearly horizontal beds record initial channel incision and subsequent bar building and migration. The relatively steep, discordant margin records the final channel cutbank. Progressive lateral decrease in the dip of beds is not expected for a symmetrical point-bar migrating under constant flow conditions.

Initially, decreasing bed dips may reflect bar asymmetry (Fig. 3) or changing bar wavelength (Fig. 2C); however in most of the storey progressive decrease in bed inclination indicates a gradual reduction in channel discharge during channel migration and that the channel accommodated lower discharges by decreasing depth but maintaining relatively constant width (Fig. 9). Lateral fining in mean grain size also suggests decreases in flow velocity. The increased abundance of planar stratification and small-scale cross-stratification adjacent to the discordant margin can be related to decreases in flow depth. The discordant nature of the channel-filling beds with the adjacent cutbank indicates cutbank retreat stopped as the channel filled. Low angle discordancies between adjacent beds within the lateral-accretion portion of the sequence can be explained by minor variations in flow intensity during different episodes of channel migration and deposition.

The point-bar model used here can predict the observed progressive lateral decrease in bedding surface inclination by progressively reducing discharge. However, the relatively constant thickness of the storey (between margins) and relatively constant grain size within laterally accreted beds is less

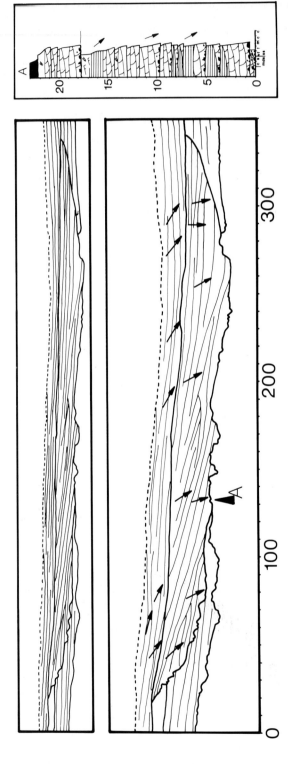

Fig. 8. Bedding geometry, palaeocurrent orientation and a sedimentological log of outcrop 3. Only one complete log through the storey was possible due to vertical nature of the outcrop. Lateral variations were observed from the base and top of the outcrop. Upper box shows bedding surfaces without vertical exaggeration, in the lower box bedding surfaces are displayed with 2 × vertical exaggeration. See caption of Fig. 4 for explanation of symbols. Scale in metres. The azimuth of the outcrop is N 90° E.

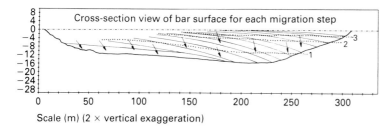
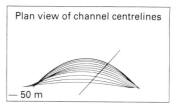

Fig. 9. Computer simulation used to interpret outcrop 3. See caption of Fig. 5 for explanation of symbols. Deposit cross-section presented as viewed in the upstream direction. In simulations channel width and bend wavelength were held constant. Channel bend sinuosity and position were varied in equal steps as shown in the plan views of the channel centreline for each migration step. Channel depth and discharge were decreased in equal steps during migration steps 4–12. Thick, solid line represents the base of the storey and is schematic.

well predicted. The model reconstructions suggest channel widths of 180 m and mean depths initially of 9 m. Sinuosity was low, bend wavelength was about 1200 m and channel discharge was initially 950 m^3 s^{-1}.

Outcrop 4

This roadcut is located within the Upper Devonian Catskill Magnafacies near Lenox, Pennsylvania, USA. This exposure is one of a sequence of roadcuts which progress from fluvial to marine over a distance of 20 km. A portion of the roadcut consisting of a single sandstone body with overlying mudstones is described here (Fig. 10). The basal erosion surface is nearly horizontal and the sandstone grades upward into the overlying mudstones. Successive beds initially steepen to the left, then gradually decrease in inclination and become more concave upward. In several locations within this sequence, adjacent beds truncate at low angles. This sequence of beds is cut by a steeper erosion surface which is inclined in the same direction as the beds below. To the left of this major disconformable erosion surface, beds initially have steeper dips again but then progressively decrease in dip and become more concave-upward. This progression is repeated twice more before the storey ends in a channel fill 50 m to the left of the bedding diagram. Palaeocurrents are generally out of the outcrop. Mean grain size of beds fines upwards. Large-scale trough cross-stratification is common lower in the storey, but small-scale cross-stratification and planar-stratification dominate the upper part of the storey. There is no abrupt change in palaeocurrent orientations or grain size across disconformable erosion surfaces.

Bedding disconformities are common within lateral-accretion sequences (Beutner et al., 1967; Allen & Friend, 1968; Bridges & Leeder, 1976; Elliot, 1976; Galloway, 1981; DeMowbray, 1983; Thomas et al., 1987). These disconformities have been interpreted as due to major discharge fluctuations (Fig. 11A) or abrupt shifts in channel position (e.g. chute cutoff, Fig. 11B). Chute cutoff will generally result in a straighter channel segment being superimposed over sediments deposited in a more sinuous channel. Therefore beds below the disconformity (i.e. deposited before the chute cutoff) will tend to have steeper dips than those deposited directly above the disconformity. If the bedding disconformity resulted from discharge fluctuation, however, beds deposited above the disconformity should be inclined at similar or steeper angles than those below the disconformity. Chute cutoff-formed bedding disconformities may also mark pronounced shifts in palaeocurrent orientation. The magnitude of this palaeocurrent shift will depend on the bend sinuosity before cutoff, the amount the bend planform migrated by translation and the outcrop orientation.

The bedding disconformities observed in the Lenox outcrop are clearly caused by major fluctuations in discharge. The initial sequence of progressively steepening beds can be related to bar migration. Vertical variation in grain size and sedimentary structures records decreased flow velocity and depth upwards on the bar surface. Restricted lateral palaeocurrent variations suggest an outcrop orientation nearly parallel with the direction of net channel migration. The following sequence of gradually decreasing bed inclinations and increasing upward concavity of beds records gradual reduction in discharge over several flood events.

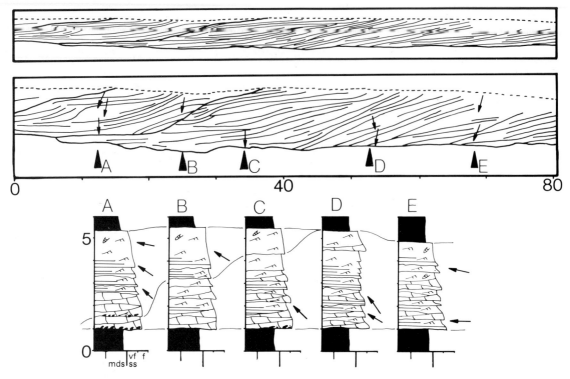

Fig. 10. Bedding geometry, palaeocurrent orientation and sedimentological logs of outcrop 3. Upper box shows bedding surfaces without vertical exaggeration; in the lower box bedding surfaces are displayed with 2 × vertical exaggeration. See caption of Fig. 4 for explanation of symbols. Scale in metres. The azimuth of the outcrop is N 20° E.

The disconformity records a major flood event and erosion of channel filling deposits. The decreasing flow sequence is repeated in beds above the disconformity. Minor disconformities within the sequence which truncate only a few adjacent beds are related to more modest discharge fluctuations. Bedding geometry within this deposit resembles sediments deposited in the modern tidal channels described by Bridges & Leeder (1976) and DeMowbray (1983).

The model describes this deposit well but cannot predict the upward concavity of channel filling beds. Reconstructions indicate that palaeochannel geometry and discharge varied in time but that palaeochannel width was up to 30 m, mean depth was up to 3.3 m, mean velocities were up to 0.39 m s^{-1} and discharge was up to 40 m^{3}s^{-1}. Bend wavelength was of the order of 500 m and sinuosity rose to 1.2 when depositing the sediments examined here.

DISCUSSION

These comparisons between point-bar model output and outcrop data suggest that bedding geometry, grain size and palaeocurrent variations within point-bar deposits can be interpreted in terms of specific cross-sections through migrating channel bends of prescribed geometry and hydraulics. Such interpretation is possible only because three-dimensional variations along modern point-bars have been well documented (i.e. as summarized here by a mathematical model). Although this study examines only point-bars, it underscores the need for a more detailed understanding of three-dimensional variations within all channel-bar types. Unfortunately, side bars with steep, down-flow dipping surfaces and braid bars (with and without steep, down-flow dipping surfaces) have only been documented very qualitatively. More study is needed to document how these types of bars

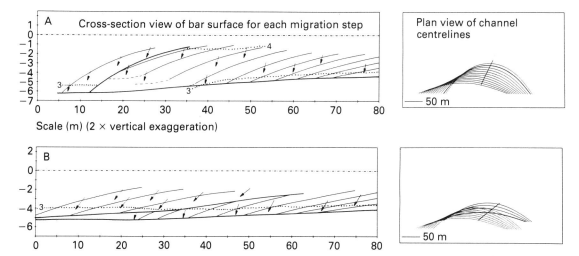

Fig. 11. Computer simulation used to interpret outcrop 4. See caption of Fig. 5 for explanation of symbols. In both simulations channel width and bend wavelength were held constant and channel bend sinuosity and position were varied in equal steps as shown in plan views of the channel centreline for each migration step. In (A) channel depth and discharge were successively decreased during migration steps 7–10, then increased to previous values for migration steps 11 and 12. In (B) dashed centreline plan views represent channel position before chute cutoff, solid lines represent channel centreline positions after cutoff. Chute cutoff is here modelled to have occurred during a single migration event.

vary in geometry and sedimentary characteristics, and how they migrate. Similarly, outcrop studies of ancient bar deposits need to document lateral deposit variations in more detail. Lateral deposit variations which are not predicted by this point-bar model must reflect channel bars of contrasting geometry, planform, migration behaviour or temporal variations in channel hydraulics. Identification of such trends may provide clues to palaeochannel bar type and palaeochannel pattern. Eventually, this type of study should lead to more general models for interpreting bedding geometry within all river-channel deposits.

SUMMARY AND CONCLUSIONS

Point-bars are complex, three-dimensional bodies which change shape and position with time as the channel migrates. Changes in point-bar shape and position are reflected in the bedding, facies and geometry of the bar deposit. Point-bar deposits need to be described in terms of lateral variation in deposit thickness, bedding geometry, variation in grain size and sedimentary structures, and palaeocurrent orientation relative to that of the outcrop plane. These variations need to be interpreted in terms of temporal and spatial variations in channel-bar geometry, facies and hydraulics, style of channel migration and outcrop orientation.

ACKNOWLEDGEMENTS

I am grateful to John S. Bridge for helpful advice and supervision during research and for introducing me to the first two outcrops presented here. I wish to thank Sharon Gabel for reviewing early versions of this manuscript and Dave Tuttle for photographic assistance. This work was supported by an NSF grant to John S. Bridge.

REFERENCES

ACKERS, P. & CHARLTON, F.G. (1970) Meander geometry arising from varying flows. *J. Hydrol.* **11**, 230–252.

ALLEN, J.R.L. & FRIEND, P.F. (1968) Deposition of the Catskill facies, Appalachian region: with notes on some other Old Red Sandstone basins. In: *Late Paleozoic and Mesozoic Continental Sedimentation, Northeastern North America* (Ed. Klein, G. deV.) pp. 21–74. Geol. Soc. Am. Spec. Paper 106.

BEUTNER, E.C., FLEUCKINGER, L.A. & GARD, T.M. (1967) Bedding geometry in a Pennsylvanian channel sandstone. *Bull. Geol. Soc. Am.* **78**, 911–916.

BRIDGE, J.S. (1976) Mathematical model and Fortran IV program to predict flow, bed topography and grain size in open-channel bends. *Computers and Geosciences* **2**, 407–416.

BRIDGE, J.S. (1977) Flow, bed topography, grain size and sedimentary structures in open channel bends: a three-dimensional model. *Earth Surf. Proc.* **2**, 401–416.

BRIDGE, J.S. (1978) Paleohydraulic interpretation using mathematical models of contemporary flow and sedimentation in meandering channels. In: *Fluvial Sedimentology* (Ed. Miall, A.D.) pp. 723–742. Can. Soc. Petrol. Geol., Calgary, Memoir 5.

BRIDGE, J.S. (1982) A revised mathematical model and Fortran IV program to predict flow, bed topography and grainsize in open-channel bends. *Computers and Geosci.* **8**, 91–95.

BRIDGE, J.S. (1984) Flow and sedimentary processes in river bends: comparison of field observations and theory. In: *River Meandering* (Ed. Elliott, M.) pp. 857–872. Proc. Rivers '83. Am. Soc. Civ. Engrs.

BRIDGE, J.S. & DIEMER, J.A. (1983) Quantitative interpretation of an evolving ancient river system. *Sedimentology* **30**, 599–623.

BRIDGE, J.S. & GORDON, E.A. (1985) Quantitative reconstructions of ancient river systems in the Oneonta Formation, Catskill Magnafacies. In: *The Catskill Delta* (Eds Woodrow, D. & Sevon, W.) pp. 163–181. Geol. Soc. Am. Spec. Paper 201.

BRIDGE, J.S. & JARVIS, J. (1982) Dynamics of a river bend: a study of flow and sedimentary processes. *Sedimentology* **29**, 499–542.

BRIDGE, J.S. SMITH, N.D., TRENT, F., GABEL, S.L. & BERNSTEIN, P. (1986) Sedimentology and morphology of a low-sinuosity river: Calamus River, Nebraska Sand Hills. *Sedimentology* **33**, 851–870.

BRIDGES, P.H. & LEEDER, M.R. (1976) Sedimentary model for intertidal mudflat channels with examples from the Solway Firth, Scotland. *Sedimentology* **23**, 533–552.

CARLTON, C.W. (1965) The relation of free meander geometry to stream discharge and its geometric implications. *Am. J. Sci.* **263**, 864–885.

DEMOWBRAY, T. (1983) The genesis of lateral accretion deposits in recent intertidal mudflat channels, Solway Firth, Scotland. *Sedimentology* **30**, 425–435.

DIETRICH, W.E. (1987) Mechanics of flow and sediment transport in river bends. In: *River Channels: Environment and Process* (Ed. Richards, K.S.) pp. 179–227. Inst. Brit. Geogr. Spec. Publ 18.

DIETRICH, W.E. & SMITH, J.D. (1983) Influence of the point bar on flow through curved channels. *Water Resources Res.* **19**, 1173–1192.

DURY, G.H. (1965) Theoretical implications of underfit streams. US Geol. Surv. Prof. Paper 452-A, pp. 1–67.

ELLIOT, T. (1976) The morphology, magnitude and regime of a Carboniferous fluvial-distributary channel. *J. Sedim. Petrol.* **46**, 70–76.

ENGELUND, F. (1974a) Flow and bed topography in channel bends. *J. Hydr. Div., ASCE* **100**, 1631–1648.

ENGELUND, F. (1974b) Experiments in curved alluvial channels. Process Report No. 34, Inst. Hydrodyn. and Hydraul. Engr., Tech. Univ. Denmark, pp. 31–36.

GALLOWAY, W.E. (1981) Depositional architecture of Cenozoic Gulf coastal plain fluvial systems. In: *Recent and Ancient Nonmarine Depositional Environments: Models for Exploration* (Eds Ethridge, F.G. & Flores, R.M.) pp. 127–155. Soc. Econ. Paleont. Miner., Tulsa, Spec. Publ. 31.

HOOKE, J.M. (1984) Changes in river meanders: a review of techniques and results of analyses. *Prog. Phys. Geogr.* **8**, 473–508.

JACKSON, R.G. (1976) Depositional model of point bars in the Lower Wabash River. *J. Sedim. Petrol.* **46**, 579–594.

NELSON, J.M. & SMITH, J.D. (1989a) Flow in meandering channels with natural topography. In: *River Meandering* (Eds Syunsuke, I. & Parker, G.) pp. 69–102. Water Resources Monograph, 12. Am. Geophys. Union.

NELSON, J.M. & SMITH, J.D. (1989b) Evolution and stability of erodible channel beds. In: *River Meandering* (Eds Syunsuke, I. & Parker, G.) pp. 321–377. Water Resources Monograph 12. Am. Geophys. Union.

SCHUMM, S.A. (1968) River adjustment to altered hydraulic regime — Murrambidgee River and paleochannels. US Geol. Surv. Prof. Paper 598, 1–58.

THOMAS, R.G., SMITH, D.G., WOOD, J.M., VISSER, J., CALVERLEY-RANGE, E.A. & KOSTER, E.K. (1987) Inclined heterolithic stratification — description, interpretation and significance. *Sedim. Geol.* **53**, 123–179.

WILLIS, B. (1989) Palaeochannel reconstructions from point bar deposits: a three dimensional perspective. *Sedimentology* **36**, 757–766.

Geometry and lateral accretion patterns in meander loops: examples from the Upper Oligocene–Lower Miocene, Loranca Basin, Spain

M. DÍAZ-MOLINA

Dpto de Estratigrafía e Instituto de Geología Económica, Universidad Complutense, 28040 Madrid, Spain

ABSTRACT

The Loranca Basin was infilled by two coalescing fluvial systems from Late Oligocene to Early Miocene times. High sinuosity rivers left meander loop deposits on the fan surfaces. Today these are preserved as exhumed point bar deposits which provide excellent three-dimensional exposures, enabling detailed examination of the internal organization and architecture of the sandstone bodies. It is possible to reconstruct the geometry of individual depositional components and thus chronicle the development of individual meander loops. Laterally stacked point bars are delimited by discordant surfaces marking changes in the displacement direction of meander bends.

Two types of point bars are distinguished on the basis of grain size distributions: type A bars display a wide range, whereas in type B bars some of the grain sizes are lacking. The vertical organization of point bar deposits was described from longitudinal or oblique sections through lateral accretion units. Type A point bars display three types of vertical bedding profiles: thickening–thinning, thickening and thinning. In contrast type B point bars grew from recurrent thickening–thinning bedding cycles.

The bed thickness of type A point bars has a positive correlation with sedimentation and channel migration rates, which in turn are controlled by channel curvature. In contrast the recurrent bedding cycles which characterize type B point bars are ascribed to lateral accretion episodes accompanied by scroll bar formation. The discordances between adjacent point bars resulted from episodic channel adjustment in response to the development of critical curvatures.

INTRODUCTION

The Loranca Basin is located in the central part of Spain (Fig. 1), within the Iberian Range. This basin was formed during Eocene (p.p.) to Late Oligocene (p.p.) times and contains continental deposits. From the Late Oligocene to the Early Miocene times the Loranca Basin was filled by two coalescing fluvial systems (Fig. 1), the Tórtola and the Villalba de la Sierra fluvial fans (Díaz-Molina *et al.*, 1989). Both fluvial systems were dominated by individual channels, and they are considered to be wet fluvial fans (following Schumm, 1977). During the Late Oligocene to the Early Miocene the Loranca Basin was connected to the Madrid Basin. In the Early Miocene, gypsum deposits extended into the basin indicating an endorheic situation.

The Tórtola fan covers an area of 2500 km², whereas the Villalba de la Sierra fan is almost twice as large, covering more than 4200 km² (Fig. 1). The Tórtola fan coalesces with the Villalba de la Sierra fan to the north, and its outline was tectonically controlled to the south-west. The Villalba de la Sierra fan coalesces with another fan to the north too. The apexes of both depositional systems were located in the Serranía de Cuenca. Tectonic deformation and erosion have altered the original continuity of the fan sediments.

A comparison of palaeocurrent distributions and fluvial characteristics has allowed the distinction of distributary and tributary areas of the basin (Díaz-Molina *et al.*, 1989). Both fans radiated downslope

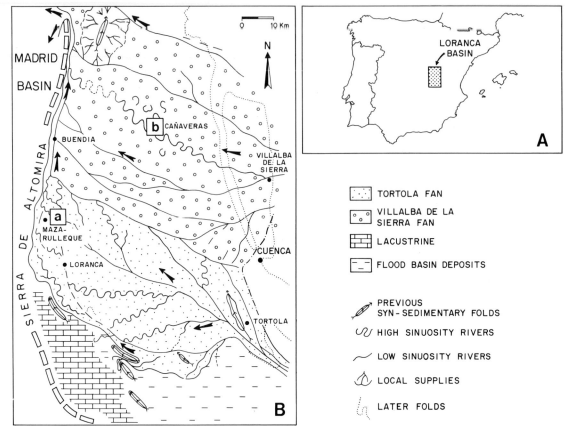

Fig. 1. (A) Location map of the Loranca Basin. (B) Palaeogeographic reconstruction of the Loranca Basin during the Late Oligocene showing the distribution of the depositional systems; a and b correspond to the studied exposures, which are illustrated in detail in Fig. 2.

from the apex areas developing a multiple channel system. There is a general downstream evolution, from braided to meandering rivers in the Tórtola fan. In contrast the Villalba de la Sierra fan transported coarser material in predominantly braided rivers. Nevertheless extensive meander loop deposits are present in the Cañaveras area (Fig. 1). Major tributaries were derived from the medial zone, where the two fans merged, and also from the western margin of the basin, along the strike of the Sierra de Altomira (Fig. 1). This western region is characterized by an increased density of low sinuosity channel deposits possibly resulting from the lateral concentration of the channels in the fluvial system (Díaz-Molina et al., 1985).

Five examples of meander loops or meander loop complexes are described below. Two of them (1 and 2) belong to the Tórtola fluvial fan and represent type A deposits, while examples 3, 4 and 5 are type B point bars found on the Villalba de la Sierra fluvial fan. The location of the exposures is shown in Fig. 2.

POINT BAR GEOMETRIES

The term 'point bar' has various usages, for example: (i) an individual lateral accretion unit (Sundborg, 1956; Leopold et al., 1964); (ii) a complex of ridges and swales within a meander loop (Allen, 1965; Jackson, 1976a); and (iii) in modern examples, the most recent unvegetated lateral accretion unit forming at present (Nanson, 1980). The term point bar is defined in the present paper as a composite bar formed by a set of comformable lateral accretion units, which in plan view correspond to a group of parallel scroll bars or ridges and swales. Thus a single point bar deposit comprises a

Accretion patterns in ancient meander loops 117

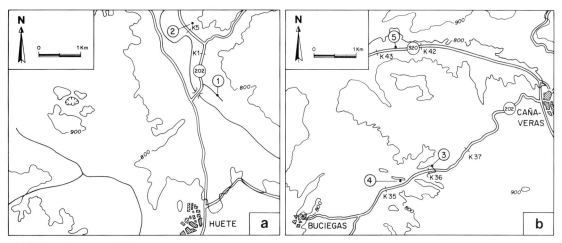

Fig. 2. Location map of the exposures.

meander loop. Two or more laterally adjacent point bar bodies constitute a meander loop complex, which is equivalent to the composite loop of Brice (1974). This simple terminology provides a framework for the analysis of the deposits.

The three-dimensional geometry of the point bar bodies has been reconstructed from the exposures (Díaz-Molina, 1978, 1979, 1984), since examples of transverse, longitudinal and intermediate sections of the meander bends are available in the outcrops. Transverse sections have sigmoidal outlines which are typical of point bar deposits (Fig. 3). Longitudinal profiles have lenticular geometries, but some distinctions can be established according to the nearness of the section to the inner bank (Fig. 3). Thus, depending on the location of the longitudinal section along the meander loop radius it varies from a perfect lenticular geometry to disconnected point bar deposits. A section in an intermediate position between the transverse and the longitudinal ones will have a convex-up profile (Fig. 3). The three-dimensional point bar geometry was simulated by Willis (1989).

Meander loop complexes are formed by the lateral juxtaposition of point bar deposits, and have a wide variety of cross-sectional geometries. They feature irregular upper surfaces and vary in width according to the number of point bars present.

MEANDER LOOP REACTIVATION SURFACES

Meander loop complexes can be divided into groups of conformable lateral accretion units, corresponding to individual point bar bodies. Each point bar has a characteristic geometry with lateral accretion units displaying unique strike and dip. The variable geometries seen between adjacent point bars demonstrate fluctuations in the channel displacement direction (Díaz-Molina, 1984). Thorne *et al.* (1985) proposed that they might represent successive major 'unidirectional' growth phases.

Although scroll bars are only preserved in exceptional cases, the top surfaces of lateral accretion units can be readily observed. Lateral accretion units have a curved outline in plan view. They form groups having specific orientations, which often were truncated by erosional surfaces.

A well-defined contact existing between adjacent point bars was referred to as a meander loop reactivation surface (Díaz-Molina, 1978, 1984). Thorne *et al.* (1985) applied the term 'set boundary' for this type of discontinuity surface. Two types of reactivation surfaces have been found, erosional and non-erosional (Díaz-Molina, 1984), and in some exposures both end-members are preserved as lateral equivalents. Though scroll-bar topography and patterns have been described in detail (Sundborg, 1956; Brice, 1974; Hickin, 1974; Nanson, 1980), surfaces delimiting point bars have received little attention. However reactivation surfaces are conspicuous features in modern floodplains.

Erosional reactivation surfaces

These are distinguished by erosion of the underlying point bar. This reactivation surface may be analogous to the erosional and slightly discordant lateral accretion surfaces distinguished by Allen

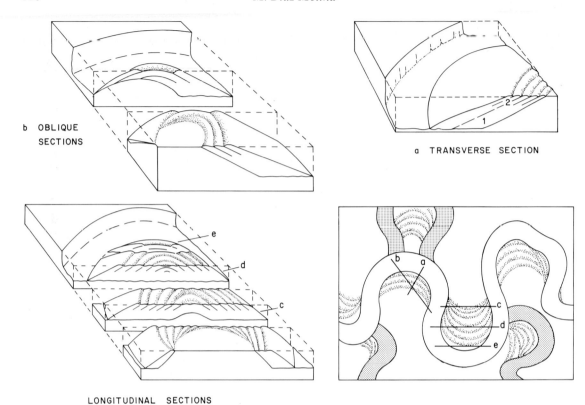

Fig. 3. Different point bar body geometries.

(1982). Field examples have been described by Allen & Friend (1968) and Elliot (1976). Elliot (1976) observed that erosion surfaces were mostly apparent in the upper two-thirds of the lower coarse member, but persisted into the trough cross-bedded sandstones and occasionally coincided with a step down in the channel base. Surfaces of this nature can also be recognized in the drawings of several other workers, for example Moody-Stuart (1966), Nami & Leeder (1978), Puigdefábregas & van Vliet

Fig. 4. Erosional reactivation surface (r.s.). The overlying point bar deposit is conformable to the erosion surface delimiting the point bars.

Fig. 5. Non-erosional reactivation surface. The overlying point bar shows an onlap on the depositional topography of the previous deposits.

(1978) and Steward (1981). Bridges & Leeder (1976) have also noticed prominent truncation surfaces in tidally influenced point bars.

The overlying point bar is either conformable with the reactivation surface (Fig. 4) or discordant. The overlying point bar is discordant when bedding onlaps the reactivation surface to form an angular unconformity.

Non-erosional reactivation surfaces

The non-erosional reactivation surface coincides with the depositional topography of the underlying point bar. The overlying point bar may be slightly or strongly discordant and it overlies the previous deposits with an onlap (Fig. 5). The onlap against the reactivation surface is interpreted as related to the progressive incorporation of the local profile to the sedimentation area over the inner bank. This type could be equivalent to the strongly discordant and extensive lateral accretion surface distinguished by Allen (1982).

TYPE A POINT BARS

Type A point bars are present on the Tórtola fan. Their thicknesses range between 0.8 and 12 m. The thinner point bar deposits are located along the western fan fringe; they are preserved as isolated meander loops comprising individual point bars. Thicker point bar deposits which frequently form meander loop complexes are preserved throughout the Tórtola fan (Fig. 1) and dominate in the north-western area. The thinner point bars were probably active during high flood events, whereas the thicker ones are considered to be associated with more actively channelled areas, where run off was relatively permanent (Díaz-Molina, 1979; Díaz-Molina et al., 1985, 1989).

A general review of the point bar bodies of the Tórtola fluvial system has been presented in earlier works (Díaz-Molina, 1978, 1979; Díaz-Molina et al., 1985). A typical point bar deposit generally consists of a fining upwards sequence of sediment grain size. Inside the point bar two units can be distinguished, a coarse grained lower member and a fine grained upper one. The lower part of the point bar profile may be formed by gravel sized materials to fine grained sands. Sedimentary structures in the lower part include: foreset cross-stratification generated by gravel bar migration, and large-scale trough and planar cross-stratification. The later displays reactivation surfaces of the kind formed by superimposed bed forms (McCabe & Jones, 1977). The upper unit consists of fine grained sandstones and siltstones. Small-scale cross-stratification can be seen internally, although preservation is often poor due to pedogenesis.

The thin point bars are similar to the Murillo point bar which was described by Puigdefábregas (1973) from the Ebro Basin. They display lateral accretion surfaces at the base of the bar deposit. In contrast, the thicker point bars always feature lateral accretion surfaces in the upper unit (Figs 4 & 5) but rarely in the lower one. Lateral accretion surfaces are recognized in the lower part of thicker point bars by the presence of silt drapes, associated with a decrease in grain size. They are eroded from the higher portions of point bars by a subsequent increase in river discharge. The preservation of fine grained laminae on lateral accretion surfaces during low stage river discharge is also described by Elliot (1976) from a Carboniferous meander belt.

The sedimentary structures generally indicate sediment transport up the point bars and are formed at acute angles to the dip of the lateral accretion surfaces. However in some point bar examples of the Tórtola depositional system, and in the downstream part of the ancient bend, sedimentary structures produced by down bar sediment migration indicate a secondary outwards directed flow. In the Wabash River, regions of outward flow are correlated with shoaling areas of point bars (Jackson, 1975). The same phenomenon occurred in three of the five bends of the Beatton River described by Hickin (1978). In the Muddy Creek the outward velocity over the point bar is related to a decrease in depth downstream (Dietrich & Smith, 1983). Thorne et al. (1985) have observed in some bends of the Fall River that the main cell of the secondary flow is directed radially outwards and that in this process the stage–width relationship has a particular importance. All these recent examples indicate that a suite of sedimentary structures can be produced by differing patterns of migration across the surface of modern point bars.

The thickness of individual lateral accretion units was obtained from the upper member of the vertical successions. Bed boundaries occurring within accretion units are defined by a decrease in grain size (see Figs 4 & 5), commonly ranging between fine sand and very fine silt. In rare examples there was little grain size contrast, beds were not identifiable and measurements could not be obtained.

The top surfaces of the point bars were usually modified by erosion and scroll bar dimensions could not be measured. Where scroll bars have been preserved they form topographic features several metres apart, similar to those indicated by Puigdefábregas & van Vliet (1978, fig. 10) in a point bar of the Lower Montañana Group.

Two meander loop complexes have been selected to examine meander loop development (examples 1 and 2). Each point bar has between two and seven lateral accretion episodes. This contrasts with the ancient meander belt described by Elliot (1976), where eight to twelve sigmoidal beds occurred between erosion surfaces. Examples 1 and 2 differ in point bar and meander loop sizes. The empirical equations of Ethridge & Schumm (1978) were used to estimate palaeochannel characteristics from point bar number 7 of example 2 (see Fig. 9) and sinuosity was calculated as 2.0 (Díaz-Molina et al., 1985). These estimations were possible because point bar 7 shows a transverse section which is the only appropriate section according to the method. In contrast example 1 illustrates point bar sections which are oblique or longitudinal to the meander loop and therefore unsuitable for morphologic reconstructions. Nevertheless, its excellent three-dimensional exposure permits a more accurate reconstruction of the meander loop.

Example 1

Example 1 (Fig. 6) comprises five adjacent point bars separated by reactivation surfaces (Díaz-Molina et al., 1985). Logs through the meander loop and four of the point bars are shown in Fig. 7. Two types of bedding sequences were found, thickening and thickening–thinning. The reconstruction of the meander loop (Fig. 7) was based on the geometry of the lateral accretion units of the point bar deposits, observed on the top of the outcrop. The reconstruction reveals a compound loop, formed by the alternate attachment of secondary arcs around the loop perimeter. Brice (1974) has observed that compound loops evolve only by the migration of individual simple loops, and moreover, the radius, length and height of most compound loops are indefinite.

Example 2

The meander loop in example 2 is 170 m wide (Fig. 8). Seven adjacent point bars are clearly distinguished (Fig. 9). Each point bar presents a longitudinal, transverse or oblique section to the successive meander bends (Fig. 3). Based on the point bar geometries a partial reconstruction of the meander loop complex has been obtained (Fig. 9). Bed thickness was measured in five point bars; point bars 1 and 2 (P.1 and P.2) in Fig. 8 have thickening–thinning trends, point bars 4 and 5 show a thinning sequence and point bar 3 a thickening sequence.

TYPE B POINT BARS

The Tórtola and Villalba de la Sierra fluvial systems

Fig. 6. Example 1. The upper part of this outcrop can be divided into bed sets. Each bed set corresponds to an individual point bar.

can be distinguished by grain size trends. Generally coarser material occurs in the Villalba fluvial fan, and there is less sand grade detritus in the meander belts exposed in the Cañaveras area (Figs 1 & 2). These differences were controlled by the lithologies of the catchment areas. The source area of the Villalba fan mainly consists of calcareous Mesozoic rocks, while other lithologies were present in the catchment area of the Tórtola fan.

Point bars are between 6 and 11 m thick in the Cañaveras area (Fig. 1). Up to 50% of the point bars consists of fine sand to silt. Point bar logs reveal either a discontinuous range of sediment sizes or a dominance of fine grained materials and the bound-

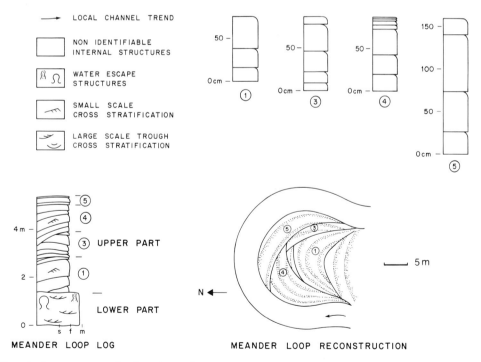

Fig. 7. Meander loop reconstruction based on point bar geometric models. The bed sequences are thickening–thinning or thickening.

Fig. 8. Example 2. Meander loop exposure containing seven adjacent point bars.

ary between the lower and upper part of the sequences is transitional. Sedimentary structures include: gravel bar deposits (trough, planar, massive and horizontally bedded framework gravel), thin strata of scattered gravels marking lateral accretion surfaces, trough and planar large-scale cross-stratification, small-scale cross-stratification and climbing ripple cross-stratification. The coarser sediments are found in the lower member of the point bar, while fines are present in any location. The vertical succession of sedimentary structures commonly shows departures from the continuous fining upwards pattern. In the upper part of the point bar bodies each lateral accretion unit is formed by

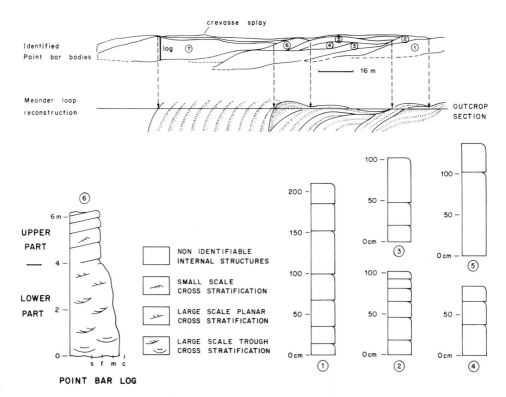

Fig. 9. Meander loop reconstruction. Bed thicknesses measured in five point bars reveal thickening–thinning, thickening and thinning bed sequences.

Fig. 10. Point bar transverse section. The exposure corresponds to P.3 in example 3 (Fig. 12). Lateral accretion surfaces exhibit a segmented profile.

superimposed cosets of climbing ripple cross-stratification. The thicker lateral accretion units may represent bar morphologies and lateral thickness variations; they correspond to preserved ridges in the upper unit of the point bar deposit. Because of the bar convex-upward topography, the following lateral accretion episode tends to smooth this local topography and is thinner or even absent on the crest of the previous deposit. These features were named 'compensation cycles' from turbidite sandstone lobes (Mutti & Sonnino, 1981).

Lateral accretion surfaces are more frequently preserved in the lower part of the point bars than in type A examples. They are conspicuous features because of the marked change in sediment size between some accretional episodes. Sedimentary units clearly delimited by lateral accretion surfaces can be observed in Figs 10 and 12. The upper outline of the body has a segmented profile and comprises several discrete ridge sections, seen from the base to top, where scroll bar morphology is preserved. Scroll bars are composite forms, showing prograding units. Two consecutive scroll bars were 10 m apart.

Sedimentary structures show the same palaeocurrent trends as those observed in type A point bars. In addition, rare examples of outward directed climbing ripple cross-stratification have been found in the upper part of point bars (example 3, Fig. 11).

Two exposures have been selected (Fig. 2). One of them has 3 km of lateral continuity and comprises several adjacent meander loop complexes. Two of these meander loops are described in examples 3 and 4. The bed thickness has been measured in the upper part of the point bars. Although there is lateral variation in bed thickness, the data represent generalized bed thickness trends.

Example 3

The exposure is 140 m wide and consists of several point bars (Fig. 12). In P.3 and P.4 the lateral

Fig. 11. Lateral accretion unit with bar morphology. Internal climbing ripples are directed outwards from the inner bank.

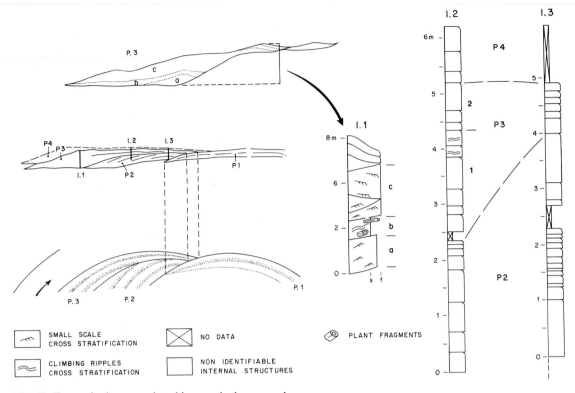

Fig. 12. Four point bars complete this meander loop complex.

accretion surfaces produce a segmented profile where ridges can be distinguished (Fig. 10). The ridges are composite bar deposits mainly formed by ripple cross-stratification (see l.1, Fig. 12).

Suitable exposures of the upper unit are seen in only three point bar exposures (Fig. 12), and these differ from previous examples. The geometry of the accretion units varies laterally. The variation is very noticeable in the thicker beds which are convex-up. Log 1 (Fig. 12) illustrates the typical vertical succession of sedimentary structures, although these have not been preserved at the top of the deposit. Along strike logs 2 and 3 are dominated by climbing ripple cross-stratification (Fig. 12). They represent the evolution of bed thickness in three different point bars. Unfortunately most of the beds cannot be traced laterally to produce a cross-section. Bedding consists of superimposed sequences displaying rough thickening–thinning, thinning or thickening trends. The thickening–thinning cycles are related

Fig. 13. Example 4, indicated with arrows. The exposure is interpreted as a point bar longitudinal section.

to the formation of the composite ridges whose morphologies have been preserved in the upper members of the point bars.

Example 4

The exposure shows a 90 m longitudinal section through a meander loop (Figs 13 & 14). Two logs, corresponding to the upstream (l.1) and downstream (l.2) parts of the bend, have been represented in Fig. 14. Both logs reveal coarse sediments interstratified with finer materials. The upstream log reveals gravel bar deposits, while lag deposits are present in the downstream bar section. The intercalation of coarse sediments high in the vertical succession of point bars has been demonstrated by Jackson (1976a) and Bridge & Jarvis (1976). Jackson (1976a) has observed this phenomenon in the upstream end of point bars, where spiral flow is not fully developed.

Bed thickness measurements were obtained in the downstream part of the bend (1.2, Fig. 14); the succession can be divided into a thickening–thinning lower sequence overlain by a group of poorly organized thin beds.

Example 5

This exposure exhibits four point bars (Figs 15 & 16), presenting a variety of sections. The planimetric reconstruction (Fig. 16) only includes point bars 2 and 3. Point bars 1 and 4 belong to lower and higher stratigraphic levels respectively. Climbing ripple cross-stratification is well preserved in the upper part of these point bars. Only one vertical section was measured through P.2, and two superimposed thickening–thinning cycles were recognized (S.1 and S.2, Fig. 16).

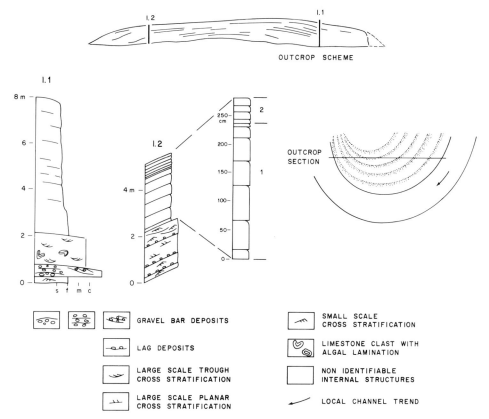

Fig. 14. Sedimentary logs and planimetric reconstruction of example 4.

Fig. 15. Example 5. The exposure consists of four adjacent or stacked point bars.

INTERPRETATION OF REACTIVATION SURFACES

In plan view the reactivation surface is a curved line separating groups of parallel scroll bars and indicating a change in the geometric evolution of the meander loop. Elliot (1976), Díaz-Molina (1978), Allen (1982) and Thomas *et al.* (1987) have suggested that they are related to high stage events or unsteady channel flow.

Leopold & Wolman (1960) have observed that the ratio of the radius of channel curvature/stream width (rm/b) of the meander bends tends towards a constant common value between 2 and 3. Bagnold

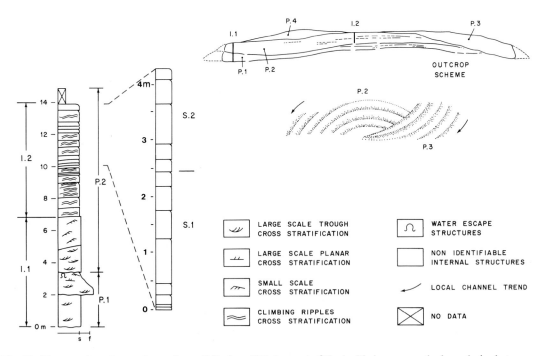

Fig. 16. Measures have been taken only partially from P.2. In most of the bedded sequence the boundaries between the lateral accretion units are not well defined.

(1960) demonstrated that the resistance to flow in a channel of a uniform cross-section falls to a sharply defined minimum within these values. On the other hand, Brice (1974) remarked that the elongation of simple loops does not continue indefinitely, and that they develop a second arc on their perimeters. Hickin (1974) studied the meandering growth-patterns of the Beatton River and demonstrated that there is a critical value for the ratio of the radius of channel curvature to channel width which exerts a control over subsequent direction and rate of lateral migration. In the Beatton River the critical value averages 2.11. When a channel bend achieves a value for r/wm of 2.0, erosion elsewhere on the bends tends to increase r/wm to values greater than 2.0. Changes in the direction of lateral erosion are a response to variations in the flow resistance due to channel curvature changes during the development of the meander. These changes in the direction of lateral erosion are accompanied by correlative changes in the direction of lateral accretion of the meander loop. According to this model the reactivation surfaces can be interpreted as markers of critical curvature thresholds.

LATERAL ACCRETION PATTERNS

Type A point bars

In the Beatton River, Hickin & Nanson (1975) have estimated that the rate of channel-bend migration reaches a maximum value where the ratio radius of channel curvature to stream width approximates 3.0. The rate of channel migration rapidly declines for bends with values of r/wm greater or less than 3. In addition, the spacing of floodplain ridges enlarges with increasing rates of bend migration.

The Beatton River and these ancient examples display different sets of data. In the modern example meander loops have been considered as a whole evolving along a continuous erosional axis (Hickin, 1974) whereas in the Loranca Basin they have been divided into successive individual point bar bodies. In the Beatton River the wavelength of the ridge–swale interval averages 11.60 m (Hickin & Nanson, 1975); however this implies a larger scale than that estimated for some entire individual point bars in the Loranca Basin (see scale on the planimetric reconstructions in Figs 7 & 9).

In these ancient meandering deposits the r and wm values cannot be obtained for an individual point bar, although in some instances it is possible to measure or estimate one of the parameters. However, changes in the spacing of floodplain ridges should be accompanied by correlative changes in the thickness of the lateral accretion episodes. The lateral accretion episodes corresponding to each scroll bar may consist of several lateral accretion units, and the variations in sedimentation rates are examined from them.

If the lateral accretion episodes pinched out towards the top of the point bars, a vertical section would show a thinning upwards trend. However, bed thickness seems to be independent of its vertical location in the point bar profile. In spite of the difficulty in contrasting the modern and ancient data sets, and given that bed thickness could be controlled by the sedimentation rate, it is still possible to compare evolving channel curvature in the present day Beatton River with bedding patterns observed in the Loranca Basin meander loops. A thickening–thinning bedding cycle suggests a decrease in the ratio of radius of channel curvature to stream width, from values higher than 3.

Type A point bars corresponding to examples 1 and 2 (Figs 7 & 9) show few lateral accretion units and a relatively high density of adjacent point bars. Both features could result from the meander loop lying close to critical values of channel curvature along the meander loop development. It is possible to reconstruct a hypothetical model to explain the spatial relationships of the thickening–thinning, thickening and thinning sequences. In Fig. 17 two point bars are represented, with point bar 2 depicting a down-current translation. Other kinds of point bar migrations could also have been considered following Jackson's models (1976b). Three logs (1, 2 and 3 in Fig. 17) exhibit different locations and bedding trends. Log 1 cuts the thinning trend of point bar 1, and the thickening sequence of point bar 2. Log 2 represents the entire thickening–thinning cycle of point bar 2. Log 3 shows a strongly discordant reactivation surface. In Log 3, the bed set corresponding to the first point bar has a thickening upwards sequence, while that belonging to the second point bar contains a thickening–thinning bed sequence and an onlap relationship with the underlying point bar.

When the erosional axis of two adjacent point bars coincides (Fig. 18) and the section is perpendicular to the erosional axis, the reactivation surface is not easily seen. The reactivation surface could only be distinguished where there was a

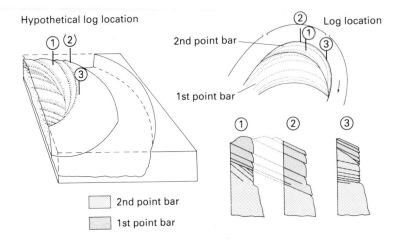

Fig. 17. Hypothetical model indicating the lateral relationships between different kinds of sequences. The surface bars correspond to lateral accretion units.

difference in bed set dip, resulting from a change in the depositional slope. However two different bedding patterns could be seen superimposed as shown in Log 1. Logs 2 and 3, seen in Fig. 18, show similarities. Both logs reflect the lateral pinch out of the second point bar. Since point bars 1 and 2 in Fig. 18 are nearly parallel, the reactivation surface between them is less noticeable here than in log 3 of Fig. 17.

Type B point bars

Type B point bars have a larger lateral development than type A. They comprise a greater number of lateral accretion units, arranged in several superimposed sequences, and showing thickening–thinning bedding trends.

Several of the characteristics displayed in example 3, e.g. a discontinuous range of sediment sizes, segmented profiles and palaeocurrents directed outwards in the upper part of the deposit (Fig. 11), have been previously described from the Beatton River by Nanson (1980). In the Beatton River scroll bars grow upstream and downstream from a point bar platform. As the platform aggrades an initial scroll bar is deposited between the centre and the channel limb of the platform. Nanson (1980) suggested that the discontinuous range of sediment sizes was responsible for the formation of ridges of sediments, which in turn would generate Langmuir circulation. The Langmuir circulation can also explain the preserved outwards directed climbing ripples in this ancient deposit. Later, the same author (Nanson, 1981) suggested that an initial

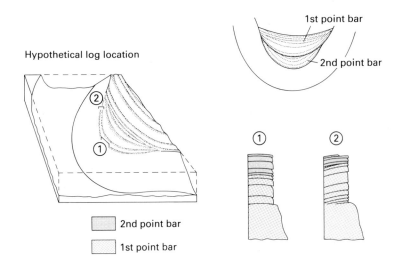

Fig. 18. The meander loop complex consists of two point bars sharing the same erosional axis. The second point bar shows an expansion, which is responsible for overlapping the earlier deposits (log 1).

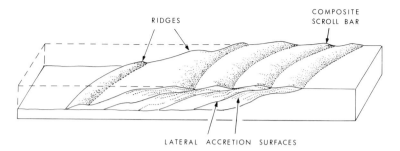

Fig. 19. Reconstruction of the accretion geometry of a point bar with segmented profile. The drawing is based on P.2 of example 3.

scroll bar may form around a stranded dead tree, acting as a nucleus for sedimentation on a point bar platform. Nevertheless, floodplain scrolls can result from the coevolution of several mechanisms.

Example 4 and P.2 of example 5 show longitudinal sections which correspond to the outer extreme of the point bar. Likewise, section (e) of Fig. 3 illustrates an identical section. In this location bedding should reflect the evolution of the rate of sedimentation before either the abandonment of the bend or the adjustment of the channel. Example 4 presents a rough thickening–thinning sequence, while the other example shows two superimposed thickening–thinning cycles and the last cycle is thinner. The bases of the point bars (example 4 and P.2 in example 5) have not been defined by reactivation surfaces and the logs represent only a part of the lateral accretion units; thus the overall bedding analysis is incomplete.

The bedding characteristics of two point bars (P.2 and P.3, Fig. 12) can be observed in the meander loop of example 3. The complete bedding trends of P.2 can be analysed by superimposing 1.2 on 1.3. In example 3 the bedding trends of the lateral accretion units are strongly dependent on the position of the ridges along the point bar profiles (Fig. 19), and the 'compensation beds' make sequential analysis difficult. However, in example 3 each composite ridge comprises a rough thickening–thinning cycle. Since in the Beatton River ridges development and aggradation originate scroll bars, thickening–thinning bedding trends of example 3 might correspond with scroll bar formation. Scroll bars might also account for the formation of bedding cycles observed within other examples of lateral accretion units.

The expected overall trend in the upper part of the point bar deposits is illustrated in Fig. 20. Ridge morphologies are not represented because they are absent from examples 4 and 5. Thickening–thinning bedding cycles correspond to scroll bar units in Fig. 20. The hypothetical reconstruction is based on the increase and subsequent decrease in migration rates as values of r/wm decline.

SUMMARY AND CONCLUSIONS

In the Loranca Basin the meander loop complexes cannot be considered as a whole evolving along a continuous axis. Instead they are formed by discrete parts (point bars), identifiable from geometric criteria. The adjacent point bars are formed by conformable lateral accretion units. Each point bar body

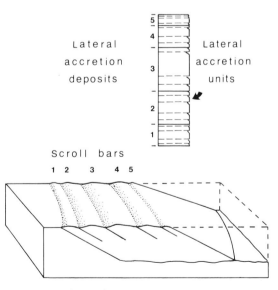

Fig. 20. Idealized point bar where scroll bar sequences are formed by thickening–thinning bedding cycles. Scroll bar cycles complete a lateral accretion succession with a thickening–thinning trend. A decrease of the ratio r/wm would explain this hypothetical model.

shows a longitudinal, transverse or oblique section through the meander loop. Transverse geometries are frequently recognized in other ancient fluvial deposits; however other kinds of sections should be expected.

The contact between laterally juxtaposed point bars is defined by a meander loop reactivation surface, which can be erosional or non-erosional and represent different phases of planimetric meander loop evolution. Similar phenomena observed in modern rivers are interpreted as resulting from channel readjustment when thresholds of critical curvature have been reached. Thus point bars delimited by two reactivation surfaces represent sedimentary evolution until a new critical curvature developed.

In type A point bars thickening–thinning bedding sequences are interpreted as the result of increasing and subsequently decreasing sedimentation rates, controlled by a fall in the ratio r/wm from values higher than 3. Thinning bedding sequences may have resulted from a decrease in r/wm values from approximately 3. Thickening sequences could represent channel evolution to a critical curvature arising from an increase in the sedimentation rate. Nevertheless the different bedding patterns could also be spatially related, for example the lateral changes might be a consequence of the translation or expansion of the point bars. Meander loop complexes of examples 1 and 2 probably formed when near-critical values of channel curvature had developed along the meander loop.

In type B point bars the evolution of the ratio r/wm could only be deduced from the megasequences formed by the superposition of the minor cycles. The superimposed cycles show a decrease in thickness in examples 4 and 5, although they are incomplete at the base, and do not represent the complete evolution of the meander loops. The very well preserved segmented profiles in P.3 and P.2 of example 3 (Fig. 10), the discontinuous range of sediment sizes and the outwards directed climbing ripples in the upper member in P.2 are closely comparable to the sedimentary characteristics of the Beatton River point bar deposits.

Thickening–thinning bedding sequences found in the upper finer members of the point bars could represent the same rhythmic sedimentation pattern which was responsible for the formation of scroll bars. Alternatively the thickening–thinning sequences of type A point bars might result from the formation of individual scroll bars. In examples 1 and 2, where a point bar section consists of only a few beds, the sedimentary response to a fall in the r/wm ratio may have coincided with the development of an isolated scroll bar on the point bar surfaces.

Bedding cycles exist in the point bars of the Loranca Basin, and their meanings have been tentatively discussed in this work. However more data are required, from modern and ancient river examples, to understand the origin of the lateral accretion units, and the meaning of the sequences formed by the lateral accretion units.

ACKNOWLEDGEMENTS

The author is grateful to the 'Comisión Interministerial de Ciencia y Tecnología' (Spain) for providing the financial support for the project PB85-0022. She thanks J. Alexander, J. Bond and C. Vondra for constructive reviews of the manuscript. The assistance of Fernando Pérez in the drafting and of Jose Luis Gonzalez-Pachón in the photography laboratory is greatly appreciated.

REFERENCES

Allen, J.R.L. (1965) A review of the origin and character of recent alluvial sediments. *Sedimentology* 5, 89–191.
Allen, J.R.L. (1982) *Sedimentary Structures. Their Character and Physical Basis.* Developments in Sedimentology 30, Vol. II. Elsevier, Amsterdam, 663 pp.
Allen, J.R.L. & Friend, P. (1968) Deposition of the Catskill Facies, Appalachian Region, with notes on some other Old Red Sandstone Basins. In: *Late Paleozoic and Mesozoic Continental Sedimentation, Northeastern North America* (Ed. Klein, G. de V.) pp. 21–74. Geol. Soc. Am. Spec. Paper 106, Denver.
Bagnold, R.A. (1960) Some aspects of the shape of river meanders. US Geol. Surv. Prof. Paper 282-E, pp. 135–143.
Brice, J.C. (1974) Evolution of meander loops. *Bull. Geol. Soc. Am.* 85, 581–586.
Bridge, J.S. & Jarvis, J. (1976) Flow and sedimentary processes in the meandering River South Esk, Glen Clova, Scotland. *Earth Surf. Proc.* 1, 303–336.
Bridges, P.H. & Leeder, M.R. (1976) Sedimentary model for intertidal mudflat channels, with examples from the Solway Firth, Scotland. *Sedimentology* 23, 533–552.
Díaz-Molina, M. (1978) *Bioestratigrafía y Paleogeografía del Terciario al este de la Sierra de Altomira.* PhD thesis, Universidad Compultense de Madrid.
Díaz-Molina, M. (1979) Características sedimentológicas de los paleocanales de la Unidad Detrítica Superior al N de Huete (Cuenca). *Estudios Geol.* 35, 241–251.

DÍAZ-MOLINA, M. (1984) Geometry of sandy point bar deposits, examples of the Lower Miocene. Tajo Basin, Spain (abstr.). In: *Int. Ass. Sediment. 5th European Regional Meeting of Sedimentology, Marseille*, pp. 140–141.
DÍAZ-MOLINA, M., ARRIBAS-MOCOROA, J. & BUSTILLO-REVUELTA, A. (1989) The Tórtola and Villalba de la Sierra fluvial fans: Late Oligocene–Early Miocene, Loranca Basin, Central Spain. *Int. Ass. Sediment. 4th Int. Conf. on Fluvial Sedimentology, Barcelona. Field Trip 7.* Servei Geològic de Catalunya, Barcelona.
DÍAZ-MOLINA, M., BUSTILLO-REVUELTA, A., CAPOTE, R. & LOPEZ-MARTINEZ, N. (1985) Wet fluvial fans of the Loranca Basin (central Spain). Channel models and distal bioturbated gypsum with chert. In: *Int. Ass. Sediment. 6th European Regional Meeting of Sedimentology, Lérida, Spain. Excursion Guide-Book*, pp. 149–185.
DIETRICH, W.E. & SMITH, J.D. (1983) Influence of the point bar flow through curved channels. *Water Resources Res.* **19**, 1173–1192.
ELLIOT, T. (1976) The morphology, magnitude and regime of a carboniferous fluvial-distributary channel. *J. Sedim. Petrol.* **46**, 70–76.
ETHRIDGE, F.G. & SCHUMM, S.A. (1978) Reconstructing palaeochannel morphologic and flow characteristics: methodology, limitations and assessment. In: *Fluvial Sedimentology* (Ed. Miall, A.D.) pp. 703–722. Can. Soc. Petrol. Geol., Calgary, Memoir 5.
HICKIN, E.J. (1974) Development of meanders in natural river channels. *Am. J. Sci.* **274**, 414–442.
HICKIN, E.J. (1978) Mean flow structure in meanders of the Squamish River, British Columbia. *Can. J. Earth Sci.* **15**, 1833–1849.
HICKIN, E.J. & NANSON, G.C. (1975) The character of channel migration on the Beatton River, Northeast British Columbia, Canada. *Bull. Geol. Soc. Am.* **86**, 487–494.
JACKSON II, R.G. (1975) Velocity–bedform–texture patterns of meander bends in the lower Wabash River of Illinois and Indiana. *Bull. Geol. Soc. Am.* **86**, 1511–1522.
JACKSON II, R.G. (1976a) Large scale ripples of the Wabash River. *Sedimentology* **23**, 593–623.
JACKSON II, R.G. (1976b) Depositional models of point bars in the Lower Wabash River. *J. Sedim. Petrol.* **46**, 579–594.
LEOPOLD, L.B. & WOLMAN, M.G. (1960) River meanders. *Bull. Geol. Soc. Am.* **38**, 125–316.
LEOPOLD, L.B., WOLMAN, M.G. & MILLER, J.P. (1964) *Fluvial Processes in Geomorphology.* W.H. Freeman, San Francisco, 522 pp.
MCCABE, P.J. & JONES, C.M. (1977) Formation of reactivation surfaces within superimposed deltas and bed forms. *J. Sedim. Petrol* **47**, 707–715.
MOODY-STUART, M. (1966) High- and low-sinuosity stream deposits, with examples from the Devonian of Spitsbergen. *J. Sedim. Petrol.* **36**, 1102–1117.
MUTTI, E. & SONNINO, M. (1981) Compensation cycles: a diagnostic feature of turbidite sandstone lobes (abstr.). In: *Int. Ass. Sediment. 2nd European Regional Meeting of Sedimentology, Bologna* (Eds Valloni, R., Colella, A., Sonnino, M., Mutti, E., Zuffa, G.G. & Ori, G.G.) pp. 120–123. Tecnoprint, Bologna.
NAMI, M. & LEEDER, M.R. (1978) Changing channel morphology and magnitude in the Scalby Formation (M. Jurassic of Yorkshire, England). In: *Fluvial Sedimentology* (Ed. Miall, A.D.) pp. 431–440. Can. Soc. Petrol. Geol., Calgary, Memoir 5.
NANSON, G.C. (1980) Point-bar and flood plain formation of the meandering Beatton River, northeastern British Columbia, Canada. *Sedimentology* **27**, 3–30.
NANSON, G.C. (1981) New evidence of scroll-bar formation on the Beatton River. *Sedimentology* **28**, 889–891.
PUIGDEFÁBREGAS, C. (1973) Miocene point bar deposits in the Ebro Basin, Northern Spain. *Sedimentology* **20**, 133–144.
PUIGDEFÁBREGAS, C. & VAN VLIET, A. (1978) Meandering stream deposits from the Tertiary of the southern Pyrenees. In *Fluvial Sedimentology* (Ed. Miall, A.D.) pp. 469–486. Can Soc. Petrol. Geol., Calgary, Memoir 5.
SCHUMM, S.A. (1977) *The Fluvial System.* John Wiley & Sons, New York, 338 pp.
STEWARD, D.J. (1981) A meander-belt sandstone of the Lower Cretaceous of southern England. *Sedimentology* **28**, 1–20.
SUNDBORG, A. (1956) The river Klarälven, a study of fluvial processes. *Geogr. Ann.* **38**, 125–316.
THOMAS, R.G., SMITH, D.G., WOOD, J.M., VISSER, J., CALVERLEY-RANGE, E.A. & KOLSTER, E.H. (1987) Inclined heterololithic stratification — terminology, description, interpretation and significance. *Sedim. Geol.* **53**, 123–179.
THORNE, C.R., ZEVENBERGEN, L.W., PITLICK, J.C., RAIS, S., BRADLEY, J.B. & JULIEN, P.Y. (1985) Direct measurements of secondary currents in a meandering sand-bed river. *Nature* **315**, 746–747.
WILLIS, B.J. (1989) Palaeochannel reconstructions from point bar deposits: a three-dimensional perspective. *Sedimentology* **36**, 757–766.

Alluvial ridge-and-swale topography: a case study from the Morien Group of Atlantic Canada

M.R. GIBLING* and B.R. RUST†

*Department of Geology, Dalhousie University, Halifax, Nova Scotia B3H 3J5, Canada; and
†Department of Geology, University of Ottawa, and Ottawa-Carleton Geoscience Centre, Ottawa, Ontario K1N 6N5, Canada

ABSTRACT

Ridge-and-swale topography (RST), a prominent feature of meander lobes in modern rivers, is generally attributed to scroll-bar accretion. An exhumed meanderbelt in the Silesian Morien Group of Nova Scotia, Canada shows four types of RST, three of which are attributed to scroll bars. *Asymmetric* ridges (type 1) contain single or stacked planar cross-sets that advanced up inclined surfaces that strike parallel to the ridge crests. Type 1a ridges show 0.5 m relief, 5 m mean spacing and low-angle (5°) inclined surfaces in sandstone. Type 1b ridges are larger in scale (1 m relief, 10 m mean spacing); their thin-bedded sandstone and siltstone show inclined surfaces that dip at up to 20°. Type 1 RST is attributed to scroll bars, and probably originated in part from the inward migration of two-dimensional dunes.

Symmetric (type 2) ridges consists of thick-bedded, trough cross-stratified sandstone with ridge-parallel palaeoflow. Type 2a ridges (up to 1.2 m of relief, 12 m average spacing) underlie scroll-bar ridges (1b), and probably represent initial scroll-bar deposits. Type 2b ridges are paired, with well-marked erosional swales, and are attributed to levees bordering chute channels.

Sandstone-rich ridges (1a, 2a, 2b) are resistant to weathering and form the most pronounced exhumed topography. Such ridges are preserved at a storey boundary within the meanderbelt deposit.

INTRODUCTION

Ridge-and-swale topography (RST) is a descriptive term for curvilinear elevations and depressions on meandering river floodplains. Other terms used for these features include accretion topography, meander scrolls and scroll bars (Lobeck, 1939; Fisk, 1952; Allen, 1965a, 1968; Nanson, 1980). Where the floodplain lies within a meander loop, RST roughly conforms to the curve of the channel and the ridges accurately locate convex banks of pre-existing meanders (Allen, 1968; Hickin, 1974). The floodplains of meandering rivers with bankfull depth of about 5–10 m generally show 10–30 m ridge spacing and ridge-to-swale relief from 0.5 m to a few metres (Table 1). Relief reaches about 4.5 m on the Mississippi floodplain (Fisk, 1947).

RST in meandering rivers has been attributed to

† Deceased.

the growth of scroll bars, which are streamwise accumulations of sediment that accrete to the point bars of meandering channels. A series of scroll bars are attached successively to the subaerial, convex bank of the meander bend (Allen, 1965a; Hickin, 1974). The form, internal stratification and genesis of scroll bars were discussed by Sundborg (1956) for the Klaralven River of Sweden, by Jackson (1976a) for the Wabash River of Indiana, and by Nanson (1980, 1981) for the Beatton River of British Columbia. In the first two rivers, scroll-bar initiation is closely associated with the inward migration of transverse bars, and the scroll bars migrate towards the convex bank until flow is diverted channelward, leaving a largely inactive swale on the landward side of the scroll bar. In the Beatton River, many scroll bars were initiated around stranded trees, and lateral migration of scroll bars

Table 1. Dimensions of alluvial ridge-and-swale forms of some modern meandering rivers; asterisk indicates estimates made from published figures

River	Ridge spacing (m)	Ridge-to-swale height (m)	References
Klaralven, Sweden	17–27	Up to 3	Sundborg (1956)
Beatton, Canada	11.6 (av.)	0.52 (av.)	Hickin & Nanson (1975) Nanson (1980)
Endrick, Scotland*	10	Up to 1.5	Bluck (1971)

was not noted, although downstream, and locally upstream, growth was observed. Vegetation colonizes the bars at low-flow stage, effectively connecting them to the adjacent floodplain (see also Fisk, 1944, p. 19). Nanson (1980) concluded that floodplain scrolls are examples of convergent evolution of landforms and can result from different mechanisms. Sundborg (1956) and Nanson (1980) examined natural sections of RST, and identified scroll-bar deposits in the ridges, thus establishing the connection between scroll bars and floodplain ridges.

Potentially, the geological record can furnish both the exhumed meanderbelt surfaces and cross-sections necessary for further investigation of RST. For example, Puigdefàbregas (1973) documented a point-bar deposit in the Spanish Miocene, in which bedforms represented by planar cross-bed sets a few decimetres thick had migrated up inclined surfaces that span the entire thickness of the deposit (1–2 m). The exhumed cross-sets formed curvilinear ridges (Fig. 1).

The present paper documents an exhumed Pennsylvanian meandering channel deposit in the Sydney Basin of Nova Scotia (Fig. 2). RST has been dissected to various levels on wavecut surfaces, and the adjacent cliffs yield cross-sections parallel to the dip of the inclined stratification (IS) and inclined heterolithic stratification (IHS) (Thomas et al., 1987) that is a prominent feature of the deposit. We describe here the form and internal structure of the exhumed ridges, and discuss their origin. We also discuss the importance of relative resistance to weathering and depth of erosion in determining exhumed topography.

GEOLOGICAL SETTING

The Sydney Mines Formation is the uppermost unit of the Morien Group of the Sydney Basin (Boehner & Giles, 1986). The formation, dated as Late Carboniferous (Westphalian D to Stephanian: Bell, 1938; Zodrow & Cleal, 1985), is about 500 m thick in onshore exposures, and consists of sandy channel deposits interbedded with grey and red mudstones, thin sandstones, coals and limestones ascribed to floodbasin deposition. Rust et al. (1987) interpreted the formation as meandering fluvial because the channel deposits show upward fining, upward decrease in the scale of cross-beds, and abundant inclined and commonly heterolithic stratification attributed to lateral accretion on point bars. One channel complex documented from extensive cliff exposures (Gibling & Rust, 1987) contains three

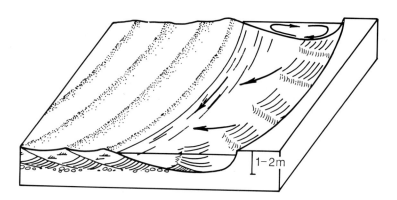

Fig. 1. Model to explain the genesis of ridge-and-swale topography in the Murillo point bar, Miocene of Spain. Helicoidal flow within the meander bend causes large-scale, two-dimensional bedforms (origin not specified) to migrate obliquely up the point-bar surface now represented by channelward-inclined stratification. (Modified from Puigdefàbregas (1973, fig. 7) and Allen (1968, fig. 3.21).)

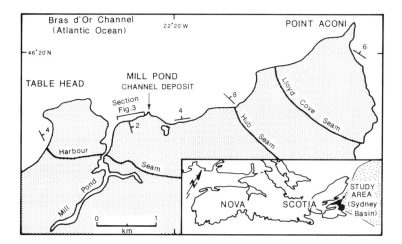

Fig. 2. Location of Mill Pond channel deposit and stratigraphic section of Fig. 3. Stippled area is underlain by the Sydney Mines Formation. Geology from Boehner & Giles (1986). Inset locates Sydney Basin in Nova Scotia, showing the basin's onshore (black) and partial offshore (stippled) extent.

erosionally based storeys up to 10 m thick with well-developed IHS. Palaeoflow direction for adjacent storeys differs by up to 90°, and the complex was interpreted as a meanderbelt body comprising a series of point-bar lenses.

The Sydney Mines Formation contains a predominantly non-marine biota (Rust *et al.*, 1987), although the presence of sparse agglutinated (marsh) foraminifera suggests localized marine (brackish) influence (Thibaudeau & Medioli, 1986).

This paper describes the Mill Pond channel deposit, located at an unnamed headland along Bras d'Or Channel where the Sydney Mines Formation dips at about 2° (Fig. 2). The exposure comprises a cliff about 5 m high and a wavecut platform up to 50 m wide at low tide. The deposit shows features noted above as indicative of meandering fluvial sedimentation.

THE CHANNEL DEPOSIT

The channel deposit exposed in the cliffs is 11.9 m thick and shows two superimposed storeys (Fig. 3). The lower storey, designated Unit 1, is up to 3 m thick; it thickens and coarsens eastward from planar-bedded sandstone and siltstone to trough cross-bedded sandstone. This lateral transition may represent a passage from marginal levee and crevasse splay deposits to channel deposits (from 'wings' to 'central body': Bersier, 1958). In easterly exposures, the top of Unit 1 shows RST with up to 1 m of relief, rendered indistinct due to amalgamation with sandstones of the upper storey.

The upper storey is 8.9 m thick and comprises three units (Fig. 3). The lowermost unit (2) mantles the uneven basal surface and contains claystone intraclasts from a few centimetres to 1.7 m long. This suggests that clay formerly capped the lower storey or lay upstream of the depositional site. Unit 3 consists of medium-grained sandstone with trough cross-sets that decrease in thickness from 20–50 cm at the base to 10–30 cm at the top. Planar and trough cross-sets, ripple cross-laminae and a few carbonate rhizoliths (Klappa, 1980) are present in the topmost metre, which shows IS. RST is well developed locally at the contact between Units 3 and 4, the ridges adding 1–2 m of strata to Unit 3. Inclined stratification passes upward into inclined heterolithic stratification where siltstone layers appear in the ridges (forming a 'composite set': Thomas *et al.*, 1987). The wavecut platform approximates the level of the Unit 3/4 contact.

Unit 4, up to 2.5 m thick, contains IHS sets of fine-grained sandstone and siltstone with planar cross-sets, ripple cross-laminae, roots and rhizoliths. The topmost 50 cm, interpreted as a palaeosol, is intensely rooted and capped by 5 cm of unlithified yellow silt overlain by overbank deposits (Unit 5).

A marked discordance in accretion and palaeoflow direction is evident between northern and southern segments of Units 3 and 4 (Fig. 4). The southern segment contains IS and IHS units that dip consistently northeastward (the accretion direction) for 105 m parallel to their dip and show southeastward palaeoflow. The northern segment contains units that dip southeastward for 85 m parallel to their dip and show northeastward palaeo-

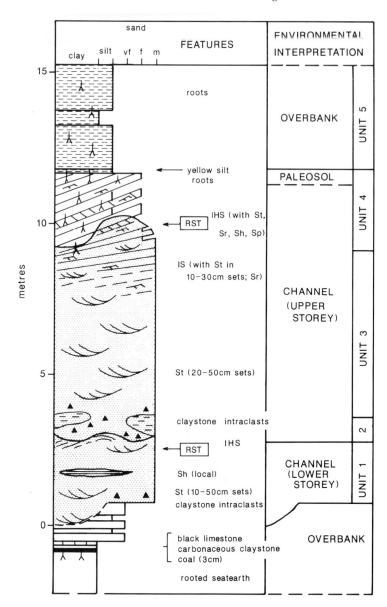

Fig. 3. Stratigraphic section of the Mill Pond channel deposit. The units indicate stratigraphic divisions referred to in the text. RST, Ridge-and-swale topography; IS and IHS, inclined and inclined heterolithic stratification, respectively (Thomas et al., 1987); St, large-scale trough cross-stratified sandstone; Sr, ripple cross-laminated sandstone; Sh, planar stratified sandstone; Sp, large-scale planar cross-stratified sandstone. (Facies codes from Miall (1978).)

flow. The segment contact runs subparallel to the palaeoflow direction of the southern segment, which is inferred to be the younger as confirmed by overlap of the southern segment on bevelled strata of the northern segment. The Unit 3/4 contact lies at a similar elevation in the two segments. We interpret the segments as juxtaposed point-bar lenses formed while the meanderbelt surface lay at a relatively uniform topographic level (Bridge & Diemer, 1983; Gibling & Rust, 1987). An abandoned-channel fill, the upper part of which comprises 3 m of rooted mudstone, at the segment contact was formed during abandonment of the southern meanderbelt segment.

The thickness of the completely preserved upper storey (Units 2–4: 8.9 m) serves as an approximation of bankfull channel depth (Leeder, 1973). As the storey contains IHS (lateral accretion) surfaces, this value is taken to represent bankfull depth at a meander bend. Bankfull width was estimated at

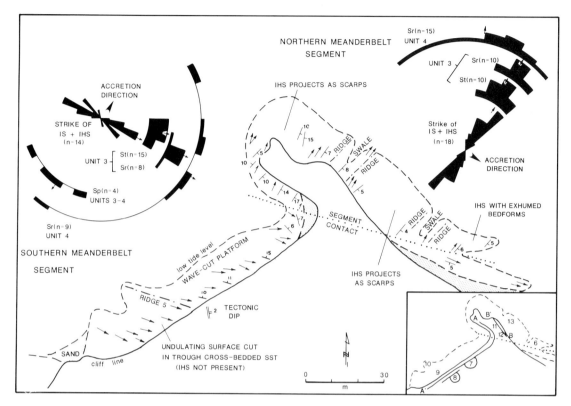

Fig. 4. Upper part (Units 3 and 4) of Mill Pond channel deposit as mapped on the wavecut platform. An erosional contact separates the strata into northern and southern meanderbelt segments with strongly contrasting palaeoflow and accretion directions. Representative orientations of palaeoflow indicators (arrows) and IS/IHS sets (dip and strike symbols) are shown. Rose diagrams show orientation data obtained from both wavecut platform and cliffs: statistical vector means are indicated by arrows on the rose diagrams. The most prominent exhumed ridges on the wavecut platform are indicated. Facies codes are explained in caption of Fig. 3. Inset shows the geographic location of other figures and sections A–A' and B–B' (Figs 7 & 11, respectively).

about 200 m using the relationship:

$$w = 6.8h^{1.54}$$

where w and h are bankfull width and depth, respectively, for channel reaches with a sinuosity >1.7 (Leeder, 1973). A large standard deviation is associated with the regression line. The IHS and IS sets cannot be utilized to determine channel width (see Allen, 1965b) because they are visible only in the topmost strata. The sets show slight curvature in planview, but the extent of the wavecut platform is insufficient to determine their radius of curvature.

Sinuosity can be estimated from the degree of palaeoflow divergence in the deposit using the relationship:

$$\text{sinuosity} = \frac{1}{1 - \left(\frac{\Theta}{252}\right)^2}$$

where Θ is the maximum angular range of the mean channel azimuth (Miall, 1976). Based on IHS and trough cross-bed vector means (Fig. 4), the two segments show a palaeoflow divergence of about 80°, which yields a sinuosity of 1.12. This is considered a low estimate because only two meanderbelt segments are represented, and because both are cut radially by the wavecut platform, giving little palaeocurrent variation. Other channel deposits in the Sydney Mines Formation exhibit up to 180° divergence in trough cross-bed vector means in adjacent storeys which, assuming that the crossbeds reflect channel trend, indicates a sinuosity of up to 2.0 in those cases.

Other channel parameters were computed using quantitative relationships determined for modern meandering rivers (Method 2 of Ethridge & Schumm, 1978). Meander wavelength was estimated at 3 km, radius of curvature at 1 km and slope at about 30 cm km^{-1}. The results must be treated with great caution because channel width, an input parameter for the computation, was derived theoretically and because broad confidence intervals apply to the equations used.

RIDGES AT THE TOP OF THE UPPER STOREY

Four types of ridge were identified in the Mill Pond deposit (Table 2, Fig. 5). They are recognized as asymmetric (type 1) or symmetric (type 2) by (i) their exhumed form and cross-sectional shape, and (ii) the predominant direction of sediment transport during ridge construction, parallel to ridge crests (symmetric) or transverse to ridge crests (asymmetric). Each type can be further subdivided. Type 1a, 1b and 2a ridges are interpreted as scrollbar deposits, whereas type 2b ridges are ascribed to levees bordering chutes on an upper point-bar surface (Table 2). Types 1a and 2b were observed only in the northern meanderbelt segment, whereas types 1b and 2a were observed only in the southern segment.

Asymmetric ridges

Asymmetric ridges (*type 1a;* Fig. 6) are well developed on the wavecut platform of the northern segment (Fig. 4). IHS surfaces, which dip consistently southeastward at 3–8° (average 5.1°, $n = 7$), bound planar cross-sets 20–50 cm thick with foresets dipping up the inclined surfaces. The tops of isolate or stacked sets form low ridges with 1.6–8 m spacing (average 4.9 m, $n = 8$) and less than 50 cm relief. The ridges are slightly curved in plan view and can be traced for 50 m across the platform. Their crestal height varies along strike.

Lenses of interstratified siltstone and medium- to very fine-grained sandstone are prominent in Units 3 and 4 in the southern segment (ridges 6–9 between the 78 and 107 m positions in the cliff face: Figs 7 & 8). Limited three-dimensional exposure shows that the lenses form part of asymmetric ridges (*type 1b*). The ridges contain IHS surfaces which separate beds 10–50 cm thick and dip consistently northeastward at 5–20° (average 11.2°, $n = 11$). The ridges show convex-upward upper surfaces and their

Table 2. Description and inferred origin of exhumed RST in the Mill Pond channel deposit

		Average spacing (m)	Relief (m)	Internal structure	Origin
Asymmetric (type 1) Cross-section asymmetric. Contain accretion surfaces with strike parallel to ridge crests. Planar cross-sets migrated up accretion surfaces. Ridges flow-parallel (as shown by trough cross-beds in strata below)	1a	4.9	0.5	Planar cross-sets (20–50 cm), minor ripple cross-lam.	Transverse bars stacked to form small scroll bars.
	1b	9.7	1–2	Planar cross-sets (max. 85 cm), with trough and ripple cross-strat. and planar strat. Convex-upward accretion surfaces.	Large scroll bars associated with transverse bars.
Symmetric (type 2) Cross-section symmetric. Accretion surfaces rare. Composed mainly of trough and ripple cross-strat.	2a	12.5	0.4–1.2	Underlie type 1b ridges on channelward side. Trough and minor ripple cross-strat.	Initial scroll-bar ridges underlying large scroll bars.
	2b	10.5	0.7	Ridges paired with strongly erosional swale. Trough cross-strat. below, planar sets (35–65 cm) above.	Levees bordering chute channels.

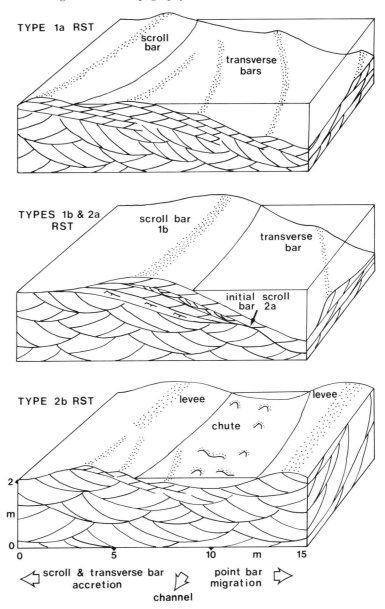

Fig. 5. Schematic diagrams to illustrate types and inferred origin of ridge-and-swale topography at Mill Pond.

contained IHS is roughly form-concordant. Individual IHS surfaces can be traced updip for 1–2 m, but ridge crests and swales differ in elevation by less than 1 m. The ridges are asymmetric in cross-section, with longer northeastern limbs, and their sandy strata overlie finer-grained strata, up to 1 m thick, in the swales on their southwestern sides (95 m position, Fig. 7). Ridge spacing averages 9.7 m.

Planar cross-sets are prominent in ridges 6–9. The sets range in thickness from 8 to 85 cm, with a mean of 28 cm ($n = 9$), and are isolate or grouped as cosets. Their cross-strata are 1–3 cm thick with carbonaceous and commonly sigmoidal partings (Fig. 9). Where measurable, the cross-strata dip southwestward up the IHS surfaces (Fig. 4). A few planar cross-sets show an apparent foreset dip down the IHS surfaces (78 m position, Fig. 7). The ridges also contain ripple cross-lamination and small

Fig. 6. Type 1a RST. Wavecut platform of the northern segment shows a series of scarps with up to 50 cm relief. The scarps are slightly curvilinear in planview, and most show a convex-upward form in cross-section. They represent bedforms that have advanced northwestward (to the left) up the IHS surfaces, clearly seen where arrowed. Letter r indicates prominent ridge crests. The water-filled trough (dashed) in the middle distance covers the segment contact (see Fig. 4) and IHS sets in the foreground dip towards the camera. Scale (middle distance) 1 m long.

(5–10 cm) trough cross-sets, and are capped locally by thin, planar-stratified sets. Roots are common in the ridges.

Symmetric ridges

Symmetric to slightly asymmetric ridges and swales (*type 2a*: ridges 1–5 and 10 of Fig. 7) are exhumed on the wavecut platform of the southern segment at a lower stratigraphic level than the type 1b ridges. Mean ridge spacing is 12.5 m, relief ranges from 0.4 to 1.2 m, and ridge sides dip at 7–11°. The ridges are traceable across the wavecut platform, a distance of 25 m, and are slightly sinuous. They consist of trough cross-stratified and ripple cross-laminated sandstone, with ridge-parallel palaeoflow, and locally contain IS which strikes parallel to the ridge crests and dips northeastward at 6–11° (average 8°, $n = 3$). Planar cross-sets were not observed. The exhumed swales commonly show a thin covering of mudstone intraclasts up to 2 cm in diameter.

Ridge 5 is seen in cross-section at the seaward

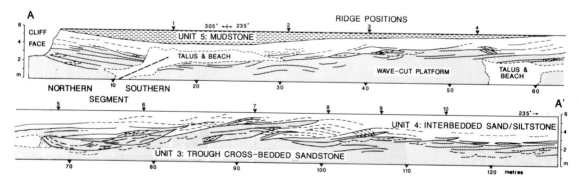

Fig. 7. Type 1b and 2a RST. Cliff face and uppermost part of wavecut platform along the line A–A' (Fig. 4), traced from a photomosaic and shown in two parts. The stippled area represents thick-bedded sandstone of Unit 3 (see Fig. 3); the Unit 3/4 contact is locally gradational and difficult to define. Note in the upper part of Unit 3 the inclined stratification (IS), which dies out downdip into trough cross-stratified, medium-grained sandstone and grades updip into IHS sets (siltstone and fine-grained sandstone) of Unit 4, shown white on the diagram. 'Ridge positions' (1–10) indicate well-defined ridges developed at the top of Unit 3 and (in one case: ridge 8) in Unit 4. Ridges 6–9 (and associated swales) belong to type 1b RST; ridges 1–5 and 10 belong to type 2a RST. Tectonic dip is about 2°. Scale (vertical = horizontal) is approximate. Continuous cliff face ends at about the 130 m position.

Fig. 8. Photomosaic of cliff face from 67–110 m position in A–A' (inset, Fig. 4). Type 1b ridges (6–9, numbers placed just above ridge crests) contain sandstone lenses associated with inclined surfaces which dip consistently to the left (NE). Ridges 6, 7 and 9 contain prominent sandstone beds whereas ridge 8 consists mainly of siltstone (compare with Fig. 7). Type 2a ridges form an undulating surface developed in trough cross-bedded sandstones of Unit 3 below the type 1b ridges. Ridge 5 at left of photograph is a good example of a type 2a ridge (see Fig. 10). Scale 1 m long.

margin of the wavecut platform (Fig. 10). The ridge, 1.2 m above the adjacent swale to the southwest, contains a scour fill 80 cm thick. The scour cuts through trough cross-bedded sandstone, and the fill consists of inclined layers of ripple cross-laminated sandstone which strike subparallel to the ridge, decrease in dip upward and curve over to form the ridge crest (Fig. 10). Flow during deposition of the fill was parallel to the ridge. The swale between ridges 5 and 6 is cut into the underlying sandstones and is partially filled by a lens of trough cross-bedded sandstone (70–82 m position in Figs 7 & 8). Ridges 1–4 show only limited cross-sectional exposure.

The relationship between types 2a and 1b ridges is illustrated by ridge 5 (type 2a) which underlies the northeastern part of ridge 6 (type 1b) (Fig. 8). Type 2a ridges also underlie the northeastern parts of type 1 ridges 7–9 (88, 95 and 108 m positions in Figs 7 & 8, where they are not numbered separately).

Fig. 9. Close-up of set of sigmoidal, planar cross-strata (S) in a type 1b ridge. The set climbs to the right (SW) up the IHS surface which forms its lower boundary. Ridge 8, 100 m position within Unit 4 (Figs 7 & 8). Scale 1 m long.

Fig. 10. Type 2a ridge (ridge 5) on wavecut platform at the 68 m position in Figs 7 & 8. The ridge runs into the cliff and consists of a sandstone channel fill composed of inclined layers with Sr. Where arrowed, the topmost inclined surfaces of the fill curve over northeastward (to the left) to form the ridge crest. The channel is incised into trough cross-beds of Unit 3. The swale fill to the right (below 1 m scale) is composed of trough cross-sets which rest on the ridge strata. Numbers indicate ridges shown in Fig. 7.

Paired ridges (*type 2b*) separated by a deep swale are present (two occurrences) in the uppermost part of Unit 3 in the northern segment (Figs 4, 11–13). Relief is 70 cm in each case, and the paired ridges are 8 and 13 m apart. The swales are cut into trough cross-bedded sandstone, show terraced sides, and are lined with thin, inclined layers of ripple cross-laminated sandstone. IHS in the topmost strata of the ridges (Figs 11 & 12) strikes parallel to the ridge crests and dips at 8–17° (average 13.1°, $n = 11$). The inclined surfaces bound sets of planar cross-strata 35–65 cm thick (average 48 cm, $n = 5$), the crests of which project from the cliffs as low ridges (Figs 11 & 12). The cross-strata are separated by thin layers of ripple cross-laminated siltstone. Cross-stratal surfaces are smooth and subparallel to

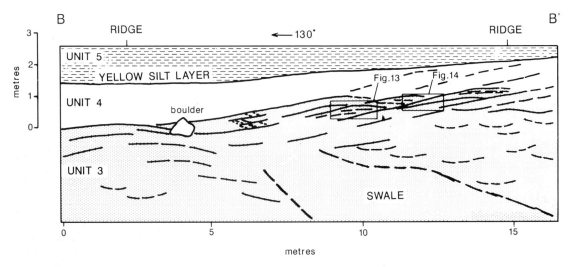

Fig. 11. Type 2b RST. Cliff face and uppermost part of wavecut platform along the line B–B' (inset, Fig. 4), traced from photomosaic (Fig. 12). The stippled area represents thick-bedded sandstone of Unit 3, the top of which shows two well-defined ridges (at 2 and 14 m positions). The lower part of the latter ridge, exposed on the wavecut surface, consists of trough cross-strata; the upper part, in the cliff face, consists of sets of planar cross-strata emplaced along IS surfaces. Scale is approximate.

Alluvial ridge-and-swale topography 143

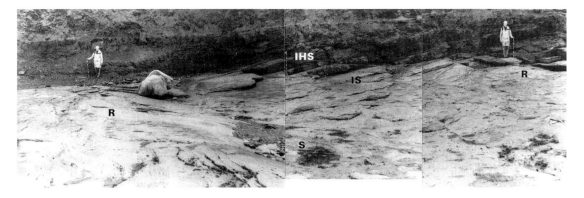

Fig. 12. Type 2b RST. Photomosaic of cliff face shown in Fig. 11. R, Type 2b ridge; S, swale; IS and IHS, inclined and inclined heterolithic stratification, respectively. Scale 1 m long.

the preserved accretion faces (form-concordant). They dip up the slope of the bounding IHS surfaces and at 15–20° to the surfaces. Some IHS surfaces bound thin (10–20 cm) lensoid beds of planar-laminated and ripple cross-laminated sandstone which show a downlapping relationship. Interbedded sets of ripple cross-laminated and planar-laminated sandstone/siltstone are present locally. The wavecut surface transects progressively lower strata away from the cliffs, and inclined units (IS) cease to be discernible 0.5–1 m below the ridge surfaces, passing downward into trough cross-stratified sandstone with cross-sets 30 cm thick.

RIDGES AT THE TOP OF THE LOWER STOREY

Five ridges at the top of the lower storey (Unit 1: Fig 3) are exposed mainly in the cliff face and show 30–100 cm relief and 4.9 m mean spacing. Thick-bedded, trough cross-stratified sandstone, conformable with the underlying strata, forms the bulk of the ridges and passes upward locally into thin-bedded sandstone and siltstone with IHS but few discernible sedimentary structures (Fig. 14). The IHS surfaces dip predominantly northward but locally southward (Fig. 14). Unit 1 is overlain by

Fig. 13. Type 2b RST. Northeastward view across wavecut surface of northern meanderbelt segment (Unit 3, Fig. 11). (The same boulder is shown in Figs 11–13.) Note in the foreground the two well-defined type 2b ridges (R) separated by a swale (S), and in the background IHS sets which dip to the right (SE) at 4–6° and are truncated by the wavecut erosion surface. Scale 1 m long.

Fig. 14. RST (types 2a and 1b superimposed) at the top of the lower storey. The ridges are founded in trough cross-bedded sandstones visible in the wavecut platform in the foreground and are thin-bedded and heterolithic in their upper parts in the cliff face. The RST is overlain by thick-bedded sandstones of the upper storey. Numbers indicate Units 1 and 2 (Fig. 3). Scale is 1 m long.

medium- to thick-bedded sandstones of the upper storey (Unit 2), the lower beds of which fill the swales. The lower parts of the ridges resemble the type 2a ridges of the upper storey in the predominance of trough cross-stratified sandstone. The upper parts of the ridges, preserved locally, resemble type 1b ridges in their thin bedding and the presence of IHS. The ridges' preserved relief is similar to that of type 2a ridges but their spacing is shorter.

ORIGIN OF RST

The ridges, with a height of up to 2 m and a spacing of up to 12.5 m (Table 2), are mesoforms as defined by Jackson (1975a, 1976a). Their interpretation is based on: (i) their dimensions, internal structure and exhumed form in comparison with modern bedforms; (ii) their stratigraphic level within the channel deposit and relation to palaeosols and roots; (iii) their palaeoflow patterns in relation to channel trend as defined by trough cross-stratal orientation in the subjacent strata; and (iv) their relationship to major stratification surfaces (Brookfield, 1977; Haszeldine, 1983; Miall, 1985), principally IS and IHS.

Type 1a

Planar cross-sets, the main component of the ridges, are similar in scale and internal structure to the transverse bars described from many point bars (Sundborg, 1956; McGowen & Garner, 1970; Singh & Kumar, 1974; Jackson, 1976a; Gustavson, 1978; Levey, 1978). The Mill Pond bedforms were small- to medium-scale, two-dimensional dunes, according to the terminology recently recommended by Ashley (1990). Sundborg (1956, pp. 270–272) noted that transverse bars with heights of 10–70 cm and wavelengths of 5–15 m are especially common on bars at high levels in the Klaralven channel. Transverse bars (30–200 cm height, 20–500 m wavelength) in the Wabash River occur most commonly in shallows near inner banks, actively migrating at depths of as little as 30 cm (Jackson, 1976a). The deeper channels of most of these modern rivers show three-dimensional dunes. Shallow-water conditions are inferred for the Mill Pond type 1a ridges, based on their high stratigraphic position and their association with thinly interbedded sets of planar and ripple cross-laminated sandstone. Similar interbedded sets occur in upper point-bar deposits (Singh, 1972; Ray, 1976), and in flood deposits where they probably reflect settling of suspended material under relatively low energy conditions (Stear, 1985).

Planar cross-sets in type 1a ridges at Mill Pond migrated up the associated IHS surfaces. Although most of the modern transverse bars noted above are oriented with crests normal to flow, their crests commonly swing inward on meander bends due to helicoidal flow and velocity differentials in the channel and the longer time available for bar migration in channelward segments of the bedforms (Sundborg, 1956; Jackson, 1976a). Sundborg suggested that longitudinal (scroll) bars in the Klaralven River are largely built up from sediment supplied by transverse bars. The close spacing of the type 1a ridges at Mill Pond (<5 m) compared with the wavelengths of many modern transverse bars, the presence of stacked cross-bed sets within some ridges and the set-bounding IHS surfaces dipping at a few degrees suggest that the ridges represent bars accreted to the upper part of the point bar in shallow water. The ridges thus are interpreted as deposits of scroll bars (unit bars: Ashley, 1990).

Type 1b

Type 1b ridges are the largest and stratigraphically highest observed at Mill Pond. Ridge tops are bioturbated by roots and locally truncated. The planar cross-strata within the ridges dip up the associated inclined surfaces so that sandy ridge strata overlie finer-grained strata within the swale on the 'updip' side of the ridge.

Type 1b ridges are attributed to scroll bars similar to those of the Wabash River, where scroll bars are up to 2 m high and contain inwardly directed avalanche sets decimetres thick with tabular to convex-up bounding surfaces (Jackson, 1976a). Swales landward of the Wabash scroll bars are underlain by trough cross-stratified sandstone and contain fine-grained fills over which ridge sands are transported during bar migration. The scroll bars are colonized by vegetation at low-flow stage. The planar sets of cross-strata probably represent associated transverse bars (they are similar in scale to cross-sets in type 1a ridges). Natural sections of ridges ascribed to scroll-bar deposition along the Klaralven River (Sundborg, 1956, fig. 47) show inwardly directed cross-sets.

The IHS surfaces, locally truncated, can be traced through 2 m of strata, but ridge-to-swale height is less than 1 m. We suggest that scroll bars about 2 m high were incorporated into the floodplain, where relief decreased due to ridge truncation and progressive swale filling before pedogenesis modified the meanderbelt surface. RST on modern alluvial floodplains is progressively buried beneath a sheet of younger overbank sediment deposited from levee progradation, from suspension during floods and from slope wash in relatively confined valleys (Sundborg, 1956; Schmudde, 1963; Bluck, 1971).

Relief of ridges on the Beatton and Endrick River floodplains is less than 1.5 m (Table 1).

Type 2a

Type 2a ridges largely consist of and overlie trough cross-stratified sandstone. Large-scale trough cross-beds have been ascribed to the downstream migration of three-dimensional dunes (Harms et al., 1963; Harms & Fahnestock, 1965). The common location of the ridges beneath the northeastern side of overlying type 1b ridges, coupled with the presence of northeastward-dipping IS, suggest that the ridges are closely connected with an early stage of scroll-bar accretion. Nanson (1980) noted that some scroll bars in the Beatton River were initiated during a single flood and continued to build as a result of convergent flow over the bar from both the main channel and the landward swale. The initial scroll bars were 20–70 cm high and contained 'flood cyclothems' (Nanson, 1980) with planar stratification and ripple-drift cross-lamination. Jackson (1976a) and Nanson (1981) stressed the importance of colonizing and stranded vegetation in initiating scroll-bar growth. Although flood cyclothems, convergent palaeoflow and vegetation were not observed in the type 2a ridges at Mill Pond, we suggest that these ridges represent initial scroll-bar deposits. Subsequent scroll-bar deposition (generating type 1b ridges) was associated with finer-grained sediments, inward sediment transport and a progressive bankward shift of the bar crest.

Type 2b

The well-defined, erosional swales between paired ridges resemble chute channels which act as conduits during flood stage on the upper levels of some modern point bars (McGowen & Garner, 1970; Levey, 1978). A levee-like ridge borders the channelward margin of a chute on the Colorado River, Texas (McGowen & Garner, 1970). The planar cross-sets associated with type 2b ridges resemble in scale those of type 1 ridges, and are interpreted as the deposits of two-dimensional dunes that migrated from the chute channel up the adjacent ridge side.

The type 2b ridges could represent modified floodplain (scroll-bar) ridges. Preferential growth of floodplain ridges was noted by Schmudde (1963) following a Missouri River flood when about 30 cm of sandy sediment was laid down on floodplain ridge crests whereas a few centimetres of silty sediment was deposited in the swales. Kesel *et al.* (1974) noted that Mississippi flood sediment deposited on ridges was coarser-grained than that deposited in the adjoining swales. However, the absence of roots and of inclined surfaces at deeper levels in the type 2b ridges suggests that they developed on a point-bar surface that had not undergone prolonged subaerial exposure.

DISCUSSION

Exhumed meanderbelt surfaces that show curvilinear bedding traces are known in Devonian to Oligocene strata (Table 3). However, few studies have included detailed descriptions of RST. An exception is the study by Puigdefábregas (1973), which also documents the relationship between RST and large-scale inclined stratification. Large-scale cross-beds with foresets dipping up associated inclined surfaces were interpreted as probable scroll-bar deposits by Nami & Leeder (1978) and Bridge & Diemer (1983) (see also Elliott, 1976, who suggested that each inclined bed represents one scroll bar). Some workers have described the differentially weathered intersection of inclined strata with the present-day erosional surface rather than truly exhumed ridges (Padgett & Ehrlich, 1976; Edwards *et al.*, 1983; Smith, 1987).

The ridges described by Puigdefábregas (1973) (Fig. 1) consist predominantly of planar cross-sets and resemble in scale and internal structure the Mill Pond type 1a ridges, ascribed to transverse and scroll bars. However, the maximum depth inferred for the Murillo channel (1-2 m) is much less than that inferred for the Mill Pond channel (9 m). In channels deeper than those inferred for Murillo, trough cross-sets should underlie the planar cross-sets, as they do at Mill Pond. Puigdefábregas (1973) did not ascribe the Murillo cross-sets to a particular bedform, nor comment on the latter's apparent inward migration.

The exhumed ridges described in several studies listed in Table 3 resemble in scale and internal structure the type 1b ridges at Mill Pond, attributed to large scroll bars. Exhumed transverse bars were described by Nijman & Puigdefábregas (1978) and Stear (1980, 1983), and inferred transverse-bar deposits bounded by inclined point-bar surfaces were described by Beutner *et al.* (1967). Chute channels were identified in the upper parts of exhumed point bars by Nijman & Puigdefábregas (1978), Bridge & Diemer (1983) and Edwards *et al.* (1983). In one case (Nijman & Puigdefábregas, 1978, fig. 15), the chute channel was bordered by a prominent ridge reminiscent of type 2b ridges. No analogues of our type 2a ridges appear to have been described. The 'scroll-bar relief' with accretionary bedding described by Puigdefábregas & Van Vliet (1978) appears to be truly exhumed. Planar cross-sets were not noted. Their apparent absence may reflect homogenization by roots but could indicate that RST formed without contribution from two-dimensional dunes.

The topography of ancient exhumed meanderbelts depends largely on differential resistance to weathering. Ridge types 1a, 2a and 2b at Mill Pond are composed predominantly of thick-bedded sandstone and form the most prominent exhumed topography. In contrast, type 1b ridges are heterolithic and thinner-bedded, and have weathered more readily. Scroll-bar deposits, analogous to type 1b ridges, in several modern rivers are sandy, and might generate resistant ridges in ancient analogues.

IS and IHS sets are present only at the top of the Mill Pond upper storey, principally in association with scroll- and transverse-bar deposits. The dip of the inclined surfaces increases from a few degrees in Unit 3 to as much as 20° in Unit 4. This increase may represent in some cases the increase in local gradient from the lateral-accretion surface of the point bar up the stoss slope to the crest of the superimposed bars. A similar upward steepening of inclined surfaces in ancient meanderbelt deposits was noted by Nijman & Puigdefábregas (1978) and Edwards *et al.* (1983) (Table 3).

The form, associated mesoforms and lithology of most point bars are related to position in the meander bend (Bluck, 1971; Jackson, 1975b, 1976b; Levey, 1978). Scroll bars are initiated at the axis or in the downstream parts of bends (Sundborg, 1956; Jackson, 1976a; Nanson, 1980). Transverse bars are especially common on large, shallow flats in the downstream parts of bends in the Klaralven River (Sundborg, 1956); those in the Wabash River are commonly found in fully developed flow zones (downstream areas) of bends (Jackson, 1976a), and they are most common in mid-bar areas of the Congaree River (Levey, 1978). Chutes occur on both upstream and downstream areas of point bars (McGowen & Garner, 1970; Levey, 1978). From this brief survey, we suggest that the Mill Pond meanderbelt segments represent mid to downstream parts of point bars, where preservation potential is high on account of the common tendency for meanders to show a downstream compo-

Table 3. Details of some exhumed meanderbelt deposits reported in the literature: Rc, radius of meander curvature, mainly estimated from curvature of exhumed ridges; IS and IHS, inclined and inclined heterolithic stratification (Thomas *et al.*, 1987); FU, fining upward; facies codes from Miall (1978); figure numbers refer to those of authors cited

Age and location	Fluvial-channel deposit	Inclined units	Exhumed ridge-and-swale topography	References
Miocene, Spain	1–2 m thick, FU	IHS extends from base to top; large-scale cross-strat. dips obliquely up IHS surfaces	Curvilinear accretion units 1–2 m wide, 0.25 m relief; spacing est. 10–13 m (fig. 5); Rc = 200 m	Puigdefábregas (1973)
Jurassic, UK	3–4 m thick, FU	IHS forms lensoid units 0.3–0.7 m thick (plate 2b) cross-sets (scroll bars?) dip up IHS surfaces in associated deposits	Curvilinear features; Rc = 77.5 m	Nami (1976) Nami & Leeder (1978)
Carboniferous, Morocco	5 m thick, FU		Curvilinear scarps with <1 m relief; Rc, not noted	Padgett & Ehrlich (1976)
Eocene–Oligocene, Spain	3 bodies, 5–10 m thick, FU, single–multistorey	Decimetre-scale IHS extends from base to top, or only in upper part of storey	Sand-rich beds form ridges, mud-rich beds form swales; curvilinear 'scroll-bar relief' on upper surfaces (figs 5B, 10, 12)	Puigdefábregas & Van Vliet (1978)
Eocene, Spain	40 m thick multistorey, 11 m topmost storey FU, with exhumed transverse bars	IS in basal part of storey (5° dip), IHS set 2 m thick at top (max. 18° dip)	Curvilinear accretion units, 1.5 m max. relief, 15 m spacing; Rc = 850 m	Nijman & Puigdefábregas (1978)
Triassic, S. Africa	Multilateral sheet sandstones, <10 m thick; St + Sh/Sl below, Sr and small-scale St above; exhumed transverse bars	IHS underlies ridges	Curvilinear ridge-and-swale, 1 m relief, est. 13–16 m spacing	Stear (1980, 1983)
Triassic, S. Africa	Single-storey deposit, est. bankfull depth 10 m (top eroded) and meanderbelt width 3 km; FU, St below, Sr/Sh above; chute channels at top	IHS underlies ridges, 16–20° dips, and extends to base of deposit; palaeoflow along strike	Curvilinear 'accretion ridges', 0.5 m relief: eroded scarps, not exhumed scroll bars; Rc = 600 m	Smith (1980, 1987)
Devonian–Carboniferous Ireland	Multistoreyed, 4–10 m inferred channel depth, commonly FU; chute deposits at top	IHS dips up to 12°; large-scale cross-sets at storey tops diverge in dip direction from mean palaeoflow (scroll-bar sets?)	Ridge-and-swale form (fig. 11B)	Bridge & Diemer (1983)
Permian, Texas	2–3 m thick, FU, chute deposits at top	IHS dips 5–8° at base, 15–25° at top	Curvilinear scarps, topography reflects differential weathering; Rc = 45–535 m	Edwards *et al.* (1983)
Pennsylvanian, Nova Scotia	Multistoreyed, upper storey 8.9 m thick, FU, exhumed scroll bars, transverse bars, chutes with levees, point-bar benches with gentle topography	IS and IHS sets av. 6° dips below, 12° above (max. 20°); transverse and scroll bars emplaced along inclined surfaces	Exhumed forms show up to 2 m relief; scroll bars 9.7 m av. spacing, transverse bars 4.9 m; Rc est. 1 km (from equations)	This paper

nent of migration (Sundborg, 1956; Jackson, 1975b, 1976b; Nanson, 1980).

The Mill Pond type 1b scroll-bar deposits with their numerous bedsets of varied type are probably the result of numerous flow events. Hickin & Nanson (1975) determined a mean periodicity of 27 years for ridge formation and incorporation into Beatton River point bars, and Sundborg (1956) estimated that successive scroll bars develop at 10–20 year intervals in the Klaralven River.

The Mill Pond channel deposit resembles the modern Wabash and Klaralven Rivers in the presence of planar cross-sets, probably originating as transverse bars, in the scroll-bar deposits. The Klaralven is 10–11 m deep (Sundborg, 1956, pp. 266–8) and the lower Wabash River about 10 m deep (Jackson, 1975b, fig. 8) at meander bends, dimensions comparable to those inferred for the Mill Pond upper storey. The Wabash River is freely meandering, except locally where it encounters bedrock and Pleistocene deposits, but the Klaralven is largely confined and thus cannot be considered closely analogous to the Mill Pond river which is inferred to have been freely meandering. The Beatton River is shallower (6–7 m bankfull depth at meander bends), lacks transverse bars, and shows predominant ripple cross-lamination and planar stratification in scroll-bar deposits. Channel migration rate is relatively slow and the channel relatively narrow on account of incision into Pleistocene deposits (Hickin & Nanson, 1975), and this, coupled with the relatively fine sediment grade, probably accounts for the absence of transverse bars (Nanson, 1980, pers. comm., 1989).

The presence of RST ascribed to scroll-bar accretion at the top of the lower storey (Fig. 3) implies relatively complete preservation of the point-bar deposits, an inference that would otherwise be difficult to make. The local abundance of claystone intraclasts at the base of the upper storey suggests that overbank strata covering the lower storey were stripped off, as were, perhaps, the upper parts of the scroll-bar deposits.

Finally, we find that the origin of scroll bars and RST, important aspects of models of meandering rivers, remains far from clear. The hydrodynamics and internal structure of scroll bars have been investigated for only three rivers and in only two cases was the relationship between scroll bars and floodplain ridges established (see Introduction). Nanson (1980) concluded that scroll bars are of multiple origin. Additionally, few studies of ancient strata have documented RST in sufficient detail to contribute to the question of its origin. We suggest that this topic is worthy of future research.

SUMMARY AND CONCLUSIONS

1. Ridge-and-swale topography (RST) exhumed at the top of a Pennsylvanian meanderbelt deposit at Mill Pond, Nova Scotia, is of multiple origin (Fig. 5, Table 2). Occurrences of RST are ascribed variously to initial and fully developed scroll bars, associated with transverse bars, and to chute-and-levee systems. The form of the exhumed meanderbelt surface depends largely on the bedding and lithology of the various deposits, which governs resistance to weathering and depth of erosion.

2. Scroll-bar deposits are partially preserved at the top of a lower storey amalgamated with overlying sandstones. The identification of scroll-bar deposits at this level indicates that the point bar was relatively completely preserved.

3. The strata are inferred to represent the downstream areas of point bars formed within channels about 9 m deep at bends. The fluvial system is comparable to the modern Wabash and Klaralven Rivers in scale and in the importance of transverse bars in scroll-bar accretion.

ACKNOWLEDGEMENTS

We thank Gerald Nanson for discussion and comments on an earlier version of the manuscript, and Frank Ethridge and an anonymous reviewer for their helpful reviews. Therese Carmody and Barbara McGoldrick assisted with typing, Ferenc Stefani and Max Perkins with photography, and David Martin with drafting. Financial assistance from the Natural Sciences and Engineering Research Council of Canada is gratefully acknowledged. M.R.G. thanks many colleagues at the University of Wollongong, NSW, where the manuscript was written, and acknowledges his great debt of gratitude to co-author Brian Rust (thesis supervisor, colleague and friend) who died on 22 June 1990.

REFERENCES

ALLEN, J.R.L. (1965a) A review of the origin and characteristics of recent alluvial sediments. *Sedimentology* **5**, 89–191.

ALLEN, J.R.L. (1965b) Sedimentation and palaeogeography of the Old Red Sandstone of Anglesey, North Wales. *Proc. Yorks. Geol. Soc.* **35**, 139–185.

ALLEN, J.R.L. (1968) *Current Ripples.* North-Holland, Amsterdam, 433 pp.

ASHLEY, G.M. (1990) Classification of large-scale subaqueous bedforms: a new look at an old problem. *J. Sedim. Petrol.* **60**, 160–172.

BELL, W.A. (1938) *Fossil Flora of Sydney Coalfield, Nova Scotia.* Geol. Surv. Can., Ottawa, Ontario, Memoir 215, 334 pp.

BERSIER, A. (1958) Sequences detritiques et divagations fluviales. *Eclog. Geol. Helvetica* **51**, 854–893.

BEUTNER, E.C., FLUECKINGER, L.A. & GARD, T.M. (1967) Bedding geometry in a Pennsylvanian channel sandstone. *Bull. Geol. Soc. Am.* **78**, 911–916.

BLUCK, B.J. (1971) Sedimentation in the meandering River Endrick. *Scot. J. Geol.* **7**, 93–138.

BOEHNER, R.C. & GILES, P.S. (1986) *Geological Map of the Sydney Basin.* Map 86-1. Nova Scotia Dept Mines Energy, Halifax, Nova Scotia.

BRIDGE, J.S. & DIEMER, J.A. (1983) Quantitative interpretation of an evolving ancient river system. *Sedimentology* **30**, 599–623.

BROOKFIELD, M.E. (1977) The origin of bounding surfaces in ancient aeolian sandstones. *Sedimentology* **24**, 303–332.

EDWARDS, M.B., ERIKSSON, K.A. & KIER, R.S. (1983) Paleochannel geometry and flow patterns determined from exhumed Permian point bars in north-central Texas. *J. Sedim. Petrol.* **53**, 1261–1270.

ELLIOTT, T. (1976) The morphology, magnitude and regime of a Carboniferous fluvial-distributary channel. *J. Sedim. Petrol.* **46**, 70–76.

ETHRIDGE, F.G. & SCHUMM, S.A. (1978) Reconstructing paleochannel morphologic and flow characteristics: methodology, limitations, and assessment. In: *Fluvial Sedimentology* (Ed. Miall, A.D.) pp. 703–721. Can. Soc. Petrol. Geol., Calgary, Memoir 5.

FISK, H.N. (1944) *Geological Investigation of the Alluvial Valley of the Lower Mississippi River.* Mississippi River Commission, Vicksburg.

FISK, H.N. (1947) *Fine-grained Alluvial Deposits and Their Effect on Mississippi River Activity.* Mississippi River Commission, Vicksburg, 82 pp.

FISK, H.N. (1952) Mississippi River Valley geology in relation to river regime. *Trans. Am. Soc. Civil Engs* **117**, 667–689.

GIBLING, M.R. & RUST, B.R. (1987) Evolution of a mud-rich meander belt in the Carboniferous Morien Group, Nova Scotia, Canada. *Bull. Can. Soc. Petrol. Geol.* **35**, 24–33.

GUSTAVSON, T.C. (1978) Bed forms and stratification types of modern gravel meander lobes, Nueces River, Texas. *Sedimentology* **25**, 401–426.

HARMS, J.C. & FAHNESTOCK, R.K. (1965) Stratification, bed forms, and flow phenomena (with an example from the Rio Grande). In: *Primary Sedimentary Structures and Their Hydrodynamic Interpretation* (Ed. Middleton, G.V.) pp. 84–115. Soc. Econ. Paleont. Miner., Tulsa, Spec. Publ. 12.

HARMS, J.C., MACKENZIE, D.B. & MCCUBBIN, D.G. (1963) Stratification in modern sands of the Red River, Louisiana. *J. Geol.* **71**, 566–580.

HASZELDINE, R.S. (1983) Descending tabular cross-bed sets and bounding surfaces from a fluvial channel in the Upper Carboniferous coalfield of north-east England. In: *Modern and Ancient Fluvial Systems* (Eds Collinson, J.D. & Lewin, J.) pp. 449–456. Spec. Publs Int. Ass. Sediment. 6. Blackwell Scientific Publications, Oxford.

HICKIN, E.J. (1974) The development of meanders in natural river-channels. *Am. J. Sci.* **274**, 414–442.

HICKIN, E.J. & NANSON, G.C. (1975) The character of channel migration on the Beatton River, northeast British Columbia, Canada. *Bull. Geol. Soc. Am.* **86**, 487–494.

JACKSON, R.G. (1975a) Hierarchical attributes and a unifying model of bedforms composed of cohesionless material and produced by shearing flow. *Bull. Geol. Soc. Am.* **86**, 1523–1533.

JACKSON, R.G. (1975b) Velocity–bed-form–texture patterns of meander bends in the lower Wabash River of Illinois and Indiana. *Bull. Geol. Soc. Am.* **86**, 1511–1522.

JACKSON, R.G. (1976a) Largescale ripples of the lower Wabash River. *Sedimentology* **23**, 593–623.

JACKSON, R.G. (1976b) Depositional model of point bars in the lower Wabash River. *J. Sedim. Petrol.* **46**, 579–594.

KESEL, R., DUNNE, K.C., MCDONALD, R.C. & ALLISON, K.R. (1974) Lateral erosion and overbank deposition on the Mississippi River in Louisiana caused by 1973 flooding. *Geology* **2**, 461–464.

KLAPPA, C.F. (1980) Rhizoliths in terrestrial carbonates: classification, recognition, genesis and significance. *Sedimentology* **27**, 613–629.

Leeder, M.R. (1973) Fluviatile fining-upwards cycles and the magnitude of palaeochannels. *Geol. Mag.* **110**, 265–276.

LEVEY, R.A. (1978) Bed-form distribution and internal stratification of coarse-grained point bars, Upper Congaree River, SC. In: *Fluvial Sedimentology* (Ed. Miall, A.D.) pp 105–127. Can. Soc. Petrol. Geol., Calgary, Memoir 5.

LOBECK, A.K. (1939) *Geomorphology.* McGraw-Hill, New York, 731 pp.

MCGOWEN, J.H. & GARNER, L.E. (1970) Physiographic features and stratification types of coarse-grained point bars: modern and ancient examples. *Sedimentology* **14**, 77–111.

MIALL, A.D. (1976) Palaeocurrent and palaeohydrologic analysis of some vertical profiles through a Cretaceous braided stream deposit, Banks Island, Arctic Canada. *Sedimentology* **23**, 459–483.

MIALL, A.D. (1978) Lithofacies types and vertical profile models in braided river deposits: a summary. In: *Fluvial Sedimentology* (Ed. Miall, A.D.) pp. 597–604. Can. Soc. Petrol. Geol., Calgary, Memoir 5.

MIALL, A.D. (1985) Architectural-element analysis: a new method of facies analysis applied to fluvial deposits. *Earth Sci. Rev.* **22**, 261–308.

NAMI, M. (1976) An exhumed Jurassic meander belt from Yorkshire, England. *Geol. Mag.* **113**, 47–52.

NAMI, M. & LEEDER, M.R. (1978) Changing channel morphology and magnitude in the Scalby Formation (M. Jurassic) of Yorkshire, England. In: *Fluvial Sedimentology* (Ed. Miall, A.D.) pp. 431–440. Can. Soc. Petrol. Geol., Calgary, Memoir 5.

NANSON, G.C. (1980) Point bar and floodplain formation of

the meandering Beatton River, northeastern British Columbia, Canada. *Sedimentology* **27**, 3–29.

NANSON, G.C.(1981) New evidence of scroll-bar formation on the Beatton River. *Sedimentology* **28**, 889–891.

NIJMAN, W. & PUIGDEFÁBREGAS, C. (1978) Coarse-grained point bar structure in a molasse-type fluvial system, Eocene Castisent Sandstone Formation, South Pyrenean Basin. In: *Fluvial Sedimentology* (Ed. Miall, A.D.) pp. 487–510. Can. Soc. Petrol. Geol., Calgary, Memoir 5.

PADGETT, G.V. & EHRLICH, R. (1976) Paleohydrologic analysis of a late Carboniferous fluvial system, southern Morocco. *Bull. Geol. Soc. Am.* **87**, 1101–1104.

PUIGDEFÁBREGAS, C. (1973) Miocene point bar deposits in the Ebro Basin, northern Spain. *Sedimentology* **20**, 133–144.

PUIGDEFÁBREGAS, C. & VAN VLIET, A. (1978) Meandering stream deposits from the Tertiary of the southern Pyrenees. In: *Fluvial Sedimentology* (Ed. Miall, A.D.) pp. 469–485. Can. Soc. Petrol. Geol., Calgary, Memoir 5.

RAY, P.K. (1976) Structure and sedimentological history of the overbank deposits of a Mississippi River point bar, *J. Sedim. Petrol.* **46**, 788–801.

RUST, B.R., GIBLING, M.R., BEST, M.A., DILLES, S.J. & MASSON, A.G. (1987) A sedimentological overview of the coal-bearing Morien Group (Pennyslvanian), Sydney Basin, Nova Scotia, Canada. *Can. J. Earth Sci.* **24**, 1869–1885.

SCHMUDDE, T.H. (1963) Some aspects of landforms on the Lower Missouri River floodplain. *Ann. Assoc. Am. Geogr.* **53**, 60–73.

SINGH, I.B. (1972) On the bedding in the natural-levee and the point-bar deposits of the Gomti River, Uttar Pradesh, India. *Sedim. Geol.* **7**, 309–317.

SINGH, I.B. & KUMAR, S. (1974) Mega- and giant ripples in the Ganga, Yamuna, and Son Rivers, Uttar Pradesh, India. *Sedim. Geol.* **12**, 53–66.

SMITH, R.M.H. (1980) The lithology, sedimentology and taphonomy of flood-plain deposits of the Lower Beaufort (Adelaide Supergroup) strata near Beaufort West. *Trans Geol. Soc. S. Africa* **83**, 399–413.

SMITH, R.M.H. (1987) Morphology and depositional history of exhumed Permian point bars in the southwestern Karoo, South Africa. *J. Sedim. Petrol.* **57**, 19–29.

STEAR, W.M. (1980) Channel sandstone and bar morphology of the Beaufort Group uranium district near Beaufort West. *Trans. Geol. Soc. S. Africa* **83**, 391–398.

STEAR, W.M. (1983) Morphological characteristics of ephemeral stream channel and overbank splay sandstone bodies in the Permian Lower Beaufort Group, Karoo Basin, South Africa. In: *Modern and Ancient Fluvial Systems* (Eds Collinson, J.D. & Lewin, J.) pp. 405–420. Spec. Publs Int. Ass. Sediment. 6. Blackwell Scientific Publications, Oxford.

STEAR, W.M. (1985) Comparison of the bedform distribution and dynamics of modern and ancient sandy ephemeral flood deposits in the southwestern Karoo region, South Africa. *Sedim. Geol.* **45**, 209–230.

SUNDBORG, A. (1956) The River Klaralven — a study of fluvial processes. *Geogr. Ann.* **38A**, 127–316.

THIBAUDEAU, S.A. & MEDIOLI, F.S. (1986) Carboniferous thecamoebians and marsh foraminifera: new stratigraphic tools for ancient paralic deposits. *Geol. Soc. Am. Annual Meeting,* San Antonio, Abstracts with Program, 771.

THOMAS, R.G., SMITH, D.G., WOOD, J.M., VISSER, J., CALVERLEY-RANGE, E.A. & KOSTER, E.H. (1987) Inclined heterolithic stratification — terminology, description, interpretation and significance. *Sedim. Geol.* **53**, 123–179.

ZODROW, E.L. & CLEAL, C.J. (1985) Phyto- and chronostratigraphical correlations between the late Pennyslvanian Morien Group (Sydney, Nova Scotia) and the Silesian Pennant Measures (South Wales). *Can. J. Earth Sci.* **22**, 1465–1473.

Processes and products of large, Late Precambrian sandy rivers in northern Norway

S.-L. RØE* and M. HERMANSEN†

*Geological Institute, Dept A, University of Bergen, 5007 Bergen, Norway; and
†Norsk Hydro A/S, PO Box 200, 1321 Stabekk, Norway

ABSTRACT

Late Precambrian fluvial sandstones of the Fugleberget Formation and Hestman Member, northern Norway are dominated by sheet-like sandstone bodies (0.7–6 m thick, with exposed length of up to 200 m). The sandstone bodies, of mainly bar origin, consist of fine-grained sandstone and are subdivided into three types. Type A sandstone bodies (simple bars) comprise either solitary sets of large-scale, concave-up and sigmoidal-shaped cross-strata, or low-angle inclined to horizontal planar-strata, or they show a combination of the cross- and planar-stratified facies. Type B sandstone bodies (mainly compound bars) are composed of superimposed or downward-climbing cross-sets. Type C sandstone bodies (composite bars) show an upstream or downstream transition from large, solitary cross- or planar-stratified sets to cosets of medium-scale cross-sets.

The simple to composite bars of the sheet-sandstone bodies are interpreted to represent short-term depositional events, possibly within one flood cycle. Over a longer time span (one to several flood cycles) they probably grew or amalgamated into macroforms several times larger in downflow and lateral dimensions than the component bars. High discharge variability (absence of vegetated source terrains), high sediment load and transport rates (current velocities much higher than critical erosion velocity) and easily erodible banks contributed to the formation of very wide, unstable, low-sinuosity rivers with a mainly braided pattern during intermediate- and low-flow stages.

INTRODUCTION

The purpose of this paper is to present a detailed analysis of part of two Late Precambrian fluvial sandstone successions from northern Norway. Despite their somewhat different stratigraphic and geographic positions, they are remarkably similar with respect to grain size, sandstone body geometry and internal facies organization, with the implication that the two river systems were of the same kind. We argue that they were broadly comparable to wide tracts of the modern Huanghe (Yellow River; Chien, 1961) and Brahmaputra (Coleman, 1969; Bristow, 1987) in terms of river pattern, channel width/depth ratios, flood velocities and macroform migration rates. Furthermore, apparent similarities and differences in stratification, vertical sequences and architectural style between deposits of the modern and ancient river systems are briefly discussed.

The described sandstones belong to the Fugleberget Formation of the Vadsø Group and the Hestman Member of the Båsnæring Formation, Varanger Peninsula (Fig. 1). Excellent exposures of the Fugleberget Formation occur at Fugleberget and on Vadsø Island. The lower part of the Hestman Member is well exposed along the coast east of Kiberg (Fig. 1).

STRATIGRAPHIC AND DEPOSITIONAL SETTING

The Late Precambrian sedimentary strata that underlie the two regions of the Varanger Peninsula, northern Norway (Fig. 1) are interpreted to have accumulated in two widely separated basins. Strata of the Barents Sea region probably attained their

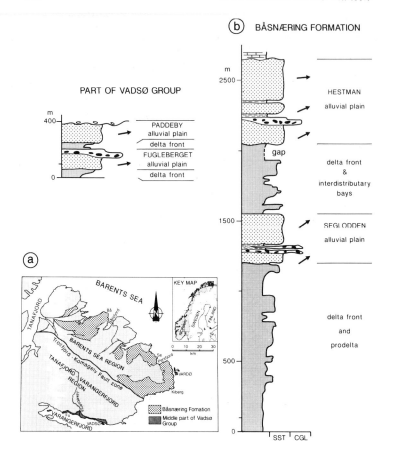

Fig. 1. (a) Map showing the middle part of the Vadsø Group (Tanafjord-Varangerfjord region) and the Båsnæring Formation (Barents Sea region) with location of the three study areas: Fugleberget (Fu), Vadsø Island and Kiberg. (b) Stratigraphic setting of the Fugleberget Formation and Hestman Member. (Modified from Siedlecka & Edwards (1980) and Siedlecka (1985).)

present position through right lateral movement along the Trollfjord-Komagelv Fault Zone in post-sedimentation time (Johnson et al., 1978; Siedlecka, 1985).

The 300 m thick middle part of the Vadsø Group of the Tanafjord-Varangerfjord region and the 3000 m thick Båsnæring Formation of the Barents Sea region (Fig. 1) comprise sediments of fluvial and deltaic origin (Hobday, 1974; Siedlecka & Edwards, 1980; Siedlecka et al., 1989). The successions of both regions each show two, overall progradational sequences (Fig. 1). The fluvial (100–600 m thick) top-components of the sequences comprise mainly sandstones. A channelized conglomeratic horizon is present in three out of the four (Fig. 1). Fine-grained floodbasin deposits occur only in three levels in the alluvial succession of the Båsnæring Formation.

The sandstone portions of the four fluvial successions have several important characteristics in common as judged from previous publications and about 1000 m of newly measured vertical profiles. These similar attributes include (Fig. 2): (i) unidirectional palaeocurrents; (ii) absence of cyclicity and vertical sedimentation time-trends; (iii) dominance of fine-grained sandstone and a general absence of argillaceous beds; (iv) dominance of medium- and large-scale cross-sets (0.2–4 m thick) interbedded with upper-stage planar strata; the cross-strata have typically concave-up or sigmoidal shapes; (v) abundant soft sediment deformation, comprising overturned cross-strata, large-scale convolution and massive beds; (vi) intraformational conglomerates, ripple cross-stratification and siltstones are rare in all successions.

These similarities between the four stratigraphically different sandstone successions suggest that they were deposited in the same type of environ-

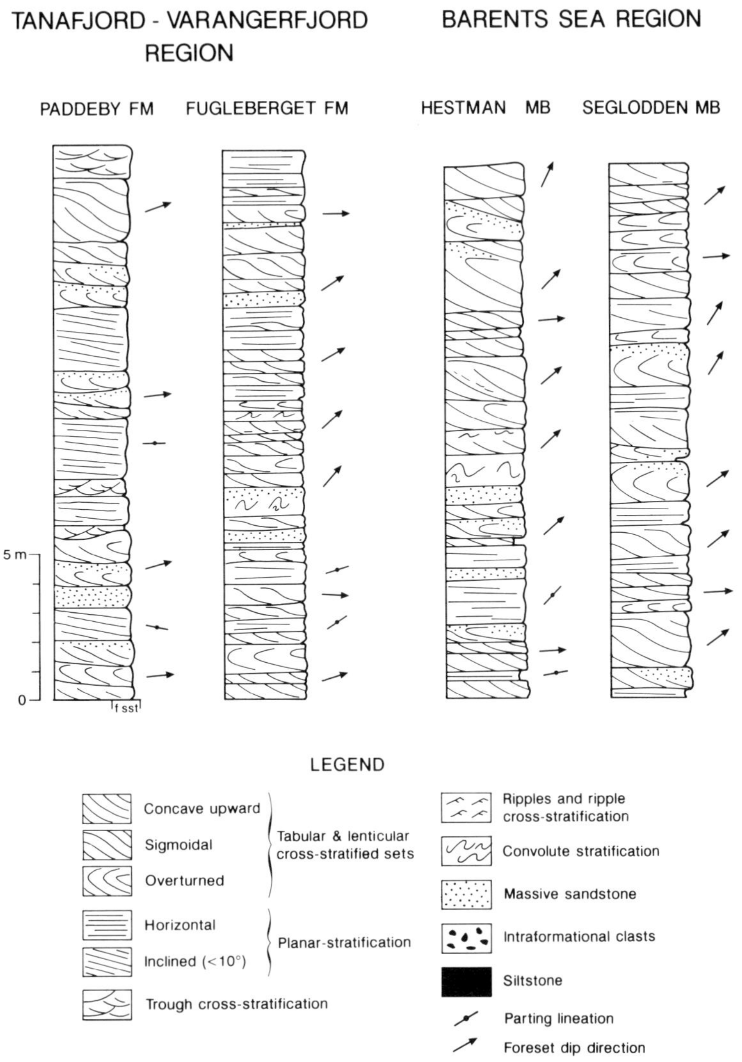

Fig. 2. Typical facies sequences of the Paddeby and Fugleberget Formations and the Hestman and Seglodden Members. Note their similar uniform fine grain sizes and stratification characteristics. The legend also applies to subsequent figures.

ment. A fluvial setting of the four successions is suggested because of their unidirectional palaeocurrents, absence of wave- or tide-generated structures and their stratigraphic context above delta front deposits (Hobday, 1974; Siedlecka & Edwards, 1980; Hjellbakk, 1987; Røe, 1987; Hermansen, 1989; Siedlecka et al., 1989). In this paper we give the first detailed account of the architecture and facies organization of sandstones present in long exposures of the Fugleberget Formation and Hestman Member and offer an interpretation of the pattern of the palaeorivers, together with their scale and dynamic behaviour.

THE SHEET SANDSTONE BODIES

Exposures of the Fugleberget and Hestman sequences are dominated by sheet-sandstone bodies of mainly bar origin. Wedge- and lenticular-shaped bodies representing channel forms and fills are subordinate components of both sequences (Fig. 3). The term 'bar', as used here, refers to the bed features that are interpreted to have formed significant positive relief on the river bed relative to the water depth and with much smaller water depth/bedform height ratios than is common for dune bedforms (e.g. Jackson, 1976; Allen, 1982, fig. 8.20;

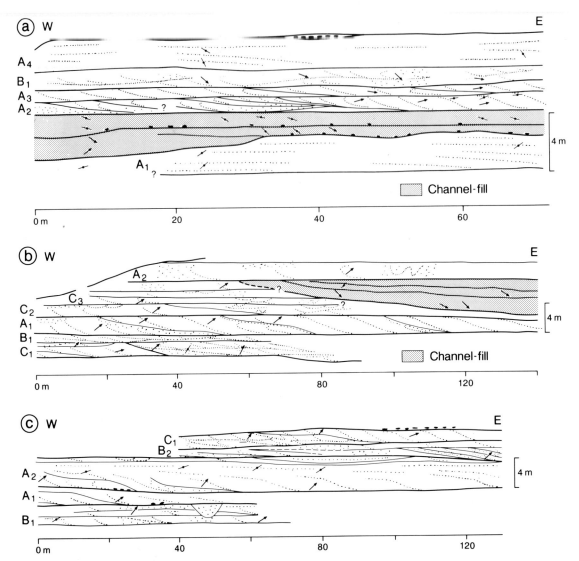

Fig. 3. Sandbody geometry and sedimentary structures in the Fugleberget Formation (a) and the Hestman Member (b, c). A_{1-4}, Type A sandstone bodies (simple bars). B_{1-2}, Type B sandstone bodies (cosets of dune cross-sets). C_{1-3}, Type C sandstone bodies (mainly composite bars). Note the dominance of sheet-sandstone bodies.

Bridge, 1985). The sheet sandstones have been subdivided into three main types.

Type A sandstone bodies

The main components of these sandstone bodies are simple bars that internally display large-scale cross-strata, planar-strata or show a combination of the two stratification types (Figs 3 & 4). These bars rest on overall planar, horizontal erosion surfaces, in places with local conglomerate-filled scours. Linguoid to straight-crested ripples, small- and medium-scale cross-stratified sets, horizontal planar-stratified sets, cross-cutting channel-fill sandstones and thin siltstones top the bars in a few places (Figs 3c (A_2) & 4d).

The cross-stratified bars (Figs 3a (A_2), 3b ($A_{1,2}$), 3c (A_1), 4a–c, 5 & 6) comprise solitary sets of cross-strata that are up to 2 m thick in the Fugleberget and up to 4 m thick in the Hestman exposures.

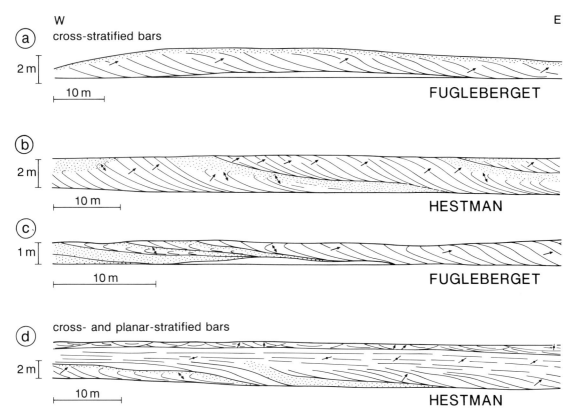

Fig. 4. Type A sandstone bodies (simple bars). Note their sigmoidal and concave-up cross-strata shapes (a–d), planar-stratified topsets (d) and reactivation surfaces associated with massive sandstone and overturned cross-strata (b–c). Trough cross-sets cap the bar in (d).

Foresets are concave-up or sigmoidal in shape with uniform dip directions and dip-angles throughout the sets (see Røe, 1987, for detailed description). Reactivation surfaces are common and the majority of these are underlain by a wedge of massive sandstone that passes with a vertical downward transition into overturned cross-strata (Figs 3b (A_1), 4b,c & 5). These reactivation surfaces are either widely spaced and separated by long segments of undeformed cross-strata (Figs 3b (A_1) & 4b) or they are so closely spaced that they bound downcurrent-dipping, deformed cross-sets (Figs 3a (A_2) & 4c).

The planar-stratified bars comprise strata that dip at low angle (less than 10°) relative to their lower bounding surfaces (Figs 3a (A_1, A_4) & 6). Parting lineations have, in general, orientations parallel or somewhat oblique to the maximum dip azimuth of the inclined strata. The planar stratified bars are up to 3 m thick in the Fugleberget and up to 6 m thick in the Hestman outcrops.

The combined cross- and planar-stratified bars include aspects of both the bar types described above. Sigmoidal and/or concave-up cross-strata are overlain by horizontal topsets (Figs 3c (A_2), 4d & 7). The topsets either overlie the cross-strata discordantly or the horizontal topset-strata pass into low-angle inclined strata that merge with the steeper inclined foresets (Fig. 7). Both occurrences are frequently encountered within the same bar. In places, cross-strata are succeeded downcurrent by low-angle inclined planar strata. Reactivation surfaces with massive and/or deformed substrata are also common in these bars (Fig. 4d).

The cross-stratified and combined cross- and planar-stratified bars are equivalent to the large solitary sets described by Røe (1987) from the

Fig. 5. Cross-stratified simple bar (Type A sandstone body, Hestman Member). Note the reactivation surface and the underlying wedge of mainly massive sandstone. Arrow points to overturned cross-strata. The photograph corresponds to part of Fig. 3b (A_1) and Fig. 4b.

Fugleberget Formation. The planar-stratified bars and the combined cross- and planar-stratified bars are similar, in terms of internal stratification, to the plane-bedded simple bars and the compound and composite-compound bars described by Allen (1983) from the Brownstone Beds. The present bars, however, contrast with those of the Brownstone Beds by being of larger scale, with horizontal bounding surfaces, and by having uniform fine grain sizes and internally consistent palaeocurrents.

Fig. 6. Low-angle inclined planar-stratified simple bar superimposed on a cross-stratified bar (Type A sandstone body, Fugleberget Formation).

Fig. 7. Cross- and planar-stratified simple bar (Type A sandstone body, Hestman Member). Note the up-slope transition from foresets to low-angle inclined topsets. Arrows point to heavy mineral rich laminae. The photograph corresponds to part of Fig. 3c (A_2) and Fig. 4d.

The internal characteristics of the cross- and combined cross- and planar-stratified bars suggest that they formed by swift, heavily suspension-laden currents in the transition to upper-stage plane-bed (e.g. Saunderson & Lockett, 1983; Røe, 1987). The common occurrence of reactivation surfaces with massive sandstone and deformed cross-strata within these bars, suggests intermittent liquefaction of the bar-fronts and downslope sediment flow as previously inferred by Jones & Rust (1983) for similar features in the Triassic Hawksbury sandstone. In the present case, the liquefaction is probably due to pressure fluctuations exerted on the bar-front by macroturbulent bursts. That the liquefaction could be caused by dumping of riverbank material on the bar-tops or stage fluctuations as suggested by Jones & Rust (1983) is not likely here because of the abundant evidence for local liquefaction within the sediment studied.

Røe (1987) suggested that the cross-stratified and combined cross- and planar-stratified bars were comparable in terms of grain size, streamwise shape, formative flow and migration rates to long (up to 450 m), low-relief bars described by Culbertson & Scott (1970) from a shallow conveyance channel to the Rio Grande. Flow over these bar-tops was shallow with velocities of 1.5 m s^{-1}. The bars advanced at rates of several tens of metres to more than 100 m per day. The planar-stratified bars probably formed at somewhat higher flow velocities or shallower depths than the cross- and combined cross- and planar-stratified ones (Allen, 1983).

The inferred flow conditions and migration rates for the bars of the Type A sandstone bodies suggest that they formed under high or intermediate discharges. Their capping, in places, by small- and medium-scale structures all showing current directions strongly oblique to that of the associated bars is probably suggestive of very shallow, relatively low-stage flow conditions.

Type B and C sandstone bodies

The Type B sandstone bodies comprise grouped, interdigitating sets of cross-strata (cosets) present throughout the length of the outcrop (Fig. 3a (B_1), 3b (B_1), 3c ($B_{1,2}$)). The Type C bodies display an upstream or downstream transition from such cosets to solitary large sets of cross-strata or low-angle inclined planar-strata (Figs 3b (C_{1-3}), 3c (C_1) & 8) similar to those already described. Low-angle inclined erosion surfaces separate the two different components of the Type C sandstone bodies. Intraformational conglomerates occur locally at the base of some of these sandstones, and they are in places capped by low-stage deposits of the same kind as those of the Type A bodies. Type B and C sandstone bodies range in thickness from 0.75 to 4 m in the Fugleberget Formation and from 1 to 6 m in the Hestman Member.

The cross-sets within the cosets, in the Type B and C sandstone bodies, are similar in scale, geometries and internal cross-strata characteristics. Sets are either lenticular or tabular in shape and have, in general, overall horizontal or downcurrent dipping set-boundaries in streamwise sections (Figs 3c ($B_{1,2}$) & 8). Cross-strata are also here typified by concave-up or sigmoidal shapes and overturned cross-strata in association with massive sandstone.

The interdigitating nature of the cross-sets within the cosets in the Type B and C sandstone bodies suggests periodic bedforms and probably dune configurations (*sensu* Ashley, 1990). The cross-strata characteristics within the sets suggest high velocities (Røe, 1987) and that the cosets probably were deposited within a short time interval. In the Brahmaputra River, sequences of dune-generated cross-sets, up to 7 m thick, often accumulate within a single day during fall in river stage (Coleman, 1969).

The majority of the Type B sandstone bodies probably represent compound bars formed by the local coalescence and vertical aggradation of dunes on wide, flat channel-floors. Where descending cross-sets are present they form components of larger compound bed features with downcurrent or oblique progradation (e.g. Banks, 1973; Allen, 1983; Miall, 1985). However, dune cosets may, in places, have been deposited on the top of simple and composite bars (e.g. Fig. 3b (C_1–B_1)) or within deeper parts of their flanking channels (e.g. Fig. 3a (B_1–A_4)) since dunes tend to occur on both these morphologically different sites in modern rivers (e.g. Jackson, 1976; Cant & Walker, 1978; Crowley, 1983; Bristow, 1989).

The Type C sandstone bodies are interpreted as composite bars on account of their up- and/or down-current transition from large solitary cross- or planar-stratified sets to cosets of dune cross-sets. Culbertson & Scott (1970) observed that dunes tend to coalesce by an increase in velocity in the transition to upper-stage plane-bed, forming the nucleus for further bar-growth similar to the situation envisaged here (e.g. Fig. 8a,b). In contrast, the downstream transition from large solitary sets to cosets of cross-stratification (Fig. 8c) suggests a decline in velocity possibly associated with lowered discharges. These interpretations imply that the com-

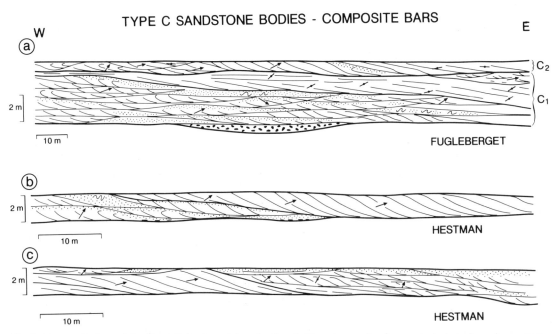

Fig. 8. Composite bars of the Type C sheet-sandstone bodies. Solitary large sets of cross-strata (a,b,c) or planar strata (a(C_1)) pass up-current (a,b) or down-current (c) into cosets of medium-scale cross-sets. Note the down-current inclined set-boundaries within the coset in (b) and (c).

posite bars could have been generated during one flood cycle although their time span was larger than that of the simple bars in the Type A sandbody.

WEDGE- AND LENTICULAR-SHAPED SANDSTONE BODIES

These sandstone bodies fill concave-up scours (up to 6 m deep) that are interpreted as former channel floors and margins (Figs 3a,b, 9 & 10). Exposed lateral dimensions are several tens of metres, and only in one outcrop are both channel margins exposed (Fig. 9). Internal characteristics of the fills, as well as their palaeocurrent trends, commonly deviate from those of the enclosing sheets. The portion of the Hestman Member studied in detail shows five channel-fill sandbodies, whereas the Fugleberget exposures have three.

The simplest and most common type of channel-fill comprises planar- and/or ripple cross-stratified very fine-grained sandstone that onlaps the channel margins. Siltstone clasts occur at the bases of these channels.

The Fugleberget sequence on Vadsø Island possesses a remarkably well exposed 120 m wide and 5 m deep channel with two distinctly different infill components (Fig. 9). The oldest component comprises a lens-shaped bank-attached body dominated by trough cross-stratified, pebbly sandstone (Fig. 9a,c). Erosion surfaces inclined towards the channel axis, locally draped with siltstone, suggest a lateral accreting sidebar involving several flood/low-stage cycles. Subsequent to the active sidebar migration, the channel was filled vertically by massive to faintly laminated fine-grained sandstone containing in places dispersed intraclasts (Fig. 9b,c). At least three sedimentation units are distinguished through their separation by thin siltstones. The attributes of these sandstones suggest hyperconcentrated, rapidly depositing flow (cf. Smith, 1986), whereas the interbedded siltstones were probably deposited from suspension. Extensive bank collapse may have been responsible for these infilling events; alternatively, sediments were introduced into the channel by overbank flooding of nearby channels.

The multistorey and heterogeneous nature of the sandstone bodies in Fig. 10a,b suggests several episodes of channel scouring and deposition. The fine material that overlies the lower erosional surfaces in each case implies an initial abandonment phase. The subsequent channel-fill comprises mainly small-scale trough cross-sets and planar-stratified sets (upper-stage) in the Fugleberget example (Fig. 10b). The Hestman channel-fill (Fig. 10a) is the only example where the cross-sets are of similar scale and have characteristics similar to those of the sheet-sandstone bodies.

The grain sizes and/or stratification types within these various channel-fills suggest, apart from the example in Fig. 10a, current velocities of smaller magnitude than those responsible for the sheet sandstones. They may represent subordinate channels within the active river (e.g. Chien, 1961; Bristow, 1987) or, alternatively, the channels and/or channel-fills formed on the alluvial plain subsequent to major river avulsion at these sites. In particular, the channel form and fill of the Fugleberget Formation (Fig. 9), with its narrow width relative to depth, probable high sinuosity and stable pattern through several flood cycles may be suggestive of the later situation.

SIGNIFICANCE OF BOUNDING SURFACES AND VERTICAL SEQUENCES

Whereas the bounding surfaces below the lenticular- and wedge-shaped sandstone bodies are the products of channel incision, the significance of the bounding surfaces between the sheet-sandstone bodies is less clear. They either represent major events related to channel migration/shifting, or minor surfaces of erosion/non-deposition prior to continued vertical aggradation of an already existing channel/bar topography.

The vertical profiles (Fig. 11) are all from exposures of sufficient length to allow a subdivision into various sandbody and bar types. Figure 11b,e corresponds to profiles from Fig. 3a,b. The profiles are dominated by simple, compound and composite bars of the sheet-like Type A, B and C sandstone bodies. In places these bars overlie intraformational siltstone conglomerates, and are capped by relatively low flow-stage deposits. There is no obvious pattern in the vertical organization of the various bar types.

Bounding surfaces between the sheet sandstones in Figs 3 & 11 are believed to represent major erosive events in the following cases:
1 where the base of a sandstone body overlies (at least partly) large channel-fill bodies, such as ob-

Fig. 9. Channel form and fill in the Fugleberget Formation. The sidebar of mainly trough cross-stratified sandstone (a,c) is onlapped by horizontal beds of mainly massive sandstone (b,c). Siltstone drapes within the two channel-fill components are marked with heavy lines.

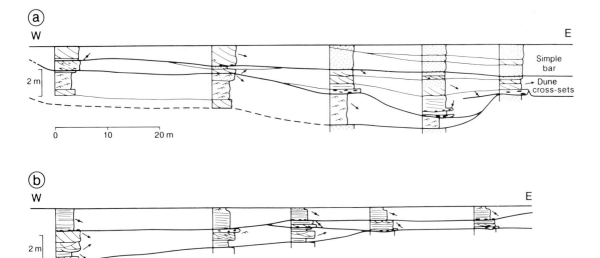

Fig. 10. Wedge-shaped sandstone bodies (channel forms and fills). (a) Hestman Member; (b) Fugleberget Formation. Note the multistorey and heterogeneous nature of the bodies and the very fine grained facies at the base of these channels; (a) corresponds to the channel-fill in Fig. 3b and (b) to that in Fig. 3a.

served for the base of the A_2 sandstones in Fig. 3a,b and Fig. 11e;

2 where relatively thick intraformational conglomerates occur, at the base of the sandstone bodies (Figs 8a & 11a,c,e); such conglomerates are, however, only locally developed along any one bounding surface, so their absence does not preclude a channel-floor origin;

3 where the boundaries separate sandstone bodies which are several metres thick and have the same internal characteristics (Fig. 11a, 11c (B_{1-2}), 11d (B_{1-3})); vertical aggradation of this magnitude, within one channel, would probably alter the flow characteristics to such an extent that the youngest deposits would appear different in stratification characteristics or set thicknesses from the underlying ones;

4 where there is a change in palaeocurrent directions between superimposed sandbodies (Figs 3a (A_3-B_1) & 11e (A_3-B_1), 11e (A_4-B_2)).

The inference that many of the bounding surfaces between sheet-sandstone bodies are major channel erosion surfaces implies that each of these superimposed sheets represents deposits within different channels. However, we suggest that a few sequences of stacked sheet-sandstone bodies accumulated within the same channel. The inferred bar-complex sequences in Fig. 11b,d all show relatively thin sandstone bodies superimposed on bed features much larger in scale. Such thinning-upward sequences may form as a consequence of a decrease in water depths upon deposition within a channel. The two simple bars, A_2 and A_3, in Figs 3a & 11e amalgamate into a poorly defined sandstone 20 m west of the section (Fig. 3a), and therefore form components of the same bar complex. In addition, sequences where bars are superimposed on dune generated cross-sets (Figs 3a (B_1-A_4) & 11e (B_1-A_4)) are common in modern rivers (e.g. Jackson, 1976; Cant & Walker, 1978; Crowley, 1983) and evident in the channel-fill sequence in Fig. 10a.

Recognition of genetic sequences of higher rank than those of the channel-bar and channel-bar-complex sequences are hampered by the general absence of fine-grained floodplain deposits within the Hestman and Fugleberget successions. However, we tentatively suggest that the multistorey sequence A_2-A_4 (Figs 3a & 11e) is the product of channel-belt aggradation occurring between major river avulsions. The sequence overlies a large fine-grained channel-fill (Fig. 10b) and the channel-form (Fig. 9) merges eastward with the horizontal bounding surface at the top of the A_4 sandstone body. The $C-A_1$ association (Fig. 11c) is also interpreted as a

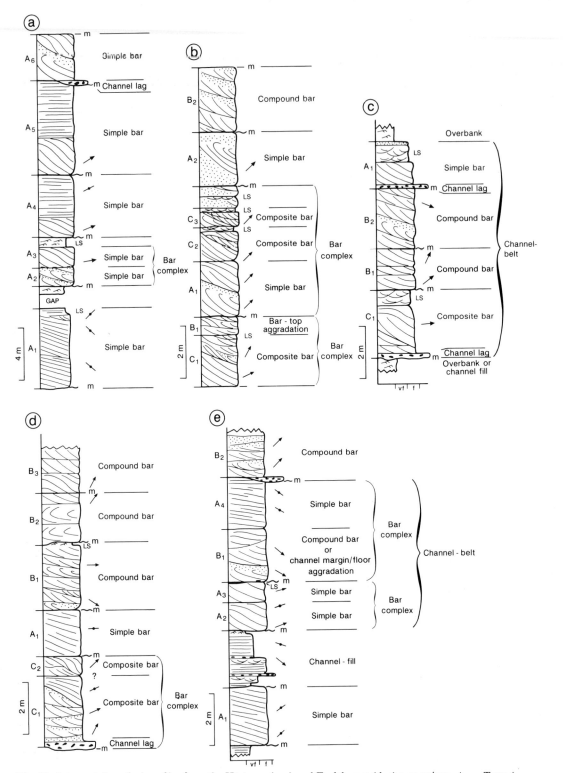

Fig. 11. Interpreted vertical profiles from the Hestman (a–c) and Fugleberget (d–e) successions. A_{1-6}, Type A sandstone bodies (simple bars). B_{1-3}, Type B sandstone bodies (cosets of dune cross-sets). C_{1-3}, Type C sandstone bodies (composite bars). The major erosion surfaces (m) separate channel sequences consisting either of solitary bars or bar-complexes. LS, relatively low flow-stage deposits.

Table 1. Flow parameters, bed-load grain size, scales and macroform migration rates in the Huanghe, Brahmaputra and Precambrian rivers; (+) inferred positive correlation

	Huanghe (Yellow River) (Chien, 1961)	Brahmaputra (Coleman, 1969; Bristow, 1987)	Present study
Parameters			
Discharge	Max. 16 000 m^3 s^{-1} (2000–5000)	Max: 77 000 m^3 s^{-1} Av: 20 000	
Velocity (flood)	2.5–3 m s^{-1}	1.3–3 m s^{-1}	> 1 m s^{-1}
Bed-load grain size	V. fine sand	Fine sand	Fine sand
Scales			
River	W: (5)–23 km $\overline{W}$ ~ 10 km	W: (3)–18 km $\overline{W}$ = 10 km $\overline{D}$ = 5 m	+
Channels	W/D: (80)–700 D_{max}: 2.5 m	W/D: (50)–500 D_{max}: 35 m	+ D_{max} > 4–6 m
Macroforms	≤ several kilometres	≤ several kilometres	+
Migration rates			
Macroforms	40–120 m d^{-1}	31 m d^{-1}	+
Dunes		70–120 m d^{-1}	+

complete channel-belt sequence since it is bounded by fine-grained overbank and/or channel-fill deposits. Notably, the rippled unit that caps this sequence forms the overbank 'wing' of a 6 m deep channel, filled with very fine-grained sandstone. This channel is incised into the channel-belt sequence 20 m east of the measured profile.

THE LATE PRECAMBRIAN RIVER SYSTEMS

The similarities in grain size, sandbody geometries and lateral facies organization between the Fugleberget and Hestman successions suggest that they represent similar river systems. We have attempted to characterize these palaeorivers in terms of their planview pattern and scales, migration rates of their channel and bar-complexes, and their discharge, velocity and stage fluctuations. The following discussion is based mainly on features of the sheet sandstones and their Precambrian context, together with data from wide tracts of the Huanghe (Yellow River) and Brahmaputra rivers (Chien, 1961; Coleman, 1969; Bristow, 1987). Both these modern rivers have fine uniform bedload grain sizes, and high flood velocities similar to those of the ancient rivers discussed here (Table 1). Furthermore, the large widths of the Huanghe and Brahmaputra allow their macroform elements (*sensu* Jackson, 1975) to wander laterally over distances of many kilometres. The absence of vegetation on the sandy alluvium of the Fugleberget and Hestman rivers suggests similar unconfined flow conditions.

The term 'bar-complex' as used here refers to sediment storage bodies much larger in lateral and downflow dimensions and with longer time spans than the component bars. Similar bed features in the Brahmaputra river are called 'bars' by Coleman (1969) and Bristow (1987).

River pattern

The low variance of palaeocurrent directions within and between superimposed sheet-sandstone bodies (Figs 3 & 11) of the Fugleberget and Hestman deposits and the general absence of argillaceous sediment throughout the 100–600 m thick successions indicate low-sinuosity, highly mobile river systems rather than meandering ones. The discussed bedform- and bar-types, and their inferred formative flow conditions, are probably consistent with the environment of relatively shallow rivers, with a mainly braided pattern during intermediate and low discharges. In addition, the Precambrian context, with its absence of vegetation, favours low bank stability within the sandy alluvium, high discharge variabilities and high sediment availability. These are conditions generally accepted as conducive to the development of braided fluvial patterns (Smith, 1976; Schumm, 1977; Carson, 1984; Smith & Smith, 1984; Bridge, 1985).

Scales and morphological elements

The inference that the bars of the Hestman and Fugleberget exposures represent short-term events (less than one flood cycle) suggests, in a braided stream setting, that on a longer time span (one to several flood cycles) they coalesced to form midchannel and sidebar-complexes (macroforms) with lateral and downstream dimensions many times those of the individual bars. Such bar-complexes scale with their associated channels (Bridge, 1985; Bristow, 1987). In the wider tracts of the Huanghe and Brahmaputra rivers, channel width/depth ratios in the order of a few hundred are common (Table 1; Chien, 1961, fig. 2; Bristow, 1987, figs 5 & 6). Fuller (1985) argued that such ratios would be expected in the absence of bank stabilizing vegetation in the Precambrian.

Minimum depths of the channels within the Fugleberget rivers were between 1 and 4 m, and within the Hestman rivers between 1 and 6 m as judged from the thicknesses of the bars, bar-complexes and channel-fill sequences (Figs 3 & 11). These depths, together with the width/depth ratios deduced above, suggest that the main channels within the rivers may have been several hundred metres wide. The larger bar-complexes probably had lateral and downstream dimensions of several hundred to a few thousand metres as suggested from the scales of their component bars (Fig. 3).

Channel and bar-complex migration rates

The fine, uniform bedload grain sizes of the Huanghe and Brahmaputra rivers relative to their high flood velocities, together with excessive sediment load, result in rapid response of the bed to changing flow conditions. Consequently, the rate of channel/bar-complex migration is in the order of several tens of metres a day to several kilometres in months or years (Table 1). Since the Precambrian rivers apparently were comparable to the modern ones with respect to their fine grain sizes, high velocities and abundant sediment load, similar high macroform migration rates are inferred.

Discharge, velocity and stage fluctuations

The high run-off rates suggested by the absence of vegetation (Schumm, 1977) imply high discharge variability and high rates of discharge decline in the Precambrian rivers. In unstable river situations such as envisaged here, the cross-sectional area of through-flow would probably respond quickly to changes in discharge thus maintaining relatively high velocities through a wide range of discharges. Maddock (1969) argued that such conditions would be favoured above a critical velocity in streams with bed and banks of incoherent unigranular material. The local presence of low-energy deposits on some of the bar-tops suggests a decline in velocities at these sites during low discharges. Their preservation potential is, however, limited because of erosion, either during successive floods or by the action of winds on emerged bar-surfaces.

In the Brahmaputra river, stage fluctuations up to 8 m have been recorded (Coleman, 1969; Bristow, 1987). This, together with the low migration rate of the river (70 m d^{-1}) as opposed to that of its channels and flats (Table 1) suggests, overall, relatively stable river banks. In the Precambrian rivers, in contrast, the probable absence of thick overbank mud deposits and the absence of vegetation imply very low bank stabilities with a tendency for the rivers to widen during an increase in discharge (Schumm, 1977; Fuller, 1985). Consequently, stage fluctuations were probably relatively small and much less in the Precambrian rivers than in the Brahmaputra.

DISCUSSION AND CONCLUSIONS

We have argued that the deposits of the Late Precambrian Fugleberget Formation and Hestman Member represent the products of wide, relatively shallow, braided rivers, broadly comparable to the Huanghe and Brahmaputra rivers in terms of channel width/depth ratios, velocities and macroform instabilities. Deposits of the Huanghe River probably comprise, on account of their grain size, depth and velocity relationships (Table 1), mainly horizontal to low-angle inclined planar-stratified sands. In contrast, the Brahmaputra River deposits are dominated by cross-stratified facies although Bristow (1989) observed that planar-stratification, in sets up to 4 m thick, occurred at the base of several bar-top sequences. In the Precambrian river deposits the amount of planar stratification varies between 5 and 70% but is, in most sections, less abundant than cross-stratification (e.g. Figs 2, 3 & 11).

Limited information from the Brahmaputra precludes detailed correlation between its channel sequences and those of the Precambrian rivers. However, the dramatic stage fluctuations in the Brahmaputra River, together with channel depths commonly in the order of 10–20 m (Bristow, 1987, figs 5 & 6), suggest that its channel-sandflat sequences comprise superimposed deposits from a variety of bedform and bar types (cf. Bristow, 1989). In the Precambrian rivers, in contrast, channels were only a few metres deep and stage fluctuations limited. Consequently, channel sequences are relatively simple and comprise in the majority of examples solitary units of simple, compound and composite bars (Fig. 11).

Despite these apparent differences in dominant stratification types and/or sequences between deposits of the two modern and two ancient river systems they probably have important architectural attributes in common. Their large widths, together with the magnitude and styles of behaviour of their macroforms, suggest that their deposits comprise sheet-sandstone bodies of at least four different scales. The first order sheet bodies scale with the channel-belt deposits and have time spans of the same duration as the avulsion rates of the rivers (several tens to several hundred years). The second order sheet bodies represent the preserved portions of individual macroforms (the second order channels and bars of Bristow, 1987) with time spans of one to several flood cycles. The third order sheet bodies form components of the macroforms deposited within one flood cycle. As evident from the Brahmaputra River (Coleman, 1969, fig. 17), single dynamic events within a flood cycle also create sandbodies with sheet geometries (fourth order sheet bodies).

The simple, compound and composite bars of the Type A, B and C sandstone bodies of the Precambrian deposits correspond to fourth and third order sheet bodies. The bar-complexes (Fig. 11b,d,e) belong to the second order sheet bodies, whereas the inferred channel-belt sequences (Fig. 11c,e) are classified as first order sheet-sandstone bodies. The many solitary occurrences of the third and fourth order sheet-sandstone bodies between channel erosion surfaces (Fig. 11) suggest, however, that their bounding surfaces, at least locally, coincide with those of the second order sheets. Furthermore, identification of surfaces bounding channel-belt sequences is, in general, hampered by the absence of fine floodplain deposits in the alluvial sediments described here. Consequently, the extensive hierarchy of bounding surfaces as proposed by Miall (1988) appears to have limited application in very wide, shallow and unstable types of river systems such as those inferred for the Late Precambrian Fugleberget and Hestman successions.

ACKNOWLEDGEMENTS

We thank Nigel Bank, Trevor Elliot, Bill Galloway and Ron Steel for comments on an earlier draft of the manuscript, and Finnmark Fylkeskommune and Nord Norges Fond ('Kometen') for financial support.

REFERENCES

Allen, J.R.L. (1982) *Sedimentary Structures: their Character and Physical Basis.* Elsevier, Amsterdam.

Allen, J.R.L. (1983) Studies in fluviatile sedimentation: bars, bar-complexes and sandstone sheets (low-sinuosity streams) in the Brownstones (L. Devonian), Welsh Borders. *Sedim. Geol.* **33**, 237–293.

Ashley, G.M. (1990) Classification of large-scale subaqueous bedforms: a new look at an old problem. *J. Sedim. Petrol.* **60**, 160–172.

Banks, N.L. (1973) The origin and significance of some downcurrent-dipping cross-stratified sets. *J. Sedim. Petrol.* **43**, 423–427.

Bridge, J.S. (1985) Paleochannel patterns inferred from alluvial deposits: a critical evaluation. *J. Sedim. Petrol.* **55**, 579–589.

Bristow, C.S. (1987) Brahmaputra River: channel migration and deposition. In: *Recent Developments in Fluvial Sedimentology* (Eds Ethridge, F.G., Flores, R.M. & Harvey, M.D.) pp. 63–74. Soc. Econ. Paleont. Miner., Tulsa, Spec. Publ. 39.

Bristow, C.S. (1989) Some observations of sedimentary structures exposed in bar tops in the Brahmaputra River, Bangladesh (abstr.). In: *4th Int. Conf. on Fluvial Sedimentology*, p. 88. University of Barcelona, Barcelona.

Cant, D.J. & Walker, R.G. (1978) Fluvial processes and facies sequences in the sandy braided South Saskatchewan River, Canada. *Sedimentology* **25**, 625–648.

Carson, M.A. (1984) The meandering–braided river threshold: a reappraisal. *J. Hydrol.* **73**, 315–334.

Chien, N. (1961) The braided stream of the Lower Yellow River. *Scientia Sinica* **10**, 734–754.

Coleman, J.M. (1969) Brahmaputra River: channel processes and sedimentation. *Sedim. Geol.* **3**, 129–239.

CROWLEY, K.D. (1983) Large-scale bed configurations (macroforms), Platte river basin, Colorado and Nebraska; primary structures and formative processes. *Bull. Geol. Soc. Am.* **94**, 117–133.

CULBERTSON, J.K. & SCOTT, C.H. (1970) Sandbar development and movement in an alluvial channel, Rio Grande near Bernardo, New Mexico. US Geol. Surv. Prof. Paper 700B, pp. 237–241.

FULLER, A.O. (1985) A contribution to the conceptual modelling of Pre-Devonian fluvial systems. *Trans. Geol. Soc. S. Africa* **88**, 189–194.

HERMANSEN, M. (1989) *Sedimentology of the lower part of the Late Precambrian Hestman Member (Barents Sea Group), Varanger Peninsula, Finnmark.* Thesis, University of Bergen.

HJELLBAKK, A. (1987) *Sedimentology of the Late Precambrian Seglodden and upper Næringselv Members, Barents-Sea Group, Varanger Peninsula, Finnmark.* Thesis, University of Bergen.

HOBDAY, D.K. (1974) Interaction between fluvial and marine processes in the lower part of the Late Precambrian Vadsø Group, Finnmark. *Norges Geol. Unders.* **303**, 39–56.

JACKSON II, R.G. (1975) Hierarchical attributes and a unifying model of bed forms composed of cohesionless material and produced by shearing flow. *Bull. Geol. Soc. Am.* **86**, 1523–1533.

JACKSON II, R.G. (1976) Largescale ripples of the Lower Wabash River. *Sedimentology* **23**, 593–623.

JOHNSON, H.D., LEVELL, B.K. & SIEDLECKI, S. (1978) Late Precambrian sedimentary rocks in east Finnmark, north Norway and their relationship to the Trollfjord–Komagelv Fault. *J. Geol. Soc. Lond.* **135**, 517–533.

JONES, B.G. & RUST, B.R. (1983) Massive sandstone facies in the Hawkesbury Sandstone, a Triassic fluvial deposit near Sydney, Australia. *J. Sedim. Petrol.* **53**, 1249–1259.

MADDOCK, T. (1969) The behavior of straight open channels with movable beds. US Geol. Surv. Prof. Paper 622A.

MIALL, A.D. (1985) Architectural-element analysis: a new method of facies analysis applied to fluvial deposits. *Earth Sci. Rev.* **22**, 261–308.

MIALL, A.D. (1988) Architectural elements and bounding surfaces in fluvial deposits: anatomy of the Kayenta Formation (Lower Jurassic), southwest Colorado. *Sedim. Geol.* **55**, 233–266.

RØE, S.-L. (1987) Cross-strata and bedforms of probable transitional dune to upper-stage plane-bed origin from a Late Precambrian fluvial sandstone, northern Norway. *Sedimentology* **34**, 89–101.

SAUNDERSON, H.C. & LOCKETT, F.P.J. (1983) Flume experiments on bedforms and structures at the dune plan bed transition. In: *Modern and Ancient Fluvial Systems* (Ed. Collinson, J.D. & Lewin, I.) pp. 49–58. Spec. Publ. Int. Ass. Sediment. 6. Blackwell Scientific Publications, Oxford.

SCHUMM, S.A. (1977) *The Fluvial System.* John Wiley, London, 331 pp.

SIEDLECKA, A. (1985) Development of the Upper Proterozoic sedimentary basins of the Varanger Peninsula, East Finnmark, North Norway. *Geol. Surv. Finland, Bull.* **331**, 175–185.

SIEDLECKA, A. & EDWARDS, M.B. (1980) Lithostratigraphy and sedimentation of the Riphean Båsnæring Formation, Varanger Peninsula, North Norway. *Norges Geol. Unders.* **335**, 27–47.

SIEDLECKA, A., PICKERING, K.T. & EDWARDS, M.B. (1989) Upper Proterozoic passive margin deltaic complex, Finnmark, N. Norway. In: *Deltas: Sites and Traps for Fossil Fuels* (Eds Whatley, M.K.G. & Pickering, K.T.) pp. 205–219. Geol. Soc., Lond., Spec. Publ. 41.

SMITH, D.G. (1976) Effect of vegetation on lateral migration of anastomosed channels of glacier meltwater river. *Bull. Geol. Soc. Am.* **87**, 857–860.

SMITH, G.A. (1986) Coarse-grained nonmarine volcaniclastic sediment: terminology and depositional process. *Bull. Geol. Soc. Am.* **97**, 1–10.

SMITH, N. & SMITH, D.G. (1984) William River: an outstanding example of channel widening and braiding caused by bed-load addition. *Geology* **12**, 78–82.

Crevasse splay sandstone geometries in the Middle Jurassic Ravenscar Group of Yorkshire, UK

R. MJØS, O. WALDERHAUG and E. PRESTHOLM

Rogaland Research Institute, PO Box 2503, N-4004 Stavanger, Norway

ABSTRACT

Two types of crevasse splay sandstones have been defined: (i) small-scale single crevasse splay sandstone lobes attached to the levee of the fluvial channel, and (ii) large-scale composite crevasse splay sandstones associated with crevasse subdelta lobes which infill interdistributary bays.

Sections through single crevasse splay sandstones in the Ravenscar Group show width/thickness ratios less than 1500 and length/thickness ratios less than 2000. Thicknesses are up to 2.5 m and widths and lengths are of similar magnitude (up to about 2000 m). The single crevasse splay sandstones thin rapidly at their outer margins and also outward from the proximal part towards the unconfined crevasse splay lobes. Sections with low width/thickness ratios (10–100), large thicknesses (1.0–2.5 m) and small widths (18–200 m) represent proximal confined crevasse splay sandstones. The geometries of single crevasse splay sandstones in Yorkshire are similar to the geometry of other crevasse splay sandstones described in the literature.

Composite crevasse splay sandstone sequences (usually between 2.5 and 6 m thick) have probably been formed by crevasse subdelta sedimentation. The extent of a composite crevasse splay sandstone level in the Gristhorpe Member is at least 20 km. In the medial to distal part of a crevasse subdelta the composite crevasse splay sequence is composed of single crevasse splay sandstones with widths of around 0.5 km. The proximal part of a crevasse subdelta contains, axially, channelized crevasse splay sandstones fringed by crevasse splay sandstone sheets.

The rock volume of crevasse splay sandstones is strongly dependent on thickness: a doubling of the thickness causes an eightfold increase in volume. The width/thickness ratio also strongly influences the volume estimate: a doubling of this ratio will cause a fourfold increase in volume.

The thickest crevasse splay sandstones are genetically connected to the top part of their fluvial feeder channel. The crevasse splay sandstones are probably in most cases in contact with a fluvial channel sandstone through a crevasse channel sandstone with a low width/thickness ratio (10–20). Sometimes, crevasse splay sandstones are cut by younger fluvial channels.

The crevasse splay sandstones within overbank sequences may place constraints on the location of fluvial channel sandstones; in particular, these constraints are useful when considering sequence architecture in fluvio-deltaic reservoirs.

The quantitative geometrical data increase our ability to make geometric predictions based on well data, and also provide constraints on the dimensions and shapes of sandbodies. Incorporation of this information into a reservoir model will reduce the uncertainty in the model and contribute to more reliable predictions of reservoir performance.

INTRODUCTION

In alluvial and deltaic sequences crevasse splay sandstones are commonly interbedded with fine grained mudrocks. Knowledge of crevasse splay sandstones and their quantitative geometries is, however, relatively meagre. In the literature, some papers have focused on the sedimentology of crevasse splay sandstones (e.g. Coleman & Gagliano, 1964; Elliott, 1974, 1986; Fielding, 1984, 1986; Guion, 1984; Smith et al., 1989), but the quantitative and architectural aspects of crevasse splay sandstones are not well known. The recent emphasis on overbank sequences is largely due to the significance of coals and palaeosols in the evaluation of alluvial and deltaic sequences (e.g. Fielding, 1984, 1985; Kraus & Bown, 1986; Behrensmeyer, 1989; Haszeldine, 1989). The crevasse splay sandstones within overbank sequences may give important information about the spatial organization of fluvial channel sandstones. Bridge (1984) has utilized this and has focused on the relation between the vertical overbank sequences and the avulsion frequency.

Crevasse splay deposits accumulate when excess discharge during floods breaches the adjacent levee. In some cases crevasse splay channels cut deeply through the levee and may tap bed load and earlier deposits in the river; thereby coarse grained material is transported into and deposited in the floodplain areas (Allen, 1964).

The crevasse splay sandstones are sheet-like or lenticular units with erosive or sharp basal contacts and the grain size and sedimentary structures often indicate an upward decrease in flow regime (Elliott, 1986). The mud and siltstones beneath a crevasse splay sandstone very often coarsen upwards (Flores, 1985). The most dominant sedimentary structures within crevasse splay sandstones are ripple and flat lamination (Allen, 1964); however, the sedimentary features are mainly determined by flow characteristics and magnitude of floods (Elliott, 1974). Plant fragments, fossils and mud or carbonate clasts are common constituents of crevasse splay sandstones (Fielding, 1984; Bown & Kraus, 1987). The crevasse channels within the crevasse splay lobe may be recognized by a distinct erosive base, fining upward trend, and content of trough and ripple lamination (Flores, 1985). The crevasse splay sandstones are normally fine to medium grained, whereas the crevasse splay channels often are slightly coarser than the crevasse splay sheets (Flores, 1985).

Crevasse splay sandstones are commonly lobe shaped but they may also be tongue shaped (Elliott, 1974), and the lobes are oriented roughly perpendicular to the fluvial feeder channel. Two mechanisms for crevasse splay sandstone deposition have been described (Elliott, 1974). The crevasse splay sandstones have either been deposited by small anastomosing reaches implying many sand lenses separated by mud and siltstone drapes, or by density current induced sheet floods that deposit sheet sandstones. The anastomosing channels are most pronounced on the large-scale crevasse subdeltas that infill the interdistributary bays. The channels enter the bay and deposit minor mouth bars (Elliott, 1974, 1986).

The main aim of this paper is to present quantitative data concerning the geometry of crevasse splay sandstones and to consider how crevasse splay sandstones relate to fluvial channel sandstones. Finally, application of the collected data in reservoir geology is discussed. The coastal cliff sections of the Ravenscar Group in Yorkshire have been selected for this study due to the similarities between the Ravenscar Group and the subsurface Ness Formation in the northern North Sea (Hancock & Fisher, 1981; Mjøs & Walderhaug, 1989).

The measurements presented in this paper have been performed by: measurement of outcrops with a 50 m steel measuring tape; determining the positions of the ends of sections and subsequent measurement of the distance between the endpoints on maps of scale 1 : 25 000; measurements on photomosaics containing scales of known length; and angular measurements performed with a theodolite.

GEOMETRICAL DATA FROM THE LITERATURE

Literature data indicate that the width/thickness ratios for crevasse splay sandstones fall in the range 150–1000 (Fig. 1). The reported thicknesses are in

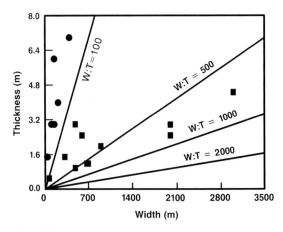

Fig. 1. Crevasse splay sandstone width and thickness data collected from the literature. (●) Crevasse channels; (■) crevasse splay lobe. In some cases, the data are based on measurements from illustrations. The data are from Baganz et al. (1975), Farrell (1987), Fielding (1986), Flores (1984), Smith (1983) and Smith et al. (1989).

the range 0.3–4.5 m, but usually less than 2 m. Crevasse splay channel sandstones have a width/thickness ratio in the range 5–60 (Fig. 1); thicknesses are up to 7 m, although usually less than 4 m.

Recent crevasse splay deposits on alluvial plains have been studied in a few cases (Coleman, 1969; Smith, 1983; O'Brian & Wells, 1986; Farrell, 1987; Smith et al., 1989). Smith (1983) noted that the geometry is lobate with several sublobe extensions along the distal margin, whereas Smith et al. (1989) show splay evolution through time from initial lobate sheet geometries towards a more mature channelized anastomosed pattern at the final stage. O'Brian & Wells (1986) and Farrell (1987) describe tongue-shaped crevasse splay sandstones, where the length is two to four times the width.

Data from crevasse subdelta and associated minor mouth bar sediments deposited in the interdistributary lower delta plain area are not included in the above mentioned data. Crevasse subdelta deposition in the modern Mississippi delta may lead to deposits up to 15 m thick, but most of the Mississippi crevasse splay sandstones are 1.5–6 m thick (Saxena, 1978).

Recent crevasse splay deposits in interdistributary areas have been described in connection with studies of the Mississippi delta (Coleman & Gagliano, 1964; Elliott, 1974; Saxena, 1978; Galloway & Hobday, 1983). The major crevasse subdeltas of the modern digitate lobe of the Mississippi delta are from 6 to 25 km long and the widths are in all instances of similar magnitude, giving a 1:1 relationship between width and length. The distal parts of crevasse subdeltas are composed of minor mouth bars deposited at the mouth of single crevasse channels. These mouth bars commonly coalesce and form a continuous sheet (Elliott, 1974, 1986). The width (alongshore) of a minor mouth bar complex deposited by a single crevasse channel in the Mississippi delta is less than 2500 m (Elliott, 1974).

THE RAVENSCAR GROUP

The Middle Jurassic Ravenscar Group is located in the Cleveland (Yorkshire) Basin, and the studied exposures are located along the Yorkshire coastal cliffs between Whitby and Yons Nab (Fig. 2). The Ravenscar Group comprises five formations (Hemingway & Knox, 1973; Fig. 2). The four lowermost formations are of Bajocian age and the uppermost of Bajocian to Bathonian age (Hancock & Fisher, 1981; Fisher & Hancock, 1985). The lowermost formation (up to 50 m thick) is the Saltwick Formation which represents delta plain sediments (Hancock & Fisher, 1981; Livera & Leeder, 1981) and consists of fluvial channel sandstones, crevasse splay sandstones and fine grained delta plain sediments. The Saltwick Formation is overlain by the Eller Beck Formation (up to 8 m thick) which is thought to have been deposited in a marine embayment (Hancock & Fisher, 1981). A return to a delta plain environment during deposition of the fluvial channel sandstones and the fine grained overbank sediments of the overlying up to 50 m thick Sycarham Member (Cloughton Formation) occurred prior to a new marine incursion which led to deposition of the Millepore Bed and the Yons Nab Beds. The Yons Nab Beds are overlain by the Gristhorpe Member (up to 30 m thick) which also comprises fluvial channel and crevasse splay sandstones and fine grained delta plain sediments. The shales, limestones and sandstones of the overlying Scarborough Formation (up to 30 m thick) were deposited in relatively open marine conditions (Livera & Leeder, 1981) or in an embayment (Hancock & Fisher, 1981). The uppermost Scalby Formation (up to 60 m thick) consists of a fluvial sheet channel complex at the base overlain by fluvial channel and crevasse splay sandstones encased in fine grained delta plain sediments (Fisher & Hancock, 1985).

The quantitative crevasse splay geometry data have been collected in the fluvio-deltaic Saltwick and Scalby Formations. A composite crevasse splay sandstone in the Gristhorpe Member has also been studied.

CREVASSE SPLAY SANDSTONES IN THE RAVENSCAR GROUP

In the literature, the relatively thin sheet sandstones in the Saltwick Formation have been interpreted as crevasse splay sandstones (Hancock & Fisher, 1981; Livera & Leeder, 1981); similar sandstones in the Scalby Formation have been identified as crevasse splays (Fisher & Hancock, 1985) or 'crevasse splays and sheet sands' (Ravenne et al., 1987).

In this study, crevasse splay sandstones were identified by their characteristic sedimentary structures, geometries and association with other facies. The sedimentary structures include current ripples,

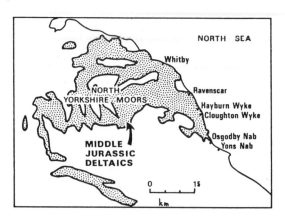

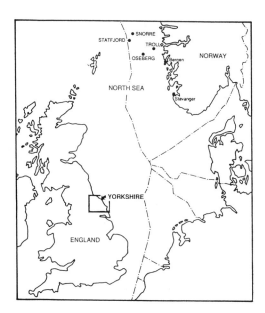

Fig. 2. Stratigraphy and location map of the Middle Jurassic Ravenscar Group of the Yorkshire Basin. The Ravenscar Group is up to 230 m thick (Hemingway & Knox, 1973) comprising the Saltwick, Eller Beck, Cloughton, Scarborough and Scalby Formations. (Modified from Ravenne *et al.* (1987).)

horizontal lamination and irregular horizontal mud drapes or laminae. Many crevasse splay sandstones appear massive without discernible sedimentary structures. Lower boundaries of crevasse splay sandstones are usually sharp or erosive and with occasional loading and gutter casts; tops commonly contain carbonized roots. Some lower boundaries are gradational and display a coarsening upward trend. Upper boundaries are often sharp, but may occasionally fine upwards. Vertical variation in grain size within the crevasse splay sandstones is usually minor. Proximal channelized parts of crevasse splay sandstones may contain high angle, planar, cross-bedding (a few decimetres high).

Quantitative data

Two-dimensional sections through crevasse splay sandstones can be classified as longitudinal (length), transverse (width) or oblique (Fig. 3). A *longitudinal* section can be recognized by lateral merging with a channel sandstone, thinning outwards from

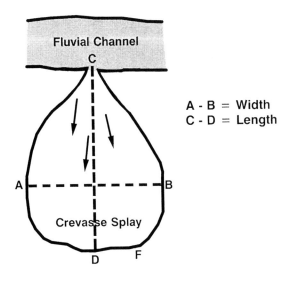

Fig. 3. Definition of length and width of a crevasse splay lobe. The width of a crevasse splay lobe is defined as the distance A–B. The length is defined as distance C–D. Sections parallel to A–B are in this paper referred to as transverse sections, and sections parallel to C–D are referred to as longitudinal sections. In some cases sections like B–F may appear to be transverse even if they in fact are nearly longitudinal.

the channel sandstone and by palaeocurrent directions parallel to the section. A *transverse* section can be recognized by thinning from the centre towards the ends of the section, by palaeocurrent directions at right angles to the section and by the absence of a channel sandstone at the ends of the section. However, sections through the outer part of a crevasse splay lobe such as B–F on Fig. 3 may appear to be transverse sections even if they are in fact nearly parallel to the length of the splay.

If the orientation of fluvial channel sandstones within a formation is known, then this can also assist in determining whether a section through a crevasse splay sandstone is longitudinal or transverse since the length axis of a crevasse splay will tend to be oriented approximately normal to the fluvial channel which fed the crevasse splay lobe.

Length, width and thickness data for crevasse splay sandstones from the Ravenscar Group are shown in Fig. 4. These data are collected from crevasse splay sandstone units which were deposited during a single flood event or period. Sandstone units separated by a major erosion surface, by a few centimetres of coal or shale or by only a surface with rootlets, are considered as two crevasse splay sandstones.

Obvious differences between widths and lengths of crevasse splay sandstones in the Saltwick Formation are not observed. The recorded widths are up to 2200 m and lengths up to 1750 m. Many of the data points in Fig. 4 represent minimum values due to smaller lateral extent of the exposure than the crevasse splay sandstones. Some of the smaller values may be the result of sections located near the edges of large lobes; however, this probably cannot explain the greater abundance of small width measurements in the Scalby Formation, since the probability distribution of the lengths of random sections through a lobe-shaped sandbody is roughly rectangular (Mjøs & Walderhaug, 1989). The abundance of relatively small width measurements in the Scalby Formation therefore probably represents real variation in crevasse splay size.

The crevasse splay sandstones thin rapidly at their outer margins and also outwards from the proximal channelized part towards the unconfined crevasse splay lobe. Within the central parts of the crevasse splay lobes thickness appears to be relatively constant and rarely varies by a factor greater than 2. Most of the measured sections through crevasse splay sandstones have thicknesses less than 1 m, but thicknesses up to 2.5 m are not uncommon (Fig. 4).

There is no linear relationship between width and thickness of the sections through crevasse splay sandstones (Fig. 4). Maximum width/thickness ratios are, however, not greater than 1500. It seems likely that a plot of maximum thickness against maximum width for crevasse splay sandstones in the Ravenscar Group would show a more linear trend, at least for the larger lobes, but since the quality of the exposures in Yorkshire is limited such a plot cannot be constructed. However, since the studied sections through crevasse splay sandstones can be considered to be randomly located within the crevasse splay lobes, it follows that the highest observed width/thickness ratios probably are closest to the maximum width/maximum thickness ratios for crevasse splay sandstones in the Ravenscar Group.

Several of the transverse sections through crevasse splay sandstones in the Saltwick Formation have very low width/thickness ratios (10–100),

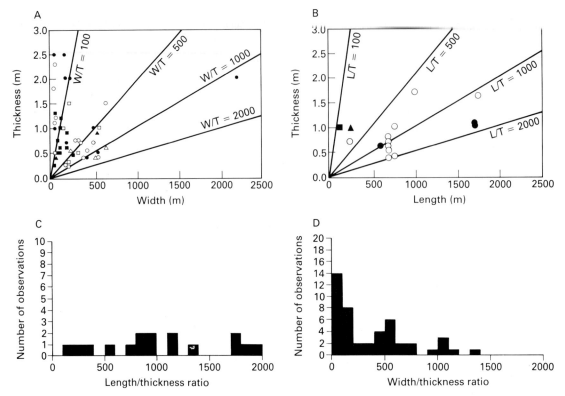

Fig. 4. Quantitative data for sections through crevasse splay sandstones from the Ravenscar Group. In (A) and (B) open symbols represent minimum values where the full extent of the crevasse splay sandstone was not observed due to termination of the exposure. (●○) Saltwick Fm.; (▲△) Gristhorpe Mb.; (■□) Scalby Fm. (A) Plot of measured widths against measured thicknesses. Points with width/thickness ratios less than 100 and thicknesses greater than 1.5 m represent sections through the proximal, channelized part of crevasse splay sandstones. (B) Plot of measured lengths against measured thicknesses. (C) Histogram of the length/thickness ratios. (D) Histogram of the width/thickness ratios.

large thicknesses (1.0–2.5 m) and small widths (18–200 m). The lower boundaries of these sections are commonly convex and seem to be erosional; the upper boundaries are in some cases also convex and may reflect bedforms forming positive features.

Length/thickness ratios do not exceed 2000 (Fig. 4), and are within the same order of magnitude as the width/thickness ratios.

Composite crevasse splay sandstones

In some cases several crevasse splay sandstone units are stacked upon each other and form a composite crevasse splay sandstone. In the Gristhorpe Member a composite crevasse splay sandstone at Cloughton Wyke has been traced laterally for 1100 m (Fig. 5). However, it probably has a much greater extent since composite crevasse splay sandstones are found at all localities along the coast where the Gristhorpe Member is exposed. The composite crevasse splay sandstone in Cloughton Wyke is 3–6 m thick and consists of sandstone units up to 2 m thick, plus some thin (decimetre) siltstones and coals in the lower part and coals up to 0.5 m thick in the upper part (Fig. 5). The individual sandstones and coals within the composite crevasse splay have a restricted lateral extent, commonly around 0.5 km. The individual crevasse splay sandstone units are often rooted in their uppermost parts, have sharp to erosive bases which may show gutter casts and also loading features when overlying coal. Sedimentary structures include current ripples, decimetre-scale cross-bedding, plane parallel lamination and more irregular horizontal

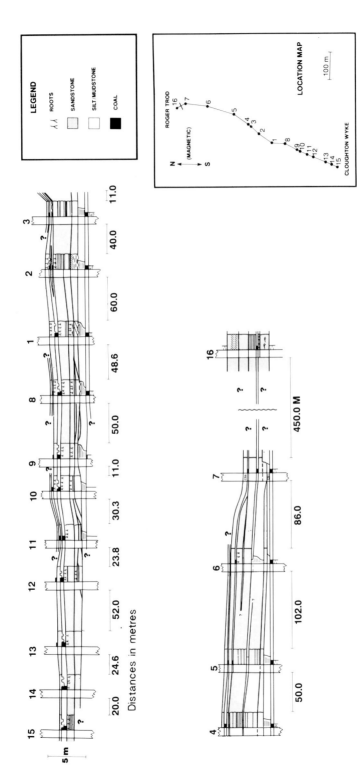

Fig. 5. Lithological correlation through a composite crevasse splay sequence between Cloughton Wyke and Roger Trod, a distance of 1060 m.

mud drapes or laminae. Apart from the characteristic crevasse splay sedimentary structures, wave ripple lamination also occurs at the base of the composite crevasse splay sandstone at Ravenscar (south). At the base of the composite crevasse splay sandstone at Cloughton Wyke the sandstone units overstep each other in a manner suggestive of lateral accretion or progradation (Fig. 5). Palaeocurrent directions determined from gutter casts (50°/230°), and the fact that some of the crevasse splay sandstone units have a distinct lensoid cross-section, suggest that the exposures in Cloughton Wyke represent a near transverse section through the composite crevasse splay sandstone.

Stacked crevasse splay sandstones are found at the same stratigraphic level at all localities where the Gristhorpe Member is exposed (Fig. 6). Palaeocurrent directions (current ripples) at Osgodby Point (Fig. 6) are towards the northwest (320°). The composite crevasse splay sandstone level is situated 8–10 m below the lower Scarborough Formation sandstone and 10–11 m above the Yons Nab Beds from Osgodby Point in the southeast towards Ravenscar in the northwest, a distance of 17 km. At Yons Nab, 2.2 km southeast of Osgodby Point, the Gristhorpe Member is much thinner and a composite crevasse splay sandstone level is located about 2 m below the Scarborough Formation and 2 m above the Yons Nab Beds.

At Ravenscar the upper mudstone and siltstone dominated part of the Gristhorpe Member thins from 10 m in the south to 5 m in the north within a distance of about 600 m. In the same area the composite crevasse splay sandstone thins southwards, in the opposite direction of the mud and siltstone unit, from 6 m to 3 m (Fig. 6). These thickness changes occur in an area where the composite crevasse splay sandstone level can be traced northwards (on photomosaics) as an interval situated 10 m above the top of a thick (24 m) fluvial channel sandstone at Ravenscar. The composite crevasse splay sandstone is not connected to this older fluvial channel sandstone; these lateral thickness changes of the composite crevasse splay sandstone may be related to the lateral changes in sedimentation style that have taken place.

The composite crevasse splay sandstone level in the Gristhorpe Member shows lateral changes in thicknesses and sedimentary characteristics (Fig. 6) and this is also to be expected since the composite crevasse splay sandstone comprises crevasse splay sandstone units with restricted lateral extent (Fig. 5).

In the lower part of the Saltwick Formation at Hayburn Wyke a composite sandstone sequence was traced laterally for 650 m (Fig. 7), but the extent is significantly larger. The sandstone units have sedimentary characteristics that are consistent with a crevasse splay origin, but this composite sandstone sequence has no rootlets, contains more mud and silt matrix and contains an erosively based massive to trough cross-bedded sandstone. Laterally this massive and trough cross-bedded sandstone interfingers with more mudstone and siltstone dominated sheet beds (Fig. 7). The upper part of the composite sandstone sequence can be traced southwards towards a fluvial channel sandstone margin at Hayburn Wyke and caps the fluvial channel infill (Fig. 7). The lower part of the composite sandstone sequence becomes less sandy southwards towards the fluvial channel margin and interfingers with mud and siltstone dominated levee sediments.

Discussion

The similar width/thickness and length/thickness ratios of the sections through crevasse splays in Yorkshire suggest that the crevasse splay sandstones are lobe-shaped. This is also in accordance with most of the literature data, although tongue-shaped and channelized geometries are reported (O'Brian & Wells, 1986; Farrell, 1987; Smith et al., 1989). The sections through crevasse splay sandstones in the Ravenscar Group with low width/thickness ratios, relatively large thicknesses and associated erosional and convex lower boundaries are probably located close to a fluvial channel margin and probably represent the channelized proximal part of a crevasse splay lobe. However, the low width/thickness ratios of many of the lensoid crevasse splay sandstones in the Scalby Formation may be related to tongue-shaped or channelized features in the distal parts of the crevasse splay. Smith et al. (1989) have described channelized features from modern crevasse splays in east-central Saskatchewan, Canada; these features develop as the splay enlarges and directs water and sediment to more localized portions of the splay margin to form a sympodial growth pattern (Fig. 8).

The composite crevasse splay sandstone level (up to 6 m thick) in the Gristhorpe Member may have a lateral extent of at least 20 km implying a width/thickness ratio possibly larger than 3000. The width/thickness ratios of the individual crevasse splay sandstone units that form the building blocks

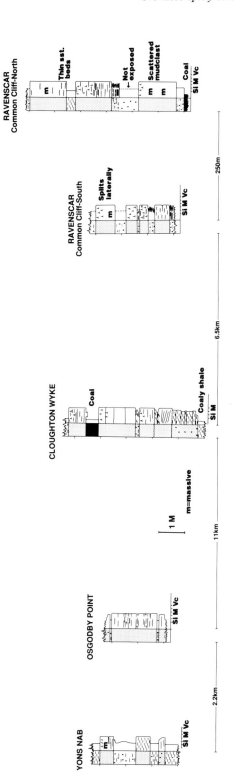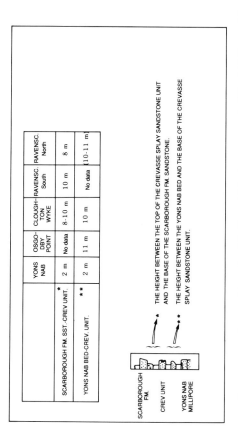

Fig. 6. Lithological logs from the composite crevasse splay sandstone level in the middle to upper part of the Gristhorpe Member. The logs are from five localities along the coastal cliffs where the Gristhorpe Member is exposed. The distance between Ravenscar in the northwest and Yons Nab in the southeast is 20 km. The vertical log scale is in metres.

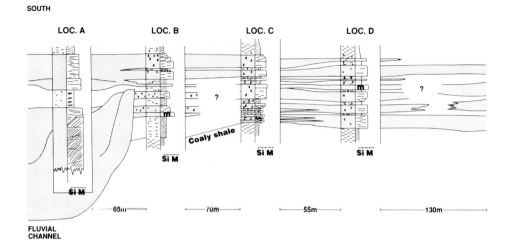

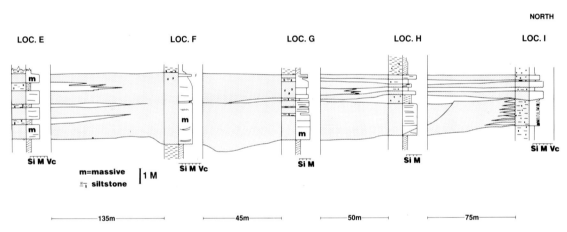

Fig. 7. Lithological correlation along a delta plain sequence at Hayburn Wyke comprising a fluvial channel sandstone in the south and a composite crevasse splay sequence to the north. The cross-bedding within the fluvial channel sandstone indicates an oblique to longitudinal section through the channel sandstone. The composite crevasse splay sequence probably exhibits a transverse section through the splay complex. Confined crevasse splay sandstones have been deposited in the axial part (loc. F, G and H) whereas sheet-shaped crevasse splay sandstones are prominent on both sides of the axial and confined crevasse splay sandstones (loc. B, C, D, E and I).

of the composite crevasse splay sandstone are comparable with what is elsewhere found in the Ravenscar Group (less than 1500, Fig. 4). The large extent of the composite crevasse splay sandstone must be the result of both lateral and vertical stacking of crevasse splay sandstones. At Ravenscar faint evidence of wave reworking at the base of the composite crevasse splay sandstone occurs, but generally wave reworking has been of minor importance and therefore cannot explain the great lateral extent of the composite crevasse splay sandstone.

The composite crevasse splay sandstone in the Gristhorpe Member (Cloughton Wyke) is interpreted by Hancock & Fisher (1981) to have been deposited as crevasse splay and sheetflood sands which periodically were capped by vegetated swamps. Palynofacies analysis of the thin coarsening upward sequence at the base of the composite

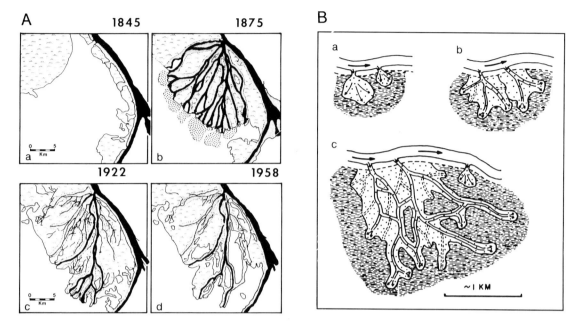

Fig. 8. Evolution of a crevasse splay from early stages to more mature stages. Examples from the recent Mississippi River delta (A) and the Cumberland Marshes in east-central Saskatchewan (B). Note the differences in scale between the large-scale crevasse subdelta (A) and the small-scale crevasse splay (B). (A) The West Bay crevasse subdelta in the interdistributary bay of the Mississippi River delta evolved from the initial levee break in 1845 to a fan-shaped crevasse splay sandstone body in 1875. Active channels are in black, inactive channels and levee deposits are densely stippled. In 1922, the subdelta was in an early stage of abandonment and a few lakes and tidal creeks had formed. Few channels were active and the active part of the crevasse subdelta had changed from fan or lobe shape in 1875 to more tongue shape in 1922. During later stages of abandonment (when subsidence was dominant, in 1958) lakes and bays were enlarging and flow in the channels was minor. (Modified from Coleman (1976).) (B) The splay evolution in the Cumberland Marshes. During early stages ((a) and (b)) small lobate splays are formed by the initial flow diversion through the levee of the trunk channel. With further growth, a sympodial growth pattern appears. Extension and rejoining of crevasse channels create an anastomosed pattern (c). Growth of long and isolated crevasse channel extensions is characteristic for the most mature stages. (Modified from Smith *et al.* (1989).)

crevasse splay sandstone indicates saline influence and deposition in a marine influenced bay (Hancock & Fisher, 1981).

The great extent of the composite crevasse splay level may be the result of either:
1 coalescing crevasse splay lobes along a trunk channel; or
2 crevasse subdelta development which in size is comparable with the crevasse subdeltas of the modern Mississippi delta.

If the crevasse sandstone units within the composite crevasse splay sandstone have a lobate geometry, then the fluvial trunk channel feeding these units must have been situated less than a kilometre or two from the exposures. This implies that the fluvial trunk channel had to flow southeastwards parallel to the coastline for at least 20 km. However, such a channel position is unlikely, although not impossible. It is therefore difficult to explain the extent of the composite crevasse splay by interpreting it as a result of coalescing crevasse splay lobes along a fluvial trunk channel.

The composite crevasse splay sandstone interval may be the result of crevasse subdelta deposition. The great extent (up to 25 km in width and length) of the crevasse subdeltas (Fig. 8A) in the Mississippi delta may indicate similarities in large-scale geometry with the composite crevasse splay sandstone in the Gristhorpe Member. The West Bay subdelta in the Mississippi delta (Fig. 8A) formed initially as a deep break in the major distributary levee during flood stage. During the next 30 years peaks of maximum discharge occurred and the crevasse subdelta prograded into the bay through a system of radial, bifurcating channels similar in plan to veins of a leaf (Coleman & Gagliano, 1964). During early stages of abandonment only a few crevasse channels were active and luxuriant plant growth and lakes increased in abundance. During the initial and rapid progradational stage the width and length of the West Bay subdelta were about the same (19 km) but the active part of the crevasse subdelta changed from lobe to tongue shaped from 1875 to 1922 (width 11 km; length 23 km). Based on Fig. 8A, the West Bay subdelta probably comprises several crevasse splay sandstones with shoestring or channelized geometries.

When comparing the small-scale crevasse splay evolution (Fig. 8B) with the crevasse subdelta evolution (Fig. 8A), similarities appear in evolution from an initial lobe geometry towards a more channelized and tongue shaped geometry at the final stage. Elliott (1974) proposed a comparable interdistributary bay model which suggested evolution from crevasse splay lobe deposition at early stages towards crevasse channel and minor mouth bar deposition at later stages. The evolution towards channelling of the crevasse splays with time may be the result of levee or alluvial ridge accretion that reduces the numbers of small-scale crevasse sheet flood lobes, and instead results in the excess discharge during floods being concentrated towards crevasse channels which cut deep into the levee. During more mature stages of crevasse splay development the favourable accommodation areas are more distant and transport of sediments towards these distant areas is most effective by development of channelized transport.

The composite crevasse splay sandstone in the Gristhorpe Member (Figs 5 & 6) is composed of crevasse splay sandstone units with widths of around 0.5 km. If these crevasse units are connected directly to the fluvial trunk channel, then they must be several kilometres long and display a shoestring morphology. Another possibility is that crevasse channels on the subdelta may feed lobe shaped extensions in a manner proposed by Fielding (1984) (Fig. 9). Relatively small channels at Cloughton Wyke cut through the mud and siltstone dominated sequence on top of the composite crevasse splay sandstone, but whether these channels are genetically related to the underlying composite crevasse sandstone is not clear.

The fluvial trunk channel or channels that have

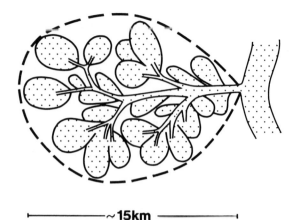

Fig. 9. Crevasse splay complex that forms a crevasse subdelta. The proximal part is dominated by channelized erosively based sandstones. The medial part is dominated by interbedded sharply based sandstones forming irregular areas whereas the distal part is dominated by single or multiple coarsening upward sequences forming lobes. (Modified from Fielding (1984).)

fed the composite crevasse splay sandstone in the Gristhorpe Member have probably been located as a north–south oriented elongated belt; this is supported by the measured palaeocurrent directions within the composite crevasse splay sandstone. The composite crevasse splay sandstone is situated stratigraphically in the middle to upper part of the regressive sequence that forms the Gristhorpe Member, and it is located some metres (c. 10 m) above the thick (24 m) fluvial channel sandstone at Ravenscar.

The composite sandstone sequence in the lower part of the Saltwick Formation at Hayburn Wyke (Fig. 7) may be of crevasse splay origin where Fig. 7 probably represents a transverse section. The confined or channelized sandstones in the axial part grade laterally into more sheet-like sandstones and siltstones. The geometry of this composite sandstone sequence at Hayburn Wyke may be the result of similar crevasse splay depositional settings as proposed by Fielding (1984) (Fig. 9), where the main crevasse channel forms the axial part fringed by crevasse splay lobes. Hancock & Fisher (1981) have documented saline influence in the Saltwick Formation at Hayburn Wyke. Based on this and the upward coarsening nature at the base, where the sequence contains thin horizontal sandstone sheets, the composite crevasse splay complex is believed to have been formed by infilling of a shallow bay or lagoon that was situated close to the fluvial channel at Hayburn Wyke (Fig. 7). The fluvial channel at Hayburn Wyke may represent the feeder channel of the composite crevasse splay complex. This suggestion is supported by the occurrence of the fluvial channel sandstone and the composite crevasse splay complex in the same stratigraphic position and by the fact that the transverse section of the composite crevasse splay is perpendicular to the longitudinal section of the Hayburn Wyke fluvial channel.

The differences in sedimentary characteristics and architecture between the composite crevasse splay sandstones in the Saltwick Formation (at Hayburn Wyke, Fig. 7) and in the Gristhorpe Member (Fig. 5), respectively, may be attributed to the different position of the sections within the crevasse subdelta. The section through the composite crevasse splay sandstone at the base of the Saltwick Formation is probably close to the fluvial feeder channel (some hundred metres) and in this proximal position the crevasse splay sediments are relatively confined compared with what would be the case in a more medial and distal position within the crevasse subdelta (Fig. 9). Consequently, the section through the composite crevasse splay sandstone in the Gristhorpe Member at Cloughton Wyke may be located in a medial position within the crevasse subdelta where the lobes with sheet geometries are more prominent (Fig. 9).

The above-mentioned data and the considerations of crevasse splay sandstone dimensions and geometries suggest great variability in types and dimensions of crevasse splay sandstones. However, the quantitative data from the Ravenscar Group place constraints upon width/thickness and length/thickness ratios for single crevasse splay sandstone units.

It seems important to differentiate between two types of crevasse splay deposition: (i) small-scale crevasse splay lobes attached to the levee or alluvial ridge of the fluvial channel, and (ii) large-scale crevasse subdelta lobes that infill bays and lagoons.

The small-scale crevasse splay lobes of the modern Mississippi delta are illustrated in Fig. 8B, where in the lower delta plain area they contribute to levee development and do not cut deep into the levee (Elliott, 1974). This type of crevasse splay lobe is also common on the modern alluvial plain in east-central Saskatchewan, Canada (Fig. 8B). Most of the quantitative crevasse splay sandstone data collected in Yorkshire reflect geometries of such small-scale crevasse splay lobes. The large-scale crevasse subdelta lobes form thick sandstones (up to 15 m) and lengths and widths are commonly larger than 10 km. The composite crevasse splay sandstones in the Ravenscar Group may have formed by crevasse subdelta deposition into embayments. The crevasse subdeltas seem to be more influenced by a channelized crevasse splay pattern than the small-scale crevasse splay lobes.

ROCK VOLUME ESTIMATES

The volumes of crevasse splay sandstones encountered in wells can be estimated from their thickness if their shape, thickness variations and width/thickness ratios are known. Volumes of crevasse splay sandstones within the Ravenscar Group can roughly be estimated on the basis of the determined thickness and width/thickness ratios as follows:

$$V = 0.25\pi a^2 t^3 + (b-1)a^2 t^3$$

where V is the crevasse splay sandstone volume, a the width/thickness ratio, b the length/width ratio, and t the sandstone thickness. The equation is valid if the thickness of a crevasse splay is relatively constant as is the case for the lobe shaped crevasse

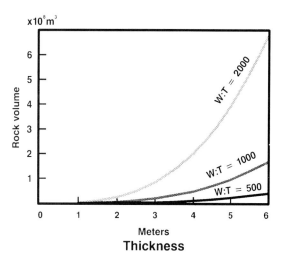

Fig. 10. Relationship between volume, thickness and width/thickness ratio for crevasse splay sandstones. The curves are valid for lobe-shaped crevasse splay sandstones whose shape can be approximated by a circle and which have a relatively constant thickness. The curves have been constructed from the equation given in the text, and for the case where $b = 1$.

splay sandstones in the Ravenscar Group. Figure 10 shows the rock volume estimate of crevasse splay sandstones with a lobe shaped geometry ($n = 1$). The volume is strongly dependent on thickness, a doubling of the thickness causing an eightfold increase in volume. The width/thickness ratio also strongly influences the result; a doubling of this ratio will give rise to a fourfold increase in volume.

The volume considerations imply that a 2.5 m thick crevasse splay sandstone has the same volume as 125 thin (0.5 m thick) crevasse splay sandstones. The composite crevasse splay sandstones have large volumes (more than 500 million m^3 in the Gristhorpe Member) due to their significant thickness and large width/thickness ratio (3000–4000?).

RELATIONS BETWEEN CREVASSE SPLAYS AND FLUVIAL CHANNELS

Crevasse splay sandstones in the Ravenscar Group can in several cases be seen to be connected to channel sandstones deposited in the fluvial channel which fed the crevasse splay. Adjacent to the fluvial channel sandstones, sections through the crevasse splay sandstones are often lenticular whereas the sheet-like shape is more apparent at some distance (a few hundred metres) from the channel. In some cases, the crevasse splay sandstone has apparently

Fig. 11. Part of a cliff section from the Saltwick Formation (at High Whitby) displaying a channel sandstone margin and an adjacent sequence with crevasse splay sandstones interbedded in overbank mud and siltstones. A channel sandstone wing comprising crevasse splay sediments is genetically connected to the upper part of the channel sandstone. These crevasse splay sediments consist of lensoid crevasse splay sandstones encased in mud and siltstones. The lensoid sandstone geometries have probably been formed within proximal crevasse channels which extended out from the adjacent fluvial feeder channel. The overbank sequence beneath the channel wing has an erosional and discordant contact to the adjacent channel sandstone. At the base of this overbank sequence a lensoid sandstone (width 25 m and thickness 2.5 m) may represent a transverse section of a proximal crevasse splay channel sandstone (see arrow). In particular, the sheet geometry of the crevasse splay sandstones is prominent within the overbank sequence on top of the fluvial channel sandstone. The cliff section is about 75–80 m high.

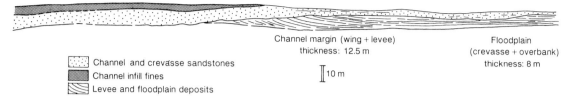

Fig. 12. Channel sandstone from the upper part of the Saltwick Formation (at Widdy Head) and its associated levee and floodplain deposits. The channel sandstone has a channel wing comprising crevasse splay sandstones located on top of the levee and floodplain sequence.

no contact with the nearby fluvial channel sandstone; the crevasse splay sandstone pinches into mud and siltstones adjacent to the channel and at a stratigraphical level near the top of the channel. This is probably a cross-sectional effect, and the crevasse splay sandstone is most likely in contact with the fluvial channel sandstone through a crevasse channel sandstone with a low width/thickness ratio (10–20). Based on the geometrical data (Fig. 4) the width of the crevasse channel sandstone connecting the crevasse splay lobe to the fluvial channel is estimated to be 1/10 to 1/100 of the maximum width of the crevasse splay. This estimate is based on the data base which also indicates that the thickness of the proximal channelized part of the crevasse splay is not more than two times the thickness of the central parts of the lobe. The low ratio between maximum crevasse channel width and crevasse splay lobe width implies low probability for a section through a crevasse splay including the crevasse channel contact with the fluvial channel. Although the crevasse splay sandstones are commonly in contact with fluvial channel sandstones, crevasse splay sandstones can also be seen to be solely in contact with the finer grained fluvial channel fill. Similar contact features are also mentioned by Galloway & Hobday (1983).

The thickest crevasse splay sandstones can commonly be correlated towards the top of the fluvial channel sandstone and thus form a channel sandstone wing (Figs 11 & 12). Similar wing features have been described from the Ebro Basin in Spain (Friend et al., 1979). In the Saltwick Formation, the overbank fines (including relatively thin crevasse splay sandstones) located below the crevasse sandstone wing may have an erosional lateral contact to the adjacent channel sandstone (Fig. 11). Moreover, the sequence of overbank sediments displays no dip away from the channel, indicating that a typical levee facies is absent adjacent to the channel.

Hence, the overbank sequence below the crevasse splay wing may not be genetically related to the adjacent fluvial channel sandstone. Based on this and indications of channel incision within the channel sandstone (Mjøs & Prestholm, in prep.), it can be argued that the fluvial channel was probably formed by relatively deep (up to 8–10 m) channel incision into pre-existing deposits (Mjøs & Prestholm, in prep.). In other cases, however, a levee facies has been recognized adjacent to the fluvial channel sandstone (Fig. 12). In Fig. 12, the levee thickness is of similar magnitude to the fluvial channel sandstone thickness. The crevasse sandstone wing overlies the levee and due to lateral channel erosion parts of the levee underlie the fluvial channel sandstone. The relation between overbank (levee and floodplain) fines and channel sandstone, shown in Fig. 12, indicates that the channel was mainly formed by levee accretion and did not incise deeply into the pre-existing deposits. In Fig. 12, the levee and crevasse sandstone wing have formed a channel margin ridge, 12.5 m thick, which thins to 8 m thickness 140 m away from the channel.

Crevasse splay sandstones are encased in mud and siltstones and erosional contacts between different crevasse splay sandstones were not observed except within composite crevasse splay sandstones (see below). However, crevasse splay sandstones were in several cases seen to be cut by younger fluvial channels.

APPLICATION OF THE DATA IN RESERVOIR GEOLOGY

The data base from Yorkshire may have several implications for the correlation of crevasse splay sandstones encountered in wells. The relevance of the data for individual subsurface reservoirs will, however, have to be evaluated in each specific case. The data base has been considered to be helpful for the description of Ness Formation reservoirs in some specific areas of the northern North Sea (Mjøs & Walderhaug, 1989).

Assuming that the data base has general relevance, the lateral extent of single crevasse splay sandstones can to a certain extent be predicted from the thickness of the crevasse splay sandstone. Since the crevasse splay sandstones seem to have length/thickness ratios lower than 2000 and since the thicknesses of the crevasse splay sandstones are relatively constant throughout most of the crevasse splay lobe, the lateral extent of a drilled crevasse splay sandstone in an imaginary well would usually be less than 2000 times the thickness of the drilled crevasse splay sandstone. However, if the crevasse splay sandstone was penetrated in the most distal part where the sandstone thins rapidly, the lateral extent could be more than 2000 times the thickness. Similarly, if the sandstone was penetrated in the proximal channelized part where the sandstone thickens significantly, the lateral extent would be less than 2000 times the thickness.

The data base indicates that there is a wide range in the size of single crevasse splay lobes, and lateral extents of 2000 times the thickness should be regarded as a maximum value. A correlation between wells therefore cannot depend on all or most of the drilled crevasse splay sandstones having lateral extents close to the maximum values encountered.

Since the data indicate only minor variation in thickness, grain size and sedimentary structures within the individual crevasse splay sandstones, it is difficult to define the well position on the crevasse splay lobe. However, the proximal channelized part of a crevasse splay sandstone may be recognized by relatively large thickness, decimetre-scale high angle cross-bedding, an erosive base and relatively mud-poor sandstones.

In simple volume calculations the thickness of crevasse splays can be set to a constant value and the geometry of the splays is commonly lobate (length between one and two times the width). Single crevasse splay sandstones are commonly less than 2.5 m thick; thicker crevasse splay sandstones are most likely composite.

When considering interconnectedness between sandbodies, single crevasse splay sandstones should not be in contact with each other; they are normally encased in mud and siltstones. Some crevasse splay sandstones (about one in five?) are cut by younger adjacent fluvial channel sandstones and most crevasse splay sandstones are probably connected to fluvial sandstones in their feeder channel.

The presence of crevasse splay sandstones places some constraints on the location of fluvial channel sandstones. A crevasse splay sandstone encountered in a well must be derived from a channel located closer than 2000 times the thickness of the drilled crevasse splay sandstone if the length/thickness ratios for the Ravenscar Group are applicable. In most cases, crevasse splay sandstones should be correlated towards the top or upper part of the fluvial channel sandstone. The contact between the fluvial channel sandstone and the crevasse splay sandstone lobe is commonly through an elongate crevasse channel sandstone and the contact area is determined by the width/thickness ratio (10–20) of the most proximal part of the crevasse channel sandstone.

Single crevasse splay sandstones within a composite crevasse splay usually exhibit a high degree of interconnectedness; in particular, amalgamation of the single crevasse sandstones is pronounced in the lower part of the composite crevasse splay sequence. Great care should be taken when correlating a single crevasse splay sandstone within the composite crevasse splay sequence between wells since the crevasse splay width/thickness ratios (Fig. 4) also place constraints on the dimension of the individual sandstones within a composite crevasse splay sequence.

Composite crevasse splay sandstone levels may have great lateral extent (at least 20 km in the Gristhorpe Member) and should show a high degree of correlation between wells. Possibly, coalescing crevasse subdeltas may form a composite crevasse splay sandstone level that could be correlated between wells in larger areas than expected by deposition from a single crevasse subdelta. However, the composite crevasse splay sandstone level is assumed to be heterogeneous due to the many crevasse splay sandstone building blocks.

SUMMARY AND CONCLUSIONS

1 The geometrical data collected in Yorkshire place constraints on the shape and dimensions of single crevasse splay sandstones.
2 Single crevasse splay sandstones may build a composite crevasse splay sequence with great lateral extent.

3 Most crevasse splay sandstone lobes are probably in contact with the fluvial feeder channel through an elongate crevasse splay channel. The thickest crevasse splay sandstones are commonly connected to the top part of a fluvial channel sandstone.
4 Crevasse splay sandstones have width/thickness and length/thickness ratios within a certain range and are connected to fluvial channels. Hence the presence of crevasse splay sandstones within the overbank sequences may place some constraints on the location of fluvial channel sandstones.
5 Application of the presented quantitative geometrical data in subsurface reservoirs may improve reservoir description and contribute to enhanced reservoir performance.

ACKNOWLEDGEMENTS

This paper is based on results from a project funded by the Norwegian State Program for Improved Oil Recovery (SPOR). We are grateful to Per Arne Bjørkum for assistance in the field and for geological support throughout the study. The manuscript benefited from reviews by R. Eschard and R.W.O. Knox.

REFERENCES

ALLEN, J.R.L. (1964) Studies in fluviatile sedimentation: six cyclothems from the Lower Red Sandstone, Anglo-Welsh Basin. *Sedimentology* **3**, 163–198.

BAGANZ, B.P., HORNE, J.C. & FERM, J.C. (1975) Carboniferous and recent Mississippi lower delta plains: a comparison. *Trans. Gulf Coast Ass. Geol. Soc.* **25**, 183–191.

BEHRENSMEYER, A.K. (1989) Overbank facies and paleosols of the Chinji Formation, Northern Pakistan. In: *Programme and Abstracts, 4th Int. Conf. on Fluvial Sedimentology, Barcelona, 2–4 October 1989*, p. 72. Servei Geologic de Catalunya, Barcelona.

BOWN, T.M. & KRAUS, M.J. (1987). Integration of channel and floodplain suites, I. Developmental sequence and lateral relations of alluvial paleosols. *J. Sedim. Petrol.* **57**, 587–601.

BRIDGE, J.S. (1984) Large-scale facies sequences in alluvial overbank environments. *J. Sedim. Petrol.* **54**, 583–588.

COLEMAN, J.M. (1969). Brahmaputra River; channel processes and sedimentation. *Sedim. Geol.* **3**, 129–239.

COLEMAN, J.M. (1976) *Deltas: Processes of Deposition, Models for Exploration*. Continuing Education Publishing Co., Champaign, 102 pp.

COLEMAN, J.M. & GAGLIANO, S.M. (1964). Cyclic sedimentation in the Mississippi River deltaic plain. *Trans. Gulf Coast Ass. Geol. Soc.* **14**, 67–80.

DREYER, T. (1990) Sandbody dimensions and infill sequences of stable, humid climate delta plain channels. In: *North Sea Oil and Gas Reservoirs — II* (Eds Buller, A.T., Berg, E. & Hjelmeland, O.) pp. 337–352. Graham and Trotman, London.

ELLIOTT, T. (1974) Interdistributary bay sequences and their genesis. *Sedimentology* **21**, 611–622.

ELLIOTT, T. (1986) Deltas. In: *Sedimentary Environments and Facies* (Ed. Reading, H.G.) pp.113–154. Blackwell Scientific Publications, Oxford, 615 pp.

FARRELL, K.M. (1987) Sedimentology and facies architecture of overbank deposits of the Mississippi River, False River Region, Louisiana. In: *Recent Developments in Fluvial Sedimentology* (Eds Ethridge, F.G., Flores, R.M. & Harvey, M.D.) pp. 111–120. Soc. Econ. Paleont. Miner., Tulsa, Spec. Publ. 39.

FIELDING, C.R. (1984) A coal depositional model for the Durham Coal measures of NE England. *J. Geol. Soc. Lond.* **141**, 919–931.

FIELDING, C.R. (1985) Coal depositional models and the distinction between alluvial and delta plain environments. *Sedim. Geol.* **42**, 41–48.

Fielding, C.R. (1986) Fluvial channel and overbank deposits from the Westphalian of the Durham Coalfield, NE England. *Sedimentology* **23**, 119–140.

FISHER, M.J. & HANCOCK, N.J. (1985) The Scalby Formation (Middle Jurassic, Ravenscar Group) of Yorkshire: reassessment of age and depositional environment. *Proc. Yorks. Geol. Soc.* **45**, 293–298.

FLORES, R.M. (1984) Comparative analysis of coal accumulation in Cretaceous alluvial deposits, Southern United States Rocky Mountain basins. In: *The Mesozoic of Middle North America*. (Eds Stott, D.P. & Colass, D.J.) pp. 373–385. Can. Soc. Petrol. Geol., Calgary, Memoir 9.

FLORES, R.M. (1985) Coal deposits in Cretaceous and Tertiary fluvial systems of the Rocky Mountain Region. In: *Recognition of Fluvial Depositional Systems and Their Resource Potential*. (Eds Flores, R.M., Ethridge, F.G., Miall, A.D., Galloway, W.E. & Fouch, T.D.) pp. 167–216. SEPM Short Course 19. Soc. Econ. Palaeont. Miner., Tulsa.

FRIEND, P.F., SLATER, M.J. & WILLIAMS, R.C. (1979) Vertical and lateral building of river sandstone bodies, Ebro Basin, Spain. *J. Geol. Soc. Lond.* **136**, 39–46.

GALLOWAY, W.E. & HOBDAY, D.K. (1983) *Terrigenous Clastic Depositional Systems, Application to Petroleum, Coal and Uranium Exploration*. Springer Verlag, New York, 423 pp.

GUION, P.D. (1984) Crevasse splay deposits and roof rock quality in the Threequarters Seam (Carboniferous) of the East Midlands Coalfield, UK. In: *Sedimentology of Coal and Coal-bearing Sequences* (Eds Rahmani, R.A. & Flores, R.M.) pp. 191–308. Spec. Publs Int. Ass. Sediment. 7. Blackwell Scientific Publications, Oxford.

HANCOCK, W.J. & FISHER, M.J. (1981) Middle Jurassic North Sea deltas with particular reference to Yorkshire. In: *Petroleum Geology of the Continental Shelf of North-west Europe* (Eds Illing, L.W. & Hobson, G.D.) pp. 186–195. Institute of Petroleum, London.

HASZELDINE, R.S. (1989) Coal review: depositional controls, modern analogues and ancient climates. In: *Deltas, Sites and Traps for Fossil Fuels* (Eds Whateley, M.K.G. & Pickering, K.T.) pp. 289–308. Geol. Soc., Lond., Spec. Publ. 41. Blackwell Scientific Publications, Oxford.

Hemingway, J.E. & Knox, R.W.O. (1973) Lithostratigraphical nomenclature of the Middle Jurassic strata of the Yorkshire basin of north-east England. *Proc. Yorks. Geol. Soc.* **39**, 161–224.

Kraus, M.J. & Bown, T.M. (1986) Paleosols and time resolution in alluvial stratigraphy. In: *Paleosols, Their Recognition and Interpretation* (Ed. Wright, V.P.) pp. 180–207. Princeton University Press, Princeton.

Livera, S.E. & Leeder, M.R. (1981) The Middle Jurassic Ravenscar Group (Deltaic Series) of Yorkshire: recent sedimentological studies as demonstrated during a field meeting, 2–3 May 1980. *Proc. Geol. Assoc.* **92**, 241–250.

Mjøs, R. & Prestholm, E. (in prep.) The geometry and organization of fluvio-deltaic channel sandstones, in the Jurassic Saltwick Formation, Yorkshire, England.

Mjøs, R. & Walderhaug, O. (1989) Sandstone geometry in fluvio-deltaic sediments of the Ravenscar Group, Yorkshire. State Program for Improved Oil and Gas Recovery (SPOR) Report (Open), 6/89, Rogaland Research Institute, Norway.

O'Brian, P.E. & Wells, A.T. (1986) A small, alluvial crevasse splay. *J. Sedim. Petrol.* **56**, 876–879.

Ravenne, C., Eschard, R., Galli, A., Mathieu, Y., Montadert, L. & Rudkiewicz, J.L. (1987) Heterogeneities and geometry of sedimentary bodies in a fluvial-deltaic reservoir. *Soc. Petrol. Engrs,* **16752**, 115–119.

Saxena, R.S. (1978) Exploration models and criteria for subsurface recognition of crevasse splay and reworked deltaic sandstone reservoirs (abstr.). *Am. Assoc. Petrol. Geol. Bull.* **62**, 560.

Smith, D.G. (1983) Anastomosed fluvial deposits: modern examples from Western Canada. In: *Modern and Ancient Fluvial Systems* (Eds Collinson, J.D. & Lewin, J.) pp. 155–168. Spec. Publs Int. Ass. Sediment. 6. Blackwell Scientific Publications, Oxford.

Smith, N.D., Cross, T.A., Dufficy, J.P. & Clough, S.R. (1989) Anatomy of an avulsion. *Sedimentology* **36**, 1–23.

Grain-size distribution of overbank sediment and its use to locate channel positions

M.J. GUCCIONE

Geology Department, Ozark Hall–#118, University of Arkansas, Fayetteville, Arkansas 72701, USA

ABSTRACT

Studies of three rivers show that overbank alluvium decreases in grain size with distance from a channel source. In a first approximation of this relationship, sand and silt fractions best illustrate the decrease in grain size of overbank sediment derived from small streams where it is deposited less than 100 m from the source. In contrast, silt and clay fractions best illustrate this decrease in grain size for a large river (the Mississippi) where the sediment is deposited at distances greater than 1 km from the source. However, the clay fraction can only be used if the accretion rate is sufficiently rapid for pedogenesis not to affect the clay content. For both small and large streams with overbank sediment from a source less than 10 km distant, the fine silt fraction best illustrates the decrease in grain size with distance.

Overbank alluvium may be used to determine the position of the channel which served as a source of that sediment. The sediment grain size and the grain size–distance plot can be used to determine the distance between a sample site and a source channel, assuming a simplistic diffusion model for sediment transport and deposition. Multiple sample sites may be used to locate the channel position.

INTRODUCTION

Valley floors of meandering rivers include both channels and flood plains. In the past, channels have been studied extensively because they are the most dynamic portion of the valley and the sediments deposited in them as point bars are the coarsest-grained and easiest to study (e.g. Allen, 1965, table IV). Though channels are extremely important parts of the valley floor, they may deposit volumetrically less sediment than is deposited on the flood plain (Boggs, 1987, figs 11.12 & 11.3). A more constant record of alluvial deposition in meandering streams is the overbank sediment deposited by vertical accretion beyond the meander belt. Because of the more complete sedimentation record of overbank sediment in comparison to channel deposits, the focus of this study is the overbank sediment that underlies the flood plain. These overbank deposits are sediment that is deposited beyond the channel by non-channelized flow. This includes natural levee and backswamp or flood-basin deposits. It does not include crevasse splay deposits, which were identified by their relatively coarse grain size and their laminated or bedded nature. Nor does it include oxbow lake or channel-fill deposits within the meander belt.

The approach for examining overbank sediment in this study is similar to the approach used for examining loess. In the field both types of sediment appear to be fine-grained, homogeneous deposits. It can be difficult to identify homogeneous silty deposits as overbank alluvium or loess where the deposit underlies a flood plain or terrace (West & Rutledge, 1987). The reason for this difficulty is the similarity in their mechanism of deposition. Both loess and overbank sediment are deposited from suspension in fluids as the velocity of that fluid decreases with distance from the source. For loess the fluid is air whilst for overbank sediment that fluid is water. Quantitative research has been conducted on loess and curves have been published showing a decrease in grain size with a logarithmic decrease in distance (Smith, 1942; West *et al.*, 1980; Rutledge *et al.*, 1985).

Similar quantitative information on grain size as a function of distance is not available for overbank sediment, though qualitatively the same relation-

Table 1. Discharge and size of the White, Buffalo and Mississippi Rivers near the study areas

	Mean annual discharge ($m^3 s^{-1}$)	Maximum discharge ($m^3 s^{-1}$)	Channel width (m)	Flood plain width (km)	Valley width (km)	Meander belt width (km)
Buffalo River[a]	29	4 474	20	0.2	0.3	NA
White River[b]	15	2 311	30	0.6	2.1	NA
Mississippi River[c]	13 430	56 076	1600	59.0	88.5	33

[a] Discharge measurements from a US Geological Survey gauging station on the Buffalo River at St Joe, Arkansas (Lamb *et al.*, 1986). This station is 50 km down valley from Erbie. The discharge is much less at Erbie than the values reported.
[b] Discharge measurements from a US Geological Survey gauging station on the White River at Fayetteville, Arkansas (Lamb *et al.*, 1986). This station is 1.9 km upstream from the study area. The discharge should be the same as the values reported.
[c] Discharge measurements from a US Geological Survey gauging station on the Mississippi River at Memphis, Tennessee (Lamb *et al.*, 1983). This station is 92 km down valley from Blytheville. The discharge at Blytheville should be less than the values reported.
NA, Not applicable.

ship has been noted many times (e.g. Allen, 1965; Brown, 1987). The purpose of this study is to quantify the grain size–distance relationship of overbank sediment that has vertically accreted upon a flood plain. Results from three different study areas are compared to examine streams with several orders of magnitude difference in size, measured in terms of discharge and channel, flood-plain and valley widths (Table 1). A simple model using the grain size–log distance curve developed in this study is proposed to determine the distance and location of a channel from multiple sample sites.

STUDY AREAS

The three streams examined in this study are located in two physiographic provinces in the south-central USA (Fig. 1). The White River and its tributary the Buffalo River are relatively small streams that are located in the Ozark Plateaus Province (Thornbury, 1965). Study areas on both rivers are along an escarpment that separates a lower plateau in the province (the Springfield Plateau) from a higher plateau (the Boston Mountain Plateau). The third river, the Mississippi River, is considerably larger and is the major river draining the interior of the North American continent. Both the White and the Buffalo Rivers are tributaries of the Mississippi River. The study area along the Mississippi River is located in the Mississippi alluvial valley (Thornbury, 1965).

Morphological, sedimentological and hydrological characteristics of these rivers are quite different. Both the White and Buffalo Rivers are entrenched meandering streams. The Buffalo River valley at

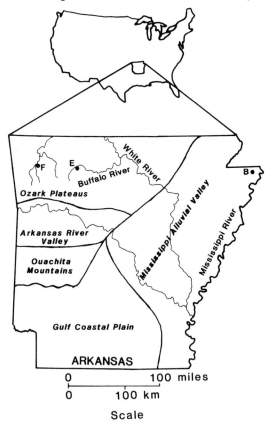

Fig. 1. Location of study areas and physiographic provinces. B, Blytheville; E, Erbie; F, Fayetteville.

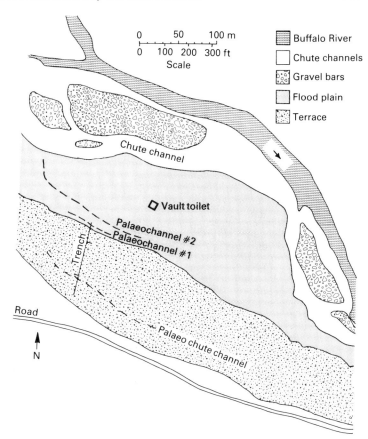

Fig. 2. Location map of study area at Erbie, Arkansas along the Buffalo River. Vertical samples were taken from four positions in the trench. Palaeochannel #1 is interpreted to be the source of lower overbank sediment and palaeochannel #2 is interpreted to be the source of upper overbank sediment beneath the terrace.

Erbie, Arkansas is comprised of a meandering channel with numerous gravel bars, a flood plain, one terrace and a colluvial fan (Fig. 2) (Guccione, 1988). Both the valley walls and valley floor are Palaeozoic bedrock (Glick, 1976). Sediment load includes clay- to gravel-sized carbonate and clastic material. The White River valley near Fayetteville, Arkansas also comprises a meandering channel with sand and gravel point bars and a flood plain, but has at least three terraces (Fig. 3) (Guccione & Rieper, 1988). Both valley walls and valley floor are Late Palaeozoic bedrock (Glick, 1976). Sediment load of the White River includes clay- to gravel-sized clastic debris.

In contrast, the Mississippi River near Blytheville, Arkansas, is a meandering stream in a wide alluvial valley. Its valley comprises a meandering channel with sandy point bars near the eastern margin of the meander belt, a flood plain along the western edge of the meander belt and multiple braided stream terraces along the western side of the valley (Fig. 4) (Saucier, 1974; Guccione, 1987).

The valley wall to the east is Palaeozoic bedrock overlain by Pleistocene loess and to the west is Tertiary and Pleistocene sediment that forms Crowley's Ridge (Haley, 1976). Pleistocene alluvium makes up the valley floor. Sediment load of the Mississippi River consists of clay- to sand-sized clastic material.

METHODS

Flood-plain sediment underlying the lowest terrace and/or the modern flood plain was examined, described and sampled at the three study areas (Fig. 1). Samples were taken from trenches in the Buffalo and White River valleys and from cores and an outcrop at the Mississippi River valley.

A total of 43 samples were analysed for grain size. Two to nine vertical samples from each flood-plain unit were analysed at each site and one to four sites were examined for each channel position. The means of all samples from each unit at each site

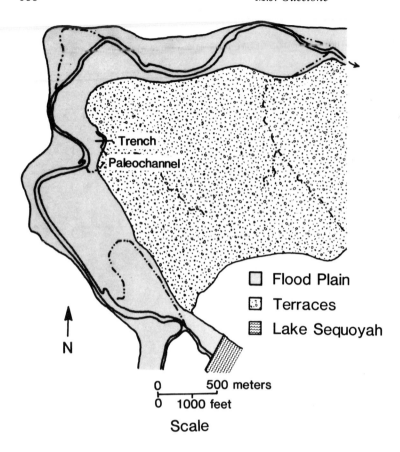

Fig. 3. Location map of the study area at Fayetteville, Arkansas along the White River. Vertical samples were taken from three positions in the trench. The palaeochannel is interpreted to be the source of overbank sediment beneath the terrace. The present channel of the White River is interpreted to be the source of overbank sediment beneath the flood plain.

were used. Samples were mechanically disaggregated with a mortar and pestle and then a water–sample slurry was agitated in a 'malt mixer'. Grain size analysis was by dry sieving the sand fractions and by pipette methods for the silt and clay fractions (Table 2) (Day, 1965). Results of the size analysis were calculated in different ways at each site. At the Buffalo and White River valleys the flood-plain sediment is relatively coarse textured, thin and highly weathered. For these samples the percentage of the sand and silt fractions was calculated on a clay-free basis to eliminate the effect of pedogenesis (West et al., 1980). In the Mississippi River valley the flood-plain sediment is fine textured and the mean sedimentation rate is so rapid, 0.06 cm yr^{-1} (Guccione, 1987), that the effect of pedogenesis is absent or minimal. Because of the dominance of silt and clay fractions, the percentage of silt and clay was calculated on a sand-free basis. In order to compare these data with those from the Buffalo and White River valleys, the Mississippi River valley data were also calculated on the clay-free basis.

RESULTS

Flood-plain deposits

Overbank sediment deposited on a flood plain was identified in the field using standard criteria (Allen, 1965; Brown, 1987). It was distinguished from channel deposits and colluvium by the general lack of gravel, the lateral fining of texture with increasing distance from the channel and the general lack of bedding, except proximal to the channel. At the natural levee, adjacent to the channel, the sediment is intermediate in texture compared to the underlying channel deposit and the lateral equivalent distal overbank deposit. It may be laminated and the ground surface is slightly higher than the surrounding flood plain. Here the deposit is relatively well

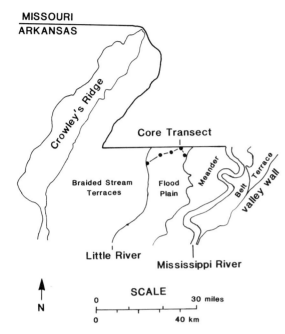

Fig. 4. Location map of the study site near Blytheville, Arkansas along the Mississippi River. Vertical samples were taken from each core and outcrop along a transect of the flood plain. The west edge of the meander belt is assumed to be the source of overbank sediment beneath the flood plain. Only samples from the eastern three sample sites were used for the grain size–distance plot because the Little River is an additional source for sediment in the western sample sites.

Table 2. Size fractions utilized in this study

Fraction	Diameter (mm)
Gravel	> 2.0
Sand	
Very coarse	2.0–1.0
Coarse	1.0–0.5
Medium	0.5–0.25
Fine	0.25–0.125
Very fine	0.125–0.0625
Silt	
Coarse	0.0625–0.016
Medium	0.016–0.004
Fine	0.004–0.002
Clay	< 0.002

drained and may have colours indicating oxidation. With increasing distance from the channel the sediment becomes finer textured and the ground surface is slightly lower. Here the deposit is poorly drained, has reduced or gleyed colours, and may be mottled.

Because overbank deposits are fine-grained it was difficult or impossible to identify multiple units derived from different channel positions in the field unless separated by a buried soil. This assessment was best done by taking multiple samples from a vertical exposure and analysing them for texture in the laboratory. If these samples had a uniform texture, it was concluded that there was only one depositional unit derived from a channel that had maintained a stable position. The grain size of overbank sediment from a stable source can be used to determine the grain size–distance relationship. The Buffalo River near Rush, Arkansas is an example of this situation (Fig. 5).

If the samples are layered with homogeneous textures within the layers and abrupt textural changes between layers, it was interpreted that the channel had avulsed and changed positions dramatically and quickly. The grain size of each of these deposits can be used to determine each grain size–distance relationship. The Buffalo River near Erbie, Arkansas, is an example of this type of situation (Fig. 5). Because of the logarithmic nature of the grain size–distance relationship, changes in channel position at Erbie are best represented in overbank sediment proximal to the channel source. It is difficult to determine if a channel position has been stable, migrating or avulsing at a distant sample site where small changes in texture record large differences in channel position.

Finally, if the samples had a gradually changing texture, it was interpreted that the source or channel had migrated. This type of overbank sediment was not used to determine the grain size–distance relationship for this study. The Buffalo River near Rush, Arkansas, is an example of this situation (Fig. 5).

Channel location

To quantify the grain-size distribution of overbank sediment, the source of that sediment or the channel position must be known. For flood-plain sites, the position of the channel is assumed to be the present channel for the youngest overbank deposit. In the White River valley there is only one overbank deposit at the study site and the present channel is assumed to be the source of the sediment (Fig. 3).

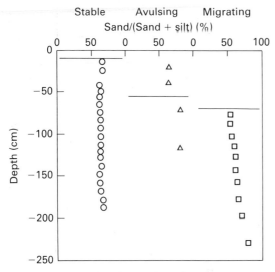

Fig. 5. Examples of vertical profiles of overbank sediment from stable, avulsing and migrating channels. A stable channel is the source of overbank #2 from Terrace 1 at the Buffalo River near Rush, Arkansas, approximately 70 km down valley from Erbie, Arkansas. It is overlain by 8 cm of a 1982 flood deposit. An avulsing channel is the source of overbank #1 and #2 from Terrace 1 at the Buffalo River near Erbie, Arkansas. The boundary between these deposits at 55 cm depth is not marked by a buried soil, indicating relatively rapid avulsion. A migrating channel is the source for overbank #1 from Terrace 1 at the Buffalo River near Rush, Arkansas. A buried soil is developed in the upper part of this deposit and clearly separates it from an overlying deposit, overbank #2.

The flood plain of the Mississippi River is larger and more complex than that of smaller bedrock-incised streams. For the Mississippi overbank sediment the channel position was assumed to be the western edge of the meander belt (Fig. 4). Abundant meander channel scars in the meander belt lead to the certain conclusion that this is a simplistic assumption. The source area has been at a multitude of locations within the meander belt during the Holocene and therefore at a variety of distances, nearly all greater than this assumed position. All overbank sediment used in this study was from the uppermost few metres of the deposit. This sediment is probably derived from the more recent meander channels on the west side of the meander belt, but probably not at the west edge. Results from this study are merely a first approximation of the grain size–distance relationship. Probably the slope of the particle size composition versus distance reported in this research is greater than the true slope.

For terrace sites in the White and Buffalo River study areas, the initial palaeochannel was assumed to be on the flood plain immediately adjacent to the terrace scarp (Figs 2 & 3). The scarp at both sites is relatively steep and was the cut bank of a palaeochannel. At Erbie in the Buffalo River valley, where two overbank deposits are present (Fig. 5), the older unit (below 55 cm depth) is assumed to be derived from a palaeochannel along the terrace scarp. There is an abrupt change in texture at 55 cm depth and the absence of a buried soil at the boundary. The upper unit is finer-grained than the underlying overbank deposit. These textural relationships indicate that the upper sediment must be from a source at a greater distance than the terrace scarp compared to the lower sediment and that the channel position must have changed rather abruptly or avulsed to another position on the flood plain. The most likely position for the younger channel is a channel swale on the flood plain that is nearest to the terrace scarp (Fig. 2).

Grain size–distance relationships

Small streams

Overbank sediment along the White and Buffalo Rivers becomes finer-grained with distance (20–100 m) from its source (Fig. 6A). Sand is an abundant size fraction in overbank sediment that underlies the flood plain and terraces of the White and Buffalo Rivers. The amount of sand decreases with distance from the source. Conversely the amount of total silt and all three subfractions of silt increases with distance (Fig. 6B–D). The clay fraction also increases with distance but it cannot be evaluated because its original abundance is modified by pedogenesis.

Large stream

Overbank sediment underlying the flood plain of the Mississippi River also becomes finer-grained with distance from its source (Fig. 7A). However, because the distance from the source is three orders of magnitude greater than that of the small streams, sand is not a major component of this sediment and only silt and clay are used to illustrate this grain size decrease. Unlike the overbank sediment of small streams, the total silt fraction of the Mississippi River overbank sediment decreases with distance from the source because it is the relatively coarse

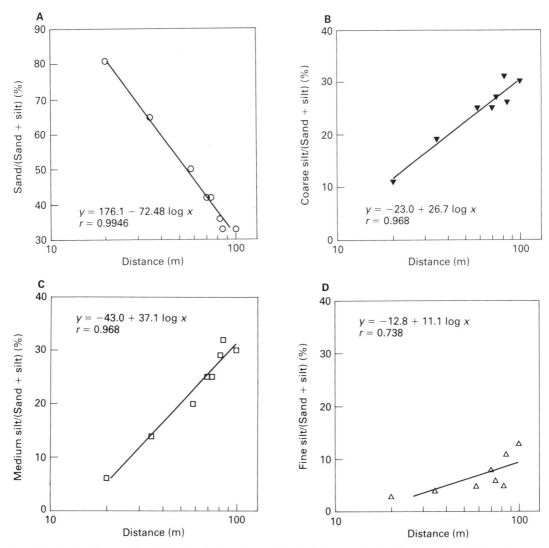

Fig. 6. Grain size–distance plot of the clay-free White and Buffalo River overbank sediment using (A) the sand fraction, (B) the coarse silt fraction, (C) the medium silt fraction and (D) the fine silt fraction.

fraction in this deposit. Within the total silt fraction, the coarse silt fraction is not very abundant and is approximately uniform with distance (Fig. 7B). Medium silt is the most abundant size fraction and it also decreases with distance. In contrast, the smallest silt fraction increases with distance, similar to the clay fraction (Fig. 7A).

All streams

Fine silt is the best size fraction to examine the grain size–distance relationship along a transect from 20 m to 10 km distant (Fig. 8). It consistently increases along this distance and its abundance is unaffected by pedogenesis. Within this distance range there is a problem with each of the other size fractions. The sand fraction decreases to a nominal amount at a short distance from its source (Fig. 6A). The coarse and medium silt fractions increase in relative abundance a short distance from the source (Fig. 6B,C) but remain constant or decrease in abundance with increasing distance from the source (Fig. 7B). The clay fraction can be affected by pedogenesis and it should only be used where the

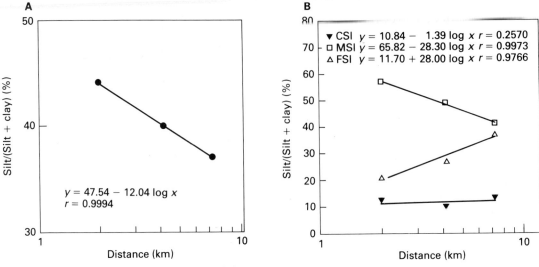

Fig. 7. Grain size–distance plot of the sand-free Mississippi River overbank sediment (Blytheville) using (A) the total silt fraction and (B) the coarse, medium and fine silt fractions.

accumulation rates are fast enough that its abundance is not affected.

Locating unknown channel positions

Assuming that the grain size–distance plots developed in this study are widely applicable or that suitable plots can be developed for an area under consideration, this relationship can be used to determine the location of unknown channel positions (Fig. 9). A single location of the channel can be determined by using two or more sample sites, if the sites are approximately perpendicular to the channel. The distance to the channel is determined using the grain size of overbank sediment at each sample site and the grain size–distance plot. Circles with radii equal to the distance between the site and the source can be drawn around each site. The point where the circles coincide is the channel location. If the sample sites are approximately parallel to the channel, two possible channel locations can be determined by using similar circles drawn around these sites. Because a stream is a linear source, a tangent drawn along all the circles will coincide with the channel position. Two tangents and therefore two channel positions are possible. A grid of sample sites that included sites approximately perpendicular and approximately parallel would allow a single channel position to be determined.

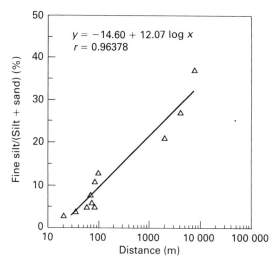

Fig. 8. Grain size–distance plot of the clay-free White, Buffalo and Mississippi River overbank sediment using the fine silt fraction.

APPLICATIONS

The applications of the grain size–distance curves for overbank sediment are numerous if the sediment can be disaggregated into its component parts and if diagenesis has not modified the original grain size of the sediment. Channel positions can be

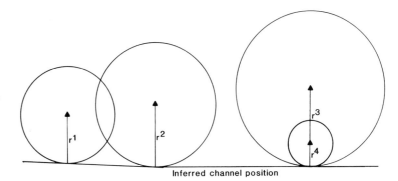

Fig. 9. Method of locating a channel position using grain size analyses of overbank sediment from multiple sample sites and the grain size–distance curve developed in this study. Triangles are the sample sites, r the radius of the circle around each sample site. The radius is equal to the distance of the site from the source. The inferred channel position is the only possible tangent of all the circles.

located using a sampling grid of a single overbank deposit where only a portion of the alluvial record is preserved or exposed, or where the lateral relationships are unclear, if there is a single channel source for the deposit. This method can be useful for petroleum, groundwater and archaeological studies. Petroleum and groundwater geologists are interested in the location of coarse channel deposits and the fluids that they may contain. Archaeologists are interested in determining the physical environment and geomorphology at the time of human occupation. By locating a shifting channel position through time, the evolution of a valley can be determined. This may also allow some inference about the mechanism of that evolution.

SUMMARY AND CONCLUSIONS

Grain size of overbank alluvium decreases with distance from a channel source. Sand and silt fractions are most useful to illustrate that relationship at short distances from the channel. Silt and clay fractions are most useful to illustrate that relationship at large distances from the channel, if accumulation rates are rapid and pedogenesis is negligible. Fine silt is the most useful size fraction to illustrate that relationship over a broad range of distances (0.02–10 km). The grain size of overbank deposits and the grain size–distance plot may be used to locate an unknown channel position.

ACKNOWLEDGEMENTS

Parts of this study were funded by the Memphis District, US Army Corps of Engineers, Contract No. DACW66-86-C-0034, the US National Park Service, Contract No. CX7029-6-0018, McClelland Consulting Engineers and the City of Fayetteville. Barbara Rieper, Michael Sierzchula and Keith Smith assisted in the field. Barbara Rieper also provided some of the size analyses. Robert H. Lafferty III made the study possible by integrating geomorphic studies with his archaeological work. C.S. Bristow and J. Best reviewed the manuscript and provided valuable comments.

REFERENCES

ALLEN, J.R.L. (1965) A review of the origin and characteristics of recent alluvial sediments. *Sedimentology* **5**, 89–191.

BOGGS, S. (1987) *Principles of Sedimentology and Stratigraphy.* Merrill, Ohio, 784 pp.

BROWN, A.G. (1987) Holocene floodplain sedimentation and channel response of the lower River Severn, United Kingdom. *Z. Geomorph. NF* **31**, 293–310.

DAY, P.R. (1965) Particle fractionation and particle size analysis. In: *Methods of Soil Analysis, Part I* (Ed. Black, C.A.) pp. 545–567. Agronomy Vol. 9. American Society of Agronomy, Inc. and Soil Science Society of America, Inc., Madison.

GLICK, E.E. (1976) Geologic map of Arkansas. In: *Geologic Map of Arkansas* (Ed. Haley, B.R.). Scale 1 : 500 000. Arkansas Geological Commission, Little Rock.

GUCCIONE, M.J. (1987) Geomorphology, sedimentation, and chronology of alluvial deposits, northern Mississippi County, Arkansas. In: *A Cultural Resources Survey Testing, and Geomorphic Examination of Ditches 10, 12, and 29, Mississippi County, Arkansas* (Lafferty III, R.H., Guccione, M.J., Scott, L.J., Aasen, D.K., Watkins, B.J., Sierzchula, M.C. & Baumann, P.F.) pp. 67–99. Mid-Continental Research Associates Report 86-5 to the Memphis District, US Army Corps of Engineers. MCRA, Lowell.

GUCCIONE, M.J. (1988) Late Quaternary history of the Buffalo River. In: *Tracks in Time: Archeology at the Elk Track Site (3NW205) and the Webb Branch Site (3NW206), Erbie Campground Project, Buffalo National*

River, Newton County, Arkansas (Lafferty III, R.H., Lopinot, N.H., Guccione, M.J., Santeford, L.G., Sierzchula, M.C., Scott, S.L., King, K.A., Hess, K.M. & Cummings, L.S.) pp. 27–43. Mid-Continental Research Associates Report 88-1 to the US National Park Service, Southwest Region. MCRA, Lowell.

Guccione, M.J. & Rieper, B. (1988) Late Quaternary history of the White River, Fayetteville, Arkansas. *The Compass* **65**, 199–206.

Haley, B.R. (Ed.) (1976) *Geologic Map of Arkansas.* Scale 1 : 500 000. Arkansas Geological Commission, Little Rock.

Lamb, T.E., Porter, J.E., Lambert, B.F. & Edds, J. (1983) Water resources data Arkansas water year 1983. US Geol. Surv. Water Data Report Ar-83-1.

Lamb, T.E., Porter, J.E., Lambert, B.F. & Edds, J. (1986) Water resources data Arkansas water year 1986. US Geol. Surv. Water Data Report AR-86-1.

Rutledge, E.M., West, L.T. & Omakupt, M. (1985) Loess deposits on a Pleistocene age terrace in Eastern Arkansas. *Soil Sci. Soc. Am. J.* **49**, 1231–1238.

Saucier, R.T. (1974) Quaternary geology of the Lower Mississippi Valley. *Arkansas Archeology Survey Research Series* **6**, 26.

Smith, G.D. (1942) Illinois loess: variations in its properties and distribution. *Univ. Illinois Agricultural Experiment Station Bull.* **490**, 139–184.

Thornbury, W.D. (1965) *Regional Geomorphology of the United States.* John Wiley & Sons, New York, 609 pp.

West, L.T. & Rutledge, E.M. (1987) Silty deposits of a low Pleistocene-age terrace in eastern Arkansas. *Soil Sci. Soc. Am. J.* **51**, 709–715.

West, L.T., Rutledge, E.M. & Barber, D.M. (1980) Sources and properties of loess deposits on Crowley's Ridge in Arkansas. *Soil Sci. Soc. Am. J.* **44**, 353–358.

Geometrical facies analysis of a mixed influence deltaic system: the Late Permian German Creek Formation, Bowen Basin, Australia

A.J. FALKNER and C.R. FIELDING

Department of Earth Sciences, University of Queensland, Queensland 4072, Australia

ABSTRACT

The Late Permian German Creek Formation outcrops along the western margin of the foreland Bowen Basin in central Queensland, Australia. The unit comprises a coarse clastic wedge enclosed below, above and to the south and east by rocks of interpreted marine shelf origin. This coarse clastic interval is interpreted to be of deltaic origin, sourced from the cratonic Australian landmass to the west, and prograding eastwards (transversely) into a marine basin. The upper part of the formation, which is extensively mined for coal, is the focus of this study.

Six facies have been identified: Facies A, thick and laterally extensive, channelized sandstone bodies, interpreted as deltaic distributary channel fills; Facies B, mostly siltstone with thin sandstone interbeds interpreted as interdistributary bay fills and tidal flat deposits; Facies C, sheet-like, mostly horizontally laminated, well sorted sandstones interpreted as foreshore deposits; Facies D, laterally extensive, tabular sandstone bodies internally dominated by hummocky cross-stratification interpreted as proximal mouth bar deposits; Facies E, heavily bioturbated, interbedded siltstone and sandstone with fining-upward graded units and occasional hummocky cross-stratification interpreted as distal mouth bar deposits; and Facies F, coals, interpreted as *in situ* mire deposits which accumulated on an extensive coastal plain. Palaeocurrent data from Facies A–C indicate bipolar east–west directed currents.

The facies assemblage is interpreted as the deposits of lower delta plain and delta front environments within a mixed influence (i.e. wave–tide–fluvial influenced) deltaic complex. The great lateral extent of the major coal seams within the German Creek Formation is difficult to explain in terms of purely sedimentary processes in the context of a lower delta plain, and may imply the role of external base level changes.

INTRODUCTION

There is an urgent need to obtain high quality data on sediment body geometry and internal heterogeneity from a broad range of sedimentary systems if our understanding of clastic depositional systems is to advance. Hydrocarbon reservoir description is one field where the need for quantitative data is critical in production planning and enhanced recovery schemes. Powerful computer programs that model hydrocarbon reservoir performance under simulated production conditions require extensive geometrical databases that have not, as yet, been generated.

The Bowen Basin open-cut coal-mines of central Queensland, Australia (Fig. 1) provide some of the best exposures of alluvial and fluvio-deltaic sequences in the world. Within the Bowen Basin coal is extensively mined along a strike length of 5–600 km exposing various parts of a Late Permian terrestrial to shallow marine sequence. Individual mines cover 10–30 km of strike and may work several different seams that provide stratigraphic correlations between mines. In addition, extensive close-spaced borehole networks drilled to prove open-cut reserves and prospective underground mines extend several kilometres down dip (Fig. 2). Since several mines exploit the same sequence, entire depositional systems may be traced from source to basin depocentre, allowing detailed reconstruction of complete systems.

This paper presents some results from a project

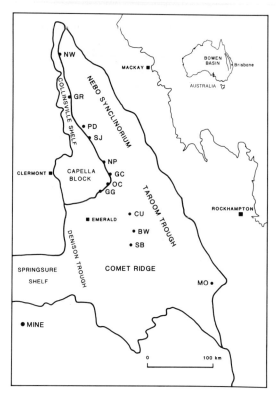

Fig. 1. Map showing location and disposition of the Bowen Basin in eastern Queensland, major tectonic elements and the locations of coal mines investigated during this study. The Bowen Basin continues to the south of the mapped outcrop limit in the subsurface, beneath the cover of the Surat Basin.

based at the University of Queensland and funded by BP Petroleum Development Ltd (London), designed to provide a detailed database on facies geometry within alluvial and fluvio-deltaic coal-bearing sequences of the Bowen Basin. Detailed data on external geometry and internal architecture of sediment bodies were gathered from controlled, orthogonal photomosaics of Bowen Basin highwalls and other large exposures. Where access allowed, typically at the ends of highwalls, at highwall failures and in mine ramps, creek diversions and railway and road cuttings, information on sedimentary structures and palaeocurrent directions was collected. This database has been supplemented by close-spaced boreholes that allow the accurate reconstruction of sediment bodies in three dimensions over considerable areas.

The depositional setting of the German Creek Formation, one of the intervals examined during the course of the project, is the subject of this paper. An overview of the entire project is given by Falkner & Fielding (1993). Detailed datasets are not published due to confidentiality constraints.

REGIONAL GEOLOGY

The Bowen Basin (Fig. 1) has been interpreted as a foreland basin, associated with convergent plate margin tectonics at the eastern edge of the ancient Gondwanan continent (Murray, 1985). The fill is of Permian to Triassic age, with dominantly marine Early Permian strata overlain by a coastal plain to alluvial plain (coal-bearing) Late Permian sequence, and a terrestrial Triassic interval. The basin initially developed as a series of graben and half-graben arising from a phase of back-arc crustal extension, which coincided with a period of arc volcanism (Fielding et al., 1990). These early sub-basins were filled by thick sequences of continental clastic sediments, which were ultimately transgressed by the sea as a phase of more even, widespread, thermal subsidence began. For much of Late Permian times, the Bowen Basin remained largely marine, with modest clastic sediment input from basin margins. Late in the Late Permian compressive deformation began, coinciding with a resurgence in arc volcanism. The marine basin was rapidly infilled, and as thrust sheets propagated from the rising orogen to the east the basin became once again continental. Sedimentation in the Bowen Basin probably ceased during Mid Triassic times, and was followed by a Mid–Late Triassic period of intense uplift and erosion (Beeston, 1986). The basin is preserved as a tectonically modified, north–south trending synclinal structure.

The early Late Permian German Creek Formation crops out in an arcuate belt from Emerald in the south to Saraji in the north, and supports the Gregory, Oaky Creek, German Creek and Norwich Park mines (Fig. 1). The German Creek Formation (Fig. 3) can be divided into a lower unit, up to 160 m thick, barren of coal, and an upper coal-bearing interval that averages 100 m thick (Fig. 4). The Formation is enclosed above and below by sequences of marine shelf origin, the Macmillan and Maria Formations respectively (Fig. 3). To the south and east the German Creek Formation interfingers with the Crocker Formation, which is interpreted as a nearshore marine shelf deposit. To the north, however, the upper German Creek Forma-

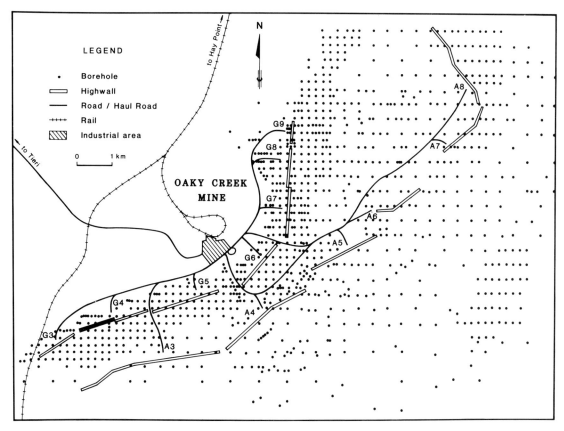

Fig. 2. Plan of the Oaky Creek mine, illustrating the extent of highwall and borehole coverage in the Bowen Basin coal mines. Further exposures are available in ramps, creeks and railway cuttings. Shaded highwall shows the location of rocks illustrated in Fig. 6.

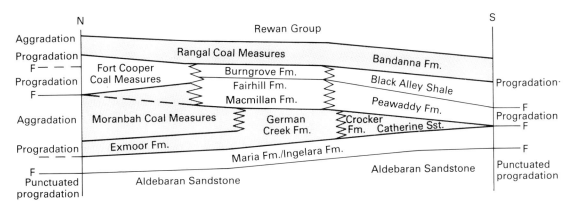

Fig. 3. Schematic north–south (axial) cross-section through the Late Permian section in the Bowen Basin. Shaded sections denote major coal-bearing intervals. Note the southward pinchout of the German Creek Formation and equivalents. F, Major flooding surface. (Modified from Draper & Balfe (1985).)

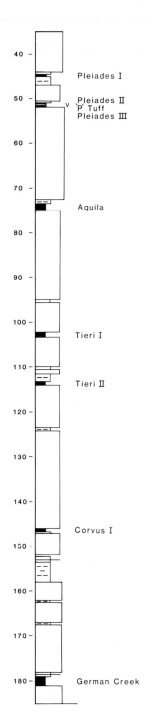

Fig. 4. Generalized graphic log of the upper German Creek Formation from a borehole at Oaky Creek mine, showing the sandstone-dominated nature of the unit and seam stratigraphy.

tion interfingers with the terrestrial Moranbah Coal Measures, with coal seams traceable continuously between the two units.

The German Creek Formation may be regarded as a coarse clastic wedge along the western limb of the Bowen Basin, enclosed to the east and south by marine sedimentary rocks. The lower part of the unit contains horizons rich in marine body fossils, and is interpreted as a complex of shallow-water deltaic and marine shelf deposits. However, the character and sequence context of the upper, coal-bearing part are suggestive of a largely deltaic plain environment.

The major economic seam of the German Creek Formation is the Lilyvale–German Creek–Dysart–Goonyella Lower seam (terminology from south to north and extending into the Moranbah Coal Measures). It has upper and lower splits and is on average 3.0 m thick, low in ash (9%) and sulphur (0.5%) and is exported as coking coal. The other seams in the unit are in stratigraphic order the Corvus, Tieri, Aquila and Pleiades seams which also split and are generally thinner and higher in ash and sulphur than the German Creek seam (Fig. 4).

Coal seams in the upper part of the Formation can be traced continuously over hundreds of kilometres indicating that the coastal plain was areally extensive. Some interseam intervals, particularly the German Creek–Corvus, are well exposed and drilled over large areas, and can be interpreted in terms of both local and regional environments of deposition. Lithofacies recognized in the present study are described and interpreted in the following section.

FACIES ANALYSIS

The German Creek Formation is sandstone dominated with subordinate thin, fine-grained sequences of interbedded siltstone and sandstone, and seams of coal (Fig. 4). Sedimentological interpretations of the upper coal-bearing section have been proposed by Phillips et al. (1985), Godfrey (1985) and Mallett et al (1987), who postulated a delta plain to marginal marine setting. The notion of a coastal environment of deposition is further supported by the unit's stratigraphic setting, enclosed conformably by sequences of interpreted offshore marine shelf origin (Fig. 3).

The lower German Creek Formation contains an abundant marine fauna and is extensively bioturbated. This, together with the nature of the observed facies assemblage, allows the tentative proposal of a mixed inner marine shelf/deltaic plain depositional environment for the interval. This study, however, concentrates on the more extensively exposed, upper coal-bearing part of the German Creek Formation. The coal-bearing section can be divided into six separate facies coded A, B, C, D, E and F. In the coal mines which together define a strike parallel strip (Fig. 1), Facies A, B and C occur predominantly in the interval between the German Creek, Corvus and Tieri seams (Fig. 4). Stratigraphically higher intervals are dominated by Facies D and E whilst the seams themselves are designated as Facies F. Facies not apparent within the outcrop of particular intervals can be identified from subsurface data down dip and along strike.

Facies A, Distributary Channel

This facies comprises erosively based, fine- to coarse-grained quartz-lithic sandstone bodies (Fig. 5) that vary from 2 to 15 m thick with an average of about 7 m (Fig. 6). The bases of these sandstone bodies are sharp and planar with modest relief (*c*.0.5 m). Rare exposures of channel margins indicate that the channels are incised into mainly Interdistributary Bay deposits (Facies B). Upper contacts may be gradational, showing a degree of interbedding with overlying facies, or sharp, sometimes with a top surface lag (Fig. 7). Grain size is usually fine to medium and mostly fines upward. Rare pebbles up to 5 cm occur either concentrated into basal or top surface lags, or scattered throughout the sandstone body. Organic material is common as impressions of logs, leafy fronds, leaves, coaly particles and rare peat rafts. Trough cross-bedding and ripple cross-lamination are ubiquitous with unimodal or bimodal palaeocurrent distributions parallel to the direction of elongation of the sandstone body (Fig. 8). Sets are generally less than 0.5 m thick but occasionally reach 3 m. Planar cross-bedding and climbing ripples are common. Combined flow and symmetrical ripples are rare and occur mostly in the upper parts of channel fills. This facies also shows varying development of flat and uneven lamination. Siltstone laminae are common while thicker layers up to 0.5 m are rare. Bioturbation comprises a restricted suite of small burrows including *Planolites, Chondrites* and *Arenicolites*. Channel plugs have been noted at two localities and comprise dark grey carbonaceous siltstone and claystone with a small proportion of fine sandstone that occurs as thin lenticular beds concentrated at the base of the plug sequence. The plugs are 3.5 and 5.0 m thick and display sharp contacts with underlying, coarse-grained channel fill sandstones a few metres thick, indicating relatively late-stage, rapid abandonment.

The internal architecture of Distributary Channel facies is dominated by lateral accretion surfaces (Fig. 5) that are generally planar and can be traced continuously from lower to upper contact. The surfaces dip at between 8° and 15° and have dip azimuths orientated at angles between 30° and 90° to palaeoflow directions. The extent of lateral accretion surfaces in the down dip direction can be used

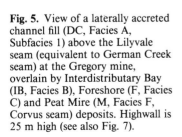

Fig. 5. View of a laterally accreted channel fill (DC, Facies A, Subfacies 1) above the Lilyvale seam (equivalent to German Creek seam) at the Gregory mine, overlain by Interdistributary Bay (IB, Facies B), Foreshore (F, Facies C) and Peat Mire (M, Facies F, Corvus seam) deposits. Highwall is 25 m high (see also Fig. 7).

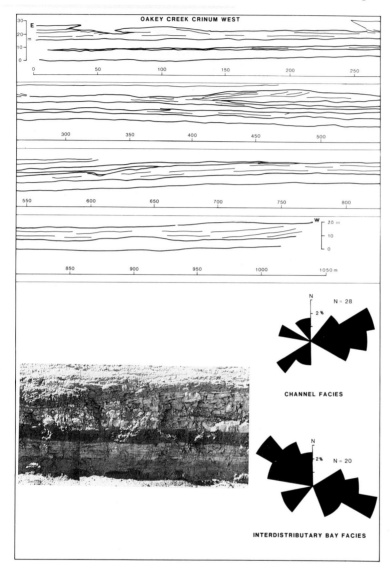

Fig. 6. Montage illustrating geometry and internal characteristics of Distributary Channel (Facies A, Subfacies 2) deposits above the German Creek seam at Oaky Creek mine (see Fig. 2 for location of highwall). The interpreted highwall map, based on a controlled photomosaic, illustrates the two-dimensional geometry of a channel body and enclosing Interdistributary Bay deposits in a direction roughly parallel to palaeoflow. The photograph illustrates a section of the map from 400 to 475 m and rose diagrams summarize palaeocurrent data for the entire highwall.

to calculate the instantaneous widths of channels using the method of Allen (1965). Rarely, entire channel fills are exposed in mine highwalls.

Three subfacies can be defined within this facies using the distinguishing criteria of lateral accretion, interlamination of sandstone and siltstone, and palaeocurrent relationships.

1 Subfacies 1 (Fig. 5) comprises laterally accreted channel fills which display unimodal palaeocurrent directions and abundant siltstone laminae. These laminae are typically 1 cm thick and occur throughout the unit except where a coarse-grained lower section is present. They are planar to slightly undulating and can be traced several metres laterally.

2 Subfacies 2 consists of laterally accreted sandstone bodies with bimodal palaeocurrent directions. Siltstone laminae are rare.

3 Subfacies 3 has been noted only at Norwich Park mine where a channel fill with an internally complex, vertically accreted fill displaying unimodal palaeoflow is exposed. Siltstone laminae are rare.

Facies A is interpreted as the active fills of distributary channels affected by tidal flows on a lower delta plain. These channels were of variable

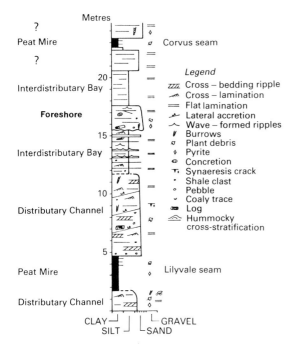

Fig. 7. Detailed graphic log of a highwall exposure at Gregory mine, illustrating the vertical sequence character of Distributary Channel (Facies A) and associated deposits.

sinuosity, from the architectural features noted above. Flows within the channels were evidently sluggish for the most part, from the dominance of cross-bedding and cross-lamination structures. Such sluggishness may be related to tidal effects, which would have retarded fluvial outflow.

The geometry of distributary sandstone bodies is profoundly affected by the position of the channel system with regard to the major avenue of sediment dispersal. Channel belts greater than 5 km wide occur near the centre of the delta at about the latitude of the Gregory mine, whereas in more marginal positions belt widths are around 1 km. Channel depths and instantaneous widths calculated from lateral accretion surfaces are not affected by channel position within the delta. Channel belts bifurcate at their mouths, their fills thicken and at German Creek mine become more incised, eroding the underlying coal seam by up to 1.5 m (Fig. 8). Channel depths vary between 2 and 15 m with an average of approximately 7 m. Instantaneous widths calculated from lateral accretion surfaces range from 90 to 170 m.

Subfacies 1 and 3 display unimodal palaeocurrent distributions, interpreted to reflect a dominance of outflow processes. Because of its complex, multistorey fill, Subfacies 3 may represent a trunk distributary channel. Subfacies 1 with its abundant siltstone partings also represents major channels whose flow was periodically held up by tidal movements. Such alternation of lithologies in sinuous channels, or inclined heterolithic stratification (IHS), is often associated with tide-influenced environments, where the inclined, fine-grained partings represent slack water accumulations deposited at the ends of tidal cycles. Substantial fluvial outflow, however, would have ensured that bedforms were aligned dominantly in the downstream direction.

Subfacies 2, which displays bipolar palaeocurrent patterns, may be interpreted as the deposits of channels whose courses were swept regularly by the tide.

Directly measured parameters for channel depth, channel width (from lateral accretion surfaces) and channel belt width from this study were compared with the dataset of Fielding & Crane (1987) to gauge channel morphology for the upper German Creek Formation. Channel width/depth ratios for Subfacies 1 range from 15 to 25, and from 10 to 16 for Subfacies 2. Instantaneous channel widths for Subfacies 3 could not be measured but a channel belt width/depth ratio of 60 was noted. By comparison of observed data with Collinson's (1978) relationship for channel depth to channel belt width in fully developed meandering channels, we infer that Subfacies 1 channels were truly meandering, Subfacies 2 channels were of moderate sinuosity and Subfacies 3 channels of low sinuosity.

Facies B, Interdistributary Bay

Sequences 1–30 m thick of dark grey siltstones interbedded with fine- to coarse-grained quartz-lithic sandstones characterize this facies (Figs 6 & 7). Typically, medium-grained siltstones comprise 70% of this facies in units 1 mm to 30 cm thick, interbedded with fine- to medium-grained sandstones up to 50 cm thick. Individual beds can be traced tens of metres and may show a rhythmic character. Sandstones display flat or ripple cross-lamination with evidence of both current and wave activity. Small-scale trough cross-bedding associated with thicker, sharply bounded sandstone beds and ripple cross-lamination, show bimodal or polymodal palaeocurrent distributions (Figs 6 & 8). Rare extraformational pebbles and siltstone clasts

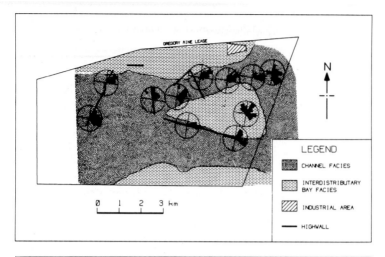

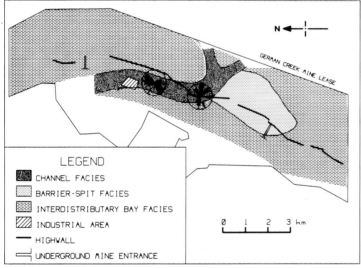

Fig. 8. Interpreted facies maps for the German Creek to Corvus interval at the Gregory and German Creek mines. The Gregory map shows bifurcation of a Distributary Channel belt deposit (Facies A, Subfacies 1) close to its mouth, and passage into Proximal Mouth Bar deposits (Facies D) at approximately the eastern limit of the mine lease. The German Creek map again shows a bifurcating Distributary Channel (Facies A, Subfacies 2) deposit close to its point of issue into the marine basin, and an associated barrier-spit sandbody developed through wave reworking and southward longshore drift of sand. Arrowheads on some palaeocurrent roses indicate dip of associated lateral accretion surfaces.

are generally less than 5 cm in diameter. This facies is commonly carbonaceous and in places contains units of carbonaceous claystones with rare thin coals. Small-scale soft sediment deformation and synaeresis cracks are common. Bioturbation is common and comprises a restricted suite of burrows, notably *Planolites* and *Chondrites*.

Facies B displays sharp or gradational vertical contacts with adjacent facies, varies from 1 to 30 m thick and can be traced laterally for up to 10 km, but is generally only traceable for a few kilometres. This facies has an irregular sheet-like geometry giving way laterally to Facies A and C.

Units of Facies B are closely associated with Facies A (Fig. 6) and are thought to represent sediment accumulation within interdistributary bays by overbank and crevasse splay deposition, and by suspension fallout from turbid waters following floods. Symmetrical ripples and top surface lags often present in this facies indicate that the interdistributary bays were affected by waves and were probably open to the sea to the east, at least in part.

Facies C, Foreshore

Facies C comprises fine- to medium-grained, sheet-like sandstone bodies 1–10 m thick that internally are dominated by flat and very low-angle lamina-

Fig. 9. View of Foreshore deposits (F, Facies C) above the German Creek seam at German Creek mine (see Fig. 8; location is just south of southernmost underground mine entry). Note flat and low-angle master bedding planes within the light-coloured, quartzose sandstone body. Geologist is 2.0 m tall.

tion (Fig. 9). Basal and upper contacts are sharp and planar except where extensive bioturbation has intermixed finer overlying sediments. Interdistributary Bay (Facies B) or coal seams (Facies F) occur above and below foreshore deposits while distributary channels (Facies A) may be associated with them laterally. Master bedding planes, where present, are horizontal or slightly inclined. The sandstone petrology of this facies contrasts with that of sands in associated distributary or interdistributary bay facies. At German Creek the sandstones of Facies C are quartz-rich and very strong due to a pervasive quartz cement; at Gregory interpreted foreshore sands are also relatively quartz rich and better sorted than sands of adjacent facies. Symmetrical (wave formed) ripples, cross-lamination, cross-bedding and pebbles are rarely visible. The sandstones are pyritic in part, often with coaly traces, and contain rare *Skolithos* burrows. Soft sediment deformation is common, particularly in occasional sandy siltstone interbeds up to 30 cm thick that are laterally persistent for hundreds of metres. Peculiar structures similar to gutter casts (cf. Whitaker, 1973) are exposed as roof rolls at Southern Colliery, German Creek. They occur as elongate troughs and bars (i.e. both negative and positive relief) averaging 50 cm deep, 2 m wide and up to 5 m long that are asymmetrical in cross-section and clearly modified by compaction. Coalified logs up to 50 cm in diameter and 5 m long are also common in the roof at Southern Colliery. They occur both *in situ*, as stumps rooted in the underlying coal seam, and as drifted logs. Orientations of roof rolls and logs are parallel to one another and perpendicular to crestline trends of symmetrical ripples (Fig. 10). Limited ripple data indicate a westerly wave propagation direction which is roughly perpendicular to palaeoshoreline trend, while the logs and roof rolls are also orientated perpendicular to the palaeocoastline (cf. Whitaker, 1973; MacDonald & Jefferson, 1985) (Fig. 8).

Facies C can be traced for several kilometres and appears to comprise lensoid sheet-like sandstone bodies, elongate parallel to palaeoshoreline trend (Fig. 8).

This facies is interpreted as foreshore deposits formed by wave reworking of channel mouth sands. An alternative interpretation of a crevasse splay origin for Facies C is rejected on the lack of abundant current generated structures and waning flow characteristics. The contrasting petrology of the sands with respect to Facies B (Interdistributary Bay) sands and the palaeogeographic setting of the German Creek example (Fig. 8) also contradict a crevasse splay origin. The orientation of wave ripple structures suggests an oblique wave approach to the palaeoshoreline, at least at German Creek mine (Figs 8 & 10). The lensoid sandstone body illustrated in Fig. 8 is interpreted specifically as a barrier-spit deposit because of its geometry and palaeogeographic setting. Similar units at Gregory are more widespread and cannot be accurately mapped and are, therefore, interpreted in more general terms as foreshore deposits. Barrier-spit facies are interpreted to arise from longshore

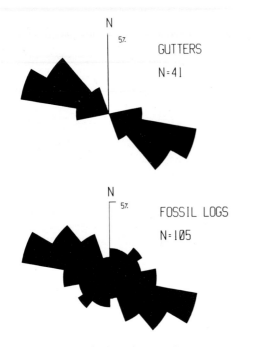

Fig. 10. Summary of orientation data from Foreshore deposits (Facies C), Southern Colliery (underground), German Creek mine. Data gathered from fossil log impressions and gutter casts in the roof of the German Creek seam (mostly collected by R. Byrnes, Capcoal). Both fossil log and gutter orientations indicate alignment perpendicular to palaeoshoreline (Fig. 8). Symmetrical ripple crest line trends (025°, 032°, 340°) are roughly parallel to the palaeoshoreline.

transfer of sand and preferential accumulation on the down-drift side of the mapped palaeochannel mouth (Fig. 8). The dominance of flat and low-angle lamination within the foreshore facies is interpreted to reflect swash–backwash processes in an uppermost shoreface–foreshore setting. Other foreshore deposits are interpreted to represent channel-mouth sands that were reworked in shoreface to foreshore environments, perhaps following the avulsion of a distributary channel to another site.

Facies D, Proximal Mouth Bar

Facies D comprises very large, sheet-like bodies of sandstone (Fig. 11). These fine- to very coarse-grained quartzose sandstones have sharp, planar, sulphur-stained bases and show thick tabular bedding. Grain size is most commonly medium-grained but extremes occur, particularly in the Gregory mine area where rounded extraformational cobbles up to 20 cm in diameter are common

in some intervals. Coarsening-upward and fining-upward trends are not developed over the entire thickness of the sandstone body, but both are noted as minor trends over intervals, commonly at the upper or lower contacts. Internally this facies displays flat, low-angle and undulating lamination and hummocky cross-stratification (Fig. 11). Siltstone layers up to 50 cm thick are common and laterally continuous for hundreds of metres. Cross-bedding, both trough and planar, and ripple cross-lamination are rare, as are small channel scours. Wave ripples are common. Bioturbation is varied and abundant, commonly as deep (30 cm), vertical, tubular *Skolithos* burrows. Other types include *Arenicolites*, *Rosselia* and *Diplocraterion*. Rare marine bivalves of the genus *Atomodesma* and non-specific brachiopods have been noted.

Sandstone bodies of this facies vary from 4 to 20 m thick and are up to 30 by 100 km in plan, elongate parallel to palaeoshoreline trend (Fig. 12).

Facies D is interpreted as proximal mouth bars of shallow water marine deltas, constructed under the influence of fluvial outflow, waves and tides. Their great lateral extent, tabular external form and internal bedding, and dominance of wave/combined flow generated structures, distinguish sandstones of Facies D from those of Facies A (Distributary Channel). Together with the occurrence of marine body fossils, these features suggest deposition in a shallow marine environment subjected to vigorous wave activity. Given the sand dominated nature of the facies, deposition in an area immediately offshore from river mouths is envisaged (i.e. proximal mouth bar). The great lateral extent of Facies D in some cases suggests that sand bodies represent the combined proximal delta front deposit of several outflowing channels.

Facies E, Distal Mouth Bar

Extensively bioturbated, interbedded siltstone and sandstone sequences intersected in core in association with hummocky cross-stratified, brachiopod-bearing sandstones (Facies D, Proximal Mouth Bar) are interpreted as Distal Mouth Bar (Facies E) deposits. Boreholes close to Gregory mine intersect sequences of Facies E (Distal Mouth Bar) up to 4 m thick, while 30 km to the east similar facies have been noted up to 60 m thick. The ratio of siltstone to sandstone in this facies is roughly 1 : 1 with graded bedding common within sharply based 5 cm units. Siltstones are generally fine- to medium-grained and commonly carbonaceous while associated sand-

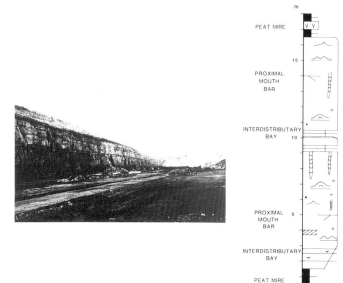

Fig. 11. Montage illustrating geometry and internal characteristics of Proximal Mouth Bar deposits (Facies D) at the German Creek mine, Aquila to Pleiades seam interval. General view shows tabular geometry of the mouth bar sandbodies, while close-up view (geologist is 1.65 m tall holding 2.0 m scale stick) and graphic log (all of same highwall) illustrate the internal characteristics of the facies. Note the short coarsening-upward sequence at the base of the Proximal Mouth Bar Facies.

stones are very fine- to medium-grained. Horizontal lamination, flaser bedding and ripple cross-lamination are common and hummocky cross-stratification is noted in thicker sandstone beds. Units of this facies are often extensively bioturbated by a varied suite of burrow types.

Heavily bioturbated, interbedded siltstones and sandstones associated with proximal mouth bar deposits are interpreted as the more distal equivalents, both seaward and laterally, to mouth bar deposits. Graded units and physical structures were probably produced mainly during storms with extensive bioturbation occurring in calmer spells.

Facies F, Peat Mires

The upper German Creek Formation contains five coal seams that can be traced continuously for hundreds of kilometres north–south (i.e. along strike). They are in stratigraphic order the German Creek, Corvus, Tieri, Aquila and Pleiades seams (Fig. 4). These coal seams split and coalesce in a relatively gradual manner with interseam intervals constant over wide areas. Seam thicknesses vary from 0.3 to 5 m, with most seams being vitrinite rich. Ash contents of the German Creek seam average below 9%, with sulphur contents of around 0.5%. Stratigraphically higher seams have ash contents between 9 and 28%, and sulphur contents from 0.6 to 3% (Zillman, 1976).

Facies F represents accumulation in long-lived, low-lying, areally extensive peat mires that were remote from clastic sediment input and only rarely received flood-borne sediment or distal, airfall volcanic ash. From their great lateral extent, the peat bodies are interpreted to have formed on an extensive coastal plain.

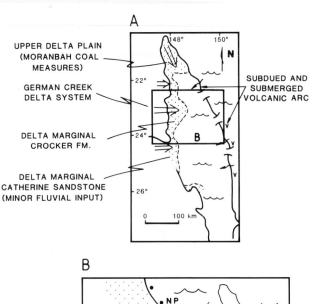
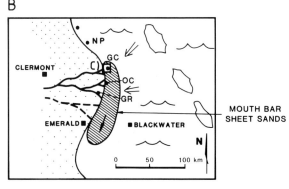
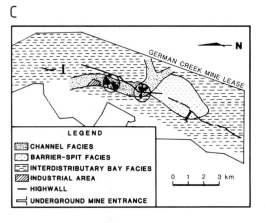

Fig. 12. Interpreted palaeogeographic maps illustrating the German Creek delta system at: (A) basinal; (B) regional; and (C) local scales. (A) Extent of the entire system and its context within the Bowen Basin. (B) Shape of the deltaic coastline (generalized) and extent of distributary mouth bar sands (hachured). Open arrows, direction of palaeowave approach; closed arrows, direction of longshore drift. NP, Norwich Park; GC, German Creek; OC, Oaky Creek; GR, Gregory mines. (C) Distribution of lithofacies immediately above the German Creek seam at the German Creek mine.

DISCUSSION

The facies assemblage as defined comprises six distinct lithofacies, with some variation within facies encompassed by subfacies. It must be noted, however, that the data set is biased by being restricted largely to mine exposures orientated along a strike-parallel (north–south) strip. Examination of subsurface and other data regionally suggests that the stratigraphic intervals examined are representative of the formation as a whole. Because the mines are arranged in a strip parallel to the depositional strike of the German Creek delta system, however, facies trends in a dip-parallel direction are less evident.

To the west of the mines, equivalent strata have been removed by erosion. Eastward, however, borehole data suggest that (i) coal seams split, wedge out and deteriorate in quality; (ii) channel deposits are poorly represented; and (iii) mouth bar deposits dominate, passing eastward into thinner, more interbedded facies with some coarsening-upward trends and marine body fossils. These trends are consistent with the notion of an easterly directed palaeoslope and drainage net.

Many of the coal seams in the upper German Creek Formation are traceable over hundreds of square kilometres parallel to strike. Given that coals are continuous beyond the extent of the German Creek delta system, some external base level control on their geometry seems likely. Analysis of possible base level shifts during the German Creek Formation accumulation is beyond the scope of this paper, but it is worthy of note that this period coincided with a major change in subsidence style in the Bowen Basin (Fielding et al, 1990).

The upper German Creek Formation may be interpreted as resulting from repeated progradation and retreat of deltas into a shallow-water marine environment. The deltas advanced eastward into the western part of the marine Bowen Basin, each time progressing a different distance before being abandoned and flooded.

The German Creek deltaic complex (Fig. 12) displays evidence of having been deposited under the combined influence of fluvial outflow, waves and tides, and may be considered as a mixed-influence delta according to the classification of Galloway (1975). Such a balance of fluvial outflow and basinal influences has been noted to characterize coastal facies assemblages throughout the Permian of the Bowen Basin (Fielding, 1989).

The situation envisaged for the German Creek delta complex is similar in some respects to that shown by the modern Burdekin River delta in northeast Australia (Coleman & Wright, 1975) and the Copper River delta of Alaska (Galloway, 1976). Both these deltas have an arcuate coastline, moulded by wave action, and funnel-shaped channels in plan view, affected by tidal activity. Both show barriers, spits and offshore shoals fashioned by wave-driven longshore drift systems.

While the delta plain portion of the German Creek delta complex can be envisaged in terms of these analogues, the delta front environment is more enigmatic. Few descriptions of delta front facies in wave-influenced deltas exist, making comparison with other sequences difficult.

Vanderburgh & Smith (1988) have shown that the progradational phase of the modern Slave River delta, a fluvial–wave interactive system in western Canada, is represented by a laterally extensive, wave-reworked sandbody internally dominated by low-angle lamination. In terms of setting, this sandbody is similar to the interpreted proximal mouth bars of the German Creek Formation. The base of the delta-front sandbody is sharp in many places, as with the German Creek Formation. The modern delta front is reported to show low, short barrier islands and shoals (Vanderburgh & Smith, 1988), again the situation envisaged for the German Creek Formation.

A more spectacular example of a wave dominated river mouth bar is given by the modern Amazon River in South America (Nittrouer et al., 1986). Proximally, this feature comprises thick, stratified sand deposits, which grade distally into interbedded sand and mud. The huge sediment wedge which has been constructed at the Amazon River mouth is attributed by Nittrouer et al. (1986) to vigorous wave reworking of the sediment load of a large river.

A possible ancient analogue to the German Creek deltaic system is given by the upper part of the Jurassic Brent Group in the northern North Sea. Livera (1989) has illustrated laterally extensive mouth bar sandbodies which accumulated on the delta front of a fluvial–wave–tide interactive delta system. These mouth bar sandbodies are both externally and internally similar to those described in this paper, as indeed is the delta plain facies assemblage. Both systems prograded into shallow water which was subjected to a vigorous wave regime, and tidal influence is evident in both cases. Classical coarsening-upward sequences are not well developed in either system.

Chan & Dott (1986) described an Eocene wave dominated deltaic succession from Oregon, which also displays some similarity to the German Creek Formation. The delta front facies of Chan & Dott (1986) is comparable to the Distal Mouth Bar facies (E) described herein, whilst their 'delta margin' facies appears similar to Facies D (Proximal Mouth Bar) of this study.

As has already been stated, the nature of delta front deposits in wave influenced deltas is poorly understood, and a need exists for more detailed descriptions. This paper places geometrical constraints on such sandbodies together with the distributary channels which supplied them.

SUMMARY AND CONCLUSIONS

A geometrical facies analysis of the early Late Permian, upper German Creek Formation, based mainly on large mine exposures, reveals that it accumulated as a coarse clastic wedge which prograded eastwards (transversely) into the marine Bowen Basin. At outcrop the unit comprises six facies that accumulated in a mixed influence (i.e. wave–tide–fluvial influenced) delta: Facies A, distributary channel fills; Facies B, interdistributary bay deposits; Facies C, foreshore deposits; Facies D, proximal mouthbar deposits; Facies E, distal mouth bar deposits; and Facies F, peat mire deposits.

Three varieties of channel fill are recognized, two of which represent fluvially dominated distributary channels and the third tidal channels. Finer-grained sediments accumulated in interdistributary bay environments. Where the sediment load of distributary channels was deposited at river mouths and reworked by waves and tides, extensive mouth bar sandbodies accumulated immediately offshore. In their proximal parts these mouth bar deposits were sharply based, giving way distally to more interbedded sand/silt and coarsening-upward facies. Discontinuous sheet sandstones also accumulated in foreshore environments, in some cases forming low-lying barrier-spits adjacent to river mouths. These did not form an effective barrier to marine influence.

This paper is one of few detailed descriptions of ancient fluvial–wave–tide interactive delta systems. It provides new data on the geometry and internal characteristics of distributary channels and distributary mouth bars in such systems.

ACKNOWLEDGEMENTS

This project was made possible through the financial support of BP Petroleum Development Ltd. BHP Utah Coal Ltd, Capcoal, Central Queensland Coal Associates and Oaky Creek Coal are thanked for allowing access to exposures and other data. Roger Byrnes (Capcoal) allowed use of his orientational data from German Creek Southern Colliery, and Anna Thompson drafted the figures. The manuscript has been improved considerably through the reviews of K. Shanley and an anonymous individual.

REFERENCES

ALLEN, J.R.L. (1965) The sedimentation and palaeogeography of the Old Red Sandstone of Anglesey, North Wales. *Proc. Yorks. Geol. Soc.* **35**, 139–185.

BEESTON, J.W. (1986) Coal rank variation in the Bowen Basin, Queensland. *Int. J. Coal Geol.* **6**, 163–179.

CHAN, M.A. & DOTT, R.H., JR (1986) Depositional facies and progradational sequences in Eocene wave-dominated deltaic complexes, southwestern Oregon. *Bull. Am. Assoc. Petrol. Geol.* **70**, 415–429.

COLEMAN, J.M. & WRIGHT, L.D. (1975) Modern river deltas: variability of processes and sand bodies. In: *Deltas: Models for Exploration* (Ed. Broussard, M.L.) pp. 99–149. Houston Geol. Soc., Houston.

COLLINSON, J.D. (1978) Vertical sequence and sand body shape in alluvial sequences. In: *Fluvial Sedimentology* (Ed. Miall, A.D.) pp. 577–588. Can. Soc. Petrol. Geol., Calgary, Memoir 5.

DRAPER, J.J. & BALFE, P.E. (1985) Late Permian stratigraphy — western Bowen Basin, Queensland. In: *Proc. 19th Symp. Advances in the Study of the Sydney Basin*, pp. 106–109. Newcastle University, Newcastle, Australia.

FALKNER, A.J. & FIELDING, C.R. (1993) Quantitative analysis of deltaic and alluvial sequences of the Bowen Basin, Australia. In: *Quantitative Description and Modelling of Clastic Hydrocarbon Reservoirs and Outcrop* (Eds Flint, S. & Bryant, I.D.) pp. 81–97. Spec. Publ. Int. Ass. Sediment. 15.

FIELDING, C.R. (1989) Controls on marine sedimentation in the southern Bowen Basin, Qld. In: *Proc. 23rd Symp. Advances in the Study of the Sydney Basin*, pp 9–14. Newcastle University, Newcastle, Australia.

FIELDING, C.R. & CRANE, R.C. (1987) An application of statistical modelling to the prediction of hydrocarbon recovery factors in fluvial reservoir sequences. In: *Recent Developments in Fluvial Sedimentology* (Eds Ethridge, F.G., Flores, R.M. & Harvey, M.D.) pp. 321–327. Soc. Econ. Paleont. Miner., Tulsa, Spec. Publ. 39.

FIELDING, C.R., GRAY, A.R.G., HARRIS, G.I. & SALOMON, J. (1990) The Bowen Basin and overlying Surat Basin. In: *The Eromanga–Brisbane Geoscience Transect: A Guide to Crustal Development Across Phanerozoic Australia in Southern Queensland* (Ed. Finlayson, D.M.) pp. 105–116. Bull. Aust. Bur. Min. Res. Geol. Geophys., Canberra, Vol. 232.

GALLOWAY, W.E. (1975) Process framework for describing the morphologic and stratigraphic evolution of deltaic systems. In: *Deltas: Models for Exploration* (Ed. Broussard, M.L.) pp. 87–98. Houston Geol. Soc., Houston.

GALLOWAY, W.E. (1976) Sediments and stratigraphic framework of the Copper River fan-delta, Alaska. *J. Sedim. Petrol.* **46**, 726–727.

GODFREY, N.H.H. (1985) Strip mining in the Moranbah and German Creek Coal Measures. In: *Bowen Basin Coal Symp.*, pp. 75–81. Geol. Soc. Aust., Brisbane, Abstr., Vol. 17.

LIVERA, S.E. (1989) Facies associations and sand-body geometries in the Ness Formation of the Brent Group, Brent Field. In: *Deltas: Sites and Traps for Fossil Fuels* (Eds Whateley, M.K.G. & Pickering, K.T.) pp. 269–286. Geol. Soc., Lond., Spec. Publ. 41.

MACDONALD, D.I.M. & JEFFERSON, T.H. (1985) Orientation studies of waterlogged wood: a palaeocurrent indicator? *J. Sedim. Petrol.* **55**, 235–239.

MALLETT, C.W., BUCKLAND, A., BONNER, G., ROBERTS, G. & SULLIVAN, D. (1987) A down dip study of geological conditions, German Creek Mine, central Bowen Basin. Commonwealth Scientific and Industrial Research Organization Site Report No. 35. CSIRO, Brisbane.

MURRAY, C.G. (1985) Tectonic setting of the Bowen Basin. In: *Bowen Basin Coal Symp*, pp. 5–16. Geol. Soc. Aust., Brisbane, Abstr., Vol. 17.

NITTROUER, C.A., KUEHL, S.A., DEMASTER, D.J. & KOWSMANN, R.O. (1986) The deltaic nature of Amazon shelf sedimentation. *Bull. Geol. Soc. Am.* **97**, 444–458.

PHILLIPS, R., GREEN, D. & MOLLICA, F. (1985) German Creek Mine. In: *Bowen Basin Coal Symp.*, pp. 243–251. Geol. Soc. Aust., Brisbane, Abstr., Vol. 17.

VANDERBURGH, S. & SMITH, D.G. (1988) Slave River delta: geomorphology, sedimentology, and Holocene reconstruction. *Can. J. Earth Sci.* **25**, 1990–2004.

WHITAKER, J.H.M. (1973) 'Gutter casts', a new name for scour-and-fill structures: with examples from the Llandoverian of Ringerike and Malmöya, southern Norway. *Norsk geol. Tidsskr.* **53**, 403–484.

ZILLMAN, N.J. (1976) Coal resources of the German Creek Formation in the Oaky Creek area, south central Bowen Basin. *J. Queensland Govt. Min.* **77**, 572–579.

Computer modelling of flow lines over deformed surfaces: the implications for prediction of alluvial facies distribution

P.J. WESTON and J. ALEXANDER

Department of Geology, University of Wales College of Cardiff, PO Box 914, Cardiff CF1 3YE, UK

ABSTRACT

Computer modelling methods can be used to predict the distribution of sedimentary facies deposited by alluvial processes. The method presented here calculates the most likely flow paths for fluid moving over modelled topographic surfaces. The floodplain topography evolves largely through tectonic activity, together with deposition and erosion. This presentation investigates the effects of changing fault plane dip, fault offset and fault orientation on the distribution of flow lines, and discusses the implications of the model results for the patterns of sedimentary facies deposited in areas of surface deformation.

INTRODUCTION

In an alluvial setting, channel position and overbank flow patterns are controlled largely by the floodplain topography, which is controlled by sedimentation, erosion and tectonic activity (Alexander, 1986). The distribution of facies in alluvial formations of the rock record is controlled by the evolving topography during the period of deposition and postdepositional deformation. Therefore, if the palaeotopography can be modelled throughout the depositional period, it is possible to predict the resulting distribution of facies.

The probable distribution and orientation of sandstone bodies deposited by alluvial systems can be predicted using a computer-generated idealized flow model, a simple version of which is presented here. This modelling technique is potentially a powerful tool for testing theories of palaeogeography and synsedimentary tectonic control of sedimentation. The method calculates the most likely flow paths for fluid moving over modelled topographic surfaces assuming that the fluid flows directly downslope. The areas with high flow densities are areas of greater probability of channel deposits and the orientation of flow lines predicts the probable orientation of channel sandstone bodies.

THE MODEL

In order to construct a flow model, it is necessary to determine the mean palaeoslope direction. The initial depositional surface is assumed to be a simple slope (Fig. 1a) whose gradient can be varied. In the examples below, a gradient of 1 in 10^4 is used, based on comparison with modern coastal plains (Baker, 1983; Bristow, 1986; and others). The surface is simulated on a computer using a 50×50 grid of square cells. Flow lines are initiated at regular intervals along the top of the sedimentary slope. The flow lines are constructed perpendicular to the contours and move downslope until they reach an edge of the grid or until the flow line is 'ponded' by a closed topographic low. On the initial, simple slope (Fig. 1a) a set of parallel lines is generated.

The depositional surface can be modified by normal faults which are assumed to be active at the time of deposition (Fig. 1b). The orientation, position, length, maximum offset and the dip of a fault(s) are input variables. Movement on a fault causes the grid cells in the region adjacent to the fault to be displaced from their original positions. The magnitude of this displacement is calculated at

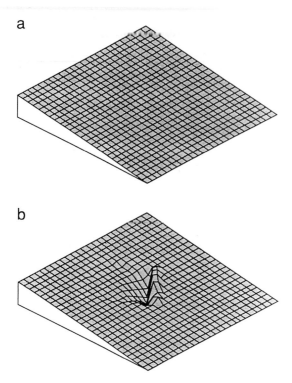

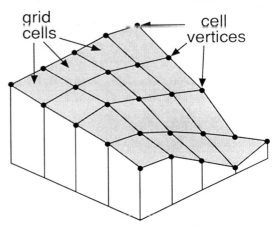

Fig. 2. Detail of deformed grid surface illustrating grid interpolation method. Interpolation of the displacement within the grid cell yields a surface which is a hyperbolic paraboloid fitting the cell vertices.

Fig. 1. Isometric projections of sedimentary surfaces. (a) Simple undeformed sedimentary surface. (b) Surface deformed by a single normal fault.

the vertices of each grid cell. Interpolation of the displacement within the grid cell yields a surface which is a hyperbolic paraboloid fitting the cell vertices exactly. This method ensures that there is no discontinuity of the surface at cell boundaries (Fig. 2) as discussed by Tetzlaff & Harbaugh (1989). The variations in magnitude of displacement along the trace of the fault and in the footwall and hangingwall of the fault are calculated using equations developed by the Fault Analysis Group at Liverpool University (Gibson et al., 1989) and are illustrated in Fig. 4. These equations may need some modification to model synsedimentary faults accurately, but detailed studies of displacement variations along synsedimentary faults are limited from the literature.

Fitting a hyperbolic paraboloid function to each grid cell permits the value of the height to be determined at any point within a grid cell. Differentiation of the height function with respect to both X and Y axes enables the surface gradients in the X- and Y-directions to be calculated uniquely at any point within the grid. In order to eliminate 'false ponding' produced during gridding of the input, the flow line program also considers the heights of the four grid vertices forming a 3×3 square of grid cells around (X,Y). A second calculation of the X and Y gradients is made. If the sense of the two sets of gradients is the same, i.e. if the local gradient reflects the more regional one, the local gradient is used. If there is a difference in the sense of the gradients then the more regional gradient is used. Simple trigonometry can be used to determine the resultant gradient at the point (X,Y) and any flow line arriving at (X,Y) will move forward in the direction of the resultant gradient (Fig. 3).

The flow lines remain parallel until they reach a region where the slope changes direction such as in an area of deformation around a fault where they are deflected according to the local gradients. Areas where flow lines become relatively closely spaced are interpreted as regions into which flow will be preferentially diverted and therefore have a greater probability of being sites of sandstone deposition. Conversely, areas where flow lines are widely spaced or non-existent are interpreted as regions away from which flow is diverted and therefore probable sites of peat accumulation or more mature palaeosol development. In reality, the relationships between flow line densities and sedimentation or erosion are more complicated than the very simple model presented here, and require further study.

This presentation sets out to discuss the effects of

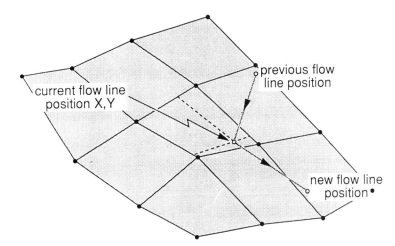

Fig. 3. Detail of the generation of the new flow line direction calculated at X,Y.

three properties of faults which control the concentration of flow lines and consequently alluvial architecture. Simple models are presented to demonstrate the influence of:
1 the dip of a fault plain,
2 the offset across a fault, and
3 the orientation of a fault with respect to the palaeoslope,
on flow line patterns and by inference probable facies distribution. Flow lines are solid and topographic contours dotted in each of the models presented.

The dip of the fault

Changing the dip of a normal fault causes the distribution of the total offset across the fault assigned to hangingwall subsidence and footwall uplift to alter (Fig. 4). Vertical faults result in equal amounts of footwall uplift and hangingwall subsidence, by analogy with elastic dislocations which intersect a free surface (Stekatee, 1958). Non-vertical faults result in asymmetric distributions of displacement magnitude. The asymmetry of displacement magnitude increases with decreasing fault dip (Fig. 4; Gibson *et al.*, 1989). The effect of faults with differing dips can be illustrated by models where the fault orientation and displacement are kept constant but the dip is varied, as in Fig. 5. In this example increasing the fault dip causes the point at which the flow lines are first deflected to occur further upslope due to the in-

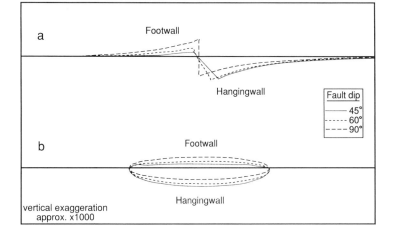

Fig. 4. Calculated surface displacement profiles for regions adjacent to faults. (a) Displacement in sections perpendicular to fault trace for faults with dips of 45°, 60° and 90°. (b) Displacement along the fault planes for faults with dips of 45°, 60° and 90°.

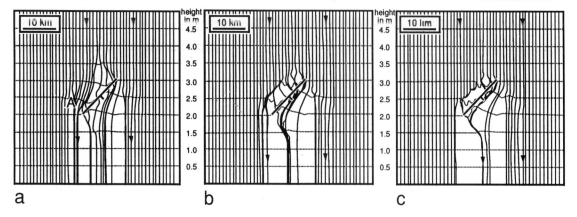

Fig. 5. Flow models for surfaces deformed by normal faults of different fault plane dips, 0.5 m offset and NE–SW orientation, downthrowing to the SE. (a) Fault plane dip 90°. (b) Fault plane dip 60°. (c) Fault plane dip 45°.

creased proportion of footwall uplift. With a fault dip of 90° (Fig. 5a), the greatest density of flow lines occurs beyond the tip of the fault (point A). A weak convergence of flow lines occurs in the hangingwall of the fault. At lower fault dips (Fig. 5b,c), there is a stronger concentration of flow lines in the hangingwall region of the fault. This simple model (Fig. 5) implies a greater axial divergence of flow in the example with lowest fault dip, thus predicting that channel sandstone body distribution and orientation will vary with fault dip.

Quarles (1953) records dips at 45° on synsedimentary faults in the Gulf coast alluvial deposits. We have used this value in the models to examine the effects of other fault properties presented below.

The offset across the fault

The pattern of flow around a fault depends on the displacement across that fault. By maintaining constant fault orientation this effect can be illustrated as in Fig. 6. In this example, as in most cases, the amount of disturbance to the flow pattern is proportional to the magnitude of fault displacement. With a small offset across the fault (0.25 m in Fig. 6a), only minor deflection of flow lines occurs. The depositional slope is sufficient for the flow lines to cross the fault scarp and continue downslope. Increasing the offset (0.5 m in Fig. 6b) causes some 'ponding' of flow lines to occur in the hangingwall of the fault. The flow lines escape close to the fault

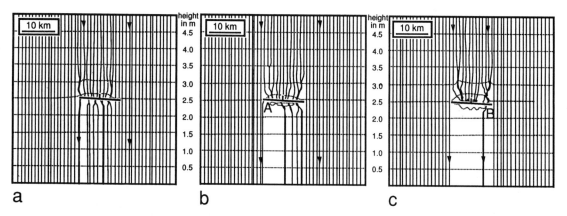

Fig. 6. Flow models for surfaces deformed by normal faults with different offsets, 45° fault plane dip and E–W orientation, downthrowing to the N. (a) Fault offset 0.25 m. (b) Fault offset 0.5 m. (c) Fault offset 1.0 m.

tips where the offset is less (A on Fig. 6b). Increasing the offset further (1.0 m in Fig. 6c) generates a closed topographic low and intensifies the 'ponding' effect such that flow lines can only continue downslope at the right tip of the fault (B on Fig. 6c).

The magnitude of fault displacement controls the extent to which the flow line pattern is changed and the extent of flow line concentration. This predicts that the distribution and orientation of alluvial channel deposits depend on the magnitude of surface displacement. The 'ponding' of flow lines may predict areas of probable lake formation.

Orientation of fault with respect to palaeoslope

The effect of a fault on alluvial processes depends on the orientation of that fault with respect to slope and source area. This effect is illustrated in Fig. 7 where five faults with the same geometry of displacement but differing orientations are modelled. In all five examples there is a strong concentration of flow lines in the hangingwall of the fault, predicting a greater probability of channelized deposits than average for the modelled area. The orientation of flow lines in the zones of high concentration mirrors the orientation of the faults. Downslope from the fault, the orientation of the fault controls the position of the region of flow line concentration. In Fig. 7a, b the flow lines concentrate downslope of the west tip of the fault (B on the maps), whereas in Fig. 7c–e the concentration is downslope of the centre of the fault.

The orientation of a fault in an alluvial setting controls the distribution of flow in the area of deformation and also the position of higher probability of flow in areas downslope of the fault.

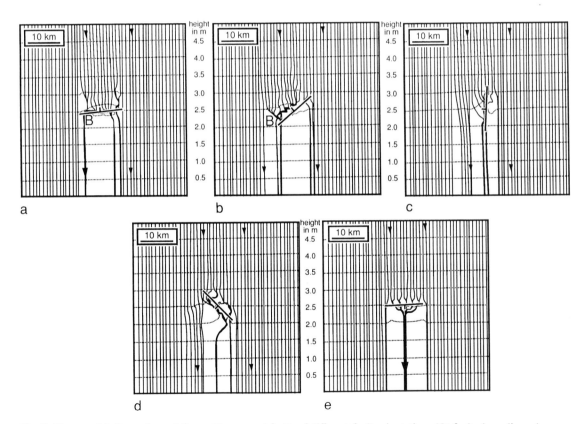

Fig. 7. Flow models for surfaces deformed by normal faults of different fault orientation, 45° fault plane dip and 0.5 m offset. (a) E–W fault trend, downthrowing N. (b) NE–SW fault trend, downthrowing NW. (c) N–S fault trend, downthrowing W. (d) NW–SE fault trend, downthrowing SW. (e) E–W fault trend, downthrowing S.

DISCUSSION

The models presented here are simplistic and should only be used to consider trends in facies distribution. The main advantage of this simplistic approach is that it considers the controls on facies in three dimensions and consequently it is a significant improvement on earlier two-dimensional simulations (e.g. Bridge & Leeder, 1979) that have tended to overemphasize the importance of axial channel systems and underplay the influence of lacustrine deposition. It is possible, using the flow line modelling methods, to construct more realistic models for understanding facies distribution.

The positions of flow line concentrations in the simple models presented here demonstrate that areas of high probability of flow, and therefore by inference channel activity, may occur on the hangingwall or footwall sides of a fault as a result of the combined effect of surface tilting and barrier formation. Flow line orientations produced by the surface deformation in these simple models indicates that although there are many situations where there is a high probability of finding axial channel systems, many combinations of fault geometry and slope produce patterns of flow with little or no axial drainage. Locally in many of these models (e.g. Fig. 5) flow lines on the hangingwall are orientated away from the fault scarp, a situation that many people find difficult to envisage but that may occur in nature.

The architecture of an alluvial suite deposited, for instance, on the hangingwall of an active fault, contrary to commonly held beliefs, is not always strongly dominated by channel deposits. There are a number of ways in which this can be explained by the use of flow line models. Firstly, fine sediment deposition may occur in a tectonically subsiding area as a result of lake formation in the topographic depression; this may be predicted by flow line 'ponding'. Lakes are very common in alluvial environments where there is active surface deformation (see Alexander & Leeder, 1986, for example) and their importance in the rock record has been largely underplayed in alluvial facies architecture models to date. It must be remembered, however, that not all closed depressions will contain lakes and that with deposition, topographic lows may be rapidly filled and erosion may counteract the topographic barrier. Consequently although the model presented here improves the prediction of sites of lacustrine facies deposition, it may overstate the effects of ponding by small topographic depressions. Secondly, other topographic features within, or upslope of, the area can have a strong influence on channel position by deflecting flow away from the topographic depression. It is very important therefore to model the topography of an area greater than that of specific interest and particular care must be taken when considering the topography in the upslope direction. Thirdly, fine sediment deposition in an area of flow line concentration may result from low sediment discharge within the channel present in the area of the hangingwall as discussed by Alexander (1986).

In most rock record examples, it is not possible to model the exact topography at any moment in time, due to limited data and the continually changing nature of topography in depositional areas. It is only possible to model topographic trends. This is of particular importance where there is periodic movement on a fault as the topography related to that fault movement will be continuously modified. It is conceivable that some faults will affect the depositional surface for limited periods of time; however surface deformation produces significant lateral variations in probability of flow over the area. Long-lived patterns of lateral variation are not produced by autocyclic processes. As the pattern of influence of deformation may be predicted, a model using that pattern will be more reliable at predicting the position and orientation of accumulations of channel bodies than a model for a situation where autocyclic processes dominate.

SUMMARY AND CONCLUSIONS

1 Computer-generated flow line models are potentially powerful predictive tools for facies distribution and a means for testing the validity of theories on synsedimentary tectonic control of facies distribution.
2 The simple models presented here go some way to demonstrate the importance of fault geometry, inclination and orientation for the pattern of flow over deforming surfaces.
3 The importance of lakes in tectonically influenced alluvial settings has been underplayed by facies architectural models to date. Flow models allow an improved prediction of probable sites of lake formation.
4 The simple models presented here suggest that the importance of axial channel systems, and chan-

nel deposit accumulations in the immediate hangingwall of faults, has been overstressed in the literature by the reliance on two-dimensional architectural simulations.

5 To understand facies distributions it is not sufficient to say that a fault was active and affecting the facies pattern: some understanding of the fault kinematics is required to understand the complex relations between surface deformation and sedimentation.

ACKNOWLEDGEMENT

We thank British Petroleum for funding this research project.

REFERENCES

ALEXANDER, J. (1986) Idealised flow models to predict alluvial sandstone body distribution in the Middle Jurassic Yorkshire Basin. *Mar. Petroleum Geol.* 3, 298–305.

ALEXANDER, J. & LEEDER, M.R. (1987) Active tectonic control of alluvial architecture. In: *Recent Developments in Fluvial Sedimentology* (Eds Ethridge, F.G., Flores, R.M. & Harvey, M.D.) pp. 243–252. Econ. Paleont. Miner., Tulsa, Spec. Publ. Soc. 39.

BAKER, V.R. (1983) Large-scale fluvial palaeohydrology. In: *Background to Palaeohydrology* (Ed. Gregory, K.J.) pp. 453–474. J. Wiley & Sons, Chichester.

BRIDGE, J.S. & LEEDER, M.R. (1979) A simulation model of alluvial stratigraphy. *Sedimentology* 26, 617–644.

BRISTOW, C.S. (1986) Brahmaputra river: channel migration and deposits. In: *Recent Developments in Fluvial Sedimentology* (Eds Ethridge, F.G., Flores, R.M & Harvey, M.D.) pp. 63–74. Soc. Econ. Paleont. Miner., Tulsa, Spec. Publ. 39.

GIBSON, J.R., WALSH, J.J. & WATTERSON, J. (1989) Modelling of bed contours and cross-sections adjacent to planar normal faults. *J. Struct. Geol.* 11, 317–328.

QUARLES, M. JR (1953) Salt-ridge hypothesis on origin of Texas Gulf Coast Type of faulting. *Bull. Am. Assoc. Petrol. Geol.* 37, 489–508.

STEKATEE, J.A. (1958) Some geophysical applications of elastic theory of dislocations. *Can. J. Phys.* 36, 192–205.

TETZLAFF, D.M. & HARBAUGH, J.W. (1989) *Simulating Clastic Sedimentation.* Van Nostrand Reinhold, 202 pp.

Geomorphic and Structural Controls on Alluvial Systems

Geomorphic and structural controls on facies patterns and sediment composition in a modern foreland basin

J.F. DAMANTI*

Department of Geological Sciences, Cornell University, Ithaca, NY 14853, USA

ABSTRACT

First-order controls on foreland-basin sedimentation can be inferred from facies patterns and sediment composition in the modern Bermejo basin of west-central Argentina. Sediment composition and texture are controlled to a great extent by drainage systems in the bounding structural provinces. The drainage networks themselves are products of complex interactions among structural style, bedrock lithology and climate.

Local drainage divides form on linear basin-bounding structures isolating portions of the sedimentary basin from sediment sources located in the interior of the mountain belt. Laterally extensive (along strike) bajada facies are produced in these areas. The distribution of bajadas is controlled by structural geometry. Sediment composition reflects the adjacent source-rock lithology.

The broad, topographically high interior of the mountain belt facilitates the formation of large integrated drainage nets that typically extend over a variety of source-rock lithologies. Point-source fans form where trunk streams fed by large drainage networks breach basin-bounding structures. Fan composition is heterogeneous reflecting the variety of lithologies exposed in the headwaters, and may contain material derived great distances (>150 km) from the depositional site.

Variability in drainage geometry along the basin margin implies non-uniform sediment flux into the basin; the larger the drainage system the higher the sediment yield. Large drainage networks also may respond quickly to renewed uplift by releasing stored sediment. Small watersheds along the basin margin can only increase their sediment yield through increased bedrock erosion, and thus sediment supply changes may lag behind episodes of uplift.

INTRODUCTION

The study of modern sedimentary basins is an important step in understanding erosion and deposition in the geological past. Yet, documentation of modern analogues for ancient aggradational basins is sparse. The Bermejo foreland basin of west-central Argentina is an actively aggrading foreland basin that provides an opportunity to evaluate the relative importance of structural style, drainage geometry and source-rock lithology as controls on the texture and composition of basin fill. This paper presents a simple descriptive study of this aggrading sedimentary basin and its bounding structural provinces. These observations form the basis for speculation about spatial and temporal variations in sediment flux.

With the advent of satellite remote sensing, we have the ability to observe surficial processes over large areas and in otherwise inaccessible, remote locations. In addition, multispectral data provide lithological information on a basinwide scale. One first step in this rapidly expanding field is to document spatial patterns of erosion, sediment dispersal and deposition in tectonically active areas, and determine the first-order controls on foreland-basin sedimentation.

Observed patterns of sediment texture and composition in the Bermejo basin raise interesting questions about the spatial and temporary varia-

* Present address: 67–71 Yellowstone Blvd YB, Forest Hills, NY 11375, USA.

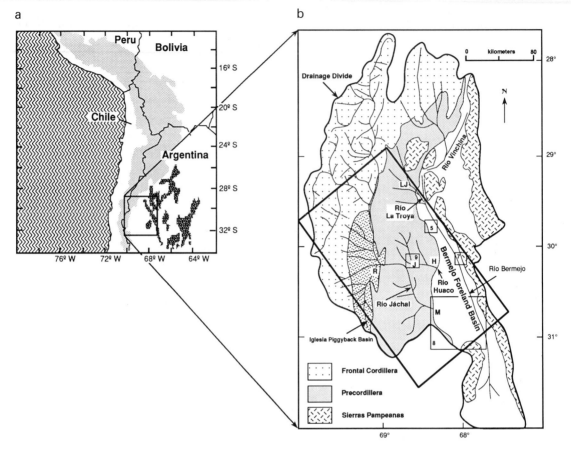

Fig. 1. (a) Location of study area in South America. Lightly shaded area has average elevation greater than 3 km. Sierras Pampeanas shown in dark shading. (b) Modern hydrographic Bermejo drainage basin. Principal river systems are labelled. R, Town of Rodeo; J, town of Jáchal; H, town of Huaco; M, town of Mogna; LJ, town of Las Juntas. Large outlined box indicates location of Fig. 2. Small numbered boxes indicate location of Figs 5, 7, 8 & 9.

bility of sediment flux. The size and shape of drainage networks supplying sediment to the basin vary spatially. This implies a spatial variability in the volume of sediment delivered by the different networks. Inasmuch as these drainage systems are evolving, so too will sediment flux patterns evolve. Stream systems typically expand by headward erosion; with the increased size of a catchment basin, there is an accompanying increase in sediment yield to a particular point in the sedimentary basin. Such observations and attendant inferences have significant implications for the development of mathematical and conceptual models of foreland-basin sedimentation, and will help guide interpretations of ancient basin fill.

GEOLOGICAL CONTEXT

The Bermejo watershed is located between approximately 27–32° S latitude and 67–70° W longitude, and occupies ~68 252 km^2 of central western Argentina (Fig. 1). The watershed includes a foreland basin which has been accumulating synorogenic sediment for the last 18 million years (Johnson *et al.*, 1986; Reynolds *et al.*, 1990). It contains up to 8 km of Cenozoic fill and is actively aggrading today (Figs 2 & 3).

The Frontal Cordillera is the main topographic expression of the Andes Mountains in the study area (Figs 1 & 2). It contains an assemblage of late Palaeozoic volcanic, plutonic and low-grade

Fig. 2. Large-format photograph taken from the Space Shuttle *Challenger*. Refer to Fig. 1 for orientation, location and principal features.

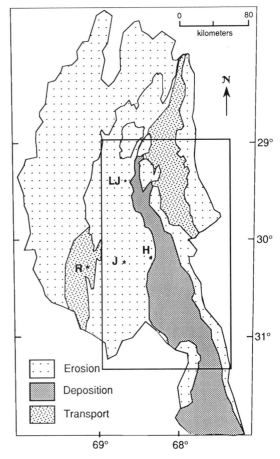

Fig. 3. Division of Bermejo watershed into zones of erosion, deposition and transport. Towns are lettered as in Fig. 1. Outlined box indicates location of Figs 4 & 6.

metasedimentary rocks and Miocene volcanics (Caminos, 1979; Maksaev et al., 1984). Movement on reverse faults with significant vertical throw has been dated as Miocene. Field relations in Chile and Argentina indicate that most of the faulting occurred between ~18.9 and 16.6 Ma (Maksaev et al., 1984; Ramos et al., in press) and that minor faulting continued between ~11 and 6 Ma in Argentina.

The Frontal Cordillera is flanked on the east by the Precordillera thrust belt (Fig. 1), a deformed wedge of Palaeozoic and Mesozoic marine clastics and carbonates, continental siliciclastics and minor volcanic rocks (Furque, 1972; Furque & Cuerda, 1979; Ortiz & Zambrano, 1981). The Precordillera has been subdivided into three segments, the Western, Central and Eastern Precordillera, based on stratigraphic and structural criteria (Ortiz & Zambrano, 1981). The Western and Central Precordillera are characterized by thin-skinned east-verging folds and faults. The Eastern Precordillera consists of west-verging structures that either involve basement or are part of a triangle zone not involving basement (Fielding & Jordan, 1988). Deformation of the Precordillera began about 20 Ma (Jordan et al., 1993) and is seismically active today (Smalley, 1988).

In the southern part of the study area, the Frontal Cordillera and Precordillera are separated by the Iglesia basin (Figs 1 & 2). This feature represents a Miocene piggyback basin (Beer et al., 1990) of the type described by Ori & Friend (1984).

The eastern boundary of the study area is marked by the Sierras Pampeanas (Figs 1 & 2), a region of reverse-fault-bounded crystalline basement blocks, generally with an east or west vergence, separated by broad basins (Caminos, 1979). The Sierras Pampeanas Ranges contain the only medium-grade metamorphic rocks within the drainage basin study area. Amphibolites, granulites, gneisses, schists and locally marbles are exposed (Furque, 1972; Caminos *et al.*, 1982)

METHODS

Sedimentary facies were digitally mapped on Thematic Mapper (TM) images using a high-resolution colour monitor and the International Imaging System (IIS) 600 software package. Area calculations were performed by digitally counting the number of pixels in a region of interest and converting to equivalent ground area in metres. Latitude and longitude coordinates were calculated from the header information provided with each image using a program written by E.J. Fielding of Cornell University. Catchment basins and river systems were digitally mapped on 1 : 1 000 000 scale and 1 : 500 000 scale topographic maps (ONC series, Defense Mapping Agency), and randomly checked for accuracy with TM.

Spectral data were collected using IIS 'region of interest' and 'statistics' commands which provide mean spectral reflectance for seven TM bands within the training sites.

GEOMORPHIC PARAMETERS

General

The Bermejo watershed can be divided into regions characterized either by aggradation (deposition), degradation (erosion) or transport (Fig. 3). Erosion is the dominant process over 73.8% of the drainage basin (50 398 km^2), and removes mass from portions of three morphotectonic regions: (i) Frontal Cordillera, (ii) Precordillera and (iii) Sierras Pampeanas (compare Figs 1 & 3).

Deposition occurs over 15.7% of the drainage basin (10 745 km^2). The principal locus of deposition is the main foreland basin (Figs 1–3). Local accumulations of sediment also are present within the thrust belt. However, because of their position within a zone of net uplift, these deposits have poor preservation potential.

Approximately 10.4% (7109 km^2) of the drainage basin appears to be neither aggrading nor eroding (Fig. 3). A thin veneer of sediment is present; however, underlying structures are clearly visible on TM images suggesting little net accumulation. River systems with their headwaters in the erosional zone pass through the transport areas and deliver sediment and water to the zone of aggradation.

Principal drainage geometries and facies products

Drainage systems in the mountains in part control the texture and composition of sediments deposited in the basin (Eisbacher *et al.*, 1974; Hirst & Nichols, 1986; Nichols, 1987). Three distinct drainage geometries exist in the study area: (i) short (~10 km), straight, low-order streams perpendicular to the basin margin; (ii) large, complex drainage nets made up of numerous low-order streams and a high-order trunk stream that flows across the thrust belt; and (iii) longitudinally draining fluvial system similar to (ii), but whose principle distributary flow is along the basin axis. Each drainage configuration produces a distinct sedimentary facies: (i) bajada; (ii) point-source alluvial fans; and (iii) fluvial floodplain, respectively (Fig. 4).

Bajada facies

Bajada deposits cover 25.7% (~2763 km^2) of the foreland basin. They consist of numerous small alluvial fans of approximately equal dimensions (Fig. 5). Fans coalesce to form an apron of sediment which is laterally extensive in one dimension (parallel to the basin-bounding structures). Perpendicular to strike, bajada deposits typically are 6–7 km in width and in all cases they are restricted to within 10 km of the basin margin, regardless of lithology or structural vergence of the source (Fig. 6). Bajada fans are fed by short, high-gradient, low-order streams emerging from local cuestas (Fig. 4). Feeder streams rarely extend more than ~10 km from the basin margin into the zone of erosion. These consequent streams form on newly uplifted basin-bounding structures, and are younger than or equal in age to those structures. This contrasts with antecedent rivers which form prior to the deformation of young structures through which they flow (Oberlander, 1985).

The composition of alluvial fan sediment can vary greatly along-strike, mimicking the variation

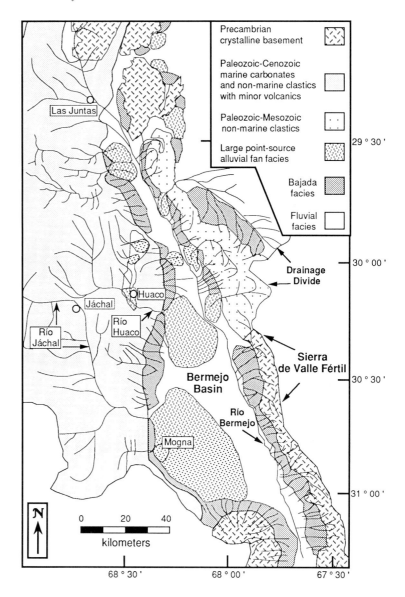

Fig. 4. Principal drainage and facies types and source rock lithologies of the Bermejo Valley. See Fig. 3 for location.

of bedrock exposures in the source (Fig. 7). Spatial variations in source-rock units often reflect structural complexities. However, bajada deposits of homogeneous composition, which are laterally extensive in the along-strike dimension, occur in areas of exposure of homogeneous source-rock lithology (Fig. 6). The key is that feeder streams are short, and thus sediment composition reflects the lithology of the source rock exposed along the adjacent basin margin.

Point-source alluvial fans

Alluvial fans, which are areally extensive in two dimensions, occur where runoff from large catchment areas is focused at points of entry to the foreland basin (Figs 1, 4 & 8). Such large fans cover 22.1% (~2371 km²) of the foreland basin. The size (from apex to toe) of point-source fans varies from ~10 to >50 km. Source-rock composition is heterogeneous and may include lithological

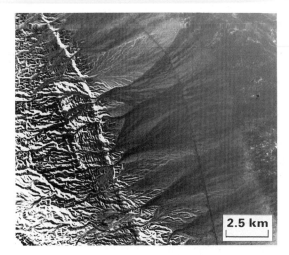

Fig. 5. Full resolution Thematic Mapper Image Bands 5-4-2 of typical bajada facies. Note numerous small fans coalescing to form bajada apron. North is to the top. See Fig. 1 for location.

units located great distances (>150 km) from the fan (Fig. 1).

Large drainage networks may form by capture of sub-basins in the case of consequent stream formation. They also may be antecedent. Antecedent rivers form when numerous streams coalesce and produce a trunk stream prior to deformation of structures through which they flow. The resultant trunk river erodes down through actively rising thrusts. Sediment composition may display complex temporal patterns and considerable blending of diverse lithologies as newly uplifted structures contribute material to the system (Damanti *et al.*, 1988; Steidtmann & Schmitt, 1988; Damanti, 1989). Some of the clasts carried by large trunk streams are derived from the interior part of the mountain belt (>150 km from the basin) and are not directly related to active structures.

Fluvial floodplain deposits

Floodplain deposits are associated with the axially draining Bermejo River and cover 52% (~5611 km^2) of the foreland basin. In most years the Bermejo River dries up south of about 31°S latitude due to evaporation and infiltration. The Bermejo is thus a closed foreland basin by virtue of its hydrographic properties. The overall width of the river floodplain is controlled by the essentially north trending Eastern Precordillera and the northwest trending Sierra de Valle Fértil. These structural provinces converge to form a triangular shaped basin in map view (Fig. 4). The floodplain is laterally extensive downstream, but its width varies from less than 10 km in the northern part of the basin to greater than 60 km in the southern part (Fig. 4). Locally, the position of the river channel is controlled by the volume of sediment influx perpendicular to the basin margin. Major deflections in the river's course occur at sites of encounters with point-source alluvial fans (Fig. 4).

The drainage configuration in the headwaters of the Bermejo River is similar to that for point-source fans (i.e. large well-integrated drainage net) (Fig. 1). However, fluvial floodplain deposits are produced, because distributary flow is along the gently southward slope of the basin axis. Poorly developed fans are reworked, and sediment is carried downstream. Considerable secondary mixing and amalgamation occur in the depositional basin as material enters the Bermejo fluvial system by transport perpendicular to the flow direction of the Bermejo River.

Rivers entering the basin perpendicular to the Bermejo River flow direction carry clasts whose composition reflects the adjacent source rock (Fig. 7). Composition may vary considerably along-strike as a function of structural complexities and lateral variations in source-rock exposure (Fig. 7). The proportion of particular detrital components in the Bermejo River varies along the direction of flow as material from separate sources is added at the boundary between fluvial floodplain and alluvial-fan deposits.

MAJOR DRAINAGE-BASIN SUBDIVISIONS

The Bermejo watershed can be divided into four main sub-basins, each draining a region of substantially different areal extent and substrate lithology. These sub-basins are informally referred to here as the Río Jáchal, Río Huaco, Río La Troya and Río Vinchina (Fig. 1). The Río La Troya drains the northern Precordillera. The Río Vinchina drains the westernmost Sierras Pampeanas and northern Precordillera. Together, these two sub-basins form the headwaters of the Río Bermejo. Neither of these drainage basins produces a well-developed alluvial fan at the point of entry into the sedimentary basin. This is probably because the rivers enter the basin

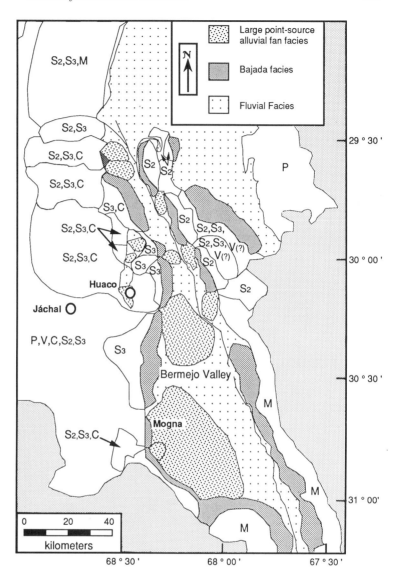

Fig. 6. Spatial distribution of drainage basins and sedimentary facies present in the Bermejo region (area indicated on Fig. 3). Patterned areas are depositional zones; unpatterned areas are source zones. Closed polygons with symbols are individual drainage basins. Lithologic symbols are as follows: S2, second-cycle sediment; S3, third-cycle sediment; C, carbonate (limestone); P, plutonic; V, volcanic. Note that source-rock composition does not control the size and shape of catchment basins.

at its apex where gradients are high (~2.4%) and evaporation and infiltration have not yet decreased discharge.

In contrast, the Río Jáchal and Río Huaco sub-basins feed the two largest alluvial fans in the basin (Figs 1 & 4). The Río Huaco has its headwaters within the Precordillera south of the village of Las Juntas. The Río Jáchal sub-basin (Fig. 1) is disproportionately larger than any other sub-basin (more than four times the size of the next largest sub-basin), and is the main system carrying material from the Frontal Cordillera to the foreland basin.

Río Jáchal drainage basin

The Río Jáchal drainage system (27 783 km^2) represents ~41% of the entire watershed. It is an extensive drainage network with headwaters reaching the continental divide (Fig. 1). Virtually all streams draining the Frontal Cordillera between 27° and 30°30′ S are diverted from their west–east flow by structural/topographical highs in the westernmost Precordillera (Figs 1 & 2). Streams originating in the Frontal Cordillera are blocked behind the thrust belt. The rivers then flow southward or

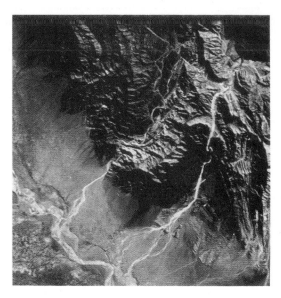

Fig. 7. Full resolution (false colour) Thematic Mapper Image Bands 5-4-2. Alluvial fans have same spectral signature as the immediately adjacent source unit. Río Bermejo is also shown truncating the alluvial fans. Composition in the river will change as it receives input from the individual fans. North is to the top. Horizontal scale approximately 15 km. See Fig. 1 for location.

Fig. 8. Full resolution Thematic Mapper Image Bands 5-4-2 of largest point-source fan in study area. North is to the top. Horizontal scale approximately 75 km. See Fig. 1 for location.

northward in the Iglesia valley until they reach Rodeo where they coalesce before crossing the Precordillera as a single river, the Río Jáchal (Figs 1 & 2).

In theory, the Río Jáchal may be antecedent to, or superposed on, the structures of the Precordillera, or it may have evolved by headward growth across the structures and subsequent capture of the Iglesia basin drainages. In the Western and Central Precordillera, the Río Jáchal cuts straight across anticlines and the hanging walls of thrust plates (Fig. 2). It is unlikely that headward erosion is a significant mechanism, given the lack of deflections in the river's course as it cuts across structures. Nor is there any evidence that the Río Jáchal is superposed on pre-existing structures. This is consistent with Furque (1972), who concluded, from extensive field mapping, that the Río Jáchal is antecedent to the structures of the Western Precordillera.

The Río Jáchal trunk stream may have formed at Rodeo in response to a transverse structural low or lateral ramp. The Iglesia piggyback basin represents the present structural low in the area around Rodeo, and a lateral ramp probably exists to the east near the city of Jáchal (R.W. Allmendinger, pers. comm, 1989). As deformation of the westernmost Precordillera occurred at this latitude, the Río Jáchal was able to erode down through the rising structures.

Farther to the east at the location of Jáchal, the present course of the Río Jáchal bends dramatically from nearly east–west to virtually north–south (Figs 1 & 2). Along this north–south stretch the river is structurally controlled. Downstream from this segment the river again trends east–west winding around structural lows (fold noses) and along faults in the Eastern Precordillera until it debouches into the Bermejo basin where it deposits a very large alluvial fan near the town of Mogna (Figs 4 & 8). The youngest rocks through which the Río Jáchal flows are ~2.5 million years old. Since the river is structurally controlled along this reach, and therefore younger than the structure, the large alluvial fan at Mogna must be <2.5 million years old.

Río Huaco drainage basin

The Río Huaco sub-basin covers a relatively large area (7104 km, or ~10% of the watershed) completely within the Central and Eastern Precordillera (Fig. 1). The upper reaches of the Río Huaco drain a structural low (the northern part of the Jáchal valley), two elevated thrust plates east and west of the Jáchal valley and a mountainous area of complex structures north of the valley. The present course of most of the Río Huaco is essentially east–west, nearly perpendicular to structural strike. In its present manifestation, the valley appears to be antecedent to the Central Precordillera (Fig. 2). Several lines of evidence suggest that this is not strictly true.

The lack of deflection of the Río Huaco valley as it crosses structural highs and its transverse east–west course, perpendicular to the thrust belt for nearly 30 km, suggest antecedence or superposition. Headward erosion and stream capture do not appear to be viable alternatives to antecedence given the straight river valley morphology and lack of evidence for young east–west trending faults (Oberlander, 1965, 1985; Furque, 1972). I conclude that the valley now occupied by the Río Huaco was formed by an antecedent river, but not by the Río Huaco.

A river with the discharge of the Río Huaco, which has a small catchment area in an arid climate, probably could not have had enough erosive power to incise the actively rising Central Precordillera in an antecedent manner. Major antecedent rivers require the development of a substantial headland drainage network in the hinterland prior to thrusting (Oberlander, 1965, 1985). Diagenetic studies and facies analyses (Beer & Jordan, 1989; Damanti & Jordan, 1989) indicate that the climate in this area was semi-arid to arid for most of the last 14 Ma. It is likely, therefore, that the river that cut the Río Huaco valley was large, perhaps related to the ancestral Río Jáchal. The Río Jáchal subsequently abandoned the antecedent valley probably in response to tilting during the Quaternary (see below).

Convincing evidence for tilting of the Jáchal valley during the Quaternary comes from TM images of alluvial fans. Each of the modern alluvial fans of the Jáchal valley (Fig. 1) experienced a southward shift of the active segment of the fan (Fig. 9). The older portions of the fans can be recognized because they are more dissected and more vegetated than are younger portions (Damanti, 1990). In virtually every fan on the east and west flanks of the Jáchal valley the older portions of the fans are to the north and the actively depositing segments are to the south. The shift in active deposition suggests that the valley has experienced a recent southward tilting. Deformation across a lateral ramp which may exist within the thrust belt at

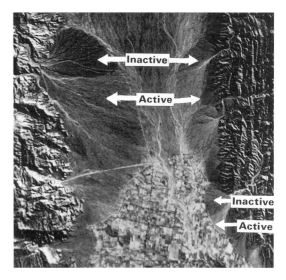

Fig. 9. Full resolution Thematic Mapper Image Bands 5-4-2 of Jáchal Valley. The southward shift of active fan segments is clearly visible. North is to the top. Horizontal scale approximately 15 km. See Fig. 1 for location.

this latitude (R.W. Allmendinger, pers. comm., 1989) could have produced such a tilting. This deformation may also be responsible for the southward deflection of the Rio Jáchal from its previous east–west course across the entire Precordillera.

CONTROLS ON FACIES PATTERNS AND SEDIMENT COMPOSITION

General

The principal controls on sediment texture and composition are: (i) source-rock lithology; (ii) structural geometry; (iii) climate; and (iv) drainage geometry. The climate throughout the Bermejo basin is arid, and therefore climate can be eliminated as a spatial variable in this study.

Individual catchment basins and their compositional makeup are illustrated in Fig. 6. These relationships demonstrate that source-rock lithology is not the primary control on gross dispersal patterns. All three geometries and facies are produced from a variety of provenances. Bajada facies are distributed around the perimeter of the basin. Point-source alluvial fans, although more abundant on the west side of the basin, are found associated with crystalline basement source rocks on the east side as well. The differences in bedrock lithology between adjacent sub-basins emphasize along-strike variability in a linear fold-thrust belt typically considered to be relatively uniform along-strike.

Drainage basin controls

The size and shape of the drainage basin are primary controls on facies and sediment composition. Source-rock lithology may affect the grain size distribution within a particular facies, but it does not control the overall depositional environment. On both the east and west sides of the basin, the highlands are formed by roughly north–south trending linear structures, which create local drainage divides separating the basin from more interior portions of the mountain belt. These local divides are formed by uplift associated with folding and thrusting, and occur on structures with both east and west vergence. The position of the bajada facies around the perimeter of the Bermejo valley demonstrates the importance of structural geometry.

Some point-source fans, on the other hand, are directly controlled by neither basin-bounding structures nor source-rock lithology. This makes the prediction of their location in ancient strata difficult. Although the size and shape of the drainage basins depend on structural style and lithology, the drainage geometry produced by the combined effects of these controls often breaches the structural trend.

Alluvial fan dimensions

Several workers (Bull, 1964; Denny, 1965; Hooke, 1968) have recognized that the area occupied by an alluvial fan increases with drainage-basin area. This relationship is expressed by the following equality:

$$A_f = C(A_d)^n$$

where C and n are empirically determined. This relationship holds for alluvial fans in the Bermejo basin, and can be expressed by $A_f = 0.78(A_d)^{0.9}$ for fans as a group (Fig. 10a). The relationship is not a function of structural province or gross bedrock lithology (Fig. 10a). However, if the fans are grouped by type into bajada and point-source alluvial fans, the data show a stronger correlation (Fig. 10b).

The bajada facies is best described by $A_f = 0.79(A_d)^{1.0}$, and the point-source fans by $A_f = 0.77(A_d)^{0.8}$. A value of $n < 1$ typically is in-

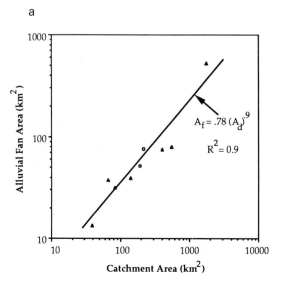

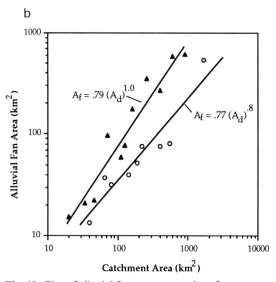

Fig. 10. Plot of alluvial fan area versus size of catchment basin. (a) (▲) Precordillera; (○) Sierras Pampeanas. There is no indication that the relationship varies according to tectonic province. (b) (○) Point-source fans; (▲) bajada. A stronger correlation is evident when fans are grouped according to type (i.e. bajada fan or point-source fan).

ferred to indicate storage of sediment in the catchment basin (Langbein & Schumm, 1958; Hooke, 1968). A value of $n = 0.8$ for point-source fans is consistent with their extensive catchment areas which provide a significant potential for sediment storage. More efficient transport of sediment from source to basin for bajada facies is suggested by the value of $n = 1.0$. Although the size of the drainage basin is independent of structural province, the largest fans clearly require extensive watersheds covering areas of considerable topographical relief. Structural style may, therefore, limit the size of alluvial fans by limiting the area affected by uplift (Fig. 4).

GEOLOGICAL IMPLICATIONS

The Bermejo drainage patterns and associated sedimentary facies have important implications for interpreting ancient strata and understanding basin-filling processes. The observations presented here also can be used to guide the further development of mathematical and conceptual models of basin evolution. The study of ancient basins from around the globe has led to conflicting interpretations and conclusions of the causal relationship between thrusting and sediment dispersal. For example, on ongoing debate concerning whether progradation of coarse-grained, basin-margin deposits occurs during periods of thrusting or quiescence (Blair & Bilodeau, 1988; Burbank et al., 1988; Heller et al., 1988; Flemings & Jordan, 1990) can be addressed with data from modern basins.

The crux of the conflict is whether times of active thrusting are indicated by the appearance of coarse-grained deposits or fine-grained deposits in a cyclic overall upward coarsening sediment package. Although Burbank et al. (1988) concluded that no single model could explain the variability among different basins, Blair & Bilodeau (1988) proposed that the onset of deposition of finer-grained sediments is a 'more consistent and more logical indicator' of tectonic activity in virtually all types of sedimentary basins (e.g. rifts, pull-apart basins *and* foreland basins). They suggest that this interpretation is consistent with: (i) the major changes in basin and source-area geomorphology dictated by cyclothems; (ii) modern examples; (iii) the disparity between the rate of response to tectonic subsidence of the environments that produce fine-grained units and those that produce coarse-grained units; (iv) the disparity between the rates of tectonic uplift and erosion; and (v) the unlikely event that a rapidly and widely dispersed coarse clastic wedge can form during active tectonic episodes (Blair & Bilodeau, 1988).

Evidence from the Bermejo basin demonstrates that these assertions are not always true. As the area of the catchment basin increases the size of the alluvial fan produced also increases. In other words, as catchment basins increase in size their total sediment yield also increases. Therefore, sediment flux must be greater for point-source fans than for bajada fans. This simple observation illustrates that, given a particular subsidence rate, the basinward extent of marginal facies is a function of sediment supply (implicit in the arguments presented in the above mentioned publications). Large point-source fans prograde out across the basin, whereas bajada fans are restricted to the basin margin. Given these relationships, it may be possible to predict the clastic-wedge response to episodic deformation as a function of facies or source-area drainage geometry.

Point-source fans extend many tens of kilometres into the foreland basin. Thus, a widely dispersed clastic wedge can form during active tectonic episodes, given sufficient sediment flux. The disparity between uplift rates and erosion rates was discussed at some length by Schumm (1963). This work focuses attention on the much higher rates of uplift relative to denudation in tectonically active regions. In the context of the argument presented here, the key parameter is the size of the catchment basin and the geometry of the transport network. Although uplift might outpace erosion locally, the net flux of sediment to a particular point in the basin could be sufficient to produce a widely dispersed clastic wedge.

Point (iii) of Blair & Bilodeau (1988) is perhaps the most important. It probably is true that progradation of bajada facies lags behind renewed uplift in the source. The small catchment areas combined with an arid climate to produce a local system which reacts slowly to tectonic activity. However the Bermejo basin example demonstrates that large point-source alluvial fans, in addition to longitudinal fluvial systems, are fed by extensive drainage networks, and also should respond quickly to subsidence. In addition, the value of $n = 0.8$, in the equation $A_f = C(A_d)^n$, suggests that a significant amount of sediment storage occurs in the headwater regions of point-source fans. This creates an even greater potential for rapid response of the clastic wedge to tectonic activity as stored sediment may be flushed during intervals of thrusting. The determining factor for a particular point in the sedimentary basin is the size of its catchment basin. This is because the size of the drainage net is the primary control on the volume of sediment delivered to a particular point in the basin.

This is one example of the many geological uncertainties that need to be addressed with data collected from modern basins. Although these data do not necessarily solve the problem, they do set boundary conditions. Such constraints from modern basins also are needed to guide mathematical models. Given the observations presented here, the next step may be to expand mathematical models to predict three-dimensional stratal geometries in a basin with spatially variable sediment influx.

SUMMARY AND CONCLUSIONS

The study of modern systems provides important information for interpreting ancient strata. It is clear that a three-dimensional perspective is required, and that proper emphasis must be placed on the role of drainage network evolution.

Whether the drainage system enters the foreland basin as a high-order trunk stream (point source) or as a series of smaller low-order streams (line source) is of great importance to the sedimentary deposits produced. The drainage geometry controls in part the facies and range of sediment composition characteristic of each facies. The direct relationship between the size of alluvial fans and catchment basin area suggests the tendency for large alluvial fans to have heterogeneous compositions, whereas sediments on small fans will tend to be less varied and reflect the composition of immediately adjacent source-rock exposures.

On a basinwide scale, the spatial distribution of lithologically distinct sediment packages is an indication of variations in source-rock distribution and, therefore, along-strike structural changes. Along-strike variability in bedrock exposure is best inferred from bajada sediments, because their composition is representative of the adjacent source rock. Sediment deposited on point-source alluvial fans is less conclusive, because their catchment areas extend over a diverse suite of lithologies.

Interior portions of the mountain belt continue to supply material to the sedimentary basin after the deformation front has migrated tens of kilometres toward the foreland, producing considerable 'noise' in the clast data, and providng a sediment source external to the thrust load. Since point sources can be associated with antecedent streams, detrital components may contain information about the evolution of several spatially separated structures through which they flow. In addition, because of their size, point-source fans are likely to be encountered in the rock record.

Although not independent of structural geometry and source-rock distribution, the drainage network is the immediate control on spatial and temporal variation in sediment composition. Even with extensive knowledge of source-rock lithologies, the significance of sediment texture and composition can be erroneously interpreted, without knowledge of the type of drainage system that produced them.

ACKNOWLEDGEMENTS

Funding for this research was provided by the National Aeronautics and Space Administration GSRP (contract NGT-50242). TM images and supporting software were purchased under an additional grant from NASA (contract NAS5–28767). I thank the members of the Cornell Andes Project, in particular Brian Isacks, Teresa Jordan, Arthur Bloom and Rick Allmendinger, for discussions that helped focus this work. I also thank Eric Fielding for his assistance with the image processor.

REFERENCES

BEER, J.A. & JORDAN, T.E. (1989) The effects of Neogene tectonism on sandy, ephemeral deposition in the Bermejo Basin, Argentina. *J. Sedim. Petrol.* **58**, 330–345.

BEER, J.A., ALLMENDINGER, R.W., FIEGUEROA, D.E. & JORDAN, T.E. (1990) Seismic stratigraphy of a nonmarine piggyback basin, Argentina. *Bull. Am. Assoc. Petrol. Geol.* **74**, 1183–1202.

BLAIR, T.C. & BILODEAU, W.L. (1988) Development of tectonic cyclothems in rift, pull-apart, and foreland basins: sedimentary response to episodic tectonism. *Geology* **16**, 517–520.

BULL, W.B. (1964) Geomorphology of segmented alluvial fans in western Fresno County, California. US Geol. Surv. Prof. Paper 437-A, pp. A1–A7.

BURBANK, D.W., BECK, R.A., RAYNOLDS, R.G.H., HOBBS, R. & TAHIRKHELI, R.A.K. (1988) Thrusting and gravel progradation in foreland basins: a test of post-thrusting gravel dispersal. *Geology* **16**, 1143–1146.

CAMINOS, R. (1979) Cordillera Frontal. *Segundo Simposio de Geología Regional Argentina* **1**, 397–454.

CAMINOS, R., CINGOLANI, C.A., HERVE, F. & LINARES, E. (1982) Geochronology of the pre Andean metamorphism and magmatism in the Andean Cordillera between latitudes 30° and 36°S. *Earth Sci. Rev.* **18**, 333–352.

DAMANTI, J.F. (1989) *Evolution of the Bermejo foreland*

basin: provenance, drainage development and diagenesis: Unpublished PHD thesis, Cornell University.

DAMANTI, J.F. (1990) Lithologic mixing in a modern foreland basin: evidence from Landsat Thematic Mapper Images. *Geology* **18**, 835–838.

DAMANTI, J.F. & JORDAN, T.E. (1989) Cementation and compaction history of synorogenic foreland basin sedimentary rocks from Huaco, Argentina. *Bull. Am. Assoc. Petrol. Geol.* **73**, 858–873.

DAMANTI, J.F., JORDAN, T.E. & BEER, J.A. (1988) Progressive thrusting and sediment dispersal patterns in the Andean foreland: 29°–31° South latitude. *Geol. Soc. Amer. Abstr. with Programs* **20**, 396.

DENNY, C.S. (1965) Alluvial fans of the Death Valley region, California and Nevada. US Geol. Surv. Prof. Paper 466.

EISBACHER, G.H., CARRIGY, M.A. & CAMPBELL, R.B. (1974) Paleodrainage pattern and late-orogenic basins of the Canadian Cordillera. In: *Tectonics and Sedimentation* (Ed. Dickson, W.R.) pp. 143–166. Soc. Econ. Paleont. Miner., Tulsa, Spec. Publ. 22.

FIELDING, E.J. & JORDAN, T.E. (1988) Active deformation at the boundary between the Precordillera and Sierras Pampeanas, Argentina, and comparison with ancient Rocky Mountain deformation. Geol. Soc. Am. Memoir **171**, pp. 143–163.

FLEMINGS, P.B. & JORDAN, T.E. (1990) Stratigraphic modelling of foreland basins: interpreting thrust deformation and lithosphere rheology. *Geology* **18**, 430–434.

FURQUE, G. (1972) Descripción geológica de la Hoja 16b, Cerro La Bolsa. Servicio Geolólico Nacional (Argentina), Boletín 125.

FURQUE, G. & CUERDA, A.F. (1979) Precordillera de La Rioja, San Juan y Mendoza. *Segundo Simposio de Geología Regional Argentina* **1**, 455–522.

HELLER, P.L., ANGEVINE, C.L., WINSLOW, N.S. & PAOLA, C (1988) Two-phase stratigraphic model of foreland-basin sequences. *Geology* **16**, 501–504.

HIRST, J.P.P. & NICHOLS, G.J. (1986) Thrust tectonic controls on Miocene alluvial distribution patterns, southern Pyrenees. In: *Foreland Basins* (Eds Allen, P.A. & Homewood, P.) pp. 247–258. Spec. Publs Int. Ass. Sediment. 8.

HOOKE, R.L. (1968) Steady-rate relationships on arid-region alluvial fans in closed basins. *Am. J. Sci.* **266**, 609–629.

JOHNSON, N.M., JORDAN, T.E., JOHNSSON, P.A. & NAESER, C.W. (1986) Magnetic polarity stratigraphy, age and tectonic setting of fluvial sediments in the eastern Andean foreland basin, San Juan Province, Argentina. In: *Foreland Basins* (Eds Allen, P.A. & Homewood, P.) pp. 63–75. Spec. Publs Int. Ass. Sediment. 8.

JORDAN, T.E., ALLMENDINGER, R.W., DAMANTI, J.F. & DRAKE, R.E. (1993) *J. of Geology* **101**, 137–158.

LANGBEIN, W.B. & SCHUMM, S.A. (1958) Yield of sediment in relation to mean annual precipitation. *Am. Geophy. Union Trans.* **39**, 1076–1084.

MAKSAEV, V., MOSCOSO, R., MPODOZIS, C. & NASI, C. (1984) Las unidades volcánicas y plutónicas del Cenozoico superior en la Alta Cordillera del Norte Chico (29–31°S): geología alteración hidrotermal y mineralización. *Revista Geológica de Chile* **21**, 11–51.

NICHOLS, G.J. (1987) Structural controls on fluvial distributary systems – the Luna system, northern Spain. In: *Recent Developments in Fluvial Sedimentology* (Eds Ethridge, F.G., Flores, R.M. & Harvey, M.D.) pp. 269–278. Soc. Econ. Paleont. Miner., Tulsa, Spec. Publ. 39.

OBERLANDER, T.M. (1965) *The Zagros Streams: A New Interpretation of Transverse Drainage in an Orogenic Zone.* Syracuse Geographical Series 1, 168 pp.

OBERLANDER, T.M. (1985) Origin of drainages transverse to structures in orogens. In: *Tectonic Geomorphology* (Eds Hack, J.T. & Morisawa, M.) pp. 155–182. The Binghampton Symposia in Geomorphology: Int. Series 15.

OPERATIONAL NAVIGATION CHART (1980) Sheet Q-26, Edition 3. Defense Mapping Agency Aerospace Center, St Louis Air Force Station, MO.

ORI, G.G. & FRIEND, P.F. (1984) Sedimentary basins formed and carried piggyback on active thrust sheets. *Geology* **12**, 475–478.

ORTIZ, A. & ZAMBRANO, J.J. (1981) La Provincia Geológica Precordillera Oriental. *Actas del Octavo Congreso Geológico Argentino* **3**, 59–74.

RAMOS, V.A., KAY, S. & PAGE, R.N. (in press) La ignimbrita Vacas Heladas y el cese del volcanismo en la Valle del Cura-Provincia de San Juan. In: *Andean Volcanism* (Ed. Coira, B.). Asociación Geológica Argentina.

REYNOLDS, J.H., JORDAN, T.E., JOHNSON, N.M., DAMANTI, J.F. & TABBUTT, K.D. (1990) Neogene deformation of the flat-subduction segment of the Argentine–Chilean Andes: chronological constraints from Las Juntas, La Rioja Province, Argentina. *Bull. Geol. Soc. Am.* **102**, 1607–1622.

SCHUMM, S.A. (1963) The disparity between rates of denudation and orogeny. US Geol. Surv. Prof. Paper 454-H, pp. H1–H13.

SMALLEY, R. (1988) *Two earthquake studies: (1) seismicity of the Argentine Andean Foreland and (2) a renormalization group approach to earthquake mechanics.* Unpublished PhD thesis, Cornell University.

STEIDTMANN, J.R. & SCHMITT, J.G. (1988) Provenance and dispersal of tectogenic sediments in thin-skinned thrusted terrains. In: *New Perspectives in Basin Analysis.* (Eds Kleinspehn, K.L. & Paola, C.) pp. 353–366. Springer-Verlag, New York.

Quaternary alluvial fans in southwestern Crete: sedimentation processes and geomorphic evolution

W. NEMEC* and G. POSTMA†

*Geological Institute (A), University of Bergen, 5007 Bergen, Norway; and
†Comparative Sedimentology Division, Institute of Earth Sciences, University of Utrecht, PO Box 80.021, 3508 TA Utrecht, The Netherlands

ABSTRACT

This is a study of the Quaternary alluvial sedimentation and its controls in the piedmont zone of the Lefka Ori limestone massif in southwestern Crete, Greece. The piedmont was sculptured by wave erosion during a major marine transgression in Pliocene time. The sea-level highstand controlled also the development (erosion base level) and early alluvial infill of large intramontane valleys, whose outlets at the mountain front are 'hanging' high above the piedmont plain. These palaeovalleys were once filled with thick gravelly alluvium, and periodically conveyed large sediment and water discharges during Quaternary times, when the highest mountain range was subject to inferred glaciations and deglaciations. The palaeovalleys have later been degraded by deep incision of narrow axial trenches. The Quaternary piedmont deposits occur as an array of coalescent alluvial-fan complexes, each comprising several generations of superimposed fans built of limestone gravel. Three main generations of fans have been distinguished as broadly isochronous depositional 'stages'.

1 The older fans (earliest Pleistocene), which have apices at altitudes of 300–400 m, are relatively small (radii of $c.1$ km), very steep (20–22° near apices) and composed of reddish debris-flow deposits. These fans were built of highly immature debris derived locally through headward erosion of mountain-front ravines, and received little or no sediment from the pre-existing intramontane valleys. Climate was probably arid to semiarid, such that the valleys stayed inactive. The fan toes were modified and cliffed by a marine incursion into the piedmont zone.

2 The younger (Pleistocene) fans, which are several tens of metres thick and constitute the bulk of the piedmont alluvium. They have apices at altitudes of 320–370 m and comprise at least five generations of variously stacked fan lobes. These fans are relatively large (radii in excess of 2 km) and steep (up to 13°), but consist of streamflow deposits. Moreover, their longitudinal morphometric profiles tend to be convex, broadly sigmoidal in shape. The deposits are relatively well-sorted, grey gravels that occur as alternating channel-lag and channel-bar sheets. This stage involved large water discharges, attributed to mountain ice-cap melting, whereby the pre-existing intramontane valleys had been activated as principal drainage and sediment-yield areas. The hanging valley outlets made the systems prograde on to the piedmont plain as relatively steep, 'cascading' fans. The preserved toes of smaller, lower-lying fans in the east show the record of deposition in a protected marine embayment.

3 The Holocene fans and trench-confined lobes, which are of minor volumetric importance and consist of grey, unconsolidated, openwork 'sieve' gravels. Their origin is attributed to ephemeral stream-flood surges issued from the degraded intramontane valleys in semiarid, possibly monsoonal, climatic conditions.

INTRODUCTION

Modern sedimentological and geomorphological literature abounds with studies of alluvial fans (see reviews by Bull, 1977; Schumm, 1977; Rachocki, 1981; Nilsen, 1982, 1985; Rust & Koster, 1984; French, 1987; Schumm *et al.*, 1987), and these specific depositional systems have attracted much interest as sensitive sedimentary recorders of piedmont or basin-margin conditions (e.g. see recent selections of studies edited by Miall, 1978; Koster & Steel, 1984; Nemec & Steel, 1988; Rachocki & Church, 1990; see also Gloppen & Steel, 1981; Gibling *et al.*, 1987; Wells & Harvey, 1987; Blair & Bilodeau, 1988; Frostick & Reid,1989; Haughton, 1989; Hill, 1989; North *et al.*, 1989; Jolley *et al.*, 1990). Various aspects of alluvial-fan development have been studied in considerable detail, and the analytical approaches adopted by researchers range from sedimentological to purely geomorphological. However, there have been relatively few detailed, integrated studies of the actual relationships between alluvial-fan sedimentology and geomorphic evolution, especially in longer-lived piedmont systems (e.g. Harvey, 1987; Muto, 1987, 1988; DeCelles *et al.*, 1991). Surficial 'snapshot' studies of modern fans lack the critical time dimension, or stratigraphic perspective, whereas the studies of ancient fans generally fail to highlight the geomorphic aspects of fan and piedmont evolution, as these are scarcely apparent in the rock record.

As pointed out by Harvey (1984, 1987), few authors have sought to document and fully understand the role of geomorphic factors in the development of particular ancient fans. Sedimentological reconstructions of alluvial fans either rely totally on the limited geomorphic observations from modern systems, often adopting them quite uncritically as stereotype models, or merely invoke geomorphic controls as a kind of 'ballpark' variable. This common approach is rather unfortunate, if not somewhat paradoxical, since researchers generally agree that the geomorphic aspects of fan and piedmont evolution are far too important to sedimentological reconstructions to be ignored or treated in a superficial way.

The present study is from a late Cenozoic piedmont in southwestern Crete, Greece. The Quaternary alluvium there comprises an array of gravelly alluvial-fan complexes (multiple fan systems), which evolved from small, debris flow-dominated cones, through relatively large, stream flow-dominated fans, to small fans built of sieve deposits. Moreover, the stream-dominated systems display quite unusual morphometric characteristics and comprise several generations of variously stacked, successive fan lobes. This study is a reconstruction of the depositional history of the alluvial-fan complexes, with emphasis on their geomorphic aspects, sedimentation processes and regional controlling factors. The study has several important implications for the sedimentological analysis of ancient alluvial fans. It shows that some of the common notions of alluvial-fan morphology as a simple function of sediment dispersal processes may not be as universal as widely asserted in textbooks and assumed in sedimentological reconstructions.

DEPOSITIONAL SETTING

Piedmont geology and geomorphology

In the study area in southwestern Crete (Fig. 1), the Quaternary alluvium occurs in a narrow coastal plain, which is a remnant of the late Cenozoic piedmont zone of the Lefka Ori (White Mts) limestone massif (Fig. 2). This is the only segment of the cliffed, retreating southern Cretan coast where a considerable portion of the Lefka Ori piedmont has been preserved.

The Lefka Ori massif, which sourced the alluvium, is the frontal part of a pre-Neogene to early Neogene thrust-pile of Mesozoic to early Tertiary limestones, ranging from metamorphic to nonmetamorphic. The nappe front remained inactive since its emplacement at the end of Miocene. The piedmont zone consists of a phyllitic-quartzitic basement (scarcely exposed) with a remnant cover of Pliocene littoral deposits, overlain unconformably by the Quaternary alluvium. The latter comprises coalescent alluvial-fan complexes built of limestone gravels. The Holocene coastal cliff cut in the piedmont deposits commonly reaches 10–20 m in height and its foot-zone shows local remnants of raised early Holocene beaches (Postma & Nemec, 1990). The cliffed coast today is subject to intense marine erosion and further bulk retreat.

The onshore geomorphology of the area is of particular importance here, as it reflects quite closely the original piedmont landscape in which the Quaternary alluvial sedimentation has taken place. The relative dryness and increasing aridity of the modern climate prevented major alterations of the landscape, and even the surfaces of the subrecent alluvial fans appear to have been altered relatively little. The study area comprises a number of distinct, narrow, coast-parallel geomorphological zones (Nixon *et al.*, 1988; see also Fig. 2).

1 The *coastal plain* (piedmont zone) is an arid, almost treeless area with an annual rainfall of less than 180 mm. This rain-shadow zone belongs to the so-called 'European Sahara', dominated by the dry

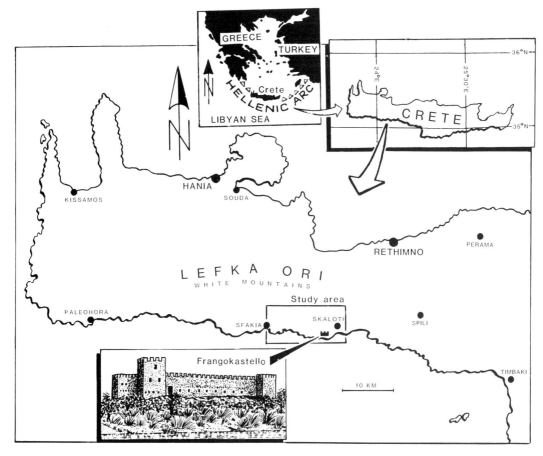

Fig. 1. Locality map of the study area in southwestern Crete, southern Greece.

northerly sirocco that carries African desert dust across the Libyan Sea.

2 *Lower mountain slopes* (foothill zone) are relatively steep, extend to an altitude of 800 m, and are dissected by the outlets of numerous intramontane valleys from which the Quaternary fans have prograded on to the piedmont plain (Figs 3 & 4). Notably, the valley outlets show 'hanging' palaeovalley mouths, at altitudes of 400–450 m, degraded by deeply incised, steeper, narrow, younger trenches whose floors at the outlets are at altitudes of c.200 m (Figs 2 & 3). This zone and coastal plain are irregularly covered with sparse phrygana and steppe vegetation, some undershrubs, minor shrubs and very sparse trees. Soil cover is patchy, poor to absent.

3 *Upper mountain slopes* (mountain zone) extend up to 1700 m, are less steep and generally less arid, but seldom have much soil cover. This zone, partly woodland, is cut by the main, intramontane segments of the valleys mentioned above; the valleys are trending north–south (Fig. 4), are V-shaped and several hundred metres deep.

4 *High mountains* comprise the highest zone of the Lefka Ori, with a peak (Mt Pahnes) at 2452 m only 8 km from the southern coast. This 'mountain desert', with an astonishing landscape of glacio-karstic origin, is quite barren of continuous vegetation. Our own brief geomorphological reconnaissance and the survey conducted by colleagues (Nixon *et al.*, 1988) indicate an earlier, presumably Pleistocene, high-altitude glaciation in the Lefka Ori. Although the exact character and history of this inferred glaciation are yet to be documented, there is much compelling geomorphic indication of probable cirque glaciers or an ice-cap with glaciers; the

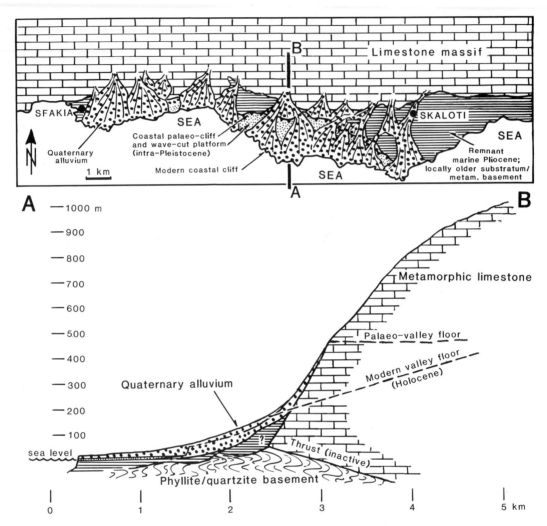

Fig. 2. Simplified geological map of the study area and a generalized cross-section, showing the occurrence of Quaternary alluvial fans in the remnant piedmont zone of the Lefka Ori limestone massif. (Note the exaggerated vertical scale in the cross-section.)

glaciers were presumably short and never descended into the lower-altitude intramontane valleys. Data on the past Quaternary climates of Crete are scarce, but glacial cirques and moraines have been found in the high part of Mt Ida, the easterly extension of the Lefka Ori massif (J. Shaw, pers. comm., 1990), and a cold, 'truly glacial' Pleistocene climate has been inferred for southern Greece (Pope & van Andel, 1984). High-altitude Pleistocene glaciations have also been documented in the adjacent southwestern Turkey (Messerli, 1967; Birman, 1968; Ardos, 1973; Butzer, 1975; Roberts, 1982, 1983). Even in the present-day warm and dry climate, the highest range of Lefka Ori has nearly permanent snow caps.

Probable Pleistocene glaciogenic features in the highest range of the Lefka Ori massif include the smooth, U-shaped profiles of many bedrock valleys. Among the more spectacular glaciokarstic features of the high Lefka Ori are also *oropedhia:* large intramontane plains in pod-like, oval or circular depressions, several kilometres across, with partial alluvial or fluvio-lacustrine infill, soils, human settlements and cultivation. The intramontane depres-

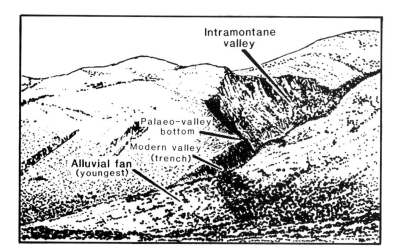

Fig. 3. Sketch of the outlet of an intramontane valley in the Lefka Ori foothill zone (for location, see Nomikiana fan complex in Fig. 4). Note the 'hanging' palaeovalley and the deeply incised younger trench. Overlay drawing from a photographic print.

sions and lower-altitude valleys probably acted as areas of temporal sediment storage during some periods of the Quaternary denudation history of the mountain range (see further discussion in text). The lower-altitude valleys (Fig. 4) show no evidence of glacier encroachment, but the mountain slopes there are of Richter type, typically periglacial (J. Shaw, pers. comm., 1990).

Tectonics and sea-level changes

The late Cenozoic sedimentary and palaeogeographic history of southern Crete involved frequent changes in relative sea level (Drooger & Meulenkamp, 1973; Fortuin, 1978; Pirazolli et al., 1982; Fortuin & Peters, 1984; Peters et al., 1985). At least some of them are thought to have been related to the tectonic movements of the Cretan block.

The island of Crete is the southernmost emerged part of the Hellenic Arc (Fig. 1, inset), an arcuate segment of the active convergent boundary between the African and Eurasian lithospheric plates in the eastern Mediterranean (Angelier et al., 1982; Horvath & Berckhemer, 1982; Ryan et al., 1982; de Boer, 1989). The compressional tectonic regime in the 'backarc' (Cretan-Aegean Sea) region to the north changed into an extensional one in late Miocene time, causing intense normal faulting and differential vertical movements (Le Pichon & Angelier, 1979; Horvath & Berckhemer, 1982; Lister et al., 1984; de Boer, 1989). While the central Aegean region generally subsided, further pulses of folding and thrusting occurred within the Hellenic Arc: the external (Cretan) non-volcanic arc emerged by the end of Miocene times, and the internal (South Aegean) volcanic arc subsequently developed at the end of Pliocene times (Mercier, 1977; McKenzie, 1978; Le Pichon & Angelier, 1979; Horvath & Berckhemer, 1982; Meulenkamp et al., 1988). At the same time, after a general regression related to the well-known Messinian 'salinity crisis', the region underwent major marine transgression (Cita, 1982; Horvath & Berckhemer, 1982).

The southern coast of Crete thus became subject to a stepwise marine transgression in Pliocene times, concurrently with some north–south tectonic extension and southerly listric faulting, whereby the front of the Miocene nappe-pile had been strongly eroded and cliffed (Fig. 2; see also Spaak, 1981; Peters et al., 1985). These latter authors have recognized Pliocene marine terraces (wave-cut platforms), in eastern Crete, as high as 450 m above the present-day sea level. Also, the associated marine facies and palaeontological data suggest progressive deepening of water during the Pliocene, with oceanic-type circulation and water depths of at least 500 m at the transgression maximum (Benson, 1972; Spaak, 1981; van Harten, 1984; Peters et al., 1985).

The Pliocene transgression apparently sculptured the piedmont and lower slopes of the Lefka Ori massif (Fig. 2), subsequently draping the drowned landscape with the littoral sands/gravels and abundant marls (see also Peters et al., 1985). Accordingly, we infer that the transgression, with a presumed late Pliocene highstand of sea level at an altitude of around 450 m (Peters et al., 1985, fig. 6), controlled also the development (erosion base

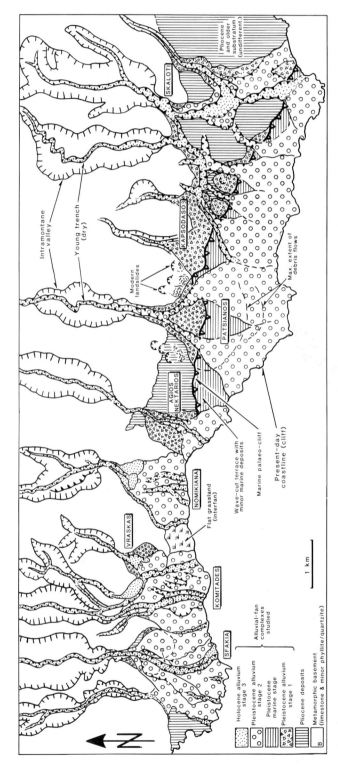

Fig. 4. Map of the Quaternary alluvial-fan complexes in the study area (cf. Fig. 2), showing their intramontane feeder valleys and three main stratigraphic divisions (depositional 'stages'). Lines with hatches indicate topographic scarps; the present-day coastline, though marked differently, is a scarp too (sea cliff).

level) and early alluvial infill of the intramontane palaeovalleys, whose outlets today are 'hanging' high above the piedmont plain (see Figs 2 & 3).

The marine transgression was followed by a more abrupt, major regression at the end of Pliocene times (Peters *et al.*, 1985), when the relative sea level dropped to somewhat below its present-day position and most of the Pliocene littoral deposits had been eroded. This relative drop in sea level, although considerable, did not equal the magnitude of the preceding sea-level rise, as indicated by the still submerged coastal palaeocliffs covered with Pliocene marls in eastern Crete (Peters *et al.*, 1985). The subsequent, Pleistocene sea-level stands with respect to the southern Cretan coast are thought to have been controlled by glacioeustatic variations, superimposed on the effects of mild tectonic uplift (Peters *et al.*, 1985). However, the actual pattern of these changes remains largely unknown. We have recognized an early Pleistocene coastal palaeocliff in the piedmont alluvium (see Fig. 2, map), some 15 m above the present-day sea level, and evidence of a later Pleistocene sea level (fan-delta deposits) around 10 m above the present-day level. Raised Holocene palaeobeaches in the study area (Postma & Nemec, 1990) and the radiometrically dated wave-cut notches in the coastal cliffs of western Crete (Pirazzoli *et al.*, 1982) further indicate Recent episodes of a northeasterly tilting of the Cretan block, some 1530 years BP (Fig. 5) and somewhat earlier. The present study of the piedmont alluvium suggests that similar episodes of gentle uplift, with an oblique easterly tilting of the piedmont, have probably punctuated also the earlier Quaternary history of the piedmont, in addition to probable glacioeustatic oscillations of sea level.

THE ALLUVIAL-FAN COMPLEXES

The Quaternary alluvium occurs as an array of coalescent fan complexes (Figs 2 & 4), each comprising several generations of variously stacked, successive fans. On the basis of conspicuous facies differences and geomorphic analysis, three main generations of fans have been distinguished as broadly isochronous depositional 'stages' (Fig. 4).
1 The older fans (?early Pleistocene), which are relatively small, very steep, and consist of debris-flow deposits.
2 The younger (Pleistocene) fans, which constitute the bulk of the piedmont alluvium, are generally

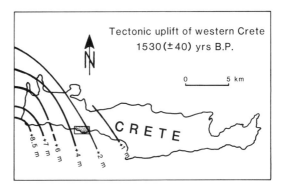

Fig. 5. Recent uplift of western Crete, inferred from radiometrically dated, wave-cut notches in the coastal cliffs of the island. (Modified from Pirazzoli *et al.* (1982).)

steep but consist of stream-flow deposits, and comprise at least five generations of superimposed fan lobes.
3 The Recent (Holocene) fans and related, trench-fill gravel lobes, which are of minor volumetric importance and consist of unconsolidated 'sieve' deposits.

The development and facies characteristics of these successive fans are the main topic of the present study.

Except for the oldest fans (stage 1), the gravelly alluvium was derived from the adjacent intramontane valleys (Fig. 4). These deep, bedrock valleys today are generally devoid of alluvium, with only sparse Recent gravel and large, isolated limestone blocks present along their axial trenches. However, the intramontane valleys are inferred to be older, late Pliocene geomorphic features (see preceding section). They were once filled with thick gravelly alluvium, and then periodically conveyed large discharges in Quaternary times. At least some of these 'hanging' palaeovalleys (Figs 2 & 3) show remnant terraces of a bedded gravelly alluvium on their walls, with preserved thicknesses of up to 30–40 m (for best-preserved example, see Fig. 9). The valleys are broadly V-shaped, with maximum widths of around 1 km, and their axial parts are deeply incised by the narrow, steeper, younger trenches (see Fig. 4 and cross-section in Fig. 2). In a plan-view, therefore, the dissected fan apices appear to be connected with the broad palaeovalley outlets by the neck-like 'breaching' segments of these narrow, incised trenches (Figs 3 & 4).

DEBRIS FLOW-DOMINATED FANS (STAGE 1)

General

These oldest fans, of latest Pliocene(?) to early Pleistocene age (time-frame uncertain), occur as small, very steep semiconical features directly at the mountain front (Fig. 4), against which they abut. Their apical parts, where preserved, are at altitudes of 300–400 m. The fans have been affected by an episode of contemporaneous marine erosion (see next section) and further modified by subsequent alluvial sedimentation (Fig. 4). The latter involved trenching, whereby the early-stage alluvial cones have been partly eroded and partly covered by the younger alluvium. Some of these cones have probably been totally destroyed or concealed (see Komitades and Nomikiana fan complexes in Fig. 4). The exposures of these older fans are thus very limited and generally poor.

The measured slopes of remnant fan-head segments are up to 20–22°, and could originally be slightly steeper on account of the later tectonic tilting of the Cretan block (see earlier section); however, the actual angular difference is unlikely to have significantly exceeded 1° and can be considered negligible (even more so for the younger fans). Longitudinal morphometric profiles, where traceable, are concave, with the cone slope decreasing to 8–9° merely a few hundred metres from the apex. Estimated fan radii (axial lengths) are around 1 km, but at least some of the cones appear to have extended their toes, by an additional few hundred metres, in the form of thin alluvial sheets. In the Patsianos fan (Fig. 4), for example, such distal 'feather-edge' alluvium comprises merely three or four gravel units, each c.1 m thick, interbedded with sandy marine deposits (described further below). Comparable distal outcrops of the other alluvial fans of stage 1 are not available.

Fan deposits

The alluvial-fan deposits of stage 1 are thick-bedded, unsorted, matrix-supported gravels with maximum clast sizes ranging from cobbles to large isolated boulders (Fig. 6). The debris is predominantly angular (Fig. 7), comprising fragments of local bedrock (metalimestone, commonly also phyllite and/or quartzite). Matrix is an unsorted mixture of finer-grained gravel, sand and a reddish-brown

Fig. 6. The oldest Pleistocene alluvium (stage 1), interpreted as debris-flow deposits, exposed in a road-cut section in the northeastern flank of Sfakia fan complex (Fig. 4). Palaeotransport direction is to the left, at 40° out of the picture. Overlay drawing from a photographic print.

mud of *terra rossa* type. In some cases, as in the Sfakia fan (Fig. 4), the matrix is rich in detrital mica derived from the phyllitic bedrock locally exposed to weathering and erosion at the mountain front. The abundance of reddened, muddy matrix and the polymict composition and distinct angularity of debris are particularly characteristic of these older fan deposits.

Bed boundaries are generally indistinct, commonly amalgamated, but the thicknesses of beds in most cases can be estimated as ranging from less than 1 m to slightly more than 2 m (Fig. 6). Beds seem to be tabular or broadly lenticular in vertical sections, are internally massive (unstratified), and their bases show little or no obvious scour. Clast fabric is disorganized, and there is also little or no clast-size grading within the beds. Large cobbles and boulders are 'floating' randomly, some in nearly

Fig. 7. Characteristic, highly immature texture of the Pleistocene alluvium of stage 1 (detail from same section as in Fig. 6). The photograph shows the basal, finer-grained portion of a cobbly gravel bed with crude, coarse-tail inverse grading. The lens cap (scale) is 5 cm.

vertical positions (Fig. 6). Beds with a flow-parallel (downfan) alignment of clast longer axes, or $a(p)$ clast fabric, are uncommon. Equally uncommon are beds with somewhat finer-grained basal portions or crude, coarse-tail inverse grading (Fig. 7).

Beds with scoured bases, more clast-supported textures and normal grading are rare. No gravel beds with tractional structures and no significant sandy interbeds have been observed within this older alluvium.

Interpretation

The alluvium of stage 1 is interpreted as debris-flow deposits. The sediment gravity flows are thought to have been cohesive, highly viscous and sluggish, of very low mobility, as indicated by their considerable clast-support competence (floating megaclasts), disorganized clast fabric (little or no pervasive shear strain) and deposition on steep slopes directly at the mountain front. The occasional beds with somewhat better-organized, $a(p)$ clast fabric imply flows that have experienced some pervasive laminar shear, albeit rather brief or transient. Beds with coarse-tail basal inverse grading represent flows whose lower parts were subject to more persistent laminar shear (whereby the largest clasts would lose support and settle from the shearing viscous mass, and thus be dropped from the flow), while the thicker upper parts behaved largely as non-shearing 'rigid plugs' (see Johnson, 1970; Naylor, 1980). The few debris-flow beds that extend some hundred metres beyond the cone toe, as in the Patsianos fan (Fig. 4), apparently represent some exceptionally mobile flows, which were more 'watery' and/or gained greater momentum by descending higher slopes.

The variation in internal bed characteristics discussed above is not accompanied by any obvious differences in the gravel texture or mineral composition. The inferred variation in the rheological behaviour of debris flows is then attributed primarily to the varied content and vertical distribution of water in the mobilized, slope-derived masses of debris. The rare beds with erosive bases and normal grading thus probably represent some water-rich sediment flows, whose movement on steep slopes involved intense shearing and possibly turbulent churning (see Lawson, 1982).

The fans are volumetrically small, developed essentially as localized prisms within a broader veneer of slope-waste debris that mantled the pre-existing bedrock topography of the mountain footzone. Local bedrock knobs, with a relief of a few tens of metres, have not even been mantled or overtopped by this alluvium (e.g. see later Figs 11 & 12). Climate was probably arid to semiarid, as indicated by the low sediment yield (small fan volumes), the strictly local derivation and extreme immaturity of debris, and the deposition solely by debris flows. The fans apparently formed through localized slope wastage and headward erosion of mountain-front ravines, due to differential weathering and retreat of the fractured mountain front. Factors that generally promote debris flows in piedmont areas and can be invoked here are the availability of water (intense rainfall) over short periods of time at irregular intervals, steep slopes with insufficient vegetation cover to prevent rapid erosion, and a source that provides a readily available, abundant detritus with a muddy matrix (Bull, 1977; French, 1987).

There is no recognizable difference, in either the character of debris or fan sizes, between the fans which are spatially associated with the intramontane valleys and those which lack such an association (Fig. 4). Apparently, very little or no debris had been derived from these pre-existing valleys at stage 1, although they probably contained an earlier, Pliocene alluvium (see earlier section). Stream-flow deposits are virtually absent in the alluvial fans of stage 1. The intramontane valleys, clearly, did not experience flood events or convey major discharges at this stage. The aridity of the late Pliocene–early Pleistocene climate and the probable early carbonate encrustation of the Pliocene valley-fill alluvium are thought to have rendered the pre-existing intramontane valleys essentially inactive.

RECORD OF A MARINE INCURSION

The alluvial fans of stage 1, in their lower (distal) parts, appear to have been eroded by marine processes. There is a coastal palaeocliff (Figs 2 & 4) that rims an erosive platform cut in the alluvium and associated marine strata (see Fig. 8, discussed below). The cliff has been strongly altered by the later processes of subaerial denudation and fluvial dissection (Fig. 4), but is still recognizable as a distinct morphological step, some 2–4 m in height. The wave-cut platform (Fig. 4) is at an altitude of 13–14 m in the eastern to central part of the study area (e.g. see 13 m level in the left-hand part of Fig. 8), but seems to be a few metres higher in the western part, where it has been largely concealed under the younger fans and further destroyed by the coastal cliff development and retreat (Figs 9 & 10).

The exact relationship between the alluvial-fan development and the marine incursion is difficult to establish on a regional scale, because of the scarcity of adequate exposures. Our observations are from the distal part of the Patsianos fan (see Fig. 4), which has been dissected by a narrow man-made trench, some 2 m deep and 300 m long, directly north of Frangokastello (Fig. 1). In the trench section (Fig. 8), the fan-derived deposits are merely 3–4 m thick, but show the record of a complex interplay of alluvial and marine sedimentation. The substratum comprises Pliocene marls, rich in burrows, planktonic Foraminifera and sandstone intercalations, whereas the sedimentary cover consists of late Holocene soils. Only the older part (stage 1) of the Pleistocene alluvium is preserved at this particular locality. The sedimentary sequence comprises (Fig. 8) the following.

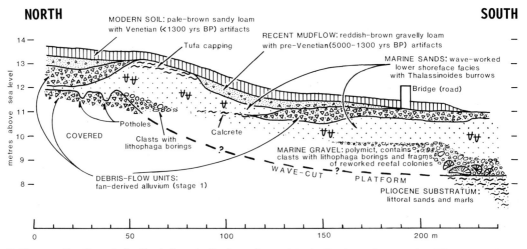

Fig. 8. Cross-section through the distal, fan-toe alluvium of stage 1 in the Patsianos fan complex (Fig. 4), showing intercalations with shallow-marine deposits. Measured, with detailed geodetic levelling, in a man-made trench north of Frangokastello (Fig. 1). Note the vertical exaggeration ($c.10 \times$) in scale.

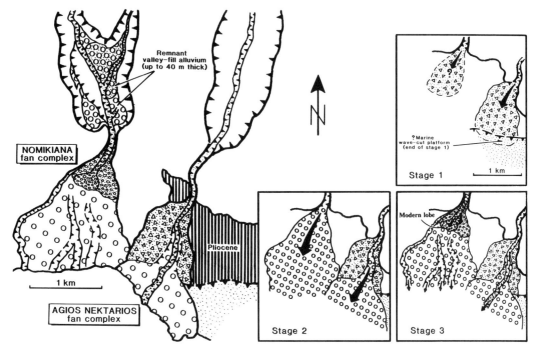

Fig. 9. Detailed plane-table map and reconstructed depositional history of the Nomikiana and Agios Nektarios fan complexes (for location, see Fig. 4). Lines with hatches are erosional topographic scarps.

1 Basal transgressive 'lag', c. 1 m thick, which is a coarse gravel unit with a sandy matrix and maximum clast sizes of up to 40–50 cm. Larger clasts show common *Lithophaga* borings. Clasts are sub- rounded to rounded fragments of metalimestone, limestone conglomerate, marine bioclastic sandstone (calcarenite) and biogenic limestone (eroded reefal colonies). The latter two components are

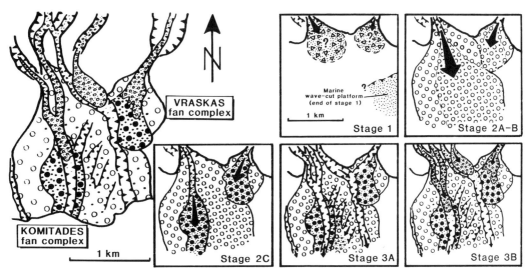

Fig. 10. Detailed plane-table map and reconstructed depositional history of the Komitades and Vraskas fan complexes (for location, see Fig. 4). Lines with hatches are erosional topographic scarps.

thought to represent the reworked cover of an earlier marine terrace, now at an altitude of c.9 m (Fig. 8), possibly analogous to the Pliocene wave-cut features described by Peters et al. (1985). The limestone conglomerate probably represents an early-indurated alluvium of stage 1, or an older (Pliocene) coastal deposit.

2 Carbonate-indurated, wave-worked sands (shoreface facies), with common gravel lenses and *Thalassinoides* burrows, intercalated with the isolated beds of unsorted debris-flow gravel. The latter are the alluvial fan-fringe deposits of stage 1 (see the distal segment of Patsianos fan in Fig. 4).

The late Holocene cover of the sequence has been dated (see details in Fig. 8) by archaeologists on the basis of pottery sherds (J. Moody, pers. comm., 1988).

The sea thus encroached on to the piedmont plain near the end of stage 1. The data (Fig. 8), though very limited, clearly indicate an interfingering of the fan-derived alluvium with marine shoreface sands, and this would qualify the coastal systems essentially as *fan deltas* (see Nemec & Steel, 1988, p. 5). The facies record further indicates oscillations from subaqueous to subaerial conditions (see calcrete and tufa horizons in Fig. 8). The deposits were eventually truncated, and the alluvial fans cliffed, when the extensive wave-cut platform developed (Fig. 4; altitude of 13 m in Fig. 8) during the sea-level highstand and subsequent retreat. The retreat was probably rapid, whereby the erosional platform emerged *en bloc* as a subaerial plain to become covered by the younger alluvium of stage 2 (Figs 2 & 4). Where this younger alluvium is absent, due to non-deposition or local later erosion, the Holocene soil cover rests directly on the abandoned marine platform (as in Fig. 8).

STREAM FLOW-DOMINATED FANS (STAGE 2)

General

The alluvium of stage 2 is gravelly and constitutes the bulk of the Quaternary fan complexes in the study area (Fig. 4). The exact age-frame and internal stratigraphy of this alluvium are uncertain, because of the scarcity of datable material. The polyphase carbonate cement of the limestone gravel is generally unsuitable for radiometric dating. The cement comprises calcrete-type amorphous phases of early-diagenetic surficial encrustations, strongly contaminated with detritus, as well as several more pervasive phases of pore-filling calcite, amorphous to crystalline. Therefore, few reliable $^{234}U/^{230}Th$ dates have been obtained.

A sample of an early-diagenetic, amorphous calcitic cement from fan lobe 2A in the Sfakia complex (see below) has been dated as older than 350 ka. The fan deposits were cemented prior to the entrenchment and deposition of subsequent fan lobe 2B, as indicated by large cavities and overhang features in the palaeotrench wall, but it remains uncertain whether this particular cement sample actually pre-dates the younger fan. More reliable U/Th dates have been obtained from the middle lobe of the younger, multiple fan 2C in the Sfakia complex. A large, fan-derived clast of a calcitic pore-fill 'flowstone' has been dated at 290 ($\pm$ 45) ka, and a similar *in situ* flowstone from a large pore infill at 79 ($\pm$ 6) ka. All these dates point to Late Pleistocene.

The alluvial fans of stage 2 comprises at least five generations of variously stacked, carbonate-encrusted lobes. They are hardly distinguishable in the eastern to central part of the study area (Fig. 4), where the successive fan lobes have apparently been superimposed upon one another by simple bulk onlap (Fig. 9), but are increasingly better manifested within the fan complexes to the west, where the development and progradation of the successive fan lobes involved considerable trenching (Figs 10 & 11). The reason for this east–west variation is discussed later in the text.

The stratigraphy of the alluvium is thus most apparent in the westernmost fan complex (Fig. 11), used here as a local reference: there are three successive entrenched fans (denoted as 2A, 2B & 2C) that overstepped one another by means of cutting and filling, and the youngest (2C) comprises at least three fan lobes stacked by simple onlap and separated by palaeosols. This gives a total of five consecutive fan lobes. The cutting and filling here involved considerable entrenchment relief (Figs 11–13); fan 2B in its 'inset' proximal segment is more than 50 m thick (base unexposed), and also the inset head of fan 2C exceeds 10 m in thickness.

It should be clear to the reader that our distinction of the component fans (stacked lobes) within their individual complexes is based on detailed mapping and geomorphic criteria, and thus the regional stratigraphic correlation of the fans is tentative and to be regarded with caution. The successive fans of stage 2, although markedly dif-

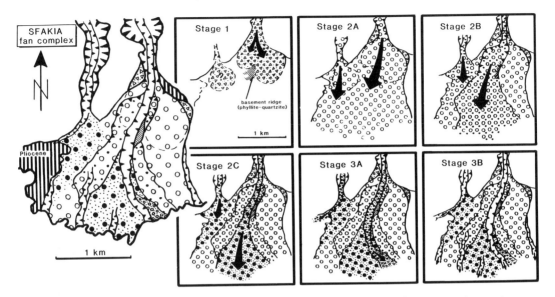

Fig. 11. Detailed plane-table map and reconstructed depositional history of the Sfakia fan complex (for location, see Fig. 4). Lines with hatches are erosional topographic scarps. Further details are shown in Figs 12 & 13.

ferent from those of stage 1 and 3, show no obvious facies differences, and the scarcity of reliable radiometric dates further renders exact chronostratigraphic correlations impossible. Our correlation relies on the geomorphic assumption that the alluvial systems, within such a relatively small area, would respond in a roughly synchronous manner to factors like climatic changes and/or sea-level fluctuations. Therefore, our correlation is only as reliable as the latter assumption itself.

The alluvial fans of stage 2 are relatively large, with axial lengths in excess of 2 km (Fig. 4). Their lowermost segments are unpreserved, due to Holocene coastal erosion, but the distal parts of the smaller eastern fans have been truncated only slightly; their toes exposed in the coastal cliff are thin and intercalated with marine facies (discussed further in text). The smaller volumes and radii (c.2 km) of the eastern fans are clearly due to the division of valley-derived discharges into two or more trenches cut in the older deposits, whereby a single intramontane valley fed simultaneously two or more fan lobes (see the multilobate Skaloti and Patsianos systems in Fig. 4).

The apices of the alluvial fans of stage 2, where not truncated or removed by entrenchment processes (Fig. 4), are at altitudes of 320–370 m (e.g. Figs 12–13). The fans, especially in the western part of the piedmont, are unusually steep as to their stream-flow origin (documented below), showing depositional slopes of up to 13° (Figs 12–14). Moreover, the longitudinal morphometric profiles of the fans tend to be convex and broadly sigmoidal in shape, with the depositional slopes ranging from 4–7° in the uppermost segments, through 10–13° in the medial segments, to 3–4° in the lower preserved segments. The smaller fans in the east (Fig. 4) are less steep; their medial, narrow, trench-confined (inset) segments have slopes of up to 8–10°, and the distal, unconfined segments have slopes of less than 2–3°.

The occurrence of the fans at the outlets of the large intramontane valleys (Fig. 4) indicates that the voluminous alluvium of stage 2 has been derived directly from these valleys. Notably, the alluvial systems of stage 1 whose headward erosion did not reach any of the intramontane valleys, as in the case of the Kapsodasos fans (Fig. 4), essentially lack the younger alluvium of stage 2. There seems also to be a direct relationship between the size of an intramontane valley and the size of the associated fan of stage 2 (e.g. see Vraskas fan complex and its neighbours in Figs 4 & 10), although the multilobate systems fed by single valleys (see Patsianos and Skaloti systems, Fig. 4) clearly deviate from the rule.

Fan deposits

The alluvium of stage 2 consists of relatively well-

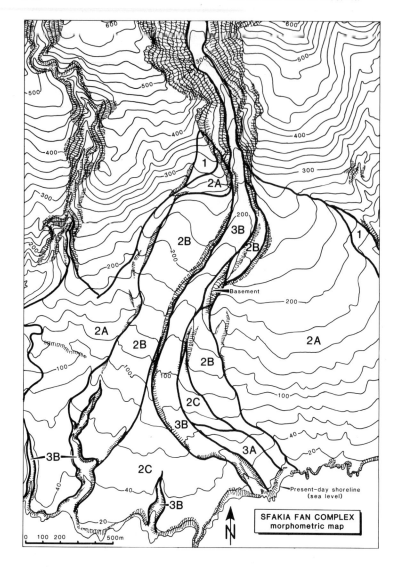

Fig. 12. Morphometric map of the Sfakia fan complex (Fig. 11). The successive fans, or alluvial 'stages', are numbered as in Fig. 11.

sorted, grey or yellowish-grey limestone gravel. Mean grain sizes vary from pebbles to cobbles, with maximum sizes in the large cobble to boulder range. Clasts are subangular to rounded, but mainly subrounded.

The gravel is generally well bedded, showing extensive, sheet-like units 0.2–1.4 m in thickness (Fig. 14). The geometry of individual sheets is tabular or broadly lenticular in sections parallel to fan radius (Fig. 15A), but tends to be more lenticular or wedge-shaped in sections perpendicular to the radius (Fig. 15B). Internally, the gravel sheets range from massive (unstratified) to parallel stratified, and commonly show vertical and/or lateral grading in clast sizes. Sandy interbeds are virtually rare; where present, they are very coarse-grained and pebbly, laterally impersistent, and merely a few centimetres thick. The only fine-grained deposits are isolated palaeosol units, described further below. These general characteristics pertain to the entire alluvium of stage 2 (Fig. 4), although our detailed observations and illustrating material are mainly from the Sfakia fan complex (Figs 11 & 12), which is the one best exposed.

The gravel sheets (Figs 14 & 15), as the basic depositional units of the alluvium, have been clas-

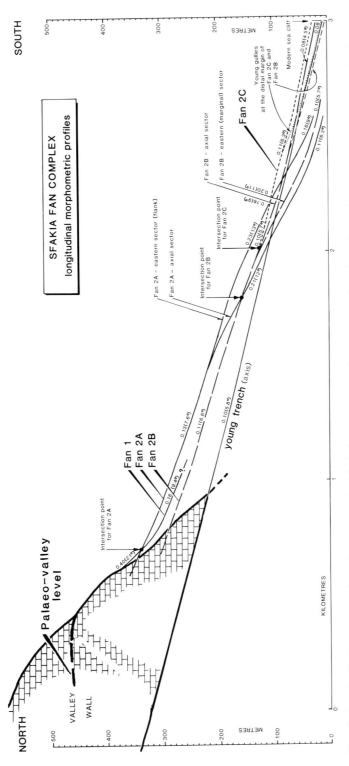

Fig. 13. Longitudinal morphometric profiles through the Sfakia fan complex and its component fan lobes (cf. map in Fig. 12 and the reconstruction in Fig. 11). Note the vertical exaggeration (c.2.5×) in scale.

Fig. 14. (A) Deposits of the proximal 'inset' part of fan 2B in the Sfakia complex (Fig. 11), as seen in the western wall of the Recent trench 3B at an altitude of $c.180$ m (Fig. 12); downfan direction is to the left, away from the viewer. (B) Deposits of the proximal part of fan 2C in the Sfakia complex, as seen in a roadcut section through the western wall of the modern trench 3B at an altitude of $c.90$ m (Fig. 12); downfan direction is to the left, slightly out of the picture. Note the steep depositional dip of the well-bedded fan deposits in both cases.

sified into two broad categories (Figs 16–18): (i) channel-lag sheets, and (ii) channel-bar sheets. The distinction is based on descriptive criteria, and the genetic labels are used for sheer clarity and convenience, to reflect our interpretation of these gravelly units in terms of a braided-stream distributary network.

Channel-lag sheets

These gravel sheets are generally coarser-grained (see clast-size histogram in Fig. 18), and characteristically have scoured, uneven bases and relatively flat tops (Figs 16–18). They typically consist of a clast-supported framework of cobble- to boulder-sized gravel (Fig. 19). Matrix is coarse sand to fine pebbly gravel, but the upper parts of some beds tend to have a pure sand matrix, which often gives the clast-supported gravel a bimodal textural appearance. In a few cases, the upper parts of the gravel units show an extremely bimodal texture, with the intersitital spaces of the coarse clast framework filled with a buff-coloured silt or silty sand, or a reddish-brown mud.

Some sheets display crude normal grading, at least on a very local scale, but others are virtually ungraded. Large clasts tend to have an $a(t)b(i)$ fabric (notation after Walker, 1975, fig. 7-2), although this imbrication pattern is not always obvious and some gravel units seem to have a chaotic, disorganized fabric.

In cross-sections perpendicular to fan radius, these erosive gravel sheets are broadly lenticular and 50–150 m wide, but their thicknesses are often highly irregular, with more than one deeper-scoured zone, some 8–15 m wide (see lower-right diagram in Fig. 18). Accordingly, gravel sheet thicknesses range from less than 0.2 m to 1 m, most often between 0.3 and 0.5 m (see thickness histogram in Fig. 18). In the downfan direction, the sheets tend to be very extensive, several hundred metres long and possibly up to 1–2 km in some cases, and it is primarily these sheets that give the fan alluvium its apparent 'continuous' bedding (Fig. 14A). Overall, the gravel sheets thus have a belt-like, or rather multiple ribbon-type, geometry.

These erosive sheets are interpreted as channel-floor lag deposits, comprising the coarsest and least mobile bed-load material deposited on a stream floor immediately after the erosion maximum of channel peak-flood phase. Their poor clast-size grading and tractional fabric suggest deposition as a

Fig. 15. Well-bedded Pleistocene alluvium of stage 2 in the Sfakia complex (Fig. 11), as seen in vertical sections parallel (A) and nearly perpendicular (B) to the fan radius. Palaeotransport direction in (A) is to the left, and in (B) is away from the viewer.

pavement or stream-floor 'armour' layer. The deposition of the successive sheets of this type (Figs 16 & 17) would then represent the armouring of newly scoured channel-floor surfaces with the coarsest material derived from the fan drainage area and through erosional reworking of the fan surface itself. The matrix of this clast-supported gravel is probably due to entrapment of some finer-grained material concurrently with the coarsest bed-load deposition, and due to subsequent *in situ* infiltration (see Carling, 1984; Frostick *et al.*, 1984). The muddy matrix in the upper parts of some sheets is analogous to the material of the loamy palaeosol units preserved within the alluvium (see distal section in Fig. 17 and further text), and is thought to have infiltrated the openwork pavement gravel of some abandoned channels as a result of erosion and redeposition of contemporaneous soils, probably by rainwash processes.

Channel-bar sheets

The gravel sheets of this type are finer-grained and generally better sorted, although their 'bulk-sample' grain-size distributions vary from unimodal to polymodal. The latter typifies units with internal stratification, where the alternating layers of finer- and coarser-grained gravel (4–10 cm thick) often have highly varied unimodal distributions. Mean grain sizes are in the coarse granule to pebble range, and maximum sizes vary from 2 to 20 cm, occasionally up to 40 cm (Fig. 17; see clast-size histogram in Fig. 18).

These gravel sheets are large elongate lenses, or lobes, some 20–50 m long (downfan direction) and 10–15 m wide. Their measured thicknesses (Fig. 17) are between 0.25 and 1.4 m, most often 0.4–0.7 m (see thickness histogram in Fig. 18). The tops of the gravel sheets tend to be convex up-

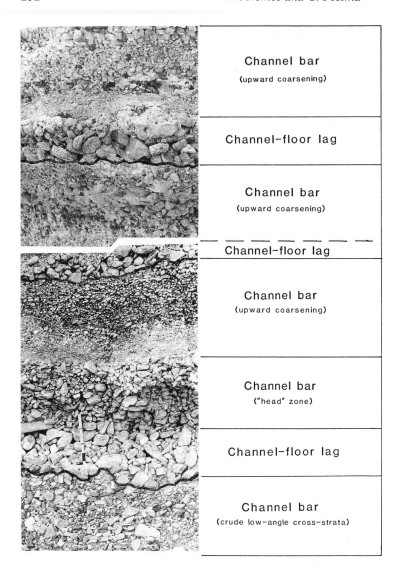

Fig. 16. Two basic types of alternating gravel sheets that constitute the alluvium of stage 2. Example from the upper segment of Sfakia fan 2B (Fig. 11). Hammer (scale) is 33 cm long.

wards, whereas their bases are relatively flat and unscoured, except where onlapping local relief of erosional or depositional origin (Fig. 15B; see also Fig. 24B).

Another characteristic attribute of these gravel sheets is their lateral fining, particularly in downfan direction, and a corresponding change in the internal structure. Irrespective of their location within a fan, the individual lobate sheets have relatively narrow, thicker and coarser-grained upstream parts, which are unstratified, more poorly sorted, typically rich in large imbricate discs (bladed or oblate cobbles and boulders) and showing little or no vertical grading (Fig. 20A). Downstream and sideways, this coarsest part of a lobate sheet, interpreted as *bar-head* zone, becomes progressively thinner and finer-grained, and shows an upward coarsening (Figs 20B & 21). The main, long 'medial' part of the sheet is dominated by better-sorted, coarsening upward gravel, usually unstratified (Fig. 22), whereas the broad downstream part, interpreted as *bar-tail* zone, is relatively thin and well sorted, distinctly finer-grained and parallel stratified, with or without an upward coarsening (Fig. 23). Plane-parallel stratification dominates the tail zones, but some sheets display low-angle cross-stratification

(Fig. 17), with strata dip azimuths parallel or oblique to the downfan direction.

These gravel sheets are interpreted as mid-channel bars (Fig. 18, upper-right diagram) deposited in shallow braided streams. Similar bars have been described from modern, gravel-bed braided streams and outwash fans, and variously referred to as 'sheet bars' (Boothroyd, 1972), 'longitudinal bars' (Boothroyd & Ashley, 1972) or 'small/unit bars' (Bluck, 1982). Analogous bars have also been recognized in some stream-dominated, coarse-gravelly ancient alluvial fans (Fernández et al., 1988; Haughton, 1989).

The coarse bar heads in the present case are thought to have built up by accretion of localized cobble/boulder 'jams', turned into loci of gravel deposition and segregation (see Bluck, 1974, 1982; Smith, 1974). Such coarse loci, from which the bars would develop as discrete lobate features, might form downstream of riffles (Bluck, 1982); as erosional remnants of dissected/reworked earlier bar lobes (Southard et al., 1984); or as local mounds and bouldery clusters related to the stream-floor armouring with coarse 'lag' material (see preceding section).

In the empirical model by Bluck (1982, fig. 12.5), the coarse bar head, concentrating large discs (least mobile clasts), moves more slowly than the downstream tail. In general, such lobate bars would be most mobile when there is still relatively little downstream clast-size segregation. As soon as the coarse gravel population emerges at the head and the finer tail evolves due to particle segregation, the bar moves more slowly and/or extends its long axis by growing downstream. The relative increase in clast size of the head increases the scale and intensity of turbulence, which causes removal of finer material out of the head and, at the same time, increases the rate of migration of the tail by redistributing some of the material there. The successive layers of finer gravel swept downstream thus produce the stratified bar tail, and the inclination of these strata reflects the downstream or lateral (flank) inclination of the accreting bar surface. The lack of high-angle cross-stratification in the present case indicates that the bars did not develop avalanching slipfaces at their fronts or flanks. This character of the bars could be due to their development in shallow, powerful streams on relatively steep fan slopes.

The coarse clasts of the bar head probably create sufficient turbulence at the water–sediment interface to allow only slightly finer particles, large enough to tolerate these turbulent conditions, to come to rest directly downstream of the head (Bluck, 1982). Similar selection of progressively smaller particles would then occur further downstream, with a corresponding decrease in the bed's 'grain roughness' and scale of near-bed turbulence, such that the clast sizes within a bar lobe would effectively decline in this direction. As the coarser, head-derived material progrades downstream, the bar develops an internal upward coarsening of its particle sizes (cf. Figs 18, 21 & 22).

As the downstream segregation of clast sizes increases, the bar head becomes relatively fixed and the overall migration rate of the bar decreases (Bluck, 1982). The bar thus becomes more 'mature', and a new bar may be attached to it or stacked upon it through bar-welding processes (see multiple bar units in Figs 15B, 17, 18 & 24B). Mature bars also tend to multiply, due to the 'chute-and-lobe' processes described by Southard et al. (1984). Once a bar stalls, its head may aggrade and the bar lobe continues to enlarge as new material is swept downstream to the lobe surface. When the downstream inclination of the bar-lobe surface reaches some critical value, the lobe deposits begin to unravel quickly. Axial incision during falling or low stages thus exposes one or two remnants of the lobe, and these elongate mounds then become loci for the formation of new bar lobes (see Southard et al., 1984, fig. 6). This process of bar multiplication, by axial dissection and lobe re-establishment, is known to cause braiding in shallow, gravel-bed streams (Southard et al., 1984; see also Boothroyd & Ashley, 1975).

There is at least one point here worth special emphasis. The extensive, coarsening upward gravel sheets (Fig. 22), if not traced laterally into their head/tail segments and unrecognized as longitudinal bars (Fig. 18, upper-right diagram), can easily be misinterpreted as the deposits of 'density-modified grain flows' (sensu Lowe, 1976). Sheet-like bed geometry, clast-supported texture and inverse grading are commonly regarded as the characteristic depositional attributes of cohesionless debris flows dominated by clast collisions and dispersive stresses (Lowe, 1976, 1982, fig. 12), in both subaqueous (Nemec et al., 1980; Lowe, 1982; Clifton, 1984; Massari, 1984; Nemec & Steel, 1984; Postma, 1984) and subaerial settings (Lowe, 1976; Allen, 1981; Nemec et al., 1984; Todd, 1989). The risk of misinterpretation is particularly high in the context

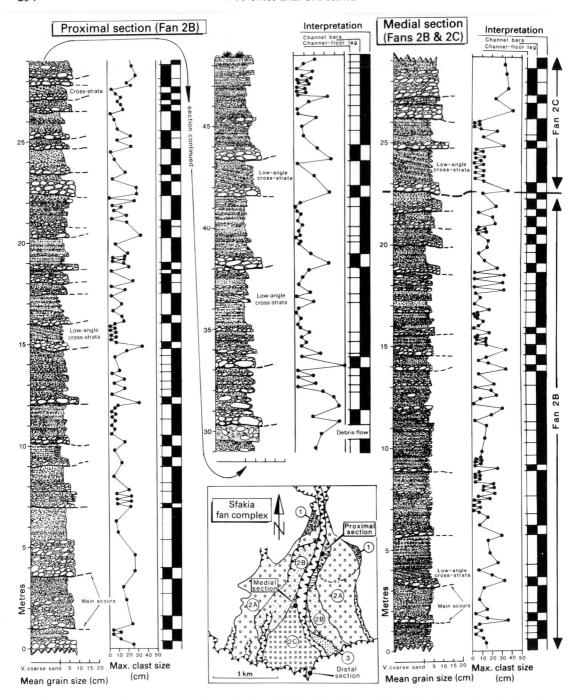

Fig. 17. Detailed vertical logs through the proximal, medial and distal parts of the Sfakia fan complex, showing the alluvium of fan lobes 2B and 2C (cf. Figs 11 & 12) and its facies interpretation. The base of fan 2B in the proximal and medial sections is unexposed, and the topmost part of fan 2C in the medial section is not shown in the log. The base of fan 2C in the distal section is unexposed (concealed by the sea), and the topmost few metres of the fan are not shown in the log.

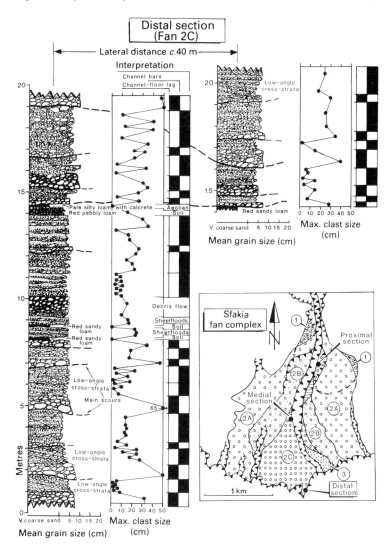

Fig. 17. *Continued*

of steep alluvial fans, as in the present case, especially when outcrops are poor or small.

The apparent clast-size grading in the present case (Fig. 22; also Figs 16 & 24B) is clearly a result of tractional 'grain-by-grain' deposition, as discussed above and as indicated by the downstream transitions into finer-grained stratified gravel (Fig. 18, top right) and common tractional $a(t)b(i)$ clast fabric. Therefore, we have used the term 'upward coarsening', rather than 'inverse grading', when describing the bar sheets, so as not to imply inadvertently any mass-flow mode of emplacement. However, it is worth emphasis that the apparent similarity between certain gravelly braid-bars and cohesionless debris-flow deposits may be confusing and lead to serious misinterpretations, especially in studies based on small outcrops or well-core samples.

Associated debris-flow deposits and palaeosols

Deposits of inferred debris-flow origin are very rare in the alluvium of stage 2 (see Fig. 17). They occur as isolated, lenticular units of matrix-rich gravel, up to 0.6–0.8 m thick, in which a matrix of sand and/or reddish mud supports 'floating' pebbles and cobbles. The debris flows are thought to have been derived locally from the fan surfaces and adjoining

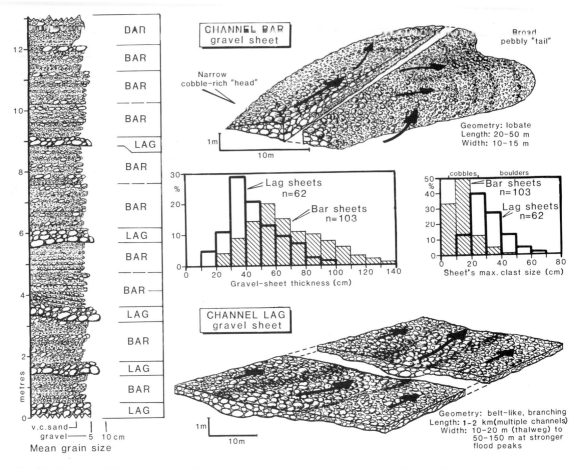

Fig. 18. Summary of the measured and interpreted characteristics of channel-lag and channel-bar gravel sheets (alluvium of stage 2). Quantitative data and estimates are from the Sfakia fan complex, which is the best exposed. The log shows, for illustration, a representative portion of Sfakia fan 2B.

slopes, probably in association with rainwash processes and due to occasional failure of undercut channel banks or other unstable local slopes. Proximal fan alluvium is no richer in debris-flow deposits than is the distal alluvium (Fig. 17), with no evidence of significant mass-flow derivation from the intramontane drainage areas.

Notably, the reddish-brown matrix of inferred debris-flow beds is analogous to the material that constitutes palaeosol units within the alluvium of stage 2 (see distal section in Fig. 17). Such units are best recognizable within the Sfakia fan complex (Fig. 11), where they separate the lower segments of the three successive, superimposed lobes of the multiple fan 2C (Figs 17 & 24A). The palaeosol units here are 0.2–0.3 m thick and laterally persistent, draping the surfaces of the lower and middle lobes, although they are locally truncated by the overlying fluvial gravels. The third, upper lobe is carbonate-encrusted and covered by a thin, patchy Recent soil.

The palaeosols typically consist of an unmottled, reddish-brown sandy loam (terminology after Fitzpatrick, 1980, fig. 4.6), commonly gravel-bearing. They often show some crude layering due to varied gravel contents, but display little or no true stratification, clast grading or imbrication. These deposits are interpreted as a *terra rossa* soil material, derived by rainwash on to abandoned fan surfaces and variously mixed with the alluvial gravel. Their substratum is commonly an openwork, coarse alluvial gravel infiltrated with a similar red loam

Fig. 19. Details of channel-lag gravel sheets. (A) Example from Sfakia fan 2B; hammer head (scale) is 12 cm long. (B) Example from Sfakia fan 2C; the white lighter (centre) is 8 cm long.

(Fig. 25). There are also some associated, thin interbeds of graded sandy gravel of inferred sheet-flood origin, attributed to rain-water runoff (Fig. 17).

In better-preserved cases, the reddish-brown loam is overlain by a pale, buff-coloured silty loam, which is similar to the modern loess-like material, blown by wind and partly rainwashed, deposited as thin local patches in the dry, wind-swept coastal zone of southern Crete. The pale palaeosol in some instances appears to be separated from the underlying red loam by a one pebble-thick gravelly horizon, which may be a deflation lag. In other cases, the pale palaeosol itself contains an admixture of coarse sand or gravel in the lower part (Fig. 26), suggesting that the deposition commenced with rainwash processes.

The pale palaeosol is unmottled, contains small scattered calcareous glaebules or larger, isolated calcrete nodules, and often shows a thin, vaguely laminated calcareous layer, or 'petrocalcic' horizon, at the top (Fig. 26). The petrocalcic layer locally comprises the entire pale-loam unit, where relatively thin (Fig. 25). These features are generally characteristic of Aridisols (Retalleck, 1988), although it is uncertain whether the petrocalcic capping here is of purely pedogenic origin, or is partly due to a secondary filtration flow of carbonate-rich solutions through or from the overlying porous gravel (Figs 25 & 26).

Fig. 20. Details of channel-bar gravel sheets. (A) Bar-head gravel, relatively coarse and ill-sorted, showing imbricate discoidal cobbles and boulders; palaeoflow direction is to the right, at 50° out of the picture; the white lighter (middle right) is 8 cm long. (B) Downstream portion of a bar-head segment, showing upward coarsening and cobble-rich top; palaeoflow direction is towards the viewer; the lens cap (centre) is 5 cm. Examples from Sfakia fan 2B (Fig. 11).

Fig. 21. Detail of a channel-bar gravel sheet, showing the transition from bar-head zone (left) to the finer-grained, coarsening-upward mid-bar zone (right). The hammer is 33 cm long. Example from Sfakia fan 2B (Fig. 11).

Fig. 22. Details of channel-bar gravel sheets, showing their medial segments, which are the transitions from bar heads to tails (cf. Fig. 18, top right). Note the upward coarsening of clast sizes. Scale: the lighter in (A) is 8 cm long; the pocket knife in (B) is 9 cm long; and the hammer in (C) is 33 cm long. Examples from Sfakia fans 2B and 2C.

The same uncertainty pertains to the carbonate cementation and apparent encrustation of the successive fans of stage 2, since there is mixed evidence of both pedogenic and secondary calcitic precipitates. The latter include flowstone-type infill of large pores and interstitial cavities, as well as some elongate, calcite-rich, white-coloured zones that run downfan and are clearly due to intrastratal filtration flow. Amorphous carbonate encrustations of pedogenic origin are indicated by the reddish, 'oxidized' colour of cement-rich, presumably calcretized fan tops that show coalescent carbonate nodules rich in host detritus, and by the associated palaeosols.

It is to be emphasized that the palaeosols and pedogenic carbonate encrustations are not developed on fan terraces or surfaces lateral to the younger lobes, but clearly *separate* the successive, superimposed fan lobes (e.g. Figs 17 & 24A). The presence of palaeosols and pedogenic encrustations on fan lobes thus indicates considerable time-breaks in the alluvial sedimentation of stage 2. Apparently, there must have been relatively long periods of non-deposition, when the intramontane drainage areas and valleys stayed inactive, while the rainwash, aeolian and pedogenic processes made their recognizable imprint on the abandoned fan surfaces.

Fig. 23. Details of channel-bar gravel sheets, showing the stratified, finer-grained, downstream portions of bar tails (cf. Fig. 18, top right). The standing figure gives scale in (A), and the hammer in (B) is 33 cm long. Examples from Sfakia fans 2C and 2B (Fig. 11), respectively.

Subaqueous fan-toe deposits

The lowest, distal segments of the alluvial fans of stage 2 are generally unpreserved, eroded due to the Holocene coastal retreat (Fig. 4). However, some of the smaller fans in the eastern part of the piedmont have been only slightly truncated and their thin toes are exposed at the top of the coastal cliff. For example, the toe of the westernmost Patsianos fan (Fig. 4), cropping out c.10 m above the present-day sea level, shows stream-derived, more or less isolated gravel units embedded in muddy shallow-marine deposits (Fig. 27). The toes of the adjoining fans in this cliff section (Fig. 14) are even thinner and less coarse-grained, but embedded in sandier marine deposits. The gravel units (Fig. 27) are interpreted as underwater channel-fill and mouth-bar deposits of the fan's distributary streams, and the fan system is inferred to have prograded into a shallow, protected, muddy marine embayment between the Frangokastello area to the west and the extended foothills to the northeast (cf. Fig. 4). This sedimentary record thus shows that the alluvial fans of stage 2 in the eastern part of the piedmont have prograded into the sea and developed *fan deltas*. It is uncertain whether the same pertains to the western, higher part of the piedmont, where the Pleistocene shoreline was probably a few kilometres further to the south.

The fan deltas were shoal-water systems, lacking steep-face mouth bars or avalanching subaqueous slopes. The nearshore zone was apparently shallow, such that nearly all fan-derived sand and much gravel effectively bypassed this zone, carried offshore by the powerful, erosive stream effluent. The coarse gravel units (Fig. 27) have clearly been emplaced by very powerful tractional currents, but are thin and embedded in distinctly fine-grained, muddy 'background' facies. The fan-delta facies appear to be more sandy in the relatively more distal, less gravelly toes of the two adjoining fans (Fig. 4). These facts support the notion of sediment 'bypass' processes. If the notion is correct, the relatively thin gravelly succession (Fig. 27) might,

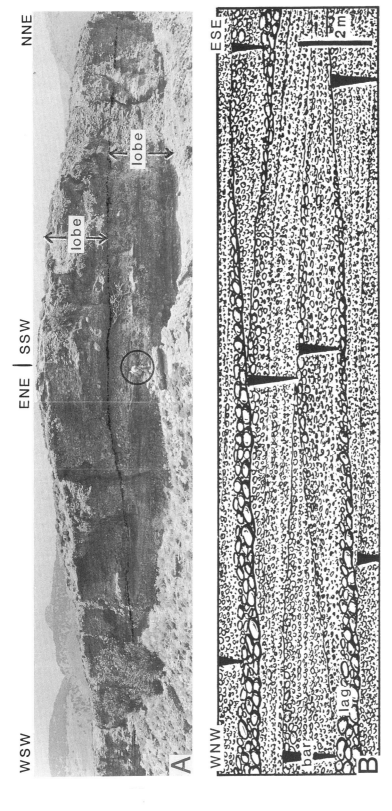

Fig. 24. (A) Alluvium of Sfakia fan 2C, as seen in the modern coastal cliff (for locality, see 'distal section' on the map in Fig. 17). The two thick packets of stream-flow gravel, each underlain by a palaeosol, are the upper lobes of fan 2C that onlapped one another with some lateral offset (see also log in Fig. 17); the standing figure (centre) gives scale. The lighter-shade rubble (modern scree) along the cliff's foot has covered the outcrop of the lower lobe of fan 2C, which is exposed further to the left and right. (B) Close-up detail of the upper fan lobe, showing the alternating channel-lag and channel-bar gravel sheets portrayed in Fig. 18. The sketch is from the far left-hand corner of the outcrop in (A); palaeoflow direction is roughly towards the viewer.

Fig. 25. Palaeosol unit between the lobes of Sfakia fan 2C (for stratigraphic position, see the lower palaeosol in 'distal section' in Fig. 17). The sequence comprises: 1, openwork channel-lag gravel infiltrated with reddish-brown sandy loam; 2, unsorted, gravelly, reddish-brown sandy loam attributed to rainwash processes; 3, faintly laminated, yellowish-white calcareous layer interpreted as petrocalcic Aridisol capping; and 4, erosive, loaded basal part of a channel-lag gravel. The binoculars (scale) are 10 cm long.

indeed, be a subaqueous nearshore equivalent of the entire alluvium of stage 2, with the 'condensed' record due to effective sediment transfer through this shoal-water bypass zone.

Discussion

The alluvium of stage 2 clearly represents a profound change in both sediment and water discharge, and in the sediment textural maturity and mode of transport. This change is ascribed to a climatic factor. The large discharges involved in the formation of the successive, coarse-grained fans of stage 2 are attributed to meltwater runoff, and ascribed to periodical melting and re-establishment of ice caps or cirque glaciers in the adjacent, high range of Lefka Ori. The pre-existing intramontane valleys had apparently been activated as principal drainage and sediment-yield areas (see Fig. 4), whereby the fans were built of relatively mature debris and solely by stream-flow processes. The high-lying valley mouths made these alluvial systems prograde on to the adjacent piedmont plain as relatively steep, 'cascading' fans, whose longitudinal morphometric profiles thus tend to be convex and broadly sigmoidal in shape (Fig. 13). All these inferences are discussed in detail below.

There is much compelling geomorphic indication and regional palaeoclimatic evidence that the high mountain range was subject to glaciation (see earlier section), although there is no evidence that the inferred high-altitude ice-caps or cirque glaciers ever descended directly into the lower-altitude, fan-feeding intramontane valleys. The latter are V-shaped in cross-sections, lack any moraine-type deposits, and locally show karstic bedrock 'towers' and precariously perched blocks. However, the sedimentary facies and inferred depositional processes of the alluvial fans of stage 2 are very similar to those of modern gravelly outwash fans (see McDonald & Banerjee, 1971; Boothroyd, 1972; Boothroyd & Ashley, 1975). The fan gravels here are elongate, lensoidal sheets deposited by sediment-laden flood water that spread out from the valley mouths and was conveyed downfan through a distributary network of shallow channels, which repeatedly divided and rejoined due to formation of low-relief longitudinal bars (Fig. 18). Steep fan gradients and high water discharges promoted the deposition of gravel, while sand and finer material largely bypassed the distributary channel networks.

Since the entire voluminous alluvium of stage 2 has been derived from the intramontane valleys (see Fig. 4), with little or no concurrent sediment derivation from the mountain-front slopes and ravines, the periods of high discharge cannot readily be attributed to factors like an absolute increase in precipitation or the piedmont's 'pluvial' climatic conditions (see also discussion by Roberts, 1983, p. 169). The five successive generations of stream-dominated fans are thus interpreted as low-altitude periglacial features related to five major periods of

Fig. 26. Palaeosol unit between the lobes of Sfakia fan 2C (for stratigraphic position, see the upper palaeosol in 'distal section' in Fig. 17). The sequence comprises: 1, unsorted, gravelly, reddish-brown sandy loam attributed to rainwash processes; 2, pale, buff-coloured silty loam with some sand and minor gravel, interpreted as wind-blown material mixed with fan-surface debris by rainwash; 3, pale silty loam with yellowish-white, faintly laminated calcareous upper part, interpreted as a wind-blown material with petrocalcic capping; and 4, channel-lag gravel. The lens cap (scale) is 5 cm.

high-altitude glacier melting, marked by enormous water runoff through the intramontane valleys.

The intramontane valleys, though progressively incised during the runoff periods, are thought to have been recharged with coarse material during the intervening periods of high-altitude glaciation, when rainfall and minor runoff (?seasonal snowmelt) probably continued to wash material into the lower-altitude valleys. The pattern of the alternating periods of major runoff and fan abandonment is well-evidenced by the stacked fan lobes capped with pedogenic encrustations and Aridisols, although it is not equally well recognizable within all of the individual fan complexes. The exact number of glacial/interglacial periods thus remains somewhat uncertain.

It is worth noting that the individual fans, as illustrated by the Sfakia complex, essentially lack those characteristic grain-size trends that have been reported from many ancient fans (e.g. Steel *et al.*, 1977; Heward, 1978) and are regarded by sedimentologists as particularly typical or diagnostic of progradational fan lobes. There is no marked down-fan fining, and the stacked fan-lobe sequences show no obvious upward coarsening or fining either (Fig. 17). This apparent lack of 'classical' grain-size signatures is attributed to valley incision and fan-head trenching processes, which clearly played an important role in fan growth at stage 2 and must have caused considerable recycling of the alluvium (e.g. Figs 11 & 13).

The unusual depositional geometry of the alluvial fans of stage 2 (Fig. 13) requires special comment. The flanks of these fans tend to be somewhat steeper than the axial sectors, and such a relationship can be expected for relatively steep fans built of coarse cohesionless material and characterized by high water discharges (Hooke & Rohrer, 1979). However, the steep depositional slopes (gradients of up to 0.21–0.23) and the sigmoidal, convex–concave longitudinal profiles of these stream-dominated fans are quite unusual. Natural alluvial fans typically have concave profiles, possibly comprising several straight or concave segments (Bull, 1961, 1964, 1972, 1977; Denny, 1967; Hooke, 1968; Schumm, 1977; Hooke & Rohrer, 1979). Although relatively steep fan slopes are known to be associated with large particle sizes, modern stream-dominated gravelly fans have gradients scarcely steeper than 0.05 and the gradients, in fact, tend to be inversely proportional to discharges (Hooke, 1968, 1972; Beaumont, 1972; Bull, 1977; Schumm, 1977). Even when dominated by very coarse gravel and involving mass flows, small fans (area of $c.1$ km^2) are reported to have steepest axial gradients of 0.10–0.15, whereas larger fans have steepest gradients below 0.10 and often below 0.05 (Lustig, 1965, p. 187, table 1; Hooke & Rohrer, 1979, fig. 4b). Braided, gravelly, modern outwash fans have steepest reported gradients of $c.0.02$ (McDonald & Banerjee, 1971, fig. 3; Boothroyd & Ashley, 1975, fig. 8). The analogous fans in the present case

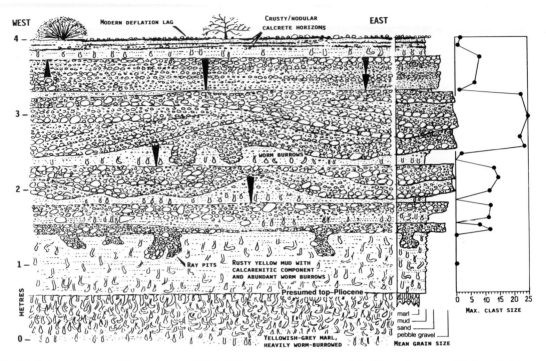

Fig. 27. Subaqueous toe of the westernmost Skaloti fan (stage 2), near the point of its coalescence with the adjoining Patsianos fan (Fig. 4), exposed at the top of Recent coastal cliff, c.11 m high. The sketch shows a small portion of a broader outcrop.

have gradients one order steeper than the latter maximum.

Tectonic steepening of fan slopes, whether syn- or post-depositional, can be precluded in the present case. The inclined, surface-parallel bedding of fan alluvium (Fig. 14) and the concentric distribution of fan thicknesses and morphometric contours (Fig. 12) show clearly that the steep fans are depositional features, not tectonic forms. There is also no indication that the thrust front of the Lefka Ori massif (Fig. 2) has ever been reactivated, on either a regional or a local scale, in post-Pliocene time. Although western Crete experienced some *en bloc* regional uplift in the Quaternary time, at least in the Holocene (Fig. 5), the structural effect of the mild northeasterly tilting would be to decrease, rather than increase, the fan surface inclinations, and the actual angular effect on fan strata would anyway be negligible (cf. Fig. 5). It does not mean, of course, that other potential effects of the regional uplift are to be ignored (see below).

As the only conceivable explanation for the excessive fan gradients, we thus infer a *cascading fan* model (Fig. 28). The alluvial fans of stage 2 were sourced from the high-lying outlets of the reactivated, 'hanging' palaeovalleys, and thus prograded on to the adjacent piedmont plain as relatively steep, 'cascading' systems, yet dominated by stream-flow processes. The conditions of high water runoff determined the character of sediment dispersal processes, while the geomorphic setting itself determined the depositional morphometry and internal architecture of the resulting fans. The sigmoidal longitudinal profile (Fig. 13) and stratal architecture (Fig. 14) thus simply reflect the adjustment of an alluvial-fan system to the pre-existing piedmont relief, namely to the excessive step, or jump, in local base level that has not been cancelled or significantly reduced during the 'dry' stage 1 (see earlier section). The stream-dominated fans advanced rapidly, and this forced them even more to adjust to a pre-existing relief.

Contemporaneous tectonic tilting, in turn, might be responsible for the observed regional variation in the actual style of fan progradation and stacking along the mountain front. In the eastern part of the piedmont (Fig. 4), the older fans of stage 1 had been deeply incised by streams debouching from the

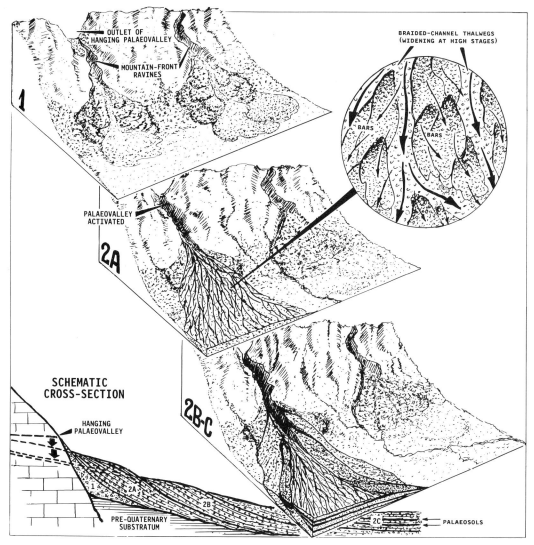

Fig. 28. Schematic model for the development of 'cascading' fans (stage 2) in the Lefka Ori piedmont. The small, debris flow-dominated fans of stage 1 failed to reduce significantly the topographic step between the hanging palaeovalleys and the piedmont plain. The stream-dominated fans of stage 2, issuing from the activated palaeovalley, thus had to adjust their morphometry to this pre-existing jump in local base level. The youngest two, unentrenched lobes of fan 2C are superimposed by bulk onlap, with some lateral offset, and are separated by palaeosols. The model pertains directly to fans in the western part of the piedmont; differences in the central to eastern part of the piedmont are discussed in the text.

mountain valleys, whereby the younger fans prograded beyond the wave-cliffed margin of the former alluvium and progressively aggraded, back-filling the trenches; the Skaloti fan system (Fig. 4) may serve as an extreme example. The steepest, medial segments of these younger fans are thus 'inset' in trenches, whereas the lower, unconfined segments show relatively gentle slopes and a simple, conformable style of vertical stacking by bulk onlap. The successive generations of stage-2 alluvium are thus hardly distinguishable in the eastern to central part of the piedmont, partly due to the scarcity of vertical sections there, although they are clearly hinted at by intrastratal pedogenic encrustations and the evidence from fan-delta sequences (see preceding section).

The style of fan stacking changes gradually towards the west (Fig. 4), where the fans of stage 1 appear to have initially been onlapped and overstepped by the younger systems with limited incision, while the progradation of the successive fans of stage 2 involved considerable fan-head trenching (Figs 9–11). The entrenchment (*sensu* Wasson, 1977) of fan heads was permanent; although followed by back-filling, each major event of trenching apparently shifted the area of deposition progressively further downslope from a previous fan head. This 'telescoping' mode of fan stacking (Denny, 1967; Bull, 1977) is most striking in the westernmost fan complex (Figs 11 & 12), where the increasing lengths of successive fan-head trenches and the associated, irreversible shifts of fan intersection points (Fig. 13) are particularly well pronounced. The fan trenches, in their upper segments, have bottoms a few tens of metres below the previous fan surface.

Fan-head trenching and back-filling may be due to source-terrain degradation and progressive diminution of sediment yield (Eckis, 1928); to an autocyclic mechanism involving intrinsic thresholds of fan aggradation (Schumm, 1977; Schumm *et al.*, 1987); to the fan-toe regime controlled by the piedmont's base level, which may promote 'headcut' processes instead of aggradation (Harvey, 1987); to climatic variation (Bull, 1964; Wasson, 1974, 1977); or to tectonic uplift relative to the piedmont's base level (Bull, 1977). Glacioeustatic sea-level changes alone cannot serve as a plausible explanation, because this factor would be expected to affect the piedmont zone uniformly (rather than differentially) and to raise the piedmont's base level during interglacial (runoff) periods, thus promoting fan aggradation rather than dissection. The fan trenching here thus cannot readily be attributed to glacioeustatic or climatic changes, since the alluvial fans in the eastern part of the piedmont and also the younger fans 2C in the western part are virtually unentrenched.

The varying style of fan stacking along the piedmont zone in the present case favours episodic tectonic tilting as the probable cause. Uplift is known to induce entrenchment, which leaves parts of the uplifted fan as paired terraces. Abnormally deep fan-head trenching, such that the intersection point shifts irreversibly downfan and overbank deposition on the adjoining fan terraces becomes impossible (cf. Figs 10 & 11), is more likely to be triggered by tectonic uplift, rather than by climatic or autocyclic factors (Hooke, 1967, p. 458). Moreover, the western fans are, indeed, several metres higher than their eastern counterparts; are deeply dissected, also by the Recent trenches (see Figs 12 & 13); and there is also the evidence of fan deltas in the east. Fan trenching related to piedmont uplift and accelerated intramontane valley incision, together with the steep fan slopes and high discharges, would cause major transfer of coarse material to the lower fan reaches, and this would explain the lack of marked downfan decrease in the mean and maximum clast sizes, as well as the lack of coarsening upward trends in the progradational fan alluvium.

Fan trenching, caused by episodic uplift, would be associated with an accelerated downcutting of the intramontane 'feeder' valley. However, the deposition farther downfan would eventually result in back-filling in the fan trench above the intersection point (Denny, 1967; Hooke, 1967), as indeed observed in the present case (Figs 10 & 11). As the back-filling shifts deposition back to the fan-apex area, the valley upfan from the apex will aggrade in order to maintain a common gradient (Hooke, 1967, 1968; Bull, 1977). This would cause recharging of the intramontane valley with debris.

The Pleistocene eastward tilting of western Crete is thought to have comprised a number of discrete episodes, comparable to the Recent one (Fig. 5). The sedimentary record indicates, indeed, that not every period of high water runoff and fan growth bore an effect of piedmont tilting. Even in the westernmost Sfakia fan complex (Fig. 4), supposedly most prone to the uplift effects, the two youngest lobes of fan 2C are stacked vertically in a non-entrenched, onlap/offset fashion, although each is underlain by an Aridisol ascribed to a glacial period of negligible water runoff. A similar style of fan stacking, though with less obvious offset, is observed in the eastern part of the piedmont, where the sea stayed closer to the mountain front and the effects of contemporaneous uplift were probably insignificant.

As regards the onlap/offset style of fan-lobe stacking, it is probably the most simple, most natural mode of longer-term fan aggradation. Longer-term hydraulic processes on alluvial fans are known to result in lateral shifts of the area of deposition, from one radial sector to another, as a result of aggradation and growing fan-gradient differentials (Denny, 1967; Bull, 1977; Schumm, 1977; Schumm *et al.*, 1987). A longer-term decline in water discharges, as during a glacial period, might raise the 'shrinking' fan surface near the apex or intersection point sufficiently to favour shifting of the distributary stream network to an adjacent lower sector with the beginning of the subsequent period of major runoff. The occurrence of palaeosols and pedogenic encrustations supports this notion. However, even such major, climatically controlled shifts in the area of deposition may not be easy to recognize if not marked by preserved palaeosols or variations in stratal geometry, as indeed is the case here with the poorly exposed eastern fan complexes (where vertical sections are small or lacking).

In summary, our analysis of the alluvium of stage 2 suggests alternating periods of fan growth and abandonment, which we relate to the periods of deglaciation and renewed glaciation of the highest range of Lefka Ori. As many as five consecutive periods of high water runoff have been recognized from the sedimentary record, and the stream-flow processes recognized from the alluvial facies are fully comparable to those described from modern, coarse-gravelly outwash fans. The geomorphic conditions of the piedmont, with 'hanging' palaeovalley outlets, are thought to have been responsible for the unusual morphometry of the stream-dominated fans of stage 2. Episodes of gentle easterly tilting of the piedmont are invoked as an additional factor, to account for the varied style of fan stacking along the mountain front. Sea level is thought to have acted as the piedmont's base level. The principal effect of glacioeustatic sea-level rises on the sedimentation in the piedmont zone would be to promote fan aggradation and trench back-filling processes. Glacioeustatic sea-level falls might be expected to promote widespread fan trenching, but probably coincided with the periods of negligible runoff. Fan entrenchment varied along the piedmont and was by no means widespread or correlatable in the present case.

HOLOCENE FANS AND TRENCH DEPOSITS (STAGE 3)

General

The alluvial fans of stage 2 apparently became subject to subsequent incision processes (*sensu* Wasson, 1977, p. 149), which marked another major change in climatic and discharge conditions. Stage 3 is dated broadly as Holocene, on account of the well-known regional evidence of a major climatic change, towards drier conditions, at the end of Pleistocene (see Roberts, 1982, 1983). This Recent stage of fan destruction involved downcutting of a variety of narrow channels that probably debouched, mainly or entirely, beyond the previous fan margins (see Fig. 4). Two types of fan channels can be distinguished (see Figs 9-12).

1 Long trenches that are cut through the fan-head zones as direct extensions of the intramontane valley-axis trenches (Fig. 4), with which they share uniform floor profiles (e.g. Fig. 13). The depths of such trenches are up to 40-70 m near fan apices, where the trench-floor gradients are around 0.10 (6°), but decrease to merely a few metres in the lower fan segments, where the floor gradients decline to 0.08-0.05 (4 to 3°) and approach the gradient of the adjoining, earlier fan surface (e.g. Figs 12 & 13). Some trenches die out in this way, by downfan shallowing, before reaching the fan margin (e.g. see Skaloti fan complex in Fig. 4).

2 Shorter gullies, up to 2-5 m deep, that emerge on fan surfaces well beyond the fan-head zones, thus shallowing and dying out in the upfan direction (Fig. 4). Their gradients (0.08-0.05) are similar to those of the lower reaches of fan trenches, only slightly steeper than the gradients of adjoining fan surfaces (Figs 12 & 13, see the lower fan segment).

Some fans have only one type of channel, or lack recognizable channels, whereas others show both channel varieties (Fig. 4). However, it is uncertain in this latter case whether the trenches and gullies formed concurrently, at strictly the same time, on a particular fan (see further discussion below).

The associated alluvial deposits of stage 3 are, overall, of minor volumetric importance. They occur as gravel lobes within the upper segments of the fan trenches, particularly near the intramontane valley outlets, and as small gravelly fans superimposed on the apical parts of the earlier fans of stage 2 (Fig. 4; see also details in Figs 9-11). Fan gullies are typically erosive, devoid of Holocene alluvium.

These Holocene alluvial gravels are unconsolidated and have not been darkened by lichen or blue-green algae, and thus are readily distinguishable in the field and easily recognizable, by white tones, on aerial photographs and satellite images. At least two main generations of the Holocene alluvium can be distinguished in some of the fans in terms of geomorphic criteria (see stages 3A and 3B in Figs 10-12), although they are practically indistinguishable in the others. The youngest gravel (stage 3B) has no organic encrustations and is light grey. There is also little organic coating on the earlier gravel (stage 3A), but just sufficient to make its grey shade somewhat darker.

There is no sign of sediment reddening or pedogenic carbonate encrustations. Significant Holocene soils, mainly rainwashed and relatively young, occur in the Frangokastello plain (Fig. 8) and as irregular, thicker patches in the foothill zone.

Deposits

The alluvium of stage 3 is a moderately well-sorted, limestone gravel of mainly coarse pebble to cobble grade, with sporadic boulders. Clasts are subangular to subrounded.

The alluvial fans built of this material are steep, in excess of 15°, and of small radius (Figs 3 & 29; see also stage 3 in Figs 9 & 10). Their surfaces are relatively smooth, showing broad, low-relief gravel lobes, characteristically with an openwork texture (Fig. 30A). Channel-and-bar features are lacking. In scarce vertical sections, such as road-cuts and other man-made scarps, the deposits show crude bedding parallel to the fan surface. Beds are lobate sheets, around 1 m in thickness, often with somewhat 'bulging', coarser-grained downfan margins. Bedding is recognizable due to the presence of vertical clast-size grading (commonly inverse or normal) and horizons of finer-grained and/or sand-enriched gravel (Fig. 30B). In the absence of such features, beds tend to be amalgamated and virtually indistinguishable. The gravel lobes show no preferential clast fabric, although either $a(t)b(i)$ or $a(p)a(i)$ orientation of large clasts may seem to prevail on a very local scale.

Similar gravel lobes, albeit more elongate, occur as isolated or multiple features on the floors of fan trenches, most commonly near the outlets of the intramontane valleys from which the trenches extend. Such in-trench gravel lobes usually show considerable erosional modification, marked by low-relief terraces (up to 1–2 m in height) and small, disconnected lobes or coarse gravel mounds. Sandy deposits occur as minor interbeds and surficial patches, apparently as an infill of local, sheltered topographic lows on a trench floor.

The alluvium of stage 3 is interpreted as *sieve deposits* and attributed to debris-laden, ephemeral stream-flood surges issued from the intramontane valleys. Sieve deposits (*sensu* Hooke, 1967, pp.

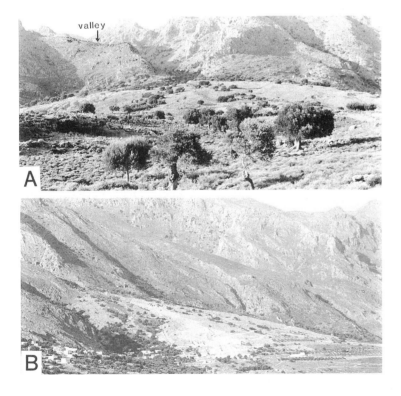

Fig. 29. Alluvium of stage 3 in the Nomikiana fan complex (see Figs 4 & 9). (A) General, frontal view of fan 3, showing the light-tone most recent lobe to the left, the light-tone previous lobe to the right, and the darker-tone older lobe in the middle. The middle sector, once cultivated, shows olive trees whose stumps have been deeply buried by the alluvium. The road scarp (dark, middle left) is c.100–150 m upslope from the fan terminus. (B) Side-view of the same fan, showing the light-tone most recent lobe in its western sector (see also stage 3 in Fig. 9). This recent lobe is less than 100 years old, according to a local farmer. Note the small radius and steep slope of the fan.

Fig. 30. Details of the youngest lobe of Nomikiana fan 3 (Fig. 29), showing sieve deposits. (A) Openwork surficial texture of fan gravel. (B) Vertical section through the uppermost portion of fan gravel, showing its crude bedding marked by sand-infiltrated horizons (darker) and clast-size grading; palaeoflow direction is to the right.

453–456) are lobes of gravel, characteristically openwork or bimodal, formed where the surficial fan material is coarse and permeable, and the fan is relatively dry, such that most of the discharging water simply infiltrates the substratum before reaching the middle or lower fan segment. Hooke (1967, p. 454), observed that because 'water passes through rather than over such [permeable] deposits, they act as strainers or sieves by permitting the water [and fines] to pass while holding back the coarse material in transport'. Stream water discharge may thus decrease drastically when flowing over a permeable surface, thereby losing competence and causing rapid deposition. In such conditions, a lobe of debris (Hooke's 'sieve lobe' or 'sieve deposit') is rapidly deposited at the point where water is unable to effect further transport. Ephemeral, gravel-laden stream floods, issued over a relatively dry, porous fan, would be particularly prone to this phenomenon.

The depositional mechanism of gravel sieve lobes (studied by Hooke, 1967, and French, 1987) is specific, and should not be confused with that of either debris flows (Johnson, 1970; Takahashi, 1981) or the inferred high-density gravel traction carpets of 'hyperconcentrated' stream floods (Todd, 1989), even though their depositional settings may be broadly similar. Sieve deposits result when high infiltration rates through the substratum decrease rapidly the amount of water available for sediment transport and cause the deposition of a highly permeable, openwork gravel lobe. Hence, a sieve lobe is not quite the same as a traction carpet supported by dispersive stresses and left *en route* by a very powerful stream along its longer course (cf. Todd, 1989). Sieve-type deposition requires the following two conditions (French, 1987, p. 157).

1 The gravel being transported must have insufficient fines to fill the interclast spaces and act as a lubricant, or friction-reducing substance, for the

rapidly dumped gravel. Otherwise, a debris flow would originate, as indeed observed on most modern fans in semiarid regions (e.g. Hooke, 1967). The deficiency of fines also ensures that once a sieve lobe is formed, the continuing flood flow effectively passes through it, rather than over it, and thus fails to take the gravel readily into traction.

2 There must be a discontinuity in the sediment transport capacity of the flow; that is, the sediment load at some point must exceed the transport capacity of the flow. This condition, ensuring rapid dumping of gravel, is most readily fulfilled by flows that are short-lived surges (flash floods) issued from an intramontane valley, and by an abrupt spreading of the flow and/or abrupt flattening of the depositional slope (as due to a pre-existing relief or preceding sieve deposition).

The position of sieve deposits on an alluvial fan is controlled by the ratio of water discharge and infiltration rate (Hooke, 1967; French, 1987). The youngest gravelly lobes will be most prone to cause sieving, and an openwork gravelly sieve lobe, once formed near the fan apex, will then act as a sieving trap for subsequent flood surges issued from the intramontane valley. Such localized, self-propagating deposition in the longer term may create steep, substantial fans, as shown by the present examples (Figs 3 & 29). When conveyed by deep fan trenches, with an older alluvium as a substratum, large flood surges may continue for some distances downfan before the loss of water is sufficient to cause bulk deposition (e.g. see stage 3 in Agios Nektarios fan, Fig. 9; stage 3B in Komitades fan, Fig. 10; and stage 3A in Sfakia fan, Fig. 11). The older alluvium may be indurated, and the infiltration rate may thus be insufficient to instigate rapid deposition within the upper trench reaches. However, some lower-magnitude floods will tend to drop their load near fan apices, due to valley widening at the outlet and the lesser volumes of flood water involved. Larger or less debris-laden floods will then dissect or remove such valley-mouth sieve deposits. This explains the range of occurrences and characteristics of the in-trench gravel lobes (see preceding description).

Fan trenches were probably eroded by flows undercharged with debris, whether due to its temporal deficiency in a given drainage area or due to sieve deposition further upslope. The dumping of gravel at the valley mouth probably reduced the sediment load of many flows below the level of their retained transporting capability, enhanced by downfan trench confinement.

During heavy floods, water passing through the porous fans probably emerged below the fan-head segments and continued downfan as surface runoff. Subsurface, filtrating flow emerging on lower fan surfaces is known to accompany sieve-type deposition (Hooke, 1967, p. 454). This would explain the origin of some of the distal gullies on the fan surfaces (see preceding description). Other gullies might be due to direct surficial runoff, channelled by pre-existing topographic irregularities, or to occasional overflow from some of the fan trenches, when still relatively shallow (cf. Fig. 10).

Discussion

The alluvium of stage 3 has apparently recorded another major change in the climatic and discharge conditions. The intramontane valleys at this stage had been subject to strong further incision, with the development of the deep, narrow, axial trenches (Figs 4, 12 & 13). They periodically conveyed large, but ephemeral, surge-type discharges that apparently left the valley bedrock floors almost bare. There was no major storage of debris within the valleys, and these sediment-starved conduits thus probably cut down rather rapidly into the fractured, karstified limestone. Where the depth of valley incision exceeded a morphological threshold, defined by the height of the adjoining, earlier fan surface, the trench has been extended far beyond the valley mouth by dissection of this surface. Otherwise, the sediment derived from the bedrock valley was deposited directly at its mouth, to aggrade the apical part of the adjoining earlier fan in an attempt to establish a common profile. The two different patterns of deposition, as observed in the present case, would then represent two alternative modes of the geomorphic adjustment of fan profile to a particular valley profile (see also Denny, 1967; Hooke, 1967; Bull, 1977; Schumm, 1977).

The coarse-grained, permeable substratum, combined with the ephemeral, 'flashy' surge-type discharges, promoted the deposition of sieve-type alluvium. Water runoff was clearly non-perennial, possibly seasonal (?monsoonal), reflecting the climatic shift towards the modern semiarid conditions. Kutzbach (1981) inferred monsoon intrusions into the Aegean region between 10 000 and 5000 years BP; if true, rare but violent summer rains may have occurred, causing large sediment yields for brief periods (see also Pope & van Andel, 1984). Sediment yield was primarily due to removal of remnant intramontane Pleistocene alluvium and the continued degradation of intramontane valleys. 'Paraglacial' effects (Church & Ryder, 1972), related to debris-laden, prewashed (devoid of fines) postglacial intramontane slopes could play an important role.

It has been suggested in the literature (Bull, 1972, p. 69; 1977, p. 236) that 'unique' source terrains, such as hinterlands composed of jointed quartzites, are the essential factor responsible for the development of alluvial fans built of sieve deposits. The present study shows that this common notion, widely reiterated in textbooks, may not necessarily be correct. The limestone terrain in the present case has apparently sourced fans of highly disparate types (cf. stages 1, 2 & 3 above). It was rather the specific combination of climatic conditions (ephemeral high discharges), porous gravelly substratum and coarse, somewhat presorted source material that promoted the deposition of abundant sieve-type alluvium in the present case.

At stage 3, the piedmont plain and its Pleistocene alluvium have been subject to strong marine erosion related to the Holocene retreat of the southern Cretan coast (Fig. 4). The alluvial-fan complexes had been cliffed, and early Holocene gravelly beaches developed against this coastal cliff (Postma & Nemec, 1990). Pulses of more recent tectonic uplift (including that shown in Fig. 5) have further raised the coastal zone by a few metres above the sea level, thus initiating the development of the modern, present-day gravelly beaches against the rejuvenated cliff that contains erosional remnants of the raised, early Holocene palaeobeaches (for details, see Postma & Nemec, 1990). As a result, the fan trenches and gullies of stage 3 are perched several metres above the present-day sea level in the coastal cliff section (see Postma & Nemec, 1990, fig. 3).

SUMMARY AND CONCLUSIONS

The southern piedmont and lower slopes of the Lefka Ori limestone massif were sculptured by wave erosion and covered with sand/gravel and abundant marls during a major marine transgression in Pliocene time. Most of these littoral clastics were subsequently eroded when the sea rapidly retreated near the end of Pliocene times. Importantly, the marine transgression, with a probable sea-level highstand $c.450$ m above its present-day position (as documented by others), controlled the development (erosion base level) and early alluvial infill of an array of intramontane valleys (Fig. 4), whose outlets are 'hanging' high above the Quaternary piedmont plain (Figs 2 & 3). These palaeovalleys were once filled with gravelly alluvium, and periodically conveyed large sediment and water discharges in Quaternary times; some of the valleys show remnant terraces of alluvial gravel on their walls, with preserved thicknesses of up to 30–40 m. The valleys have later been degraded, with the incision of narrow, axial Holocene trenches (Fig. 4), whose floors at the mountain front are at altitudes of $c.200$ m (Figs 3 & 13).

The Quaternary piedmont alluvium occurs as an array of closely spaced, coalescent alluvial-fan complexes, each comprising several generations of variously stacked, successive fans. Their deposits are limestone gravels, texturally immature to mature and largely indurated. Three main generations of alluvial fans have been distinguished as broadly isochronous depositional 'stages' (Fig. 4), and their development may be summarized as follows.

1 The oldest fan (?latest Pliocene–early Pleistocene) are relatively small, with radii up to 1 km, very steep (20–22° near apices and 8–9° near the toes) and composed of debris-flow deposits. Fan apices are at altitudes of 300–400 m. These fans were built of unsorted, angular debris derived through localized slope wastage and headward erosion of mountain-front ravines, and developed as local cones within a broader veneer of slope-waste (colluvial) debris aprons in the mountain foot-zone (see stage 1 in Fig. 28). These fans apparently received little or no sediment from the pre-existing intramontane valleys; some of the fans are not even spatially related to the latter. Climate was probably arid to semiarid, such that the intramontane valleys stayed inactive. There was a marine incursion on to the piedmont plain, whereby the alluvial fans distally interfingered with littoral clastics, and were eventually eroded and cliffed by wave action.

2 The younger (Pleistocene) fans are relatively large, with radii greater than 2 km (distal segments unpreserved), several tens of metres thick, and constitute the bulk of the piedmont alluvium. Fan apices, where uneroded, are at altitudes of 320–360 m. These alluvial systems comprise at least five generations of variously stacked fan lobes, encrusted with pedogenic carbonates and some capped by Aridisols. The successive lobes are best recognized in the western part of the piedmont, where they prograded with considerable entrenchment. The fans are generally steep (up to 13°), but consist of stream-flow deposits. Moreover, their longitudinal morphometric profiles tend to be convex, broadly sigmoidal in shape (e.g. Fig. 13). The deposits are relatively well-sorted gravels that occur as broadly lenticular units interpreted as alternating channel-lag sheets and channel-bar sheets (Fig. 18), comparable to deposits of modern periglacial outwash fans described by other authors.

This stage (2) of piedmont sedimentation apparently involved large water discharges, attributed to periodical melting of mountain ice-caps or high-altitude cirque glaciers, whereby the pre-existing intramontane valleys were activated as principal drainage and sediment-yield areas, and the fans were constructed solely by stream-flow processes. The high-lying valley outlets made the alluvial systems prograde on to the piedmont plain as relatively steep, 'cascading' fans (Fig. 28). During the inferred glacial periods, the alluvial fans were abandoned and subject to rainwash, pedogenic and aeolian processes, while the intramontane valleys were being recharged with debris. Episodes of regional tectonic uplift, with a gentle easterly tilting of the piedmont, are inferrred to explain the differential style of fan stacking along the piedmont zone.

3 The youngest (Holocene) stage involved dissection of the earlier fans by long, valley-related trenches and minor distal gullies. The Recent fans and trench-confined gravel lobes are of minor volumetric importance. They consist of unconsolidated sieve deposits, attributed to ephemeral stream-flood surges issued from the intramontane valleys. The deposition of sieve-type alluvium was promoted by the permeable alluvial substratum and ephemeral discharges, probably due to torrential rains associated with a monsoonal or semiarid climate. The depth of intramontane valley incision, relative to the adjoining fan-head surface, determined the actual style of sieve deposition — whether as steep, unconfined fans, or as in-trench gravel tongues.

The alluvial-fan complexes thus comprise a very wide range of fan types, successively superimposed upon one another in various ways. Our detailed reconstruction shows how the geomorphic setting, varied climatic conditions and probable episodes of tectonic uplift have controlled the development of these fan complexes. The study as a whole demonstrates that alluvial-fan systems are, indeed, highly sensitive recorders of piedmont conditions. In our discussion of the successive fans, we have pointed at several important implications of this study for the sedimentological analysis and reconstruction of ancient alluvial fans. Some of the general implications are as follows.

1 Stream-dominated fans in some geomorphic settings may have depositional slopes in excess of 10–12°, much steeper than commonly perceived by sedimentologists for such systems.

2 Longitudinal braid-bar gravels may resemble those deposited by cohesionless debris flows (or gravelly 'density-modified grain flows'), and can easily be confused with the latter deposits in the stratigraphic record.

3 Accordingly, ancient stream-dominated ('wet') fans may easily be misinterpreted as mass flow-dominated ('dry') systems, with serious consequences to a basin analysis.

4 Fan-head trenching processes tend to obliterate grain-size trends in fan lobes, such as downfan fining and upward coarsening, which are regarded by many sedimentologists as particularly characteristic or diagnostic of progradational fans.

5 Fan entrenchment may not be a correlative feature, even for closely adjacent fans and relatively small piedmont areas.

6 Although it is generally true that larger stream-dominated fans are related to larger drainage areas (Schumm, 1977; Kostaschuk et al., 1986; Schumm et al., 1987), the application of this geomorphic rule to the stratigraphic record requires much caution. Large intramontane valleys may give rise to small fans if the valley-derived discharge happens to be split into two or more minor trenches within the proximal piedmont zone. The present study also emphasizes that the aforementioned geomorphic rule does not apply to fans built by different alluvial processes (cf. stages 2 and 3 in the present case); fan size is apparently far more dependent on the climatic conditions of drainage and the character of transport processes, than on the drainage area itself.

7 Alluvial 'sieve' deposition is not uniquely related to a particular source-rock type, as mistakenly perceived by many sedimentologists, but appears to be conditioned by a combination of factors, including climate (ephemeral high discharges), favourable geomorphic setting (well-pronounced intersection points or valley mouths), porous substratum and coarse, somewhat presorted or prewashed source material.

ACKNOWLEDGEMENTS

The field project was financed by a research grant from the Royal Geological Society, London, to G. Postma (while at the University of East Anglia), and was further supported with a travel grant from Bergen University to W. Nemec. Carbonate samples were collected in the field and radiometrically dated by Dr Peter Rowe of the University of East Anglia, Norwich. Geodetic measurements were made by Mr David Feltham (Norwich). Dr Jennifer Moody (University of Minnesota) and Dr Lucia Nixon (Queen's University at Kingston) kindly provided their archaeological expertise in the field. Professor R. Craig Kochel (Southern Illinois University) and Dr Simon Todd (Bristol University) reviewed the manuscript and gave helpful critical comments.

REFERENCES

ALLEN, P.A. (1981) Sediment and processes on a small stream-flow dominated, Devonian alluvial fan, Shetland Islands. *Sedim. Geol.* **29**, 31–66.

ANGELIER, J., LYBÉRIS, N., LE PICHON, X., BARRIER, E. & HUCHON, P. (1982) The tectonic development of the Hellenic Arc and the Sea of Crete: a synthesis. In: *Geodynamics of the Hellenic Arc and Trench* (Eds Le Pichon, X., Augustithis, S.S. & Mascle, J.). *Tectonophysics* **86**, 159–196.

ARDOS, M. (1973) Observations sur la structure et le géomorphologie de la montagne de Hacibaba à l'ouest de Karaman (Anatolie Centrale). *Rev. Geogr. Inst., Univ. of Istanbul* **14**, 119–130.

BEAUMONT, P. (1972) Alluvial fans along the foothills of the Elburz Mountains, Iran. *Palaeogeog. Palaeoclim. Palaeoecol.* **12**, 251–273.

BENSON, R.H. (1972) Ostracods as indicators of threshold depth in the Mediterranean during the Pliocene. In: *The Mediterranean Sea: A Natural Sedimentation Laboratory* (Ed. Stanley, D.J.) pp. 63–72. Dowden, Hutchinson & Ross, Stroudsburg.

BIRMAN, J.H. (1968) Glacial reconnaissance in Turkey. *Bull. Geol. Soc. Am.* **79**, 1009–1026.

BLAIR, T.C. & BILODEAU, W.L. (1988) Development of tectonic cyclothems in rift, pull-apart, and foreland basins: sedimentary response to episodic tectonism. *Geology* **16**, 517–520.

BLUCK, B.J. (1974) Structure and directional properties of some valley sandur deposits in southern Iceland. *Sedimentology* **21**, 533–554.

BLUCK, B.J. (1982) Texture of gravel bars in braided streams. In: *Gravel-bed Rivers* (Eds Hey, R.D., Bathurst, J.C. & Thorne, C.R.) pp. 339–355. John Wiley & Sons, New York.

BOOTHROYD, J.C. (1972) *Coarse-grained Sedimentation on a Braided Outwash Fan, Northeast Gulf of Alaska*. Tech. Report No. 6-CRD. Coastal Research Division, University of South Carolina, Columbia, 127 pp.

BOOTHROYD, J.C. & ASHLEY, G.M. (1975) Processes, bar morphology, and sedimentary structures on braided outwash fans, northeastern Gulf of Alaska. In: *Glaciofluvial and Glaciolacustrine Sedimentation* (Eds Jopling, A.V. & McDonald, B.C.) pp. 193–222. Soc. Econ. Palaeont. Miner., Tulsa, Spec. Publ. 23.

BULL, W.B. (1961) Tectonic significance of radial profiles of alluvial fans in western Fresno County, California. US Geol. Surv. Prof. Paper 424-B, pp. 182–184.

BULL, W.B. (1964) Geomorphology of segmented alluvial fans in western Fresno County, California. US Geol. Surv. Prof. Paper 352-E, pp. 89–129.

BULL, W.B. (1972) Recognition of alluvial-fan deposits in the stratigraphic record. In: *Recognition of Ancient Sedimentary Environments* (Eds Hamblin, W.K. & Rigby, J.K) pp. 63–83. Soc. Econ. Paleont. Miner., Tulsa, Spec. Publ. 16.

BULL, W.B. (1977) The alluvial-fan environment. *Prog. Phys. Geogr.* **1**, 222–270.

BUTZER, K.W. (1975) Patterns of environmental change in the Near East during late Pleistocene and early Holocene times. In: *Problems in Prehistory: North Africa and the Levant* (Eds Wendorf, F. & Marks, A.E.) pp. 389–410. Southern Methodist University Press, Dallas.

CARLING, P.A. (1984) Deposition of fine and coarse sand in an open-work gravel bed. *Can. J. Fish. Aquat. Sci.* **41**, 263–270.

CHURCH, M. & RYDER, J.M. (1972) Paraglacial sedimentation: a consideration of fluvial processes conditioned by glaciation. *Bull. Geol. Soc. Am.* **83**, 3059–3072.

CITA, M.B. (1982) The Messinian salinity crisis in the Mediterranean: a review. In: *Alpine–Mediterranean Geodynamics* (Eds Berckhemer, H. & Hsü, K.) pp. 113–140. Geodyn. Ser. Am. Geophys. Union/Geol. Soc. Am. 7. Am. Geophys. Union, Washington.

CLIFTON, H.E. (1984) Sedimentation units in stratified resedimented conglomerate, Paleocene submarine canyon fill, Point Lobos, California. In: *Sedimentology of Gravels and Conglomerates* (Eds Koster, E.H. & Steel, R.J.) pp. 429–441. Can. Soc. Petrol. Geol., Calgary, Memoir 10.

DE BOER, J.Z. (1989) The Greek enigma: is development of the Aegean orogen dominated by forces related to subduction or obduction? *Mar. Geol.* **87**, 31–54.

DECELLES, P.G., GRAY, M.B., RIDGWAY, K.D., COLE, R.B., PIVNIK, D.A., PEQUERA, N. & SRIVASTAVA, P. (1991) Controls on synorogenic alluvial-fan architecture, Beartooth Conglomerate (Palaeocene), Wyoming and Montana. *Sedimentology* **38**, 567–590.

DENNY, C.S. (1967) Fans and pediments. *Am. J. Sci.* **265**, 81–105.

DROOGER, C.W. & MEULENKAMP, J.E. (1973) Stratigraphic contributions to geodynamics in the Mediterranean area: Crete as a case history. *Bull. Geol. Soc. Greece* **10**, 193–200.

ECKIS, R. (1928) Alluvial fans in the Cucamonga district, southern California. *J. Geol.* **36**, 111–141.

FERNÁNDEZ, L.P., AGUEDA, J.A., COLMENERO, J.R., SALVADOR, C.I. & BARBA, P. (1988) A coal-bearing fan-delta complex in the Westphalian D of the Central Coal Basin, Cantabrian Mountains, northwestern Spain: implications for the recognition of humid-type fan deltas. In: *Fan Deltas: Sedimentology and Tectonic Settings* (Eds Nemec, W. & Steel, R.J.) pp. 286–302. Blackie, London.

FITZPATRICK, E.A. (1980) *Soils — Their Formation, Classification and Distribution.* Longman, London, 473 pp.

FORTUIN, A.R. (1978) Late Cenozoic history of eastern Crete and implications for the geology and geodynamics of the southern Aegean area. *Geol. Mijnbouw* **57**, 451–464.

FORTUIN, A.R. & PETERS, J.M. (1984) The Prina Complex in eastern Crete and its relationship to possible Miocene strike-slip tectonics. *J. Struct. Geol.* **6**, 459–476.

FRENCH, R.H. (1987) *Hydraulic Processes on Alluvial Fans.* Elsevier, Amsterdam, 256 pp.

FROSTICK, L.E. & REID, I. (1989) Climatic versus tectonic controls of fan sequences: lessons from the Dead Sea, Israel. *J. Geol. Soc. Lond.* **146**, 527–538.

FROSTICK, L.E. LUCAS, P.M. & REID, I. (1984) The infiltration of fine matrices into coarse-grained alluvial sediments and its implications. *J. Geol. Soc. Lond.* **141**, 955–965.

GIBLING, M.R., BOEHNER, R.C. & RUST, B.R. (1987) The Sydney Basin of Atlantic Canada: an upper Paleozoic strike-slip basin in a collisional setting. In: *Sedimentary Basins and Basin-forming Mechanisms* (Eds Beaumont, C. & Tankard, A.J.) pp. 269–285. Can. Soc. Petrol. Geol., Calgary, Memoir 12.

GLOPPEN, T.G. & STEEL, R.J. (1981) The deposits, internal structure and geometry in six alluvial fan–fan delta bodies (Devonian–Norway) — a study in the significance of bedding sequences in conglomerates. In: *Recent and Ancient Nonmarine Depositional Environments: Models for Exploration* (Eds Ethridge, F.G. & Flores, R.) pp. 49–69. Soc. Econ. Paleont. Miner., Tulsa, Spec. Publ. 31.

HARVEY, A.M. (1984) Aggradation and dissection sequences on Spanish alluvial fans: influence on morphological development. *Catena* **11**, 289–304.

HARVEY, A.M. (1987) Alluvial fan dissection: relationships between morphology and sedimentation. In: *Desert Sediments: Ancient and Modern* (Eds Frostick, L.E. & Reid, I.) pp. 87–103. Geol. Soc., Lond., Spec. Publ. 35.

HAUGHTON, P.D.W. (1989) Structure of some Lower Old Red Sandstone conglomerates, Kincardineshire, Scotland: deposition from late-orogenic antecedent streams? *J. Geol. Soc. Lond.* **146**, 509–525.

HEWARD, A.P. (1978) Alluvial fan sequence and megasequence models, with examples from Westphalian D–Stephanian B coalfields, northern Spain. In: *Fluvial Sedimentology* (Ed. Miall, A.D.) pp. 669–702. Can. Soc. Petrol. Geol., Calgary, Memoir 5.

HILL, G (1989) Distal alluvial fan sediments from the Upper Jurassic of Portugal: controls on their cyclicity and channel formation. *J. Geol. Soc. Lond.* **146**, 539–555.

HOOKE, R.L. (1967) Processes on arid-region alluvial fans. *J. Geol.* **75**, 438–460.

HOOKE, R.L. (1968) Steady-state relationships on arid-region alluvial fans in closed basins. *Am. J. Sci.* **266**, 609–629.

HOOKE, R.L. (1972) Geomorphic evidence for Late-Wisconsin and Holocene tectonic deformation, Death Valley, California. *Bull. Geol. Soc. Am.* **83**, 2073–2098.

HOOKE, R.L. & ROHRER, W.L. (1979) Geometry of alluvial fans: effects of discharge and sediment size. *Earth Surf. Proc.* **4**, 147–166.

HORVATH, F. & BERCKHEMER, H. (1982) Mediterranean backarc basins. In: *Alpine–Mediterranean Geodynamics* (Eds Berckhemer, H. & Hsü, K.) pp. 141–173. Geodyn. Ser. Am. Geophys. Union/Geol. Soc. Am. 7. Am. Geophys. Union, Washington.

JOHNSON, A.M. (1970) *Physical Processes in Geology.* Freeman, Cooper & Co., San Francisco, 577 p.

JOLLEY, E.J., TURNER, P., WILLIAMS, G.P., HARTLEY, A.J. & FLINT, S. (1990) Sedimentological response of an alluvial system to Neogene thrust tectonics, Atacama Desert, northern Chile. *J. Geol. Soc. Lond.* **147**, 769–784.

KOSTASCHUK, R.A., MACDONALD, G.M. & PUTNAM, P.E. (1986) Depositional processes and alluvial fan–drainage basin morphometric relationships near Banff, Alberta, Canada. *Earth Surf. Proc.* **11**, 471–484.

KOSTER, E.H. & STEEL, R.J. (EDS) (1984) *Sedimentology of Gravels and Conglomerates.* Can. Soc. Petrol. Geol., Calgary, Memoir 10.

KUTZBACH, J.E. (1981) Monsoon climate in the early Holocene: climate experiment with the Earth's orbital parameters for 9000 years ago. *Science* **214**, 59–61.

LAWSON, D.E. (1982) Mobilization, movement and deposition of active subaerial sediment flows, Matanuska Glacier, Alaska. *J. Geol.* **90**, 279–300.

LE PICHON, X. & ANGELIER, J. (1979) The Hellenic Arc and Trench system: a key to the neotectonic evolution of the eastern Mediterranean area. *Tectonophysics* **60**, 1–42.

LISTER, G.S., BANGA, G. & FEENSTRA, A. (1984) Metamorphic core complexes of Cordilleran type in the Cyclades, Aegean Sea. *Geology* **12**, 221–225.

LOWE, D.R. (1976) Grain flow and grain flow deposits. *J. Sedim. Petrol.* **46**, 188–199.

LOWE, D.R. (1982) Sediment gravity flows: II. Depositional models with special reference to the deposits of high-density turbidity currents. *J. Sedim. Petrol.* **52**, 279–297.

LUSTIG, L.K. (1965) Clastic sedimentation in Deep Springs Valley, California. US Geol. Surv. Prof. Paper 352-F, pp. 1–192.

MCDONALD, B.C. & BANERJEE, I. (1971) Sediments and bed forms on a braided outwash plain. *Can. J. Earth Sci.* **8**, 1232–1301.

MCKENZIE, D. (1978) Active tectonics of the Alpine–Himalayan belt: the Aegean Sea and surrounding regions. *Geophys. J.R. Astr. Soc.* **55**, 217–254.

MASSARI, F. (1984) Resedimented conglomerates of a Miocene fan-delta complex, Southern Alps, Italy. In: *Sedimentology of Gravels and Conglomerates* (Eds Koster, E.H. & Steel, R.J.) pp. 259–278. Can. Soc. Petrol. Geol., Calgary, Memoir 10.

MERCIER, J. (1977) L'arc égéen, une bordure déformée de la plague euroasiatique. *Bull. Soc. Géol. Fr.* **19**, 663–672.

MESSERLI, B. (1967) Die eiszeitliche und die gegenwartige Vergletscherung im Mittelmeeraum. *Geogr. Helvetica* **22**, 105–228.

MEULENKAMP, J.E., WORTEL, M.J.R., VAN WAMEL, W.A., SPAKMAN, W. & HOOGERDUYN STRATING, E. (1988) On the Hellenic subduction zone and the geodynamic evolution of Crete since the late Middle Miocene. *Tectonophysics* **146**, 203–215.

MIALL, A.D. (Ed.) (1978) *Fluvial Sedimentology.* Can. Soc. Petrol. Geol., Calgary, Memoir 5.

MUTO, T. (1987) Coastal fan processes controlled by sea level changes: a Quaternary example from the Tenryugawa system, Pacific coast of central Japan. *J. Geol.* **95**, 716–724.

MUTO, T. (1988) Stratigraphical patterns of coastal-fan sedimentation adjacent to high-gradient submarine slopes affected by sea-level changes. In: *Fan Deltas: Sedimentology and Tectonic Settings* (Eds Nemec, W. & Steel, R.J.) pp. 84–90. Blackie, London.

NAYLOR, M.A. (1980) The origin of inverse grading in muddy debris flow deposits — a review. *J. Sedim. Petrol.* **50**, 1111–1116.

NEMEC, W. & STEEL, R.J. (1984) Alluvial and coastal conglomerates: their significant features and some comments on gravelly mass-flow deposits. In: *Sedimentology of Gravels and Conglomerates* (Eds Koster, E.H. & Steel, R.J.) pp. 1–31. Can. Soc. Petrol. Geol., Calgary, Memoir 10.

NEMEC, W. & STEEL, R.J. (Eds) (1988) *Fan Deltas: Sedimentology and Tectonic Settings.* Blackie, London, 464 pp.

NEMEC, W., PORĘBSKI, S.J. & STEEL, R.J. (1980) Texture and structure of resedimented conglomerates: examples from Ksiąz Formation (Fammenian–Tournaisian), southwestern Poland. *Sedimentology* **27**, 519–538.

NEMEC, W., STEEL, R.J., PORĘBSKI, S.J. & SPINNANGR, Å. (1984) Domba Conglomerate, Devonian, Norway: process and lateral variability in a mass flow-dominated, lacustrine fan-delta. In: *Sedimentology of Gravels and Conglomerates* (Eds Koster, E.H. & Steel, R.J.) pp. 295–320. Can. Soc. Petrol. Geol., Calgary, Memoir 10.

NILSEN, T.H. (1982) Alluvial fan deposits. In: *Sandstone Depositional Environments* (Eds Scholle, P.A. & Spearing, D.R.) pp. 49–86. Am. Assoc. Petrol. Geol., Memoir 31.

NILSEN, T.H. (1985) *Modern and Ancient Alluvial Fan Deposits.* Van Nostrand Reinhold, New York, 372 pp.

NIXON, L., MOODY, J. & RACKHAM, O. (1988) Archeological survey in Sphakia, Crete. *Classical Views* **32**, 159–173.

NORTH, C.P., TODD, S.P. & TURNER, J.P. (1989) Alluvial fans and their tectonic controls. *J. Geol. Soc. Lond.* **146**, 507–508.

PETERS, J.M., TROELSTRA, S.R. & VAN HARTEN, D. (1985) Late Neogene and Quaternary vertical movements in eastern Crete and their regional significance. *J. Geol. Soc. Lond.* **142**, 501–513.

PIRAZZOLI, P.A., THOMMERET, J., THOMMERET, Y., LABOREL, J. & MONTAGGIONI, L.F. (1982) Crustal block movements from Holocene shorelines: Crete and Antikythira (Greece). In: *Geodynamics of the Hellenic Arc and Trench* (Eds Le Pichon, X., Augustithis, S.S. & Mascle, J.). *Tectonophysics* **86**, 27–43.

POPE, K.O. & VAN ANDEL, T.H. (1984) Late Quaternary alluvium and soil formation in the southern Argolid: its history, causes and archeological implications. *J. Archeol. Sci.* **11**, 281–306.

POSTMA, G. (1984) Mass-flow conglomerates in a submarine canyon: Abrioja fan-delta, Pliocene, southeast Spain. In: *Sedimentology of Gravels and Conglomerates* (Eds Koster, E.H. & Steel, R.J.) pp. 237–258. Can. Soc. Petrol. Geol., Calgary, Memoir 10.

POSTMA, G. & NEMEC, W. (1990) Regressive and transgressive sequences in a raised Holocene gravelly beach, southwestern Crete. *Sedimentology* **37**, 907–920.

RACHOCKI, A.H. (1981) *Alluvial Fans.* John Wiley & Sons, New York, 161 pp.

RACHOCKI, A.H. & CHURCH, M.J. (Eds) (1990) *Alluvial Fans: A Field Approach.* John Wiley & Sons, Chichester, 455 pp.

RETALLECK, G.J. (1988) Field recognition of paleosols. In: *Paleosols and Weathering through Time: Principles and Applications* (Eds Reinhardt, J. & Sigleo, W.R.) pp. 1–20. Spec. Pap. Geol. Soc. Am. 216. Blackwell Scientific Publications, Oxford.

ROBERTS, N. (1982) Lake levels as an indicator of Near East palaeoclimates: a preliminary appraisal. In: *Palaeoclimates, Palaeoenvironments and Human Communities in the Eastern Mediterranean Region in Later Prehistory* (Eds Bintliff, J.L. & van Zeist, W.) pp 235–267. British Archeological Reports, International Series No. 133. Blackwell Scientific Publications, Oxford.

ROBERTS, N. (1983) Age, palaeoenvironments, and climatic significance of late Pleistocene Konya Lake, Turkey. *Quater. Res.* **19**, 154–171.

RUST, B.R. & KOSTER, E.H. (1984) Coarse alluvial deposits. In: *Facies Models*, 2nd edn (Ed. Walker, R.G.) pp. 53–69. Geosci. Can. Reprint. Ser. 1. Geological Association of Canada, Toronto.

RYAN, W.B.F., KASTENS, K.A. & CITA, M.B. (1982) Geological evidence concerning compressional tectonics in the eastern Mediterranean. In: *Geodynamics of the Hellenic Arc and Trench* (Eds Le Pichon, X., Augustithis, S.S. & Mascle, J.) *Tectonophysics* **86**, 213–242.

SCHUMM, S.A. (1977) *The Fluvial System.* John Wiley & Sons, New York, 338 pp.

SCHUMM, S.A., MOSLEY, M.P. & WEAVER, W.E. (1987) *Experimental Fluvial Geomorphology.* John Wiley & Sons, New York, 413 pp.

SMITH, N.D. (1974) Sedimentology and bar formation in the Upper Kicking Horse River, a braided outwash stream. *J. Geol.* **82**, 205–224.

SOUTHARD, J.B., SMITH, N.D. & KUHNLE, R.A. (1984) Chutes and lobes: newly identified elements of braiding in shallow gravelly streams. In: *Sedimentology of Gravels and Conglomerates* (Eds Koster, E.H. & Steel, R.J.) pp. 51–59. Can. Soc. Petrol. Geol., Calgary, Memoir 10.

SPAAK, P. (1981) Earliest Pliocene paleoenvironments of western Crete, Greece. *Proc. R. Nether. Acad. Sci.* **B-84**, 189–199.

STEEL, R.J., MAEHLE, S., NILSEN, H., RØE, S.L. & SPINNANGR, Å. (1977) Coarsening-upward cycles in the alluvium of Hornelen Basin (Devonian), Norway: sedimentary response to tectonic events. *Bull. Geol. Soc. Am.* **88**, 1124–1134.

TAKAHASHI, T. (1981) Debris flow. *Ann. Rev. Fluid Mech.* **13**, 57–77.

TODD, S.P. (1989) Stream-driven, high-density gravelly traction carpets: possible deposits in the Trabeg Conglomerate Formation, SW Ireland and some theoretical considerations of their origin. *Sedimentology* **36**, 513–530.

VAN HARTEN, D. (1984) A model of estuarine circulation in the Pliocene Mediterranean based on new ostracod evidence. *Nature* **312**, 359–361.

WALKER, R.G. (1975) Conglomerate: sedimentary structures and facies models. In: *Depositional Environments as Interpreted from Primary Sedimentary Structures and Stratification Sequences* (Eds Harms J.C. *et al.*) pp. 133-158. Short Course Notes No. 2. Soc. Econ. Paleont. Miner., Tulsa.

WASSON, R.J. (1974) Intersection point deposition on alluvial fans: an Australian example. *Geogr. Ann.* **56-A**, 83-93.

WASSON, R.J. (1977) Catchment processes and the evolution of alluvial fans in the lower Derwent valley, Tasmania. *Z. Geomorph. NF* **21**, 147-168.

WELLS, S.G. & HARVEY, A.M. (1987) Sedimentation and geomorphic variations in storm-generated alluvial fans, Howgill Fells, northwest England. *Bull. Geol. Soc. Am.* **98**, 182-198.

Palaeogeomorphological controls on the distribution and sedimentary styles of alluvial systems, Neogene of the NE of the Madrid Basin (central Spain)

A.M. ALONSO ZARZA, J.P. CALVO *and* M.A. GARCÍA DEL CURA

UEI Petrología, Inst. Geología Económica, Facultad de C. Geológicas, Universidad Complutense, 28040 Madrid, Spain

ABSTRACT

A large variety of alluvial systems were developed in the NE of the Madrid Basin during the Miocene. The different alluvial morphologies as well as the distribution of facies within each of the systems provide an opportunity to evaluate the influence of the palaeomorphology of the basin margins and other factors in the construction of the alluvial deposits.

Two different kinds of depositional systems are described: (i) major fluvial distributary systems (averaging 50 km^2 in extent), and (ii) minor fans (2–10 km^2) and/or slope-scree deposits. The largest distributary fluvial systems (Alarilla, Jadraque, Baides and Tajuña) represented the points of highest sediment input into the basin. The location of these major systems was controlled by the palaeomorphology of the margins, particularly in areas where large linear depressions were developed in relation to structural lineations. The sedimentation in the fluvial distributary systems was dominated by braided streams associated with more or less broad alluvial plains.

In constrast, minor fans and slope deposits were located along more regular areas of the palaeoreliefs and their deposition was mainly by debris flows. Gravel fill channels occur occasionally in this type of alluvial system.

The mosaic of alluvial depositional systems in the northeastern part of the Madrid Basin during the Miocene is interpreted in the light of their location and depositional styles. The role of both intrinsic and extrinsic factors in the development of the alluvial systems is also discussed.

INTRODUCTION

The Madrid Basin forms the major part of the Tajo Basin, one of the three large continental basins developed in Spain during the Tertiary. It was initially linked in the east to the Intermediate Depression but the uplift of the Altomira Range during the late Oligocene to early Miocene led to the division of the Tajo Basin in two separated confined areas (Fig. 1).

The Madrid basin is an intracratonic basin with a complex deformational history that formed as a result of differential compressive strains from the north and south against a rigid lithospheric block (the Hesperic Massif) during Alpine movements (Alvaro *et al.*, 1979; Portero & Aznar, 1984). The basin has an extent that exceeds 10 000 km^2. The basin margins are commonly bounded by faults that display different styles: high-angle reverse faults are the common feature in the northern margin whereas both the southern and northeastern margins were bounded by less active normal faults. The eastern margin, the Altomira Range, was very active at the end of the Palaeogene and early Miocene but became quite stable after that time.

Miocene deposits constitute most of the sedimentary materials that outcrop at present in the Madrid Basin (Fig. 1). They are, in general, horizontally stratified. The thickness of the Miocene deposits does not exceed 300 m in outcrop but the real thickness of the Miocene sequence reaches up to 1000 m in central parts of the basin, as demonstrated by deep drilling and geophysical information (Racero, 1988).

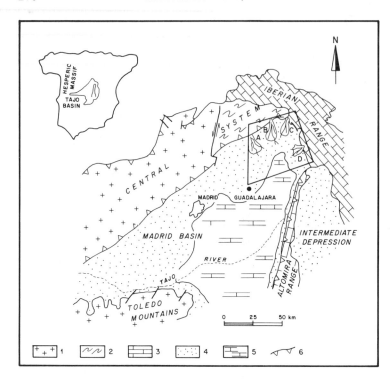

Fig. 1. Geographical and geological setting of the Madrid Basin with indication of the major fluvial distributary systems in the northeastern part of the basin. A, La Alarilla; B, Jadraque; C, Baides; D, Tajuña. 1, Granitic and high-metamorphic rocks; 2, low-metamorphic rocks; 3, sedimentary rocks of the Iberian and Altomira Ranges; 4, Miocene alluvial terrigenous sediments; 5, Miocene lacustrine sediments; 6, high-angle reverse faults and thrusts. Palaeogene formations are not drawn in view of their small outcropping extent.

The general Miocene succession of the Madrid Basin has been divided into three main units on the basis of vertical sequential changes and lithostratigraphic characteristics (Junco & Calvo, 1983). In central parts of the basin the Lower Unit is mostly made up of evaporite deposits. These are overlain by an alternating sequence of lacustrine carbonates, mudstones and gypsum which constitute the Intermediate Unit (Calvo et al., 1989). The age of this unit spans middle Aragonian (Daams & Freudental, 1981) to early Vallesian. Both the evaporite deposits of the Lower Unit and the overlying sediments of the Intermediate Unit grade laterally into thick clastic sequences near the margins of the basin. Finally, a thin (10–50 m) sequence (Upper Unit) composed of clastics and carbonates overlies the Intermediate Unit.

This paper focuses on the Miocene alluvial deposits outcropping in the NE margin of the Madrid Basin. The clastic complexes in this area were fed by the Central System low-grade metamorphic rocks and by the Iberian Range sedimentary rocks at the N and NE, respectively (Fig. 1). Four major fluvial distributary systems (*sensu* Nichols, 1987; Friend, 1989) and some minor alluvial fans are the dominant features of the marginal Miocene facies.

The purpose of this paper is to describe size, location, internal structure and sedimentary facies of these alluvial systems as well as to analyse their relationships with laterally associated deposits along the basin margin. The different depositional styles of the alluvial systems are discussed in terms of palaeomorphological and structural controls of the surrounding reliefs.

The palaeomorphology of the basin margin is divisible into: (i) areas of the basin margin with large structurally controlled depressions, and (ii) areas lacking such lineations. In the first case fluvial distributary systems were developed. In the other situations only small fans and slope-scree deposits are present.

THE ALLUVIAL SYSTEMS OF THE NE MADRID BASIN

The Miocene alluvial deposits largely extend in the NE area of the Madrid Basin where they are quite well exposed. These alluvial sediments outcrop in sequences that reach up to a maximum of 300 m in thickness and exhibit a typical red-brown colour.

From west to east a number of major fluvial

distributary systems have been distinguished. The location of the major alluvial systems is related to linear depressions trenched into the mountain reliefs that formed the basin margin. These linear depressions were fault controlled. However, each of the major alluvial systems developed under different conditions, in terms of tectonic activity of the surrounding reliefs, lithology of the catchment area and local palaeomorphology of the basin margin. The average areal extent of these systems is about 50 km^2. The development of all these alluvial systems was not isochronous. This has been deduced from the lithostratigraphic logs of the alluvial sequences (Fig. 2) as well as by correlation with sequences containing micromammal localities (Sesé et al., 1990). Thus, most of these major alluvial systems were active during the early Miocene (Lower Unit) but some of them became inactive during the middle Miocene (Intermediate Unit). This is the case of the Alarilla, which is located in the northeastern part of the area (Fig. 1). In contrast, the other major alluvial systems (Jadraque, Baides, Tajuña) continued to develop until the lower Vallesian (top of the Intermediate Unit) (Fig. 2).

Between these major fluvial distributary systems other minor alluvial systems (2–10 km^2) occur and they have their own distinct alluvial architecture along extensive zones of the basin margin. These minor systems are represented by telescoped small fans and wedges of slope-scree deposits.

In the following paragraphs we describe the main sedimentological features of the alluvial complexes occurring in the NE Madrid Basin and discuss the extrinsic and intrinsic factors that controlled their development.

La Alarilla fluvial distributary system

Coarse Miocene deposits in this area discordantly overlie carbonate and mudstone deposits of Palaeogene age. The deposits of the Alarilla system are exposed with a maximum thickness of 60 m in the marginal areas, although in distal zones more than 100 m of mudstones with interbedded fine sands can be recognized. The vertical evolution of the alluvial deposits shows a clear fining and thinning upward sequence.

Within this fluvial distributary system, proximal and distal zones can be easily distinguished. Facies assemblages in the proximal zones are mainly made up of conglomerates composed dominantly of metamorphic clasts, though some carbonate and sandstone clasts of Mesozoic and Palaeogene origin can also be recognized. The facies associations in these areas are varied and include the following.

1 Tabular conglomerate beds. Two different types are present. The first type consists of massive, usually imbricated gravels. They are arranged in tabular beds whose thickness is up to 0.8 m and lateral extent about 30 m (Fig. 3). The other type is represented by thicker bodies (up to 2–3 m) that show fining upward trend and occasional imbrication. Its width reaches 50 m. Internal erosive surfaces are often present. The mean size of the clasts is 15 cm. These facies assemblages correspond respectively to A and B of Fig. 4a.

2 Multistorey conglomerate and sandstone bodies. The thickness of these bodies commonly reaches up to 3 m. The conglomerates occur at the bottom of the bodies and show Gm, Gp and Gt lithofacies (Miall, 1977, 1978). The top of these conglomerates is usually well defined and displays a convex up geometry (Fig. 4a, C). The small depressions are occupied by fine gravels and sands which show different sedimentary structures, such as horizontal lamination and trough and planar cross-bedding. One of the most striking features of these bodies is the occasional occurrence of large Palaeogene clasts (70–80 cm).

3 Two different types of channels are present. The first type corresponds to multistorey channels displaying very steep erosive bases, filled with cross-stratified gravels (Fig. 4a, D). The thickness of these channels is about 0.5–1 m. Similar channels have been recognized in Triassic gravels of the Iberian Ranges by Ramos & Sopeña (1983) and Ramos et al. (1986). They refer to these structures as transverse fill cross-stratification. The other type of channel is represented by small multistorey gravel channels whose thickness is close to 2 m and whose width does not exceed 3 m. The infill of this second type of channel consists of vertically accreted, horizontally bedded gravel units (Fig. 4a, E).

4 Massive, brown to orange, silty mudstones occur interbedded with the gravel and sandstone bodies. The frequency of this lithofacies increases towards the top of the sequence as well as towards the south. These mudstones are frequently mottled and contain carbonate nodules (Fig. 4a, F).

In distal areas the facies assemblages change rapidly to others composed mainly of mudstones with intercalations of coarse detrital beds as well as carbonates. The facies assemblages recognized in

Fig. 2. Sedimentary logs of Miocene proximal alluvial successions showing broad environmental interpretation and correlation with distal fan and lacustrine areas. 1, Gravels; 2, sands; 3, mudstones; 4, limestones; 5, planar cross-bedding; 6, trough cross-bedding; 7, rippled sands; 8, nodular sandstones; 9, root-traces; 10, mottling.

Neogene alluvial systems, Madrid Basin 281

Fig. 3. La Alarilla fluvial distributary system. Tabular bodies consisting of Gm imbricated facies that represent the initial stage of longitudinal bars. Silts and clays are locally interbedded with the gravels.

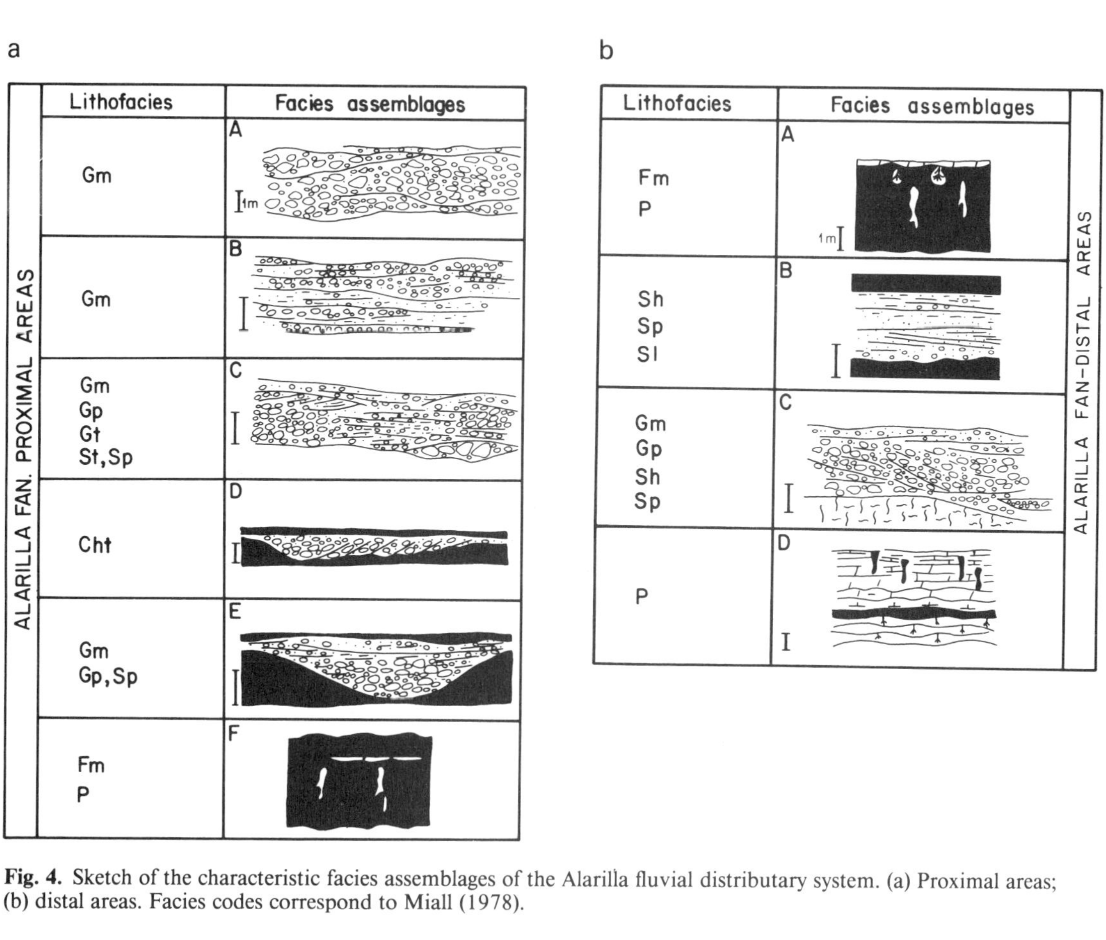

Fig. 4. Sketch of the characteristic facies assemblages of the Alarilla fluvial distributary system. (a) Proximal areas; (b) distal areas. Facies codes correspond to Miall (1978).

these areas are sketched in Fig. 4b. The most common facies are mottled mudstones, sandstone channels, gravel ribbons and nodular carbonates.

The characteristics and distribution of facies assemblages in the Alarilla system characterize a broad alluvial plain (Vos, 1975), reaching 4 km in width, in which braided channels were quite stable. In this alluvial plain the sedimentation took place in a complex of bars and channels. In the distal areas, flood plain deposits were dominant over channel or lacustrine ones. An idealized sketch of the distribution of these deposits is given in Fig. 5.

In the proximal areas several types of bars have been recognized: (i) unit bars or diffuse gravel sheets (Smith, 1974; Bluck, 1976; Hein & Walker, 1977) which represent the initial stage of growth of larger bars (Fig. 4a, B); (ii) morphologically well-defined larger bars showing different stages of growth correspond to longitudinal bars mainly formed by Gm lithofacies (Fig. 4a, C); (iii) the sands and gravels with planar or trough cross-stratification occurring at the top of the multistorey bodies of association 2 (Fig. 4a, C) may be interpreted as transverse bars formed in lower stages of the stream when the longitudinal bars were emergent (Rust, 1984).

Between these bar complexes as well as in less active areas of the alluvial plain the two abovementioned types of channel were located. Channels displaying transverse fill cross-stratification were placed in active areas of the alluvial plain whereas deeply incised channels are associated with less active lateral areas of this plain.

The sedimentation of clay and silt deposits took place in the flood plain area during periods of high flooding as well as at the top of the bars and channels during less active stages of stream flow.

Towards the south this upper alluvial plain grades to the lower alluvial plain (Vos, 1975). In this area

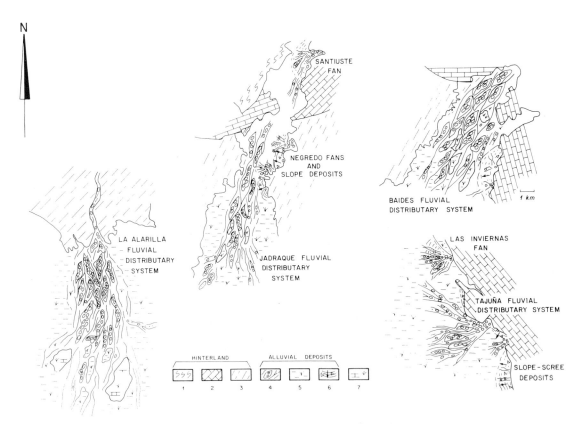

Fig. 5. Interpreted depositional models for the major alluvial complexes in the NE of the Madrid Basin. 1, Shales, quartzites and metavolcanics; 2, Mesozoic (mainly carbonates); 3, Palaeogene (carbonates and terrigenous); 4, gravels and sands; proximal and medial alluvial deposits; 5, silts and clays; distal alluvial and overbank deposits; 6, slope-scree deposits and minor fans; 7, ponds.

the channel deposits of the alluvial plain are progressively less connected. Flood plain deposits are dominant although some shallow braided channels are also present (assemblage of Fig. 4b, C). The development of small ponds on the floodplain is indicated by the presence of the above-mentioned palustrine carbonates.

Clast compositions indicate that the sources of this fluvial distributary system were mainly the Precambrian and Palaeozoic rocks of the Central System, although the fan is located at the foot of the Palaeogene relief, where the streams diverged in a radial network. This situation fits well with the general model proposed by Schumm (1977, 1981), who differentiates three major areas of alluvial development (drainage, transport and sedimentation zones). The area described in the Alarilla system corresponds to the third of these zones. The transport area is represented by a deeply incised, probably fault controlled, N-S depression which was cut into the easily erodible Palaeogene materials that limit the coarse fan facies. Quite similar situations have been recognized by several authors involved in the analysis of alluvial fan systems (Vessel & Davies, 1981) and fan deltas (Roberts, 1987).

The Jadraque system

The Jadraque system consists of the Jadraque fluvial distributary system and a number of minor coalesced alluvial fans disposed at its eastern side (Fig. 5). The Jadraque fluvial distributary system shows a palaeoflow of 180–210°, whereas minor coalescing alluvial fans are arranged approximately normal to it and palaeoflow was toward the main system.

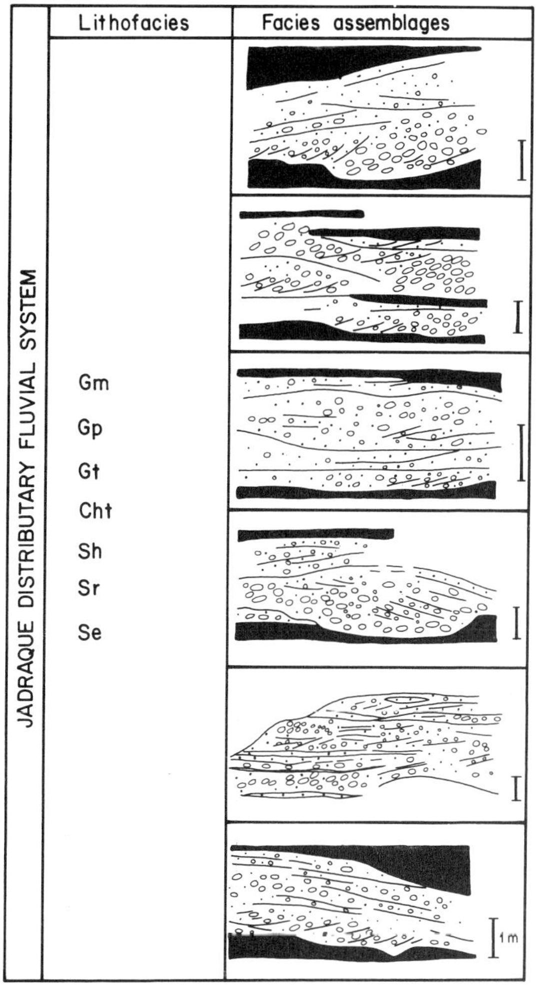

The Jadraque fluvial distributary system

This is a longitudinal, non-radial system that appears entrenched in a broad Palaeogene palaeovalley. The deposits that form this complex can be observed for a distance of nearly 12 km and a width of about 3 km. The maximum measured thickness of these deposits is 70 m. They form a fining–thinning upward sequence.

Facies assemblages of this system show some differences compared to those of La Alarilla. Most of the recognized facies consist of gravel and sand ribbons (Friend et al., 1979; Friend, 1983) interbedded with red silts and clays (Fig. 6).

Characteristically the gravel and sand ribbons show erosive bases and multistorey infill made up of massive gravels, cross-bedded gravels and sands, the sands also horizontally laminated and rippled (Fig. 7). The gravels are in general dominant over the sand deposits. The thickness of these bodies ranges between 1.2 and 5 m. Less common facies assemblages occurring in this area consist of ribbons filled with transverse cross-stratification units. In addition, thin sandstone channel-fills (0.5–1 m of maximum thickness) occur interbedded with red, massive, often bioturbated silts.

Overall, these assemblages characterize a braided fluvial system dominated by gravel deposits. The multistorey ribbons illustrate single periods of

Fig. 6. Sketch of the facies assemblages recognized in the Jadraque fluvial distributary system.

Fig. 7. Complex gravel ribbon consisting of Gm facies in the lower part and Gp facies in the top. These types of ribbons are interbedded, in the Jadraque distributary fluvial system, with red mudstones.

coarse clastic accumulation within the fluvial network, although the multistorey character of these bodies clearly indicates reactivation of the earlier deposit. The deposition of the gravels and sands took place in bars and channels. The bars were mainly longitudinal bars as demonstrated by the usual occurrence of Gm facies. However the common occurrence of the cross-bedded units at the top of the ribbons can be interpreted as either the result of transverse bars or the modification of these longitudinal bars in stages of lower energy (Bluck, 1976). The sands located at the top of the ribbons are also considered to be indicative of lower energy stages, representing the modification stages of the top of some bars.

The channels would be located between the bar complexes. These channels are either massive or transverse infilled channels (Ramos & Sopeña, 1983). Both types of channel fills are of minor importance relative to the bar complexes. On the other hand sandstone channels interbedded with overbank red silts are considered to result from slightly confined flows over the flood plain.

Santiuste and Negredo–Cendejas minor fans

This group includes several alluvial fan systems (Santiuste and Negredo) as well as some poorly developed slope-scree deposits. These systems are disposed as a narrow belt over the Mesozoic and Palaeogene reliefs (Fig. 5).

The Santiuste fan encroaches an E–W linear depression trenched in Mesozoic rocks. Its length is less than 1 km and the width is about 300 m. The maximum measured thickness of the alluvial sequence is 52 m.

In this fan the proximal areas consist of stacked conglomerate beds in which Gm facies are widely represented, although Gms facies can also be recognized. Imbricated clasts are rare. At the top of some of the conglomerates thin beds of coarse sands and gravels with diffuse cross-bedding are present. In addition, fine deposits, mainly composed of bioturbated sands and silts, occur interbedded with the conglomerates. The thickness of these finer-grained deposits ranges between 0.1 and 1.5 m. Down-system these sequences grade into silts and clays with rare interbedded conglomerate bodies. Mean size of the clasts in the conglomerates varies from 20–30 cm in the proximal areas to 8–10 cm in the distal ones.

The Negredo–Cendejas system consists of three small fans with apices embayed in Palaeogene rocks. Between the different apices slope-scree deposits were developed. Fans and slope-scree deposits coalesced easterly, displaying a telescoped architecture. Slope-scree deposits, wedge to tabular bodies composed mainly of massive gravels, are placed covering the Mesozoic or Palaeogene palaeoreliefs. In contrast to the fans they do not show a cone morphology.

Facies assemblages in these minor systems are

somewhat varied and include: (i) massive beds of matrix-supported breccia, and (ii) horizontally bedded gravels showing massive structure within the single beds. Both types of coarse-grained deposits exhibit well-developed pisolithic and laminated carbonate crusts.

From the facies analysis of the Santiuste and Negredo–Cendejas alluvial deposits we can deduce that the largest fan (e.g. the Santiuste fan) shows different facies assemblages from those that form the smaller fans and, undoubtedly, from the slope-scree deposits. In general, the larger the fan the higher the frequency of clast-supported conglomerates. These clast-supported conglomerates may be interpreted as deposited in ephemeral braided streams, although they could also be envisaged as clast-supported debris-flows. These deposits are more usual in the Santiuste fan as well as in distal areas of some of the Negredo fans. However, matrix-supported gravels will indicate transport and depositional processes dominated by debris and/or gravity flows. These are dominant in the proximal areas of the Negredo fans and they also characterize most of the slope-scree deposits.

In the studied deposits most of the intrinsic factors that influence the occurrence of debris-flows instead of stream dominated systems, i.e. drainage basin size, source rocks, lithology of the catchment area, etc. (Harvey, 1989; Wells & Harvey, 1987), might be quite similar except for the size of the drainage basin. The drainage area was larger in the Santiuste fan (dominated mainly by stream flows) than in any of the Negredo systems. However, the occurrence of channel deposits between debris-flows, as it is observed in Negredo, could result from more humid periods which favour channel processes (McArthur, 1987) or from the decrease of mass-processes in the distal areas of the fans (Hooke, 1967).

The Baides fluvial distributary system

The Baides fluvial distributary system is located in the junction between the Central System and the Iberian Ranges. It spreads out through a rather rectangular depression which is bounded by the Huermeces reverse fault in the north, the Baides anticline in the west and the Iberian Ranges to the east (Fig. 5). The alluvial deposits spread out 210–230° to the south grading into finer-grained sediments.

The movement of the Huermeces fault during the late Oligocene and probably throughout the early and middle Miocene was the major control for the development of the Baides system.

The Miocene deposits that form the fan are discordant over Palaeogene rocks. Miocene alluvial deposits, 140 m thick, show a gradual change from 20° dip at the bottom to nearly horizontal at the top.

Three minor sedimentary sequences can be distinguished within the complete alluvial succession (Fig. 2). The lower one corresponds (by correlation with adjacent areas of the basin) to the Lower Miocene Unit. The Intermediate Unit is formed of two thinning–fining upward sequences. The facies assemblages that form the sequences are, in general, quite similar.

The Baides fluvial distributary system has a width close to 3 km and the length is about 10 km. The arrangement of facies assemblages and sedimentary sequences defines a clear proximal to distal zonation. The sedimentary characteristics of these zones are as follows (Fig. 8).

1 Proximal areas. These areas consist mainly of gravel bodies which display two different morphologies.

(a) Gravel bodies showing convex up morphologies (Fig. 8, A). They show multistorey gravel infill, imbrication and occasional inverse grading. The thickness reaches up to 4 m in the central parts of the bodies but decreases to approximately 1 m in the sides. Width of these bodies varies from 300 to 500 m.

(b) Tabular gravel beds formed mainly of Gm facies and topped frequently by coarse sands (Fig. 8, B).

2 Medial areas. In these areas two different associations can be described. The first consists of multistorey gravel sheets, mostly composed of Gm facies. Thickness of the simple sheets is about 3 m. These beds show different morphologies, either flat base and convex top (Fig. 8, C) or erosive base and flat top (Fig. 8, D). Lenses of cross-stratified sands are usually interbedded with the gravel bodies.

The second assemblage occurring in these medial areas includes sands, red silts and nodular carbonates with some lenses of gravels (Fig. 8, E). They are well-defined tabular beds with irregular convex surfaces at the top. They are often bioturbated and occasionally show pisolithic crusts.

3 The distal areas are characterized by mottled silts and clays with intercalated conglomerate and sand filled channels (Fig. 8, F).

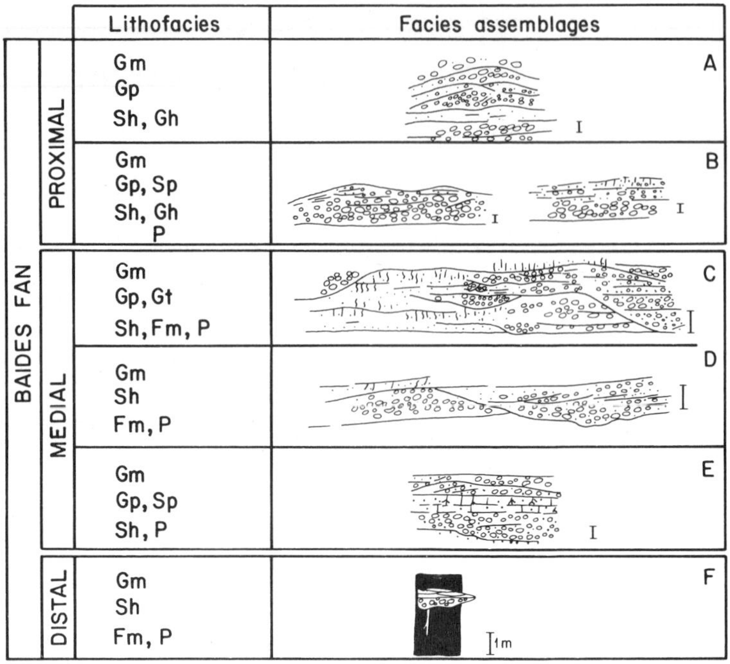

		Lithofacies	Facies assemblages
BAIDES FAN	PROXIMAL	Gm, Gp, Sh, Gh	A
		Gm, Gp, Sp, Sh, Gh, P	B
	MEDIAL	Gm, Gp, Gt, Sh, Fm, P	C
		Gm, Sh, Fm, P	D
		Gm, Gp, Sp, Sh, P	E
	DISTAL	Gm, Sh, Fm, P	F

Fig. 8. Sketch of the facies assemblages of the Baides fluvial distributary system.

The occurrence and location of these facies define a fluvial distributary system dominated by braided streams. In most proximal areas it may be characterized as a non-confined 'close braided' system (Corrales et al., 1986) or as an 'upper braided alluvial plain' (Vos, 1975). The arrangement of facies in the sheets indicates the development of a broad channel belt in which channels and longitudinal bars were the main morphological elements (Fig. 5).

The dominance of sands and silts in the medial areas and the interbedding of these sediments with gravel beds are indicative of a less interconnected pattern of the channels. Bars and channel complexes were also represented. Their interrelationships characterize a 'loose braided' system (Corrales et al., 1986) or 'lower braided alluvial plain' (Vos, 1975). Furthermore some areas placed in higher topographic levels over the channel belt were submitted to periodical flooding allowing the deposition of finer-grained sediments. These higher zones were modified by pedogenic processes related to the activity of subaerial plants, as noted by the occurrence of root moulds. There is a decrease of the density of channels from medial to distal areas in the Baides fan.

Finally, both the vertical evolution of the Miocene sedimentary sequences and the geometrical relationships observed in the Baides fluvial distributary system suggest that its development took place in different stages related to the uplift of the Huermeces reverse fault. The movement of this fault probably took place throughout the lower to middle Aragonian.

Fluvial distributary and other minor alluvial systems related to the Iberian Ranges

The alluvial deposits that are considered here form a rather narrow fringe, about 20 km in length, that flanks the Mesozoic reliefs extending from the junction between the Central System and the Iberian Range in the north until the northernmost side of the Altomira Range in the south (Fig. 1).

The alluvial fans that constitute this complex are usually embayed in linear depressions trenched in Mesozoic carbonate reliefs. Elsewhere, slope-scree deposits and minor fans are located along the steep margins of these reliefs.

The Tajuña fluvial distributary system

The length of the Tajuña system is about 5 km. The entrenchment of its apex is related to a NW–SE reverse faulting. Alluvial deposits that outcrop in the most proximal zone lie discordantly on a calcareous crust which was developed on Mesozoic carbonates. The maximum measured thickness of the deposits reaches up to 120 m. Areas of the system are as follows.

1 Apex and proximal areas. The apical area shows a complex structure, the distribution of the fanhead coarse deposits being controlled by the palaeomorphology of the adjoining reliefs. A few tens of metres downslope from the apex, the proximal deposits are confined within a narrow linear depression where they form a vertical succession of 120 m in thickness composed of two fining–thinning upward sequences.

These sequences are made up of tabular conglomerate beds, averaging 3 m in thickness, each one composed of Gm facies, whether imbricated or not. Gravels and coarse sands display cross-stratification. Fine sands and silts are commonly recognized at the top of the conglomerate beds (Fig. 9).

2 Medial areas. These areas are located away from the zone under the direct influence of the aforementioned palaeovalley. The width of the fan deposits is larger because they are not confined within the mountain relief. Two different parts, distinguished by distinct palaeocurrent directions and alluvial sedimentary styles, can be recognized within the medial areas. These parts correspond to the northern and southern parts of the fan body, which expand in N 60° W and N 120° W directions, respectively (Fig. 5).

In the northern area the deposits of the fan consist of red clays and silts with interbedded gravel channel-fills (Fig. 9). The channels appear to have been incised into the overbank lutites. They show a fining upward trend and are characterized by erosive bases, overlain by massive gravels and gravels and sands with planar and trough cross-stratification. Rippled sands can be present at the top of the channel bodies. The mean thickness of these bodies is 2–3 m. Typical width is about 100 m.

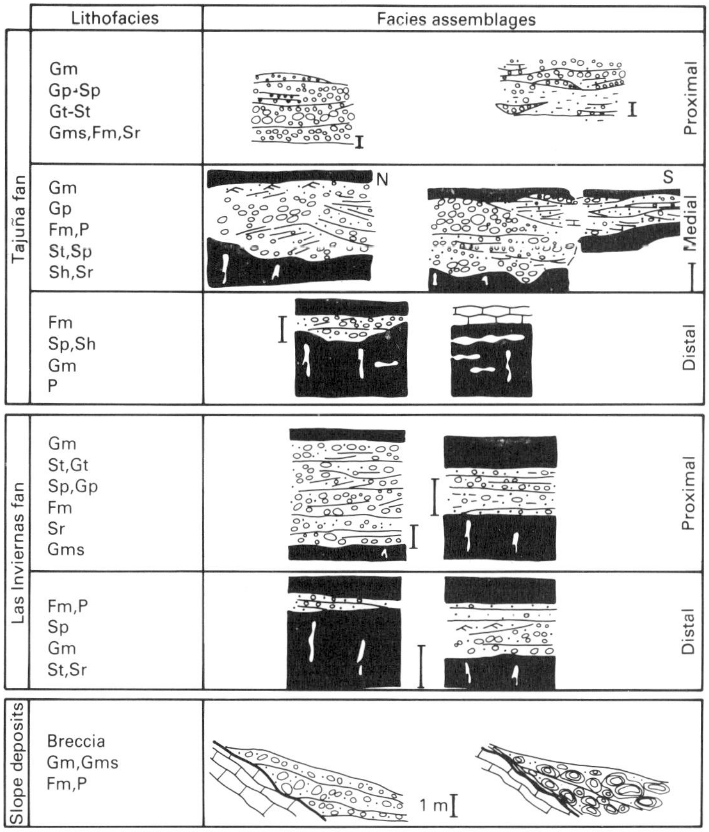

Fig. 9. Facies assemblages of Tajuña–Las Inviernas complexes, showing the presence of coated and non-coated gravels in the slope deposits.

In the southern areas the sedimentation is characterized by sheets of gravel bodies intercalated between red silts and clays. The width of each sheet reaches up to 300 m. Thickness varies between 1 and 5 m. Most commonly these sheets are formed of crudely bedded gravels, with large-scale planar cross-stratification, and sands with planar or trough cross-bedding. These assemblages are occasionally cut by channelized bodies composed of imbricated gravels. The maximum thickness of these bodies is usually seen in the central zone of each sheet, where the bases of the gravel units are sharply erosive (Fig. 10). Lateral parts of the sheets can be equated to the wings of Friend et al. (1979). They consist mainly of coarse sands and gravels showing planar cross-stratification, occasionally cut by minor channels.

3 Distal areas. Clay and silt deposits are dominant in the distal areas of this system. Down-system they include marl levels. The general facies association in these zones includes:

(a) thick levels (8–10 m) of red silts and clays which contain carbonate nodules, rhizoliths and mottling;

(b) medium- to fine- grained sandy beds locally showing cross-stratification, the thickness of these beds is about 0.5–1 m and their width about 30 m;

(c) imbricated gravel channels about 1.5 m in thickness; they show erosive bases and normal grading;

(d) nodular marls that grade laterally to pedogenically mottled limestones.

The apical or 'fan core facies' (Boothroyd & Nummedal, 1978), mainly made up of Gm facies and lesser amounts of Gp, St, Sp and Sr lithofacies, suggest that the deposition was due to stream dominated flows. The sedimentary facies and sequences are quite similar to the Scott model (Miall, 1977, 1978), characterized by active deposition in the longitudinal bars and channels of a braided alluvial system. In our case, the braided system was initially entrenched in a narrow palaeovalley and ceased being confined away from the foot of the Iberian Range. The radial expansion of the system was accompanied by differentation into two areas, this differentation being controlled by the different palaeomorphology of the reliefs where the alluvial facies were previously confined.

Thus, on the northern side the abrupt slope of the Mesozoic rocks, partially related to structural lineations, favoured the development of more stable and confined channels in this zone. In the southern area the deposition took place on a flat surface allowing the generalized lateral spreading of the flows. As a consequence gravel and sand sheets developed in this area. The sheets represent the deposits of a broad channel belt (Friend, 1983) in which the varied facies assemblages suggest that the channels were quite connected (Bridge, 1985). The deposits of the inner parts of the sheets were mainly accumulated in longitudinal bars that were fre-

Fig. 10. Section of the central part of a sheet corresponding to the medial facies of the Tajuña fluvial distributary system (southern area). The sheet is interbedded with red mudstones.

quently cut by channels. On the contrary, the deposition in the outer parts of the sheets took place mainly as sandy bars migrating along the channel belt.

The spreading of fans in their medial areas seems to be a common feature (Van der Meulen, 1986) that is commonly accompanied by the development of broader channels showing a clear decrease in their depth. In the Tajuña fan the expansion of the fan is the result of the loss of entrenchment of the main channel.

Distal areas represent the alluvial plain deposits in which only small channels were active. Both marls and mottled carbonates show the interfingering between the alluvial plain facies and the lake margin (Calvo et al., 1989).

Las Inviernas fan

This fan is not very relevant in terms of volumes of clastic influx to the basin. However it is quite well exposed and the good preservation of the initial morphology makes easy the study of the fan facies distribution and permits a detailed analysis of the sedimentological model (Alonso Zarza et al., 1990).

The length of the fan is less than 2 km. The maximum thickness of the exposed deposits (which spread out in an ENE–WSW direction) is 70 m. As in the Tajuña fluvial distributary system, two sequences can be differentiated.

The older deposits of the fan are located in the south and consist of erosive gravel bodies with thicknesses between 1.4 and 3 m. These gravels are quite well stratified and are formed of Gm, Gt, St, Sp and Sr lithofacies. Usually they show normal grading. These deposits are interbedded with red silts and clays and grade laterally to a sequence made up of mudstones which constitute the predominant lithofacies in the lower levels at the northern side of the fan. The younger coarse deposits of the fan are mainly located over these mudstones so indicating a progressive shifting of the major distributaries. These coarse deposits consist mainly of Gms facies that display occasional imbrication.

Towards the distal and lateral areas the dominant facies consist of mottled and bioturbated mudstones with interbedded lenses of sands and gravels.

Both vertical and horizontal variations of facies assemblages of Las Inviernas fan are indicative of changes of the alluvial depositional style. Initially a large braided complex developed in the southern area. Later on, this braided complex evolved to an unconfined alluvial complex where sedimentation was dominated by sporadic debris flows.

Minor fans and slope deposits

The area between the apices of the Tajuña and Las Inviernas alluvial systems (Fig. 5) is occupied by minor fans that were related to small depressions formed in the Mesozoic relief as well as by slope-scree deposits. The development of minor fans and slope-scree deposits is also a common feature in other places along the margin that the Iberian Ranges forms in the NE part of the Madrid Basin. Thus, both minor fans and slope-scree deposits built up a laterally continuous narrow clastic wedge along the basin margin between the major systems (Fig. 5). The structure of these minor fans will not be described because of their similarity with the minor systems of the Jadraque area.

The slope-scree deposits show a different sedimentary pattern compared to the minor fans. They are in a different morphological position, characterized by linear, dipping Mesozoic relief. The thickness of the Miocene slope-scree deposits reaches up to 60 m in exceptionally thick wedges. In these deposits three different facies have been recognized.

1 Mud-supported breccia. This is a characteristic deposit which covers the Mesozoic palaeorelief in much of the area. The thickness of the breccia is about 1.5 m. Clasts are surrounded by pisolithic crusts (Fig. 9) and the entire deposit is covered by a laminar crust of calcrete type (Alonso Zarza et al., 1992).

2 Gravels arranged parallel to the margin front in wedge-shaped bodies. The thickness can reach 2 m. The facies consist typically of non-imbricated massive gravels. Minor imbricated gravel channels and bodies of mixed gravel and sand with planar cross-bedding can occasionally be observed. The gravels are usually poorly sorted.

3 Both facies (1 and 2) grade distally into sandy lutites which interfinger with the distal facies of the alluvial fans.

Morphologically the slope-scree deposits can be defined as alluvial 'aprons' in the sense of Bates & Jackson (1980). The deposits over the mountain slope were the result of episodic and ephemeral flows which locally were ordered in braided ephemeral channels (Ballance, 1984). These flows caused the deposition of thick clastic wedges in areas close to the mountain front. Distally only lutites were deposited.

DISCUSSION

In dry regions, variations in the stratigraphy and morphology of fans have been considered to be the result of tectonic and climatic controls (Heward, 1978; Steel & Aasheim, 1978; Talbot & Williams, 1979; Frostick & Reid, 1989; among others). However, autocyclic processes, such as entrenchment of the fan stream, are also quite important in the development of the fans (McCraw, 1968; Went et al., 1988; Harvey, 1989).

In the NE margin of the Madrid Basin the development of the fans was mainly controlled by extrinsic factors. It has been previously noted (Calvo et al., 1989) that climate was fairly similar throughout the area studied, during the early and middle Miocene. Thus differences between the alluvial depositional systems cannot be explained by a different climatic regime. Facies assemblages as well as palaeoecological data, furnished by mammal faunas, strongly suggest that the sedimentation took place under a warm, semiarid, seasonally contrasted climatic regime throughout the study area. However a slight change towards more humid conditions has been detected in the upper Aragonian and lower Vallesian (Sesé et al., 1990).

Tectonism is another important extrinsic factor that potentially controlled the fan stratigraphy. The influence of tectonics is demonstrated by the recognition of synsedimentary uplift in the basin margins, for instance the Huermeces fault which controlled the Baides fluvial distributary system. Moreover, tectonics controlled the topograhy and/or morphology of the basin margins (Harvey, 1989). This was the most important factor that affected the lateral and vertical evolution of the alluvial deposits as well as their size.

Two different morphological situations are recognized: (i) areas in which more or less linear depressions were developed in the surrounding reliefs, and (ii) areas in which the palaeorelief had a more regular morphology, lacking the linear depressions. In the first case large fluvial distributary systems were developed, whereas minor fans and slope-scree deposits were located on the regular slopes.

With regard to the major alluvial systems each of them may represent a different morphological situation in relation to the area and/or morphology of the depression where they are located.

The Alarilla system is considered to be a braided alluvial plain located in a low-relief area extending far from the main mountain front (Central System). The connection between the mountain front and the sedimentation zone was realized through channels which were oriented according to N–S structural lineations cut into the Palaeogene rocks. In this context, the broad alluvial plain was developed at the foot of the palaeorelief allowing the development of a broad channel belt.

A rather different situation is found in the Jadraque system where the channel belt is confined within a broader (3–4 km) palaeovalley cut into the Palaeogene rocks. In this case a large non-radial distributary system was formed in which gravel and sand ribbons were interbedded with overbank deposits.

The characteristics of the Baides fluvial distributary system, dominated by gravel sheets, confirm our idea that broad depressions enabled wide alluvial plains to be developed. Contrarily narrower depressions, as in the Jadraque system, are dominated by ribbon complexes. The isochronous development of both types of gravel bodies within the same area may be explained by the different morphology of the basin margin. This morphology was ultimately controlled by tectonism.

The facies assemblages and distribution of the slope-scree deposits and minor fans are, in general, rather similar along the basin margins studied. Gravity induced flows and occasionally ephemeral and shallow braided channels originated the two types of deposits (minor fans and slope-scree) along the relatively non-incised reliefs of the basin margins.

In short, the main factors that controlled the development of the fans were extrinsic factors, and particularly the morphology of the areas where the different systems were located. Most of the alluvial systems are classified as non-entrenched systems, as confirmed by the pedostratigraphy of the fans (Wright & Alonso Zarza, 1990). Palaeosol sequences are quite typical of non-entrenched fans in which the least mature palaeosols are located in the fan head due to higher rates of sedimentation (McCraw, 1968).

In the minor fans and slope-scree deposits studied the intrinsic factors were important in determining the facies assemblages, which were controlled by the dominant sediment transport and depositional processes. In the fans these processes fluctuate between debris flows and fluvial processes (Harvey, 1989).

SUMMARY AND CONCLUSIONS

A number of different types of alluvial deposits have been studied in the NE margin of the Madrid Basin. Four major fluvial distributary systems and several minor fans and slope-scree deposits (either directly associated with the major systems or located between them) have been recognized.

The shape and sedimentary sequences of the alluvial systems were mainly controlled by the morphology of the basin margin, which was ultimately controlled by large scale to local structural features of the margin.

The major fluvial distributary systems were related to valleys eroded in the mountain relief, whereas minor fans and slope-scree deposits developed along steep slopes of the mountain front. Thus, the encroaching of the fluvial distributary systems was not related to intrinsic factors but essentially to the morphology of the basin margin.

The transport and depositional processes leading to the formation of major fluvial distributary systems, minor fan complexes and slope-scree deposits show distinct differences. In the larger systems the sedimentation was mainly realized through stream flows that were organized in braided channel patterns. In contrast, mass flows were dominant in slope deposits and minor fans where intrinsic factors, mainly the size of the drainage basin, affected the ratio between channel/debris flow deposits.

The analysis of the alluvial systems along the NE margin of the Madrid Basin provides evidence of the geological complexity of this margin and permits a detailed view of the palaeogeography of the studied area during the Miocene.

ACKNOWLEDGEMENTS

The authors would like to thank Dr Ordoñez and Dr Hoyos for their suggestions and helpful criticism. Dr De Vicente made very helpful comments on the structural features of the basin margin. We wish also to express our gratitude to Dr J.P.P. Hirst and an anonymous referee for criticism and constructive reviews of the manuscript. This work was financed by the Project PR-84-0078-CO2: 'Evolución geológica de la Cuenca media y alta del Tajo', CSIC-CAYCIT.

REFERENCES

ALONSO ZARZA, A.M., CALVO, J.P., GARCÍA DEL CURA, M.A. & HOYOS, M. (1990) Los sistemas aluviales miocenos del borde noreste de la Cuenca de Madrid: sector Cifuentes–Las Inviernas (Guadalajara). *Rev. Soc. Geol. España* **3**, 213–229.

ALONSO ZARZA, A.M., WRIGHT, V.P., CALVO, J.P. & GARCIA DEL CURA, M.A. (1992) Soil–landscape and climatic relationships in the Middle Miocene of the Madrid Basin. *Sedimentology* **39**, 17–35.

ALVARO, M., CAPOTE, R. & VEGAS, R. (1979) Un modelo de evolución geotectónica para la cadena Celtibérica. Libro Hom. Prof. Solé Sabarís. *Acta. Geol. Hisp.* **14**, 174–177.

BALLANCE, P.F. (1984) Sheet-flow-dominated gravel fans of the non-marine middle Cenozoic Simmler Formation, Central California. *Sedim. Geol.* **38**, 337–359.

BATES, R.L. & JACKSON, J.A. (1980) *Glossary of Geology.* Am. Geol. Inst., Falls Church, Virginia, 748 pp.

BLUCK, B.J. (1976) Sedimentation in some Scottish rivers of low sinuosity. *Trans. R. Soc. Edinburgh: Earth Sci.* **69**, 425–456.

BOOTHROYD, J.C. & NUMMEDAL, D. (1978) Proglacial braided outwash: a model for humid alluvial fan deposits. In: *Fluvial Sedimentology* (Ed. Miall, A.D.) pp. 641–668. Can. Soc. Petrol. Geol., Calgary, Memoir 5.

BRIDGE, J.S. (1985) Paleochannel patterns inferred from alluvial deposits: a critical evaluation. *J. Sedim. Petrol.* **55**, 579–589.

CALVO, J.P., ALONSO ZARZA, A.M. & GARCIA DEL CURA, M.A. (1989) Models of marginal lacustrine sedimentation in response to varied depositional regimes and source areas in the Madrid Basin (Central Spain). *Palaeogeog. Palaeoclimatol. Palaeoecol.* **70**, 199–214.

CORRALES, I., CARBALLEIRA, J., FLOR, G., POL, C. & CORROCHANO, A. (1986) Alluvial systems in the northwestern part of the Duero Basin (Spain). *Sedim. Geol.* **47**, 149–166.

DAAMS, R. & FREUDENTHAL, M. (1981) Aragonian: the stage concept versus Neogene Mammals zones. *Scripta Geol.* **62**, 1–17.

FRIEND, P.F. (1983) Towards the field classification of alluvial architecture or sequence. In: *Modern and Ancient Fluvial Systems* (Eds Collinson, J.D. & Lewin, J.), pp. 345–354. Spec. Publs Int. Ass. Sediment. 6. Blackwell Scientific Publications, Oxford.

FRIEND, P.F. (1989) Space and time analysis of river systems, illustrated by Miocene systems of the Northern Ebro Basin in Aragon (Spain). *Rev. Soc. Geol. España* **2**, 55–64.

FRIEND, P.F., SLATER, M.J. & WILLIAMS, R.C. (1979) Vertical and lateral building of sandstone bodies, Ebro basin, Spain. *J. Geol. Soc. Lond.* **136**, 39–46.

FROSTICK, L.E. & REID, I. (1989) Climatic versus tectonic controls of fan sequences; lessons from the Dead Sea, Israel. *J. Geol. Soc. Lond.* **146**, 527–538.

HARVEY, A.M. (1984) Debris flows and alluvial deposits in

Spanish quaternary alluvial fans: implications for fan morphology (Eds Koster, E.M. & Steel, R.J.) pp. 23–132. Can. Soc. Petrol. Geol., Calgary, Memoir 10.

HARVEY, A.M. (1989) The occurrence and role of arid zone alluvial fans. In: *Arid Zone Geomorphology* (Ed. Thomas, D.G.H.) pp. 136–158. Belhaven Press, London, 372 pp.

HEIN, F.J. & WALKER, R.G. (1977) Bar evolution and development of stratification in the gravelly braided Kicking Horse River, British Columbia. *Can. J. Earth Sci.* **14**, 562–570.

HEWARD, A.P. (1978) Alluvial fan sequence and megasequence models: with example from Westfalian D–Stephanian B coalfields, Northern Spain. In: *Fluvial Sedimentology* (Ed. Miall, A.D.) pp. 669–702. Can. Soc. Petrol. Geol., Calgary, Memoir 5.

HOOKE, R.L. (1967) Processes on arid region alluvial fans. *J. Geol.* **75**, 438–460.

JUNCO, F. & CALVO, J.P. (1983) Cuenca de Madrid. In: *Geología de España*, Vol. II, pp. 534–543. IGME, Madrid.

MCARTHUR, J.L. (1987) The characteristics, classification and origin of the late Pleistocene fan deposits in the Cass Basin, Canterbury, New Zealand. *Sedimentology* **34**, 459–471.

MCCRAW, J.D. (1968) The soil pattern of some New Zealand alluvial fans. *9th Int. Congress Soil Sci. Trans.* **4**, 631–640.

MIALL, A.D. (1977) A review of the braided-river depositional environment. *Earth Sci. Rev.* **13**, 1–62.

MIALL, A.D. (1978) Lithofacies types and vertical profile models in braided river deposits: a summary. In: *Fluvial Sedimentology* (Ed. Miall, A.D.) pp. 597–604. Can. Soc. Petrol. Geol., Calgary, Memoir 5.

NICHOLS, G.J. (1987) Structural controls on fluvial distributary systems — the Luna system, northern Spain. In: *Recent Developments in Fluvial Sedimentology* (Eds Ethridge, F.G., Flores, R.M., & Harvey, M.D.) pp. 269–277. Soc. Econ. Paleont. Miner., Tulsa, Spec. Publ. 39.

PORTERO, J.M. & AZNAR, J.M. (1984) Evolución morfotectónica y sedimentación terciarias en el Sistema Central y cuencas limítrofes (Duero y Tajo). In: *I Congreso Español de Geología, Segovia*, Vol. III, pp. 253–263.

RACERO, A. (1988) Consideraciones acerca de la evolución geológica del margen NW de la Cuenca del Tajo durante el Terciario a partir de los datos de subsuelo. In: *II Congreso Geológico de España, Granada, Simposios*, pp. 213–222.

RAMOS, A. & SOPEÑA, A. (1983) Gravel bars in low-sinuosity streams (Permian and Triassic, Central Spain). In: *Modern and Ancient Fluvial Systems* (Eds Collison, J.D. & Lewin, J.) pp. 301–312. Spec. Publs Int. Ass. Sediment. 6. Blackwell Scientific Publications, Oxford.

RAMOS, A., SOPEÑA, A. & PEREZ ARLUCEA, M. (1986) Evolution of Buntsandstein fluvial sedimentation in the northwest Iberian Ranges (Central Spain). *J. Sedim. Petrol.* **56**, 862–875.

ROBERTS, H.H. (1987) Modern carbonate–siliciclastic transitions: humid and arid tropical examples. *Sedim. Geol.* **50**, 25–65.

RUST, B.R. (1984) Proximal braidplain deposits in the Middle Devonian Molbaie Formation of eastern Quebec, Canada. *Sedimentology* **31**, 675–695.

SCHUMM, S.A. (1977) *The Fluvial System.* John Wiley & Sons, New York, 388 pp.

SCHUMM, S.A. (1981) Evolution and response of the fluvial system, sedimentological implications. In: *Recent and Ancient Nonmarine Depositional Environments: Models for Exploration* (Eds Ethridge, H.G. & Flores, R.M.) pp. 19–29. Soc. Econ. Palaeont. Miner., Tulsa, Spec. Publ. 31.

SESE, C., ALONSO ZARZA, A.M. & CALVO, J.P. (1990) Nuevas faunas de micromamíferos del Terciario Continental del NE del la Cuenca de Madrid (prov. Guadalajara). *Estudios Geol.*, 433–451.

SMITH, N.D. (1974) Sedimentology and bar formation in the Upper Kicking Horse River, a braided outwash stream. *J. Geol.* **82**, 205–223.

STEEL, R.J. & ASSHEIM, S.M. (1978) Alluvial sand deposition in a rapidly subsiding basin (Devonian, Norway). In: *Fluvial Sedimentology* (Ed. Miall, A.D.) pp. 385–412. Can. Soc. Petrol. Geol., Calgary, Memoir 5.

TALBOT, M.R. & WILLIAMS, M.A.J. (1979) Cyclic alluvial fan sedimentation on the flanks of fixed dunes, Jangari, Central Niger. *Catena* **6**, 43–62.

VAN DER MEULEN, S. (1986) Sedimentary stratigraphy of Eocene sheetflood deposits, southern Pyrenees, Spain. *Geol. Mag.* **123**, 167–183.

VESSEL, R.K. & DAVIES, D.R. (1981) Non-marine sedimentation in an active fore-arc basin. In: *Recent and Ancient Nonmarine Depositional Environments: Models for Exploration* (Eds Ethridge, F.G. & Flores, R.M.) pp. 31–45. Soc. Econ. Paleont. Miner., Tulsa, Spec. Publ. 31.

VOS, R.G. (1975) An alluvial plain and lacustrine model for the Precambrian Witwatersrand deposits of South Africa. *J. Sedim. Petrol.* **45**, 480–493.

WELLS, S.G. & HARVEY, A.M. (1987) Sedimentologic and geomorphic variations in storm-generated alluvial fans, Howgill Fells, Northwest England. *Bull. Geol. Soc. Am.* **98**, 182–198.

WENT, D.J., ANDREWS, M.J. & WILLIAMS, B.P.J. (1988) Processes of alluvial fan sedimentation, Basal Rozel Conglomerate Formation, La Tête des Houges, Jersey, Channel Islands. *Geol. J.* **23**, 75–84.

WRIGHT, V.P. & ALONSO ZARZA, A.M. (1990) Pedostratigraphic models for alluvial fan deposits: a tool for interpreting ancient sequences. *J. Geol. Soc. Lond.* **147**, 8–10.

Alluvial-fan sedimentation along an active strike-slip fault: Plio-Pleistocene Pre-Kaczawa fan, SW Poland

K. MASTALERZ* and J. WOJEWODA†

*Institute of Geological Sciences,
University of Wrocław, Cybulskiego 30, 50-205 Wrocław, Poland; and
†Department of Geology, Adam Mickiewicz University, Strzałkowskiego 5/7, 60-854 Poznań, Poland

ABSTRACT

Plio-Pleistocene syntectonic sedimentation along the Sudetic Marginal Fault (SMF), SW Poland, is reflected in the architecture and internal structures of the Pre-Kaczawa alluvial fan. The early phase of sedimentation, related to the dip-slip activity of the fault, resulted in progradation of the alluvial fan onto the downthrown Fore-Sudetic Block. The succeeding strike-slip movements brought about a shingling arrangement of successive fan lobes and the development of deformational structures in the alluvial fan and interfingering proglacial deposits. The cross-cutting relationships define the following sequence of deformations: (i) soft sediment convection deformations: (ii) planar normal faults and shear zones; and (iii) listric faults. The occurrence and orientation of planar faults and shear zones are related to the left-lateral strike-slip shear couple along the SMF. The displacement of the fan apex with respect to the supplying river outlet, the shingling arrangement of successive fan lobes and a characteristic suite of deformations seem to define useful criteria in distinguishing alluvial fans developed in strike-slip regimes.

INTRODUCTION

Syntectonic sedimentation has attracted the attention of sedimentologists who know many examples of alluvial fans developed along tectonically active morphological escarpments in various tectonic settings (Blissenbach, 1954; Bluck, 1964; Bull, 1968; reviews in: Miall, 1984; Collinson, 1986; Mitchell & Reading, 1986; Nemec & Steel, 1988). Sediments of such fans often have an abundance of deformational structures (e.g. Sieh, 1978a, b; Sieh & Jahns, 1984). However, only architectural features of alluvial fans (shape, extent, arrangement, facies organization) are commonly considered as indicative of a tectonic regime (e.g. Hooke, 1972; Steel et al., 1977; Crowell, 1982; Biddle & Christie-Blick, 1985; Alexander & Leeder, 1987; Colella, 1988; see also Nemec & Steel, 1988). To our knowledge no comprehensive report exists, as yet, describing associations of structural features which could be regarded as diagnostic of tectonic regimes in alluvial-fan settings.

The present authors have been intrigued by the position and form of the Pre-Kaczawa alluvial fan attached to the Sudetic Marginal Fault (SMF) near Złotoryja, where the Kaczawa River enters the Fore-Sudetic Block (Fig. 1). The alluvial-to-lacustrine sediments exposed close to the fault display a variety of deformational structures pointing to the synsedimentary tectonic activity of the depositional area. The fan shape is compound, showing three arcuate segments of its contour line (Fig. 1B). The complexity of the fan structure is also suggested by a dispersal pattern of palaeocurrent indicators. The fan apex position, northeastwards of the river intersection with the SMF (Fig. 1B), seems to suggest a strike-slip component of the displacements along the fault. To date, however, the SMF has been considered as a high-angle normal fault (or fault zone) with a throw of up to 300 m (Oberc & Dyjor, 1969; Oberc, 1972).

† Present address: Institute of Geology, Adam Mickiewicz University, Maków Polnych 16, 61-606 Poznań, Poland.

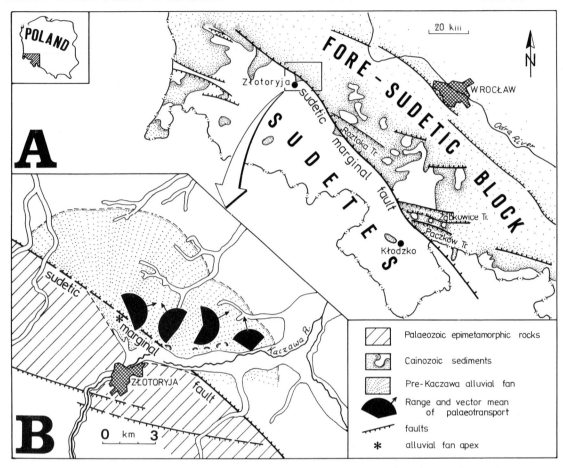

Fig. 1. Schematic map of the Sudetes and Fore-Sudetic Block showing (A) distribution of Cenozoic deposits and (B) sketch map of the Pre-Kaczawa alluvial fan.

The purpose of this paper is a description of deformational structures associated with the Pre-Kaczawa fan evolution, and to discuss their relation to the synsedimentary strike-slip activity of the SMF. The study of the type, orientation and succession of deformations allows the determination of the palaeotectonic regime, sense of fault movements and time relations between tectonic activity and sedimentation. Another important aim of this paper is to suggest some criteria, including structural criteria, useful in distinguishing alluvial fans developed in strike-slip regimes.

GEOLOGICAL SETTING

The SMF, possibly still active (Gierwielaniec & Woźniak, 1983), is one of the most conspicuous features in the topography of SW Poland. It extends as far as 150 km from SE to NW and separates the uplifted Sudetes from the downthrown Fore-Sudetic Block (Fig. 1A). The fault was suggested to have originated in the Palaeogene (Cloos, 1922; Oberc, 1955; Walczak, 1966). However, the oldest sediments of the Fore-Sudetic Block, which display a close relation to the SMF in respect of facies distribution and thickness, are of Pliocene age. The first stage of fault development was attributed to the formation of large flexural bending and gradual uplifting of the Sudetes, which restricted the area of sedimentation of the Poznań Formation (Upper Miocene–Lower Pliocene) (Oberc & Dyjor, 1969; Oberc, 1972). Oberc & Dyjor (1969) concluded that the fault had not formed until Late Pliocene, when the final uplift of the Sudetes took place, resulting in alluvial-fan sedimentation along the SMF. This type of sedimentation lasted until the Pleistocene glaciation (Dyjor, 1966).

The SMF crosscuts several structural units of different age and history (Oberc, 1966, 1972). Late Proterozoic(?)–Early Palaeozoic to Cretaceous complexes are exposed in the Sudetes, only locally being covered by Tertiary sedimentary and volcanogenic rocks. In contrast, most of the Fore-Sudetic Block is covered with up to 200 m thick Cenozoic strata (Oberc & Dyjor, 1969). The thickest sedimentary sequences infill E–W trending synsedimentary grabens, oblique to the SMF (Fig. 1A). The Oligocene to Upper Miocene succession of the Fore-Sudetic Block comprises sandy to argillaceous deposits with coal seams, acccumulated in alluvial-to-lacustrine settings, including temporary peat bogs (Dyjor, 1966; Oberc & Dyjor, 1969). The overlying Poznań Formation comprises predominantly argillaceous deposits settled in the ancient Poznań Lake in mid-west Poland. The Upper Pliocene Gozdnica Formation rests unconformably on the older deposits and displays facies distribution obviously influenced by the activity of the SMF (Dyjor, 1966; Oberc & Dyjor, 1969). The thickest and the coarsest-grained deposits occur close to the fault zone; they become thinner and finer-grained, and interfinger with lacustrine facies of the recessing Poznań Lake further northwards.

The Quaternary deposits are from 13 to 18 m thick on the Fore-Sudetic Block, only locally attaining 50 m (Ruhle & Mojski, 1965). They consist mainly of fluvioglacial gravels and sands, and of the Saalian and Elsterian tills. The uppermost part of the Pleistocene cover comprises loess and loess-like loams. Holocene fluvial deposits are restricted to valley floors.

SEDIMENTARY FACIES

Several gravel pits located close to the SMF near Złotoryja (Fig. 1B) expose sediments of the Pre-Kaczawa alluvial fan. The sections attain 5–15 m in thickness and exhibit the upper portions of the fan sequence covered with glacigenic deposits (Fig. 2).

Alluvial-fan facies (AF)

Alluvial-fan facies comprise yellowish-coloured, predominantly coarse-grained sediments. The grain framework is entirely composed of local Sudetic rocks. The directional structures suggest northeastward palaeotransport (Fig. 2). The sediment accumulated within shallow channels shifting across the stream-dominated fan surface, and as widespread sheets outside channels. The general coarsening-upward trend of the sequence suggests a progradational style of the fan development.

Sand bar/sheetflood deposits consist of cross- and horizontally stratified sands and pebbly sands (St, Sh and Gt: facies symbols as defined in Miall's (1978) classification). Medium- to fine-grained, ripple cross- and parallel-laminated sands (Sr and Sl) are subordinate.

Gravel bar/sheet deposits comprise fine- to medium-grained (MPS 2–6 cm) gravels organized into 5–15 cm thick sets of trough cross-bedding (Gt). Less frequent are massive sets with pebble imbrication (Gm). These deposits form thick fining-upward sequences with sandy interlayers at the top (Fig. 3A).

Gravel-bed channel deposits form medium- to coarse-grained (MPS 6–12 cm) gravel beds showing erosive bases (Fig. 3B). They are underlain by channel lags with pebble imbrication and scarce intraclasts (Fig. 3C). The sediment is organized into thick (15–60 cm) sets, massive or with fining-upward grain size trend (Gm) and, less frequently, showing trough cross-bedding (Gt). Some sets display crude horizontal stratification.

Proglacial lacustrine–deltaic facies (L–D)

Interlayered lacustrine and deltaic, muddy to gravelly deposits contain an admixture of northerly derived (Scandinavian) pebbles. The directional structures indicate NE palaeotransport (Fig. 2). The lacustrine interlayers formed due to the submergence of the fan in proglacial (ice-marginal?) lake waters.

Lacustrine deposits consist of silts and muds which form 5–15 cm thick homogeneous or normally graded beds (Fm). Some are bipartite with thicker silty lower parts and thinner clayey–muddy tops. Thinly laminated beds (Fl) are also common. Deposits of this facies form 50–100 cm thick, composite layers of relatively large lateral extent or restricted to abandoned channels. These deposits are abundant in soft sediment deformation structures (Figs 3D & 4).

Deltaic lobe deposits include cross- and horizontally stratified sands and gravels (St, Sh and Gt)

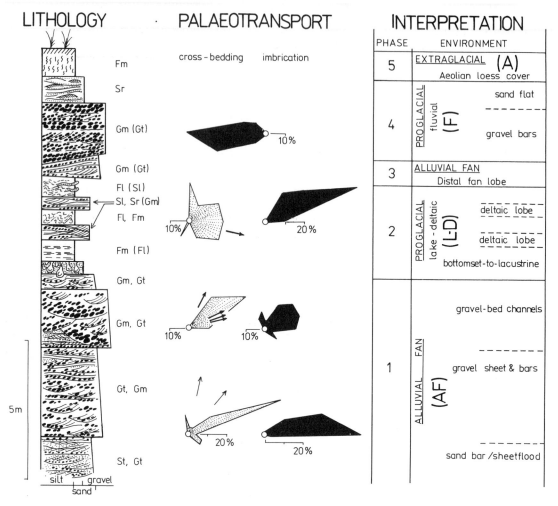

Fig. 2. Section showing generalized features of the Pre-Kaczawa alluvial fan and the Pleistocene, glacier-related deposits. Thin arrows, trough cross-set axes; thick arrows, channel axes.

interlayered with ripple cross-laminated, medium to fine sands (Sr). The beds comprise numerous thin (2–5 cm) sets of irregular shape in cross-section. Some gravel beds contain imbricated pebbles.

Proglacial fluvial facies (F)

Proglacial fluvial facies include brownish-coloured gravels and sands. These deposits contain considerable amounts of northerly derived pebbles (Scandinavian granitoids and metamorphic rocks) and show western palaeotransport directions (Fig. 2). Deposits of this facies accumulated as in-channel bars and as sand sheets on a gently inclined sand flat of a fluvioglacial braided outwash.

Gravel bar deposits are organized into 5–20 cm thick sets of medium-grained (MPS 4–8 cm), massive gravels (Gm) or those showing distinct cross-bedding (Gt). Sorting is moderate to poor, but some sets show good sorting and sometimes an open grain framework.

Sand flat deposits include medium-grained and, less commonly, fine- or coarse-grained sands which display ripple cross-lamination. Horizontal lamination occurs less frequently. Individual sets are 1–6 cm thick.

Sedimentation along an active strike-slip fault

Fig. 3. Pre-Kaczawa alluvial-fan facies. (A) Cross-bedded fine-grained conglomerates with thin horizontally stratified sandy interlayers; gravel bar/sheet deposits. (B) Gravel-bed channel deposits showing a distinct concave-up base. (C) Channel lag gravels with pebble imbrication and large imbricated intraclasts; gravel-bed channel deposits. (D) The lowermost packet of the lacustrine–deltaic deposits; note the soft-sediment deformations.

Extraglacial aeolian facies (A)

Loess-like loam consists of yellowish, non-stratified to crudely laminated, silty deposit of high porosity. It is in sharp contact with the underlying fluvioglacial deposits and reaches 1–2 m in thickness (Fig. 2). It accumulated from airborne silt winnowed from the periglacial zone. Locally, loess cover was slightly displaced by downslope creep.

DEFORMATIONAL STRUCTURES

Convection structures

The lacustrine–deltaic part of the Pre-Kaczawa fan section shows three laterally continuous horizons of soft-sediment deformation structures (Figs 2, 3d & 4). Deformations occur in 0.3–1.0 m thick sheets of fine-grained material, separated from each other by

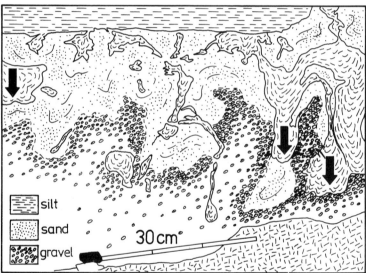

Fig. 4. Soft-sediment deformation structures — convection deformations in alluvial-fan-to-lacustrine transition. Note the irregular and drop-shaped forms sunk in coarse-grained substratum and gravelly pillars invading upwards.

coarser-grained, undeformed sediment. The sheets have sharp and planar top boundaries, while their bottom contacts vary from sharp and planar to gradational and wavy. The geometry of deformations varies within and between the horizons. The lowermost horizon is characterized by diversified three-dimensional structures; the irregular, pseudonodular, columnar and mushroom or drop shaped (Fig. 4) being the most common. The two higher horizons display much more regular, often symmetrical convolutions ranging from 0.1 to 0.5 m in width.

The authors suggest that the deformational structures resulted from sediment liquefaction triggered by seismic shocks. The deformations have apparently developed due to convection in fine-grained lacustrine sediments displaying high liquefaction potential. Remnants of original fining-upward sequences in the horizons containing contortions seem to exclude simple sediment loading as the physical process able to bring about the deformation. The geometry of the structures and the flat position of the interlayered undeformed sediments

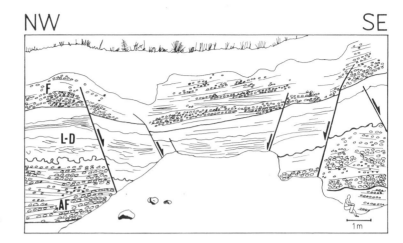

Fig. 5. Planar normal faults forming an asymmetrical graben. Gravel pit near Złotoryja (redrawn from photograph).

also preclude a significant influence of slope processes such as sediment creep and/or slumping. Similar deformations reported from lacustrine deposits in active tectonic areas have been interpreted as seismically induced (Selley *et al.*, 1963; Sims, 1973, 1975; Begin, 1975; Weaver, 1976; Muir, 1984; Hempton & Dewey, 1983; Hempton *et al.*, 1983; Seilacher, 1984).

Planar normal faults and extensional grabens

The alluvial-fan gravels and overlying lacustrine–deltaic sediments are cut by normal faults which form systems of asymmetrical grabens and half-grabens (Fig. 5). Fault surfaces are planar dipping toward south and north (Fig. 6) and they die out upward within the fluvioglacial sediments. The dips vary between 50° and 85°. The faults usually do not deform the sediment they cut. Only locally are fault-drag zones (up to 30 cm wide) or thin zones (up to 10 cm) of homogenized sediment in direct contact with the faults.

The grabens are up to several metres wide and up to 80 cm deep. They trend from east to west, obliquely to the SMF. Their spatial relation to the fault is typical of extensional fractures (e.g. Tchalenko, 1970; Jaroszewski, 1972; Ramsay & Huber, 1983) and in agreement with the left-lateral sense of displacement on the major fault (cf. Naylor *et al.*, 1986; Ingersoll, 1988; Sylvester, 1988; Valle *et al.*, 1988; Groshong, 1989). Similar extensional features have been reported from areas of recent

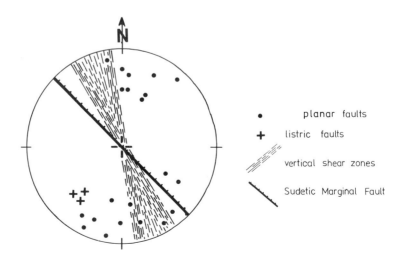

Fig. 6. Orientation of planar deformational structures of the Plio-Pleistocene sequence near Złotoryja; normal to planes of planar and listric faults in projection on the upper hemisphere, Schmidt net.

continental sedimentation close to active strike-slip zones (Clark, 1972; Clark et al., 1972; Sieh, 1978a,b, 1984; Sieh & Jahns, 1984).

Vertical shear zones

Within the alluvial-fan and lacustrine–deltaic facies there occur planar, steeply dipping (exceeding 80°) zones with a characteristic tripartite structure (Fig. 7). They reach 30–50 cm in width. The axial parts of shear zones contain partly homogenized sediment. On both sides the grain framework shows distinct effects of sediment collapsing into the centre of a zone (Fig 7). The degree of deformation gradually decreases outwards. Sediments on both sides of these zones do not show evidence of vertical displacements.

The orientation of shear zones, slightly oblique to the SMF (Fig. 6), corresponds to low-angle P-shears (cf. Tchalenko, 1970; Valle et al., 1988). Similar vertical planar structures observed in close proximity to synsedimentary strike-slip faults have been reported as shear features (Wojewoda, 1987).

Listric normal faults

The surfaces of the listric normal faults are inclined to south-southwest (Fig. 6). The southern sides of these faults are downthrown 1–3 m (Fig. 8). The downthrown blocks do not show considerable deformation except in near-fault surfaces where there are weakly fault-dragged beds and partly homogenized sediments.

The orientation of the listric faults suggests that the Kaczawa River Valley played an important role in their development. They must have formed as rotational slides related to the unstable northern slope of the valley incised into the alluvial-fan-to-proglacial deposits. Similar slope processes are often triggered due to seismic events (e.g. Foster & Karlstrom, 1967; Seed, 1968).

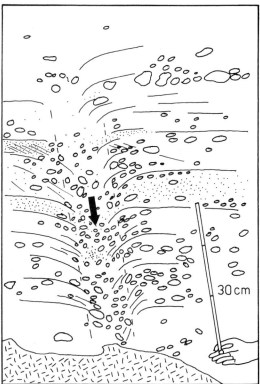

Fig. 7. Vertical shear zone in alluvial-fan conglomerates. Note the well-defined deformation of grain framework that resulted from sediment collapse into the shear zone.

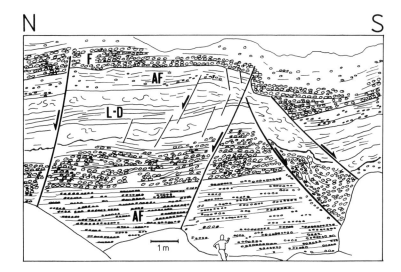

Fig. 8. Extensional listric and planar normal faults cutting the sequence of alluvial-fan-to-fluvioglacial deposits; gravel pit near Złotoryja. Note the rotational character of downthrown blocks on the listric faults (redrawn from photograph).

Succession of deformations

The cross-cutting relationships permit a reconstruction of the temporal succession of the deformations (Fig. 9). The horizons of convection structures (C) are the oldest. These horizons are separated by undeformed sediments, and hence the convection deformations must have been syndepositional and, most likely, seismically induced. Extensional faults (E) and vertical shear zones (S) cut the horizons of contortions. The authors interpret both as being associated with regional shear couple along the SMF and synchronous with the fluvioglacial sediments into which they die out. The relations between shear zones and extensional faults are not clear enough to solve the problem of their temporal order. The listric faults (L) associated with rotational slides on the northern slope of the Kaczawa valley are the youngest. They are related to the Late Pleistocene phase of the fan incision, as evidenced by the undisturbed cover of loess-like loams above these faults.

SUMMARY AND CONCLUSIONS

The above-described sediments and deformational structures document the Plio-Pleistocene depositional history of the area located close to the morphological, fault-related margin of the Sudetes, near Złotoryja. They also reflect synsedimentary tectonic activity of the basement and suggest left-lateral strike-slip displacements along the SMF. Syntectonic sedimentation along this fault started in the Pliocene, as reflected by the fault-related facies distribution of the Gozdnica Formation. The Pre-Kaczawa alluvial-fan deposits represent the continuation of the near-marginal facies of the Gozdnica Formation.

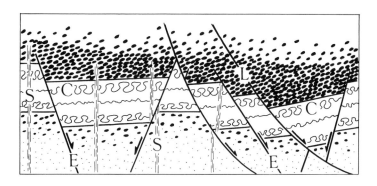

Fig. 9. Generalized cross-cutting relationships between deformational structures in the Plio-Pleistocene sequence near Złotoryja. C, Contortions; E, extensional fractures; S, shear zones; L, listric faults.

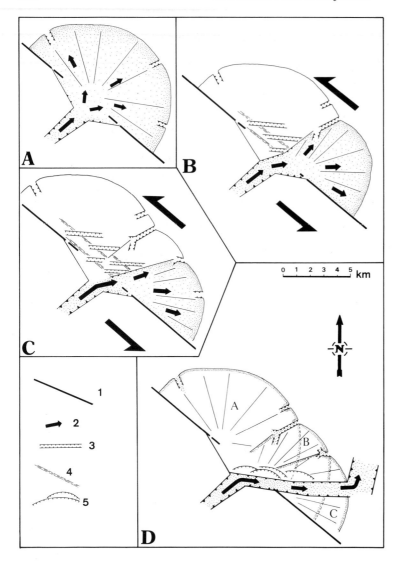

Fig. 10. Evolution of the Pre-Kaczawa alluvial fan (for explanation see text). 1, Sudetic Marginal Fault; 2, sediment dispersal; 3, extensional grabens; 4, shear zones; 5, listric faults.

During the incipient growth of the Pre-Kaczawa alluvial fan (Fig. 2, phase 1), dip-slip movement along the SMF was dominant, as suggested by the significant thickness (10–20 m) of the fan deposits, the progradational style of the depositional system and the symmetrical radial shape of the oldest fan lobe (Fig. 10A). Afterwards, the fan was offset to the west and submerged under a proglacial (terminoglacial?) lake (Fig. 2, phase 2). After recession of the lake, alluvial-fan deposition continued (phase 3). However, the successive fan lobes were also displaced northwestwards (Fig. 10B, C). The resulted shingled arrangement (cf. Crowell, 1982) of the fan lobes suggests a sinistral strike-slip displacement along the SMF.

Deposition on the Pre-Kaczawa alluvial fan was significantly restricted during the following glaciation. The progradational fan system was replaced by an aggradational fluvioglacial outwash (Fig. 2, phase 4). At that time the SMF was still active. The orientation of extensional faults and shear zones suggests that the fault kept the same left-lateral mode of displacement.

Although it was assumed that two glaciations took place in the study area during the Pleistocene (Ruhle & Mojski, 1965; Ruhle, 1973), the present authors tend to exclude the possibility of a glacigenic origin for the reported deformations. During the Pleistocene the area was located just between the uplifted Sudetes and the postulated front of a continental glacier. At that time the Pre-Kaczawa fan system constituted an elevated form particularly sensitive to glacitectonic compression. The lack of compressional features in the discussed sediments seems to contradict an overriding by a glacier as a potential deformational mechanism.

The accumulation area of the Pre-Kaczawa alluvial fan was tectonically active during the Plio-Pleistocene with the SMF as the main active structure. It acted not only as a dip-slip fault but also as an important strike-slip zone. The features described from the Pre-Kaczawa alluvial fan suggest that the following criteria may be useful for distinction of alluvial fans developed along active strike-slip margins:

1 an offset of the alluvial-fan apex in relation to the position of the outlet of the supplying river;
2 the shingling arrangement (cf. Crowell, 1982) of the successive fan lobes, the younger lobes growing backward with respect to the sense of tectonic movement in the basement;
3 the occurrence and orientation of structures typical of shear couple along strike-slip faults (e.g. planar normal faults and extensional grabens, shear zones).

ACKNOWLEDGEMENTS

The authors are grateful to Paweł Aleksandrowski for discussion in the field and critical suggestions.

REFERENCES

ALEXANDER, J. & LEEDER, M.R. (1987) Active tectonic control on alluvial architecture. In: *Recent Developments in Fluvial Sedimentology* (Eds Ethridge, F.G., Flores, R.M. & Harvey, M.D.) pp. 245–252. Soc. Econ. Paleont. Miner., Tulsa, Spec. Publ. 39.

BEGIN, Z.B. (1975) Palaeocurrents in the Plio-Pleistocene Samra Formation (Jerycho region, Israel) and their tectonic implications. *Sedim. Geol.* **14**, 191–218.

BIDDLE, K.T. & CHRISTIE-BLICK, N. (1985) Glossary — strike-slip deformation, basin formation, and sedimentation. In: *Strike-slip Deformation, Basin Formation, and Sedimentation* (Eds Biddle, K.T. & Christie-Blick, N.) pp. 375–386. Soc. Econ. Paleont. Miner., Tulsa, Spec. Publ. 37.

BLISSENBACH, E. (1954) Geology of alluvial fans in semiarid regions. *Bull. Geol. Soc. Am.* **65**, 175–189.

BLUCK, B.J. (1964) Sedimentation of an alluvial fan in southern Nevada. *J. Sedim. Petrol.* **34**, 395–400.

BULL, W.B. (1968) Alluvial fans. *J. Geol. Educ.* **16**, 101–106.

CLARK, M.M. (1972) Collapse fissures along the Coyote Creek fault. US Geol. Surv. Prof. Paper, 787, pp. 190–207.

CLARK, M.M., GRANTZ, A. & RUBIN, M. (1972) Holocene activity of the Coyote Creek fault as recorded in sediments of Lake Cahuilla. US Geol. Surv. Prof. Paper 787, pp. 112–130.

CLOOS, H. (1922) *Der Gebirgsbau Schlesiens und die Stellung seiner Bodenschatze.* Gebr. Borntr., Berlin, 107 pp.

COLELLA, A. (1988) Gibert-type fan deltas in the Crati Basin (Pliocene–Holocene, southern Italy). In: *Excursion Guidebook International Workshop on Fan Deltas 1988* (Ed. Colella, A.) pp. 19–77. Grafica Cosentina, Cosenza, 152 pp.

COLLINSON, J.D. (1986) Alluvial sediments. In: *Sedimentary Environments and Facies* (Ed. Reading, H.G.) pp. 20–62. Blackwell Scientific Publications, Oxford, 615 pp.

CROWELL, J.C. (1982) The Violin Breccia, Ridge Basin, southern California. In: *Geologic History of Ridge Basin, Southern California* (Eds Crowell, J.C. & Link, M.H.) pp. 89–98. Publ. Pac. Sec. Soc. Econ. Paleont. Miner., Los Angeles.

DYJOR, S. (1966) Młodotrzeciorzędowa sieć rzeczna zachodniej części Dolnego Śląska (Late Tertiary drainage system of the western part of the Lower Silesia). In: *Z Geologii Ziem Zachodnich* (Ed. Ober, J.) pp. 275–318. PWN, Worcław.

FOSTER, H.L. & KARLSTROM, T.N.V. (1967) Ground breakage and associated effects in the Cook Inlet area, Alaska, resulting from the March 27, 1964, earthquake. US Geol. Surv. Prof. Paper 543-F, 28 pp.

GIERWIELANIEC, J. & WOŹNIAK, J. (1983) Ocena współczesnej aktywności uskoku sudeckiego brzeżnego w świetle archiwalnych materiałów niwelacyjnych (Estimation of the Recent activity on the Sudetic Marginal Fault). In: *Proc. III Symp. Modern Neotectonic Movements of the Earth Crust in Poland, Wrocław 1981*, pp. 109–123, Ossolineum, Wrocław.

GROSHONG, R.H. JR (1989) Half-graben structures: balanced models of extensional fault-bend folds. *Bull. Geol. Soc. Am.* **101**, 96–105.

HEMPTON, M.R. & DEWEY, J.F. (1983) Earthquake-induced deformational structures in young lacustrine sediments, East Anatolian Fault, southeast Turkey. *Tectonophysics* **98**, 7–14.

HEMPTON, M.R., DUNNE, L.A. & DEWEY, J.F. (1983) Sedimentation in an active strike-slip basin, southeastern Turkey. *J. Geol.* **91**, 401–412.

HOOKE, R.L. (1972) Geomorphic evidence for Late-Wisconsian and Holocene tectonic deformation, Death Valley, California. *Bull. Geol. Soc. Am.* **83**, 2073–2098.

INGERSOLL, R.V. (1988) Tectonics of sedimentary basins. *Bull. Geol. Soc. Am.* **100**, 1704–1719.

JAROSZEWSKI, W. (1972) Drobnostrukturalne kryteria tektoniki obszarów nieorogenicznych na przykładzie północno-wschodniego obrzeżenia mezozoicznego Gór Świętokrzyskich (Mesoscopic structural criteria of tectonics of non-orogenic areas: an example from the north-eastern Mesozoic margin of the Świętokrzyskie Mountain). *Studia Geol. Polon.* **38**, 3–215.

MIALL, A.D. (1978) Lithofacies types and vertical profile models in braided river deposits: a summary. In: *Fluvial Sedimentology* (Ed. Miall, A.D.) pp. 597–604. Can. Soc. Petrol. Geol., Calgary, Memoir 5.

MIALL, A.D. (1984) *Principles of Sedimentary Basin Analysis.* Springer-Verlag, Berlin, 490 pp.

MITCHELL, A.H.G. & READING, H.G. (1986) Sedimentation and tectonics. In: *Sedimentary Environments and Facies* (Ed. Reading, H.G.) pp. 471–519. Blackwell Scientific Publications, Oxford, 615 pp.

MUIR, S.G. (1984) *Holocene deformed sediments of the southern San Joaquin Valley, Kern County, California.* Unpublished PhD thesis, Univ. S. California.

NAYLOR, M.A., MANDL, G. & SIJPESTEIJN, C.H.K. (1986) Fault geometries in basement-induced wrench faulting under different initial stress states. *J. Struct. Geol.* **8**, 737–752.

NEMEC, W. & STEEL R.J. (Eds) (1988) *Fan Deltas: Sedimentology and Tectonic Settings.* Blackie, Glasgow, 444 pp.

OBERC, J. (1955) Wpływ budowy geologicznej na morfologię w regionie bardzkim (De l'influence de la structure geologique sur la morphologie de la region de Bardo). *Czas. Geogr.* **26**, 339–362.

OBERC, J. (1966) Ewolucja Sudetów w świetle teorii geosynklin (Evolution of the Sudetes in the light of geosyncline theory). *Proc. Inst. Geol.* **47**, 1–92.

OBERC, J. (1972) *Budowa Geologiczna Polski, IV. Tektonika, 2. Sudety i obszary przyległe* (Geological Structure of Poland, IV. Tectonics, 2. The Sudetes and Surrounding Areas). Wyd. Geol., Warszawa, 307 pp.

OBERC, J. DYJOR, S. (1969) Uskok sudecki brzeżny (Sudetic Marginal Fault). *Biul. Inst. Geol.* **236**, 41–142.

RAMSAY, J.G. & HUBER, N.I. (1983) *The Techniques of Modern Structural Geology, Vol. 1: Strain Analysis.* Academic Press, London, 307 pp.

RUHLE, E. (1973) Stratygrafia czwartorzędu Polski (Quaternary stratigraphy of Poland). In: *Metodyka Badań Osadów Czwartorzędowych* (Ed. Ruhle, E.) pp. 31–78. Wyd. Geol., Warszawa.

RUHLE, E. & MOJSKI, J.E. (1965) *Geological Atlas of Poland. Stratigraphic and Facial Problems. Fasc. 12 - Quaternary.* Wyd. Inst. Geol., Warszawa.

SEED, H.B. (1968) Landslides during earthquakes due to liquefaction. *J. Soil Mech. Found. Div.* **94**, 1053–1122.

SEILACHER, A. (1984) Sedimentary structures tentatively attributed to seismic events. *Mar. Geol.* **55**, 1–12.

SELLEY, R.C.D., SHERMAN, D.J., SUTTON, J. & WATSON, J. (1963) Some underwater disturbances in the Torridonian of Skye and Raasay. *Geol. Mag.* **100**, 224–243.

SIEH, K.E. (1978a) Slip along the San Andreas Fault associated with the great 1857 earthquake. *Bull. Seismol. Soc. Am.* **68**, 1421–1448.

SIEH, K.E. (1978b) Prehistoric large earthquakes produced by slip on the San Andreas Fault at Pallet Creek, California. *J. Geophys. Res.* **83**, 3007–3939.

SIEH, K.E. (1984) Lateral offsets and revised dates of large prehistoric earthquakes at Pallet Creek, southern California. *J. Geophys. Res.* **89**, 7641–7670.

SIEH, K.E. & JAHNS, R.H. (1984) Holocene activity of the San Andreas Fault at Wallace Creek, California. *Bull. Geol. Soc. Am.* **95**, 883–896.

SIMS, J.D. (1973) Earthquake-induced structures in sediments of Van Norman Lake, San Fernando, California. *Science* **182**, 161–163.

SIMS, J.D. (1975) Determining earthquake recurrence intervals from deformational structures in young lacustrine sediments. *Tectonophysics* **29**, 141–152.

STEEL, R.J., MAEHLE, S., NILSEN, H., ROF, S.L. & SPINNANGR, A. (1977) Coarsening-upward cycles in the alluvium of Hornelen Basin (Devonian) Norway: sedimentary response to tectonic events. *Bull. Geol. Soc. Am.* **80**, 1124–1134.

SYLVESTER, A.G. (1988) Strike-slip faults. *Bull. Geol. Soc. Am.* **100**, 1666–1703.

TCHALENKO, J.S. (1970) Similarities between shear zones of different magnitude. *Bull. Geol. Soc. Am.* **81**, 1625–1640.

VALLE, B., COUREL, L. & GELARD, G.-P. (1988) Synsedimentary and syndiagenetic tectonic indicators in the Blanzy-Montceau Stephanian strike-slip basin (Massif Central). *Bull. Soc. Geol. France*, **8**, 529–540.

WALCZAK, W. (1966) Sudety Kłodzkie i ich przedpole (The Sudetes near Kłodzko and foreland). In: *Guidebook, 9th Meeting Pol. Soc Geogr., Wrocław* (Eds Golachowski, S., Jahn, A. & Nalczak, W.) pp. 15–25, Publ. Wrocław University, Wrocław.

WEAVER, J.D. (1976) Seismically-induced load structures in the basal Coal Measures, *South Wales Geol. Mag.* **113**, 535–543.

WOJEWODA, J. (1987) Sejsmotektoniczne osady i struktury w kredowych piaskowcach niecki śródsudeckiej (Seismotectonic deposits and structures in the Cretaceous sandstones of the Intra-Sudetic Basin). *Prz. Geol.* **408**, 169–175.

Present-day changes in the hydrologic regime of the Raba River (Carpathians, Poland) as inferred from facies pattern and channel geometry

B. WYŻGA

Nature and Natural Resources Protection Research Center, Polish Academy of Sciences, ul. Lubicz 46, 31-512 Kraków, Poland

ABSTRACT

Changes in the channel geometry and facies pattern of the Raba River, a mountain gravel-bed stream of the temperate zone, have been analysed to reconstruct transformations of the hydrologic regime during the last 200 years.

The river straightened, became shallower and widened during the nineteenth century. Point bar deposits showing a diversified facies pattern were then replaced with and overlain by very poorly sorted channel bar deposits of a low-sinuosity river. These changes are attributed to a marked increase in bed load due to the repeated occurrence of flash floods of high magnitude.

Channel degradation and coarsening of bed material during the twentieth century have resulted from increased stream power (channelization effect) and reduced sediment supply, due to variations in basin management and a change in flood hydrographs. A tendency to meander has reappeared but, as point bar deposits are much coarser now, the nineteenth century aggradational phase has had a persistent effect upon the river.

INTRODUCTION

During recent historical time, great changes have taken place in the channels and alluvial plains of the main Carpathian tributaries to the Vistula River (Klimek & Starkel, 1974). At first, until the second half of the nineteenth century, there occurred fast vertical accretion of the alluvial plains. During the present century, channel degradation and formation of lower and lower terraces have been observed.

Changes in river geometry result from variations in hydrologic regime (Schumm, 1969). These variations may be triggered by modifications in the amount of run-off, flood waves, sediment yield and suspended load to bed load ratio (Schumm, 1968), due to either climatic modifications (Hjulström, 1949) or human activities (Gregory, 1987).

This study deals with changes in the geometry of the Raba River channel and concomitant changes in the facies pattern of channel sediments during the last 200 years. These changes illustrate the response to the varying hydrologic regime of a mountain gravel-bed river of the temperate zone.

RIVER CHARACTERISTICS AND STUDY METHODS

The Raba River drains the northern slopes of the Western Carpathians (Fig. 1A). This river is characterized by great variability of water stage and discharge (Punzet, 1969). Such a hydrologic regime results from the low retention potential of the flysch bedrock, high relief (extreme elevations 1310 and 180 m a.s.l.) and the predominant north-western exposition of the slopes, favouring high precipitation from oceanic air-masses arriving from this direction. Floods occur usually during the summer, and snow melting results in prolonged freshets in March and April.

The investigations have been carried out in a

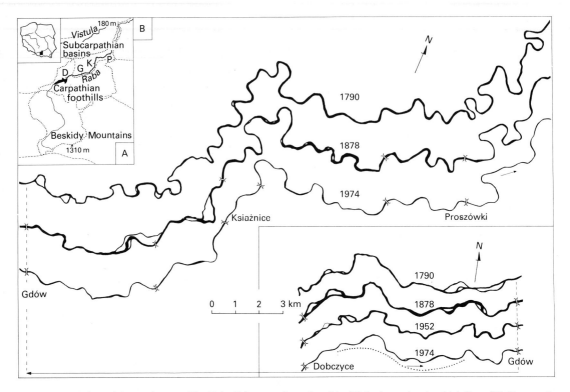

Fig. 1. (A) Location of the study area. The Raba River reach analysed in (B) is shown by the thick line. (B) Changes in the course of the Raba River channel during the last 200 years. Extent of sampling channel sediments for textural analysis is indicated by the dotted line.

Raba River reach extending downstream of Dobczyce, including the middle (within the Carpathian Foothills) and lower (within the Sandomierz Basin) river course (Fig. 1A). A comparison of maps dated at 1790, 1855, 1878, 1901, 1932, 1952 and 1974 allows for evaluation of the changes in channel pattern, width and sinuosity. On the basis of repeated measurements made by the Hydrologic Survey, the twentieth-century changes in the shape and depth of the channel at water-gauge cross-sections have been reconstructed. The vertical position and shape of the older channels have been inferred from hand-auger drillings in parts of the alluvial plain, dated on the basis of the cartographic data. Facies pattern of sediments of the known age has been studied in river cutbanks and walls of gravel pits.

A part of the middle river course between Dobczyce and Gdów, a few kilometres long (Fig. 1B), has been chosen for a detailed textural analysis of channel sediments. The average river slope is 0.00235 there. Sediment samples, $c.10$ dcm^3 in volume, have been collected from upper parts of gravel bars. These parts of the wetted perimeter of a perennial gravel-bed stream are supposed to be most sensitive to changes in water conditions and sediment transfer in the channel. The samples represent different lithological types, whose frequency of occurrence in the sediments of different age has been estimated by eye. The subpavement material was sampled in the present-day channel so as to use fabric and grain-size distribution of the sediment to make inferences on depositional conditions during the passage of flood-wave crests. The sediments sampled were sieved through sieves spaced at $1/3\phi$ intervals.

NINETEENTH-CENTURY EVOLUTION

Changes in channel geometry

Late in the eighteenth century the Raba River channel at the mountain reach, upstream of

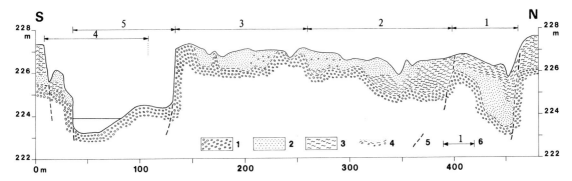

Fig. 2. Cross-section of the present-day channel and the left-side alluvial plain of the Raba River at Fałkowice (3 km upstream of Gdów). **1**, Gravels; **2**, sands; **3**, muds; **4**, boundary between substratum gravels and top stratum fines; **5**, erosional surfaces; **6**, extent of the alluvial-plain elements dating from (top): 1, the eighteenth century; 2, first half of the nineteenth century; 3, second half of the nineteenth century (fragment); 4, the 1950s; 5, present-day channel.

Dobczyce, was of a braided type and reached up to 250 m in width. At the Dobczyce–Gdów reach, the river flowed in a relatively straight (sinuosity index SI = 1.14), single-thread or anastomosing channel (Fig. 1B). The channel was triangular in cross-section (Fig. 2); its bankfull width amounted to 50 m whereas the width/depth ratio was $c.12$ (Fig. 3). Downstream of Gdów, the river meandered (SI = 1.63), although there was a tendency for the channel to straighten through meander cut-off in the uppermost part of this reach (Fig. 1B).

The nineteenth century Raba River evolution was characterized by an increase in the width and a reduction in the depth of the channel, the latter being progressively aggraded (Fig. 2). This trend attained its climax during the second half of the nineteenth century. From 1790 until that time, the channel widened approximately 2–3 times (up to 80–200 m) in the middle river course, and 1.5–2 times (up to 80–170 m) in the lower one (Fig. 1B). Filling up the thalweg with a sediment layer 1.5–2 m thick, and a rise in gravel bar elevation, contributed to channel aggradation in the middle river course (Figs 2 & 3). This decrease in bankfull depth and the considerable widening of the river caused its width/depth ratio to increase by several times.

Alterations in channel pattern and river sinuosity (Fig. 1B) accompanied the dramatic change in channel shape. At the Gdów–Książnice reach, the

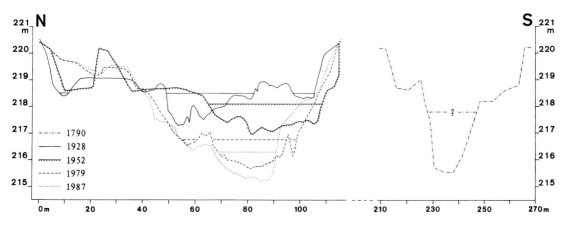

Fig. 3. Changes in channel cross-section of the Raba River at the Gdów water-gauge station between 1790 and 1987. Elevation in metres a.s.l. of mean annual stage is marked for each year. The cross-section was surveyed in the twentieth century, and suggested for 1790 channel on the basis of the gravel bar position.

river shortened by 20% in the period 1790–1878. Consequently, the channel was straight or braided in the middle course. In the lower course, downstream of Książnice, river shortening through meander cut-off amounted merely to a few per cent. In the upper part of this reach, however, numerous mid-channel bars developed giving rise to a complex braided–meandering pattern.

Sediment changes

The alluvia of recognized eighteenth-century age crop out occasionally. A cutbank in the uppermost part of the lower river course, just downstream of Książnice, reveals the following:
1 massive or crudely horizontally stratified pebble gravels, 2 m thickness visible, interpreted as bar platform deposits (terminology according to Bluck, 1971); they are followed upwards by
2 sands and gravelly sands showing large-scale trough cross-stratification, 2 m thick. They form complexes a dozen or so metres wide, and are separated by inclined beds of massive silty sands overlain by fine sands with climbing-ripple lamination. The sediments originated under conditions of lower-regime bed roughness (Harms & Fahnestock, 1965), and are interpreted as supraplatform bar deposits. The internal variability of the sediments reflects successive stages of the lateral accretion of a point bar and the presence of broad scroll ridges and narrow scroll swales on the bar.
The overlying overbank deposits, 2 m thick, consist of alternating sand and mud layers.

The adjacent channel sediments of late nineteenth-century age (Fig. 4) truncate channel-fill deposits. The base of the former is situated $c.0.75$ m above the present-day mean water level. Massive pebble gravels, 2.5 m in thickness, are covered here by horizontally stratified sandy gravels, 1.5 m thick. Transport and deposition of the latter took place under conditions of upper-regime bed roughness (Fahnestock & Haushild, 1962; Harms & Fahnestock, 1965). The sandy gravels are interpreted as the deposits of the upper part of a channel bar. The overlying overbank deposits, $c.0.75$ m thick, consist of parallel laminated sands.

A cutbank located 4 km upstream of Gdów reveals changes in the facies pattern, accomplished in the middle river course during the nineteenth century (Fig. 5). Massive or crudely horizontally stratified cobble gravels, 1.1 m thickness visible (bar platform deposits), are followed here by a

Fig. 4. Structure of the Raba River alluvial plain from the end of the nineteenth century seen in the cutbank at Siedlec (1 km downstream of Książnice). Description of the sequence in the text. Note that the sandy overbank deposits contribute little to the alluvial-plain formation. 25-cm scale on levelling rod. The present-day bar platform gravels in the foreground.

complex of finer-grained sediments, 1–1.2 m thick. These comprise: (a) very coarse sands with trough cross-stratification; (b) gravelly sands with planar cross-stratification; and (c) massive pebble gravels forming cut-and-fill structures a few metres wide and 0.7 m deep. The sediments of the complex are characterized by poor sorting and a large scatter of mean grain size (Fig. 6).

An investigation of the present-day Raba River point bars (see Fig. 8B) allowed the described complex to be recognized as supraplatform bar deposits of a meandering river, and the facies distinguished to be attributed to mesoenvironments of the bar (cf. Gustavson, 1978). Type 'a' represents deposits of a longitudinal ridge, and originated due to migration of dunes through a bar zone covered with shrubs or grass. Type 'b' is recognized as deposits of a transverse bar, and type 'c' as gravel-plug deposits, filling a chute or an inner channel. Tranquil flows were the dominant conditions under which the supraplatform bar deposits (types 'a' and 'b') originated. Location of the sediments near a historical boundary of villages, running arcuately across the valley floor, also points to a meandering river (older than 1790) as their depositional environment.

The supraplatform bar deposits are truncated and covered with massive pebble gravels, 0.8–1 m thick (Fig. 5). They are very poorly sorted and coarser than a majority of the underlying sediments. The largest clasts found in the pebble gravels are $c.2$–4 times coarser than those of the latter. The massive

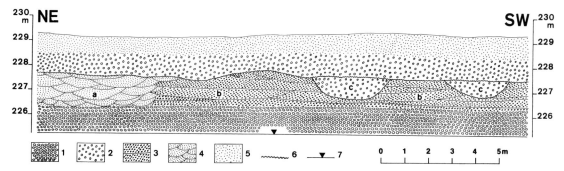

Fig. 5. Structure of the Raba River alluvial plain seen in the cutbank at Stadniki (4 km upstream of Gdów). 1, Massive or crudely horizontally stratified cobble gravels; 2, massive pebble gravels; 3, planar cross-stratified gravelly sands; 4, trough cross-stratified very coarse sands; 5, massive fine sands; 6, major erosional surfaces; 7, mean water table. Explanation of letter symbols in the text.

pebble gravels originated under conditions of upper-regime bed roughness. These are only the conditions of rheologic bed stage that allow either a thick sediment layer to rapidly accrete or separate grains to be deposited at random there (Moss, 1972), thus preventing development of traction structures. These sediments may be associated with the channel from the second half of the nineteenth century. They are interpreted as the channel bar deposits of a shallow, aggrading, low-sinuosity river that periodically transported great volumes of bed load under conditions of shooting flow.

It was under conditions of a relatively narrow channel and concentrated flow that normally loose gravels (with a gravel mode dominating over sand, like the present-day sample D; see Figs 11D & 12D), were deposited by this river. Such gravels occupy the lower part of the field of the nineteenth-century sediments on the plot of mean size vs standard deviation (Fig. 6). Dilated framework gravels (see Fig. 11A) originated in wider channel reaches, the flow being diverged among mid-channel bars. These deposits consist of nearly equal proportions of gravel and sand (see Figs 12A & 13A), and are grouped in the upper part of the nineteenth-century field of the Mz/δ_I plot (Fig. 6). Most likely, reduced stream power and flow turbulence caused some of the sand carried in suspension to settle and be introduced into a rheologic layer in such reaches.

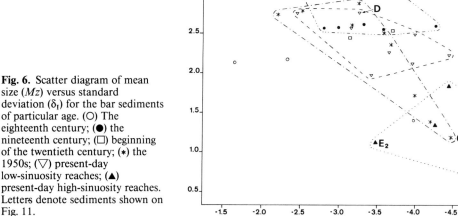

Fig. 6. Scatter diagram of mean size (Mz) versus standard deviation (δ_I) for the bar sediments of particular age. (○) The eighteenth century; (●) the nineteenth century; (□) beginning of the twentieth century; (∗) the 1950s; (▽) present-day low-sinuosity reaches; (▲) present-day high-sinuosity reaches. Letters denote sediments shown on Fig. 11.

Interpretation

The changes in the channel geometry characteristic of the nineteenth-century Raba evolution may be attributed to growth in discharge and an increased proportion of bed load in the total river load (cf. Schumm, 1969). Considering a gravel-bed river like the Raba, it seems reasonable to link changes in channel geometry with alterations in flood discharges rather than mean annual discharge. Therefore, it can be inferred that the nineteenth-century Raba River channel was shaped by flood flows greater than those of the eighteenth-century channel.

Two causes for the growth in flood magnitude and/or frequency are likely. It may be assumed that the high humidity of the final part of the Little Ice Age (Pfister, 1980) contributed greatly to the frequent occurrence of heavy floods in mountain rivers of Central Europe at that time. Age determinations of the debris flows from the Alpine altitudinal belt of the nearby Tatra Mts point to a great intensity of meteorological and hydrologic phenomena during the nineteenth century (Kotarba, 1989). The other reason for the growth in flood discharges was a diminishing retention potential of the basin due to human activity. Progressive slope deforestation and growth in the density of cart-tracks and, especially, the introduction of potato cultivation late in the eighteenth century, all resulted in the growing intensity of the surface and concentrated run-off (Klimek, 1987). Moreover, the removal of leats, ponds and weirs, abundant in the Carpathian valleys during the Middle Ages, favoured high velocities of flood-wave transmission. This rapid run-off of rain water resulted in formation of flood waves with great peak discharges and short time-bases.

The transformation of a high- or moderate-sinuosity channel into a low-sinuosity one progressed downstream with time. This process, distinct during the nineteenth century, must have started earlier in the mountain river reach. The change in channel sediments of the Raba River from those deposited by tranquil flows to those laid down by shooting flows, accomplished during the nineteenth century, is indicative of upstream-controlled aggradation (Teisseyre, 1985). The intensification of agricultural activity in the mountains of the basin must have caused increased sediment supply to the channel (Łajczak, 1988). This resulted in the widening and shallowing of the channel, increasing the ability of the river to carry bed load (Carling, 1983), thus restoring the balance between sediment supply and transporting capacity of the river (Bagnold, 1977). Throughout the nineteenth century, the Raba River was being transformed from a mixed load into a bed load stream in the middle course, and from a suspended load into a mixed load stream in the lower course (cf. Schumm, 1968).

The channel system of the Raba River developed during the nineteenth century was adjusted for conveying high discharges. Flood waves passing through such a high-flow channel are not altered, and their peak discharges are sustained (Burkham, 1976). Waves of these flash floods originated in the mountain part of the basin and carried huge volumes of bed-material load. After the peak of such a flood wave had passed, the sediment was rapidly deposited, and subjected to little or no reworking. High bed shear stresses produced by the flood waves (high discharges, steepened river slope) resulted in the transport of particles much coarser than those transported within the previous low-flow channel. Little difference in bed shear stresses generated within the wide, shallow channel was the reason for marked homogeneity of the sediments of that time.

TWENTIETH-CENTURY RIVER EVOLUTION

Changes in channel geometry

Channelization works began in 1904 in the lower and middle course of the Raba River. They consisted of channel straightening through meander cut-off (Fig. 1B), channel narrowing by groins and the lining of concave banks by gabions and rip-rap. These works resulted in the shortening of the channel downstream of Książnice by 15%.

Channel downcutting commenced in the middle river course in 1908, and in 1913 in the lower one (Wyżga, 1991). Multi-year variations of the lowest annual water stages at the Gdów and Proszówki gauge stations (Fig. 7) reveal the timing of the process. The degradation resulted in channel downcutting by c.2 m at Proszówki (lower course), and by c.3 m at Gdów (middle course) (Figs 3 & 7).

During the first half of the century, the braided channel was transformed into a single channel in the middle river course (Fig. 3). This single-

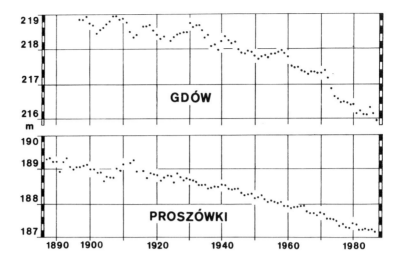

Fig. 7. Changes in the lowest annual stage of the Raba River since the end of the nineteenth century at the Gdów and Proszówki stations.

thread channel showed a conspicuous tendency to meander. The cessation of river-control works during, and immediately after, the Second World War allowed the river to increase its sinuosity in the Dobczyce–Gdów reach by 14% (Fig. 1B). The channelization works were resumed late in the 1950s and they resulted in shortening of the river by 9% in the Dobczyce–Gdów reach and by 4% downstream.

Channel narrowing accompanied the straightening and downcutting of the river during the twentieth century, its rate having reached its peak in the last thirty years (Figs 1B & 3). Nowadays, the channel width amounts to 60–120 m in the middle course and 50–80 m in the lower one. The degradation and narrowing of the channel resulted in a considerable change of its width/depth ratio. At Gdów, it diminished from $c.40$ in 1928 to 20 nowadays (Fig. 3).

The present-day channel in the middle river course comprises reaches of low (dominant) and high sinuosity. Reaches accompanied by side bars of little morphological and sedimentological activity prevail within the former, braided reaches (Fig. 8A) being of secondary importance there. High-sinuosity reaches originate due to fast bank retreat and formation on the convex bank of a point bar, being characterized by the presence of a transverse gravel bar with a distinct brinkpoint and a slip-face (Fig. 8B) (cf. Hickin, 1969). In such cases the existing side bar is incorporated into the point bar as an initial bar platform (Fig. 9).

Fig. 8. Overall view of the Raba River channel in braided (A) and high-sinuosity (B) reaches. In the latter, three depositional mesoenvironments within the supraplatform bar are distinguished: a, longitudinal gravel ridge; b, transverse gravel bar; and c, inner channel.

Fig. 9. Point bar origination due to formation of a transverse bar. Progradation of the transverse-bar front brings about the depositionally inactive side bar to be included into the point bar as an initial bar platform.

Sediment changes

Numerous mud and sand bodies embedded in gravels are the most conspicuous features of the sedimentary record in the middle Raba River course from the beginning of the twentieth century (Fig. 10). Larger bodies, from the middle part of the profile, are recognized as the infills of shallow braids; smaller ones, occurring in its higher part, as the infills of chutes that dissected a gravel bar.

The shifting of active zones within a shallow, wide channel caused some braids to become abandoned and filled with fines. These fines were eroded and covered with channel bar gravels with the repeated approach of a thalweg.

Masssive, normally loose gravels predominate, and dilated framework and openwork gravels are of secondary importance within the bar deposits (Fig. 10). All the sediments are very poorly sorted but vary highly in mean grain-size (Fig. 6).

The point bar deposits of the sinuous river from the 1950s are highly variable texturally. Normally loose and filled underloose (Fig. 11B) gravels originated in a bar head environment. Both these types, as well as the dilated framework and openwork (Fig. 11C) gravels were deposited at bar tails, each type being of nearly equal importance there. Sorting improvement and mean grain-size increase, observed when shifting from dilated framework to openwork gravels on the Mz/δ_I plot (Fig. 6), reflect decreasing percentage of the sand mode in the sediments (Figs 12B, C & 13B,C).

The clustering of the present-day deposits in two distinct fields on the Mz/δ_I plot (Fig. 6) reflects the differentiation of the river into low- and high-sinousity reaches. The former are typified by normally loose (Fig. 11D) and filled underloose gravels, showing significant sand content (Figs 12D and 13D) and very poor sorting (Fig. 6).

Mechanisms of transport and deposition acting in high-sinuosity reaches result in the formation of transverse bar(s), within which cross-stratified gravels originate (Figs 8B & 9) (cf. McGowen & Garner, 1970; Gustavson, 1978).

The present-day supraplatform bar deposits of the middle Raba River are openwork gravels (Fig. 11E). They contain little or no sand (Figs $12E_1$ & $13E_1, E_2$) and show poor to fair sorting (Fig. 6). The gravels exhibit perpendicular normal grading when viewed in section, owing to reverse tangential grading developed on foresets (Fig. 11E).

Two mechanisms may be responsible for the effective sorting of bed material on the transverse bar. One mechanism operates on the surface of a supraplatform bar and results in winnowing of sand from bed-material load. As sand is being set in suspension, the sediment reaching a bar brinkpoint and deposited immediately below it (Fig. 11E) has only slight sand content (Fig. $13E_2$).

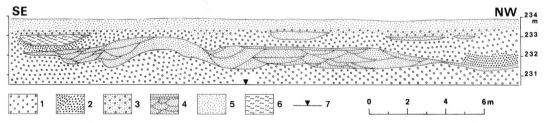

Fig. 10. Structure of the Raba River alluvial plain seen in the cutbank at Winiary (4 km downstream of Dobczyce). 1, Normally loose pebble gravels; 2, openwork pebble gravels; 3, dilated framework pebble gravels; 4, trough cross-stratified and parallel laminated medium- to very coarse-grained sands; 5, massive fine sands; 6, muds; 7, mean water table.

Fig. 11. Photographs showing changes in texture and structure of the Raba River bar gravels between the second half of the nineteenth century and the present. (A) Massive, dilated framework gravels (second half of the nineteenth century); (B) filled underloose gravels, visible gravel imbrication (the 1950s); (C) normally graded, openwork gravels, clasts dip downflow (the 1950s); (D) massive, normally loose gravels (present-day mid-channel bar deposits); (E) normally graded, planar cross-stratified, openwork gravels originating on the foresets of the present-day transverse bar in a high-sinuosity reach. 25-cm scale on levelling rod. Visible paving of the bar surface.

The second one, the overpassing mechanism, causes segregation of grains moving down a bar slip-face (Fig. 6). This results in unimodal gravels (Figs $12E_1$ & $13E_1$) to be formed in the lower part of the slip-face (Fig. 11E). The mechanism consists of preferential movement of coarser particles over a relatively smooth bedlayer of smaller grains, owing to a greater fluid drag exerted on the former, and to their greater momentum conditioning a comparatively straight downstream path and limiting the

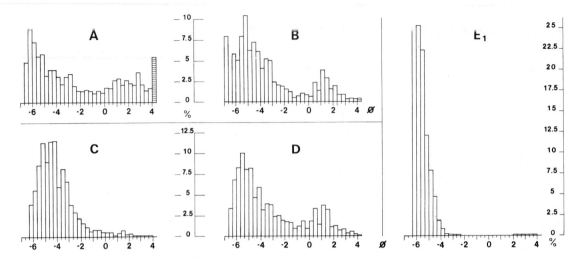

Fig. 12. Histograms of the grain-size distribution (in weight frequency percentage) of the sediments shown on Fig. 11. Hachured areas represent percentage of undivided fines.

number of their bed contacts (Carling & Glaister, 1987).

Pavement horizons and imbricated gravels have not been found in the channel bar deposits from the

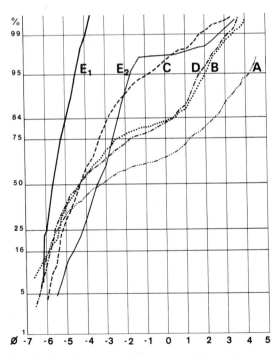

Fig. 13. Cumulative probability curves of the sediments shown on Fig. 11.

second half of the nineteenth century. Such structures occur occasionally in deposits dated at the beginning of the present century, but are more frequent in the point bar deposits from the 1950s. Nowadays, the channel bed is paved on its whole width (Fig. 11E) and shows numerous pebble clusters.

Openwork gravels from the twentieth century show great changes in the thickness of cross-sets and in the foreset angle of deposition. Cross-set thickness, reflecting the height of a transverse bar-front, changed from 5–15 cm in the sediments from the beginning of the century to $c.30$ cm in those from the 1950s and 60–100 cm nowadays. The foreset angle increased during that period from 20–25° to 30–35°.

Interpretation

The twentieth-century sedimentation changes strongly indicate a reduction in the fine sediment load of the Raba River during that period. This is indicated by the increasing contribution of openwork gravels in the bar structures, by the growing degree of bar surface paving and by the steepening angle of openwork-gravel foresets.

The third change invoked is known to reflect the reduced percentage of sand in the supplied sediment (Carling & Glaister, 1987). The reduction of sand content in the bar deposits directly points to a smaller amount of sand carried in the bed load.

Pavement formation requires considerable turbulence, the intensity of the latter being inversely related to the suspended-sediment concentration of a river (Beckinsale, 1972).

There were changes in basin management which caused the progressive reduction in sediment supply during the present century. They comprised alterations in some agricultural practices, i.e. introduction of contour ploughing and terracing of slopes, as well as construction of check-dams in the headwaters (Wyżga, 1991).

If a channel is not constrained, the river adjusts to the lowered sediment load through an increase in sinuosity (Schumm, 1968). The shortening and narrowing of the Raba River, due to channelization works, increased the stream power and brought the river into disequilibrium. Thus, channel degradation was induced as a compensating mechanism.

The channel downcutting of the middle Raba River was temporarily stopped at the mid-century, as a result of the sinuosity increase that followed destruction of regulation structures. During the last thirty years the repeated channelization and changes in run-off have caused the radical reorganization of facies pattern and further degradation in this reach, both the river responses being much faster now. Following the rise in mean annual discharge by 30%, mean annual peak discharge was reduced more than twice, and some growth in duration of flows higher than the mean took place then (Wyżga, 1991). It means that high flash floods gave way either to flattened and more prolonged flood waves or to frequent but low ones.

The sediment yield of the basin must have been reduced with the lowering of peak discharges, thus joining the effect of man's control of bed-material load in the headwaters. At the same time, the longer duration of flows higher than the mean facilitated the erosion of the bed.

Apart from the overall decrease in sediment load, the percentage of bed load in the total load of the river was reduced during the present century. As a result, the middle Raba River changed from a bed load stream into a mixed load one. The main causative factors were: (i) the coarsening of bed material, (ii) the elimination of overloose gravels from bed material in favour of closely packed gravels, (iii) the increase in channel depth, and (iv) the fall in peak discharges. All these tended to reduce the bed load dimensions through a decrease either in bed shear stresses ((iii), (iv): Bagnold, 1977), or in particle susceptibility to entrainment ((i), (ii), (iii): Baker & Ritter, 1975; Church, 1978; Carling, 1983).

SUMMARY AND CONCLUSIONS

The growth in the frequency/magnitude of floods and the increase in bed load were the main factors responsible for transformation of the Raba River during the nineteenth century, including straightening, shallowing and widening of the river channel. Point bar deposits showing a diversified facies pattern were then replaced with and overlain by fairly uniform, very poorly sorted channel bar deposits of a low-sinuosity river. The marked size increase of the coarsest clasts of the bar deposits testifies to the growth in river competence brought about by these flash floods.

The rise in mean annual flood and sediment delivery is attributed to the high humidity of the final part of the Little Ice Age, and to the lowering of the retention potential of the basin due to human activity.

The increased stream power (effect of channelization) and reduced sediment supply (due to some variations in basin management), during the twentieth century, resulted in progressive outwashing of finer grains from bed material and in channel degradation. The great reduction in peak discharges and the prolongation of flood flows has led to a marked intensification of the two processes during the last thirty years.

The river responded to the abating sediment supply and the altered water run-off conditions with a decrease in bed load percentage and reorganization of the facies pattern of the bar deposits. The channel was transformed from a multithalweg into a single-thread one. However, as the increased river competence and channel shortening caused grains of a given grade to be moved into a more distal position during the nineteenth-century high-flow phase, the point bar deposits are now much coarser than during the pre-nineteenth-century low-flow period (Fig. 6).

The lateral migration of the river during the high-flow phase resulted in the removal of the former, thick mud–sandy overbank sediments (Fig. 2). Instead, a relatively thin blanket of sands was formed over the aggraded channel sediments. As the nineteenth-century alluvial plain was subsequently deeply dissected, the present-day river banks are composed nearly exclusively of cohesionless material, mostly gravels or sandy gravels (Figs 2, 4, 5 & 10). Such composition of the banks facilitates their erosion and exerts a major control on the river slope by feeding the channel with coarse debris (cf. Lane, 1955).

ACKNOWLEDGEMENTS

The author wishes to thank Professor Kazimierz Klimek and Drs Witold Zuchiewicz and Szczepan Porębski for helpful criticism of the manuscript and improvement of the English translation of the paper. Thanks are also due to Dr P.A. Carling and an anonymous referee for their critical comments on the manuscript.

REFERENCES

BAGNOLD, R.A. (1977) Bed load transport by natural rivers. *Water Resources Res.* **13**, 303–312.

BAKER, V.R. & RITTER, D.F. (1975) Competence of rivers to transport coarse bedload material. *Bull. Geol. Soc. Am.* **86**, 975–978.

BECKINSALE, R.P. (1972) The effect upon river channels of sudden changes in load. *Acta Geogr. Debrecina* **10**, 181–186.

BLUCK, B.J. (1971) Sedimentation in the meandering River Endrick. *Scot. J. Geol.* **7**, 93–138.

BURKHAM, D.E. (1976) Effects of changes in an alluvial channel on the timing, magnitude, and transformation of flood waves, Southeastern Arizona. US Geol. Surv. Prof. Paper 655-K, pp. 1–25.

CARLING, P.A. (1983) Threshold of coarse sediment transport in broad and narrow natural streams. *Earth Surf. Proc. Landf.* **8**, 1–18.

CARLING, P.A. & GLAISTER, M.S. (1987) Rapid deposition of sand and gravel mixtures downstream of a negative step: the role of matrix-infilling and particle overpassing in the process of bar-front accretion. *J. Geol. Soc. Lond.* **144**, 543–551.

CHURCH, M. (1978) Palaeohydrological reconstructions from a Holocene valley fill. In: *Fluvial Sedimentology* (Ed. Miall, A.D.) pp. 743–772. Can. Soc. Petrol. Geol., Calgary, Memoir 5.

FAHNESTOCK, R.K. & HAUSHILD, W.L. (1962) Flume studies on the transport of pebbles and cobbles on a sand bed. *Bull. Geol. Soc. Am.* **73**, 1431–1436.

GREGORY, K.J. (1987) River channels. In: *Human Activity and Environmental Processes* (Eds Gregory, K.J. & Walling, D.E.) pp. 207–235. John Wiley & Sons, Chichester.

GUSTAVSON, T.C. (1978) Bed forms and stratification types of modern gravel meander lobes, Nueces River, Texas. *Sedimentology* **25**, 401–426.

HARMS, J.C. & FAHNESTOCK, R.K. (1965) Stratification, bed forms, and flow phenomena (with an example from the Rio Grande). In: *Primary Sedimentary Structures and Their Hydrodynamic Interpretation* (Ed. Middleton, G.V.) pp. 84–115. Soc. Econ. Paleont. Miner., Tulsa, Spec. Publ. 12.

HICKIN, E.J. (1969) A newly-identified process of point bar formation in natural streams. *Am. J. Sci.* **267**, 999–1010.

HJULSTRÖM, F. (1949) Climatic changes and river patterns. *Geogr. Ann.* **31**, 83–89.

KLIMEK, K. (1987) Man's impact on fluvial processes in the Polish Western Carpathians, *Geogr. Ann.* **69A**, 221–226.

KLIMEK, K. & STARKEL, L. (1974) History and actual tendency of flood-plain development at the border of the Polish Carpathians. *Abhandl. Akad. Wiss. Göttingen* **29**, 185–196.

KOTARBA, A. (1989) On the age of debris flows in the Tatra Mountains. *Stud. Geomorph. Carp. Balcan.* **23**, 139–152.

LAJCZAK, A. (1988) Impact of various land use on the intensity of sediment runoff in the Polish Carpathians' flysch catchments. In: *Interpraevent 1988 — Graz* (Eds Fiebiger, G. & Zollinger, F.) Vol. 3, pp. 131–165. Tagungspubl. Herausgeber, Klagenfurt.

LANE, E.W. (1955) The importance of fluvial morphology in hydraulic engineering. *Proc. Am. Soc. Civ. Engrs.* **81**, (745), 1–17.

McGOWEN, J.H. & GARNER, L.E. (1970) Physiographic features and stratification types of coarse-grained point bars: modern and ancient examples. *Sedimentology* **14**, 77–111.

Moss, A.J. (1972) Bed-load sediments. *Sedimentology* **18**, 159–219.

PFISTER, C. (1980) The climate of Switzerland in the last 450 years. *Geogr. Helvetica* **35**(5), 15–20.

PUNZET, J. (1969) Hydrological characteristics of the River Raba (English summary). *Acta Hydrobiol.* **11**, 423–477.

SCHUMM, S.A. (1968) River adjustment to altered hydrologic regimen — Murrumbidgee River and palaeochannels, Australia. US Geol. Surv. Prof. Paper 598, pp. 1–65.

SCHUMM, S.A. (1969) River metamorphosis. *J. Hydr. Div., ASCE* **95**, 255–273.

TEISSEYRE, A.K. (1985) Recent overbank deposits of the Sudetic valleys, SW Poland. Part I. General environmental characteristics (with examples from the upper River Bóbr drainage basin) (English summary). *Geol. Sudetica* **20**(1), 113–170.

WYŻGA, B. (1991) Present-day downcutting of the Raba River channel (Western Carpathians, Poland) and its environmental effects. *Catena* **18**, 551–566.

Alluvial Stratigraphy

A revised alluvial stratigraphy model

J.S. BRIDGE *and* S.D. MACKEY

Department of Geological Sciences, State University of New York, PO Box 6000, Binghamton, NY 13902-6000, USA

ABSTRACT

Quantitative models of alluvial stratigraphy have been widely used to interpret the architecture of ancient alluvium. However, with few exceptions, these interpretations have tended to be too simplistic because the complexity of the controls on alluvial architecture has not been appreciated. The alluvial stratigraphy model of Bridge & Leeder has been improved to more accurately simulate alluvial depositional processes and to predict more aspects of alluvial architecture. The revised model predicts: (i) compacted mean floodplain deposition rate; (ii) proportion, connectedness and multistorey character of channel-belt deposits; (iii) the distribution and dimensions of channel sandstone bodies; and (iv) nature of avulsion-related sequences and cyclicity in overbank deposits, including the potential for palaeosol development. Theory and model simulations demonstrate that these architectural features are controlled to varying extents by at least: (i) channel-belt width/floodplain width; (ii) channel depth; (iii) channel-belt deposition rate; (iv) across-floodplain variation in deposition rate; (v) mean avulsion period; (vi) depth of burial (affecting compaction); and (vii) tectonic tilting of the floodplain. The relationships between these controlling variables and alluvial architecture features are complex, and the effects of one controlling variable cannot be considered in isolation from the others. In order to interpret ancient alluvium using this model, it is desirable to observe as many aspects of alluvial architecture as possible. Ideally, this requires large outcrops, closely spaced borehole logs, seismic data, absolute age dating, estimates of burial depths and knowledge of modern fluvial processes.

INTRODUCTION

An understanding of the nature and distribution of channel-belt sandstone bodies within finer-grained overbank deposits (alluvial architecture) is important because channel-belt sandstones are commonly sources of water, oil, gas, gold and other metals. Overbank deposits commonly seal hydrocarbon reservoirs and may contain coals. Understanding of alluvial architecture has greatly improved as a result of the qualitative and quantitative modelling of Allen (1965, 1974, 1978, 1979), Leeder (1978) and Bridge & Leeder (1979). These models have been used recently to interpret the large-scale stratigraphy of ancient alluvial deposits (e.g. Allen & Williams, 1982; Behrensmeyer & Tauxe, 1982; Bridge & Diemer, 1983; Blakey & Gubitosa, 1984; Gordon & Bridge, 1987; Kraus & Middleton, 1987; Shuster & Steidtman, 1987).

Such interpretations have tended to be too simplistic because the complexity of the controls on alluvial architecture has not been fully appreciated. For instance, some authors have emphasized the role of subsidence (deposition) rate (Kraus & Middleton, 1987; Shuster & Steidtman, 1987; Jordan *et al.* 1988), or channel-belt size/floodplain width (Nichols, 1987) without considering the influence of other controlling factors. Bridge & Leeder (1979) found that major controls on some aspects of alluvial architecture are: ratio of channel-belt width/floodplain width, channel-belt deposition rate, avulsion frequency and tectonic tilting of the floodplain. However, they did not explicitly consider the important effects of depth of burial (affecting compaction) and the across-floodplain variation in overbank deposition rate, nor did they consider other aspects of alluvial architecture considered here.

This paper concerns modifications to, and use of, Bridge & Leeder's (1979) computer simulation model. The model has been improved by the implementation of more accurate mathematical descriptions of depositional processes and by consideration of additional aspects of alluvial architecture not included in the original model. A list of symbols employed can be found in the Appendix. We have also developed theoretical expressions for mean floodplain aggradation rate, channel-belt proportion and channel-belt connectedness which are compared with values simulated in the revised model. Model limitations are briefly discussed, and additional improvements to the model are suggested.

MODEL DESCRIPTION

In the original model, a floodplain of finite width, W, is occupied by a single active channel-belt of width, w, formed by a channel with maximum bankfull depth, d (see Fig. 1). Within the channel-belt, the channel is free to migrate laterally and may exhibit any type of pattern (e.g. meandering, braided) with the channel-belt width approximating the maximum amplitude of meander bends (single curved channel) or bankfull width (braided channel). Deposits within the channel-belt are considered to be mainly sand and/or gravel whereas adjacent overbank deposits are considered to be predominantly silt and clay. Channel-belts are aggraded at a constant aggradation rate, a. Adjacent floodplain surfaces are aggraded at a rate, r, controlled by an exponent, b, that describes the rapidity with which deposition rate decreases with increasing distance, z, from the edge of the channel-belt.

Immediately prior to an avulsion, fine-grained overbank deposits are compacted using Baldwin's (1971) composite porosity–depth function for argillaceous sediment. If selected, tectonic tilting of the floodplain surface is simulated by changing the relative elevations of the floodplain margins. The number of tectonic events and associated movements (direction and magnitude) is determined stochastically. The elevation of the floodplain surface is recalculated incorporating the effects of differential aggradation, compaction and tectonic tilting. The channel-belt is then relocated to the position of lowest elevation on the floodplain surface (i.e. avulsion). If there is more than one possible location, the point nearest to the currently active channel-belt is chosen. It is assumed that the point of avulsion occurs far enough upstream to allow free migration of the channel to any point on the floodplain surface. When the new channel-belt is initially established, the underlying floodplain deposits are eroded to a depth equal to the bankfull depth of the channel. Then the cycle of aggradation, compaction, tectonism and avulsion is repeated until the desired number of avulsions is reached. The time between avulsive events, t, is determined by Monte Carlo sampling from a presumed distribution with a fixed mean avulsion period.

A two-dimensional stratigraphic cross-section (oriented normal to the valley axis) is plotted illustrating coarse-grained channel-belt deposits, fine-grained overbank deposits and time lines representing the compacted floodplain surface at the time an avulsion occurs. Information on the thickness and lithology of each avulsion-bounded depositional increment for a series of reference verticals across the floodplain section is used to calculate the areal proportion of coarse-grained channel-belt deposits (channel deposit proportion) and the proportion of horizontal contact between channel-belts (connectedness ratio).

Modifications to the algorithms used in the

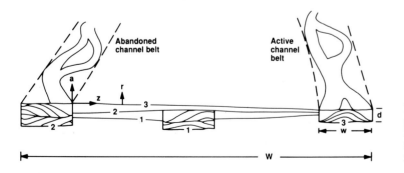

Fig. 1. Definition diagram for alluvial stratigraphy model. Numbers refer to floodplain surfaces and channel-belt margins immediately following an avulsion.

Table 1. Architectural parameters calculated by the revised FPSM program; results of experiments relating sandstone body dimensions and frequency distributions to various architectural parameters (Bridge & Mackey, 1993)

GENERAL

Channel-deposit proportion (areal proportion of channel-belt deposits)[a]
Vertical channel-deposit proportion and its lateral variation[a]
Connectedness ratio (proportion of horizontal contacts)[a]
Proportion of connected channel-belts
Mean floodplain deposit thickness
Mean floodplain aggradation rate
Absolute age of depositional surfaces (time lines)[a]
Vertical cyclicity of overbank and channel-belt deposits

SANDSTONE BODY DIMENSIONS

Individual sandstone bodies
Width
Mean thickness
Mean sandstone thickness
Sandstone proportion
Width/mean thickness
Width/mean sandstone thickness
No. of channel-belts per sandstone body

Total simulated section
Mean sandstone body width
Mean of mean thickness
Mean of mean sandstone thickness
Mean sandstone proportion
Mean width/mean of mean thickness
Mean width/mean of mean sandstone thickness
Mean no. of channel-belts per sandstone body

FREQUENCY DISTRIBUTIONS

Individual sandstone bodies
Width
Thickness
Thickness/mean CB thickness
Sandstone thickness/mean CB thickness
Width/mean thickness
Width/mean sandstone thickness

Total simulated section
Width
Thickness
Width/floodplain width
Mean thickness/mean CB thickness
Mean sandstone thickness/mean CB thickness
Width/mean thickness
Width/mean sandstone thickness

New parameters are in **bold**.
[a] Calculated in original model (Bridge & Leeder, 1979).

model, which are described herein, include: (i) improved description of the rate of decrease of overbank deposition rate with increasing distance from the channel-belt; (ii) compaction of coarse-grained channel-belt deposits; and (iii) compaction of a simulated section to a specified depth of burial. Many new parameters are calculated in the revised model (see Table 1) including: (i) overall mean floodplain deposition rate; (ii) thickness and deposition rate of avulsion-bounded sequences in overbank deposits; (iii) statistics on the width and thickness of single and connected channel-belts (channel-belt sandstone bodies); and (iv) statistics on the width and thickness of connected overbank deposits (shale bodies). Additional modifications to the model include the ability to vary channel-belt width and bankfull depth *during* a simulation, new algorithms to calculate channel deposit proportion and the connectedness ratio, less restrictive input requirements, and output routines that more clearly present results from the model (see Mackey & Bridge, 1992).

Numerous experiments have been performed using the revised model to evaluate the effects on alluvial architecture of varying channel-belt width, overbank sedimentation (b exponent), channel-belt aggradation rate, bankfull depth and mean avulsion period. In each simulation experiment, a stratigraphic section was produced by running the model for fifty avulsions. A list of input parameters used in these experiments is given in Table 2. In these and subsequent simulations, realistic values of input parameters are chosen based on observations from natural environments (full discussion in Bridge & Leeder, 1979; Bridge & Mackey, 1993). Results of these experiments are presented in the following sections, with the exception of sandstone body dimensions and statistics which are discussed in a separate companion paper (Bridge & Mackey, 1993). A further description of the revised Floodplain Simulation Model is given in Mackey & Bridge (1992).

FLOODPLAIN AGGRADATION

The time-averaged aggradation rate, r, at any distance, z, from the edge of the channel-belt was given empirically by Bridge & Leeder (1979) as:

$$r = a(z+1)^{-b} \quad (1)$$

where a is the channel-belt deposition rate and b

Table 2. Input data for experiments varying channel-belt width, overbank deposition rate, channel-belt aggradation rate, bankfull depth of channel, mean avulsion period, depth of burial and tectonic tilting; the selection of initial input parameters is discussed in Bridge & Leeder (1979) and Bridge & Mackey (1993)

Standard case
Number of avulsions: 50
Floodplain width: 10 000 m
Channel-belt width: 1000 m[a]
Bankfull depth: 5.0
Mean avulsion period: 1000 years
Avulsion exponent: 0.4
Overbank function: 1
Aggradation rate: 0.02 m yr^{-1}
b exponent: 5.0
Tectonic option: disabled

Suite 1 experiments (varying overbank deposition rate, channel-belt width)
Channel-belt width: 200, 600, 1000, 2000, 4000 m[a]
b exponent: 0.0, 0.2, 0.5, 1.0, 1.5, 3.0, 5.0, 10.0, 20.0, 30.0, 50.0

Suite 2 experiments (varying channel-belt aggradation rate, channel-belt width)
Channel-belt width: 200, 600, 1000, 2000, 4000 m[a]
Channel-belt aggradation rate: 0.001, 0.002, 0.005, 0.01, 0.015, 0.02, 0.03, 0.05, 0.1 m yr^{-1}

Suite 3 experiments (varying channel bankfull depth, channel-belt width)
Channel-belt width: 200, 600, 1000, 2000, 4000 m[a]
Bankfull depth: 2.0, 4.0, 5.0, 6.0, 8.0, 10.0, 12.0, 14.0, 16.0, 18.0, 20.0 m

Suite 4 experiments (varying mean avulsion period, channel-belt width)
Channel-belt width: 200, 600, 1000, 2000, 4000 m[a]
Mean avulsion period: 100, 250, 500, 1000, 2000, 3500, 5000 years

Suite 5 experiments (varying overbank deposition rate)
b exponent: 0.0 to 50.0 by 1.0

Suite 6 experiments (varying channel-belt aggradation rate)
Channel-belt aggradation rate: 0.1 to 0.01 by 0.01; 0.01 to 0.001 by 0.001; and 0.001 to 0.0001 by 0.0001 m yr^{-1}

Suite 7 experiments (varying mean avulsion period)
Mean avulsion period: 100 to 1000 by 50; 1000 to 5000 by 100; and 5000 to 10 000 by 1000 years

Suite 8 experiments (varying depth of burial, channel-belt width)
Channel-belt width: 200, 600, 1000, 2000, 4000 m[a]
Depth of burial: 0, 50, 100, 250, 500, 1000, 2000, 4000 m

Suite 9 experiments (varying mean tectonic period)
Mean tectonic period: 60, 80, 100, 150, 200, 300, 400, 500, 750, 1000 years
Tectonic exponent: 0.4
Relative vertical movement: 0.5 m
Std dev. vertical movement: 0.5 m

[a] Channel-belt widths held constant for each experiment.

determines the rate of decrease of overbank deposition with distance from the active channel-belt (Fig. 1). However, the dimensions in this equation are clearly incorrect. We have now replaced equation (1) with a more desirable dimensionless version, i.e.

$$\frac{r}{a} = (1 + z/z_m)^{-b} \quad (2)$$

where z_m is the maximum floodplain distance from the edge of the channel-belt. Equation (2) has the advantage over equation (1) of having r/a varying from 1 to 2^{-b}, so that maximum and minimum values of r can easily be specified and used to estimate b. Another possible model of overbank deposition rate (see Crane, 1982) is:

$$r = a(e)^{-bz/z_m} \quad (3)$$

such that values of r/a range from 1 to e^{-b}. Figure 2 shows that equations (2) and (3) give similar results.

Natural data to determine realistic values of b for equations (2) and (3) are very sparse. The data of Kesel et al. (1974, p. 463) suggest that b values of 5 to 10 are appropriate for both the power and exponential functions for a single flood on the Mississippi River (Fig. 3A). Theoretical and field results from Pizzuto (1987, pp. 304, 309) suggest that b values of 0.5 to 1.8 (power function, Fig. 3B) and 0.35 to 1.4 (exponential function) are appropriate for Brandywine Creek, Pennsylvania. As the theoretical deposition rates r and a are averaged over hundreds to thousands of years in the model, it may be inappropriate to assign values of b based solely on data from a single flood (e.g. Kesel et al., 1974). The natural data from Pizzuto (1987) and Bridge & Leeder (1979, p. 620) suggest that b values may actually decrease as the time period over which deposition occurs increases.

In the model, mean floodplain deposit thickness is calculated by averaging the thickness (in metres)

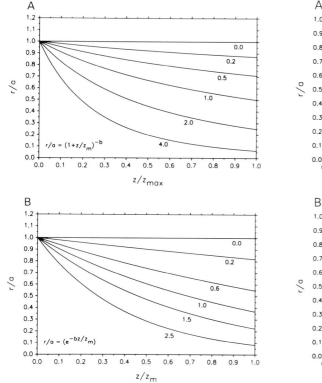

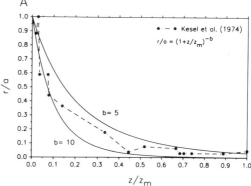

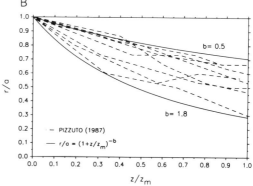

Fig. 2. Variation of overbank deposition rate (r/a) with distance from edge of channel-belt (z/z_m) for increasing values of b. (A) Power function (equation (2)). (B) Exponential function (equation (3)).

Fig. 3. Comparison of power function (equation (2)) with natural data. (A) Kesel *et al.* (1974) for a single flood on the Mississippi River. (B) Pizzuto (1987) from Brandywine Creek, Pennsylvania.

of vertical reference columns across the entire generated section. The mean floodplain aggradation rate, $\bar{r}$, is calculated by dividing the mean floodplain deposit thickness by the total time elapsed to generate the section. These parameters include both overbank and channel-belt deposits and are *compacted* values. Theoretical (uncompacted) values of $\bar{r}$ (below) are also calculated by the model for comparison with simulated results.

The theoretical mean floodplain deposition rate, including channel-belt and overbank deposits averaged across the floodplain width, can be calculated for two distinct cases. In the first case, the channel-belt is located next to the floodplain margin and is expected to yield a minimum value of $\bar{r}$. In the second case, the channel-belt is in the centre of the floodplain and is expected to give a maximum value of $\bar{r}$. In reality, as the channel-belt occupies various positions on the floodplain, the actual $\bar{r}$ will lie between the two anticipated extreme cases.

For the first case, the mean deposition rate calculated from equation (2) is:

$$\bar{r} = \frac{1}{W}\left[\int_0^w a\,dz + \int_0^{W-w} \frac{a}{(1+z/z_m)^b}\,dz\right] \quad (4a)$$

where W is floodplain width and w is channel-belt width. When evaluated, equation (4a) becomes:

$b \neq 1$:

$$\bar{r} = \frac{a}{W}\left[w + z_m \frac{\left(1+\frac{W-w}{z_m}\right)^{1-b}}{1-b} - \frac{z_m}{1-b}\right] \quad (4b)$$

$b = 1$:

$$\bar{r} = \frac{a}{W}\left[w + z_m \ln\left(1+\frac{W-w}{z_m}\right)\right] \quad (4c)$$

As z_m equals $(W-w)$, these equations reduce to:

$b \neq 1$:

$$\bar{r} = \frac{a}{W}\left[w + \frac{(W-w)}{1-b}(2^{1-b} - 1)\right] \quad (5a)$$

$b = 1$:

$$\bar{r} = \frac{a}{W}[w + 0.693(W - w)] \quad (5b)$$

For the second case, $z_m = (W - w)/2$, and the solutions for $\bar{r}$ are exactly the same. This is another advantage of using equation (2) instead of (1), because the calculated values of $\bar{r}$ using equation (1) are not the same for both cases described above. The solutions for $\bar{r}$ using equation (3) are also identical for the first and second cases, and are given by:

$$\bar{r} = \frac{a}{W}\left[w - \frac{(W-w)}{b}(e^{-b} - 1)\right] \quad (6)$$

Equations (5) and (6) indicate that $\bar{r} \to a$ as $b \to 0$, and that $\bar{r} \to a(w/W)$ as b becomes very large (see Fig. 4A and discussion below).

The values of $\bar{r}$ derived from equations (5) and (6) are based on uncompacted rates (even though the deposits will have undergone some compaction near the surface). However, with increasing depth of burial, compaction will significantly reduce the thickness of the floodplain sequence, decreasing the apparent deposition rate. Theoretical values of $\bar{r}$ also do not take into account reduced values of r over buried channel-belts that form topographic highs on the floodplain surface. The result is an overestimation of $\bar{r}$ compared with simulated values.

Figure 4 illustrates how compacted $\bar{r}/a$ varies with W, w, a, d, t and b in the simulation model. The variation of $\bar{r}/a$ with avulsion period, t, and bankfull channel depth, d (not expected from equation (5)), is due to the effects of compaction. At

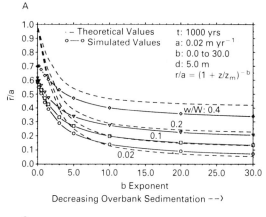

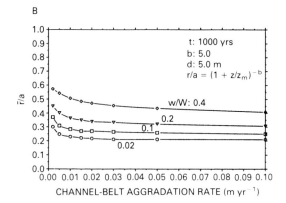

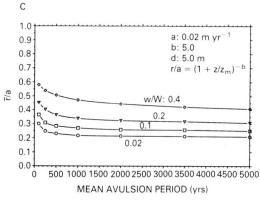

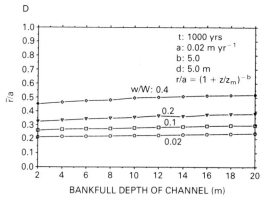

Fig. 4. Simulated variation of compacted $\bar{r}/a$ with: (A) overbank sedimentation, b; (B) channel-belt aggradation rate, a; (C) mean avulsion period, $\bar{t}$; and (D) bankfull depth of channel, d, for varying values of w/W. Simulated $\bar{r}/a$ values for various values of b are compared with theoretical values from equation (5).

small values of t and a, and large values of d, large proportions of overbank deposits are eroded and replaced by the same amount of channel-belt deposits. As the proportion of overbank deposits decreases, the compacted value of $\bar{r}/a$ increases as channel-belt deposits compact less readily than overbank deposits in these experiments. Also, in Fig. 4A, simulated $\bar{r}/a$ values would equal 1 for b equal to 0 if it were not for the effects of compaction. Note that theoretical values of $\bar{r}/a$ using equation (5) are closest to those simulated in the model when the proportion of overbank fines is low (large b values) (Fig. 4A).

CHANNEL DEPOSIT PROPORTION

Channel deposit proportion, CDP, is defined as the areal proportion of channel-belt deposits relative to total area of the generated section. The ratio of channel deposit thickness to total thickness in a single reference vertical is the vertical channel deposit proportion. By summing these values for each reference vertical across the section, and correcting for overestimation due to channel-belt 'edge effects', the channel deposit proportion for the entire section is calculated. CDP values range from 0.0 (all overbank deposits) to 1.0 (all channel-belt deposits). Theoretical values of channel deposit proportion (below) are also calculated by the model for comparison with simulated results.

Within the cross-section considered in Bridge & Leeder's (1979) model, the total area, A_t, of deposits accumulated during the average time between avulsions is:

$$A_t = W\bar{r}\bar{t} \quad (7)$$

where W is the width of the floodplain, $\bar{r}$ is mean aggradation rate and $\bar{t}$ is the mean avulsion period. Some of these deposits will be eroded by the channel when it moves to a new location on the floodplain. However, the material removed is subsequently replaced during initial aggradation of the channel-belt.

The area of channel-belt deposits, A_c, that accumulates over the same time period, $\bar{t}$, is:

$$A_c = w(a\bar{t} + d) - wdp \quad (8)$$

where d is bankfull channel depth and p is the proportion of the subsequent channel-belt(s) which erode(s) into the channel-belt under consideration. The channel deposit proportion is therefore:

$$CDP = \frac{A_c}{A_t} = \frac{w(a\bar{t} + d) - wdp}{W\bar{r}\bar{t}} \quad (9)$$

The value of p is dependent on the value of CDP. If $CDP = 0$, $p = 0$, by definition because there are no channel-belt deposits. If CDP is less than approximately 0.5, most channel-belts are not eroded by subsequent overlying channel-belts and p remains zero. In this case, equation (9) is identical to Leeder's (1978) equation for channel deposit proportion (equation (10a) below). If $CDP = 1$, $p = 1$ because there are no overbank deposits to separate channel-belts. Thus a possible theoretical model for the variation of p and CDP is:

$p = 0$:

$$CDP = \frac{w(a\bar{t} + d)}{W\bar{r}\bar{t}} \quad (10a)$$

$p = 2(CDP) - 1$:

$$CDP = \frac{wa\bar{t} + 2wd}{W\bar{r}\bar{t} + 2wd} \quad (10b)$$

Equation (10) correctly predicts that $CDP \to 0$ when $w \to 0$, that $CDP \to 1$ when $w \to W$ and $a \to \bar{r}$, and that $CDP \to 1$ as b in equations (2) and (3) becomes large because $\bar{r}/a$ approximately equals w/W. Equation (10) clearly indicates that the proportion of channel-belt deposits in alluvium is dependent upon channel-belt aggradation rate, mean floodplain deposition rate, avulsion period, channel depth, channel-belt width and floodplain width (Fig. 5). Equation (10a) (Leeder, 1978) appears to work best for CDP values less than 0.5–0.6 (i.e. low values of p). However, Fig. 5 shows that equation (10b) is a reasonable approximation for CDP under most conditions, i.e. $p = 2(CDP) - 1$.

Bridge & Leeder (1979) emphasized the influence of w/W, a and $\bar{t}$ on CDP. In particular, CDP increases dramatically with w/W. An increase in a or $\bar{t}$ with other variables held constant causes a decrease in CDP. Equation (10) shows that for low values of at/d (say < 1), a small change in at/d will cause a large change in CDP. For large values of at/d (say > 1), a small change in at/d will result in a small change in CDP (Bridge & Leeder, 1979, p. 634). The exponent b in equations (2) and (3) has a dramatic effect on CDP through its influence on $\bar{r}$ (Fig. 5). Thus, high values of b (low values of $\bar{r}/a$) will result in high values of CDP and, conversely, low values of b (high values of $\bar{r}/a$) will result in low values of CDP.

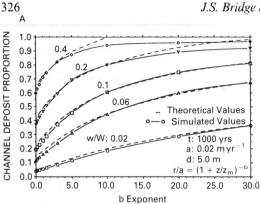
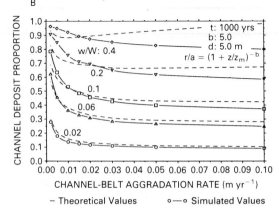
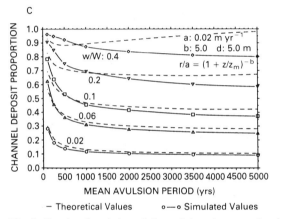
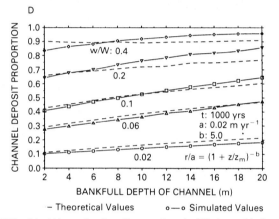

Fig. 5. Simulated variation of channel deposit proportion (*CDP*) with: (A) overbank sedimentation, b; (B) channel-belt aggradation rate, a; (C) mean avulsion period, t; and (D) bankfull depth of channel, d, for varying values of w/W. Simulated *CDP* values are compared with theoretical values from equation (10b).

CONNECTEDNESS

The connectedness ratio, *CR*, is defined as the width of horizontal contact between connected channel-belts over the total width of all channel-belts in the section. After correction for 'edge effects' at the channel-belt margins, the horizontal distance of connected portions of channel-belts is summed for all reference verticals in the section. This distance divided by the total width of channel-belts in the section is the connectedness ratio. *CR* values range from 0.0 (isolated channel-belts) to 1.0 (all channel-belts are connected along their entire width). The initial (basal) channel-belt is not included in the connectedness calculation as it is unconnected by definition.

A theoretical criterion for connectedness is based on a comparison of the amount of overbank deposition that occurs over a channel-belt following its abandonment and the amount of erosion that occurs when the channel returns to the same area. The amount of overbank sediment accumulation (r_o) on an abandoned channel-belt is:

$$r_o = \left(\frac{W\bar{r} - wa}{W - w}\right) t \cdot N_a \quad (11a)$$

$$N_a = n\left(\frac{W}{w}\right) \quad (11b)$$

where n_a is the number of avulsions before return of the channel-belt. For normal (non-tilted) sections, the value of N_a cannot exceed W/w and n is less than unity. Bridge & Leeder's (1979) model suggests that n has an average value close to 0.6 for a w/W of 0.1. When a channel returns to an area, previous deposits are eroded by an amount d. Thus, a criterion for connectedness is:

$$d > \frac{(W\bar{r} - wa)}{W - w} tn\left(\frac{W}{w}\right) \quad (12a)$$

$$c = \frac{(W - w)d}{(W\bar{r} - wa) tn\left(\frac{W}{w}\right)} \quad (12b)$$

Equation (12) suggests that connectedness will increase as both W/w and t decrease, or with increasing d and b (i.e. $\bar{r}$ decreases). Thus connectedness is expected to behave in a similar way to channel deposit proportion (Bridge & Leeder, 1979) (Fig. 6). The connectedness criterion, c (equation (12b), can vary from zero to ∞, whereas the connectedness ratio (CR) of Bridge & Leeder (1979) varies from 0 to 1. When all channel-belts are in contact with one another ($c \geq 1$), CR is expected to have a value of the order of 0.5. Thus as an approximation, $c = CR/(1 - CR)$. Figure 6E indicates that there is reasonable agreement between c and $CR/(1 - CR)$, and that n ranges from approximately 0.25 for large values of w/W to 1.0 for smaller values of w/W.

EFFECTS OF COMPACTION

Bridge & Leeder (1979) did not explicitly consider the effects of compaction on CDP or other architectural parameters. In their study, values of CDP were averaged over the entire simulated sequence even though CDP clearly varies with depth due to compaction. In the original program, Bridge & Leeder (1979) utilized the composite shale compaction curve of Baldwin (1971) to simulate the compaction of fine-grained overbank deposits. This curve approximates empirical data (within a 10% envelope) from fifteen different published sources for 'normally' compacted shales from different depositional environments (Baldwin & Butler, 1985). A polynomial expression for this curve was derived by Bridge & Leeder (1979):

$$P(y) = 0.78 - 0.043 \ln(y + 1) - 0.0054 [\ln(y + 1)]^2 \quad (13)$$

where y is burial depth in metres and P is porosity. The solid thickness, T_s, for a given sediment layer is:

$$T_s = \int_{y_1}^{y_2} [1 - P(y)] dy \quad (14)$$

where y_1 is the upper surface and y_2 is the base of the layer to be compacted (Perrier & Quiblier, 1974). Inserting equation (13) into equation (14) and integrating for depth results in the expression:

$$T_s = 0.19y + (y + 1) \ln(y + 1) \\ \times [0.032 + 0.0054 \ln(y + 1)]\Big|_{y_1}^{y_2} \quad (15)$$

This expression is used to compact fine-grained overbank deposits in the model. We are currently examining alternative porosity–depth expressions that may more accurately reflect variation in grain size and effects of pedogenic processes inherent in fine-grained alluvial deposits (e.g. Sclater & Christie, 1980; Jones & Addis, 1985; Wilson & McBride, 1988; Xiao & Suppe, 1989).

For burial depths less than 500 m, compaction of sandstone bodies is minimal (Perrier & Quiblier, 1974; Chilingarian & Wolf, 1975; Baldwin & Butler, 1985). The original model of Bridge & Leeder (1979) did not compact coarse-grained channel-belt deposits. However, it may be necessary to simulate sequences greater than 500 m thick and to compact simulated sections to a given depth of burial. We have incorporated an algorithm for the compaction of coarse-grained channel-belt deposits in the revised model. The porosity–depth curve for sandstone from Sclater & Christie (1980) can be expressed (Baldwin & Butler, 1985) as:

$$P(y) = 0.49(e)^{-y/3700} \quad (16)$$

where $P(y)$ is porosity at a given burial depth, y, in metres. Converting $P(y)$ to solid fraction and integrating for depth yields an expression for solid thickness, T_s:

$$T_s = y + 1813(e)^{-y/3700}\Big|_{y_1}^{y_2} \quad (17)$$

Expression (17) is used to compact coarse-grained channel-belt deposits in the revised model. Other sandstone porosity–depth curves have been published and are also being evaluated for use in the model (e.g. Chilingarian & Wolf, 1975; Sclater & Christie, 1980; Scherer, 1987). The model is capable of generating thick alluvial sequences (up to 200 avulsions) and contains an option that allows the generated section to be compacted to a specified depth of burial. These improvements facilitate the

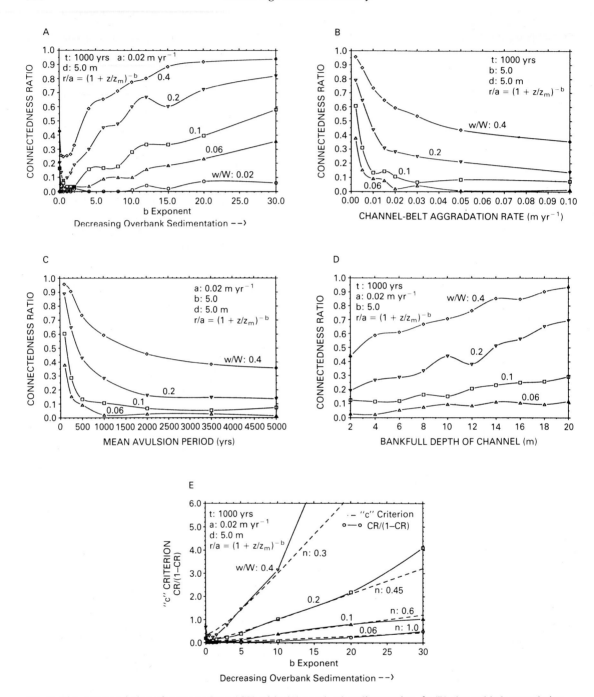

Fig. 6. Simulated variation of connectedness (CR) with: (A) overbank sedimentation, b; (B) channel-belt aggradation rate, a; (C) mean avulsion period, $\bar{t}$; and (D) bankfull depth of channel, d, for varying values of w/W. In (E), the connectedness criterion, c, from equation (12) is compared with $CR/(1 - CR)$.

comparison of model results with compacted outcrop or subsurface data.

As depth of burial and degree of compaction of overbank deposits increase, $\bar{r}/a$ will decrease while CDP increases (Fig. 7). Figure 7 shows that the largest changes in $\bar{r}/a$ and CDP occur in the upper 300 m of section. Below 1000 m, both $\bar{r}/a$ and CDP slowly decrease with increasing depth of burial. Values of $\bar{r}$ decrease by 20–30% at a 1000 m burial depth and up to 50% at a 4000 m burial depth (Fig. 8). CDP increases markedly from the surface to approximately 300 m, then levels off and begins to gradually decline at burial depths greater than 1000 m. The increase in CDP at shallow depths is due to a rapid reduction in thickness of overbank fines due to compaction. Below 300–500 m, channel-belt sandstone deposits begin to compact offsetting the diminishing effect of compaction of overbank fines.

Theoretical values of CDP calculated with equation (10) will be overestimated unless values of $\bar{r}$ and a are the properly compacted values. By contrast, values of $\bar{r}_o$ and a used in the connectedness criterion (equation (12)) must be near surface values for those portions of the sequence between an active channel-belt and an underlying abandoned channel-belt.

In sections where the initial CDP is high, compaction due to burial would cause only a minor increase in CDP and decrease in $\bar{r}$. In contrast, if the uncompacted CDP is small, burial will result in an increase in CDP, and significant decrease in $\bar{r}$. The importance of this result is that in an ancient sequence with a relatively low compacted $\bar{r}$ and CDP, the initial (uncompacted) deposition rates could be equal to, or greater than, the rates for a sequence with higher compacted $\bar{r}$ and CDP. Clearly, the depth of burial of ancient sequences must be known before they can be compared with the model.

MULTISTOREY CHARACTER

In alluvial sequences a 'storey' is defined as the deposit of a single channel bar and adjacent channel fill (Friend *et al.*, 1979, and others). The term 'multistorey' strictly means more than one storey, although it is commonly restricted to the case of vertically superimposed storeys. The inexplicit term 'multilateral' is used by some workers to describe the occurrence of storeys or channel-belts laterally adjacent to each other (e.g. Marzo *et al.*, 1988). In reality, all connected storeys or channel-belts will have some degree of overlap in both vertical and horizontal directions, and the term 'multistorey' with a descriptive qualifier is sufficiently explicit to describe the varieties possible.

Multistorey character within a *single* channel-belt deposit is created when the rate of channel migration within the aggrading channel-belt is large enough to cause superposition of channel bars and fills before the channel-belt is abandoned (Figs 9 & 10a). A measure of the multistorey character within a channel-belt is the ratio of aggraded channel-belt thickness to single storey thickness (i.e channel

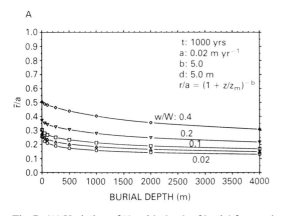
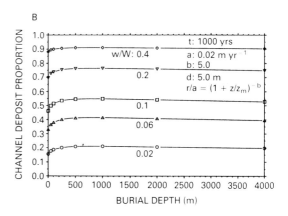

Fig. 7. (A) Variation of $\bar{r}/a$ with depth of burial for varying values of w/W. (B) Variation of CDP with depth of burial for varying values of w/W.

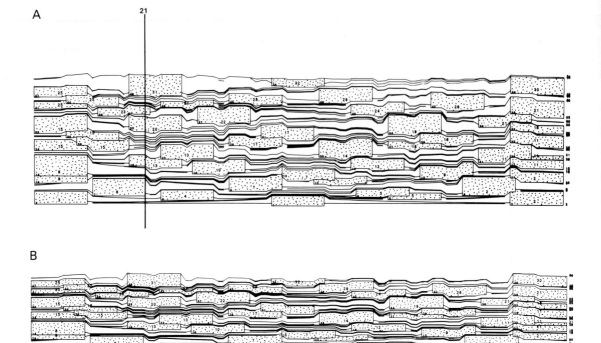

Fig. 8. Effect of compaction as burial depth increases. (A) Simulated section at the surface (standard case). (B) Same section buried to a depth of 4000 m. Both fine-grained overbank deposits and coarse-grained channel-belt deposits are compacted in the revised model. Vertical line marks the location of the reference vertical described in Fig. 11A. Parameters for these sections are: (A) w/W: 0.10; a: 0.02 m yr^{-1}; CDP: 0.46; CR: 0.12; and (B) w/W: 0.10; a: 0.02 m yr^{-1}; CDP: 0.49; CR: 0.12.

depth):

$$\frac{d+at}{d} = 1 + \frac{at}{d} \qquad (18)$$

By definition, multistorey channel-belts occur if $at/d > 0$. However, such character may only be obvious for larger values of at/d. As the value of expression (18) increases, the channel deposit proportion, CDP, decreases due to deposition of thick overbank deposits (see equation (10) and Bridge & Leeder, 1979). For $at/d < 1$, small increases in at/d cause large decreases in CDP. For $at/d > 1$, small increases in at/d cause minor decreases in CDP. As connectedness behaves in a similar way to CDP, increases in at/d result in decreasing connectedness.

If individual channel-belts are connected, another type of multistorey character can occur (Figs 9 and 10b). In this case, low at/d values result in single storey channel-belts in a sequence with thin overbank deposits and high values of CDP and connectedness. It is therefore expected that single storey channel-belts will tend to be connected, whereas individual channel-belts with well-developed multistorey character will tend to be unconnected. At high values of connectedness, the amount of horizontal overlap between channel-belts is at a maximum, giving the *appearance* of a single channel-belt deposit with multistorey character. It is clearly crucial to be able to distinguish these different types of multistorey character in ancient alluvium, as they will respond in different ways to changes in the controlling variables.

The development of these two types of multistorey character has a marked influence on the widths and thicknesses of sandstone bodies composed of one, or more, channel-belts. This

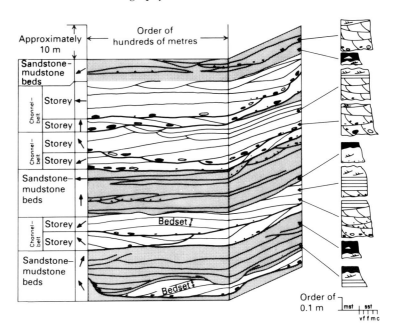

Fig. 9. Definition of multistorey character. (Modified from Gordon & Bridge (1987).)

important aspect of alluvial architecture is addressed in a separate paper (Bridge & Mackey, 1993).

SEQUENCES IN OVERBANK DEPOSITS

The number of avulsions which occur before a channel-belt returns to a given area, N_a, in the absence of a preferred direction of tilting, is $n(W/w)$, where n ranges from 0.3 to 1 (Bridge, 1984; this paper). If the mean avulsion period is $\bar{t}$, the return period, t_r, is:

$$t_r = \bar{t} n \left(\frac{W}{w} \right) \quad (19)$$

If W/w is of the order of 10 and $\bar{t}$ is of the order of 1000 years, the channel-belt return period will be of the order of 10^4–10^5 years.

It may be possible to recognize avulsive events in overbank deposits (e.g. Elliot, 1974; Bridge, 1984; Farrell, 1987). The initiation of an avulsion may be recorded by a regional disconformity or erosion surface which may have a channelized form (Behrensmeyer, 1987; Smith et al., 1989), and which may be overlain by relatively coarse-grained deposits associated with major overbank flooding. Subsequent overlying deposits are associated with overbank flow from the new channel-belt. These deposits may have recognizably different grain sizes, structures and palaeocurrent indicators from those underneath. If the new channel-belt is further away than the previous channel-belt, the new overbank deposits will be generally finer-grained and individual flood deposits will be thinner. Assuming that the degree of soil maturation increases as deposition rate decreases, soils will generally be more mature with increasing distance from the active channel-belt (e.g. Leeder, 1975; Bown & Kraus, 1987; Kraus, 1987). This will be more obvious when soils are developed on abandoned levee or crevasse splay deposits, as opposed to floodplain deposits which may have already developed a mature soil profile. Note that floodplain soils will generally tend to become more mature as a decreases and b increases, all other controlling factors being constant.

If the new channel-belt moved closer to a given area, overbank deposits may be coarser-grained and individual flood deposits may be thicker. Coarsening-upward sequences occur due to development and progradation of levee and crevasse splay systems (e.g. Elliot, 1974; Farrell, 1987). However, fining-upward and coarsening-upward sequences may be superimposed upon the avulsion controlled sequence, due to channel migration,

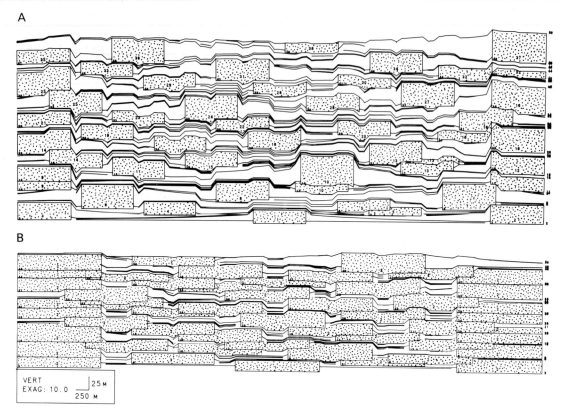

Fig. 10. (A) Simulated section showing sandstone bodies formed by *single* channel-belts. Multistorey character is due to the migration and superposition of channel bars and fills within a single channel-belt. (B) By reducing the value of at/d and/or increasing w/W, channel-belts become superimposed producing a different type of multistorey character. Parameters for these sections are: (A) w/W: 0.10; a: 0.03 m yr^{-1}; *CDP*: 0.44; *CR*: 0.09; and (B) w/W: 0.16; a: 0.015 m yr^{-1}; *CDP*: 0.74; *CR*: 0.47.

growth and abandonment of levee segments and crevasse splays within an established channel-belt. Distinguishing these different types of fining-upward or coarsening-upward sequences requires detailed vertical and lateral studies of overbank deposits.

Figure 11A shows that there are groups of approximately three to four avulsion-related sequences in which the deposits become progressively thicker (coarser) or thinner (finer). These sequences are expected to represent a time period of the order of 10^3–10^4 years. When a channel-belt avulses into an area underlain by overbank deposits, the underlying deposits are eroded to the maximum depth of the channel. If the maximum channel depth ranges from 1 to 10 m, and the average floodplain deposition rate is of the order of 0.1–1 mm yr^{-1}, the time loss represented by the eroded section is of the order of 10^3–10^4 years.

EFFECTS OF TECTONIC TILTING

Tectonic tilting of the floodplain in a direction normal to palaeoflow is not included in the theoretical models for channel deposit proportion and connectedness developed previously. Varying the degree of tectonic tilting does not have a significant effect on the *cross-section averaged* values of channel deposit proportion and connectedness. However, tilting *does* reduce the effective width of the floodplain. As a result, channel deposit proportion and connectedness increase *locally* on the downthrown side of the section and decrease on the upthrown side of the section (Fig. 12) (see also Bridge & Leeder, 1979; Alexander & Leeder, 1987).

On the downthrown side of the floodplain, the proportion of overbank deposits is small, but the avulsion-related sequences are relatively thick due

A revised alluvial stratigraphy model 333

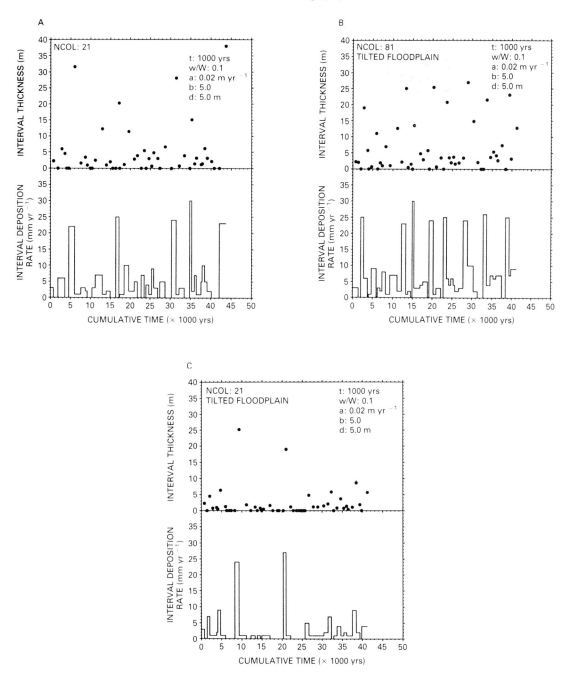

Fig. 11. Variation through time of compacted deposition rate and deposit thickness for a representative vertical through a simulated section. The oldest deposits are to the left in these plots. Points of avulsion are represented by steps in the histograms. The sections used are illustrated in Figs 8A & 12. (A) Untilted section (standard case). Channel-belt deposits are represented by peaks where deposition rates are greater than 20 mm yr^{-1}. Zero deposition rates immediately to the left of these peaks represent erosion of underlying overbank deposits by the channel-belt. Note the thickening and thinning trends of successive avulsion-bounded overbank sequences. (B) Downthrown vertical within a tilted section. (C) Upthrown vertical within a tilted section. Deposition rates and interval thicknesses *increase* in the downthrown vertical and *decrease* in the upthrown vertical.

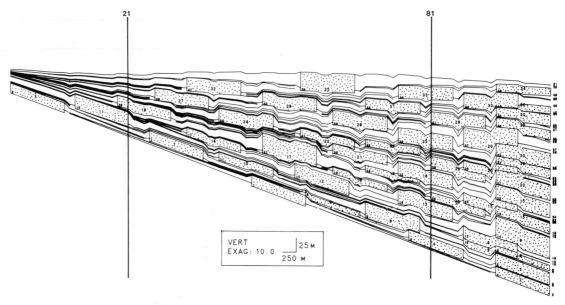

Fig. 12. Section illustrating the effects of floodplain tilting. Tilted section produced with a mean tectonic period of 300 years. Note that channel deposit proportion and connectedness increase *locally* on the downthrown side of the section and decrease on the upthrown side of section. The vertical lines mark the location of the reference verticals described in Fig. 11B,C. Section parameters are w/W: 0.1; a: 0.02 m yr^{-1}; CDP: 0.47; CR: 0.12.

to close proximity to the active channel-belt (Fig. 11B). On the upthrown side of the floodplain, successive channel-belts are separated by long periods of time, and overbank deposition rates are relatively low due to the increased distance from the channel-belt (Fig. 11C). Here, approximately three to four avulsion-related sequences are separated by approximately two to three avulsion-related sequences with very low deposition rates, representing in the order of 10^3–10^4 years. It is likely that these avulsion-related sequences would be very difficult to recognize as such. Furthermore, they are likely to include particularly mature soils.

CONCLUDING DISCUSSION

In order to interpret ancient alluvium using this model, it is desirable to observe all of the aspects of alluvial architecture discussed above in rocks of known ages in different regions of the ancient alluvial basin. Ideally this requires large outcrops, closely spaced borehole logs, seismic data, absolute age dating, estimates of burial depths and knowledge of modern fluvial processes. For instance, the width, thickness, proportion and connectedness of channel-belt deposits, and their multistorey character may be estimated from large outcrops and closely spaced borehole logs. Floodplain widths and the presence of marginal faults may be determined by regional geological mapping and seismic data. Determination of deposition rates requires absolute age dating and knowledge of burial depth (for decompaction). Other input variables can be estimated from knowledge of modern fluvial processes.

Potential problems include factors such as: failure to distinguish different types of multistorey character; confusion of overbank sandstone beds with main channel-belt sandstone beds; inaccurate lithostratigraphic correlations using borehole data; and use of an inappropriate modern analogue. Even if the alluvial architectural parameters can be described adequately and input variables are suitably constrained, the model will not be capable of simulating an ancient sequence if the fluvial system is not modelled completely or correctly. The relationships between controlling variables and alluvial architecture are complex and the effects of one controlling variable cannot be considered in isolation from the others.

Many aspects of the current model could be improved upon. For instance, the model could be improved by replacing the composite compaction curves of Baldwin (1971) and Baldwin & Butler

(1985) with compaction curves based on (very rare) data from modern and ancient alluvial sequences. Many of the input parameters are clearly interdependent (e.g. aggradation rate and avulsion period, channel-belt width and bankfull channel depth). This interdependency is not explicitly stated in the model because the processes *controlling* these parameters are not considered. Desirable improvements to the model include:

1 channel-belt width and bankfull channel depth as a function of channel pattern, discharge, slope and sediment supply;
2 more than one coeval channel-belt on the floodplain;
3 texture of overbank deposits and effect of floodplain topography on overbank aggradation rate;
4 soil and coal development as a function of aggradation rate;
5 improved grain size dependent compaction models for alluvial deposits;
6 avulsion frequency linked to flood magnitude, spatially varying aggradation rate, local floodplain slope and tectonism;
7 periodic degradation superimposed on aggradation;
8 development of a three-dimensional model.

ACKNOWLEDGEMENTS

This research is supported by grants from Conoco, Inc. and ARCO Oil and Gas Company. We are grateful for their encouragement and support of this research. This manuscript benefited from the comments of Tim Cross, Mike Leeder and Jan Alexander.

APPENDIX: LIST OF SYMBOLS

a	Channel-belt deposition rate (m yr^{-1})
A_c	Area of channel-belt deposits (m^2)
A_t	Total area of deposits (m^2)
b	'b' exponent
c	Connectedness criterion
CDP	Channel deposit proportion
CR	Connectedness ratio
d	Maximum bankfull depth of channel (m)
n	Channel-belt return coefficient
N_a	Number of avulsions for channel-belt to return to same location
p	Proportion of channel-belts eroding underlying channel-belts
$P(y)$	Porosity (%) at depth y (m)
r	Time-averaged overbank deposition rate (m yr^{-1})
$\bar{r}$	Mean floodplain deposition rate (m yr^{-1})
$\bar{r}/a$	Dimensionless mean floodplain deposition rate
r_o	Overbank sedimentation rate on an abandoned channel-belt (m yr^{-1})
t	Avulsion period (years)
$\bar{t}$	Mean avulsion period (years)
t_r	Mean channel-belt return period (years)
T_s	Solid thickness (m)
w	Width of channel-belt (m)
W	Width of floodplain (m)
w/W	Dimensionless channel-belt width
y_1	Depth to top of layer to be compacted (m)
y_2	Depth to base of layer to be compacted (m)
z	Distance from edge of channel-belt (m)
z_m	$W - w$, width of floodplain minus width of channel-belt (m)
z/z_m	Dimensionless distance from edge of channel-belt

REFERENCES

ALEXANDER, J. & LEEDER, M.R. (1987) Active tectonic control on alluvial architecture. In: *Recent Developments in Fluvial Sedimentology* (Eds Ethridge, F.G., Flores, R.M. & Harvey, M.D.) pp. 243–252. Soc. Econ. Paleont. Miner., Tulsa, Spec. Publ. 39.

ALLEN, J.R.L. (1965) A review of the origin and characteristics of Recent alluvial sediments. *Sedimentology* 5, 89–101.

ALLEN, J.R.L. (1974) Studies in fluviatile sedimentation. Implications of pedogenic carbonate units, Lower Old Red Sandstone, Anglo-Welsh outcrop. *J. Geol.* 9, 181–208.

ALLEN, J.R.L. (1978) Studies in fluviatile sedimentation: an exploratory quantitative model for the architecture of avulsion-controlled alluvial suites. *Sedim. Geol.* 21, 129–147.

ALLEN, J.R.L. (1979) Studies in fluviatile sedimentation: an elementary model for the connectedness of avulsion-related channel sand bodies. *Sedim. Geol.* 24, 253–267.

ALLEN, J.R.L. & WILLIAMS, B.P.J. (1982) The architecture of an alluvial suite: rocks between the Townsend Tuff and Pickard basal Tuff Beds (Early Devonian), Southwest Wales. *Phil. Trans. R. Soc. Lond. B* 297, 51–89.

BALDWIN, B. (1971) Ways of deciphering compacted sediments. *J. Sedim. Petrol.* 41, 293–301.

BALDWIN, B. & BUTLER, C.O. (1985) Compaction curves. *Bull. Am. Assoc. Petrol. Geol.* 69, 622–626.

BEHRENSMEYER, A.K. (1987) Miocene fluvial facies and vertebrate taphonomy in northern Pakistan. In: *Recent Developments in Fluvial Sedimentology* (Eds Ethridge,

F.G., Flores, R.M. & Harvey, M.D.) pp. 169–176. Soc. Econ. Paleont. Miner., Tulsa, Spec. Publ. 39.

BEHRENSMEYER, A.K. & TAUXE, L. (1982) Isochronous fluvial systems in Miocene deposits of northern Pakistan. *Sedimentology* **29**, 331–352.

BLAKEY, R.C. & GUBITOSA, R. (1984) Controls of sandstone body geometry and architecture in the Chinle Formation (Upper Triassic) Colorado Plateau. *Sedim. Geol.* **38**, 51–86.

BOWN, T.M. & KRAUS, M.J. (1987) Integration of channel floodplain suites, I. Developmental sequence and lateral relations of alluvial paleosols. *J. Sedim. Petrol.* **57**, 587–601.

BRIDGE, J.S. (1984) Large-scale facies sequences in alluvial overbank environments. *J. Sedim. Petrol.* **54**, 583–588.

BRIDGE, J.S. & DIEMER, J.A. (1983) Quantitative interpretation of an evolving ancient river system. *Sedimentology* **30**, 599–623.

BRIDGE, J.S. & LEEDER, M.R. (1979) A simulation model of alluvial stratigraphy. *Sedimentology* **26**, 617–644.

BRIDGE, J.S. & MACKEY, S.D. (1993) A theoretical study of fluvial sandstone body dimensions. In: *The Geological Modelling of Hydrocarbon Reservoirs* (Eds Flint, S. & Bryant, I.D.) pp. 213–236. Spec. Publs Int. Ass. Sediment. 15.

CHILINGARIAN, G.V. & WOLF, K.H. (1975) *Compaction of Coarse-grained Sediments, I.* Developments in Sedimentology 18a. Elsevier, Amsterdam, 552 pp.

CRANE, R.C. (1982) *A computer model for the architecture of avulsion controlled alluvial suites.* Unpublished PhD thesis, University of Reading.

ELLIOT, T. (1974) Interdistributary bay sequences and their genesis. *Sedimentology* **21**, 611–622.

FARRELL, K.M. (1987) Sedimentology and facies architecture of overbank deposits of the Mississippi River, False River region, Louisiana. In: *Recent Developments in Fluvial Sedimentology* (Eds Ethridge, F.G., Flores, R.M. & Harvey, M.D.) pp. 111–120. Soc. Econ. Paleont. Miner., Tulsa, Spec. Publ. 39.

FRIEND, P.F., SLATER, M.J. & WILLIAMS, R.C. (1979) Vertical and lateral building of river sandstone bodies, Ebro Basin, Spain. *J. Geol. Soc. Lond.* **136**, 39–46.

GORDON, E.A. & BRIDGE, J.S. (1987) Evolution of Catskill (upper Devonian) river systems. *J. Sedim. Petrol.* **57**, 234–249.

JAMES, C.S. (1985) Sediment transfer to overbank sections. *J. Hydr. Res.* **23**, 435–452.

JONES, M.E. & ADDIS, M.A. (1985) Burial of argillaceous sediments. *Mar. Petroleum Geol.* **3**, 243–255.

JORDAN, T.E., FLEMING, P.B. & BEER, J.A. (1988) Dating thrust fault activity by use of foreland-basin strata. In: *New Perspectives in Basin Analysis* (Eds Kleinspehn, K.L. & Paola, C.) pp. 307–330. Frontiers in Sedimentary Geology. Springer-Verlag, New York.

KESEL, R.H., DUNNE, K.C., MCDONALD K.R. & SPICER, B.E. (1974) Lateral overbank deposition on the Mississippi River in Louisiana caused by 1973 flooding. *Geology* **1**, 461–464.

KRAUS, M.J. (1987) Integration of channel and floodplain suites, II. Vertical relations of alluvial paleosols. *J. Sedim. Petrol.* **56**, 602–612.

KRAUS, M.J. & MIDDLETON, L.T. (1987) Contrasting architecture of two alluvial suites in different structural settings. In: *Recent Developments in Fluvial Sedimentology* (Eds Ethridge, F.G., Flores, R.M. & Harvey, M.D.) pp. 253–262. Soc. Econ. Paleont. Miner., Tulsa, Spec. Publ. 39.

LEEDER, M.R. (1975) Pedogenic carbonates and flood sediment accretion rates: a quantitative model of alluvial arid-zone lithofacies. *Geol. Mag.* **112**, 257–270.

LEEDER, M.R. (1978) A quantitative stratigraphic model for alluvium, with special reference to channel deposit density and interconnectedness. In: *Fluvial Sedimentology* (Ed. Miall, A.D.) pp. 587–596. Can. Soc. Petrol. Geol., Calgary, Memoir 5.

MACKEY, S.D. & BRIDGE, J.S. (1992) A revised FORTRAN program to simulate alluvial stratigraphy. *Computers and Geosciences* **18**, 119–181.

MARZO, M., NIJMAN, W. & PUIGDEFABREGAS, C. (1988) Architecture of the Castissent fluvial sheet sandstones, Eocene, South Pyrenees, Spain. *Sedimentology* **35**, 719–738.

NICHOLS, G.J. (1987) Structural controls on fluvial distributary systems — the Luna system, northern Spain. In: *Recent Developments in Fluvial Sedimentology* (Eds Ethridge, F.G., Flores, R.M. & Harvey, M.D.) pp. 269–277. Soc. Econ. Paleont. Miner., Tulsa, Spec. Publ. 39.

PERRIER, R. & QUIBLIER, J. (1974) Thickness changes in sedimentary layers during compaction history: methods for quantitative evaluation. *Bull. Am. Assoc. Petrol. Geol.* **58**, 507–520.

PIZZUTO, J.E. (1987) Sediment diffusion during overbank flows. *Sedimentology* **34**, 301–317.

SCHERER, M. (1987) Parameters influencing porosity in sandstones: a model for sandstone porosity prediction. *Bull. Am. Assoc. Petrol. Geol.* **71**, 485–491.

SCLATER, J.G. & CHRISTIE, P.A.F. (1980) Continental stretching: an explanation of the post-mid-Cretaceous subsidence of the central North Sea Basin. *J. Geophys. Res.* **85**, 3711–3739.

SHUSTER, M.W. & STEIDTMANN, J.R. (1987) Fluvial-sandstone architecture and thrust-induced subsidence, northern Green River Basin, Wyoming. In: *Recent Developments in Fluvial Sedimentology* (Eds Ethridge, F.G., Flores, R.M. & Harvey, M.D.) pp. 280–285. Soc. Econ. Paleont. Miner., Tulsa, Spec. Publ. 39.

SMITH, N.D., CROSS, T.A., DUFFICY, J.P. & CLOUGH, S.R. (1989) Anatomy of an avulsion. *Sedimentology* **36**, 1–23.

WILSON, J.C. & MCBRIDE, E.F. (1988) Compaction and porosity evolution of Pliocene sandstones, Ventura Basin, California. *Bull. Am. Assoc. Petrol. Geol.* **72**, 664–681.

XIAO, H. & SUPPE, J. (1989) Role of compaction in listric shape of growth normal faults. *Bull. Am. Assoc. Petrol. Geol.* **73**, 777–786.

Quantified fluvial architecture in ephemeral stream deposits of the Esplugafreda Formation (Palaeocene), Tremp-Graus Basin, northern Spain

T. DREYER

Norsk Hydro Research Centre, Sandsliveien 90, N-5020, Bergen, Norway

ABSTRACT

The Esplugafreda Formation belongs to the Tremp Group of the Tremp-Graus Basin, and consists of coarse-grained ephemeral stream deposits interbedded with dominantly red-coloured floodplain fines. Six coarse member and two fine member facies associations have been recognized. For each facies association, analysis of depositional environment were combined with geometrical studies. The most common coarse member deposits are strongly to moderately incised sand- and gravel-bodies with low width (W)/thickness (T) ratios ('ephemeral ribbons'), and multistorey lobe-like sandbodies with intermediate W/T ratios (stacked sheetflood deposits). Ephemeral stream sandbodies with high W/T ratios (sheet-like braided stream deposits) are only present at the top of the formation. Factors which encouraged the formation of mainly ribbon-like coarse member deposits include proximity to source area, channel incision due to upwarping of the basin margin, and a mean rate of sediment supply equal to or only slightly higher than the accommodation rate. In the multistorey ephemeral ribbon-bodies, it can be demonstrated that the W/T ratio decreases as the number of storeys increases. Three megacycles, formed by variations in sediment supply, subsidence rate and palaeoslope gradients, have been recognized in the Esplugafreda Formation and the overlying Claret Formation. These show a marked fining upwards tendency, with the most mature palaeosols present in the upper part. From the basal to the middle parts of the lower two megacycles, there is an upwards increase in the number and lateral extent of sandbodies. The upper megacycle is dominated by fine member deposits, the only notable channel unit being a laterally extensive multistorey sand- and gravel-body at the base.

The fine member deposits are characterized by the influence of pedogenic processes. Between sedimentation episodes, these processes created 'pause-planes' at exposed sandy surfaces within the inactive channels. In the interchannel areas, variegated palaeosols formed. Mature palaeosols mainly occur at the top of megacycle 1 in proximal parts, but become widespread in areas where clastic supply was less frequent. The mature palaeosols are traceable over distances of several kilometres, and represent promising correlation-horizons.

INTRODUCTION

A growing number of studies concerned with fluvial sedimentology have embraced the idea of looking beyond the 'vertical profile approach' to the wealth of data present in the lateral dimension (i.e. Miall, 1985). The term fluvial architecture has become associated with this shift of focus, stating the need for sedimentary models where two or three-dimensional arrangements of facies types and lithologies are specified. The architectural models include quantitative data about sandbody dimensions, continuity of mudstones, and relative positioning of the different types of sedimentary bodies. This kind of information is valuable in the reconstruction of ancient fluvial systems. Moreover, since considerable amounts of hydrocarbons are contained within fluvial reservoirs, application of quantitative, outcrop-derived data in reservoir modelling may significantly improve the quality of such models.

In the Triassic of the North Sea, fluvial sequences

are widespread (Steel & Ryseth, 1990). The main hydrocarbon accumulations are found in the Rhaetian Upper Lunde Formation (e.g. Nystuen *et al.*, 1989), which contains several hundred metres of sandstones and mudstones deposited by ephemeral streams in an extensional basin setting. Optimal production planning from the Upper Lunde Formation requires a thorough understanding of the fluvial architecture of this heterogeneous formation. As a first step in this direction, studies of well-exposed ephemeral stream sequences similar to the Upper Lunde Formation have been initiated. In this paper, an outline of the fluvial architecture of one of these, the Esplugafreda Formation of the Spanish Pyrenees, is given. The Esplugafreda Formation is a mudstone-dominated sequence, and is thought to be an analogue of those reservoir zones in the Upper Lunde Formation with the lowest net/gross ratio (Nystuen *et al.*, 1989).

By definition, ephemeral streams are characterized by short periods of flow, following local and intense rainfall, and alternating with long periods in which the channel is dry (Picard & High, 1973). Most ephemeral streams probably experience several flows of varying magnitude each year (Mabutt, 1977). Ephemeral floods tend to carve out minor 'desert-valleys' termed wadis or arroyos. These dry-landscape features have a varied morphology (e.g. Glennie, 1972; Picard & High, 1973; Mabutt, 1977; Talbot & Williams, 1979), although narrow and quite deep ravine-like shapes seem to be most common proximally in semiarid settings. Distally, at desert margins, the near absence of vegetation, the presence of an easily erodible sandy substratum, and the desiccated nature of most muddy beds, favour a lower degree of channel confinement. This tends to make the ephemeral streams wider and shallower (Glennie, 1972).

Traditionally, ephemeral stream deposits have been described as alternations between reddish shales and quite thin, flat-based, and laterally extensive sandstones and conglomerates (e.g. McKee *et al.*, 1967; Williams, 1971; Steel, 1974; Tunbridge, 1981; Hubert & Hyde, 1982; Graham, 1983). This kind of succession, mainly involving sandy sheetflood-deposits interbedded with oxidized floodplain fines, have often been regarded as typical of ephemeral stream sedimentation (Miall, 1977). Recently, however, several studies (Allen *et al.*, 1983; Stear, 1983; Tunbridge, 1984; Turner, 1986) have indicated the need for more sophisticated depositional models, emphasizing the variability rather than the uniformity of these systems. In the present study, the diversity of sandbodies produced by ephemeral streams will be demonstrated, and related to temporal and spatial changes in depositional style.

GEOLOGICAL SETTING, FACIES TYPES AND CYCLICITY IN THE ESPLUGAFREDA FORMATION

Introduction

The Tremp Group (Fig. 1) contains widespread continental sediments deposited during an overall regressive–transgressive cycle of late Cretaceous (Maastrichtian) to uppermost Palaeocene (Ilerdian) age (Puigdefábregas & Souquet, 1986). In the study area (Fig. 2), four formations have been distinguished in this group (Cuevas, 1989). The subject of this paper, the Esplugafreda Formation, is the second youngest of these formations (Figs 1 & 3), and consists entirely of alluvial sediments. The basal contact (towards the Talarn Formation) is characterized by the development of very mature palaeosols (see below), indicating a break in sedimentation. Across this boundary, there is a reddening of mudstones, and calcite-material of the Microcodium type (e.g. Klappa, 1978) appears. The boundary between the Esplugafreda and Claret Formations is a gradual one, defined by a colour change from mainly red to more mottled (grey–yellow–purple) mudstones (Cuevas, 1989). In the Aren area, a widespread conglomeratic body with large clasts is present near the base of this transition zone, and the formation boundary is defined at the upper boundary of this conglomerate (Fig. 3). The Claret Formation has an overall transgressive development (Cuevas, 1989), and is conformably overlain by the marine Alveoline Limestone Formation (Lower Eocene, Puigdefábregas & Souquet, 1986).

The Tremp Group accumulated in an incipient foreland basin which started to form in the Palaeocene or even latest Cretaceous times as a response to convergent plate motions in the Bay of Biscay (Puigdefábregas & Souquet, 1986). Excellent outcrop conditions in the Tremp–Graus Basin (Fig. 2) allow detailed reconstruction of depositional conditions and geometries. Most of the data have been collected from the Aren outcrop (Fig. 2). This locality was situated close to the northeastern margin of the foreland basin throughout the Palaeocene (Puig-

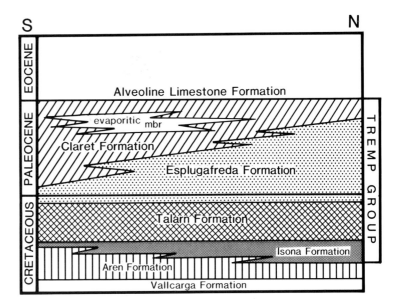

Fig. 1. Cretaceous to Lower Eocene stratigraphy of the Tremp–Graus Basin, Northern Spain. (Modified from Cuevas (1989).) Note the southward thinning of the Esplugafreda Formation.

defábregas & Souquet, 1986; Cuevas, 1989), and the studied sequence thus contains proximal clastics shed from the rising Pyrenean massif. The exposure in Gurp (Fig. 2) is located southeast of the Aren area, and contains more distal alluvial deposits. Palaeocurrent directions measured on crossbeds and gutter casts from both areas imply deposition by rivers flowing towards the south and southwest. Near Aren, the Esplugafreda Formation can be followed in the east–west direction for almost 5 km. Hence, this outcrop (Fig. 4) offers a view of the formation in a section more or less normal to the main palaeocurrent direction. This gives an excellent opportunity to measure the lateral dimensions of the different alluvial sediment bodies.

The Esplugafreda Formation gradually thins to the south, and is replaced by the Claret Formation (Cuevas, 1989, see Fig. 1). The grain size and frequency of channels also decrease southwards, and channelbodies tend to become more sheet-like in this direction. At the southern margin of the Tremp–Graus Basin, the Esplugafreda Formation is less than 100 m thick and virtually devoid of channels. In the northwards-thinning Claret Formation, evidence for marine influence is most notable in the southern areas (Cuevas, 1989). Palaeographically, it seems likely that the Esplugafreda and Claret Formations form the proximal and distal parts of an extensive ephemeral stream-clay playa complex (see below).

By studying the sediment accumulation rate for the upper part of the Tremp Group, some conclusions regarding external controls on sedimentation can be drawn. The Esplugafreda and Claret Formations were deposited over a period which approximately corresponds to the Palaeocene epoch (Fig. 1, Cuevas, 1989). Thus, deposition of these formations lasted about 9 m.y. Knowing that the formations have a maximum compacted thickness of about 400 m in Aren, the estimated mean sediment accumulation rate over this period, 0.04 m/1000 years, is relatively low. Although sedimentation rates measured over long periods of time tend to underestimate the true rates, the calculated value must still be considered to fall well below the average rate for ancient fluvial sedimentation (Sadler, 1981). The numerous palaeosols (below) present in the studied sequence are probably associated with significant time gaps, and this may partly explain the low sedimentation rate (Bown & Kraus, 1987). This rate primarily depends upon the relationship between sediment supplied to the basin and the accommodation (space available for sediment to fill; Posamentier & Vail, 1989) at any specific location (Fig. 5). Given the low mean sediment accumulation rate and the fact that a rather uniform continental sedimentation pattern persisted in the study area throughout most of the Palaeocene, it follows that the ratio between sediment supply and accommodation must have been

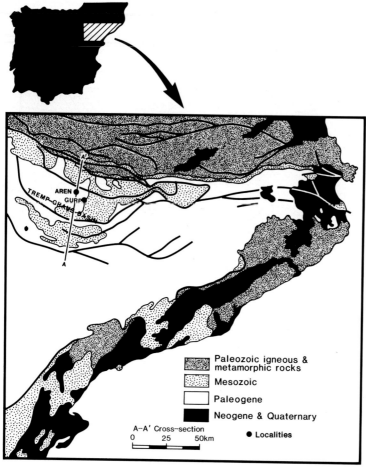

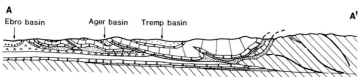

Fig. 2. Location of the Tremp–Graus Basin with study localities in the southern part of the Spanish Pyrenees. (Modified from Cámara & Klimowitz (1985).)

equal to or higher than 1 (Fig. 5). Only towards the top of the Claret Formation, where a transgressive trend is present, this ratio seems to have fallen below 1. It also follows that generation of accommodation in the continental part of the basin must have taken place at a low mean rate. Had accommodation been generated at a high mean rate throughout the Palaeocene epoch, an equally high rate of sediment supply would have been required to maintain continental conditions. This situation, which would have given the deposits of the Esplugafreda and Claret Formations a sediment accumulation rate well above average, is contradicted by the rate of 0.04 m/1000 years presented above.

The mean rate of sediment supply must also have been low when the Esplugafreda and Claret Formations were deposited. If the sediment influx had been significantly higher than the accommodation, a prograding sequence probably characterized by numerous interconnected channelbodies (Fig. 5, left drawing) would have formed. This kind of architecture can only be seen at the top of the Esplugafreda Formation. In the Claret Formation, there is evidence for a slow transgression (Cuevas,

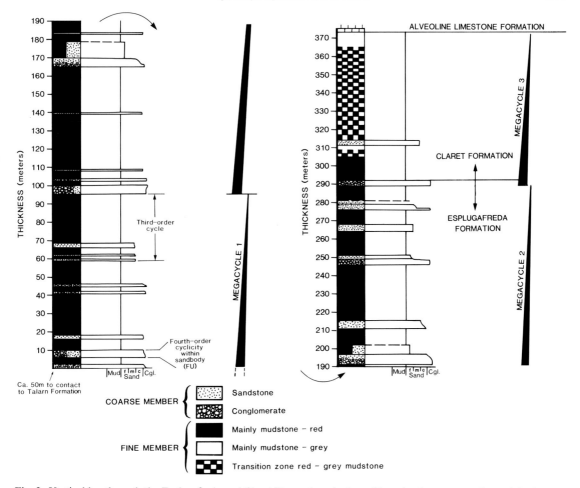

Fig. 3. Vertical log through the Esplugafreda and Claret Formations in Aren. Note the three megacycles and the low sand/shale ratio.

1989), indicating that the sediment supply at this stage was insufficient to keep pace with the increase in accommodation. In summary, the low sediment accummulation rate estimated for the Esplugafreda and Claret Formations indicates that this sequence formed in a setting characterized by a low rate of sediment supply and limited accommodation. These conclusions regarding external controls on sedimentation are important for understanding the development of the various sandbody geometries present in this fluvial sequence.

Outline of fluvial architecture

The Esplugafreda Formation consists of two major lithological units, here termed the *coarse member* and the *fine member* (Figs 3 & 4, Table 1). Fine member deposits constitute about 80% of the formation. However, at more or less well-defined levels within the formations, the percentage of coarse member deposits may be as high as 40–80% (see below).

The coarse member is defined as sandstones (fine to very coarse-grained) and conglomerates which are found in erosively based bodies of high to moderate lenticularity. The interpreted sedimentary processes for the various coarse member facies (Table 1) suggest deposition in bedload-dominated channels or lobes that periodically extended across the floodplain. In the Tremp Group as a whole, eight facies associations have been recognized in the coarse member deposits (Fig. 6). Six of these are

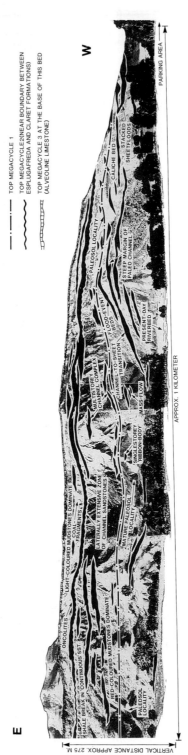

Fig. 4. Overview of the main sedimentological features of the Esplugafreda and Claret Formations in the western part of the Aren exposure. Coarse member deposits in black. Note the megacycles, the varied sand- and gravel-body geometries and the dominance of fine member deposits.

present in the Esplugafreda Formation, and will be discussed in detail below.

The fine member constitutes all the sediments surrounding the coarse member (Fig. 4). For the most part, this member contains mudstones and siltstones, but minor amounts of sandstone (mostly very fine to fine grained) and even gravel occur. As is evident from Table 1, depositional environments of the fine member range from proximal to distal well-drained floodplain deposits.

Three megacycles, 100–175 m thick, have been recognized in the studied sequence. These have the following characteristics (Figs 3 & 4).

Megacycle 1

The base of this megacycle corresponds to the boundary between the Talarn and Esplugafreda Formations (Cuevas, 1989). Megacycle 1 is approximately 150 m thick in Aren (Fig. 3) and 125 m thick in Gurp. The basal 10–20 m are dominated by fine member deposits (building blocks FM1 and FM2, see below), and the lowermost few metres contains very mature palaeosols, especially in the Gurp exposure. This indicates a break in sedimentation, and the base of megacycle 1 should perhaps be placed above this marked discontinuity (cf. base of megacycle 2). The lower to middle parts of megacycle 1 are also dominated by fine member deposits, but there is a marked increase in the number of coarse member bodies. In Aren, this stratigraphic level mostly contains building blocks FM1, C4 and C5, whereas building blocks FM2 and C4 dominate in Gurp. The grain-size and percentage of coarse member bodies are generally highest in Aren.

The percentage of coarse member deposits remains unchanged in the upper to middle parts (20% in Aren, 10% in Gurp), but there is a gradual vertical change from the ribbon-like building blocks C4/C5 to the sheetflood-dominated building blocks C2/C3 (Fig. 6). A pronounced decrease in mean grain-size accompanies this change in sandbody geometry. The uppermost 30–50 m are characterized by a dominance of mature palaeosols in both localities (building block FM2), thus completing the overall fining upwards tendency of the megacycle.

Megacycle 2

Megacycle 2 is about 190 m thick in Aren and 160 m thick in Gurp, and is similar to megacycle 1 in all aspects concerning lateral and vertical changes in grain-size, palaeosol maturity and percentage of coarse member deposits. However, the zone of mature palaeosols capping megacycle 1 is absent in megacycle 2 (Fig. 4). The base of this megacycle is taken where the mature palaeosols of building block FM2 is replaced vertically by building blocks FM1, C3 and C4 (see below for description and definition of building blocks). As in megacycle 1, these building blocks dominate in the lower to middle parts of the megacycle, whereas the sheetflood-dominated building blocks C2 and C3 replace C4 and C5 in the upper to middle parts.

Megacycle 3

The base of this megacycle coincides with the base of the sheet-like sandbody present at the top of the Esplugafreda Formation (Cuevas, 1989). This

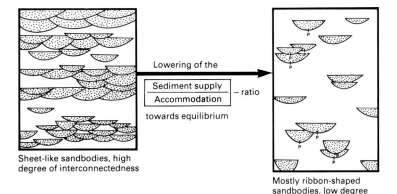

Fig. 5. The effects of changes in the ratio between sediment supply and accommodation on sandbody interconnectedness. P, Palaeochannel axis. Note the differences in the sand/shale ratio and sandbody geometry in the two scenarios.

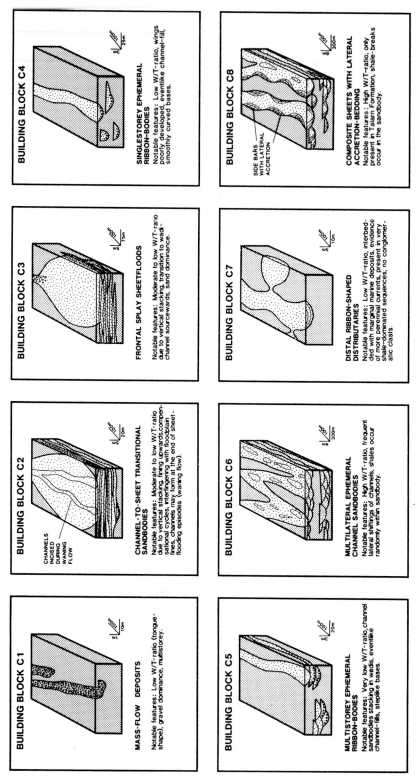

Fig. 6. Schematic illustration of the coarse member building blocks (facies associations) in the ephemeral stream deposits of the Tremp Group. Building blocks 1–6 are present in the Esplugafreda Formation, and are discussed in the text. Building blocks C7 and C8 have only been recognized in the Claret and Talarn Formations, respectively.

Table 1. Facies types in ephemeral stream sedimentary bodies and corresponding floodplain deposits, Esplugafreda Formation

Facies[a]	Characteristics	Depositional processes
G1	Massive or horizontally bedded conglomerate, often with deep basal scour. Beds tend to be 5–65 cm thick, and are arranged into bedsets up to 5 m thick. FU common both for individual beds and bedsets. Clast supported. Some CU units. Boulder to gravel-size clasts.	Stacked or single conglomeratic sheetfloods, sometimes bedload deposits in channels. FU indicates waning flow, and the occasional CU units may reflect deposition on rising flood.
G2	Poorly sorted and unstratified conglomerate with both extra- and intra-formational clasts. Limited basal erosion. Matrix-supported, occurs in a few 1 m thick lenticular beds. Ungraded or inverse grading in lower centimetres. Up to 30 cm large clasts, no imbrication.	Mass-flow deposits (mostly debris flows or density-modified grainflows). The lenticular shape indicates that the mass-flows may have accumulated in narrow depressions.
G3	Crossbedded conglomerate with well-developed imbrication. Maximum clast size 15 cm, varying clast size in neighbouring foresets. Beds 0.3–1.25 m thick, crude normal grading. Solitary sets most common.	In-channel longitudinal bars and small transverse bars deposited under high-energy flow conditions.
G4	Clast-supported massive conglomerate filling up to 2 m deep scours. Beds 0.1–1.1 m thick and very lenticular. Crude normal grading, max. clast size 20 cm. Abundant flutes at bases. Poor sorting.	Scour-fill deposits at the base of major and minor channels. Dominance of scour-fills towards central parts of sandbodies implies strongest degree of channelization here.
S1	Massive to diffusely horizontally bedded coarse-grained to pebbly sandstone. Normal graded beds 0.1–1.35 m thick. S1 beds are thickest, coarsest and most common in lower and central parts of sandbodies and at the margins of gravel-dominated bodies.	Rapid deposition from both channelized and unconfined high-energy flows. Deposition occurred in the initial phase of infilling in sandbodies and in the final phase of infilling in gravelbodies.
S2	Fine grained, very fine grained or muddy beds that occur as discontinuous sheets or lenses in coarse member bodies. Mudcracks, red or grey colours, immature palaeosol developments. 1 cm to 1.5 m thick beds, transitional to fine member beds at margins of sandbodies.	Suspended load fallout formed in channels or on sheetflood lobes at the end of depositional events. Often modified by bio- and pedoturbation. The thicker beds may represent clay plugs.
S3	Sheet-like beds displaying horizontal lamination in medium- to coarse-grained beds. Up to 20 beds stacked in 0.5–6 m thick intervals. Root-traces in upper parts, present near the top of FU sequences. Beds are normal graded and dominate in building block C3.	Deposits from short-lived, unconfined flows (sheetfloods). Vertical stacking caused by upslope entrenchment (see text).
S4	Strongly pedoturbated beds of fine-grained sandstone. Beds show FU, are sheet-like and 0.1–0.5 m thick. Present at top of sandbodies.	Sheetflood deposits formed by low- to moderate-energy flows and subjected to modification by soil processes.
S5	Medium- to very coarse-grained sandstone displaying cosets of moderate- to large-scale crossbedding. Present at sandbody margins and interfingers with fine member deposits. Beds are 0.5–1.8 m thick and tend to contain mud-drapes between foresets. Crossbed-dips vary from 7 to 17° and the foresets are tangential.	Lateral relations imply deposition at the margins of sheet flood lobes, probably as ephemeral flows entered shallow floodplain ponds (small 'frictional' Gilbert deltas?).

continued on p. 346

Table 1. *Continued*

Facies[a]	Characteristics	Depositional processes
S6	Crossbedded pebbly to medium-grained sandstone. Cosets and simple sets of limited lateral extent. Beds are 0.1–1.35 m thick and most common in central parts of sandbodies. Both trough and tabular XB, dips mostly to the south. Decreasing grain-size and set-size towards sandbody tops. Strong basal erosion.	These crossbedded sandstones reflect the migration of bedforms along the bottom of channels. Also infilling of scours.
F1	1–10 m thick units in which pebbly to very fine-grained sandstones fine up into grey, yellow and light red mudstone. Individual beds are 0.1–0.95 m thick and have strongly pedoturbated bedtops. Facies F1 forms 'wings' to channels and pinches out laterally into facies F5. Horizontal lamination and current ripples common.	Sediments deposited at the interface between channels/lobes and the floodplain. Represent proximal crevasse splays and suspended load overbank material.
F2	Laterally discontinuous intervals of grey and silty fine-grained sandstone. Coal fragments common and oncolites moderately common. Facies F2 occurs close to channels and tends to be 0.3–1.1 m thick.	Suspended load material deposited in minor ponds close to active channels. Grey colours and coals indicate that reducing conditions existed in the ponds.
F3	2–20 cm thick granule/conglomerate beds occurring within the fine member mudstones. Massive sheet-like beds showing normal grading and erosive bases. Max. clast size 10 cm.	Sheet-like 'lag' deposits formed when gravelly high-energy floods mainly bypassed muddy substratum on proximal parts of the floodplain.
F4	Moderately pedoturbated beds of very fine-grained sandstone and silt. Light red to greenish grey colours, some relict stratification. Mottling, sharp bases and diffuse tops. 5–25 cm thick beds.	Distal parts of crevasse splay deposits, moderately influenced by soil processes (forms C horizons in palaosols).
F5	Mudstones with caliche nodules in units up to 15 m thick. Red colour dominates, also brown and orange. Facies F5 is segregated into diffuse horizons displaying some colour variation. Moderate degree of caliche development. No primary structures preserved.	Overbank material deposited some distance away from channels. All original features overprinted by soil processes. Consists of immature palaeosols deposited on a well-drained semi-arid floodplain.
F6	Mudstones with large caliche concretions which often have coalesced into hardpans. Some beds of gypsum, well-developed colour banding and mottling. Reddish and purple colours dominate. Palaeosol horizons recognizable in units up to 5m thick. Together with F5/F4 form sequences more than 100 m thick. Large lateral extent, can be traced laterally and vertically into facies F6. Only present in stratigraphic intervals where channels are scarce.	Distal floodplain overbank material, strongly oxidized. All original features overprinted by mature palaeosol formation (mainly Aridisols w/vertic features). Accumulated in areas that were cut off from sediment supply for long periods.

[a]G,S, Coarse member deposits (gravel and sand); F, fine member deposits.

(gravel- and) sandbody belongs to building block C6, and is most prominently developed in Aren. Apart from this multistorey–multilateral sandbody, the rest of megacycle 3 consists of fine member deposits belonging to the Claret Formation. Megacycle 3 is approximately 80 m thick in Aren, and is transgressively overlain by marine sediments of the Alveoline Limestone Group. In Gurp, the minimum thickness is 90 m, and the top is erosively overlain by Oligocene alluvial fan conglomerates (Cuevas, 1989).

Criteria for ephemeral stream interpretation

Prior to discussing the details of these fluvial deposits, it is convenient to review some features which provide insight into the climatic and discharge conditions that existed when the Esplugafreda Formation was formed. Studies of facies, petrography and large-scale fluvial architecture suggest that the alluvial deposits of the Esplugafreda Formation were formed in an ephemeral stream environment. Listed below are the main points in favour of such an interpretation.

1 The fundamental element in most gravel- and sandbodies in the Esplugafreda Formation (Fig. 6) is stacked, thin to moderately thick (0.1–1.7 m) beds in which normal grading is the most noticeable feature (e.g. Fig. 7). Their bases are strewn with intraclasts, but only in lower parts of sandbodies are signs of deep scour common (Table 1). Internally, these beds are dominantly massive to parallel bedded (Fig. 7). Together, these features indicate deposition from episodic, shallow, high-energy stream- or sheetfloods (Picard & High, 1973; Tunbridge, 1981, 1984; Stear, 1983). These beds rarely display transitions to other sedimentary structures upwards, suggesting rapidly declining flow power. Additional indications of short-lived flow events are the lack of reactivation surfaces and the paucity of crossbedding within these beds.

2 The climate during deposition of the Esplugafreda Formation was semi-arid, as indicated by the presence of well-oxidized floodplain fines, and the common occurrence of caliche and gypsum beds. Numerous studies in present-day semi-arid settings suggest that rivers in such environments tend to be ephemeral, transporting sediment only in short 'bursts' of flow associated with periods of intense rainfall (e.g. Picard & High, 1973; Mabbut, 1977; Talbot & Williams, 1979).

3 Pause planes (Fig. 8) may be defined as surfaces within sediments which signify a break in sedimentation. In fluvial settings, pause planes are expected to form during periods of little or no discharge. They are thus very useful in distinguishing between ephemeral and perennial fluvial deposits, since the former type experiences periods of 'drying up' far more often than the latter. In the Esplugafreda Formation, pause planes are common, supporting

Fig. 7. Stacked, normally graded beds of facies G1. This bed type is the fundamental element of most sand- and gravelbodies in the Esplugafreda Formation. It is thought that they represent the deposits of individual channelized to unchannelized ephemeral flow events.

348 T. Dreyer

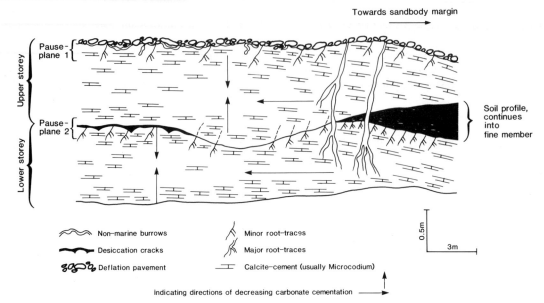

Fig. 8. Field sketch highlighting the characteristics of two pause-planes in a building block C2 sandbody in Aren. The six features shown in the legend are regarded as indicative of breaks in sedimentation.

the notion of flashy and ephemeral flow conditions.
4 The formation has a very immature mineralogical composition, being dominated by unstable carbonate rock fragments. The textural maturity of the sediments is also very low (Table 1). Carbonate rock fragments were derived from caliche developments on the surrounding floodplain, from the source area (Cuevas, 1989), and from redistribution of pedogenic carbonate (*Microcodium*) that formed during breaks in sedimentation (Fjellbirkeland, 1990). Up to 80% of the clasts in some sandbodies may consist of unbroken or slightly fractured *Microcodium* rosettes. Since the rosette is regarded as an unstable form (Fjellbirkeland, 1990) this strongly suggests that the *Microcodium* clasts were subjected to very short transport distances. These features imply ephemeral stream processes. In perennial watercourses, grain transportation/abrasion would occur more frequently, and the stability of calcite would be reduced due to the higher degree of exposure to freshwater. Hence, deposits with higher compositional and textural maturities and more fractured *Microcodium* grains would be expected to form as a result of perennial flow.
5 Levee and crevasse splay deposits are rarely developed along the perimeter of Esplugafreda Formation channels. This might suggest that the channel stability was relatively low (Galloway, 1981), and/or that the channels were too short-lived to build up extensive channel-margin deposits.

FACIES-RELATED BUILDING BLOCKS: THE LINK BETWEEN DEPOSITIONAL PROCESSES AND GEOMETRY

Overview

A problem commonly encountered in sedimentological studies is the lack of integration between qualitative (i.e. interpretation of depositional processes) and quantitative (i.e. dimensions, petrophysical values) geological data. To overcome this problem, the concept of 'building blocks' (Dreyer, 1990) is applied. A building block is defined as a body of rock (e.g. channel, crevasse splay, mouthbar) distinct from other bodies of rock on the basis of specific geometrical, petrophysical and spatial properties. In the context of quantified fluvial models, the building blocks can be viewed as individual rock units which in combination define the full fluvial architecture. Building blocks have many features in common with architectural elements (i.e. Miall, 1985), and the term is intended as an alternative that can be applied in studies where the petrophysical aspects of sedimentary heterogeneities need to be highlighted.

As discussed above, the geometry of ephemeral stream deposits may vary considerably. In this study, it was discovered that the Esplugafreda Formation contained six coarse member (Fig. 6) and two fine member building blocks. The six coarse member building blocks represent the deposits of single or multiple palaeochannels/palaeolobes (see below). In the latter case, we are dealing with composite bodies formed by superpositioning of palaeochannels, or by deposition in an interconnected, multilateral channel network. For every sandbody, the width/maximum thickness (W/T) ratio was recorded. In multistorey bodies, the width and maximum thickness of the entire composite deposit were measured, and the number of stories were noted. A W/T plot of all measured ephemeral stream sedimentary bodies is shown in Fig. 9A. From this figure, it is obvious that there is a very large spread in the W/T ratio of ephemeral stream sandbodies. The line on the plot shows the mean W/T ratio, but due to the large scatter of data-points it was impossible to obtain an optimal fit. The line is included for illustration purposes only, and it should *not* be used in calculations of subsurface ephemeral stream sandbody dimensions. For this purpose, the W/T plots for individual building blocks (Fig. 9B–F) may be applied, provided the sandbody type can be properly identified. The dominance of sedimentary bodies with low W/T ratios (*ribbons* according to Friend *et al.*, 1979) is also evident from Fig. 9. This contrasts with many previously published accounts of ancient ephemeral stream deposits, where sandbodies with mostly high W/T ratios (sheets) have been described or postulated (e.g. Steel, 1974; Tunbridge, 1981, 1984; Hubert & Hyde, 1982; Graham, 1983). However, Friend *et al.* (1979), Allen *et al.* (1983) and Stear (1983) have shown that ancient ephemeral stream sandbodies can have a wide range of morphologies, varying from very narrow ribbons to kilometre-wide sheets. This conclusion is supported both by the present study (Fig. 9) and studies of modern ephemeral streams (Mabutt, 1977, his fig. 45). Causes for this variation will be discussed below.

C1: mass-flow deposits

Only three occurrences of building block C1 have been discovered in the Esplugafreda Formation, all in megacycle 1 in Aren. The sole facies constituent is G2 (Table 1), interpreted to be mass-flow deposits (usually cohesive debris flows, as suggested by the substantial amount of muddy matrix). Building block C1 contains the coarsest clasts in the Esplugafreda Formation. The largest of these (up to 30 cm in diameter) are commonly present at bedtops or in the upper half of beds, indicating the influence of shear stress pushing these clasts upwards through the flows during emplacement. The mostly ungraded and disorganized appearance of these beds might be due to rapid deposition caused by a break in slope (Nemec & Steel, 1984). This, together with the very low W/T ratio of these beds and the semi-arid setting, might imply deposition in narrow, confined trenches running down hillslopes (e.g. the arroyos of Fig. 10, cf. Glennie, 1972). The multistorey nature of the C1 gravelbodies indicates superpositioning of successive mass-flows, or that the flows were subjected to several surges (Nemec & Steel, 1984). Significantly, the mass-flow deposits occur in gravelbodies where stream-flood deposits are absent. In the fluvial gravelbodies, mass-flow deposits may have existed, but only in a phase prior to reworking by stream-flood processes.

The presence of subaerial mass-flow deposits in semi-arid areas is often indicative of relatively high slope gradients and periods of heavy rainfall (e.g. Nemec & Steel, 1984). The strong lenticularity of building block C1 in the east–west direction implies transport from the incipient Pyrenees to the north.

C2: channel-to-sheet transitional sandbodies

Deposits of this kind (Fig. 6) are very common in megacycles 1 and 2 of the Esplugafreda Formation. They tend to be the most common coarse member body in the upper parts of these megacycles, and contain a large variety of facies (G1–G4 and S1–S6, Table 1) relatable to deposition in both channelized and unchannelized parts of ephemeral stream systems. The vertical arrangement of facies in these sandbodies records a progressive decrease in the degree of channelization (DeLuca & Eriksson, 1989), accompanied by a lowering of the energy level in the ephemeral flows. In the lowermost metre, dominance of very coarse-grained, lenticular beds of facies G4 and S4 suggests high-energy conditions in which scours were excavated and then rapidly infilled by poorly stratified gravel and sand. Above this basal zone, beds become more sheet-like

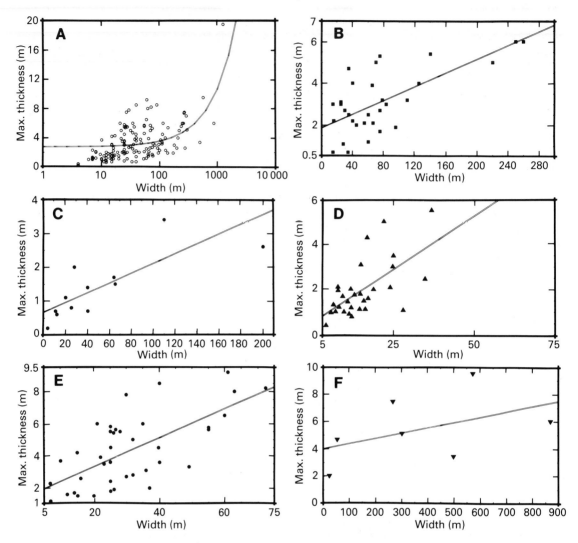

Fig. 9. Plots of maximum sandbody thickness vs width for coarse member bodies in the Esplugafreda Formation. The lines illustrate the mean W/T relationships. (A) All coarse member bodies. The correlation coefficient (r) for the 'best-fit' line is in this case 0.17, indicating that no generalizations regarding W/T ratios should be made from this pooled data-set. Instead, this plot emphasizes the large spread in dimensions of ephemeral stream channelbodies. (B) Channel-to-sheet transitional bodies (C2), $r = 0.55$. (C) Frontal splay sheetfloods (C3), $r = 0.75$. (D) Single-storey ephemeral ribbon-bodies (C4), $r = 0.40$. (E) Multistorey ephemeral ribbon-bodies (C5), $r = 0.44$. (F) Multilateral ephemeral channel sandbodies (C6), $r = 0.22$. The intermediate values of the correlation coefficients in (B) to (F) indicate that there is a certain degree of correlation between maximum thickness and sandbody width for most of the individual building blocks.

and less coarse-grained. High-energy, sheetflood-related facies like G1, S3 and S5 (at sandbody margins) are most common in this interval. Sheet-like, strongly pedoturbated beds dominate at the top of these sandbodies. The facies assemblage here (S2 and S3), along with the finer grain-size and the small bed thickness, suggest deposition from low-energy sheetfloods. Sometimes, ribbon-shaped channels similar to those in building block C4 cut into these topmost beds. This might be due to channel incision during the waning stages of sheet-flooding, when the volume of water in the system

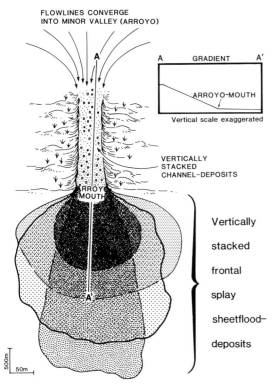

Fig. 10. Plan view of two major sedimentary features of the Esplugafreda Formation: the channel-like arroyo and the downslope sheetflood lobes. Note dominance of vertical stacking.

was insufficient to maintain flow over the entire lobe (Fig. 6). The stacking of sheetfloods on top of channels was caused either by a broadening of the arroyo by sheetflooding, or by a sourcewards retreat of the depositional system shown in Fig. 10 (cf. Schumm & Hadley, 1957).

The lateral transitions between C2 sandbodies and their surrounding fine member deposits may either be gradual (lateral fining) or abrupt. The sandbodies which terminate abruptly do so by the simultaneous pinchout of sheet-like facies S3/S4 beds, commonly over a distance of less than 20 m. An incised palaeochannel margin was found in the lower parts of some of these bodies. Where incision can be seen, the simultaneous pinchout of sheet-like beds is probably related to infilling of arroyos (Fig. 10). In the cases where incision appears to be lacking, the C2 sedimentary bodies were perhaps formed by unconfined sheetfloods which repeatedly covered more or less the same area (see discussion under building block C3).

A W/T plot for building block C2 is given in Fig. 9b. The best-fit line on this diagram has a low correlation coefficient, indicating the scatter in the geometry of this building block. For the most part, these sandbodies have W/T ratios ranging from 20 to 40. These W/T values are low compared to similar data published elsewhere for sheetflood-dominated deposits (see below).

C3: frontal splay sheetfloods

Deposits belonging to building block C3 are common in megacycles 1 and 2 (Fig. 4). They consist of sandbodies where as many as 20 thin (0.1–1.3 m) and laterally extensive beds are stacked on top of each other (Figs 6 & 10). These beds belong to facies S3, S4, S1 and S5, indicating deposition from unconfined sheetfloods (Table 1). The beds are either amalgamated, or separated by centimetre to decimetre thick beds of facies S2 (reddish fines). In vertical sections, sandbodies tend to be dominated by facies S3 in lower parts, while upper parts mostly contain facies S4. This suggests a gradual reduction in discharge through time, perhaps attributable to infilling and gradual abandonment of the channels which fed many of the sheetflood lobes (Fig. 10). The margins of these sandbodies display the same abrupt or gradual transitions into the fine member as building block C2.

The W/T plot for building block C3 (Fig. 9c) indicates that the mean W/T ratio for these stacked sheetflood sandbodies varies from 20 to 55. Earlier studies of ephemeral sheetflood deposits have rarely produced data quantifying their dimensions, but a review of the literature on this subject (e.g. McKee *et al.*, 1967; Steel, 1974; Hubert & Hyde, 1982; Tunbridge, 1984) indicates that W/T ratios exceeding 100 are common. However, it is important to notice that in virtually every case, authors have concentrated on the dimensions of *single* sheetflood-produced beds, whereas in this study, the dimensions of sandbodies containing stacked sheetflood deposits were recorded. As shown in Fig. 11, the W/T ratio for single-storey sheetflood-dominated sandbodies is higher than the W/T ratio for composite sandbodies of this kind.

The data in Fig. 11 also indicate that there is an inverse relationship between the W/T ratio and the number of stories (beds) in the C2 and C3 sandbodies. This observation is of vital importance for reservoir modelling, and may be related to vertical stacking of sheetfloods which over a period of time

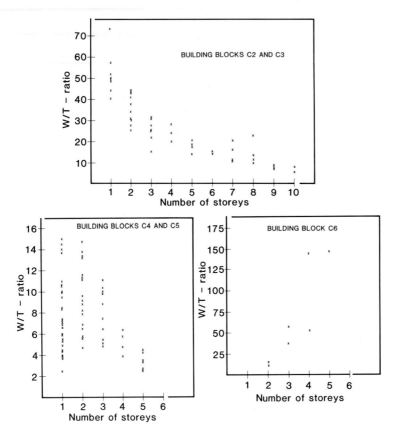

Fig. 11. Relationship between W/T ratio and number of storeys for various ephemeral stream coarse member building blocks. Building blocks C2 and C3 correspond to frontal splays and channel-to-sheet transitional sandbodies, respectively. Building blocks C4 and C5 represent single-storey and multistorey ribbons, whereas building block C6 corresponds to multilateral ephemeral channel sandbodies. See text for discussion.

covered approximately the same area. Optimal conditions for creating such deposits are present in basins with:

1 a low ratio of sediment supply vs accommodation (promotes vertical stacking);

2 entrenched watercourses (arroyos) which are able to capture the sporadic discharge. In a semi-arid setting, high bank stabilities due to concentration of vegetation near arroyos may also promote entrenchment (e.g. Turner, 1986).

In Fig. 10, it is illustrated that flow events in such a setting will generate a mixture of channelized and unchannelized deposits. These deposits are spatially related via a transitional zone (the arroyo mouth). Since the arroyo mouth may remain stationary for a prolonged period of time (e.g. Mabutt, 1977), successive sheetflood beds deposited downslope can be expected to cover roughly the same portion of the floodplain (Fig. 10). As shown by Schumm & Hadley (1957), arroyos can also regress sourcewards by headcutting in their proximal parts. Such a sourcewards retreat may generate fining- and thinning-upwards motifs in the sheetflood deposits, features commonly observed in building block C2/C3 deposits. These modes of vertical stacking promote the generation of sheetflood-dominated sandbodies with relatively low W/T ratios (Fig. 9).

C4: single-storey ephemeral ribbon-bodies

This building block is common in the lower parts of megacycles 1 and 2 (Fig. 4). The channel-fills contain subequal amounts of sand and gravel, usually arranged in decimetre-thick, normally graded units (Fig. 7) deposited by episodic flash-floods (Table 1, facies G1 and S1). The coarsest material is normally concentrated in scour-fills at the channel-base (facies G4, Table 1), but apart from this, overall upwards fining is rarely found. Crossbedding (mainly facies S6, Table 1) is moderately common. The C4 bodies tend to have smoothly curved lower bases and sharp tops, with the infill commonly being conformable to the basal curvature (Fig. 12, showing stacked ribbons of building block C5).

Fig. 12. Multistorey ephemeral ribbon-body (building block C5) encased in fine member deposits, Gurp. Note extremely low W/T ratio of sandbody and the oblique vertical stacking of storeys. The change in fine member colour from deep red (dark in this photo, below base of channel) to yellowish-red (lighter colours lateral to channel) is related to a decrease in palaeosol maturity. This colour change corresponds to the boundary between megacycles 1 and 2. Scale bar only applicable to foreground.

'Wings' (e.g. Friend et al., 1979) are poorly developed. The single-storey ephemeral ribbon-bodies tend to be isolated units incised into the floodplain fines of building blocks FM1 and FM2.

The presence of several ephemeral flood-units with intervening pause-planes (Fig. 8) in the single-storey channelbodies argues against a cut-and-fill origin for these deposits. Infilling periods of several hundreds to thousands of years is considered more likely, since a substantial amount of time is likely to have elapsed during pause-plane generation and between flooding events. This is also supported by the lower than average sedimentation rate presented above. These channels may have formed minor arroyo-like features where active sedimentation only took place during sporadic discharge episodes.

Figure 9D illustrates the W/T relationship for the single-storey ephemeral ribbon-bodies. Mean W/T ratios of less than 10 is indicated for all points along the best-fit line. It must be noted, however, that the low correlation coefficients of these lines imply that there is a substantial amount of scatter in the data. The sandbodies of building block C4 may be classified as ribbons *sensu* Friend et al. (1979). The formation of ribbon-like sandbodies is strongly influenced by the degree of channel entrenchment in the fluvial system. In the Aren exposure, a large number of entrenched channelbodies occur, especially in the lower parts of megacycles 1 and 2. The incision of these channels may be related to readjustment of stream equilibrium profiles in response to tectonic uplift. This will be discussed below in the section dealing with the tectonosedimentary model.

C5: multistorey ephemeral ribbon-bodies

This building block is illustrated in Figs 6, 12 & 13. Apart from the multistorey character and step-like bases (Figs 6 & 12) of these sedimentary bodies, they are similar to those of building block C4 in all aspects. Building block C5 is thus interpreted as vertically stacked, ribbon-like channel deposits. By analogy with building block C4, most of the C5 sedimentary bodies may have formed as a result of ephemeral flow events in major, entrenched channels (arroyos, Fig. 10). In Fig. 9E, a plot of the W/T relationship for building block C5 is presented. The mean W/T ratios, as read from the best-fit line, are varied but always low (3–13, see Fig. 12). The number of storeys in these bodies varies from two to five.

As demonstrated in Fig. 11, there is an inverse relationship between the number of storeys and the W/T ratio for the C5 sandbodies (two stories or more). This is due to the dominantly vertical stacking of storeys in this building block. The lateral distance between palaeochannel axes (P in Fig. 5) in successive storeys tend to be small, a feature which enhances the impression of vertical stacking (Fig. 12). Several factors may have influenced this dominantly vertical mode of sandbody construction. First, the postulated low values of the ratio

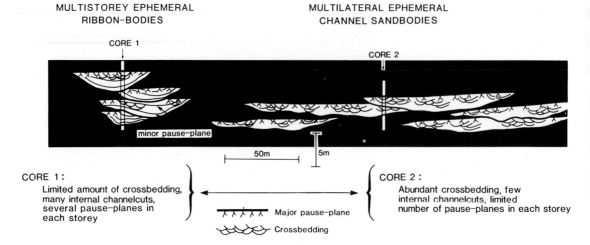

Fig. 13. Comparison of features seen in hypothetical cores through sheet-like multilateral (building block C6) and multistorey ribbon-like (building block C5) ephemeral stream sandbodies.

between sediment supply and accommodation promoted vertical rather than lateral accretion (Fig. 5). Further, stabilization of banks by vegetation and abundant suspended load deposits is likely to have discouraged lateral channel migration. The formation of channel-within-channel sandbodies may also have been encouraged by the postulated temporary high gradients in the study area (see below). In upslope parts of semiarid fluvial settings, most of the runoff from the sporadic and ephemeral flood events tends to be captured by arroyos (Picard & High, 1973; Mabutt, 1977). These tend to have steep margins that remain in a fixed position during the time it takes to fill them in, thus promoting vertical aggradation. Observations from modern arroyos (Schumm & Hadley, 1957; Picard & High, 1973; Mabutt, 1977; Talbot & Williams, 1979) indicate that these dry-landscape features usually have a width which is less than twenty times larger than the depth. This corresponds well with the W/T ratios from sand- and gravelbodies of building blocks C4 and C5.

C6: multilateral ephemeral channel sandbodies

Building block C6 contains approximately equal amounts of sand and conglomerate. The dominant facies are G1, G3, G4, S1 and S6 (Table 1), indicating deposition from both channelized and unchannelized flows of varying magnitude and duration. C6 occurs in both the Aren and Gurp exposures, seemingly at the same stratigraphic level (just below the transition to the Claret Formation, Fig. 3). However, the grain-size is coarser in Aren, and the degree of interconnectedness between channels is higher. Building block C6 is only present in the lower parts of megacycle 3 (Figs 3 & 4), and the base of the lowest C6 body corresponds to the base of this megacycle. This base shows considerable erosional relief, and appears to lie at approximately the same level when traced along depositional strike (Fig. 14) and dip (Aren to Gurp). However, by reference to correlatable palaeosol horizons (Fig. 14), there is a slight tendency for the base of megacycle 3 to become younger towards the east.

Within the C6 bodies, there appears to be two hierarchical levels of channel-like forms. At the lowest level, individual cross-cutting lenses, tens of metres wide and 0.5–2 m thick, are present. These basic units are the only ones which commonly display fining upwards tendencies. The lenses are separated by *minor* scour surfaces (limited downcutting, few intraclasts, limited lateral continuity). On a higher hierarchical level, the connected,

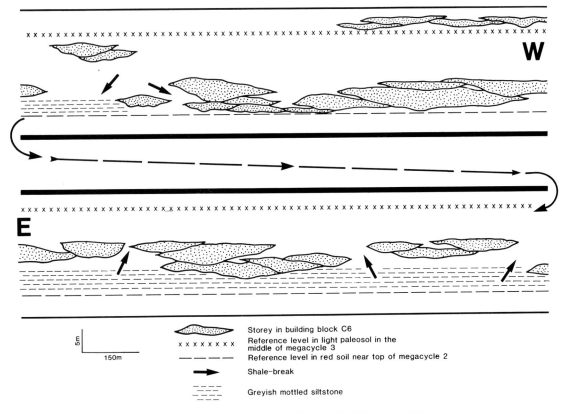

Fig. 14. Sketch of the extent of multilateral ephemeral channelbodies (building block C6) along the Aren exposure. The section is normal to palaeocurrent. Note the local clusters of storeys (separate and superimposed braid-belt deposits).

channel-like lenses form tabular bodies with relatively high W/T ratios (the storeys on Fig. 14). The top of separate storeys normally contain a moderately mature pause-plane (Fig. 8). When traced laterally, individual storeys link up to form multilateral bodies in about 75% of the cases. The connection points between two storeys are normally located near the margins of both sandbodies (Fig. 14), thus justifying the term multilateral rather than multistorey. *Major* erosion surfaces separate the storeys (moderate downcutting, abundant intraclasts, high lateral continuity). These multilateral bodies have the highest W/T ratios in the Esplugafreda Formation (Figs 9F & 11).

Individual storeys of building block C6 were probably deposited in a network of shallow, interconnected braided streams (braid-belts, e.g. Galloway, 1981; Stear, 1983). The lenses at the lowest hierarchical level might represent deposits formed in local segments of the braid-belt (channel-fills, longitudinal bar deposits, sheetflood deposits, see the facies interpretation in Table 1). The high degree of connectedness between storeys gives the C6 bodies a sheet-like geometry which contrasts with the more ribbon-like geometry of the other coarse member building blocks (Fig. 6). This contrast might be due to a lowering of the ratio between sediment supply and accommodation (see Fig. 5 and discussion below).

When planning the production strategy for fluvial reservoirs, it is important to distinguish between ribbon-like and sheet-like coarse member bodies (Fig. 6). From Fig. 13, it is evident that the facies arrangement in these deposits is very similar, thus making it problematical to make the distinction between these building blocks in cores. However, some criteria for differentiation exist. Pause-planes (Figs 8 & 13) are believed to be most abundant in the ephemeral ribbon-like sandbodies, since these bodies for the most part formed in slowly infilling arroyos. Further, it might be argued that slightly more perennial flow conditions existed as the braided stream deposits of building block C6 were laid down. Hence, crossbedding ought to be more dominant in these deposits than in the ribbon-like channels (e.g. Tunbridge, 1984). This is illustrated in Fig. 13, and can also be deduced from the common occurrence of facies G3 and S6 in building block C6. A third feature which might be used to distinguish between ribbon- and sheet-like ephemeral channel deposits in cores is the density of internal channelcut surfaces. In *equally* thick channel sandbodies of the two types, one would expect a random core to penetrate the highest number of internal channelcuts in ribbon-like sandbodies. This is due to the vertical rather than lateral buildup of these bodies, which leads to a stacking of successive channelcut surfaces more or less on top of each other (Fig. 13).

FM1: proximal floodplain deposits

The two fine member building blocks of the Esplugafreda Formation will be described in less detail than the coarse member building blocks (Fig. 6). For building block FM1, the facies assemblage of F1, F2, F3 and F4 (Table 1) suggests deposition by overbank sedimentation (crevasse splays, sheetfloods, levees) and infilling of shallow, ephemeral ponds. Because sedimentation took place close to active channels, soil-forming processes did not last long enough for well-developed palaeosols to form (Bown & Kraus, 1987). Building block FM1 is common in all three megacycles, and may contain up to 40% bedload material close to palaeochannels. However, the bedload content decreases rapidly away from the coarse member deposits. Together with building block FM2, it may form up to 100 m thick sequences of alternating proximal and distal (palaeosol-dominated) floodplain sediments.

The floodplain deposits that formed close to the coarse member bodies form relatively thick fining-up sequences, characterized by alternating mudstones and thin gravel and sandstone beds. There is usually a darkening (reddening) of mudstones and a thinning of the levee/crevasse 'package' away from the channel. Measured lateral extents of forty-four crevasse beds in a direction normal to the channels' long axes (presumably along length sections of the crevasses) showed a variation from 7.5 to 117 m (mean 28.7 m). There was no clearcut relationship between thickness and length of these beds. The limited lengths of crevasse beds indicate event-like deposition from ephemeral overbank floods that quickly became depleted in bedload material. Along their length, these deposits display a transition from facies F1 (proximal crevasses, Table 1) to facies F4 (distal crevasses). Eventually, they peter out into the palaeosols of building block FM2.

FM2: distal floodplain deposits

The deposits of this building block account for about 60% of the sediment in the Esplugafreda Formation. Palaeosols (facies F5 and F6, Table 1) are by far the most common components of building block FM2. Work on the Esplugafreda palaeosols is still in progress (Fjellbirkeland, 1990), and only a preliminary overview will be given here.

The palaeosols of the Esplugafreda Formation are characterized by a strong red pigmentation (oxidized iron components), with subordinate amounts of grey, olive, purple and yellow colours. Clay and silt are the dominant grain-sizes, and X-ray diffraction analysis reveals that the clay fraction mostly consists of smectite (Fjellbirkeland, 1990). Beds and nodules of caliche and gypsum are also common. At stratigraphic levels with limited amounts of sandbodies, individual palaeosols are thick and have well-developed horizons displaying differences in colour, texture, chemical composition and degree of mottling/pedoturbation (Fjellbirkeland, 1990). A gypsiferous purple horizon is usually present at the top of these thick palaeosols. By analogy with other palaeosols-bearing sequences (Atkinson, 1986; Bown & Kraus, 1987; Lehman, 1989; Wright, 1989), it is thought that intervals with pronounced horizontality represent the most mature palaeosols of the Esplugafreda Formation. Mature palaeosols are by far the most laterally continuous depositional element in the Esplugafreda Formation. Some of them can be traced along the entire Aren and Gurp outcrops (up to 5 km). Hence, the mature palaeosols represent the most reliable correlation horizons within this type of fluvial system.

Other characteristics of the palaeosols include rhizocretions, extensive colour mottling, the presence of peds and cutans, evidence for argilliturbation and vertic features such as cracks (due to shrinking and swelling of the expandable clays), and the presence of pedogenic calcium carbonate (*Microcodium*). All the palaeosol features are elaborated upon in Fjellbirkeland (1990), and their relation to soil-forming processes are exhaustively treated by Wright (1989), Atkinson (1986) and Bown & Kraus (1987). The listed characteristics of the palaeosols in the Esplugafreda Formation suggest that they have features in common with both the Aridisols and the Vertisols of the USDA soil-classification system (USDA, 1975). Moreover, the most immature palaeosols in the sequence may be classified as Entisols.

The negative correlation between sandbody content and palaeosol maturity is perhaps most clearly seen when comparing the Aren and Gurp exposures. In Aren, the sand- and gravel-bodies constitute about 20% of the Esplugafreda Formation. Mature palaeosols occur sporadically at this locality, and most notably in the sand-poor interval at the top of megacycle 1 (Fjellbirkeland, 1990; Fig. 4). The more distal Gurp exposure (Fig. 15) apparently received less coarse clastic material, as witnessed by the low sandbody content (about 10%). As a possible effect of this, the fine member deposits at Gurp are characterized by a higher frequency of mature palaeosols than in Aren. This might be related to variations in the length of time available for soil formation (Atkinson, 1986). In Gurp, the limited influx of coarse clastic 'contaminants' frequently enabled palaeosols to reach mature stages of development (Bown & Kraus, 1987), whereas in Aren, the higher degree of clastic contamination at this proximal locality often arrested the palaeosol development. In this context, it is especially noteworthy that the most mature palaesols in Aren occur at a stratigraphic level almost totally devoid of coarse member deposits (top of megacycle 1, Figs 4 & 15). At first glance, the proposed relationship between distality and palaeosol maturity seems to be at odds with the sequence of lateral variations in palaeosol development given by Atkinson (1986). For a similar environmental setting, he suggested that the most mature palaeosols were developed in proximal areas, whereas the distal areas were characterized by a lower degree of soil maturity. However, Atkinson (1986) explained the basinward decrease in palaeosol maturity by postulating that coarse clastic material was more effectively distributed across the floodplain in the distal areas. He maintained that the main reasons for this was a downslope increase in channel density and discharge, and a corresponding downslope decrease in channel stability. In this study, however, a downslope *decrease* in channel density and discharge is postulated. Hence, on the Esplugafreda floodplain, the distribution of coarse clastic material is likely to have decreased in the distal direction. This trend is opposite to that of Atkinson (1986) and probably accounts for the difference between the downslope palaeosol maturation sequences in the two formations.

The downslope reduction in channel stability (entrenchment) proposed by Atkinson (1986) can be documented in the Esplugafreda Formation as well. This prevented widespread distribution of coarse clastic material on the proximal parts of the floodplain, but apparently was of less importance for the development of mature palaeosols in distal areas than the downslope decrease in channel density and discharge.

DEPOSITIONAL MODEL FOR THE ESPLUGAFREDA FORMATION

Tectonic influence on megacycle development

The spatial arrangement of the eight building blocks in the Esplugafreda Formation determines the architecture of this fluvial sequence. As described above, the studied sequence was divided into three megacycles (Figs 3 & 4) displaying specific vertical building block transitions. Moreover, the sedimentary features of the three megacycles change laterally from the proximal exposure in Aren to the more distal exposure in Gurp. In this section, this megacycle development will be related to variations in factors controlling the evolution of the fluvial system.

The described megacycle development, highlighted by vertical changes in sandbody geometry, grain-size and palaeosol maturity, may have been controlled by oscillations in climate, by eustatic sea-level changes, or by variations in tectonic activity. The latter explanation is preferred for the following reasons.
1 Tertiary deposits of the Spanish Pyrenees have been significantly influenced by compressional tectonics (e.g. Plaziat, 1981; Atkinson, 1986; Puigdefábregas & Souquet, 1986). A number of tectonosedimentary cycles have been defined (Puigdefábregas & Souquet, 1986), related to periods of active thrusting, thrust unloading and corresponding changes in base-level.
2 Soil-forming processes are sensitive to changes in climate, and palaeosols formed during periods of

alternating wet and dry conditions are likely to exhibit characteristics of both states of drainage (e.g. Lehman, 1989). If periodic changes in climate generated the observed megacycles in the Esplugafreda Formation, it is thus likely that this would have been manifested by vertical variations in palaeosol types. However, all palaeosols in the Esplugafreda Formation appear to have formed under a similar state of floodplain drainage (mainly Aridisols, see above).

3 All coarse member bodies in the Esplugafreda Formation are dominated by sedimentation units relatable to ephemeral flow conditions (see above). This suggests that the entire formation was deposited under similar (semiarid) climatic conditions.

4 The strong degree of entrenchment documented for many coarse member bodies (arroyos) indicate that the rivers had to readjust their equilibrium profiles. Such a readjustment may occur in response to increased palaeoslope gradients or a change in base-level. Both these factors might be attributable to tectonic activity (see below).

Four-stage tectonosedimentary model

Assuming the megacycles in the studied sequence were formed in response to variations in thrust-related tectonic activity, a four-stage model (Fig. 15) is proposed to account for the observed lateral and vertical changes in coarse member types, palaeosol maturity, grain-size and sand/shale ratio. In *stage A*, synorogenic deposition of coarse-grained material took place across the northern margin of the ancestral Tremp–Graus Basin. This margin was characterized by a high-gradient depositional slope due to (i) upwarping during thrust emplacement, and (ii) compensational subsidence which created a topographic low some distance basinward of the area of thrust loading. The high gradient, together with an initial (synorogenic) high rate of subsidence (McLean & Jerzykiewicz, 1978), promoted the formation of vertically accreted, narrow and strongly incised channels. Moreover, it caused initial bypassing of clastics to areas beyond the slope break. Eventually, this bypassing lowered the gradient; deposition was initiated in the channels, and sediments corresponding to building blocks C4 and C5 formed. The length of time in which channels at basin margins were entrenched depends partly upon the existence of a lag time between upwarping/thrust-load emplacement and the onset of *basin margin* subsidence. DeLuca & Eriksson (1989) demonstrate the effect of such a lag time on the fluvial architecture of the Upper Triassic Chinle Formation, whereas other authors maintain that subsidence commences simultaneously with thrust-load emplacement (i.e. negligible lag time, Heller *et al.*, 1988). The flexural rigidity of the lithosphere (e.g. Tankard, 1986) is likely to have a profound effect on the duration of the lag time (i.e. high lithospheric rigidity promotes long lag times). In the Esplugafreda Formation, the dominance of entrenched channels at the base of megacycles 1 and 2 might suggest that there was a considerable lag time between upwarping and basin margin subsidence. Stage A deposits are thought to be present in the lower parts of megacycles 1 and 2, and are interpreted as sediments formed in response to episodes of active thrusting (Heller *et al.*, 1988). In keeping with the model in Fig. 15, the proximal deposits in Aren are coarser grained, have a higher sand/shale ratio, and contain less mature palaeosols and more multistorey ribbon-bodies (building block C5) than the more distal accumulations in Gurp. Moreover, the coarse-member bodies in Aren have the highest degree of incision into their surrounding fine member deposits.

In *stage B*, active thrusting ceased, and the supply of coarse-grained sediment decreased as the hinterland relief was worn down. As argued by Heller *et al.* (1988), the subsidence in foreland basins diminishes exponentially away from the thrust-load, thus limiting the basinward dip of the depositional slope. In addition, the continued downslope accumulation of clastics has by now significantly reduced the palaeoslope gradient. As a result of these changes, coarse member deposits of a more sheet-like and less incised nature (mainly building blocks C2 and C3) are likely to have formed during this transitional phase. Some of these may be attached to ribbon-like arroyo-deposits upslope, as illustrated in Fig. 10. Deposits of this type are present in middle to upper parts of megacycles 1 and 2. Compared to Aren, these stratigraphic intervals in Gurp have more well-developed palaeosols, lower sand/shale ratios, and a higher ratio of sheetflood-dominated relative to ribbon-like coarse member bodies. This again indicates the more distal position of the Gurp exposure.

In *stage C* (Fig. 15), the depositional system entered the postorogenic phase, and the palaeoslope became negligible. Sediment supply virtually ceased, mainly due to the elimination of hinterland relief. This promoted the formation of mature

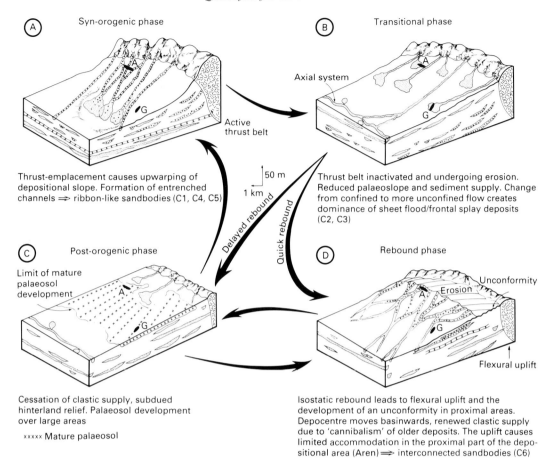

Fig. 15. Palaeogeographic evolution of parts of the northern margin of the Tremp-Graus Basin. Four tectonically related phases of deposition are invoked to explain the observed variations in sandbody types, palaeosol maturity, sand/shale ratio and grain-size. The black areas denoted A and G illustrate the positions of the Aren and Gurp exposures. See text for discussion.

palaeosol sequences (building block FM1) over considerable parts of the studied area. In Aren, stage C deposits are only present at the top of megacycle 1. In Gurp, however, stage C palaeosols occur interbedded with stage B deposits in the upper parts of megacycles 1 and 2, and they dominate in megacycle 3. In the idealized developmental sequence (Fig. 15), stage C may be succeeded by stage A, provided another episode of thrusting is initiated. This sequence of developments can be documented by the transition from stage C to stage A deposits from the top of megacycle 1 to the basal parts of megacycle 2. Stage C may also be succeeded by stage D (rebound phase), but this transition has not been documented in the study area.

Stage D, the rebound phase, was characterized by renewed coarse clastic sedimentation, but with the depocentre now occupying a more basinward position (Fig. 15). This phase corresponds to the postorogenic phase of Heller *et al.* (1988), and the renewed uplift is caused by isostatic rebound of the lithosphere as a response to removal of the tectonic load. This uplift was less 'dramatic' than the basin margin upwarping invoked in stage A, and is unlikely to have generated palaeoslope gradients of comparable magnitude. Moreover, accommodation will decrease towards the uplifted (source) area, culminating in an area of erosion in the most proximal parts, where the rate of uplift exceeds the rate of subsidence. Hence, the proximal areas of deposition will in this rebound phase have limited palaeoslopes, and a sediment supply which is likely to exceed the accommodation (Fig. 5). As a consequence, sheet-like, interconnected fluvial channel deposits of the braided stream type (building block C6) will tend to form proximally. This type of deposit occurs at the base of megacycle 3 in Aren. In Gurp, braided stream deposits are also present at this level, but the lower degree of interconnectedness between individual channel deposits suggests deposition in a more distal setting (with a lower ratio of sediment supply to accommodation). Stage D may either succeed stages C or B, depending on the time-span between thrust-belt inactivation and the onset of isostatic rebound. In the study area, only a transition from stage B deposits at the top of megacycle 2 to stage D deposits at the base of megacycle 3 has been observed. Hence, the time-span between thrust-belt inactivation and rebound was too short to enable the formation of mature palaeosols at this stratigraphic level. Alternatively, stage C deposits may have been eroded. The rebound phase is likely to be succeeded by stage C deposits (Fig. 15), as observed stratigraphically upwards in megacycle 3 in the studied area.

The proposed tectonosedimentary model involves alternations between deformation periods and tectonically quiescent periods in a compressional setting. These alternations were probably controlled by the structural evolution of *local* tectonic elements such as thrust faults and their lateral ramps. Hence, the Palaeocene tectonosedimentary model suggested for the studied area should not uncritically be applied to time-equivalent sequences elsewhere in the south Pyrenean foreland basin.

SUMMARY AND CONCLUSIONS

1 The Esplugafreda Formation of the Tremp-Graus Basin consists of *ephemeral stream* deposits formed in a semiarid climate. The streams occupied the northern margin of an incipient foreland basin characterized by a low average rate of sediment supply and limited accommodation.
2 There is considerable *variability* in the morphology of ephemeral stream sandbodies. Between five and ten different types of sandbodies (*building blocks*) may be generated in ephemeral stream environments, and failure to take this into account in reservoir modelling will lead to oversimplifications and an improper understanding of reservoir communication.
3 Three *megacycles* have been recognized in the studied sequence. These formed in response to local changes in palaeoslope gradient, sediment supply and subsidence pattern, possibly reflecting deposition during synorogenic, transitional, postorogenic and rebound phases in the studied parts of the south Pyrenean foreland basin (Fig. 15). Each phase generated sedimentary sequences with specific geometrical and lithological properties.
4 Apart from the braided stream deposits of building block C6, all coarse member bodies of the Esplugafreda Formation are characterized by vertical accretion and limited W/T ratios. This can be related to a number of factors, including the low ratio of sediment supply vs accommodation, the tendency for deep incision on a high-gradient palaeoslope, the ability of arroyos to dominate local drainage for prolonged periods, and presence of bank stabilizers like cohesive mud and vegetation.
5 *Palaeosols* dominate the fine member deposits of the Esplugafreda Formation. For the most part, the palaeosols reflect deposition in a semiarid climate (Aridisols with vertic features). The most mature palaeosols in the sequence were formed at sites which were removed from clastic contamination for prolonged periods of time. These deposits have considerable lateral extents, and thus serve as ideal marker horizons for correlations on the reservoir scale. Due to a downslope reduction in the supply of coarse clastic material, the frequency of mature palaeosols increase in the distal direction.
6 Multistorey ribbon-shaped ephemeral channels (building block C5) and sheetflood-dominated sandbodies (building blocks C2 and C3) feature prominently in the Esplugafreda Formation. An *inverse* relationship between the number of storeys and the W/T ratio in these building blocks has been demonstrated. As a consequence, the dimensions of multistorey sandbodies should not be calculated by adding the widths of the individual bodies. Instead, the more conservative W/T relationships of Fig. 9B–F should be applied.
7 The formation of *pause-planes* (with early diagenetic cement) is common in ephemeral systems, and contribute to reduce the net pay in deposits of this kind (flow-barriers). The pause-planes (Fig. 8) are well suited to distinguish between perennial and ephemeral fluvial sediments.

ACKNOWLEDGEMENTS

Thanks are due to H. Fjellbirkeland, J.L. Cuevas, L.

Mercade and R. Gundesø for assistance and cooperation in the field. I have also benefited from the collaboration with C. Puigdefábregas and M. Marzo, who have introduced me to the splendour and scientific challenges of the Tertiary deposits in the Spanish Pyrenees. Thorough reviews of earlier versions of this paper by A. Miall, R. Eschard and A. Ryseth are gratefully acknowledged.

REFERENCES

ALLEN, P.A., CABRERA, L., COLOMBO, F. & MATTER, A. (1983) Variations in fluvial style on the Eocene–Oligocene alluvial fan of the Scala Dei Group, SE Ebro Basin, Spain. *J. Geol. Soc. Lond.* **140**, 133–146.

ATKINSON, C.D. (1986) Tectonic control on alluvial sedimentation as revealed by an ancient catena in the Capella Formation (Eocene) of Northern Spain. In: *Palaeosols – Their Description and Intrepretation* (Ed. Wright, V.P.) pp. 139–179. Blackwell Scientific Publications, Oxford.

BOWN, T.M. & KRAUS, M.J. (1987) Integration of channel and floodplain suites, 1. Developmental sequences and lateral relationships of alluvial palaeosols. *J. Sedim. Petrol.* **57**, 587–601.

CAMARA, P. & KLIMOWITZ, J. (1985) Interpretacion geodinamica de la vertiente centro-occidental surpirenaica. *Est. Geol.* **41**, 391–404.

CUEVAS, J. L. (1989) *La Formacion Talarn: estudio estratigrafico y sedimentologico de las facies de un sistema aluvial en el transito Mesozoico–Cenozoico de la Conca de Trempt.* MSc thesis, University of Barcelona.

DELUCA, J.L. & ERIKSSON, K.A. (1989) Controls on synchronous ephemeral- and perennial-river sedimentation in the middle sandstone member of the Triassic Chinle Formation, northeastern New Mexico, USA. *Sedim. Geol.* **61**, 155–175.

DREYER, T. (1990) Sandbody dimensions and infill sequences of stable, humid climate delta plain channels. In: *North Sea Oil and Gas Reservoirs* (Eds Buller, A., Berg, E., Hjelmeland, O., Kleppe, J., Torsæter, O. & Aasen, J.O.) pp 331–357. Graham & Trotman, London.

FJELLBIRKELAND, H. (1990) *Pedogenese og tidlig diagenese under semi-aride forhold: Esplugafreda Formasjonen (Paleocen) i Tremp-Graus Bassenget, Spanske Pyreneer.* MSc thesis, University of Bergen.

FRIEND, P.F., SLATER, M.J. & WILLIAMS. R.C. (1979) Vertical and lateral building of river sandstone bodies, Ebro Basin, Spain. *J. Geol. Soc. Lond.* **106**, 36–46.

GALLOWAY, W.E. (1981) Depositional architecture of Cenozoic Gulf coastal plain fluvial systems. In: *Recent and Ancient Nonmarine Depositional Environments: Models for Exploration* (Eds Ethridge, F.G. & Flores, R.M.) pp. 127–155. Soc. Econ. Paleont. Miner., Tulsa, Spec. Publ. 31.

GLENNIE, K.W. (1972) Permian Rotliegendes of Northwest Europe interpreted in light of modern desert sedimentation studies. *Bull. Am. Assoc. Petrol. Geol.* **56**, 1048–1071.

GRAHAM, J.G. (1983) Analysis of the Upper Devonian Munster Basin, an example of a fluvial distributary system. In: *Modern and Ancient Fluvial Systems* (Eds Collinson, J.D. & Lewin, J.) pp. 473–483. Spec. Publs Int. Ass. Sediment. 6.

HELLER, P.L., ANGEVINE, C.L., WINSLOW, N.S. & PAOLA, C. (1988) Two-phase stratigraphic model of foreland-basin sequences. *Geology* **16**, 501–504.

HUBERT, J.F. & HYDE, M.G. (1982) Sheet-flow deposits of graded beds and mudstones on an alluvial sandflat–playa system: Upper Triassic Blomidon redbeds, St Mary's Bay, Nova Scotia. *Sedimentology* **29**, 457–474.

KLAPPA, C.F. (1978) Biolithogenesis of *Microcodium*: elucidation. *Sedimentology* **25**, 489–522.

LEHMAN, T.M. (1989) Upper Cretaceous (Maastrichtian) palaeosols in Trans-Pecos Texas. *Bull. Geol. Surv. Am.* **101**, 188–203.

MCKEE, E.D., CROSBY, E.J. & BERRYHILL, H.L., JR (1967) Flood deposits, Bijou Creek, Colorado, June 1965. *J. Sedim. Petrol.* **37**, 829–851.

MCLEAN, J.R. & JERZYKIEWICZ, T. (1978) Cyclicity, tectonics and coal: some aspects of fluvial sedimentology in the Brazeau-Paskapoo Formations, Coal-Valley area, Alberta, Canada. In: *Fluvial Sedimentology* (Ed. Miall, A.D.) pp. 441–468. Can. Soc. Petrol. Geol., Calgary, Memoir 5.

MABUTT, J.A. (1977) *Desert Landforms.* MIT Press, Cambridge, 340 pp.

MIALL, A.D. (1977) A review of the braided river depositional environment. *Earth Sci. Rev.* **13**, 1–62.

MIALL, A.D. (1985) Architectural element analysis: a new method of facies analysis applied to fluvial deposits. *Earth Sci. Rev.* **22**, 261–308.

NEMEC, W. & STEEL, R.J. (1984) Alluvial and coastal conglomerates: their significant features and some comments on gravelly mass-flow deposits. In: *Sedimentology of Gravels and Conglomerates* (Eds. Koster, E.H. & Steel, R.J.) pp. 1–31. Can. Soc. Petrol. Geol., Calgary, Memoir 10.

NYSTUEN, J.P., KNARUD, R., JORDE, K. & STANLEY, K.O. (1989) Correlations within the Triassic to Lower Jurassic sequence at the Snorre field, Northern North Sea area. In: *Correlation in Hydrocarbon Exploration* (Ed. Collinson, J.D.) pp. 273–289. Graham & Trotman, London.

PICARD, M.D. & HIGH, L.R., JR (1973) *Sedimentary Structures of Ephemeral Streams.* Developments in Sedimentology 17. Elsevier, Amsterdam, 223 pp.

PLAZIAT, J.C. (1981) Late Cretaceous to Late Eocene paleogeographic evolution of southwest Europe. *Palaeogeog. Palaeoclimatol. Palaeoecol.* **36**, 263–320.

POSAMENTIER, H. & VAIL, P. (1989) Eustatic controls on clastic deposition, 2. Sequence and system tract models. In: *Sea-level Changes – an Integrated Approach* (Eds Wilgus, C. et al.) pp. 125–154. Soc. Econ. Paleont. Miner., Tulsa, Spec. Publ. 42.

PUIGDEFÁBREGAS, C. & SOUQUET, P. (1986) Tectono-sedimentary cycles and depositional sequences of the Mesozoic and Tertiary from the Pyrenees. *Tectonophysics* **129**, 173–203.

SADLER, P.M. (1981) Sediment accumulation rates and the completeness of stratigraphic sections. *J. Geol.* **89**, 569–584.

SCHUMM, S.A. & HADLEY, R.F. (1957) Arroyos and the semiarid cycle of erosion. *Am. J. Sci.* **255**, 161–174.

STEAR, W.M. (1983) Morphological characteristics of ephemeral stream channel and overbank splay sandstone bodies in the Permian Lower Beaufort Group, Karoo Basin, South Africa. In: *Modern and Ancient Fluvial Systems* (Eds Collison, J.D. & Lewin, J.) pp. 405–420. Spec. Publs Int. Ass. Sediment. 6. Blackwell Scientific Publications, Oxford.

STEEL, R.J. (1974) New Red Sandstone floodplain and piedmont sedimentation in the Hebridean Province, Scotland. *J. Sedim. Petrol.* **44**, 336–357.

STEEL, R.J. & RYSETH, A. (1990) The Triassic–Early Jurassic succession in the northern North Sea: megasequence stratigraphy and intra-Triassic tectonics. In: *Tectonic Movements Responsible for Britain's Oil and Gas Reserves* (Ed. Hardman, R.F.P.) pp. 139–168. Geol. Soc., Lond., Spec. Publ. 55.

TALBOT, M.R. & WILLIAMS, M.A.J. (1979) Cyclic alluvial fan sedimentation on the flanks of fixed dunes, Janjari, Central Niger. *Catena* **6**, 43–62.

TANKARD, A.J. (1986) On the depositional response to thrusting and lithospheric flexure: examples from the Appalachian and Rocky Mountain basins. In: *Foreland Basins* (Eds Allen, P.A. & Homewood, P.) pp. 369–392. Spec. Publs Int. Ass. Sediment. 8. Blackwell Scientific Publications, Oxford.

TUNBRIDGE, I.P. (1981) Sandy high-energy flood sedimentation — some criteria for recognition, with an example from the Devonian of SW England. *Sedim. Geol.* **28**, 79–95.

TUNBRIDGE, I.P. (1984) Facies model for a sandy ephemeral stream and clay playa complex; the Middle Devonian Trentishoe Formation of North Devon, UK. *Sedimentology* **31**, 697–715.

TURNER, B.R. (1986) Tectonic and climatic controls on continental depositional facies in the Karoo Basin of Northern Natal, South Africa. *Sedim. Geol.* **46**, 231–257.

USDA (1975) *Soil Taxonomy.* US Dept of Agriculture Handbook No. 436, 754 pp.

WILLIAMS, G.E. (1971) Flood-deposits of the sand-bed ephemeral streams of Central Australia. *Sedimentology* **17**, 1–40.

WRIGHT, V.P. (1989) Paleosol recognition. In: *Paleosols in Siliciclastic Sequences.* Postgraduate Research Institute for Sedimentology Short Course Notes No. 001, Reading University.

Architecture of the Cañizar fluvial sheet sandstones, Early Triassic, Iberian Ranges, eastern Spain

J. LÓPEZ-GÓMEZ and A. ARCHE

Instituto de Geología Económica, CSIC-UCM,
Facultad de Geología, Universidad Complutense, 28040 Madrid, Spain

ABSTRACT

The Cañizar Sandstones (Early Triassic) of the SE Iberian Ranges display several distinct braided stream facies, including channels, bars and sandflats. Nine individual facies and three main facies associations have been defined and their three-dimensional architecture related to auto- and allocyclic processes.

The Cañizar Sandstones were deposited in a narrow semigraben or highly asymmetric graben (the Iberian Basin) trending NW–SE in what is now central Spain, by a fluvial system flowing SE towards the Tethys Sea.

Six major intervals separated by bounding surfaces have been recognized in the area, and are believed to be caused by spasmodic activity of the SW boundary fault system.

INTRODUCTION

The Iberian Ranges are part of the Alpine mountain belt of Spain, trending approximately NW–SE from central Spain to the Mediterranean coast. The Calatayud-Teruel tertiary basin separates two parallel linear segments called the Aragonian Branch (to the NE) and the Castilian Branch (to the SW), but near the Mediterranean they merge into a single, very complex fold belt. Both branches have a similar structure: a tightly folded Hercynian basement overlain by unmetamorphosed, moderately deformed sediments ranging in age from Early Permian to Miocene (Fig. 1).

This paper presents the results of a detailed sedimentological study of the Cañizar Sandstones Formation (a.C.Fm.; López, 1985), the upper part of the Buntsandstein facies in the SE Castilian Branch (Provinces of Cuenca and Valencia) (Fig. 1). In this area, the Triassic sediments rest unconformably on highly deformed Ordovician to Carboniferous rocks and show the classic Germanic trilogy of Buntsandstein, Muschelkalk and Keuper facies, although the central carbonates are divided in two parts by a wedge of gypsum and red marls.

The Buntsandstein of the SE Iberian Ranges consists of four continental red bed formations (Fig. 2), with thickness ranging from 0 to 760 m. The lowermost formation, the basal breccias (b.b) consists of quartzite and slate heterometric breccias deposited by debris flows as scree on the Palaeozoic basement. They have little lateral extension.

The Boniches Conglomerates Formation (c.B.) (Fig. 2) consists of conglomerates formed by subangular to subrounded pebbles with abundant sandy matrix and silica and clay cements; they were deposited as longitudinal and lateral bars in shallow braided fluvial channels with marked high and low-flow stages, forming a series of coalescent alluvial fans along the SW border of the basin.

The Alcotas Mudstones and Sandstones Formation (l.a.A.) consists of red mudstones with some carbonate pedogenic beds and many lenticular multilateral, multistorey sandstones and conglomerates, deposited as flood plain and channel-fill deposits in a fluvial system evolving from braided to high sinuosity channel pattern.

The Cañizar Sandstones, which form the focus of this paper, consist of pink to white, medium to fine sandstones with occasional conglomerate and mud-

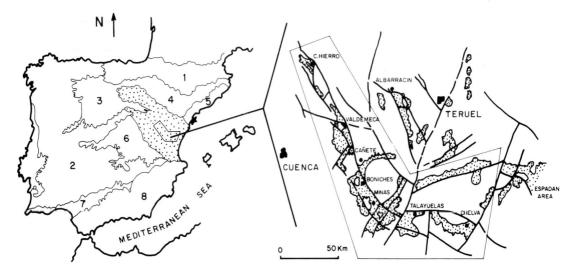

Fig. 1. Principal tectonic units of the Iberian Peninsula. Dotted area represents the Iberian Ranges. The research area is shown on the right. 1, Pyrenean Ranges; 2, Iberian Massif; 3, Duero Basin; 4, Ebro Basin; 5, Catalonian Ranges; 6, Tajo basin; 7, Guadalquivir Basin; 8, Betic Ranges.

stone beds, forming six multilateral, multistorey sandstone sheet complexes, the main architectural units which build up the sheet-like sandstone formation. Palaeocurrents point systematically towards the SE with small dispersion, and the sandstones have been interpreted as braided-stream deposits. Maximum thickness is 210 m. There is a clear decrease in grain size and an increase in the proportion of mudstone beds towards the Mediterranean.

The base of the Cañizar Sandstones is a sharp scoured surface, marking the beginning of a fluvial to shallow marine depositional sequence; it probably also represents a hiatus after the sedimentation of the previous Late Permian depositional sequence (Arche & López, 1989).

Continental sedimentation came to an end after the deposition of the Cañizar Sandstones in most of the SE Iberian Ranges, and there was a period of subaerial exposure and weathering. The region close to the Mediterranean coast (Provinces of Castellón and Valencia) (Fig. 1), however, continued subsiding and accumulating sediments.

The Tethys transgression over the eastern margin of the Iberian Ranges started with the siliciclastic Eslida Mudstones and Sandstones Formation (l.a.E.) in the SE region, followed by the shallow marine Muschelkalk carbonates that, after several alternatives covered the Ranges at the beginning of the late Triassic (Fig. 2). General references to the Permian and Triassic deposits of the Iberian Ranges are Ramos (1979), Sopeña (1979), López (1985), Pérez-Arlucea (in press) and Sopeña et al. (1988).

THE CAÑIZAR SANDSTONES: COMPOSITION AND AGE

The petrological composition of the Cañizar Sandstones falls into the arkose class of Folk (1968) although subarkoses and quartzarenites are also present, especially to the SE. The Cañizar Sandstones have a long and complex diagenetic history, with early clay cements followed by younger silica cements. Heavy minerals are of typical supermature assemblages, with tourmaline (40–90%), zircon (6–65%) and rutile (1–3%).

The first few metres of the formation consist usually of quartzite conglomerates. Pebble lags and stringers are frequent in the NW, but disappear near the Mediterranean coast, as well as most of the detrital feldspars and rock fragments.

Palynological assemblages found near the top of the formation have an Anisian age (Doubinger et al., 1990), although it is probable that the base is still Scythian (Lower Triassic) or even Thuringian.

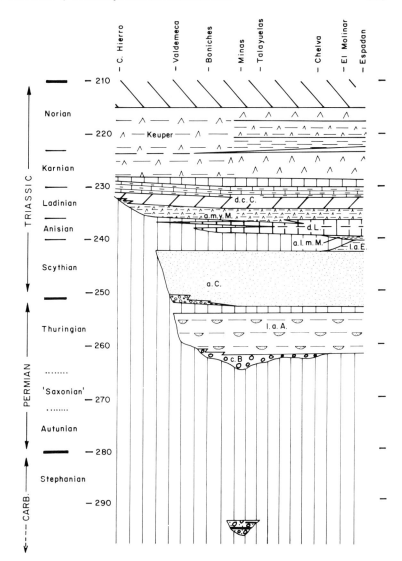

Fig. 2. Time and geographic range of the Late Permian–Late Triassic Formations in the research area. Vertical lines show periods of no sedimentation. On the upper part of the figure are located the names of the main localities (see also Fig. 1). Time scale on the left. Abbreviations: c.B., Boniches Conglomerates Fm.; l.a.A., Alcotas Mudstones and Sandstones Fm.; a.C., Cañizar Sandstones Fm. (focus of this paper); l.a.E., Eslida Mudstones and Sandstones Fm.; a.l.m.M., Marine Clays, Mudstones and Marls Fm.; d.L., Landete Dolomites Fm.; a.m.y.M., Mas Clays, Marls and Gypsum Fm.; d.c.C., Cañete Dolomites and Limestones Fm.

FLUVIAL FACIES ARCHITECTURE

Sixteen complete sections of the Cañizar Sandstones have been studied in detail in the type area of the formation (Fig. 2); seven Major Bounding Surfaces (MBS) (1 to 7 from base to top) have been identified, separating six multilateral, multistorey sandstone sheet complexes (m.m.s.s.c.) (Fig. 3); each one is a complex of multilateral fluvial channels limited by minor bounding surfaces. We use the term 'multilateral sand body' according to the original definition by Potter (1967): 'a sand body that consists of a complex of laterally coalescing deposition cycles or channel systems', and we have followed the nomenclature and analytical techniques of Bluck (1974, 1976, 1980), Leeder (1978), Allen (1978, 1983) and Miall (1983, 1985, 1988a,b).

The general aspect of these fluvial deposits is shown in detail by three profiles (I to III, Figs 4, 5 & 6) taken from the type area of the formation near Cañete (Cuenca) (Fig.1). They show the six sandstone sheet complexes (A to F, from base to top), the MBS, some of the minor bounding surfaces, the lenticular shape of the storeys and the two scales of channel-filling bodies: large channel systems 5.5–

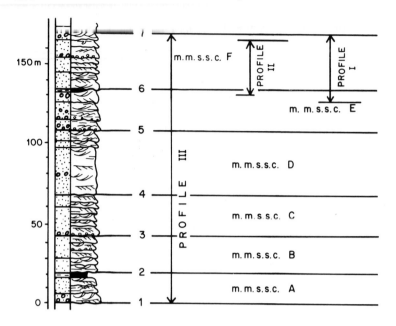

Fig. 3. General section for the research area. Six multilateral, multistorey sandstone sheet complexes (m.m.s.s.c.) are shown, separated by Major Bounding Surfaces (MBS) (from 1 to 7). Vertical arrows show the m.m.s.s.c. studied in each profile in Figs 4, 5 & 6.

11 m thick and more than 300 m long and individual simple channel deposits 1.5–4 m thick and 25–180 m long.

Profile I (Fig. 4), 12.2 km W of Cañete, is 38 m thick and 590 m long, striking E 40° S (almost parallel to the main palaeocurrents) and N 50° E and shows the upper part of the formation. Profile II (Fig. 5), 12.5 km W of Cañete, is 20 m thick and 140 m long, striking approximately perpendicular to the main palaeocurrents (N 20° E) and shows the upper part of the formation. Profile III (Fig. 6), 11.5 km W of Cañete is 140 m thick and 980 m long, striking N 10° E, almost perpendicular to the main palaeocurrents, and shows most of the formation, from the base almost to the top (see also Fig. 3). Heavy lines in Fig. 6 indicate the MBS.

Major bounding surfaces

A threefold hierarchy of bounding surfaces was envisaged by Allen (1983) for the fluvial facies, modified afterwards by Miall (1988a,b), among others. Miall expanded the original classification to a fourfold and later to a sixfold hierarchy.

The MBS (Fig. 6, heavy lines) belong to the third order surfaces of Allen (1983), the sixth order surfaces (in Fig. 5, 3rd, 4th, 5th and 6th order surfaces are also differentiated) of Miall (1988a,b), and the 'Major Surfaces' of Bridge & Diemer (1983). They delineate laterally extensive bodies up to 18 m thick and several kilometres long with a width/depth ratio in excess of 200, formed by a series of lenticular storeys (thin lines in Fig. 5) interpreted as individual channel fills and accretionary surfaces.

Palaeocurrents have been measured at 106 stations using channel bases, intersections of troughs, parting lineations and scour marks; some of the results are shown in Fig. 10, and will be discussed in the following section.

Analysis of sandstone bodies

Field observations have shown the different sandstone units to consist of nine elementary facies that might occur only once or be repeated vertically many times.

Elementary facies description

Facies a. Sandy conglomerates and pebbly sandstones with planar and trough cross-stratification (Fig. 7a)

Yellow to white sandy conglomerates and pebbly sandstones; sets 0.3–0.5 m high and up to 4 m wide

Architecture of Triassic fluvial sheet sandstones

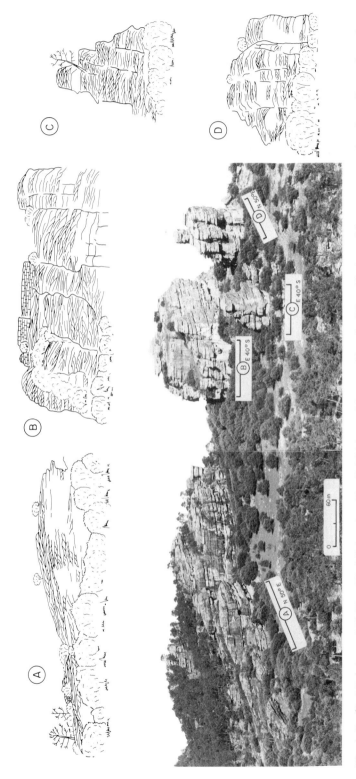

Fig. 4. Profile I. It is located 12.2 km W of Cañete (see Fig. 1). Thickness: 38 m; total length: 590 m. Horizontal lines represent minor bounding surfaces. See also Fig. 3 for location of this profile. There are two main orientations in the profile: N 50° E and E 40° S. The later is almost perpendicular to the main palaeocurrent directions.

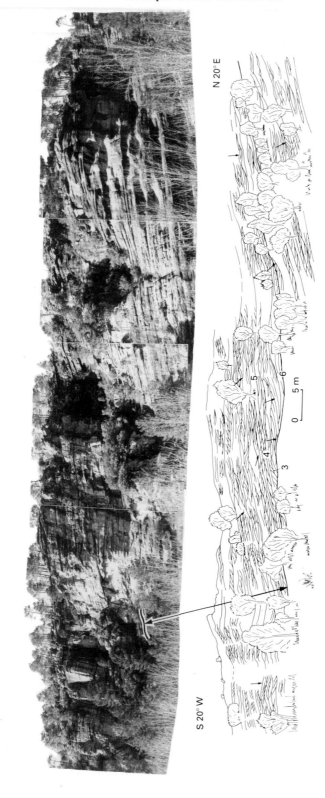

Fig. 5. Profile II. It is located 12.5 km W of Cañete (see Fig. 1). Thickness: 20 m; length: 140 m. Thin lines represent minor bounding surfaces (m.b.s.). The heavy line represents the Major Bounding Surface (MBS) number 6 which is separating multilateral and multistorey sandstone sheet complexes (m.m.s.s.c.) E and F. Arrows represent average palaeocurrents. This profile represents the upper part of m.m.s.s.c. D (see Fig. 3). Numbers in the centre of the sketch indicate the order of the bounding surface according to Miall's classification (1988a, b).

Architecture of Triassic fluvial sheet sandstones

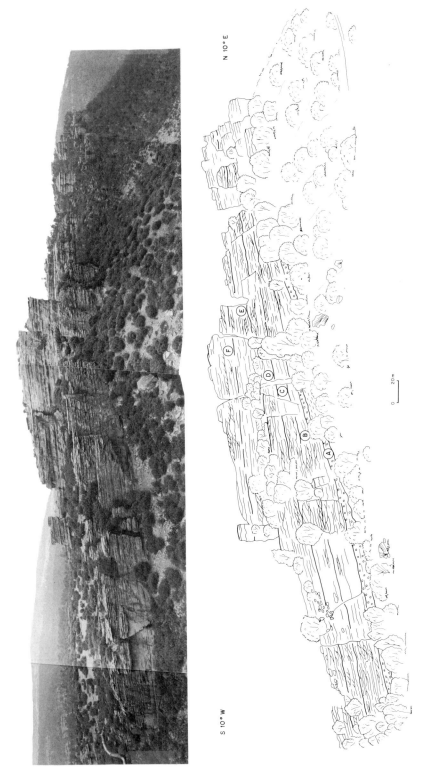

Fig. 6. Profile III. It is located 11.5 km W of Cañete (see Fig. 1). Thickness: 140 m; length: 980 m. Heavy lines represent Major Bounding Surfaces (MBS), thin lines represent minor bounding surfaces (m.b.s.). A–F: multilateral, multistorey sand sheet complexes (m.m.s.s.c.). The sketch is made almost perpendicular to the profile.

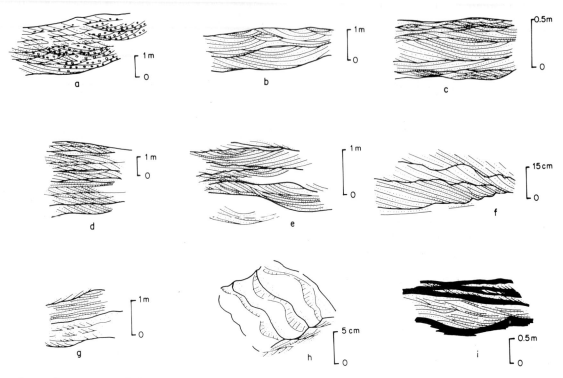

Fig. 7. Nine elementary facies differentiated in the a.C. Fm. See text for description and interpretation.

dipping up to 25°; some low-angle, tabular sets. Some reactivation surfaces; many cosets display fining-upwards trend, but not all. Thickness up to 2.4 m.

Facies b. Medium-scale trough cross-bedded sandstones (Fig. 7b)

Trough cross-bedded red sandstones, sets up to 1 m thick and 12 m wide; coarse to medium grain size, with isolated quartzite pebbles and clay chips at the base of the troughs. Sets regularly stacked in cosets bounded by scoured bases and tops. Thickness: 0.7–1.3 m

Facies c. Small-scale trough cross-bedded sandstones with rippled intervals (Fig. 7c)

Trough cross-bedded red sandstones, sets up to 0.65 m thick and 3.5 m wide; coarse to medium grain size; fining-upwards laminae very frequent. Sets regularly stacked in cosets capped by rippled intervals. Thickness: 0.5–0.85 m.

Facies d. Large-scale planar tabular cross-bedded sandstones (Fig. 7d)

Tabular cross-bedded red sandstones, sets up to 0.45 m thick, with a very flat, basal scoured surface. Cosets formed by vertical stacking of up to three sets, sometimes capped by a thin, parallel laminated interval at the top. Distinct fining-upwards trend from coarse to fine grain size. Occasional cut-and-fill structures at the top, with palaeocurrents almost perpendicular to the ones in the main sandstone body. Thickness: 0.6–1.4 m

Facies e. Large-scale planar cross-bedded sandstones with reactivation surfaces (Fig. 7e)

Tabular cross-bedded red sandstones, sets up to 0.8 m thick and 6 m wide, with irregular, scoured base. Cosets formed by vertical stacking of large sets passing into much smaller bedforms, even ripples, with many convex or flat reactivation surfaces. Distinct fining-upwards trend. Thickness: 0.1–0.3 m.

Facies f. small-scale planar tabular cross-bedded sandstones (Fig. 7f)

Planar tabular cross-bedded red sandstones, sets 0.1–0.35 m thick, and up to 6 m long. Cosets formed by vertical stacking of two to five sets with flat, scoured base; medium to fine grain size, fining-upwards trend. Thickness: 0.5–1.2 m

Facies g. Parallel laminated sandstones (Fig. 7g)

Parallel laminated, red sandstones; flat, scoured base or forming part of cut-and-fill structures; medium to fine grain size. Usually forming lenticular bodies up to 8 m wide. Thickness: 0.4–0.7 m.

Facies h. Current-rippled sandstones (Fig. 7h)

Yellow to red fine sandstones with millimetric current-ripples, sometimes with stoss and lee sides preserved; usually above trough cross-bedded sets, with transitional base and erosive top. Rare, probably due to low preservation potential. Thickness: up to 0.15 m.

Facies i. Massive mudstones and current-rippled sandstones (Fig.7i)

Massive red mudstones with isolated millimetric lenticular sandstones with current ripples; sometimes they form mud drapes along laminae of cross-bedded sets. Sharp, non-erosive base and erosive top. Unusual, probably due to low preservation potential. Thickness: up to 0.35 m.

Organic remains

There are numerous plant remains scattered all through the formation, although they are usually comminuted, poorly preserved fragments. They are preserved as silicified pieces of wood, carbonized fragments or ferruginous impressions. Pollen and spore assemblages have been recovered near the top of the formation (Doubinger *et al.*, 1990). There are also rootlet levels associated with Facies h and i.

All remains are of conifers and ferns, characteristic of a warm, seasonal climate with long dry periods.

Sandy bedforms and elementary facies interpretation

The sedimentological interpretation of the different facies and facies associations of the Cañizar Sand-stones point to a braided, bed-load fluvial system. The main geomorphic elements are *channels* (major and minor), *bars* (individual and composite), similar to the transverse bars of Ore (1963) and Smith (1970), and *sandflats*, sensu Cant & Walker (1978).

Bars migrate along the deepest parts of the channels and as they prograde downcurrent and aggrade vertically, they can reach the water surface. The emergent nucleus can grow, forming a sandflat; bars and sandflats can be deeply modified during falling and low-flow periods.

Bars, either individual or composite, cannot be considered as isolated bedforms, and should be related always to the main and inner channels. According to Allen (1983), 'sand bar' means an upstanding barrier of sediments emplaced across the current in a channel; the term can be used either as description or interpretation of a bedform.

The elementary facies are interpreted as follows.

Facies a

Usually related to Facies g, forming a–g associations. The lateral association of sandstones and conglomerates and the frequent coarsening-upwards facies associations are interpreted as the result of the migration of transverse bars with a coarse 'head' migrating over a finer 'tail' in low sinuosity shallow channels (Bluck, 1976; Ramos & Friend, 1982).

Internal scours are interpreted as bar modification during repeated high flow–low flow cycles; the pebbles may have been reworked from Late Permian and/or Late Carboniferous conglomerates found in the same basin (see Fig. 2) or from an area far to the NW, in the Iberian Massif.

Facies b

Usually related to Facies c, forming b–c associations. It is interpreted as the result of the progradation of large megaripples in the deepest parts of the channels during high flow periods. Similar to Facies A of Cant & Walker (1978).

Facies c

Usually related to Facies b, forming b–c associations and also to Facies e and f. It is interpreted as the result of the migration of small-scale bedforms on the lee side and the top of large dunes. As their orientation is not always parallel to the main sandstone body, they can be interpreted as falling-stage bedforms, when the crests of the dunes and the sandflats cause local diversions of the main flow (Collinson, 1970; Fernández & Dabrio, 1985).

Facies d

Usually related to Facies b, c, f and h forming b–d–f–h or b–d–c–h associations. It is interpreted as the result of migration of large transverse, straight crested bars, with lee faces at high angles with the main flow (Smith, 1970), cutting across sand flats towards the deepest parts of the adjacent channels. If they become inactive during aggradation, they grade vertically into Facies c (Cant & Walker, 1978).

Facies e

Usually related to Facies c and h, forming e–c–h associations; sometimes related to Facies b. It is interpreted as the result of migration of large bars with simultaneous longitudinal progradation and vertical aggradation and comparable to the DA element of Miall (1988a,b); longitudinal progradation takes the form of downcurrent descending tiers within large cosets and simultaneous vertical accretion is proved by some foresets grading upwards into horizontal topsets without any break (Haszeldine, 1983).

Reactivation surfaces are common and are related to the falling and low-flow stages, when extensive reworking of the high-flow bedforms take place (Jones, 1977; Jones & McCabe, 1980).

Facies f

Similar to the Facies d, but with smaller foresets. Usually associated with Facies d and h, forming d–f–h and b–d–f–h associations. Comparable to Facies D of Cant & Walker (1978). It is interpreted as the result of migration of straight-crested megaripples in shallow waters at the top of large sandflats or inside almost infilled channels (Harms *et al.*, 1975).

Facies g

Usually associated with the base of Facies b and a. Interpreted as upper flow regime deposits during rising-flow conditions (Simon *et al.*, 1965).

Facies h

Usually found at the top of most of the facies associations when fully preserved. Interpreted as low-regime ripples migrating over larger bedforms when a channel is almost totally infilled or during low-flow stages (Allen, 1968).

Facies i

Uncommon. Always at the top of most of the facies associations and capped by extensive scoured surfaces marking the re-establishment of the active channels. Interpreted as vertical accretion deposits in slough channels or overbank deposits in small, non-channelized parts of the alluvial plain (Smith, 1970; Cant,1978; Bluck, 1980).

Red mudstones can be found forming drapes along foresets of sandy bedforms; they are considered as sediment falling out of suspension during low-flow periods in the deepest parts of the channels.

Clay chips are locally abundant either as basal lags or along laminae of cross-bedded sets, and blocks up to 0.6 m long have been observed occasionally. The abundance of these clasts towards the SE indicates great discharge fluctuations and reworking of slough channel deposits and mudstone laminae.

FACIES ASSOCIATIONS AND THE EVOLUTION OF THE FLUVIAL STYLE

Each multilateral, multistorey, lenticular sheet sandstone complex results from the amalgamation of different channels and elementary and composite bedforms. Figure 8 shows an example of a detail from the D m.m.s.s.c. with some of its facies associations. In Fig. 9, the left part represents all the observed associations between facies, while in the right part the heavy lines indicate the more frequent associations between facies and the fine lines indicate the less frequent ones.

According to Friend (1983), there are two dominant overall geometries of sedimentary bodies in fluvial deposits: lenticular-bedded (multistorey) sheet sandstones, dominated by longitudinal progradation; and tabular bodies, dominated by lateral accretion. The Cañizar Sandstones are composed of lenticular sand bodies of different scales, represent-

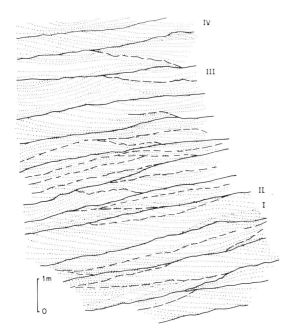

Fig. 8. Field sketch from part of the D m.m.s.s.c. Lines represent minor bounding surfaces (m.b.s.). I and IV represent deep channel fill by large transverse bars and II and III represent small-scale bars in the shallow troughs located in the upper part of those bars. Most of the megaforms represent transverse bars constituting sandflats. Palaeocurrents are almost perpendicular to the observer.

ing channels, bars and sandflats of a sandy braided fluvial system (Rust, 1978).

Facies associations

The Cañizar Sandstones cannot be described in terms of a single facies association, although elementary facies tend to be associated in a few characteristic successions (Figs 8 & 9). Facies b, d, e and f are by far the most abundant (trough cross-bedded, large and medium planar-tabular cross-bedded and planar cross-bedded sandstones). They are related to high energy flows and form the three main facies associations, although field sections usually show only truncated ones.

The b–c–(h) facies association (Fig. 9) is interpreted as the result of migration of sinuous transverse megaripples or simple bars in the deepest parts of the main channels; they can form the core of more complex composite bars and sandflats. It shows little or no vertical grading and small palaeocurrent dispersion, but there is a marked decrease in the size of sets. Where the association is fully preserved, current rippled sandstones (Facies h) represent the final stage of the infilling of a channel, but usually this part is missing due to erosion during later reactivations. The channel facies association can be up to 3.5 m thick and 60 m wide.

The (b)–d–f–h facies association (Figs 8 & 9) is interpreted as the result of the migration of straight-crested megaripples in half-infilled channels, forming sandflats. Lateral accretion was important in these bedforms, as indicated by reactivation surfaces, sometimes continued along the topset, passing laterally into small-scale, lenticular bodies with small-scale trough cross-stratification, interpreted as shallow inner or transverse chute channels crossing the main structure at wide angles, as the divergent palaeocurrents in the lenticular bodies and the main sandstone body indicate. Similar facies associations have been described by Cant & Walker (1978) in the South Saskatchewan River and by Ramos et al. (1986) and Pérez-Arlucea & Sopeña

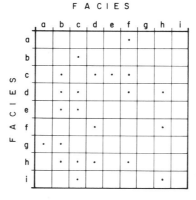
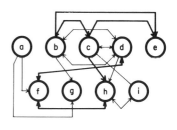

Fig. 9. The facies associations found in the a.C. Fm. are indicated in the left part of the figure. In the right part of the figure heavy lines indicate the most common facies associations, while thin lines indicate less common associations. See also Fig. 11 for the typical location of each facies and their associations.

(1986) in the Buntsandstein Facies of the NW Iberian Ranges. The facies association can be 3 m thick and 120 m long.

The b–e–c–h facies association (Figs 8 & 9) is interpreted as the result of migration of large straight-crested and sinuous megaripples forming complex, large bedforms comparable to the Platte River bars of Crowley (1983) or the 'composite bars' of Allen (1983) and comparable to the DA element of Miall (1988a,b). Foresets consist of fining-upwards laminae, characteristic of grain-flow processes. The small-scale planar cross-stratification represents minor bedforms migrating over the top sets of the sandflats. This facies association can be compared to the 'mixed facies' of Cant & Walker (1978) and can be 4.5 m thick and 150 m wide.

Palaeocurrents

Palaeocurrents are always unidirectional and point consistently to the SE and S (Fig. 10). The dispersion is always smaller than 90° and rarely exceeds 60°; the largest has been found between high-flow and low-flow structures at the top of the sandflat facies associations.

Drainage was parallel to the axis of the Iberian Basin, and there is no evidence of transverse streams feeding into the main fluvial system, although this possibility cannot be totally ruled out.

N–S palaeocurrents in an area east of Teruel (Fig. 1) reflect probably a clockwise rotation of a block during Alpine movements, caused by important dextral strike-slip faults, and not a substantial change of the drainage pattern.

Fluvial styles

The seven MBS divide the formation into six m.m.s.s.c. (A–F, Fig. 11) and can be associated with two main cycles, A to E and F. There is a clear fining-upwards tendency for each complex and for each major cycle.

The successive fluvial styles found in the Cañizar Sandstones are sketched in Fig. 12. MBS have a probable tectonic origin, as discussed in the next section, but normal channel shifting in a braided

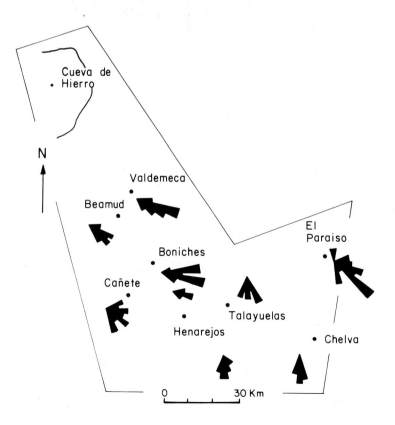

Fig. 10. Palaeocurrent diagrams for the principal zones of the research area. Each diagram shows a statistical result of the measure of the whole section. A vertical and statistical evolution of the palaeocurrents is also shown in Fig. 11. Deviations toward the S in Henarejos, Talayuelas and Chelva zones are conditioned by the Alpine tectonics.

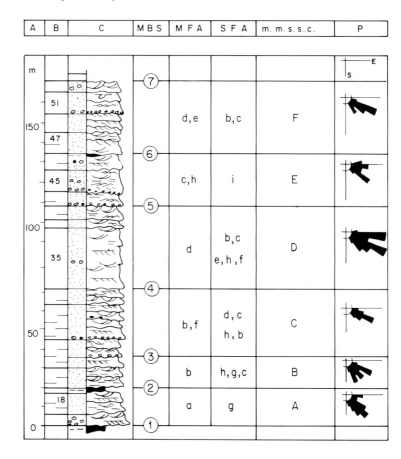

Fig. 11. Sketch of the general column (A, B and C) and the main distribution of the Major Bounding Surfaces (MBS), main facies associations (MFA), secondary facies assciations (SFA) and multilateral, multistorey sandstone sheet complexes (m.m.s.s.c.). P indicates the average of palaeocurrents in each m.m.s.s.c.

system and cycles of 'dry' and 'rainy' periods can explain the lenticular, multilateral structure of the Cañizar Sandstones.

The lower cycle (A–E) can be subdivided in two parts: A and B are dominated by channel-fill sequences and only occasional small sandflats, with frequent high-energy structures and small palaeocurrent dispersal, with channel aggradation resulting from sinuous megaripples migrating and growing along the deepest parts of the channels as the main process (Fig. 12A)

The upper part of the lower cycle (C, D and E, Fig. 12B) shows a progressive increase of complex, amalgamated forms as sandflats and composite bars, the result of many accretionary and erosive episodes. As the main channels are infilled, there is a vertical transition from large megaripples to smaller cross-bedded sets and current ripples. Lateral progradation of the sandflats is important in some cases, with reactivation surfaces at high angles with internal structures. Another important feature is the presence of shallow minor transverse channels (chute channels) cutting across the upper part of the sandflats that are infilled by sets of planar-tabular and small-scale trough cross-stratification. There are a few slough channels with vertically aggraded silts and clays.

The upper cycle (F, Fig. 11) starts with a sudden increase of energy at the base of complex F, reflected in the development of extensive sandflats and composite bars with fining-upwards sequences and simultaneous longitudinal progradation and vertical accretion; these sequences fill wide, shallow channels with a width/depth ratio in excess of 150. There are very few traces of slough channels. Comparable facies were described by Campbell (1976).

Reactivation surfaces are common in the Cañizar Sandstones; they can be convex surfaces with no or few associated minor structures or more complex, irregular surfaces covered by small-scale trough cross-stratified sandstones. According to Collinson (1970), Jones (1977) and Blodget & Stanley (1980),

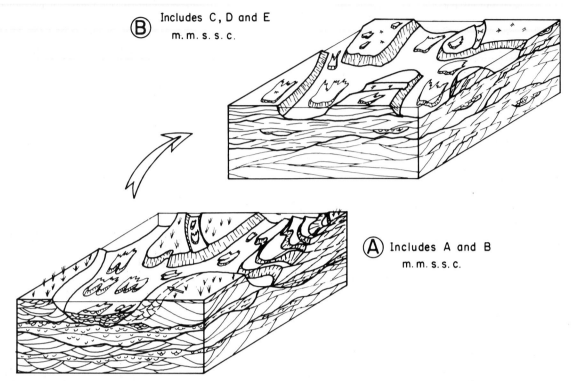

Fig. 12. Sketch of the main fluvial styles found in the a.C. Fm. The lower part (A), involving A and B m.m.s.s.c. (see also Fig. 11), represents channel-fill sequences with occasional small sandflats, low palaeocurrent dispersion and sinuous megaripples migrating and growing along the deepest part of the channels. The upper part (B) involves C, D and E m.m.s.s.c. and shows a progressive increase of complex, amalgamated forms such as sandflats resulting from many accretionary and erosive episodes. F m.m.s.s.c. represents a sudden renewal of high-energy conditions.

they are formed by extensive reworking of sandflats and bars during flood to low-flow transitions (Haszeldine, 1983).

Very similar facies associations with fining-upwards grading can be formed sometimes by low- and high-sinuosity rivers, as discussed by Moody-Stewart (1966), Miall (1977) and Puigdefábregas & Van Vliet (1978), especially when substantial lateral accretion takes place. However, the three-dimensional lenticular geometry, narrow palaeocurrent dispersion, very high ratio of channel sandstone to muds and clays (>95%), high width/depth ratio of the channels (well above 20 and probably up to 25 in some cases), total absence of point bars and evidence for simultaneous longitudinal progradation and vertical accretion, point to an overall pattern of braided channels and sandflats for the A, B and F complexes, comparable to some modern rivers like the South Saskatchewan (Cant & Walker, 1978) and Platte (Crowley, 1983), evolving to smaller systems like the Cimarron River (Shelton & Noble, 1974) and the Durance and Ardeche rivers (Doeglas, 1962). Another possible analogue is the South Saskatchewan model of Miall (1978). The equivalent facies of the Cañizar Sandstones (Rillo de Gallo Sandstones) to the NW of the Iberian Ranges were interpreted broadly in the same way by Ramos *et al.* (1986) and Pérez-Arlucea & Sopeña (1986).

There is also evidence of a NW–SE proximal–distal evolution of the fluvial system; there is a general decrease in the total thickness of the facies associations to the SE, and mud drapes along the foresets of the megaripples become frequent, especially in the middle part of the formation. The conglomerates at the base pinch out towards the SE, between Henarejos and Chelva (Fig. 10).

There is also a petrological evolution, from arkosic sandstone to more mature quartzarenites, as the unstable minerals are abraded during transport (López, 1985).

TECTONIC AND CLIMATIC CONTROLS

If a cycle may be defined as a particular sedimentary motif and cyclicity as its vertical repetition, we must conclude that there is no obvious cyclicity in the Cañizar Sandstones; however, there is a distinct pattern in the vertical evolution of the formation.

It is widely accepted nowadays that a combination of autocyclic processes (active channel migration, etc.) and allocyclic processes (local tectonics, changes in sea level, precipitation, etc.) can explain the accumulation of fluvial sediments (Beerbower, 1964). There is evidence of both types of controls in the Cañizar Sandstones.

Climatic control

The presence of numerous reactivation surfaces can be easily explained by alternating dry and rainy periods with large modifications of the stream discharge and modifications of the high-flow structures during the low-flow periods.

Palynological studies show an evolution towards warmer and drier environmental conditions from Late Permian to Late Triassic in the Iberian Ranges (Doubinger *et al.*, 1990). During the Early Triassic there was enough precipitation for conifers and ferns to thrive on the banks of the streams, but there are many forms characteristic of climates with long dry seasons. A combination of upland and lowland floras could explain the coexistence of these contrasting environments.

Tectonic control

The Cañizar Sandstones were deposited in a narrow, extensional basin, the Iberian Basin, during the final stages of a period of rift phasetectonic subsidence (Sopeña *et al.*, 1988) or, as has been suggested by Arche & López (1989), at the beginning of the thermal (flexural phase) subsidence period.

The underlying sedimentary sequence (Boniches Conglomerates and Alcotas Mudstones and Sandstones) was deposited in endoreic semi grabens, with predominant lateral drainage, during the initial stages of the tectonic subsidence. There were at least two isolated basins separated by the Cueva de Hierro-Tramacastilla High (López, 1985; Pérez-Arlucea & Sopeña, 1986) (Fig. 13).

A general tilt towards the SE caused a radical change in the drainage pattern to a longitudinal model, with a single fluvial system flowing towards the SE and the Tethys, similar to Model F of Miall (1981).

A period of non-sedimentation and erosion followed the end of the first megasequence and predated the beginning of the sedimentation of the Cañizar Sandstones; this surface represents an un-

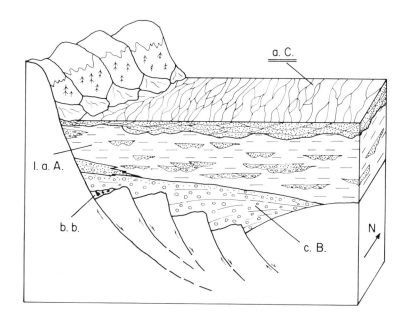

Fig. 13. Block-diagram showing a hypothetical evolution of the Buntsandstein facies SE of the Iberian Ranges: b.b., c.B. and l.a.A. Formations represent the first macrosequence; a.C. Formation the beginning of the second macrosequence (see also Fig. 2). The lower one was deposited in endorcic semigrabens with predominant lateral drainage while the second one changed the general drainage pattern to a longitudinal model with a single fluvial system flowing towards the SE. A change in the pattern of subsidence occurred between both macrosequences.

conformity (Pérez-Arlucea, in press; Pérez-Arlucea & Sopeña, 1986) to the NW, but is only a paraconformity in the SE Iberian Ranges, marked by soil profiles, bleached levels and mineralized horizons (Co, Ba), proof of long subaerial exposure prior to the sedimentation of the Cañizar Sandstones.

The MBS (comparable to third order surfaces of Allen, 1983, and sixth order surfaces of Miall, 1988a,b) can be traced laterally for more than 60 km and can probably be correlated over most of the SE Iberian Ranges. The different A to F complexes show changes in internal structures and slightly different palaeocurrents.

The MBS are related to the activity of NW–SE graben boundary faults; the streams flowed parallel to the main axis of the basin and it is very doubtful that the active channel belt occupied permanently the whole width of the basin floor during the deposition of the Cañizar Sandstones (60–80 km).

Isopachs show a maximum near the SW border of the basin (Fig. 14), thinning towards the NE, although few reliable data are available for this part of the basin. The overall geometry is a semigraben or highly asymmetric graben, and interplay between uplift of the footwall, subsidence of the hanging wall and rate of sedimentation can explain the formation of laterally extensive MBS (Fig. 15).

The activity of the faults was probably spasmodic; during quiescent periods the active channel belt migrated laterally, sweeping the whole basin (Bluck, 1974, 1980; Miall, 1977). Renewed fault activity would depress the hanging wall, especially near the fault plain, tilting the surface of the alluvial plain and causing a sudden switch to the SW of the active channel belt. Another period of slow lateral migration would follow (Fig. 15); the scoured contact between successive cycles will be one of the MBS and the successive channel fillings step up laterally, younging towards the NE, although there is no palaeontological control up to now to prove or dismiss this hypothesis. Some reactions of fluvial systems for this type of tectonic activity have been discussed by Alexander & Leeder (1987), and Marzo et al. (1988) show the migration of active fluvial belts under compressional regimes.

The overall fining-upwards motif of each cycle

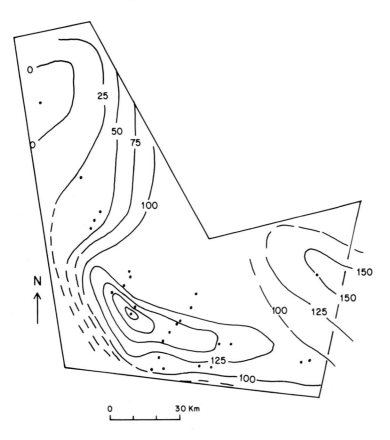

Fig. 14. Isopachs for the a.C. Formation. A maximum is clearly seen near the SE border of the basin. For locations see also Fig. 10.

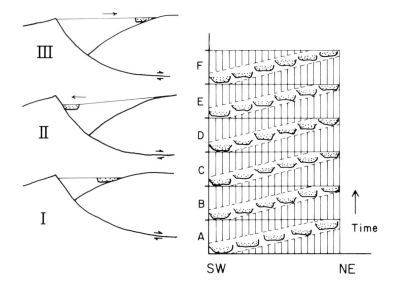

Fig. 15. Hypothetical explanation of the lateral formation of the Major Bounding Surfaces (MBS) as the result of interplay between uplift of the footwall, spasmodic subsidence of the hanging wall and migration of the active alluvial belt. A–F represent the seven m.m.s.s.c. See text for details.

can be explained by the slow decrease of gradient and transport power due to net accumulation of sediments during the quiescent periods.

Vertical aggradation was very slow during the sedimentation of the Cañizar Sandstones due to reduced subsidence (<15 m per m.y.). The fluvial system had time to adjust its longitudinal profile almost to total equilibrium, so transport was the dominant process. Sedimentation was probably related to important flood periods, but most of the deposits were afterwards reworked; the muds never had a chance to be preserved and the facies associations are always incomplete and amalgamated. The Cañizar Sandstones are a good example of sedimentation related to slow subsidence.

SUMMARY AND CONCLUSIONS

The facies analysis, palaeocurrent pattern and organic remains of the Cañizar Sandstones point to a braided river fluvial system with dominant bedload transport. They are divided into six multilateral, multistorey sandstone sheet complexes with large-scale lenticular internal structures separated by seven MBS. They were deposited in a graben or semigraben limited by NW–SE normal lystric faults.

Palaeocurrents point consistently to the SE and S, indicating a longitudinal drainage of the basin with only minor lateral sources.

The three main facies associations are interpreted as main channel transverse bars, composite bars and sandflat complexes, sometimes with lateral accretion. Internal structure is dominated by sheet-like units, several kilometres wide.

The presence of numerous reactivation surfaces suggests a climate with 'dry' and 'rainy' seasons and large discharge fluctuations and reworking of the original bars and sandflats.

MBS and large-scale cyclicity are related to the spasmodic activity of the SW graben boundary fault system and subsequent migration of the active fluvial plain.

A proximal–distal evolution of the fluvial system is marked by petrological changes of arkoses to quartzarenites, decrease of the total thickness of the facies associations and increased amount of clay drapes, clay chips and slough channel deposits. The basal conglomerates pinch out towards the SE.

Sedimentation took place during a period of slow subsidence, and the sediments were reworked extensively by the migrating channels; muds were not preserved and the remains of the channel facies association were amalgamated.

ACKNOWLEDGEMENTS

The research of the authors was financed by the CICYT Research Grants 452 and PB 0322. This work has benefited from discussions and comments in the field with Graham Evans. We thank the reviewers A.D. Miall and R. Knox for improving this paper with their comments. Thanks to Beatriz Bartolomé for typing the manuscript.

REFERENCES

ALEXANDER, J. & LEEDER, M.R. (1987) Active tectonic control on alluvial architecture. In: *Recent Developments in Fluvial Sedimentology* (Eds Ethridge, F.G., Flores, R.M & Harvey, M.D.) pp. 243–252. Soc. Econ. Paleont. Miner., Tulsa, Spec. Publ. 39.

ALLEN, J.R.L. (1968) *Current Ripples, their Relation to Patterns of Water and Sediment Motion.* North Holland, Amsterdam, 433 pp.

ALLEN, J.R.L. (1978) Studies in fluvial sedimentation: an exploratory quantitative model for the architecture of avulsion-controlled alluvial suites. *Sedim. Geol.* **21**, 129–147.

ALLEN, J.R.L. (1983) Studies in fluvial sedimentation: bar complexes and sandstone sheets (low-sinuosity braided streams) in the Brownstones (L. Devonian), Welsh Borders. *Sedim. Geol.* **33**, 237–293.

ARCHE, A. & LÓPEZ, J. (1989) *Fluvial Sedimentation during the Early Phase of Rifting in the Southeastern Iberian Ranges.* 4th Int. Conf. on Fluvial Sedimentology Field Trip Guide, 88 pp. Servei Geologic, Barcelona.

BEERBOWER, J.R. (1964) Cyclothems and cyclic depositional mechanisms in alluvial plain sedimentation. *Kansas Geol. Surv. Bull.* **169**, 31–42.

BLODGETT, R.H. & STANLEY, K.O. (1980) Stratification bedforms and discharge relations of the Platte braided system, Nebraska. *J. Sedim. Petrol.* **50**, 139–148.

BLUCK, B.J. (1974) Structure and directional properties of some valley Sandur deposits in Southern Iceland. *Sedimentology* **21**, 533–544.

BLUCK, B.J. (1976) Sedimentation in some Scottish rivers of low sinuosity. *Trans. R. Soc. Edinburgh Earth Sci.* **69**, 425–456.

BLUCK, B.J. (1980) Structure, generation and preservation of upward fining braided stream cycles in the Old Red Sandstone of Scotland. *Trans. R. Soc. Edinburgh Earth Sci.* **71**, 29–46.

BRIDGE, J.S. & DIEMER, J.A. (1983) Quantitative interpretation of an evolving ancient river system. *Sedimentology* **30**, 599–623.

CAMPBELL, C.V. (1976) Reservoir geometry of a fluvial sheet sandstone. *Bull. Am. Assoc. Petrol. Geol.* **60**, 1009–1020.

CANT, D.J. (1978) Development of a facies model for sandy braided river sedimentation: comparison of the South Saskatchewan River and the Battery Point Formation. In: *Fluvial Sedimentology* (Ed. Miall, A.D.) pp. 627–639. Can. Soc. Petrol. Geol., Calgary, Memoir 5.

CANT, D.J. & WALKER, R.G. (1978) Fluvial processes and facies sequences in the sandy braided South Saskatchewan River, Canada. *Sedimentology* **25**, 625–648.

COLLINSON, J.D. (1970) Bedforms of the Tana River, Norway. *Geog. Ann.* **52A**, 31–56.

CROWLEY, K.D. (1983) Large-scale bed configurations (macroforms), Platte River Basin, Colorado and Nebraska: primary structures and formative processes. *Bull. Geol. Soc. Am.* **94**, 117–133.

DOEGLAS, J. (1962) The structure of sedimentary deposits of braided rivers. *Sedimentology* **1**, 167–190.

DOUBINGER, J., LÓPEZ, J. & ARCHE, A. (1990) Pollen and spores from the Permian and Triassic sediments of the SE Iberian Ranges, Cueva de Hierro (Cuenca) to Chelva-Manzanera (Valencia-Teruel) Region, Spain. *Rev. Paleobotany Palynology* **66**, 25–45.

FERNANDEZ, J. & DABRIO, C.J. (1985) Fluvial architecture of the Buntsandstein facies red beds in the Middle to Upper Triassic (Ladinian–Norian) of the southeastern edge of the Iberian Meseta (Southeastern Spain). In: *Aspects of Fluvial Sedimentation in the Lower Triassic Buntsandstein of Europe* (Ed. Mader, D.), pp. 411–435. Lecture Notes in Earth Sciences 4. Springer-Verlag, Berlin.

FOLK, R.L. (1968) *Petrology of Sedimentary Rocks.* Hemphill's Bookstore, Austin, 170 pp.

FRIEND, P.F. (1983) Towards the field classification of alluvial architecture or sequence. In: *Modern and Ancient Fluvial Systems* (Eds Collinson, J.D. & Lewin, J.) pp. 345–354. Spec. Publs Int. Ass. Sediment. 6. Blackwell Scientific Publications, Oxford, 575 pp.

HARMS, J.C., SOUTHARD, J.B., SPEARING, D.R. & WALKER, R.G. (EDS) (1975) *Depositional Environments as Interpreted from Primary Sedimentary Structures and Stratification Sequences.* Soc. Econ. Paleont. Miner., Tulsa, Short Course Notes 2, Dallas, 161 pp.

HASZELDINE, R.S. (1983) Descending tabular cross-bed sets and bounding surfaces from a fluvial channel in the Upper Carboniferous coalfield of Northeast England. In: *Modern and Ancient Fluvial Systems* (Eds Collinson, J.D. & Lewin, J.) pp. 449–456. Spec. Publs Int. Ass. Sediment. 6. Blackwell Scientific Publications, Oxford, 575 pp.

JONES, C.M. (1977) Effects of varying discharge regimes on bedform sedimentary structures in modern rivers. *Geology* **5**, 567–570.

JONES, C.M. & MCCABE, P.J. (1980) Erosion surfaces within giant fluvial cross-beds of the Carboniferous in Northern England. *J. Sedim. Petrol.* **50**, 613–620.

LEEDER, M.R. (1978) A quantitative stratigraphic model for alluvium with special reference to channel deposits' density and interconnectedness. In: *Fluvial Sedimentology* (Ed. Miall, A.D.) pp. 587–596. Can. Soc. Petrol. Geol., Calgary, Memoir 5.

LÓPEZ, J. (1985) *Sedimentología y Estratigrafía de los materiales pérmicos y triásicos del sector SE de la Rama Castellana de la Cordillera Ibérica entre Cueva de Hierro y Chelva (Provincias de Valencia y Cuenca)* Seminarios de Estratigrafía 11. U. Complutense, Madrid.

MARZO, M., NIJMAN, W. & PUIGDEFABREGAS, C. (1988) Architecture of the Castissent fluvial sheet sandstones, Eocene, South Pyrenees, Spain. *Sedimentology* **35**, 719–738.

MIALL, A.D. (1977) A review of the braided river environment. *Earth Sci. Rev.* **13**, 1–62.

MIALL, A.D. (1978) Lithofacies types and vertical profile models in braided river deposits: a review. In: *Fluvial Sedimentology* (Ed. Miall, A.D.) pp. 597–604. Can. Soc. Petrol. Geol., Calgary, Memoir 5.

MIALL, A.D. (1981) Alluvial sedimentary basins: tectonic settings and basin architecture. *Geol. Soc. Can. Spec. Paper* **23**, 1–33.

MIALL, A.D. (1983) Basin analysis of fluvial sediments. In: *Modern and Ancient Fluvial Systems* (Eds Collinson, J.D. & Lewin, J.) pp. 279–286. Spec. Publs Int. Ass. Sediment. 6. Blackwell Scientific Publications, Oxford, 575 pp.

MIALL, A.D. (1985) Architectural-element analysis: a new method of facies analysis applied to fluvial deposits. *Earth Sci. Rev.* **22**, 261–308.

MIALL, A.D. (1988a) Facies architecture in clastic sedimentary basins. In: *New Perspectives in Basin Analysis* (Eds Kleinspehn, K.L. & Paola, C.) pp. 67–82. Springer-Verlag, Berlin, 453 pp.

MIALL, A.D. (1988b) Reservoir heterogeneities in fluvial sandstones: lessons from outcrop studies. *Bull. Am. Assoc. Petrol. Geol.* **72**, 682–697.

MOODY-STUART, M. (1966) High- and low-sinuosity stream deposits with examples from the Devonian of Spitsbergen. *J. Sedim. Petrol.* **36**, 1102–1113.

ORE, H.T. (1963) Characteristics of deposits of rapidly aggrading streams. *Wyo. Geol. Soc. Guidebook, Guillete*, pp. 195–201.

PEREZ-ARLUCEA, M. (in press) *Estratigrafía y Sedimentología del Pérmico y Triásico en el sector Molina de Aragón-Albarracín (Provincias de Guadalajara y Teruel).* Seminarios de Estratigrafía 13. U. Complutense, Madrid.

PEREZ-ARLUCEA, M. & SOPEÑA, A. (1986) Estudio sedimentológico del Saxoniense y del Buntsandstein entre Molina de Aragón y Albarracín (Cordillera Ibérica). *Cuad. Geol. Ibérica* **10**, 117–150.

POTTER, P.E. (1967) Sand bodies and sedimentary environments. *Bull. Am. Assoc. Petrol. Geol.* **51**, 337–365.

PUIGDEFÁBREGAS, C. & VAN VLIET, A. (1978) Meandering stream deposits from the Tertiary of the Southern Pyrenees. In: *Fluvial Sedimentology* (Ed. Miall, A.D.) pp. 469–486. Can. Soc. Petrol. Geol., Calgary, Memoir 5.

RAMOS, A. (1979) *Estratigrafía y paleogeografía del Pérmico y Triásico al Oeste de Molina de Aragón.* Seminarios de Estratigrafía 6, U. Complutense, Madrid, 313 pp.

RAMOS, A. & FRIEND, P.F. (1982) Upper Old Red Sandstone sedimentation near the unconformity at Arbroath. *Scot. J. Geol.* **18**, 297–315.

RAMOS, A., SOPEÑA, A. & PEREZ-ARLUCEA, M. (1986) Evolution of the Buntsandstein fluvial sedimentation in the North-West Iberian Ranges, Central Spain. *J. Sedim. Petrol.* **56**, 862–875.

RUST, B.R. (1978) Depositional models for braided alluvium. In: *Fluvial Sedimentology* (Ed. Miall, A.D.) pp. 605–625. Can. Soc. Petrol. Geol., Calgary, Memoir 5.

SHELTON, J.W. & NOBLE, R.L. (1974) Depositional features of a braided meandering stream. *Bull. Am. Assoc. Petrol. Geol.* **58**, 742–752.

SIMONS, D.B., RICHARDSON, E.V. & NORTON, C.F. (1965) Sedimentary structures formed by flow in alluvial channels. In: *Primary Sedimentary Structures and Their Hydrodynamic Interpretation* (Ed. Middleton, G.V.) pp. 34–52. Soc. Econ. Paleont. Miner., Tulsa, Spec. Publ. 12.

SMITH, N.D. (1970) The braided stream depositional environment: comparison of the Platte River with Silurian clastic Rocks, North-central Appalachians. *Bull. Geol. Soc. Am.* **81**, 2993–3014.

SOPEÑA, A. (1979) *Estratigrafía del Pérmico y Triásico del Noroeste de la provincia de Guadalajara.* Seminarios de Estratigrafía 5, U. Complutense, Madrid, 329 pp.

SOPEÑA, A., LOPEZ, J., ARCHE, A., PEREZ-ARLUCEA, M., RAMOS, A., VIRGILI, C. & HERNANDO, S. (1988) Permian and Triassic rifts of the Iberian Peninsula. In: *Triassic-Jurassic Rifting* (Ed. Manspeizer, W.) pp. 757–785. Elsevier, Amsterdam, 998 pp.

Effects of relative sea-level changes and local tectonics on a Lower Cretaceous fluvial to transitional marine sequence, Bighorn Basin, Wyoming, USA

E.P. KVALE* and C.F. VONDRA†

*Indiana Geological Survey, 611 N. Walnut Grove, Bloomington, IN 47405, USA; and
†Department of Geological and Atmospheric Sciences, Iowa State University, Ames, IA 50011, USA

ABSTRACT

The upper part of the Himes Member ('Greybull interval') of the Lower Cretaceous Cloverly Formation, in the Bighorn Basin of northern Wyoming is non-marine to transitional marine in origin and formed during an Early Cretaceous marine transgression. Three distinct channel-fill types are recognized in the upper Himes sequence and can be interpreted to represent deposition in: (i) a large, fluvially dominated straight channel (upper estuary), (ii) a mud-dominated mixed fluvial–tidal channel (middle estuary), and (iii) a small, high sinuosity fluvial channel (small tributary). The dominant sediment source for sands in the upper Himes was to the east, possibly the craton. The Sykes Mountain Formation conformably overlies the upper Himes and records a change from transitional marine to fully marine conditions upwards. In the lower part, the Sykes Mountain Formation is characterized by sheet sandstones, interpreted to be storm deposits, that are interbedded with mudstones. Locally, the sheet sandstones are cut by shoreward-directed (easterly directed) tidal channel bodies deposited in a lower estuary. The 'Greybull interval' overlies a regional unconformity that formed during an Early Cretaceous (late Aptian to early Albian) relative sea-level drop. Relief on the unconformity is as much as 23 m. Sea-level rise initiated the formation of estuaries with local tectonic effects influencing the placement of some of these.

INTRODUCTION

Although the effects of base level changes on coastal and marine sequences are well documented (e.g. Haq et al., 1987), such effects on non-marine sedimentation are poorly understood. Aubrey (1989) showed that eustasy can play an important role in the development of non-marine unconformities by demonstrating that a late Albian (98 Ma) type 1 sea-level drop caused the development of a sub-Dakota Sandstone unconformity in the southeastern part of the Colorado Plateau. A subsequent sea-level rise resulted in the aggradation of Dakota channels that back-filled valleys with alluvial sediment. This paper will discuss the nature of a non-marine to transitional marine depositional sequence in the Bighorn Basin of northern Wyoming (Fig. 1) associated with another Early Cretaceous (late Aptian to early Albian) relative sea-level drop and subsequent rise. This sequence comprises the upper part of the Himes Member of the Cloverly Formation (Fig. 2) (herein informally referred to as the 'Greybull interval' to conform with common usage) and the lower part of the overlying Sykes Mountain Formation. The Greybull interval illustrates the local complexity of transgressive aggradational sequences in non-marine to transitional marine deposits and demonstrates the effects of local tectonics on sedimentary facies development.

GENERAL GEOLOGY

The Lower Cretaceous Cloverly Formation in the Bighorn Basin, Wyoming consists mostly of fluvial sediments deposited in a foreland basin east of the

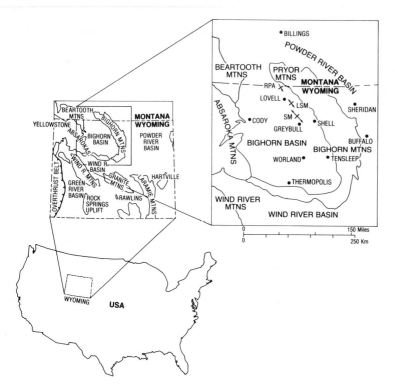

Fig. 1. Map of Wyoming and the study area. RPA, Red Pryor Anticline; LSM, Little Sheep Mountain; SM, Sheep Mountain.

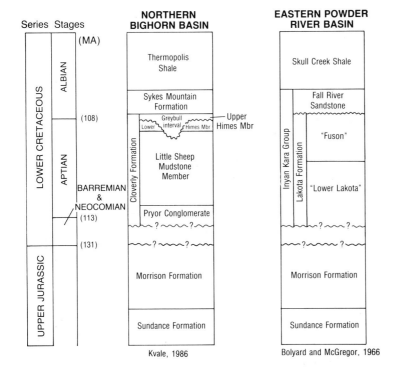

Fig. 2. Stratigraphic correlation chart for the Lower Cretaceous of northern Wyoming.

developing Sevier fold-thrust belt. Sediment source areas included the fold-thrust belt in eastern Idaho, arc volcanic vents in central Idaho, intraforeland basin uplifts and volcanic vents, and the craton (Moberly, 1960; Furer, 1970; Kvale & Beck, 1985; DeCelles, 1986; Kvale, 1986). Moberly (1960) recognized three members of the Cloverly Formation (Fig. 2). In ascending order they are: (i) the Pryor Conglomerate, (ii) the Little Sheep Mudstone Member and (iii) the Himes Member. Overlying the Himes Member conformably is the transitional marine to marine Lower Cretaceous Sykes Mountain Formation.

The Little Sheep Mudstone is part of an extensive clay (bentonitic) playa mudflat deposit that interfingers with arenaceous deposits, some conglomeratic, of ephemeral and perennial streams, that flowed to the north or to the east (Kvale, 1986; May et al., 1990). The Himes Member can be separated into two informal units, a lower Himes and an upper Himes (Kvale & Vondra, 1985a; Kvale, 1986). The lower Himes is a volcaniclastic-rich, low sinuosity channel sequence that drained western volcanic highlands, possibly in the northern Yellowstone/Absaroka Mountian region (Kvale, 1986) and is entirely restricted to the northern end of the Bighorn Basin. The upper Himes (Greybull interval) unconformably overlies the lower Himes or the Little Sheep Mudstone Member where the lower Himes is absent, and consists of channel-fill deposits of well-sorted quartz arenites or interbedded mudstones and sandstones. Channel deposits interfinger laterally with interchannel mudstones and thin sheet sands. Greybull interval streams generally flowed to the west in the Bighorn Basin from an apparent cratonic source and represent a significant gradient reversal from the underlying deposits of the rest of the Cloverly Formation (Kvale, 1986).

The Sykes Mountain Formation consists of sheet sandstones and interbedded mudstones that grade upwards into the marine black shales of the Thermopolis Shale. While many workers have considered the Greybull interval–Sykes Mountain Formation contact to be disconformable (e.g. Moberly, 1960; Winslow & Heller, 1987), Kvale & Vondra (1985a) and Kvale (1986) suggested that the contact is mostly conformable. Haun & Barlow (1962) likewise considered the Greybull interval to be more closely related to the overlying Sykes Mountain Formation and suggested that the Greybull interval not be included with the Cloverly Formation.

The idea that the Greybull interval is unconformably overlain by the Sykes Mountain Formation does not appear to be based on any direct evidence, but rather on an earlier interpretation that the Greybull interval was entirely non-marine in origin, and hence should be separated from the overlying, predominantly marine Sykes Mountain Formation by an unconformity. This contact has been correlated to the Powder River Basin to the east and considered to be equivalent to the Fall River Sandstone–Lakota Formation contact (e.g. Moberly, 1962). Greybull interval-like sandstone bodies have been identified in the Powder River Basin and placed in the upper part of the non-marine Lakota Formation by some workers (e.g. Gott et al., 1974), while others have apparently included these sandstones in the basal part of the transitional marine to marine Fall River Sandstone (e.g. Campbell & Oaks, 1973). We suggest that a major regional unconformity exists at the base of the Greybull interval rather than at the contact with the Sykes Mountain Formation and agree with the premise of Haun & Barlow (1962) that the Greybull interval should be included as a basal unit of the Sykes Mountain Formation and not part of the Cloverly Formation.

SUB-GREYBULL UNCONFORMITY

The sub-Greybull interval unconformity is a major regional unconformity that can be traced throughout northern Wyoming. It is present at the base of the Fall River Sandstone in the Powder River and Williston basins (Gott et al., 1974; Ballard et al., 1983). The age of the unconformity is considered to be latest Aptian to earliest Albian (Weimer, 1984; Heller & Paola, 1989). The unconformity manifests itself in the northern part of the Bighorn Basin as a palaeodrainage system with 23 m of local relief incised into the non-marine rocks of the Cloverly Formation. This drainage system occurs generally north of a line that extends east and west through the town of Greybull (Fig. 1) and extends at least into the northwestern edge of the Powder River Basin. Based on the occurrence of comparable sandstone channels in the same stratigraphic interval in the Powder River Basin, a similar drainage system may have existed along the eastern edge of the Powder River Basin (Bolyard & McGregor, 1966), but its relationship to the Bighorn Basin system remains unknown.

The Greybull interval attains a maximum thickness of 25 m in an area north of the town of Shell (Fig. 1) and thins rapidly to the south to a thickness of 2 m or less along the eastern edge of the basin south of Worland.

CHANNEL-FILL SEQUENCES

Channel-fill sequences were either incremental (consisting of multilateral, stacked channels) or continual (consisting of deposits from a single channel system). Locally channels continued to aggrade even after the sub-Greybull channels were filled. Three distinct types of channel-fill complexes are present within the Greybull interval. Two of the three types of complexes fill the major sub-Greybull channels. The three channel-fill types are summarized in Fig. 3.

The most obvious of the major channel-fill types is a thick (usually 15 m or more), elongate channel complex filled with well-sorted, fine-grained quartz arenites that were deposited within a low sinuosity, fluvial-dominated system (Moberly, 1960; Shelton, 1972; Stone, 1983; Kvale & Vondra, 1985b). Overbank sandstone deposits associated with this channel type are thin (usually less than 1 m) laterally continuous sandstones interbedded with non-resistant argillaceous siltstones.

The second type of major channel-fill sequence consists of massive, poorly exposed, muddy fill deposits. Locally, these deposits reflect an overall fining upward complex of stacked and multi-lateral channels filled with interbedded mudstone and sandstone. These channel-fills exhibit well-developed lateral accretion surfaces indicative of lateral barform migration. Individual channel thicknesses (based on barform thicknesses) are of the order of 3–5 m and originated in a mixed fluvial–tidal environment.

The third type of channel body is a small, 1–3 m thick, fining upward, ripple laminated sandstone channel complex that possesses lateral accretion surfaces indicative of barform migration in a high sinuosity fluvial channel complex. Relatively thin mudstones characterize the overbank deposits associated with this channel type.

CHANNEL TYPE

	Large, sandstone-dominated	Interbedded sandstone and mudstone-dominated	Small, sandstone-dominated
Estimated channel width	350–1100 m (width of sandbody)	Unknown	13 m (average) ($W_c = 1.5d$)*
Estimated channel depth (D)	up to 25 m	3–5 m	2–3 m
Brief description	Single channel-fill sequence dominated by large-scale planar and trough crossbeds that are strongly unidirectional.	Multistorey channel sequence of inclined heterolithic mudstones and v. f. grained sandstones.	Multistorey channel sequence of inclined homolithic sandstones.
Interpretation	Sandy, low sinuosity, fixed channel deposits. Fluvial-dominated upper estuary deposits.	Muddy channel deposits. Mixed fluvial-tidal middle estuary deposits.	High sinuosity (meandering) fluvial channel deposits.

* Equation from Ethridge and Schumm (1978)

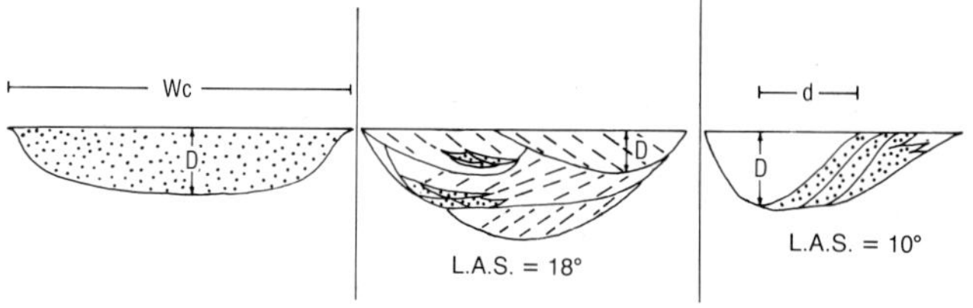

L.A.S. = 18°   L.A.S. = 10°

Fig. 3. Summary of major channel types in the Greybull interval.

Fig. 4. Large-scale planar foresets (Sp) separated by counter-current ripples (Sr). Rose diagram reflects 135 palaeocurrent measurements made in vicinity of this outcrop.

Large sandstone-dominated channels (type I)

The large sandstone-dominated channel bodies are well exposed in cliff faces along the flanks of Laramide uplifts in the northern reaches of the Bighorn Basin. These are fluvially dominated, elongate sandstone bodies that are broadly wing-shaped and deeply incised into underlying bedrock (up to 23 m). They vary in width from 350 to 1100 m. Maximum thickness of these channel bodies is 25 m in the Beaver Creek area (north of Shell) and the Sykes Mountain area (south of Pryor Mountain), but they thin to as little as 12 m westward into the basin. The sandstone bodies appear to be fixed channel systems that filled continuously after initial incision into the underlying strata. This interpretation is supported by the lack of multistorey-fill sequences typical of the other two channel types.

The base of the channel is defined by an erosional surface that is nearly horizontal to broadly concave up. The surface is lined with a concentration of large goethite-coated rip-up clasts (up to 10 cm in diameter) of argillaceous siltstones and/or argillaceous fine-grained sandstone. Rare, pebble-sized chert clasts derived from the reworking of the underlying Little Sheep Mudstone Member are locally present.

Primary structures are dominated by stacked sequences of very broad trough to tabular sets of large-scale planar foresets (Sp lithofacies of Miall, 1978) (Fig. 4). The foresets are straight with dips ranging from 26 to 33° and exhibit a distinct grain separation resulting in alternating coarse and fine laminae. Hunter (1985) interpreted this type of foreset lamination to be formed by laterally extensive avalanching over a large slipface. Oversteepening or overturning in the down-flow direction is common in the planar foresets. Toesets of counter-current ripples are locally abundant and intertongue with individual foresets. Large-scale trough cross-beds (St lithofacies) are distinctly subordinate to the Sp lithofacies in the lower sets but dominate in the upper few metres of the channel sands and nearer the channel margins.

Over 1000 palaeocurrent indicators were measured in this study. Most measurements were of large-scale trough axes and dip directions of large-scale planar foresets. For any one outcrop, palaeocurrent indicators are strongly unidirectional with deviation from the mean often not exceeding 20° in midchannel-body exposures. Within channel-body interiors, individual cross-bed sets range from 20 to 100 cm in thickness and attain lengths of 30–40 m, indicating uniform water depths. Near the channel margins the foresets are marked by uniform clay drapes (Fm lithofacies) a few centimetres thick that are often traceable onto bar tops (Fig. 5). Several clay-draped foresets may occur within one bedform and are commonly replaced with goethite.

Tongue-shaped sand-flow structures of subaqueous origin are locally present on 20 cm high foresets of three-dimensional dunes along the west margin of a channel at Red Pryor anticline, south of Pryor Mountain (Fig. 6). The dunes are linguoid in shape with 4 m maximum width and composed of fine-

Fig. 5. Goethite-replaced mudstone (arrows) draping barform. Vertical scale is 1 m.

grained sandstone. Similar sand-flow features have been reported by Buck (1985) from tidal sands of the lower Greensand (Lower Cretaceous) of southern England, by Hunter (1985) in modern fluvial deposits and by Hunter (1977) from aeolian deposits. The Greybull sand-flows are heavily cemented with goethite and appear to be related to a destabilization of the crest of the avalanche face of the three-dimensional dune during waning flow conditions after bedload transport of sediment had ceased. Sand flows are superimposed on pre-existing ripple fans occurring at the base of the dunes and show no evidence of having been reworked. No evidence of these small tongue-shaped flows is preserved in the internal laminae of the dunes, further supporting the interpretation that they reflect waning flow conditions.

The channel margins and overbank deposits of the large, sandstone-dominated channels consist of interbedded sandstones and non-resistant argillaceous siltstones. The sandstones are quartz arenites, moderately well sorted to moderately sorted, fine to

Fig. 6. Tongue-shaped sand flows formed on the avalanche face of a large three-dimensional dune. Note detachment point above flows. Also note flows impinging on ripple fan at the base of the barform. Bar top is reworked by discontinuous three-dimensional ripples. 15 cm ruler for scale.

very fine grained that commonly fine upwards to argillaceous siltstones or arenaceous siltstones. The siltstones are locally organic-rich, fissile, and, in places, totally replaced by goethite and are generally quite thin (1–5 cm) compared to the interbedded sandstones (2–45 cm). At one locality, small sand-filled mud cracks occur in the siltstone, indicating subaerial exposure. The predominant primary structure in both the sandstones and siltstones is ripple lamination. The corresponding lithofacies are Sr and Fl (laminated mudstones), respectively. Climbing ripples indicating high sedimentation rates are also preserved. The upper part of the sandstones commonly displays flaser bedding that gives way to lenticular bedding in the siltstone, suggesting a waning flow sequence. The sandstone bodies have sharp basal contacts that are only slightly irregular. This coupled with upward fining sequences and the uniform thickness and lateral persistence of the sand bodies over tens of metres suggests deposition of bed load material on very flat surfaces.

Trace fossils are rare, of low diversity and are preserved most commonly along the channel margin and in the overbank sandstones. The most common is an escape burrow (fugichnia) (Fig. 7), 2 mm in diameter, that provides additional evidence of rapid deposition of the sandstones. *Planolites* and *Arenicolites* (?) are also locally present.

The character of the overbank lithofacies is essentially the same from proximal to distal positions except that the Fl facies increases in proportion distally so that the Fl : Sr ratio becomes greater than 1.

Fossils are uncommon within the Greybull interval. However, crocodilian dermal scutes and teeth plus fragments of turtle carapace (identified by Brent Breithaupt, Geological Museum, University of Wyoming) were found in mudstones associated with the 'wing' of a large, sandstone-dominated channel fill sequence. In a similar setting, dinoflagellates, indicating brackish water conditions (E. Robertson, per. comm., 1986), were found in mudstones associated with the upper part of a channel fill along the south flank of Rose Dome, north of Sheep Mountain.

The excellent exposures in the study area allow for detailed mapping of channel margin facies (pinchouts) over an extensive area. Individual channel bodies can be traced and mapped continuously over a distance of over 30 km. The subparallel channel pinchout trends and the unimodal character of palaeocurrent indicators, along with the absence of lateral accretion surfaces and pronounced upward fining, indicate that these sandstones were deposited in an essentially straight-channel fluvial-dominated system. The presence of dinoflagellates suggests that the deposition of at least the upper part of the channel-fill sequence experienced some marine influence.

Fig. 7. Vertical escape burrows (fugichnia). Jacob's Staff marked off in 20 cm intervals.

Interbedded sandstone and mudstone-dominated channels (type II)

The second type of major channel-fill consists of non-resistant interbedded sandstones and mudstones that comprise stacked muddy barforms that fill multistory and multilateral channel complexes. These barforms are characterized by well-developed lateral accretion surfaces (Fig. 8). The lateral accretion surfaces have a maximum dip of 18° and are marked by couplets of thinly interbedded fine- to very fine-grained sandstones and mudstones.

The sandstones of the sandstone and mudstone-dominated channels are typically lenticular to wedge shaped non-resistant bodies a few centimetres thick along the channel margins but grade into resistant sheet-type bodies towards the channel interior. The sandstones are mostly massive, but faint ripple lamination is present in some sets. In some areas, channel interiors are dominated by resistant fine-grained large-scale trough cross-bedded quartz arenites with set thickness of the order of 20–40 cm indicating deposition in large, sinuous-crested to discontinuous-crested sand dunes. The tops of these sets are commonly marked by finely disseminated fragments of coalified plant remains. The mudstones are non-resistant and mostly massive, but faint ripple lamination is rarely present indicating that at least some of the mudstones were transported or reworked as bedload material. Small-scale slump features are present along channel margins. The entire interval is bioturbated by small diameter (less than 1 cm) vertical and horizontal burrows.

Locally the stacked muddy bar sequence fines upwards to a reddish brown, argillaceous, green mottled mudstone that contains dinoflagellates indicative of brackish water conditions (E. Robertson, pers. comm., 1986). These mudstones are gradational with the overlying Sykes Mountain Formation.

Whereas individual muddy barform deposits are 3–5 m thick, the complex fills a channel that locally may be incised more than 20 m through the lower Himes deposits and into the underlying Little Sheep Mudstone Member. Channel widths are uncertain but may be comparable to type I channels. Most overbank deposits associated with these channels are thin mudstones. Although generally poorly exposed, excellent examples of these types of channel-fills occur near Alkali Creek, east of Sheep Mountain.

The inclined accretionary deposits closely resemble the inclined heterolithic strata described by Smith (1987). Smith showed that steeply dipping (greater than 10°) lateral accretion surfaces (LAS) with couplets of thin sandstones and mudstones of relatively uniform thicknesses appear to be typical

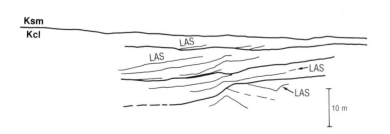

Fig. 8. Photograph and interpretive drawing of four stacked sandstone- and mudstone-dominated channels at Alkali Creek, east of Sheep Mountain. The upper three channels exhibit lateral accretion surfaces (LAS) that dip from right to left. LAS in the lowermost channel dip towards the viewer. The Sykes Mountain Formation (Ksm) contains abundant remains of brackish water bivalves at this locality. Arrow indicates sample site that produced dinoflagellates, also indicative of brackish water conditions.

of point bars formed in tidally influenced channels. De Mowbray (1983) noted similar surfaces in tidal creeks in Scotland. Rahmani (1988) has described similar deposits from an Upper Cretaceous estuarine tidal channel in Canada. Conversely, Taylor & Woodyer (1978) and Jackson (1981) have described inclined sand and mud deposits from low gradient rivers draining mud-rich basins far inland from the sea. However, based on the presence of the dinoflagellates and the gradational upper contact with the Sykes Mountain Formation, which clearly was marine to transitional marine in origin, a tidally influenced channel interpretation is considered more likely.

Small, sandstone-dominated channels (type III)

The third type of channel-fill is preserved in interflow areas that occur between the major sub-Greybull channels. This type is interpreted to represent deposition from highly sinuous fluvial channels. The channels occur within channel belts composed of erosive based, multistorey and multilateral channel complexes (Fig. 9). Incision of these channel bodies into the underlying sub-Greybull strata is generally less than 2 m. Channel bodies fine upwards from fine grained sandstones to siltstones, are typically 1–3 m thick and are characterized by lateral accretion surfaces dipping at 10°. Outcrop mapping of these channels indicates that these systems were meandering, thus suggesting a point bar origin for these surfaces. In contrast to the sandstone and mudstone-dominated channels, the lateral accretion deposits of the small, sandstone-dominated channels consist of thick sandstones separated by relatively thin and discontinuous mudstones. Along the west flank of Little Sheep Mountain, north of Sheep Mountain, scroll bars are

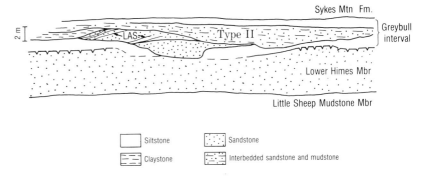

Fig. 9. Photograph and interpretation of multistorey, small sandstone-dominated channels (type III) along the south flank of Sheep Mountain. Note the 2.5 m thick clay-plug formed left of the sandstone that exhibits lateral accretion surfaces (LAS) dipping from right to left. The lowest channel is incised into the 'lower' Himes volcaniclastic-rich sandstone. An interbedded sandstone- and mudstone-dominated channel (type II) downcuts into the stacked type III channels on the right. Also note the vertisolic-like features at the contact between the Greybull interval and the 'lower' Himes. Very little is known about these, but they appear where Greybull interval illitic and kaolinitic mudstones overlie the 'lower' Himes bentonitic mudstones. The features probably reflect the formation of a soil horizon associated with the unconformity between the 'upper' and 'lower' Himes units. View is looking east.

Fig. 10. Scroll bars formed on top of a small, sandstone-dominated channel body west of Little Sheep Mountain. Photograph by D. Uhlir.

preserved on the upper surface of a sandstone body (Fig. 10). The sandstones are mostly ripple laminated and confined to the point bar complex. Locally, massive clay plugs, capped by organic-rich mudstones, are present suggesting channel cutoff and oxbow lake development. Overbank deposits are generally not laterally extensive and are composed of argillaceous siltstones.

Gymnosperm and angiosperm(?) leaf and stem fragments are locally associated with the finer grained mudstones (Scott Wing, Smithsonian Institution, pers. comm., 1989). Maceral studies completed on the overbank sediments suggest a non-marine origin for these deposits (E. Robertson, pers. comm., 1986). Lateral accretion facies that lack mudstones or contain mudstones of irregular thicknesses deposited on gently sloping surfaces (less than 10°) may be more typical of point bars formed in fluvial or upper delta plain settings (Smith, 1987).

SYKES MOUNTAIN FORMATION

The Sykes Mountain Formation conformably overlies the Greybull interval and consists of dark siltstones and interbedded darkly stained, ferruginous, fine- to very fine-grained sandstones. The sandstones form sheet deposits that are slightly erosive and laterally extensive for literally hundreds of metres, yet rarely exceed 1 m in thickness. Locally, it is at the base of the lowest of these sheet sandstones that the contact between the Greybull interval and the Sykes Mountain Formation is commonly drawn even though, lithologically, the mudstones below the sheet sandstones are the same as those above them. Dominant primary structures within the sandstones include parallel bedding with primary current lineation (PCL) on bedding plane surfaces, and isolated scour-fills manifested as low angle large-scale trough cross-bedding (Fig. 11). Burrowing is intense in the upper few centimetres of the sandstones and on upper bedding plane surfaces resulting in the destruction of primary structures in the upper parts of the sandstones (Fig. 12). Upper surfaces of the sandstone often show evidence of reworking by wind- or wave-generated currents resulting in the formation of oscillation ripples, interference ripples or straight-crested asymmetrical ripples that may bifurcate. Separating the sheet sandstones are sequences of interbedded mudstone and fine-grained sandstone that are typically lenticular to wavy bedded and dominated by hummocky cross-stratification (Fig. 13). These deposits are commonly bioturbated by large sand-filled burrows up to 2 cm across and several centimetres long. Some of the large burrows may have been generated by pelecypods that commonly occur as shell lags at the base of the sandstones. The pelecypods include *Aphrodina*(?), *Pinna*-like forms and *Leptosolen*(?). The Sykes Mountain forms are considered to have lived in marine to brackish water conditions

Fig. 11. Stacked sets of parallel-bedded sandstones from the lower part of the Sykes Mountain Formation. Note the low-angle trough scour on the right side of the photo. The lens cap is 5 cm in diameter.

(D. Hattin, Indiana University, pers. comm., 1989). Other traces include fugichnia traces identical to those found along the margins of the large, sandstone-dominated channels lower in the section.

Coarse- to medium-grained channel sandstones are commonly incised into the interbedded sandstones and mudstones of the Sykes Mountain Formation. Maximum channel dimensions are not known, but several channel complexes tens of metres across and 2 m thick are present just above the Sykes Mountain/Cloverly contact in the Sykes Mountain area, northeast of the town of Lovell. The channel-fills are dominated by large-scale trough cross-stratification that indicates dominant palaeoflow was in a general easterly direction, opposite to the general flow direction of the Greybull channels. Soliman (1988) reported bipolar cross-stratification (herringbone) in some of these channel deposits and suggested that they were tidally generated. The overall trend of the Sykes Mountain Formation is one of a gradational fining upwards to the marine black shale sequence of the Thermopolis Forma-

Fig. 12. Close-up of Fig. 10. Note the intense burrowing in the upper part of the sandstones. Also note the escape burrows in the lower part. Escape burrows in the Sykes Mountain Formation are very similar to those found in the Greybull interval (cf. Fig. 7).

Fig. 13. Hummocky cross-stratification in the Sykes Mountain Formation. 15 cm ruler for scale.

tion. The Sykes Mountain as a formation becomes much finer grained from the Sheep Mountain area (west) to the outcrops flanking the Bighorn Mountains (east).

EFFECTS OF RELATIVE SEA-LEVEL CHANGE

Maximum depth of incision into the strata underlying the Greybull interval bedrock is about 23 m. The timing of this incision is not clear because of the general lack of biostratigraphic markers for either the Cloverly or Sykes Mountain formations. The Thermopolis Shale is considered to be mid-Albian in age based on the foraminifer *Haplophragmoides gigas* (Eicher, 1960, 1962; Moberly, 1960). Volcanic tuffs present in the upper part of the Little Sheep Mudstone in the Bighorn Basin reveal radiometric ages of 115 ± 9 to 129 ± 16 m.y. BP (Chen, 1989). Thus, incision of the sub-Greybull interval channels may or may not correspond to a worldwide sea-level drop of about 40 m that occurred during early Albian time (Haq *et al.*, 1987). If the incision was caused solely by the early Aptian drop in sea level, the lack of 40 m of relief at the base of the Greybull interval may indicate that the Bighorn Basin was significantly inland, perhaps as much as 100 km more during the time of downcutting. The conformable relationship with the overlying transitional marine Sykes Mountain Formation suggests that deposition within the Greybull interval may have occurred in response to the subsequent sea-level rise and resulting marine transgression.

The relationship between the three channel-fill types is complex. However, their relative ages can be demonstrated (Fig. 14). Type I and type II channel deposits, even though lithologically different, fill channels of comparable dimensions with depths as much as 23 m. Whereas fills within the sandstone-dominated type I channels appeared to have been continuous, the presence of stacked mud-dominated barforms in type II deposits suggests a more incremental filling. Where type I and type II channel-fills can be observed together, type II interbedded sandstone and mudstone deposits are cut by type I deposits, suggesting that type II deposits are somehow related to type I deposits (Fig. 15). Type III channels are demonstrably older than either type II or type I channels by underlying these deposits or their overbank equivalents. The dinoflagellates associated with types I and II channel-fills suggest some marine influence, whereas the fill associated with the smaller type III channels apparently is totally non-marine in origin.

Kvale (1986) discussed the evidence that much of the Greybull interval was deposited within an estuary system. Recently, Rahmani (1988) described a tripartite style of sand-to-mud-to-sand fill from a Late Cretaceous estuarine tidal channel complex in Alberta, Canada. The Canadian system was associ-

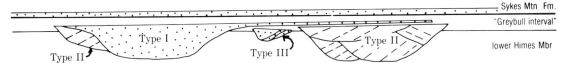

Fig. 14. Diagrammatic reconstruction illustrating the relative ages of the major channel types within the Greybull interval. Not drawn to scale. Type I: large, sandstone-dominated channels. Type II: interbedded sandstone- and mudstone-dominated channels. Type III: small, sandstone-dominated channels.

ated with a meso- to macro-tidal embayed shoreline. The sand in the lower part of the Canadian estuary was derived from a nearshore source and was transported inland by tidal and longshore processes. The middle estuarine deposits are muddy and reflect deposition from the turbidity maximum zone at the landward edge of an estuarine salt wedge. The upper part of the Canadian estuarine deposits are sandstones deposited from fluvial-dominated processes. Similar tripartite estuarine systems have been described from modern and ancient examples by Oomkens & Terwindt (1960), Karvonen (1989) and Dalrymple et al. (1990). Thus, within such a system the sandstone and mudstone-dominated channel-fills of the Greybull interval that exhibit the inclined heterolithic strata represent middle estuarine deposits, the large sandstone-dominated channel-fills represent straight-channel upper estuarine deposits, and the small sandstone- dominated channels are likely high sinuosity fluvial channels that were perhaps tributaries to the upper estuary. The sandy, lower estuarine deposits are preserved in the lower part of the Sykes Mountain Formation, which is more open marine in character.

The sandstone and mudstone-dominated channel-fills have been found in outcrop only in the Sheep Mountain area. In some cases it can be demonstrated that these channel-fills are downcut by 10–15 m thick large, sandstone-dominated channels (Fig. 15). One explanation for this is that the sandstone and mudstone-dominated channels represent the middle estuarine deposits of large, sandstone-dominated channels to the east (possibly those north of Shell). A relatively minor lowering of base level or pause during a late Aptian to early Albian sea-level rise allowed downcutting of the estuary channels in the Sheep Mountain area and a progradation of the upper estuarine sandstones over the middle estuarine muds (Fig. 16). A continuation of sea-level rise eventually resulted in the deposition of marine rocks over the Greybull interval. Another possible explanation is that fluvial discharge increased within the estuary (possibly forced

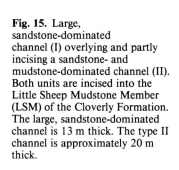

Fig. 15. Large, sandstone-dominated channel (I) overlying and partly incising a sandstone- and mudstone-dominated channel (II). Both units are incised into the Little Sheep Mudstone Member (LSM) of the Cloverly Formation. The large, sandstone-dominated channel is 13 m thick. The type II channel is approximately 20 m thick.

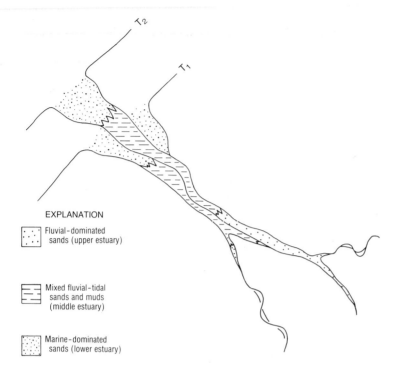

Fig. 16. Drawing illustrating the estuarine facies relationships caused by a relative drop in sea-level (T2) during an overall transgressive sequence. (Modified from Karvonen (1989).)

by an increase in rainfall) causing the progradation of the fluvial-dominated part of the upper estuary to migrate into the middle part of the estuary.

TECTONIC EFFECTS

The major incised channels in the Greybull interval are confined to regions north of what is now the area of maximum uplift in the modern Bighorn Mountains. Palaeocurrent analyses of the large, sandstone-dominated channel-fills show that these channels converge and diverge along what are now major structural lineaments associated with Laramide block uplifts (Fig. 17) (Kvale, 1986). Specifically, these channels are confined to areas delineated by the Tongue River lineament defined by Hoppin & Jennings (1971) and along the structural break between the Bighorn Mountains and the Pryor uplift. In both areas, the large sandstone-dominated channels form multilateral channel-sandstone bodies at least 20 m thick. From these points the channels 'splayed out', flowing west to southwest, and are preserved as mostly isolated channel bodies laterally separated from one another by several kilometres of relatively thin interchannel material. It appears on the basis of palaeocurrent data and field mapping that all of the thick sandstone-dominated Greybull interval channels within the Bighorn Basin can be extrapolated back to these two areas. Comparable channels are not present elsewhere along the west flank of the Bighorn Mountains.

Studies along the east flank of the Bighorn Mountains reveal that these channels likewise appear to be confined to areas north of the region of maximum uplift. Palaeocurrent measurements of large sandstone-dominated channels near the eastern reach of the Tongue River lineament reveal that flow was directed westward along the lineament. Detailed palaeocurrent data are not yet available for channels north of the Pryor uplift, but Shelton (1972) reported that sandstone-dominated channels extending northwestward from the northwest corner of the Powder River Basin into southern Montana apparently flowed westward north of the Pryor Mountains. Some of these channels must have been diverted south into northern Wyoming along the Pryor–Bighorn Mountain lineament.

The occurrence of large multilateral sandstone-dominated bodies in specific regions along the flanks of the Bighorn and Pryor mountains indi-

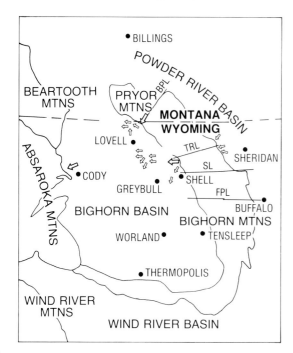

Fig. 17. Generalized palaeocurrent trends of the large, sandstone-dominated channels. Each arrow indicates an outcrop and its dominant palaeoflow direction. Large arrows indicate where three or more channels have coalesced. Trends were obtained by measuring over 1000 palaeocurrent indicators as well as mapping channel pinch-outs. Lineament positions (except for BPL) are from Hoppin & Jennings (1971). BPL, Bighorn–Pryor Lineament; TRL, Tongue River Lineament; SL, Shell Canyon Lineament; FPL, Florence Pass Lineament.

cates that the channels were areally restricted. The proximity of the outcrops to major lineations suggests that differential movement along the lineations during late Aptian to early Albian restricted the lateral development of the drainage system. Farther out into the Bighorn Basin, structural controls were not as pronounced and the drainage pattern developed over a broader area. Thus, while worldwide sea-level changes induced the incision of a major drainage system, local tectonics influenced the position of the channel system.

SUMMARY AND CONCLUSIONS

A combination of changes in relative sea level and local tectonics controlled the emplacement of a major Early Cretaceous palaeodrainage system in north-central Wyoming. Sea-level rise caused back-filling of the drainage system and the formation of a tripartite estuarine system similar to that described by other workers. Continued sea-level rise resulted in the deposition of the sand-dominated, lower estuarine Sykes Mountain Formation over the muddy, middle estuarine and sandy, upper estuarine Greybull interval.

ACKNOWLEDGEMENTS

Field and laboratory work for this study were made possible by grants from the National Geographic Society, Marathon Oil Company and Phillips Petroleum Company. We wish to thank David Uhlir, Nazrul Khandaker, James Beer, Karen Noggle, Lee Suttner and Peter DeCelles for helpful discussions in the field. Fossils were identified by E.B. Robertson, Phillips Petroleum Co. (dinoflagellates), Don Hattin, Indiana University (pelecypods), Scott Wing, Smithsonian Institution (plant remains) and Brent Breithaupt, University of Wyoming (reptilian remains). We also acknowledge the Indiana Geological Survey and Michael McKinney for assistance in preparing the line drawings and photographs used herein.

REFERENCES

AUBREY, W.M. (1989) Mid-Cretaceous incision related to eustasy, southeastern Colorado Plateau. *Bull. Geol. Soc. Am.* **101**, 443–449.

BALLARD, W.W., BLUEMLE, J.P. & GERHARD, L.C. (1983) Northern Rockies/Williston basin region. Correlation of Stratigraphic Units in North America (COSUNA) chart, one sheet. Am. Assoc. Petrol. Geol., Tulsa.

BOLYARD, D.W. & MCGREGOR, A.A. (1966) Stratigraphy and petroleum potential of Lower Cretaceous Inyan Kara Group in northeastern Wyoming, southeastern Montana, and western South Dakota. *Bull. Am. Assoc. Petrol. Geol.* **50**, 2221–2244.

BUCK, S.G. (1985) Sand-flow cross strata in tidal sands of the lower Greensands (Early Cretaceous), southern England. *J. Sedim. Petrol.* **55**, 895–906.

CAMPBELL, C.V. & OAKS, R.Q., JR (1973) Estuarine sandstone filling tidal scours, Lower Cretaceous Fall River Formation, Wyoming. *J. Sedim. Petrol.* **43**, 765–778.

CHEN, Z.Q. (1989) *A fission track study of the terrigenous sedimentary sequences of the Morrison and Cloverly Formations in the Northeastern Big Horn Basin, Wyoming.* Unpublished Master's thesis, Dartmouth College.

DALRYMPLE, R.W., KNIGHT, R.J., ZAITLIN, B.A. & MIDDLETON, G.V. (1990) Dynamics and facies model of a macrotidal sand-bar complex, Cobequid Bay–Salmon River Estuary (Bay of Fundy). *Sedimentology* **37**, 577–612.

DeCelles, P.G. (1986) Sedimentation in a tectonically partitioned, nonmarine foreland basin: the Lower Cretaceous Kootenai Formation, southwestern Montana. *Bull. Geol. Soc. Am.* **8**, 975–985.

De Mowbray, T. (1983) The genesis of lateral accretion deposits in recent intertidal mudflat channels, Solway Firth, Scotland. *Sedimentology* **30**, 425–435.

Eicher, D.L. (1960) *Stratigraphy and micropaleontology of the Thermopolis shale.* Peabody Museum of Natural History, Yale University Bulletin 15.

Eicher, D.L. (1962) Biostratigraphy of the Thermopolis, Muddy and Shell Creek Formations. In: *17th Annual Field Conference Guidebook*, pp. 72–93. Wyoming Geological Association, Casper.

Ethridge, F.G. & Schumm, S.A. (1978) Reconstructing paleochannel morphologies and flow characteristics: methodology, limitations and assessment. In: *Fluvial Sedimentology* (Ed. Miall, A.D.) pp. 703–721. Can. Soc. Petrol. Geol., Calgary, Memoir 5.

Furer, L.C. (1970) Petrology and stratigraphy of nonmarine Upper Jurassic–Lower Cretaceous rocks of western Wyoming and southeastern Idaho. *Bull. Am. Assoc. Petrol. Geol.* **54**, 2282–2302.

Gott, G.B., Wolcott, D.E. & Bowles, C.G. (1974) Stratigraphy of the Inyan Kara Group and localization of uranium deposits, southern Black Hills, South Dakota and Wyoming. US Geol. Surv. Prof. Paper 763.

Haq, B.U., Hardenbol, J. & Vail, P.R. (1987) Chronology fluctuating sea levels since the Triassic. *Science* **235**, 1156–1166.

Haun, J.D. & Barlow, J.A., Jr (1962) Lower Cretaceous stratigraphy of Wyoming. In: *17th Annual Field Conference Guidebook*, pp. 15–22. Wyoming Geological Association, Casper.

Heller, P.L. & Paola, C. (1989) The paradox of Lower Cretaceous gravels and the initiation of thrusting in the Sevier orogenic belt, United States Western Interior. *Bull. Geol. Soc. Am.* **101**, 864–875.

Hoppin, R.A. & Jennings, T.V. (1971) Cenozoic tectonic elements, Bighorn Mountain region, Wyoming–Montana. In: *23rd Annual Field Conference Guidebook* (Ed. Renfro, A.R.), pp. 39–45. Wyoming Geological Association, Casper.

Hunter, R.E. (1977) Basic types of stratification in small eolian dunes. *Sedimentology* **24**, 361–387.

Hunter, R.E. (1985) Subaqueous sand-flow cross-strata. *J. Sedim. Petrol.* **55**, 886–894.

Jackson II, R.G. (1981) Sedimentology of muddy fine-grained channel deposits in meandering streams of the American middle west. *J. Sedim. Petrol.* **51**, 1169–1192.

Karvonen, R.L. (1989) The Ostracode Member, east-central Alberta: an example of an estuarine valley fill deposit. In: *Modern and Ancient Examples of Clastic Tidal Deposits – A Core and Peel Workshop* (Ed. Reinson, G.E.) pp. 105–116. Can. Soc. Petrol. Geol., Calgary.

Kvale, E.P. (1986) *Paleoenvironments and tectonic significance of the Upper Jurassic Morrison/Lower Cretaceous Cloverly formations, Bighorn Basin, Wyoming,* Unpublished PhD thesis, Iowa State University.

Kvale, E.P. & Beck, R.A. (1985) Thrust controlled sedimentation patterns of the earliest Cretaceous nonmarine sequence of the Montana–Idaho-Wyoming Sevier foreland basin. *Geol. Soc. Am. Abstracts with Programs* **17**, 636.

Kvale, E.P. & Vondra, C.F. (1985a) Depositional environments of the Lower Cretaceous Cloverly Formation, Bighorn Basin, Wyoming. *Soc. Econ. Paleont. Miner. Abstracts Annual Midyear Meeting*, **2**, 53.

Kvale, E.P. & Vondra, C.F. (1985b) Upper Jurassic–Lower Cretaceous transitional marine and fluvial sediments in the Bighorn Basin. In: *Field Guidebook to Modern and Ancient Fluvial Systems in the United States.* (Eds Flores, R.M. & Harvey, M.) pp. 33–44. 3rd Int. Fluvial Conference. Colorado State University Press, Fort Collins.

May, M.T., Suttner, L.J. & Kvale, E.P. (1990) Evidence of complex age relations and dispersal patterns in Lower Cretaceous conglomerates, Wind River Basin, Wyoming. *Geol. Soc. Am. Abstracts with Programs* **22**, A322.

Miall, A.D. (1978) Lithofacies types and vertical profile models in braided river deposits: a summary. In: *Fluvial Sedimentology* (Ed. Miall, A.D.) pp. 597–604. Can. Soc. Petrol. Geol., Calgary, Memoir 5.

Moberly, R., Jr (1960) Morrison, Cloverly and Sykes Mountain Formations, northern Bighorn Basin, Wyoming and Montana. *Bull. Geol. Soc. Am.* **71**, 1137–1176.

Moberly, R., Jr (1962) Lower Cretaceous history of the Bighorn Basin, Wyoming. In: *17th Annual Field Conference Guidebook*, pp. 94–101. Wyoming Geological Association, Casper.

Oomkens, E. & Terwindt, J.H.H. (1960) Inshore estuarine sediments in the Haringvliet (Netherlands). *Geol. Mijnbouw* **39**, 701–710.

Rahmani, R.A. (1988) Estuarine tidal channel and near-shore sedimentation of a Late Cretaceous epicontinental sea, Drumheller, Alberta, Canada. In: *Tide-influenced Sedimentary Environments and Facies* (Eds deBoer, P.L. *et al.*), pp. 433–471. D. Reidel, Norwell.

Shelton, J.W. (1972) Trend and depositional environment of Greybull sandstone, northern part of the Pryor Uplift. In: *21st Annual Field Conference Guidebook* (Ed. Lynn, J.) pp. 75–79. Montana Geological Society, Billings.

Smith, D.G. (1987) Meandering river point bar lithofacies models: modern and ancient examples compared. In: *Recent Developments in Fluvial Sedimentology* (Eds Ethridge, F.G., Flores, R.M. & Harvey, M.D.) pp. 83–91. Soc. Econ. Paleont. Mineral., Tulsa, Spec. Publ. 39.

Soliman, H.E.A. (1988) *Stratigraphy and sedimentology of Lower Cretaceous Sykes Mountain Formation, Bighorn Basin, Wyoming.* Unpublished PhD thesis, Iowa State University.

Stone, D.S. (1983) The Greybull sandstone pool (Lower Cretaceous) on the Elk Basin thrust-fold complex, Wyoming and Montana. In: *Rocky Mountain Foreland Basins and Uplifts* (Ed. Lowell, J.D.) pp. 345–356. Rocky Mountain Association Geologists, Denver.

Taylor, G. & Woodyer, K.D. (1978) Bank deposition in suspended-load streams. In: *Fluvial Sedimentology* (Ed. Miall, A.D.) pp. 257–275. Can. Soc. Petrol. Geol., Calgary, Memoir 5.

WEIMER, R.J. (1984) Relation of unconformities, tectonics, and sea-level changes, Cretaceous of Western Interior, USA. In: *Interregional Unconformities and Hydrocarbon Accumulation* (Ed. Schlee, J.S.) pp. 7–35. Am. Ass. Petrol. Geol., Tulsa, Memoir 36.

WINSLOW, N.S. & HELLER, P.L. (1987) Evaluation of unconformities in Upper Jurassic and Lower Cretaceous nonmarine deposits, Bighorn Basin, Wyoming and Montana, USA. *Sedim. Geol.* **53**, 181–202.

Structural and climatic controls on fluvial depositional systems: Devonian, North-East Greenland

H. OLSEN* and P.-H. LARSEN†

The Geological Survey of Greenland, Øster Voldgade 10, DK-1350 Copenhagen K, Denmark

ABSTRACT

Depositional system analysis of continental Devonian deposits in North-East Greenland unravels the complex simultaneous controls of tectonism and climate. The analysis is a combination of sedimentological and structural investigations. Sediment bodies are defined as depositional systems on sedimentological criteria, and mapped by a combination of field mapping and computer assisted stereoscopic studies of vertical aerial photographs. Vertical and lateral facies changes and thickness variations of the sediment bodies are related to the geological structures exposed in the present-day outcrops.

The Upper Devonian Kap Graah and Celsius Bjerg Groups exposed in Gunnar Andersson Land in North-East Greenland are subdivided into three well defined units, termed chronosomes, separated by event-stratigraphic horizons. The chronosome boundaries define times of major palaeogeographic reorganization.

Chronosome I reflects fluvial systems transverse to the basin margins that are separated by a huge aeolian sand sea or erg occupying the basin. This palaeogeography was controlled by a climatic shift towards drier conditions, compared with pre-chronosome I conditions. Chronosome II also reflects two transverse fluvial systems, but the erg was replaced by a longitudinal southward flowing meandering river. This palaeogeography was controlled by tectonic activity both within and outside the basin. The latter caused an increase in relief of the source terrain and thus increased local precipitation. Chronosome III reflects a broad alluvial plain draining towards the north, i.e. in the opposite direction compared with chronosome II. This dramatic change was controlled by tectonic uplift of the southern part of the basin.

INTRODUCTION

The major controls on continental depositional systems are tectonic activity and climate. Although intensively studied for many decades, no easy solution to the problem of distinguishing between the two factors has been put forward, so far. However, the present paper demonstrates that an integrated sedimentological and structural study is of vital importance in approaching the problem.

Extremely well exposed continental sediments of the Upper Devonian Kap Graah and Celsius Bjerg Groups, North-East Greenland (Nicholson & Friend, 1976; Olsen & Larsen, 1993) formed the basis for a depositional system analysis (Olsen, 1993) in the sense of Miall (1990) and Galloway (1989), in combination with a structural geometric analysis (Larsen, 1990a,b,c,d). Depositional systems are three-dimensional assemblages of process-related facies that record major palaeogeomorphic basin elements. Facies association analysis was used to define individual depositional systems, which in turn were mapped in the field using vertical aerial photographs (1:50 000). The structural and sedimentological work was combined with computer assisted photogrammetric studies of vertical aerial photographs (1:150 000) at the Geological Survey

* Present address: Department of Geology and Geotechnical Engineering, Building 204, Danmarks Tekniske Højskole, DK-2800 Lyngby, Denmark.
† Present address: Maersk Oil and Gas A/S, Esplanaden 50, DK-1263 Copenhagen K, Denmark.

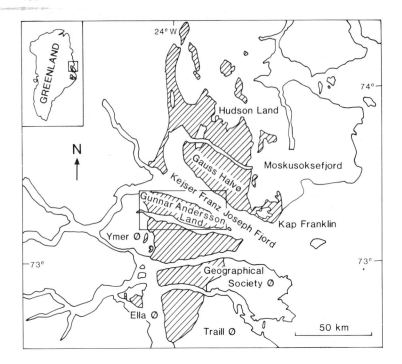

Fig. 1. Geographical setting and simplified geological map, showing outcrop areas of Devonian sediments in North-East Greenland. The present study was carried out in Gunnar Andersson Land.

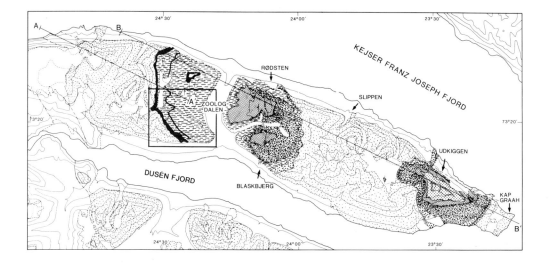

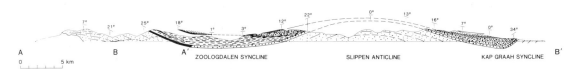

Fig. 2. Detailed geological map from Gunnar Andersson Land, showing outcrops of the formations in the Kap Graah Group and the Celsius Bjerg Group. The location of Fig. 11 is indicated. A composite cross-section through the study area is shown below the main diagram. Ornamentation corresponds to Fig. 3. (After Larsen (1990a) and Olsen & Larsen (1993).)

of Greenland. The study area (Figs 1 & 2) was visited during three successive field seasons, allowing integration of analytical work and fieldwork.

In the following the sedimentology and structural aspects are described from the E-W trending peninsula of Gunnar Andersson Land, which provides a cross-section of the Devonian basin in North-East Greenland. The Late Devonian basin evolution is illustrated and the underlying factors controlling the interrelationships of the sediment bodies and their depositional environments are discussed.

Stratigraphic nomenclature is based on Olsen & Larsen (1993).

GEOLOGICAL SETTING

The Precambrian and Caledonian structural evolution of East and North-East Greenland has been summarized recently by Henriksen & Higgins (1976), Higgins & Phillips (1979) and Henriksen (1985). The basement of the Devonian basin can be subdivided into two major structural elements: a Lower Proterozoic or older crystalline shield overlain by a thick Upper Proterozoic to Ordovician sedimentary cover sequence. The crystalline shield underwent deformation and migmatization during the Grenville orogeny (1200–900 Ma) and became the basement for the Upper Proterozoic to Ordovician sedimentary succession. Caledonian deformation affected both the crystalline shield and the sedimentary cover sequence, and it was followed by basin initiation in the late Middle Devonian due to an extensional collapse of the overthickened Caledonian crustal welt associated with left-lateral strike-slip deformations (Larsen & Bengaard, 1991). The basin developed very rapidly, accumulating more than 8 km of continental deposits in various depositional environments (e.g. Bütler, 1935, 1955; Friend et al., 1983; Larsen et al., 1989; Larsen, 1990a,b,c,d; Olsen, 1990, 1993). During deposition of the sediments the basin underwent compression, leading to a series of deformations that seem to have influenced the basin fill geometry, sedimentological facies distribution and drainage patterns (Larsen, 1990b,c,d; Olsen, 1993).

LITHOSTRATIGRAPHIC UNITS

General

The Devonian continental succession studied in Gunnar Andersson Land (Fig. 3) is more than 1500 m thick. It includes deposits from two major stratigraphic units, the Kap Graah Group and the Celsius Bjerg Group (Nicholson & Friend, 1976; Friend et al., 1983; Olsen & Larsen, 1993). These rocks are underlain by the Sofia Sund Formation of the Kap Kolthoff Group (Olsen & Larsen, 1993). The Sofia Sund Formation is not included in this detailed study, and will only briefly be described. It is composed of pebbly, medium- and coarse-grained, trough cross-bedded sandstones. The formation represents huge, often fan-shaped, sandy braidplain segments. Towards the centre of the basin these segments lost their identity to form an extensive braidplain river, draining longitudinally southwards.

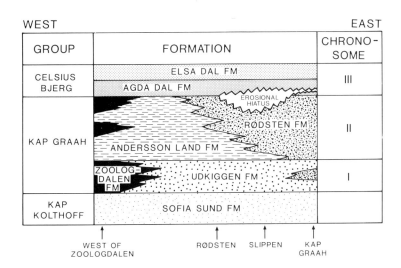

Fig. 3. Stratigraphic architecture of the Gunnar Andersson Land succession represented by a schematic W–E cross-section. The lithostratigraphic units included in the study are grouped in chronosomes delimited by event-stratigraphic horizons. For position of localities, see Fig. 2.

404 H. Olsen and P.-H. Larsen

The formations comprising the Kap Graah Group and the Celsius Bjerg Group in the Gunnar Andersson Land area (Fig. 3) will be described in turn below. The basis for the lithostratigraphic subdivision of the rocks is the concept of depositional system analysis, previously defined. Accordingly, individual formations represent individual identifiable depositional systems.

Udkiggen Formation

Description

This formation is composed of fine- to coarse-grained sandstone. It consists of two main facies associations: a cross-bedded facies association (most common) and a flat-bedded facies association.

Trough cross-bedding dominates the cross-bedded facies association, and individual sets generally range in thickness from $c.10$ cm to 5 m and between 10 and 30 m in width. Trough length commonly exceeds 50 m and may appear tabular in limited outcrops. The trough cross-bedding frequently occurs in tabular cosets 2–20 m thick (Fig. 4A). Trough set thickness commonly increases upwards within a coset from less than 1 m to 2–5 m (Fig. 4B). Laterally such cosets may grade into solitary tabular or gutter-shaped sets up to 20 m thick

Fig. 4. Udkiggen Formation. (A) Tabular coset of trough cross-bedded sandstone, 2.5 m thick, deposited by crescentic dunes, probably forming a small draa. The flat-bedding was deposited as a dry aeolian sand-sheet. (B) Coset of trough cross-bedding exhibiting upward thickening of trough set thickness. The thin trough sets are here locally deformed by soft-sediment deformation. The coset represents a draa, embedded in flat-bedded interdune (interdraa) deposits. Rucksack and hammer for scale. (C) Tabular coset of trough cross-bedding grading laterally downwind (left) into a solitary tabular cross-set, 20 m thick. This cross-set is overlain by a thin interdune deposit and another tabular cross-set. It is underlain by interdune deposits. The photograph is taken from a helicopter. (D) Low-angle and horizontally laminated sandstone with a single trough cross-set. These sediments were deposited on an aeolian sand-sheet with small crescentic dunes. Hammer for scale.

(Fig. 4C). More rarely trough- to wedge-shaped cross-bedding occurs as single sets less than 1m thick associated with the flat-bedded facies association (Fig. 4D). The foresets are composed of well sorted sandflow, grainfall and climbing translatent strata (cf. Hunter, 1977), and they possess maximum dips up to $c.30°$ towards the west-southwest (Fig. 5).

The flat-bedded facies association contains a wide range of sedimentary structures: (i) low-angle or horizontally laminated fine-grained sandstone beds with translatent strata (Fig. 4D) (abundant); (ii) wind ripple formsets (starved ripples) of medium- or coarse-grained sand (common); (iii) irregular adhesion lamination (rare to common); (iv) cross-lamination and medium-scale trough cross-bedding, commonly with intra- and extraclasts (rare); (v) symmetrical ripple formsets (rare). This facies association is closely associated with and grades laterally into single cross-sets and rare cosets, generally less than 1 m thick.

When the cross-bedded facies association dominates, tabular cosets or locally single sets up to 20 m thick commonly alternate with 0.5–3 m thick units of the flat-bedded facies association. Where the flat-bedded facies association dominates, no regular vertical sequence pattern is observed.

Interpretation

An aeolian dune origin of the cross-bedded facies association is indicated by the strata types and structural aspects. The facies association was mainly deposited by crescentic dunes forming part

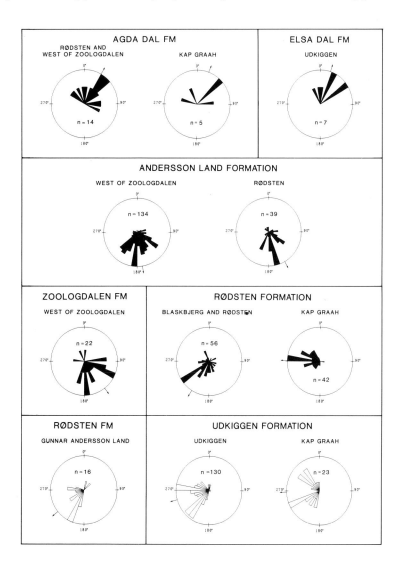

Fig. 5. Palaeoccurrent rose diagrams (linear plots) for the studied formations at different localities. Vector means and number of measurements are indicated. Fluvial currents are indicated by solid plots, winds are indicated by open plots.

of larger transverse compound dune ridges or draas (e.g. Breed & Grow, 1979; Clemmensen & Abrahamsen, 1983). The draas were only locally associated with slipfaces, indicated by the segments of tabular cross-bedding lateral to thick trough cosets. The common thickening upwards of trough sets within the cosets may indicate a larger dune height on the upper part of the draa leeside, or it may reflect a lower preservation potential of dune cross-sets formed on the low-angle dipping lower leeside of the draa, or both. The small single cross-sets associated with the flat-bedded facies association represent the migration of small isolated crescentic dunes.

The assemblage of sedimentary structures in the flat-bedded facies clearly indicates that this association is also mainly aeolian. The low-angle and horizontal lamination and starved ripples represent aeolian deposition in a relatively dry, low-relief sand-sheet or interdune environment (Fryberger et al., 1979, 1983; Ahlbrandt & Fryberger, 1981; Kocurek, 1981; Kocurek & Nielson, 1986; Lancaster & Teller, 1988). Temporary wetting is indicated by adhesion lamination (adhesion warts and ripples), cross-lamination and cross-bedding of fluvial origin, and symmetrical ripple forms of wave origin (e.g. Ahlbrandt & Fryberger, 1981; Kocurek, 1981; Kocurek & Nielson, 1986; Lancaster & Teller, 1988; Olsen et al., 1989). When the flat-bedded facies association regularly alternates with thick beds of the cross-bedded facies association it is interpreted as interdune deposits. When the flat-bedded facies association dominates it is interpreted as sand-sheet deposits.

The formation as a whole is interpreted as an erg dominated by draas of transverse crescentic dune ridge type migrating towards the west-southwest. The draas were mainly devoid of slipfaces and separated by interdune flats with associated small barchanoid dunes. Locally and/or at certain times aeolian sand-sheets dominated associated with small dunes or draas. In this respect the erg exhibits many similarities to the eastern Rub' al Khali in Saudi Arabia (Breed & Grow, 1979). The consistent dip direction of the dune foresets probably indicates trade winds from the east-northeast.

Zoologdalen Formation

Description

This depositional system is composed of alternating siltstones (to silty mudstones) and very fine- to fine-grained (and rarely medium-grained) sandstones (Fig. 6A). Siltstones generally dominate the exposed part of this formation, forming between 40 and 100% in different stratigraphic levels.

The siltstones are generally flat-bedded with parallel lamination, cross-lamination, bioturbation and desiccation cracks. Locally calcrete horizons are observed in the siltstones. The sandstones fine upwards and are trough cross-bedded, cross-laminated and parallel laminated. Most of the sandstone bodies are less than 1m thick, single or composed of thinner individual fining upward sandstone beds, but may reach up to 5 m in thickness. Such thicker sandstone bodies are invariably multistorey and/or multilateral (Fig. 7A). Desiccation cracks and rootlet horizons are common in the top of sandstone bodies as well as between individual beds in a sandstone body. The overall aspect of the sandstone bodies is sheet-like (Fig. 7B). Bases of sandstone bodies and beds are commonly erosive and intraclast strewn, but rarely cut down more than a few centimetres. Rare exceptions are the thick (2–5 m) multistorey sandstone bodies, where individual storey bases may exhibit downcutting with a storey scale relief of 0.5–2.5 m, though the base of a multistorey sandstone body is generally flat. Palaeocurrents in this formation are towards the southeast (Fig. 5). The ratio of sandstone to siltstone decreases in a downcurrent direction, and multistorey/multilateral sandstone bodies are very rare in the furthest downcurrent exposures.

Interpretation

The formation can readily be interpreted as fluvial, and deposited by an ephemeral stream system. In general aspects it resembles the modern Markanda River terminal fan (Mukerji, 1976; Parkash et al., 1983) and ancient examples of inferred terminal fan origin (Graham, 1983; Tunbridge, 1984; Olsen, 1987). The general sheet-like nature of the deposits and the lack of erosional downcutting suggest accretion on broad, poorly channelized alluvial plains. The multibed sandstone bodies can be interpreted as a series of vertical accretion deposits during successive floods in broad channels with relatively long residence times. The single sandstone beds were probably deposited in channels with residence times in the order of flood duration. The thick sandstone bodies represent the main distributary channels on the terminal fans. Their multistorey

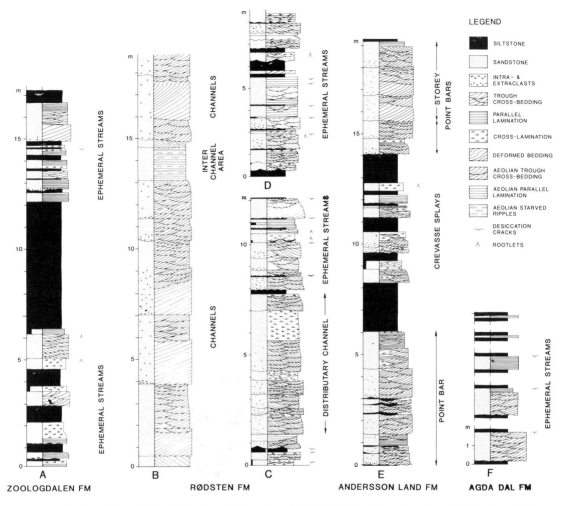

Fig. 6. Graphical logs from the fluvial dominated formations. (A) Zoologdalen Formation, deposited by a terminal fluvial system (locality: west of Zoologdalen). (B) Proximal deposits from the terminal fluvial system of the Rødsten Formation (Kap Graah locality). (C & D) Thin-bedded facies association of the Rødsten Formation: (C) intermediate deposits (Blaskbjerg), and (D) distal deposits (Rødsten). (E) Meandering river and flood basin deposits from the Andersson Land Formation (west of Zoologdalen). (F) Ephemeral stream deposits from the Agda Dal Formation (west of Zoologdalen).

and multilateral aspects resemble sandstone bodies described from Lower Jurassic desert streams (Olsen, 1989) and are likewise interpreted as braided channel deposits. The siltstones were mainly deposited as overbank deposits but some were deposited as channel fills. Calcretes indicate well drained alluvial plains and occasional sediment starvation (e.g. Leeder, 1975).

The downcurrent scarcity of distributary channel deposits and of sandstones in general indicates a medial to distal position on the fan represented in the outcrop area. The fan system is interpreted to have a source in the highlands bordering the Devonian basin in the west as indicated by the southeastward palaeocurrents.

Rødsten Formation

Description

The Rødsten Formation is the most heterogenic depositional system in the area. It is composed of

Fig. 7. Zoologdalen Formation. (A) Multistorey sandstone body, 2 m thick, interpreted in terms of a main distributary channel in a terminal fan system. (B) Multibed sandstone body interpreted as stacked channel deposits in a channel characterized by ephemeral flow. The shape of individual beds as well as the whole sandstone body is sheet-like, with only local downcutting, indicating a very wide channel compared with the depth of the channel. Lens cap is 5 cm.

two main facies associations, a cross-bedded facies association and a thin-bedded facies association, and two facies associations of only local importance, the conglomeratic facies association and the variously stratified facies association.

The cross-bedded facies association is dominated by sandstones with cosets of 10–30 cm thick trough sets, associated with 0.5–2 m thick wedge-shaped cross-sets, which occur singly (Figs 6B & 8A). Rarely, parallel lamination with parting lineation occurs, though locally it forms up to 30%. The sandstones are moderately to well sorted medium- to coarse-grained. Extraformational small pebbles and granules form less than 5%. The sandstones occur in broadly channel-shaped storeys (several tens to hundreds of metres wide) with limited local relief. Individual channel-shaped storeys are 1–4 m thick and exhibit fining upward lithology. Also, trough set height commonly decreases upwards through a storey. The wedge-shaped cross-sets usually occupy the lowest portion of a storey. The sets are sometimes traceable up- or downcurrent into trough cross-bedded cosets with inclined set boundaries. The medium- to coarse-grained sandstones are associated with fine- to medium-grained sandstone beds up to 2 m thick with parallel lamination and trough cross-bedding dominated by translatent strata (Fig. 6B). These beds generally form less than 10% of the cross-bedded facies association.

The second major deposit type, the thin-bedded facies association, is highly varied. It is dominated by trough cross-bedded fine- to medium-grained

Fig. 8. Rødsten Formation. (A) Cross-bedded facies association from the proximal part of the system, composed of trough- and wedge-shaped cross-sets. Palaeocurrent is into the rock face. Ten-centimetre bar for scale. Locality: Kap Graah. (B) Thin-bedded facies association, composed of decimetre to metre scale sandstones interbedded with siltstones, deposited in the distal part of the terminal fluvial system. The upper 10 m of the cliff face is composed of aeolian trough cross-bedding of the variously stratified facies association. The cliff face is c.50 m high facing obliquely to the palaeocurrent. Locality: Rødsten. (C) Tabular beds of the conglomeratic facies association from the Kap Graah locality. Palaeocurrent is into the cliff face. The cliff is c.15 m high. (D) Variously stratified facies association with aeolian trough cross-bedding. Figure for scale. The sediments are interpreted as interchannel deposits. Locality: Blaskbjerg.

sandstones, associated with cross-laminated and parallel laminated fine-grained sandstones. In general the facies association is characterized by fining upward sandstone beds, 0.2–1.5 m thick, interbedded with desiccation cracked siltstones (Fig. 6D). In the downcurrent part of the formation (Rødsten and Blaskbjerg) the siltstone beds are centimetres to decimetres thick and form c.10% of the facies association. In the upcurrent part (Kap Graah) siltstones exclusively occur as millimetres thick partings. The overall aspect of the sandstone beds is sheet-like (Fig. 8B). Bases of beds are erosive but local relief rarely exceeds a few centimetres. Desiccation cracks and rootlets are commonly observed in the tops of the sandstone beds. The palaeocurrent spread in thin-bedded successions is generally high, locally up to 180°. The thin bedding of sheet sandstones is interrupted by thicker (1–5 m) tabular sediment bodies of stacked channel-shaped fining upward storeys, sharing the same general palaeocurrent direction (within 90°). These sediment bodies are mainly composed of trough cross-bedded medium-grained sandstones (Fig. 6C). The sediment bodies form up to 40% of the facies association in the upcurrent part. Sandstones with translatent strata, grainfall strata and starved ripple strata generally form 5–10% of the facies association. These stratification types usually occur as decimetre thick parallel laminated beds, and trough sets less than 0.5 m thick occur rarely, either singly or in cosets up to 1 m in thickness.

The conglomeratic facies association is composed of pebbly conglomerates and sandstones with

10–90% extraformational clasts. The sediments are generally poorly sorted and dominated by scour-and-fill structures with massive fills. Parallel lamination and cross-bedding are rarely developed. The facies association is bedded in 0.2–1.5 m thick tabular beds (Fig. 8C). Bases of beds are always erosive and usually planar. Individual beds are normally graded (abundant), non-graded (common) and inversely graded (rare). Commonly the bases of pebbly sandstone beds are covered by a pebbly lag deposit. The conglomeratic facies association is developed as a localized coarsening to fining upward wedge up to 30 m thick in the Kap Graah area, wedging out within less than 10 km in a downcurrent, westward direction (Nicholson & Friend, 1976) as well as transversely to the general current direction. The wedge of exclusively conglomeratic facies is enclosed in and gradational to sediments of the thin-bedded facies association.

The last facies association in the Rødsten Formation, the variously stratified facies association, is dominated by fine- to medium-grained sandstones with translatent strata, grainfall strata and starved ripple strata forming 60–85% of the facies association. These fine- to medium-grained sandstones are mainly composed of trough sets 0.5–1 m, and rarely up to 2.5 m, thick forming cosets up to 15 m thick (Fig. 8B and D). The rest of these fine- to medium-grained sandstones are composed of flat-bedded parallel lamination. Medium-grained trough cross-bedded and parallel laminated sandstones with intraformational and rare extraformational clasts form the remaining 15–40% of the facies association. These rocks occur as channel-fill bodies 0.2–1 m thick. The assemblage of sedimentary structures resembles the Udkiggen Formation. In contrast to that formation, however, only rarely is a sequential regularity of alternating trough cosets and flat-bedding observed, and intraclast-rich sandstones are more abundant. The facies association usually exhibits an unsystematic interbedding of the different subfacies. The variously stratified facies association forms a succession more than 200 m thick in the upcurrent locality, Kap Graah. In the downcurrent localities (Rødsten, Blaskbjerg) it occurs as rare 10–15 m thick units.

Interpretation

The cross-bedded facies association is interpreted as stacked deposits of fluvial channels. A sandy braided stream environment is envisaged, composed of channels with dunes, and of foreset macroforms associated with both slipfaces and leesides with descending dunes, and rare plane beds. Thin aeolian interchannel deposits, dominated by translatent strata, commonly escaped channel erosion. The facies association is comparable with modern braided river deposits (Cant & Walker, 1978) and ancient examples described by Allen (1983), Kirk (1983) and Haszeldine (1983a, b).

The thin-bedded facies association is of typical ephemeral stream origin resembling modern deposits from Australia (Williams, 1971), Sinai (Sneh, 1983) and India (Parkash et al., 1983). Ancient examples have been described by Steel & Aasheim (1978), Graham (1983), Clemmensen et al. (1989) and Olsen (1987, 1989). The streams were poorly defined channels and sheetfloods, resulting in the general tabular aspect of the thin sandstone beds. The thicker sandstone bodies can be interpreted as the deposits of relatively persistent fluvial channels. The stacking of storeys within these sediment bodies reflects either deposition in component channels within a main channel or successive flood infillings of a major channel. The abundance of thin sandstone beds with aeolian stratifications in the thin-bedded facies association indicates common aeolian reworking.

The conglomeratic facies association is interpreted in terms of small longitudinal gravel bars and associated fluvial channels collectively forming broad, low-relief, bar complexes (e.g. Boothroyd & Ashley, 1975; Boothroyd & Nummedal, 1978; Nemec & Steel, 1984; Miall, 1985; Collinson, 1986) or chutes and lobes (Southard et al., 1984).

The variously stratified facies association is mainly of aeolian origin as evidenced by the strata types (cf. Hunter, 1977). Small aeolian dunes dominated, associated with interdunes and sand-sheets. The lack of thick successions with systematic interbedding of dune cosets and interdunes indicates that aeolian bedforms were mainly isolated or occurred as small dune fields in contrast to the erg envisaged for the Udkiggen Formation. The high proportion of fluvial deposits (sandstones with intraformational clasts) indicates common inundation of the area and thereby hampering of major dune or draa construction.

Palaeocurrents in the fluvial deposits are generally to the west. Palaeowinds in the aeolian sandstones are to the southwest (Fig. 5).

The formation represents an alluvial system with associated aeolian interfluvial areas. The system is characterized by a rapid downcurrent change in facies association types, probably indicating a rapid downcurrent decrease in gradient. The two main facies associations represent the proximal (cross-bedded facies association) and medial to distal part of the system (thin-bedded facies). The proximal part was characterized by large braided channels with long residence time and perennial flow, associated with interchannel areas exposed to aeolian activity (variously stratified units). The medial part of the system was characterized by a range of fluvial channels, from large distributary channels with a long residence time and fluctuating flow conditions to small ephemeral channels and sheetfloods with residence times in the order of flood duration. In the distal part of the system only ephemeral streams and sheetfloods with flood duration residence times existed. The thin aeolian beds, associated with the medial and distal part of the system, indicate short-term reworking of the fluvial sand between successive floods.

The variously stratified facies association is interpreted in terms of interchannel areas dominated by aeolian processes. The thick succession of this facies association in the proximal area at Kap Graah indicates large stable interchannel areas in the proximal part of the system. In contrast, the medial to distal facies at Rødsten and Blaskbjerg are associated with only thin and rare interchannel deposits indicating restricted interchannel areas and/or limited residence times. Palaeowind directions indicate trade winds from the northeast.

The formation as a whole is interpreted as a terminal distributary system with major 'feeder channels' in the proximal part with associated interchannel areas. Downcurrent these channels divided into multiple distributary channels with relatively long residence times and associated small ephemeral flood channels and sheetfloods. Further downstream the distributary channels disappeared and ephemeral streams and sheetfloods prevailed. Channel switching or temporal incision of channels in the proximal area resulted in prolonged interchannel conditions in a part of the proximal area (thick succession of the variously stratified facies association). However, because of the downcurrent multiple division of channels such switching or incision had only limited influence on the downcurrent part of the system, and interchannel areas were rapidly reoccupied by the fluvial streams.

The system is comparable to modern single entry terminal fan systems described from India (Parkash et al., 1983) and Sudan (Abdullatif, 1989), although the inferred Devonian system had multiple entries. In this respect parallels may be drawn to glacial outwash plains in Iceland (Boothroyd & Nummedal, 1978) and wadibelts in Libya (Glennie, 1970). Comparable ancient examples have been described by Graham (1983), Tunbridge (1984) and Olsen (1988a).

The conglomeratic wedge is interpreted as proximal alluvial fan deposits. The proximal fan environment graded downcurrent into ephemeral streams (thin-bedded facies association) forming the distal part of the alluvial fan. Collectively the conglomeratic facies association and the associated thin-bedded facies association form a $c.120$ m thick coarsening to fining upward alluvial fan unit at the base of the thickest part of the Rødsten Formation (see chronosome II in later section) at Kap Graah. The alluvial fan is thought to have fed the terminal fan system further westward.

Andersson Land Formation

Description

This formation is composed of alternating sandstones and siltstones. Tabular bodies of fine- to medium-grained sandstones (with rare fine pebbles), 2–15 m thick, form 25–75% of the formation. The sandstones occur both as single-storey (Fig. 9A) and as multistorey bodies (Fig. 9C). Each storey is underlain by a horizontal erosion surface with only limited relief. Internally the storeys are composed of fining upward bedsets separated by inclined ($c.7°$) erosion surfaces. The bedding, composed of trough cross-sets, parallel and cross-lamination and siltstone partings, is parallel to these inclined surfaces and palaeoflows are directed perpendicular to the dip of the surfaces. Commonly, but not always, individual storeys exhibit a gross fining upwards (Fig. 6E). Palaeocurrents are generally towards the south (Fig. 5). The tabular sandstone bodies exhibit cyclic variations in thickness, with cycles of the order of 100 m thick and superimposed smaller cycles 20 m thick (Olsen, 1990).

The tabular sandstones are embedded in siltstones and silty mudstones with associated very fine- to fine-grained broad lenses of sandstone less than 1 m thick (Fig. 9A). Desiccation cracks and rootlet horizons are common in these finer grained

Fig. 9. Andersson Land Formation. (A) A tabular sandstone body (single-storey), 7 m thick, deposited as a point bar in a meandering river, is embedded in flood basin siltstones with thin crevasse splay sandstones. Palaeocurrent is parallel to the cliff face, away from the observer. (B) Crevasse splay sandstone with intricately interbedded parallel lamination, cross-lamination and trough cross-bedding. Pencil for scale. (C) Multistorey sandstone body deposited by two point bars. The two storeys are indicated. Epsilon cross-bedding is observed in the upper storey, dipping towards the right. Palaeocurrent is towards the observer. Figure for scale at the base of the sandstone body. All photographs are from localities west of Zoologdalen.

deposits. The sandstone beds are mostly composed of intricately interbedded parallel laminated, cross-laminated and low-angle trough cross-bedded sandstone (Fig. 9B). Commonly a few centimetres to decimetres of sand with translatent strata occur in the tops of these sandstone lenses.

Interpretation

The tabular sandstone bodies resemble inferred ancient point bar deposits (e.g. Allen, 1965; Moody-Stuart, 1966; Beutner *et al.*, 1967; Leeder, 1973; Puigdefábregas, 1973; Puigdefábregas & van Vliet, 1978; Bridge & Diemer, 1983). The lateral accretion bedding, envisaged from the relationship of inclined bedding and palaeocurrents, justifies a similar interpretation for the present deposits. They are accordingly interpreted as point bars in a meandering river system. The gross fining upwards of many storeys is a result of the helicoidal flow pattern in meander loops. Vertical grain-size distributions differing from the general rule probably indicate the existence of more than one spiral vortex in the meander loops (e.g. Jackson, 1976; Bridge, 1977). Fining upwards of individual component beds is interpreted as flood controlled (cf. Bridge & Diemer, 1983). The cyclic variations in thickness of the tabular sandstone bodies are interpreted to reflect cyclic variations in channel size caused by Milankovitch type climatic changes (Olsen, 1990). The siltstones and silty mudstones are typical flood basin deposits, and the associated thin sandstone beds are interpreted as crevasse splay deposits comparable to modern (Farrell,

1987) and ancient examples (Bridge, 1984). Translatent stratification (cf. Hunter, 1977) in the tops of thin sandstone beds indicates aeolian reworking of the splays. The meandering river drained the basin longitudinally along the basin axis.

Agda Dal Formation

Description

The Agda Dal Formation was studied in the field at Kap Graah, on Rødsten and on the mountain immediately west of Zoologdalen. At Kap Graah the Agda Dal Formation is composed of medium-grained sandstones dominated by trough cross-bedding and parallel lamination. The sandstones occur as broadly channel-shaped storeys 2–4 m thick. On Rødsten the formation is composed of fine- to medium-grained sandstones in tabular beds 0.2–1 m thick. The sandstone beds fine upwards and are dominated by trough cross-bedding, parallel lamination and cross-lamination. Siltstones in beds centimetres to decimetres thick are interbedded with the sandstones and exhibit desiccation cracks. Siltstones form only 10% of the rocks at Rødsten. In contrast to the varied facies of the Rødsten Formation, no aeolian sandstones are observed in the Agda Dal Formation at this locality. West of Zoologdalen the formation is composed of 70% siltstones with associated very fine- to coarse-grained tabular sandstone beds 0.1–1.6 m thick. The sandstones exhibit fining upwards beds and parallel lamination, trough cross-bedding and minor cross-lamination (Fig. 6F). Desiccation cracks abound on top of sandstone beds. Palaeocurrents are to the north-northeast (Fig. 5).

Interpretation

The formation is interpreted as being composed of braided stream (at Kap Graah) and ephemeral stream deposits (at Rødsten and west of Zoologdalen). Within Gunnar Andersson Land insufficient data were collected to indicate whether the Agda Dal Formation was deposited by terminal fluvial systems, as in the Rødsten Formation. The facies spectrum observed in the Rødsten Formation is also recognized in the Agda Dal Formation. This may indicate a similar type of alluvial system. However, the silt content and the 'ephemerality' decrease towards the east, combined with a weakly developed change from palaeocurrents to the north-east to a more northerly direction (Fig. 5). This may indicate that transverse terminal fluvial systems, dominated by ephemeral flow conditions and connected to the western basin margin, occupied the main part of the basin floor, with a gradual transition to a longitudinal perennial braided system in the eastern part of the area. No signs of a transverse system connected to the eastern basin margin occur in Gunnar Andersson Land. An alternative interpretation is an extensive and asymmetrical longitudinal alluvial plain with a general northward palaeoslope. The discharge thus varied across the alluvial plain with a maximum along an axis in the east. This latter alternative is preferred. The most remarkable feature of the Agda Dal Formation is the palaeoslope direction, which is almost opposite to the direction of preceding fluvial systems, indicating a drainage reversal.

Elsa Dal Formation

Description

The deposits of this formation crop out only on Udkiggen and near Kap Graah and are composed of fine- to medium-grained well sorted sandstones. They are dominated by trough cross-bedding associated with parallel lamination and rare cross-lamination. Siltstone occurs as intraclasts. The exposures are generally poor but in outcrops north and south of Gunnar Andersson Land (Olsen, 1993) the sandstones are seen to be composed of small, 1–3 m thick, grossly fining upward or non-graded storeys locally displaying lateral accretion bedding (Fig. 10). Palaeocurrents are to the north-northeast (Fig. 5).

Interpretation

The formation is fully alluvial. The local occurrence of lateral accretion deposits indicates a fluvial system with local meandering reaches and/or flank accretion on braid bars. Modern examples have been described by Shelton & Noble (1974), Schwartz (1978) and Bristow (1987). Ancient examples have been described by Allen (1983) and Olsen (1988b). The deposits are interpreted in terms of a braidplain with perennial braided–meandering channels. The drainage direction was similar to the underlying Agda Dal Formation.

Fig. 10. Elsa Dal Formation. Lateral accretion bedding parallel to channel margin. To the right of this channel and above it vertical accretion bedding is observed. The lateral accretion bedding grades into vertical accretion bedding to the left of the photograph. Channel margin and top of channel unit are indicated. The base is just below the grass cover. Palaeoflow into the cliff face. Hammer for scale at the channel margin. Locality: Gauss Halvø, north of Gunnar Andersson Land.

ALLUVIAL ARCHITECTURE OF MEANDER BELT DEPOSITS (ANDERSSON LAND FORMATION)

Reconnaissance studies on the mountain to the west of Zoologdalen (Fig. 11) indicated systematic variations in the Andersson Land Formation. In order to examine these variations a detailed investigation of the density of sandstone bodies (number of tabular sandstone bodies per 100 m sediment thickness) was carried out along a section on the mountain ridge through a thickness of about 900 m (Fig. 12).

The density of sandstone bodies decreases radically, from 14.8 in the western/lowermost part of the section to 7.9 in the eastern/uppermost part. Because the section cuts obliquely through the succession it cannot be demonstrated from this section alone whether this trend reflects lateral or vertical variations in the succession. Therefore a comparative count was made in a c.200 m thick ravine section stratigraphically equivalent to the western/lowermost part of the ridge section but laterally east of the ridge outcrop (Figs 11 & 12). In comparison, the sandstone-body density decreases in an eastward direction from 14.8 to 12.7 along a bed-parallel distance of about 2.5 km perpendicular to the inferred general drainage direction of the meandering river.

According to Bridge & Leeder (1979), such a lateral eastward decrease in sandstone body density may indicate a westward tilting of the flood plain perpendicular to the channel, which may also account for the observed systematic decrease in sandstone-body density along the oblique section on the mountain ridge. The decrease is, however, larger in the mountain ridge section than would be expected from the lateral change within the lowermost part of the succession, i.e. the sandstone-body density along the mountain ridge is lower (8.9) than in the ravine section (12.7) at the same lateral position (perpendicular to the bedding) (Fig. 12).

This could point towards an increased rate of flood plain tilting during the depositional history of the formation (Bridge & Leeder, 1979; Alexander & Leeder, 1987). An alternative explanation in terms of a general increase in the aggradation rate is ruled out by independent evidence for a uniform aggradation rate. Throughout the succession, Milankovitch type cycles of uniform thickness (c.100 m) have been shown to exist (Olsen, 1990).

According to Alexander & Leeder (1987) and Leeder & Alexander (1987), two styles of lateral channel migration may occur due to transverse (to river drainage) flood plain tilting: instantaneous downtilt avulsion or slow downtilt migration by preferred downtilt erosion leaving behind convex-uptilt meander loop cut-offs. The actual process in the Andersson Land Formation was investigated in a detailed study of epsilon cross-beds in the meander belt point bar sandstones. Excellent exposures in parts of the succession made it possible to calculate the original point bar orientation and the associated palaeocurrent directions (Olsen, 1988c). Point bar orientation was calculated by measuring

Fig. 11. Vertical aerial photograph of the mountain west of Zoologdalen, displaying distinct sandstone ridges of individual tabular sandstone bodies. The bedding dips towards the east. The sections where the counting of sandstone bodies was carried out are indicated (along the mountain ridge and in a ravine). (Photograph copyright Geodaetic Institute, Denmark A.200/87; route 888 no. M1264.)

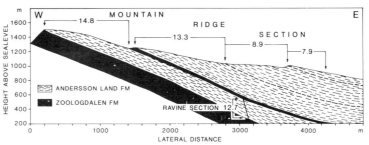

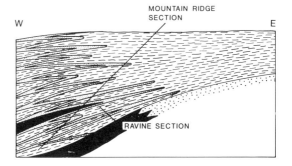

Fig. 12. Density of sandstone bodies (number of tabular sandstone bodies per 100 m sediment thickness) in the two measured sections in the Andersson Land Formation (see location in Fig. 11). The sections are projected into a W–E plane. The lower drawing is a schematic presentation of the systematic variation in the density of sandstone bodies caused by the evolution of the Slippen anticline to the east.

the mean orientation of parting planes in parallel laminated parts of epsilon cross-beds, with reorientation of these values allowing for the structural dip. Twenty-six epsilon cross-beds were measured in this way (Fig. 13). A total of twenty epsilon cross-beds indicate clockwise flow around point bars (seen from above) and thus deposition in convex-eastward meander loops. Only three point bars were deposited in convex-westward meander loops, whereas three epsilon cross-beds revealed no palaeocurrent indications. Most of the measured epsilon cross-beds represent the uppermost storey in their sandstone bodies, and thus probably record the orientation of meander loops prior to avulsion or

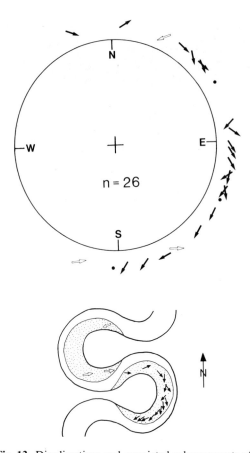

Fig. 13. Dip directions and associated palaeocurrents of 26 epsilon cross-beds (point bars) from the Andersson Land Formation. The position of individual arrows (centre of arrow) and points indicates the dip direction of individual epsilon cross-beds (reorientated allowing for the structural dip), and the directions of the arrows indicate the palaeocurrent means for the individual epsilon cross-beds. Points indicate that no palaeocurrent directions were measured. The lower sketch indicates the depositional process. Epsilon cross-beds represented by open arrows were deposited on point bars in convex-westward meander loops, whereas epsilon cross-beds represented by bold arrows were deposited on point bars in convex-eastward meander loops. The dominance of point bars from convex-eastward meander loops indicates preferred cut-offs towards the east due to westward tilting of the flood plain.

cut-off. Avulsion would, according to Leeder & Alexander (1987), lead to epsilon cross-beds showing no preferred dip direction, whereas lateral downtilt migration would result in a preferred uptilt dip of epsilon cross-bedding from convex-uptilt meander loops. The observed asymmetric dip distribution towards the east accordingly indicates

repeated downtilting towards the west associated with slow downtilt migration of the river.

In conclusion, epsilon cross-bed orientations and lateral variations in sandstone-body density collectively point towards repeated and increased transverse westward flood plain tilting during deposition of the Andersson Land Formation. As will be shown in the following sections, the tilting was caused by the formation of the so-called Slippen anticline.

PALAEOGEOGRAPHIC BASIN EVOLUTION AND UNDERLYING CONTROLS

Stratigraphic architecture

Schultz (1982) proposed the term *chronosome* for a rock body delimited by marker horizons or by well defined biostratigraphic horizons having local chronostratigraphic or event-stratigraphic significance. A chronosome, as defined by us, is a complex of all contemporaneous depositional systems in the basin, and is delimited by time planes or unconformities. The chronosome philosophy is closely related to sequence stratigraphy (e.g. van Wagoner et al., 1988) established for shallow marine settings. Chronosomes are comparable to system tracts, i.e. chronosome boundaries define times of major palaeogeographic reorganization. The definition of chronosome boundaries as time planes is solely based on sedimentological criteria, i.e. there is unlikely to be a direct sedimentological genetic relationship between rocks below and above a chronosome boundary. Since completion of this paper, an event stratigraphy has been established for the entire basin fill, and the term chronosome replaced by the term *complex* (Olsen, 1993).

In the Gunnar Andersson Land succession, three chronosomes have been defined (Fig. 3). Because the lithostratigraphy is based on a depositional system analysis, the schematic cross-section in Fig. 3 also indicates the palaeogeography at any time in the depositional history of the Gunnar Andersson Land succession.

Structural setting

The sedimentation in the Devonian basin in East Greenland was accompanied by a series of synsedimentary deformation phases, the Hudson Land phases 1–4 of Bütler (1935). The deformations, which comprised folding and faulting, led to several angular unconformities within the stratigraphic column. The Hudson Land phases, according to Bütler (1935), only affected the northern part of the Devonian basin and were not considered to have had any influence in the Gunnar Andersson Land area and in the areas farther south. Following the Hudson Land phases the entire Devonian basin was folded and deformed in Early Carboniferous time during the Ymer Ø phase of Bütler (1935). Wide, open N–S trending folds developed in Gunnar Andersson Land, referred to as the Zoologdalen syncline, the Slippen anticline and the Kap Graah syncline by Bütler (1935, 1955) (Fig. 14). However, recent investigations in Gunnar Andersson Land have shown that these structures initiated much earlier than Early Carboniferous and thus evolved simultaneous to the deposition of the Late Devonian sediments in the area (Larsen, 1990b,c,d).

In the following the Late Devonian basin evolution represented by the Gunnar Andersson Land succession will be described. Each of the three palaeogeographical basin configurations represented by the three chronosomes (Fig. 3) is discussed in relation to structural activity and climatic variations.

Chronosome I

Chronosome I is marked by a rapid change from the basin-wide Sofia Sund Formation of the Kap Kolthoff Group to a threefold subdivision into the Zoologdalen, the Udkiggen and the Rødsten Formations (Fig. 3). It is dominated by the Zoologdalen Formation in the west and the Udkiggen Formation in the eastern and central parts. The Rødsten Formation is present only as a thin wedge in the easternmost part of the chronosome. The palaeogeography reconstructed from this chronosome indicates a huge erg (Udkiggen Formation) covering the main part of the basin in the Gunnar Andersson Land region, and also outside this region (Olsen, 1993; Olsen & Larsen, 1993). A large terminal fluvial system (Zoologdalen Formation) is developed in the western part of the basin, whereas a restricted terminal fluvial system (Rødsten Formation) is situated in the east (Fig. 15, chronosome I).

The remarkable change from a braid-plain depositional environment (Sofia Sund Formation) to an erg and terminal fluvial systems is explained by a climatic shift towards drier conditions. No struc-

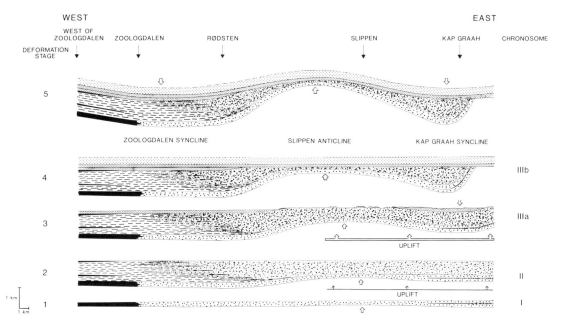

Fig. 14. The structural evolution of Gunnar Andersson Land illustrated by five successive cross-sections parallel to the cross-section in Fig. 2, but with an exaggerated vertical scale. See text for explanation. Ornamentation corresponds to Figs 2 and 3.

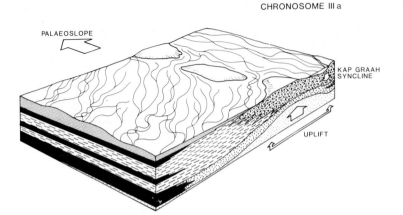

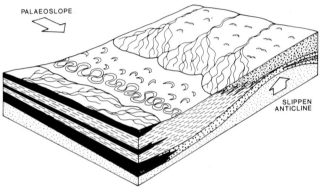

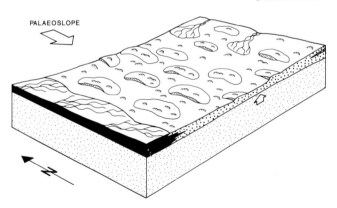

Fig. 15. Block diagrams illustrating the palaeogeography and underlying tectonic evolution during deposition of the three chronosomes. Chronosome I: terminal fluvial systems draining into the basin from the margins (Rødsten and Zoologdalen Formations) separated by an erg (Udkiggen Formation). Chronosome II: the eastern terminal fluvial system (Rødsten Formation) progrades westwards and the erg is replaced by a meander belt (Andersson Land Formation). Chronosome III a: an alluvial plain composed of mainly ephemeral streams with perennial streams restricted to the easternmost part. During continued deposition across the whole area (b in Fig. 14, not shown here) a dominance of perennial streams occurred, but the overall configuration of the alluvial plain was maintained.

tural controls seem to have been involved. The component braidplain segments of the Sofia Sund Formation were transformed into terminal systems. Due to the limited precipitation the fluvial systems terminated close to the basin margins and conditions were favourable for the development of an erg in the central part of the basin.

The thickness of chronosome unit I is greatest at Kap Graah (350 m) and Rødsten (290 m), whereas the thickness in the intervening area is less (220 m at Slippen). These variations are interpreted as the result of embryonic evolution of the Slippen anticline (Fig. 14, stage 1). A reduced thickness (210 m) of the chronosome west of Zoologdalen is interpreted as the result of later compaction of the siltstone/mudstone dominated Zoologdalen Formation. The embryonic evolution of the Slippen anticline appears to have no influence on the distribution of depositional systems within the chronosome unit.

Chronosome II

Chronosome II is characterized by the disappearance of the Udkiggen Formation and the appearance of the Andersson Land Formation (Fig. 3). The Andersson Land Formation interfingers with the Rødsten Formation to the east, and with the Zoologdalen Formation to the west. The diachronous boundary between the Rødsten and Andersson Land Formations is successively displaced towards the west (Fig. 3). At Rødsten c.100 m of the Andersson Land Formation is succeeded by several hundred metres of the Rødsten Formation, whereas west of Zoologdalen the Andersson Land Formation forms more than 1000 m (Fig. 14, stage 2) and the Rødsten Formation is absent. In contrast, the diachronous boundary between the Zoologdalen Formation and the Andersson Land Formation does not show any general inclination. Accordingly, the relative position of these two depositional systems was not changed systematically.

The palaeogeographic reconstruction from chronosome II suggests an axial meandering river (Andersson Land Formation) separating a westward prograding terminal fluvial system in the east (Rødsten Formation) from an essentially stationary terminal fluvial system in the west (Zoologdalen Formation) (Fig. 15). The streams in the transverse fluvial systems probably terminated before they reached the axial meandering river, but during extreme floods they may have drained into the axial system.

Chronosome II shows conspicuous thickness variations across the basin. A sequence of approximately 1150 m accumulated west of Zoologdalen, at Rødsten c.1050 m was deposited, whereas only 725 m was deposited in the Kap Graah area (Fig. 14, stage 2). Accordingly, uplift of the eastern part of the basin must have occurred to the west of Slippen. The upper part of chronosome II at Slippen is removed by erosion and it is, accordingly, not possible to demonstrate whether the Slippen anticline was evolving at the same time, but it seems likely. No deformation in the western part of the basin has been detected at this stage (Fig. 14, stage 2).

The expansion of the terminal fluvial system of the Rødsten Formation and the establishment of an axial meandering river as a replacement of the Udkiggen erg collectively point towards a climatic change to more humid conditions than observed in chronosome I. However, from evidence outside Gunnar Andersson Land it is shown that aeolian conditions continued to dominate in the northeastern part of the basin (Olsen, 1993). On the basis of the increasing tectonic activity and uplift of the eastern basin margin we suggest that a higher relief in the source areas might have resulted in increased local precipitation and thus expansion of the transverse fluvial system of the Rødsten Formation. The presence of a coarsening to fining upward alluvial fan sequence at the base of the chronosome at Kap Graah also indicates increased relief associated with fault activity along the eastern basin margin, at least at the onset of chronosome II deposition. Increased local precipitation north of the basin is also thought to have resulted in water transport through the part of the basin studied, by the axial meandering river of the Andersson Land Formation. A regional climatic change towards more humid conditions is therefore ruled out. The climate during deposition of chronosome II was, however, not constant, as shown by regular variations in the size of the meandering river, caused by relatively short-term (20 000–100 000 years) astronomically forced climatic cycles (Olsen, 1990).

The gradual westward displacement and narrowing of the meander belt during deposition appear, chronosome II to have been controlled by uplift of the eastern part of the basin (see previous section about flood plain tilting). The initially wide meander belt was successively forced westward and narrowed due to its constriction between the relatively stationary terminal fluvial system of the Zoologdalen Formation to the west and the rising anticline to the east (Fig. 14, stage 2).

Both terminal fluvial systems of this chronosome (Rødsten Formation and Zoologdalen Formation) kept pace with the deformation, and sediments were brought into the basin from, respectively, east and west. Rivers flowing from the east had to cross the evolving Slippen anticline. As the thickness of the Rødsten Formation increases west of Slippen the area between Slippen and Kap Graah must have acted as an area of both sedimentation and sediment bypass to the areas further west.

Chronosome III

Chronosome III is marked by the destruction of the lateral threefold subdivision of the stratigraphic architecture (Fig. 3). The Agda Dal Formation forms the lower part of the chronosome and the Elsa Dal Formation forms the upper part. Both formations are extensive, covering the total width of the exposed part of the basin in this area.

The palaeogeography consists of an extensive alluvial plain (Fig. 15, chronosome III). Initially, ephemeral stream processes characterized the main part of the depositional system with perennial streams restricted to the easternmost part (Agda Dal Formation). Perennial streams characterized the whole area during the late phase of the chronosome (Elsa Dal Formation). The reorganization of the palaeogeography from the chronosome II set-up to the chronosome III set-up was associated with a reversal of palaeoslope to a northward direction, and the basin axis was displaced to the Kap Graah syncline in the easternmost part of the area.

A prolonged uplift of the eastern parts of the basin (Fig. 15, chronosome III) in association with initiation of downfolding of the Kap Graah syncline caused erosion on the crest of the Slippen anticline, together with erosion on the eastern flank of the Kap Graah syncline. This led to a discontinuous unconformity between the Kap Graah and Celsius Bjerg Groups (Fig. 3). Today the basal conformable beds of the Agda Dal Formation are preserved in the core of the Kap Graah syncline, and at Rødsten and west of Zoologdalen where sedimentation probably continued without any major hiatus. On the crest of the Slippen anticline the thickness of the Rødsten Formation has been reduced from approximately 500–700 m originally to less than 200 m (Fig. 14, stage 3).

Little information is available of deformation stage 4 (Fig. 14) as sediments of the upper part of the Agda Dal Formation and the Elsa Dal Formation are only preserved unconformably overlying the eroded core of the Kap Graah syncline. At Rødsten and west of Zoologdalen nothing but the basal parts of the Agda Dal Formation are preserved. In Fig. 14, stage 4, a uniform thickness of c.500 m for the upper part of the Agda Dal Formation and the Elsa Dal Formation is inferred in the whole area because such a thickness is known for the formations outside Gunnar Andersson Land (Larsen, 1990b,c,d; Olsen & Larsen, 1993). This may be a simplification because the folding continued during stage 5, and it is likely that folding influenced the stratigraphic thickness also during stage 4.

Sedimentation during the transition from chronosome II to III was strongly influenced by the tectonic activity (stage 3). Across the basin the sedimentation was controlled by folding and uplift, and along the basin axis it was controlled by a palaeoslope reversal. The transition is interpreted to result from the passing of a tectonic threshold, when the palaeoslope was reversed due to basin inversion to the south. The cessation of the former threefold subdivision of the stratigraphic architecture, where the basin margins formed the site for major sediment input, and the development of an extensive alluvial plain across the whole basin, may also indicate an expansion or widening of the basin in an E–W direction. Evidence outside Gunnar Andersson Land supports this proposal, as sediments of the Celsius Bjerg Group transgress the earlier formed, fault controlled basin margins, e.g. in the Gauss Halvø area to the north (Nicholson & Friend, 1976; Larsen, 1990a).

The change in palaeogeography from chronosome II to III can hardly be explained in terms of climatic changes, which would affect the characteristics of these fluvial systems but not the palaeoslope and position of basin axis. However, climatic change towards more humid conditions could explain the transition from the Agda Dal Formation to the Elsa Dal Formation. The Elsa Dal Formation was laid down by fluvial systems devoid of ephemeral characteristics, in contrast to the Agda Dal Formation, while the general northward palaeoslope was maintained (Olsen, 1993). Chronosome III corresponds in general to the 'second phase of deposition in the Kap Graah Group' (Nicholson & Friend, 1976) and 'unit 8' (Friend et al., 1983). These authors also invoke tectonism in order to explain the transition from chronosome II to III (as defined here).

Continued folding during stage 5 (Fig. 14) postdates the sedimentation of the Devonian formations described above. The folding during this evolutionary stage may be related to the post-Devonian Ymer Ø deformation *sensu stricto* of Bütler (1935), but the present study indicates a close relationship between the Hudson Land and Ymer Ø phases in general.

SUMMARY AND CONCLUSIONS

The integrated sedimentological and structural study of Upper Devonian sediments presented here provides evidence for both structural and climatic controls on the depositional systems. The controls acted simultaneously. The major changes in the palaeogeography and thus basin fill composition occurred due to (i) a major climatic change towards dry conditions, followed by (ii) basin margin faulting and uplift and accelerating basin floor folding, and finally by (iii) a culmination in structural activity leading to a basin inversion in the south and a palaeoslope reversal in Gunnar Andersson Land. Whereas (i) and (iii) are purely climatic and structural, respectively, (ii) is a structural episode causing a change in the climate (increased local precipitation).

This paper demonstrates the importance of multidisciplinary studies in the analysis of complex continental basins.

ACKNOWLEDGEMENTS

We are grateful to P. Friend and R. Steel, who contributed with valuable reviews. H. Olsen would like to thank the Carlsberg Foundation for financial support. Thanks also to N. Turner, who typed the manuscript, and J. Halskov, B. S. Hansen, J. Lautrup and B. Thomas, who completed the figures. The paper is published with approval of The Geological Survey of Greenland.

REFERENCES

ABDULLATIF, O.M. (1989) Channel fill and sheet-flood facies sequences in the ephemeral terminal river Gash, Kassala, Sudan. *Sediment. Geol.* **63**, 171-184.

AHLBRANDT, T.S. & FRYBERGER, S.G. (1981) Sedimentary features and significance of interdune deposits. In: *Recent and Ancient Nonmarine Depositional Environments: Models for Exploration* (Eds Ethridge, F.G. & Flores, R.M.) pp. 293-314. Soc. Econ. Paleont. Miner., Tulsa, Spec. Publ. 31.

ALEXANDER, J. & LEEDER, M.R. (1987) Active tectonic control on alluvial architecture. In: *Recent Developments in Fluvial Sedimentology* (Eds Ethridge, F.G., Flores, R.M. & Harvey, M.D.) pp. 243-252. Soc. Econ. Paleont. Miner., Tulsa, Spec. Publ. 39.

ALLEN, J.R.L. (1965) The sedimentation and palaeogeography of the Old Red Sandstone of Anglesey, North Wales. *Proc. Yorks. Geol. Soc.* **35**, 139-185.

ALLEN, J.R.L. (1983) Studies in fluviatile sedimentation: bars, bar-complexes and sandstone sheets (low-sinuosity braided streams) in the Brownstones (L. Devonian), Welsh Borders. *Sediment. Geol.* **33**, 237-293.

BEUTNER, E.C., FLUECKINGER, L.A. & GARD, T.M. (1967) Bedding geometry in a Pennsylvanian channel sandstone. *Bull. Geol. Soc. Am.* **78**, 911-916.

BOOTHROYD, J.C. & ASHLEY, G.M. (1975) Process, bar morphology and sedimentary structures on braided outwash fans, north-eastern Gulf of Alaska. In: *Glaciofluvial and Glaciolacustrine Sedimentation* (Eds Jopling, A.V. & McDonald, B.C.) pp. 193-222. Soc. Econ. Paleont. Miner., Tulsa, Spec. Publ. 23.

BOOTHROYD, J.C. & NUMMEDAL, D. (1978) Proglacial braided outwash: a model for humid alluvial fan deposits. In: *Fluvial Sedimentology* (Ed. Miall, A.D.) pp. 641-668. Can. Soc. Petrol. Geol., Calgary, Memoir 5.

BREED, C.S. & GROW, T. (1979) Morphology and distribution of dunes in sand seas observed by remote sensing. In: *A Study of Global Sand Seas* (Ed. McKee, E.D.) pp. 253-302. US Geol. Surv. Prof. Paper 1052.

BRIDGE, J.S. (1977) Flow, bed topography, grain size and sedimentary structure in open channel bends: a three-dimensional model. *Earth Surf. Processes* **2**, 401-416.

BRIDGE, J.S. (1984) Large-scale facies sequences in alluvial overbank environments. *J. Sediment. Petrol.* **54**, 583-588.

BRIDGE, J.S. & DIEMER, J.A. (1983) Quantitative interpretation of an evolving ancient river system. *Sedimentology* **30**, 599-623.

BRIDGE, J.S. & LEEDER, M.R. (1979) A simulation model of alluvial stratigraphy. *Sedimentology* **26**, 617-644.

BRISTOW, C.S. (1987) Brahmaputra River: channel migration and deposition. *Recent Developments in Fluvial Sedimentology* (Eds Ethridge, F.G., Flores, R.M. & Harvey, M.D.) pp. 63-74. Soc. Econ. Paleont. Miner., Tulsa, Spec. Publ. 39.

BUTLER, H. (1935) Die Mächtigkeit der kaledonischen Molasse in Ostgrönland. *Mitt. Naturforsch. Ges. Schaffhausen* **12** (3), 17-33.

BUTLER, H. (1955) Das variscisch gefaltete Devon zwischen Duséns Fjord und Kongeborgen in Zentral-Ostgrönland. *Medd. Groenl.* **155**. 131 pp.

CANT, D.J. & WALKER, R.G. (1978) Fluvial processes and facies sequences in the sandy braided South Saskatchewan River, Canada. *Sedimentology* **25**, 625-648.

CLEMMENSEN, L.B. & ABRAHAMSEN, K. (1983) Aeolian stratification and facies association in desert sediments, Arran basin (Permian), Scotland. *Sedimentology* **30**, 311-339.

CLEMMENSEN, L.B., OLSEN, H. & BLAKEY, R.C. (1989) Erg-margin deposits in the Lower Jurassic Moenave Formation and Wingate Sandstone, southern Utah. *Bull. Geol. Soc. Am.* **101**, 759-773.

COLLINSON, J.D. (1986) Alluvial sediments. In: *Sedimentary Environments and Facies* (Ed. Reading, H.G.) pp. 20-62. Blackwell Scientific Publications, Oxford.

FARRELL, K.M. (1987) Sedimentology and facies architecture of overbank deposits of the Mississippi River, False River region, Louisiana. In: *Recent Developments in Fluvial Sedimentology* (Eds Ethridge, F.G., Flores, R.M. & Harvey, M.D.) pp. 111-120. Soc. Econ. Paleont. Miner., Tulsa, Spec. Publ. 39.

FRIEND, P.F., ALEXANDER-MARRACK, P.D., ALLEN, K.C.,

NICHOLSON, J. & YEATS, A.K. (1983) Devonian sediments of East Greenland. VI. Review of results. *Medd. Groenl.* **206**(6), 96 pp.

FRYBERGER, S.G., AHLBRANDT, T.S. & ANDREWS, S. (1979) Origin of sedimentary features, and significance of low angle eolian 'sand sheet' deposits, Great Sand Dunes National Monument and vicinity, Colorado. *J. Sediment. Petrol.* **49**, 733–746.

FRYBERGER, S.G., AL-SARI, A.M. & CLISHAM, T.J. (1983) Eolian dune, interdune, sand sheet, and siliciclastic sabkha sediments of an offshore prograding sand sea, Dhahran arla, Saudi Arabia. *Bull. Am. Ass. Petrol. Geol.* **63**, 280–312.

GALLOWAY, W.E (1989) Genetic stratigraphic sequences in basin analysis I: Architecture and genesis of flooding-surface bounded depositional units. *Bull. Am. Ass. Petrol. Geol.* **73**, 125–142.

GLENNIE, K.W. (1970) *Desert Sedimentary Environments.* Developments in Sedimentology 14. Elsevier, Amsterdam, 222 pp.

GRAHAM, J.R. (1983) Analysis of the Upper Devonian Munster Basin, an example of a fluvial distributary system. In: *Modern and Ancient Fluvial Systems* (Eds Collinson, J.D. & Lewin, J.) pp. 473–483. Spec. Publ. Int. Ass. Sediment. 6. Blackwell Scientific Publications, Oxford.

HASZELDINE, R.S. (1983a) Fluvial bars reconstructed from a deep, straight channel, Upper Carboniferous coalfield of northeast England. *J. Sediment. Petrol.* **53**, 1233–1248.

HASZELDINE, R.S. (1983b) Descending tabular cross-bed sets and bounding surfaces from a fluvial channel in the Upper Carboniferous coalfield of north-east England. In: *Modern and Ancient Fluvial Systems* (Eds Collinson, J.D. & Lewin, J.) pp. 449–456. Spec. Publ. Int. Ass. Sediment. 6. Blackwell Scientific Publications, Oxford.

HENRIKSEN, N. (1985) The Caledonides of central East Greenland 70°–76°N. In: *The Caledonide Orogen –Scandinavia and Related Areas* (Eds Gee, D.G. & Stuart, B.A.). John Wiley & Sons, Chichester.

HENRIKSEN, N. & HIGGINS, A.K. (1976) East Greenland Caledonian fold belt. In: *Geology of Greenland* (Eds Escher, A. & Watt, W.S.) pp. 183–246. Geological Survey of Greenland, Copenhagen.

HIGGINS, A.K. & PHILIPS, W.E.A. (1979) East Greenland Caledonides – a continuation of the British Caledonides. In: *The Caledonides of the British Isles – Reviewed* (Eds Harris, A.L, Holland, E.H. & Leake, B.E.) pp. 19–32. Geol. Soc., Lond., Spec. Publ. 8.

HUNTER, R.E. (1977) Basic types of stratification in small eolian dunes. *Sedimentology* **24**, 361–387.

JACKSON II, R.G. (1976) Depositional model of point bars in the Lower Wabash River. *J. Sediment. Petrol.* **46**, 579–594.

KIRK, M. (1983) Bar developments in a fluvial sandstone (Westphalian 'A'), Scotland. *Sedimentology* **30**, 727–742.

KOCUREK, G. (1981) Significance of interdune deposits and bounding surfaces in eolian dune sands. *Sedimentology* **28**, 753–780.

KOCUREK, G. & NIELSON, J. (1986) Conditions favorable for the formation of warm-climate aeolian sand sheets. *Sedimentology* **33**, 795–816.

LANCASTER, N. & TELLER, J.T. (1988) Interdune deposits of the Namib Sand Sea. In: *Eolian Sediments* (Eds Hesp, P. & Fryberger, S.G.). *Sediment. Geol.* **55**, 91–107.

LARSEN, P.-H. (1990a) *Geological map (1:100 000) of the Devonian basin, North-East Greenland.* Internal Report, Grønlands Geologiske Undersøgelse, Copenhagen, 8 pp., 3 maps.

LARSEN, P.-H. (1990b) *Geological, structural contour and isopach maps (1:50 000) of the Upper Devonian Celsius Bjerg Group on Eastern Gauss Halvø, North-East Greenland.* Internal Report, Grønlands Geologiske Undersøgelse, Copenhagen, 6 pp., 12 maps.

LARSEN, P.-H. (1990c) *Structural contour and isopach maps (1:50 000) of the Upper Devonian Kap Graah Group on western Gauss Halvø, North-East Greenland.* Internal Report, Grønlands Geologiske Undersøgelse, Copenhagen, 6 pp., 6 maps.

LARSEN, P.-H. (1990d) *The Devonian basin in East Greenland. Status of structural studies, June 1990.* Internal Report, Grønlands Geologiske Undersøgelse, Copenhagen, 54 pp.

LARSEN, P.-H. & BENGAARD, H.-J. (1991) Devonian basin initiation in East Greenland: a result of sinistral wrench faulting and Caledonian extensional collapse. *J. Geol. Soc. Lond.* **148**, 355–368.

LARSEN, P.-H., OLSEN, H., RASMUSSEN, F.O. & WILKEN, U.G. (1989) Sedimentological and structural investigations of the Devonian basin, East Greenland. *Rapp. Grønlands Geol. Unders.* **145**, 108–113.

LEEDER, M. (1973) Sedimentology and palaeogeography of the Upper Old Red Sandstone in the Scottish Border Basin. *Scott. J. Geol.* **9**, 117–144.

LEEDER, M. (1975) Sedimentology and palaeogeography of the Upper Old Red Sandstone in the Scottish Border Basin. *Scott. J. Geol.* **9**, 117–144.

LEEDER, M.R. & ALEXANDER, J. (1987) The origin and tectonic significance of asymmetrical meander belts. *Sedimentology* **34**, 217–226.

MIALL, A.D. (1990) *Principles of Sedimentary Basin Analysis*, 2nd edn. Springer-Verlag, New York, 668 pp.

MIALL, A.D. (1985) Architectural-element analysis: a new method of facies analysis applied to fluvial deposits. *Earth Sci. Rev.* **22**, 261–308.

MOODY-STUART, M. (1966) High and low-sinuosity stream deposits, with examples from the Devonian of Spitsbergen. *J. Sediment. Petrol.* **36**, 1102–1117.

MUKERJI, A.A. (1976) Terminal fans of inland streams in Satlej-Yamuna Plain, India. *Z. Geomorphol. N.F.* **20**, 190–204.

NEMEC, W. & STEEL, R.J. (1984) Alluvial and coastal conglomerates: their significant features and some comments on gravelly mass-flow deposits. In: *Sedimentology of Gravels and Conglomerates* (Eds Koster, E.H. & Steel, R.J.) pp. 1–31. Can. Soc. Petrol. Geol., Calgary, Memoir 10.

NICHOLSON, J. & FRIEND, P.F. (1976) Devonian sediments of East Greenland V. The Central sequence, Kap Graah Group and Mount Celsius Supergroup. *Medd. Groenland* **206**, 117 pp.

OLSEN, H. (1987) Ancient ephemeral stream deposits: a local terminal fan model from the Bunter Sandstone Formation (L. Triassic) in the Tønder-3, -4 and -5 wells, Denmark. In: *Desert Sediments: Ancient and Modern*

(Eds Frostick, L. & Reid, I.) pp. 69–86. Geol. Soc., Lond., Spec. Publ. 35.

OLSEN, H. (1988a) Sandy braidplain deposits from the Triassic Skagerak Formation in the Thisted-2 well, Denmark. *Dan. Geol. Unders. Ser. B* **11**, 26 pp.

OLSEN, H. (1988b) The architecture of a sandy braided–meandering river system: an example from the Lower Triassic Solling Formation (M. Buntsandstein) in W-Germany. *Geol. Rundsch.* **77**, 797–814.

OLSEN, H. (1988c) Asymmetric epsilon cross-bed dips – a sign of basin fill inhomogeneity? Devonian in East Greenland. In: *Abstr. Vol., BSRG Special Meeting, London*, 2 pp.

OLSEN, H. (1989) Sandstone-body structures and ephemeral stream processes in the Dinosaur Canyon Member, Moenave Formation (Lower Jurassic), Utah, USA. *Sediment. Geol.* **61**, 207–221.

OLSEN, H. (1990) Astronomical forcing of meandering river behaviour: Milankovitch cycles in Devonian of East Greenland. *Palaeogeogr. Palaeoclim. Palaeoecol.* **79**, 99–115.

OLSEN, H. (1993) Sedimentary basin analysis of the continental Devonian basin in North-East Greenland. *Bull. Grønlands Geol. Unders.* (in press).

OLSEN, H. & LARSEN, P.-H. (1993) Lithostratigraphy of the continental Devonian sediments in North-East Greenland. *Bull. Grønlands Geol. Unders.* (in press).

OLSEN, H., DUE, P.H. & CLEMMENSEN, L.B. (1989) Morphology and genesis of asymmetric adhesion warts – a new adhesion surface structure. *Sediment. Geol.* **61**, 277–285.

PARKASH, B., AWASTHI, A.K. & GROHAIN, K. (1983) Lithofacies of the Markanda terminal fan, Kurukshetra district, Haryana, India. In: *Modern and Ancient Fluvial Systems* (Eds Collinson, J.D. & Lewin, J.) pp. 337–344. Spec. Publs Int. Ass. Sediment. 6. Blackwell Scientific Publications, Oxford.

PUIGDEFÀBREGAS, C. (1973) Miocene point-bar deposits in the Ebro Basin, Northern Spain. *Sedimentology* **20**, 133–144.

PUIGDEFÀBREGAS, C. & VAN VLIET (1978) Meandering stream deposits from the Tertiary of the Southern Pyrenees. In: *Fluvial Sedimentology* (Ed. Miall, A.D.) pp. 469–486. Can. Soc. Petrol. Geol., Calgary, Memoir 5.

SCHULTZ, E.H. (1982) The chronosome and supersome: terms proposed for low-rank chronostratigraphic units. *Bull. Can. Petrol. Geol.* **30**, 29–33.

SCHWARTZ, D.E. (1978) Hydrology and current orientation analysis of a braided–meandering transition: the Red River in Oklahoma and Texas, USA. In: *Fluvial Sedimentology* (Ed. Miall, A.D.) pp. 231–255. Can. Soc. Petrol. Geol., Calgary, Memoir 5.

SHELTON, J.W. & NOBLE, R.L. (1974) Depositional features of a braided–meandering stream. *Bull. Am. Assoc. Petrol. Geol.* **58**, 742–752.

SNEH, A. (1983) Desert stream sequences in the Sinai Peninsula. *J. Sediment. Petrol.* **53**, 1271–1280.

SOUTHARD, J.B., SMITH, N.D. & KUHLE, R.A. (1984) Chutes and lobes: newly identified elements of braiding in shallow gravelly streams. In: *Sedimentology of Gravels and conglomerates* (Eds Koster, E. & Steel, R.J.) pp. 51–59. Can. Soc. Petrol. Geol., Calgary, Memoir 10.

STEEL, R.J. & AASHEIM, S. (1978) Alluvial sand deposition in a rapidly subsiding basin (Devonian, Norway). In: *Fluvial Sedimentology* (Ed. Miall, A.D.) pp. 385–413. Can. Soc. Petrol. Geol., Calgary, Memoir 5.

TUNBRIDGE, I.P. (1984) Facies model for a sandy ephemeral stream and clay playa complex; the Middle Devonian Trentishoe Formation of North Devon, UK. *Sedimentology* **31**, 697–715.

VAN WAGONER, J.C., POSAMENTIER, H.W., MITCHUM, R.M., VAIL, P.R., SARG, J.F., LOUTIT, T.S. & HARDENBOL, J. (1988) An overview of the fundamentals of sequence stratigraphy and key definitions. In: *Sea Level Changes: An Integrated Approach* (Eds Wilgus, C.K., Hastings, B.S., Posamentier, H.W., van Wagoner, J.C., Ross, C.A. & Kendall, C.G. St. C.) pp. 39–42. Soc. Econ. Paleont. Miner., Tulsa, Spec. Publ. 42.

WILLIAMS, G.E. (1971) Flood deposits of the sand-bed ephemeral streams of central Australia. *Sedimentology* **17**, 1–40.

Alternating fluvial and lacustrine sedimentation: tectonic and climatic controls (Provence Basin, S. France, Upper Cretaceous/Palaeocene)

I. COJAN

Ecole Nationale Supérieure des Mines de Paris, Centre de Géologie Générale et Minière, 35 rue St Honoré, 77300 Fontainebleau, France

ABSTRACT

The continental sediments that accumulated in Provence during the Upper Cretaceous/Palaeocene are organized in several similar sequences that evolve from fluvial to lacustrine facies. A first attempt to analyse the parameters governing such a sequential arrangement persisting during some 30–40 m.y. is based on a detailed analysis of the two best preserved sequences (Rognacian–Vitrollian).

The horizons of detrital sediments (quartz sand channel bodies, bars or sand sheet) developed in a braided system network which spread on a silty floodplain. Sediment supply was dominantly fine grained and water stage fluctuations were common. Sand bodies are the result of cumulative deposits of flood episodes; high stages are characterized by channel fill deposits, and low stage by build-up of sand bars and development of algal mats or oncolites. The lacustrine carbonate muds precipitated in shallow water depth lakes. Base level fluctuations controlled the sequential arrangement, and the carbonate muds were often subjected to subaerial exposure. Sediment supply was probably low as testified by the long periods of non-deposition, the intense bioturbation of the sediments, and the low detrital content of the lacustrine deposits.

The basin palaeogeography is governed by two fault systems: an E-W oriented one that defined the drainage pattern, and a NNE–SSW one that controlled the lateral variations of facies. The braided system was, from time to time, bound by locally derived conglomerates which accumulated along the E-W oriented fault system and are evidence of tectonic pulses. The subsidence rate was fairly low (some 300 m over approximately 15 m.y.) and suggests at least a local compressional deformation. Based on the lithofacies arrangement, a model involving thrust and strike-slip displacement along the E-W trending system may be suggested.

From the two analysed sequences, the lithofacies assemblages are governed by a combination of climatic influence controlling base level fluctuations as well as extensive pedogenesis, and of tectonic activity influencing sediment influxes as well as the evolution of depositional environments. Periods of tectonic quiescence and an average low rate of sediment influx allowed climatic influences on sediment to develop and be preserved. Such a balance, between braided fluvial and shallow lacustrine environments, was maintained over such a long period because the low rate of sediment influx roughly matched the subsidence rate. In this context, the environmental fluctuations linked with climatic or tectonic changes were only of limited amplitude.

INTRODUCTION

The Upper Cretaceous/Palaeocene continental deposits in Provence mainly correspond to fluvial deposits and lacustrine sediments. The continental formation, which may exceed 2000 m in thickness, includes a wide range of lithologies: siltstones, sandstones, conglomerates, breccias, lacustrine carbonate mudstones and dark organic-rich shales. These deposits are organized in several large-scale sedimentary sequences, which begin with detrital sediments and end with lacustrine deposits. Some proximal fan breccias may be interbedded in the sequence.

Repetition of similar sequences throughout a period of some 30–40 m.y. was permitted by a palaeogeographic evolution which maintained a continental environment of deposition fluctuating from floodplain to lake during this interval of time.

Parameters which controlled the spatial arrangement of the varied lithofacies will be considered in a detailed analysis of the lithofacies assemblages. Two sequences which are preserved over the widest area have been selected. They roughly correspond to the Rognacian and Vitrollian continental stages, and are up to 400 m thick. These formations outcrop in E–W oriented synclines whose axes are parallel to the major fault zones. The main synforms are the Aix en Provence (Aix) and Rians–Salernes structures (Fig. 1).

Basinwide relationships and associations of lithofacies are determined from a network of cross-sections and detailed mapping. Identification of key marker beds, such as characteristic pedogenic horizons, are utilized to establish precise local or even regional correlations in these poorly stratigraphically controlled formations, as it is often the case in continental basins.

GEOLOGICAL SETTING AND STRATIGRAPHY

During Late Cretaceous and Early Tertiary times, southern France was characterized by a rapidly evolving palaeogeography which exposed this region to continental sedimentation. The E–W trending Pyreneo-Provençal Gulf, which opened on an Atlantic-Aquitain Gulf, emerged almost entirely by the end of the Upper Cretaceous. Deformations of the Pyrenees displaced the axis of this continually subsiding North Pyrenean trough (Plaziat, 1981; Freytet & Plaziat, 1982).

In Provence, the area emerged at the end of the Santonian and continental deposits accumulated during Late Cretaceous and Palaeocene times (Durand et al., 1984). This formation has been classically divided into five regional stages: Valdonnian, Fuvelian (Campanian), Begudian, Rognacian (Maastrichtian) attributed to the end of the Cretaceous, and Vitrollian to the Dano-Montian (Babinot & Freytet, 1983; Babinot & Durand, 1984). A recent study on palaeomagnetism (Westphal & Durand, 1990) introduced a new interpretation of the Upper

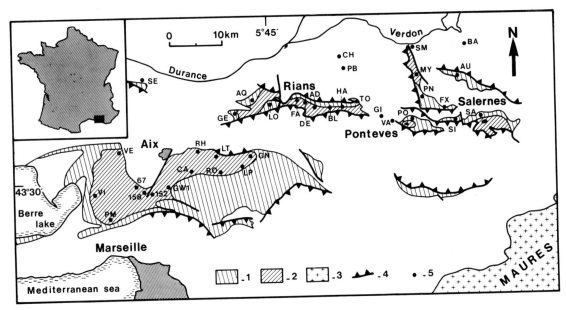

Fig. 1. Sketch map of the continental Upper Cretaceous/Tertiary outcrops in Provence and location of the sections. 1, Campanian/Maastrichtian; 2, Palaeocene/Eocene; 3, Permian plus metamorphic rocks; 4, thrust; 5, measured section. Aix area, sections: CA, Canet; GN, Gentil; LP, Les Prés; LT, Le Trou; PM, Pennes Mirabeau; RH, Roques Hautes; RO, Rousset; VI, Vitroes; VE, Ventabren; cores: GW1, 67, 152, 158. Rians-Salernes area sections: AD, Adret; AQ, Aqueduc; AU, Aups; BA, Bauduen; BL, Blacasse; CH, Chaberte; DE, Désidère; FA, Fabresse; FX, Fox Amphoux; GE, Gerle; GI, Gigeri; HA, Haut Adret; LO, Louvière; MY, Montmeyan; PB, Petite Bastide; PO, Pontéves; PN, Petit Nans; SA, Salernes; SE, St Estève-Janson; SI, Sillans; SM, St Meme; TO, Touars; VA, Varage.

Cretaceous chronostratigraphy: Valdonnian, being of normal polarity, is still of Santonian age whereas Fuvelian and Begudian roughly correspond to Campanian, and Rognacian to Maastrichtian (Berggren et al., 1985). Although stratigraphy is not the purpose of this paper, it highlights one of the key problems in the Provence Basin analysis, namely that the stratigraphic framework is based on a sparse fossil record, which almost certainly contains environment-specific fossil assemblages.

The lithostratigraphy of the Upper Cretaceous/Palaeocene sediments was described in detail by de Lapparent (1938) and Durand et al. (1984) and this provided the formations for basin analysis. Rapid lateral facies changes, as well as thickness variations due to synsedimentary deformations, caused difficulties in interpreting the depositional history of these poorly stratigraphically controlled formations (Babinot & Durand, 1984; Durand & Nury, 1984).

LITHOFACIES AND DEPOSITIONAL ENVIRONMENTS

Though Provençal lithostratigraphy is fairly well known, beside some determinations of clast origin, fluvial sediments have received little attention and palaeocurrent pattern is almost unknown in Provence. Lacustrine carbonates and pedogenic features have been extensively described and interpreted in the continental basins of southern and southwestern France by Freytet & Plaziat (1982).

In this context, before any investigation of the sedimentation controls, reconstruction of the fluvial network is necessary. Characterization of the drainage pattern, the water stage fluctuations and the rate of sediment supply is undertaken by adopting the classification of Miall (1978a) to describe the fluvial facies. It gives an approach of fluvial dynamics through the use of facies models defined from distinctive associations of standard lithofacies, geometry of sedimentary units and current orientations.

Based on the vertical arrangement of the deposits in the large-scale sequences, three major lithofacies can be recognized: floodplain, lacustrine and proximal alluvial fan. The floodplain sediments are dominated by siltstones with sandstone intercalations, and the lacustrine one by limestones with some shaly intervals (Fig. 2). The proximal alluvial fan deposits are characterized by breccia sheets.

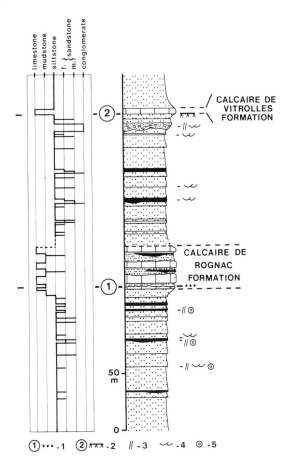

Fig. 2. Rousset section (Aix, see Fig. 1 for location of section RO). 1, Key-horizon no. 1; 2, key-horizon no. 2; 3, tabular crossbeds; 4, trough crossbeds; 5, oncolites.

Floodplain lithofacies

These sequences are dominated by red siltstones, comprising roughly 80% of the stratigraphic section (Fig. 2). No sedimentary structures were observed in this facies where invertebrate burrows and pedogenic carbonate nodules are common. It can be proposed that it corresponds to floodplain deposits (Sigleo & Reinhardt, 1988).

Distinctly channelized bodies of medium- to coarse-grained quartz sandstone cut through this fine-grained facies (Fsc of Miall, 1981). Some of them are isolated bodies (facies Ss) between 2 and 3 m wide and up to 3 m thick. These channel-fills display sharp or erosive bases and slightly undulating tops, and present some interbedded lenses of gravels and large-scale crossbedding. Other channel

Fig. 3. Channel-fill sandstone sequence. Units with planar tabular crossbedding are 80 cm thick (Rians).

sandstones are characterized by tabular geometry (Fig. 3). Owing to the general orientation of the outcrops, which are slightly oblique to palaeocurrents, the lateral extension of these sand bodies could not be defined. These channels are often infilled by one or two stage sequences and the lower part of the sequence is commonly characterized by facies Gm deposited as scour fill or gravel lag lens or by unstratified sandstone (Smu) representing channel fill (Fig. 4). The coarsest intraclasts are dominantly oncolites (up to 0.20 m in diameter) suggesting long intervals of low discharge during which algal mats grew in channels and pools (Freytet & Plaziat, 1982) around various nuclei (such as pebbles, fragments of bone or of dinosaur egg shell). During high water stage, the oncolites were transported and deposited by strong unidirectional currents suggesting that facies Gm accumulated during that interval of time.

These coarse facies (Gm and Smu) are usually overlain by stratified sands. Some are thin to thickly bedded sandstone with planar tabular crossbeds dipping 10 to 20° (facies Sp). Set thickness varies from 0.30 to 0.80 m and basal surfaces may contain scattered pebbles (Fig. 4). They are interpreted as transverse bars. Others display broad and shallow scour channels filled with lithofacies St. In these fine-grained sandstones, small-scale troughs are common (0.15–0.80 m wide and 0.03–0.10 m deep); when grain size is coarser, larger troughs are developed. Sequences are sometimes capped by ripple drift cross-laminae (facies Sr), often of a few centimetres in wavelength with crests less than 2 cm high.

These channel sequences are interpreted as a transition from longitudinal bars (facies Gm) to channel deposits (facies St) and then to transverse bars (facies Sp). The low dispersion in the set orientation is interpreted as an indication of vertical rather than lateral accretion (Bluck & Haughton, 1989). In the more distal reaches, there are tabular crossbedded sandstone bodies, 1–2 km long, which have been interpreted as sand flats (Cant & Walker, 1978).

Although coarse-grained sediments are rare (except for the oncolites), some sequences are dominated by gravel deposits (Fig. 2): the Poudingue de la Galante Formation (Aix area) and the Poudingue des Tours Formation (Rians region). In both of these conglomeratic sequences, a diverse range of clast lithologies reflects their derivation from distant extrabasinal sources (quartz, Palaeozoic sandstones, Mesozoic limestones, etc.) as well as within basin sources (carbonate, sand and silt clasts). The predominant lithofacies type is made of massive-to-poorly stratified subplanar gravel sheets (Gm and Gms) with common basal scours (Fig. 5). They display vertical stacking of channel units generally less than 2 m thick, dominated by facies Gms, Gm, Gt, Gp and Sp. Well-rounded clasts, up to 0.30 m in size and randomly distributed in a silty matrix are characteristic of an unsorted, matrix-supported lithofacies (facies Gms, Fig. 5), interpreted as debris flow deposits. These units are laterally persistent (around 500 m) but do not exceed 1–2 m in thickness. Lithofacies Gm, Gp and Gt are observed in channel-fill. These sequences often display pebble lags at their erosional base. Crude clast imbrication is the most common structure observed in the coarse-grained deposits (Gm) (Fig. 5). A predominance of the lithofacies Gt, Gp and Sp is observed in finer-grained sediments (Fig. 6). These two types of crossbeds (trough and planar) reflect sand bar morphology and may be interpreted as indicative of a change in the discharge regime. During peak discharge, in-channel

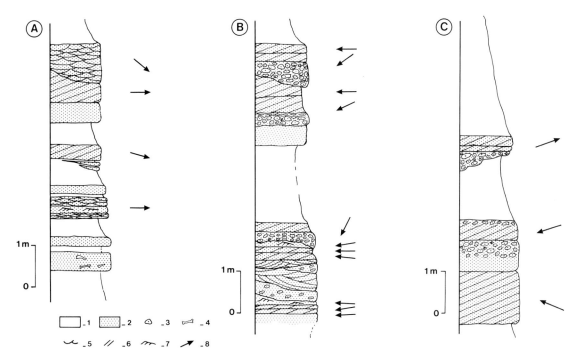

Fig. 4. Detailed section through sandstone units: (A) Rians (AQ); (B) Ponteves (PO); (C) Aix (RO). Section locations are shown on Fig. 1 and detailed section locations are shown on Fig. 12 for (A), (B) and on Fig. 11 for (C). 1, Siltstone; 2, sandstone; 3, oncolites; 4, bones; 5, trough crossbeds; 6, planar crossbeds; 7, ripples; 8, measured crossbed azimuths.

scouring and the migration of sinuous-crested dunes results in the formation of trough crossbedding, while during falling stages and lower discharge events, migration of straight-crest ripples and dunes builds up planar foresets (Fig. 7, Hein & Walker, 1977). Palaeocurrent data and the affinity of clast composition with the 'Maure Massif' lead to the interpretation of these conglomeratic episodes as outer fan deposits with conglomeratic beds.

Considering the fluvial lithofacies assemblages, the sequences are typical of a braided river system (Miall, 1977, 1978a) and may be related to the S. Saskatchewan type. Variability in grain size reflects differences in provenance (large clasts from the Maure Massif, fine-grained sands from reworking of sediments) and/or water stage fluctuations (large oncolites suggesting long periods of low discharge). The depositional system can thus be described as a distal braided system in which siltstones represent over 80% of the sediment thickness (Fig. 2). The sand facies assemblages appear to correspond to superimposed channel-fill and bar

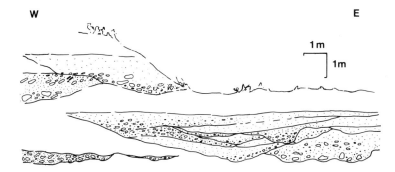

Fig. 5. Field sketch of part of the conglomeratic sequence Poudingue de la Galante Formation (Aix). Channelized bars are composed of several unsorted conglomerates (Gms). Lags at the bottom of the channels as well as clast imbrications are common features in these deposits. White, siltstone; dotted, sandstone.

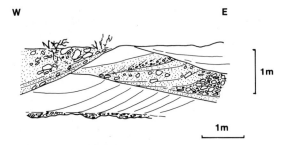

Fig. 6. Field sketch of a conglomeratic unit of the Poudingue de la Galante Formation (Aix). Trough crossbedding in a bar. White, fine-grained sandstone; dotted, medium- to coarse-grained sandstone; palaeocurrent slightly oblique to the drawing.

sedimentation cycles reflecting water stage variations.

Lacustrine lithofacies

The fluvial fine-grained sediments gradually evolve laterally and vertically into the lacustrine facies. Two main lacustrine episodes developed within this series: the Calcaire de Rognac and the Calcaire de Vitrolles Formations (Fig. 2). Even though these lakes extended over several hundred square kilometres, sediments are typical of relatively shallow-water lakes and palustrine environments.

The Calcaire de Rognac Formation lithofacies are characterized by dark, organic-rich carbonate mudstone containing abundant freshwater ostracods, gastropods and plant debris. Some beds bear evidence of sporadic subaerial exposures with associated pedogenic features, i.e. dissolution, *in situ* brecciation, mineralized mudcrack fills, etc. (Freytet & Plaziat, 1982). The lacustrine sequence, ranging from a few metres to 70 m in thickness, is organized into several limestone units separated by lignitic shales and/or silt and sandstones (Fig. 8). Transition between the limestone units and the overlying silty/shaly sediments is abrupt and may even be erosive. The sequence base is characterized by fine-grained detrital sediments accumulated under subaerial conditions or in the water (sands, silts, organic-rich shale, lignite), then chemical precipitation dominated and the lacustrine limestone deposited. Among the widespread occurrences of subaerial exposures, a ferruginous weathering horizon developed on top of one of these limestone units. It has been used as the first regional key marker in the basin analysis. This horizon is present

Fig. 7. Lower discharge events and build-up of a transverse bar. Note pebble lags, tabular planar crossbed sets in the sandstone and the sharp contact surface with the upper unsorted gravel bed (Poudingue de la Galante Formation (Aix)).

over the entire study area and is expressed as an iron encrusted hardground in the more proximal areas, changing distally toward a ferrugineous breccia. In the distal part, the weathered profile extends 40 cm into the sediments and has cemented erosional scour marks. Other limestone unit top surfaces exhibit some weathering, probably under less extreme conditions. The asymmetrical sequential arrangement of facies in the Calcaire de Rognac Formation leads to the proposal that the evolution of the lacustrine system was largely governed by base level variations.

The Calcaire de Vitrolles Formation, with a maximum thickness of 20 m, mainly corresponds to a palustrine episode. The mottled carbonated (sometimes dolomitic) mudstones show many ped-

Fig. 8. Lower part of the first main lacustrine episode: Calcaire de Rognac Formation (Rousset). The limestone unit (12 m thick) is abruptly interrupted and overlain by dark, organic-rich shaly material. Fluvial channel-fills occur before the rapid but progressive transition toward the next lacustrine limestone deposits.

ological features such as recrystallization in voids, root traces, desiccation cracks, etc. Over the entire study area, this episode was preceded by a mature or super-mature calcimorphic soil. In this palaeosoil, the isotubular root systems are surrounded by small calcareous nodules, or by irregular shaped concretions reflecting the coalescence of smaller nodules. Presence of palygorskite has been identified over the basin in connection with these facies (Sittler, 1965; Cornet, 1974). In association with palaeosoils and lacustrine episodes, palygorskite is generally related to an arid or semi-arid climate (Callen, 1984; Singer, 1984; Wright, 1986). This calcimorphic pedogenesis will be used later as the second regional key marker horizon.

These two lacustrine episodes are characteristic of shallow water depth lakes often submitted to temporary subaerial exposures. A low rate of detrital sediment supply led to the formation of nearly pure chemical rocks and enabled a deep climatic imprint to be made on the sediment during long periods of subaerial exposure.

Proximal alluvial fan deposits

The proximal alluvial fan facies consist of a series of transverse carbonate breccia flows that interdigitate into the silty sediments of the major longitudinal floodplain.

Erosional features such as channels or large scours are quite common, and a plot of current directions reflects a fan-shaped dispersion. Each of the conglomeratic wedges is 3–4 m thick and consists of several breccia sheets exhibiting numerous erosional scours at their base (Fig. 9). The breccia bodies are parallel unstratified beds (Gms), which are, apart from protruding blocks, flat topped. Unsorted, angular carbonate clasts up to 0.60 m in diameter float in a shaly carbonate matrix. These breccia sheets often fill up the channel scours and spread laterally. At the proximal break of slope, the silty material is nearly absent and the clast sheets are stacked on top of one another. Basinwards, breccia flows interfinger with the silt deposits, and shifts in the channel axis distribution are observed. The silty intervals (about 10 m thick) show pedogenic features such as diffusive carbonate concretions and burrowing. Some of the channel migrations were probably a response to more favourable gradients resulting from differential compaction (Fig. 10). The lens-shaped sandstone bodies thinning away from the mountain front

Fig. 9. Proximal fan breccia sheets (Aix). Note the stacking of several sheets in the bar and erosional scours. Bar thickness is 3.5 m.

reflect continued uplift of the mountain during fan sedimentation. Because later thrusting incorporated the fan apices, they were never preserved, thus the sequences can only be estimated to some 50 m in thickness with a lateral extension of a few kilometres.

The transverse pulses of these coarse-grained sediments, interbedded with the fine-grained floodplain sediments, may be interpreted as a response to tectonic acitivity characterized by long intervals of quiescence during which no breccias sheet are present in the lithologic succession.

PALAEOGEOGRAPHIC RECONSTRUCTION

As biostratigraphy does not allow precise correlations, nor highlights lateral lithofacies variations, which is often the case in fluvial basins with a lack of fossils, correlations are based on key marker horizons or beds. Key marker horizons 1 and 2 (Fig. 2) are present over the whole study area and are used to delimit the regional correlations (Figs 11 & 12). Locally, detailed analysis is per-

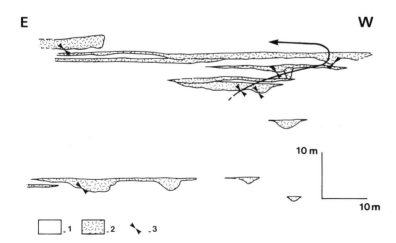

Fig. 10. Breccia sheets and some channel migration governed by differential compaction (present configuration) (Aix, section LT). 1, Siltstone; 2, breccia; 3, channel axis direction.

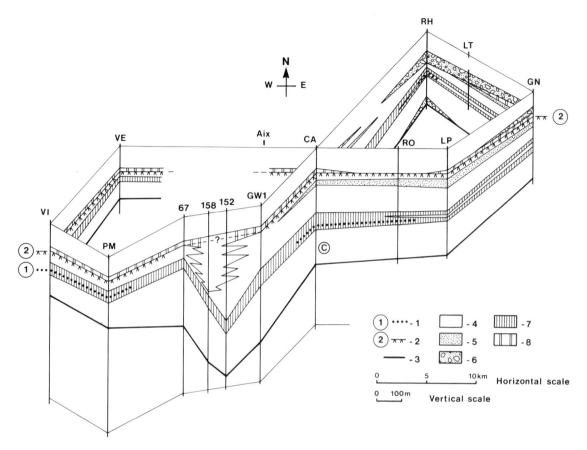

Fig. 11. Lithofacies fence diagram of the Rognacian–Vitrollian series in Aix en Provence region. Note the influence of differential subsidence on formation thicknesses along an E–W axis. Key marker horizon no. 2 is considered as horizontal in this diagram and horizontal scale has been palinspatically restored. 1, Key-horizon no. 1; 2, key-horizon no. 2; 3, approximate Begudian–Rognacian boundary. Fluvial deposits: 4, siltstone and sandstone; 5, Poudingue de la Galante Formation. Proximal fans: 6, carbonate clast breccia. Lacustrine limestones: 7, Calcaire de Rognac Formation; 8, Calcaire de Vitrolles Formation.

formed by using correlatable surfaces of a lesser extent.

Fluvial system anatomy

Fluvial pattern was reconstructed by measurements of current marks, dip of bed sets or of ripples, and clast imbrications. The drainage network involved streams flowing in parallel valleys and opposite directions as shown on Fig. 13b. These longitudinal rivers (Rians and Salernes valleys) drained into a major system (Aix valley) (Fig. 13d) and did not flow, as previously believed, toward the west in a valley parallel to the Aix one.

The fluvial deposits, restricted to the Aix region nearly up to the Rognacian, then started to accumulate also in the northern area. At the same time, several proximal alluvial fans, composed of carbonate breccias, built up along basin borders (Figs 11 & 12). The spread of the fluvial deposits, associated with tectonic pulses, may have been recording a change in the basin subsidence rate.

In the Rians syncline, the fluvial network is flowing toward the east with very little current dispersion (Fig. 13b). The quartz-rich sandstones were probably derived from the reworking of sand-rich Lower Cretaceous rocks outcropping in the north (Cotillon, 1984). In the Salernes zone, the westward system finds its major sources in the 'Maure Massif' (Fig. 13b). The exact connection

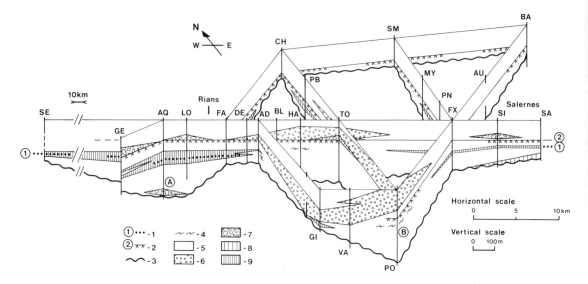

Fig. 12. Lithofacies fence diagram of the Rognacian–Vitrollian series in Rians–Salernes region. Rognacian sediments lay unconformably on the Upper Jurassic or Lower Cretaceous basement. Key marker horizon no. 2 is considered as horizontal in this diagram and horizontal scale has been palinspatically restored. 1, Key-horizon no. 1; 2, key-horizon no. 2; 3, unconformity; 4, reworked lacustrine limestone. Fluvial deposits: 5, siltstone and sandstone; 6, Poudingue des Touars Formation. Proximal fans: 7, carbonate clast breccia. Lacustrine limestones: 8, Calcaire de Rognac Formation; 9, other formations. The Calcaire de Vitrolles Formation is not pictured because, when present, its thickness does not exceed a metre or two (sections AQ, LO, GE, FA, SI).

between these two tributaries and the main westward drainage in the Aix region cannot be precisely determined due to the lack of outcrops in this area. Reconstructions show a sharp bend, probably initiated by diapirism linked to evaporitic Triassic sediments (de Lapparent, 1938; Angelier, 1974).

The dominance of fine-grained sediments in the fluvial system may reflect the lithology of source materials, long distance transport and/or low flow energies. In Provence, lithology of the source material is the main parameter as the Rians tributary could only be sourced by fine-grained material, and as Permian sediments, the dominant source rocks of the westward tributaries, would also have rapidly resulted in fine-grained sediments.

In the late Upper Maastrichtian/Early Palaeocene, outer fan sediments flowing from the east invaded the northern and southern valleys; these deposits occur first in the Aix region (Poudingue de la Galante Formation). Major units of this formation accumulated over a roughly 15 m thick interval situated some 20 m below the key marker horizon 2 (Figs 2 & 11). Similar deposits (Poudingue des Touars Formation) spread into the Rians area, where they overlie the key marker horizon 2 (Fig. 12). Undoubtedly, the Poudingue de la Galante Formation was transported along the southern axis. As regards to the Poudingue des Touars Formation, two courses can be proposed, along either the southern or northern valley. Absence of any deposits of this type in the Salernes area, where only fine-grained sediments are present, would favour the first alternative. Whatever the transit axis of the Poudingue des Touars Formation, westward-flowing coarse sediments invaded the Rians tributary, inducing a local reversal of the drainage (Figs 12 & 13c). Presence of carbonate boulders, limited to the Conglomerat des Touars Formation in the Rians area, indicates local erosion of topographic heights (presence of Triassic and Jurassic clasts, Angelier, 1974). These two conglomeratic formations are interpreted as corresponding to a shift in the distribution of outer fan lobes, enabling in this low topography context a temporary invasion of a valley by sediments transported in a direction opposite to the valley slope. A starving of the Aix valley in sediment supply from the northern tributaries probably followed these conglomeratic episodes.

The observed lithofacies assemblages are more influenced by the type of sediment load available

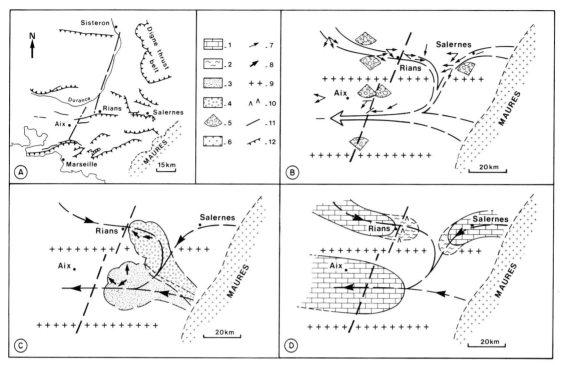

Fig. 13. (A) Structural setting of Provence. (B–D) Schematic maps illustrating the various depositional areas (in order to restore the original location of the observation points, the map has been stretched along a N–S axis of about 30 km, taking into account the Tertiary deformation). (B) Fluvial network. It has been mainly reconstructed from current measurements in the sandstone units below the Calcaire de Rognac Formation. (C) Poudingue de la Galante and Poudingue des Touars Formations. Geometry of the outer fan is reconstructed from clast imbrications and lithofacies sequences evolution). (D) Lacustrine episodes. This scheme corresponds closely to the Calcaire de Rognac Formation. Calcaire de Vitrolles Formation had a similar but restricted setting. In Aix region, the palustrine environment was probably composed of several small lakes, in Rians–Salernes area lacustrine deposits do not exceed a few metres in thickness when present. 1, Lacustrine limestone; 2, slumped lacustrine material; 3, P. Galante Formation; 4, P. Touars Formation; 5, proximal breccias; 6, 'Maures Massif' (Permian plus metamorphic rocks); 7, current measurements in sandstones; 8, clast imbrications in conglomerates; 9, divide; 10, diapirism; 11, fault; 12, thrust.

than by long transport distances. Sequences and sedimentary structures reflect a dominant low water stage with a limited flow energy allowing the development of large bars. All these features are characteristic of a low energy fluvial plain, whose width is estimated as 10 km in the tributaries and 20–30 km in the Aix region.

Extension of the lakes

From the available outcrops, no succession of perennial lacustrine or even palustrine conditions has been observed. The interfingering of fluvial and lacustrine facies is in fact present during the two main lake episodes only (Calcaire de Rognac and Calcaire de Vitrolles Formations) (Cojan & Beaudoin, 1989). Identification of two key marker horizons, each of them present within or close to one of the main lake deposits, offers the possibility of drawing a palaeogeographic sketch of the lake distribution (Figs 11, 12 & 13d). During each of these episodes, several lakes formed in the area. Local topographic relief, generating natural dams, confined these lakes and delimited the areas of carbonate sedimentation (Cojan, 1989a). In some sections, lake deposits, which are interbedded or overlain by proximal alluvial fan deposits, seem to have been partly related to tectonic pulses (Reading, 1980). The internal organization of the lacustrine facies shows a strong response to base level changes recorded by abrupt boundaries, extensive pedogenesis and asymmetric sequences.

SEDIMENTATION CONTROLS

Tectonic setting

The basin evolution was in fact controlled by two fault systems: an E–W oriented system that defined the general pattern of the fluvial network, and a NNE–SSW trending system that played a more local role, controlling rapid lateral variations in facies as well as some deformation of the basin floor (Figs 11 & 12).

The structure of the drainage network was controlled by an E–W trending fault system. Several proximal fans, restricted in size, developed along these longitudinal, E–W oriented faults, and built up transverse bodies which interfingered with the major longitudinal floodplain deposits (Figs 11, 12 & 13b). These locally derived conglomerates and breccias, of limited lateral extension, are a diagnostic feature of sedimentation in fault-bound basins. Begudian and Early Tertiary fans have an angular unconformity between them (Chorowicz & Ruiz, 1984), indicative of uplift pulses accompanied by rotation, contemporaneous with the sedimentation (Riba, 1976). Evidence of long lithofacies successions without transverse proximal fan deposits suggests that there were long periods of tectonic quiescence between tectonic pulses.

A tendency for the depocentre to migrate through time parallel to the E–W faults is shown on isopach and isobath maps established from cores drilled in the Fuvelian lignitic rich formations (Glintzboeckel, 1980). An asymmetry in the basin floor also developed with a major slope component towards the northern fault system.

Another structural control on sedimentation is linked to the NNE–SSW transverse fault system. In the northern area (Rians), it resulted in the growth of a topographic height governing lake extensions (Fig. 12). This divide can be explained by diapiric doming (Cojan, 1989a), already identified during Jurassic times (Arnaud & Monleau, 1979). A significant role of a transverse fault system of similar orientation is also revealed in the Aix valley; it delimited blocks subsiding at different rates as shown for example during the fluvial episode between the two main lacustrine formations (Fig. 11).

This structural setting has existed in Provence since the Jurassic and leads to the proposal that the fault systems governing the continental sedimentation were probably basement structures (Fig. 13a).

The characteristics of the lithofacies associations do not exactly fit with a syncline/anticline formation contemporaneous with sedimentation (Babinot & Durand, 1984) or with a normal fault-bounded basin with periods of rapid subsidence. Small-scale alternating zones of compression (synsedimentary anticlines) and extension (diapiric structures) could therefore have resulted from movement of side stepping faults (Miall, 1978b; Reading, 1980; Sylvester, 1988). The presence of transverse proximal wedges on both southern and northern reaches of the floodplain and synsedimentary rotations within the fans leads to the proposal that some thrusting occurred during this period, and might suggest a strike-slip deformation.

Climatic influence

The reconstruction of climate based upon the fossil record (palaeobotany and mollusc fossils) and the clay minerals clearly shows tropical affinities (Sittler & Millot, 1964). Transition from a warm climate (tropical in Begudian) into a subtropical one (probably cooler) was proposed from the pollen assemblages characteristic of in the lower part of the Calcaire de Rognac Formation (Ashraf & Erben, 1986).

As rainfall was seasonal, rivers were subjected to stage fluctuations with alternating periods of erosion and deposition which are reflected in the channel-fill sequences. The shallow lakes were therefore subjected to regular shrinkage causing subaerial exposures and associated plant colonization and pedogenic modifications. Fluctuations in the lake-shore across the relatively flat topography led to extensive weathering of the upper part of the lacustrine facies and even erosion (when the base level change was large enough). Sharp boundaries between the facies suggest that response to climatic changes was rapid. Organic-rich shaly/silty sediments correspond to a 'drowning' period when the lake expanded and deepened before carbonate precipitation could catch up.

Under this relatively hot climate, temporary non-depositional conditions allowed the development on a regional scale of pedogenic horizons such as key marker horizons 1 and 2 which may have been favoured by climatic shifts. Comparison with other SW European basins (Plaziat, 1981; Simo & Puigdefabregas, 1985; Fondecave-Wallez et al., 1988) leads to the proposal that these regional key marker horizons are an echo of large amplitude sea level changes (Cojan, 1989b) inducing changes in climatic conditions leading to extended pedogenic periods.

SUMMARY AND CONCLUSIONS

Reconstruction of the fluvial network during Rognacian and Vitrollian times provides a new insight into the regional palaeogeography. The northern deposits (Rians–Salernes area), which were usually considered as a lateral equivalent of the southern ones (Aix area), in fact correspond to upstream deposits and provide more proximal lithofacies sequences than in the Aix region. The fluvial system is characterized by tributaries flowing in valleys parallel to the main one. Except during the time of invasion by distal fan sediments (Poudingue de la Galante and Poudingue des Touars Formation), sediment supply of small grain size only was available. Lithofacies assemblages evidence long periods of low discharge and/or low water stage, during which transverse sand bars (or even sand flats) built up. Lacustrine lithofacies sequences also show a strong dependence on base level changes, and very little influx of detrital sediments. From these observations it can be proposed that the average sediment influx was low. That little sediment has been deposited is evidenced by the low rate of subsidence during this period (300 m during 15 m.y. without taking into account compaction).

Lithofacies analysis and palaeogeographic reconstructions of these two sequences leads to the proposal that the repetition of these large-scale sequences in Provence throughout Upper Cretaceous/Palaeocene times, a period extending around 30–40 m.y., resulted from variations in the following factors: sediment supply, stage fluctuations, tectonic activity and climatic conditions. Any change attributed to one or a combination of the above-mentioned factors led to the observed sequential arrangement: changes in sediment supply and water stages governed the distribution of coarse sediments and the sedimentary structures; changes in tectonic activity governed the evolution of depositional environments and sediment supply; changes of climatic conditions affected water stage in the fluvial system and base level in the lacustrine environment, and generated long periods of pedogenesis.

When stratigraphy can be better ascertained, a comparison of the Provence sequences with regional and/or global trends, such as the tectonics affecting western Europe and climatic evolution, will be possible.

ACKNOWLEDGEMENTS

I am grateful to Professor B. Beaudoin for the constructive discussions during this research project. D. Smith and P. Anadon provided helpful suggestions in reviewing the manuscript.

REFERENCES

ANGELIER, J. (1974) L'évolution continentale de la Provence septentrionale au Crétacé terminal et á l'Eocène inférieur: la gouttière de Rians-Salernes. *Bull. Bur. Rech. Geol. Minières* **2**, 65–83.

ARNAUD, M. & MONLEAU, C. (1979) Etude de l'évolution d'une plateforme carbonatée: exemple de la Provence au Jurassique. Thèse Doctorat d'Etat, U. de Provence, Marseille.

ASHRAF, A.D. & ERBEN, H.K. (1986) Palynogische Untersuchungen an der Kreide/Tertiar Grenze West-Mediterraner Regionen. *Palaeontographica Abt. B, Palaeophytologie* **200** (1–6), 111–163.

BABINOT, J.F. & DURAND, J.P. (1984) Crétacé supérieur fluvio-lacustre. In: *Synthèse Géologique du Sud-Est de la France* (Ed. Debrabant-Passart, S.) pp. 362–371. Mem. Bur. Rech. Geol. Minières Fr. 125.

BABINOT, J.F. & FREYTET, P. (EDS) (1983) Colloque sur les étages Coniacien à Maastrichtien. Marseille, 26–28 septembre 1983. *Géologie Méditerranéenne,* **10** (3–4), 245–268.

BERGGREN, W.A., KENT, D.V., FLYNN, J.F. & VAN COUVERING, J.A. (1985) Cenozoic geochronology. *Bull. Geol. Soc. Am.* **96**, 1407–1418.

BLUCK, B.J. & HAUGHTON, P.D.W. (1989) Recognition and structure of thalweg deposits in bedload dominated channels. In: *4th International Conference on Fluvial Sedimentology*, (Eds Marzo, M. & Puigdefábregas, C.). Generalitat de Catalunya, Barcelona.

CALLEN, R.H. (1984) Clays of the palygorskite–sepiolite group: depositional environment, age and distribution. In: *Palygorskite–Sepiolite Occurrences, Genesis and Uses* (Eds Singer, A. & Galan, E.) pp. 1–38. Elsevier, Amsterdam.

CANT, D.J. & WALKER, R.G. (1978) Fluvial processes and facies sequences in the sandy braided South Saskatchewan River, Canada. *Sedimentology* **25**, 625–648.

CHOROWICZ, J. & RUIZ, R. (1984) La Sainte Victoire (Provence): observations et interprétations nouvelles. *Géologie de la France* **4**, 41–55.

COJAN. I. (1989a) Structure diapirique controllant la sédimentation. Séries continentales de Provence (Rians-K/T). Publ. ASF 10, ASF, Paris.

COJAN, I. (1989b) Discontinuités majeures en milieu continental. Proposition de corrélation avec des évènements globaux (Bassin de Provence, S. France, Passage

Crétacé/Tertiaire). *C.R. Acad. Sci. Paris* **309** (Sér. II), 1013–1018.

COJAN, I. & BEAUDOIN, B. (1989) Interfingering of terrigeneous fluvial and lacustrine limestones (K/T Boundary, France). In: *Abstracts, 28th Int. Geol. Congress, Washington DC* **1**, 311.

CORNET, C. (1974) Sur l'existence d'un niveau repère à attapulgite au sein de remplissage des fossés Nord-Varois (Provence). *C.R. Acad. Sci. Paris* **278** (Sér. II), 809–811.

COTILLON, P. (1984) Crétacé inférieur. In: *Synthèse Géologique du Sud-Est de la France* (Ed. Debrabant-Passart, S.) pp. 362–371. Mém. Bur. Rech. Geol. Minieres Fr. 125.

DURAND, J.P. & NURY, J. (1984) Paléocène–Eocène. In: *Synthèse Géologique du Sud-Est de la France (Ed. Debrabant-Passart, S.)* pp. 425–429. Mém. Bur. Rech. Geol. Minieres Fr. 125.

DURAND, J.P., GAVIGLIO, P., GONZALES, J.F. & MONTEAU, R. (1984) Upper Cretaceous, Paleocene and Eocene Fluvio-lacustrine Sediments in the Arc Syncline (Aix-en-Provence District). *5th IAS European Meeting, Marseille, Fieldtrip Guidebook*.

FONDECAVE-WALLEZ, M.J., SOUQUET, P. & GOURINARD, Y. (1988) Synchronisme des séquences sédimentaires du comblement fini-crétacé avec les cycles eustatiques dans les Pyrénées centro-méridionales (Espagne). *C. R. Acad. Sci. Paris* **307** (Sér. II), 289–293.

FREYTET, P. & PLAZIAT, J.C. (1982) Continental Carbonate Sedimentation and Pedogenesis. Late Cretaceous and Early Tertiary of Southern France. *Contributions to Sedimentology, Schweizerbart'sche Verlagsbuchhandlung, Stuttgart*, 12, 213 pp.

GLINTZBOECKEL, C. (1980) Le gisement de charbon du bassin de l'Arc (Houillères de Provence). Reconnaissance de l'extension du gisement. *Rev. Industrie Minérale* (Suppl.) **Juin**, 41–53.

HEIN, F.J. & WALKER, T.C. (1977) Bar evolution and development of stratification in the gravelly, braided, Kicking Horse River, British Columbia. *Can. J. Earth Sci.* **14**, 562–570.

LAPPARENT, A.F. DE (1938) Etudes géologiques dans les régions provençales et alpines entre le Var et la Durance. *Bull. Serv. Carte Géol. Fr.* **40**, 198.

MIALL, A.D. (1977) A review of the braided river depositional environment. *Earth Sci. Rev.* **13**, 1–32.

MIALL, A.D. (1978a) Lithofacies types and vertical profile models in braided rivers: a summary. In: *Fluvial Sedimentology* (Ed. Miall, A.D.) pp. 537–564. Can. Soc. Petrol. Geol., Calgary, Memoir 5.

MIALL, A.D.(1978b) Tectonic setting and syndepositional deformation of molasse and other non-marine paralic sedimentary basins. *Can. J. Earth Sci.* **15**, 1613–1632.

MIALL, A.D. (1981) *Analysis of Fluvial Depositional Systems*. Am Assoc. Petrol. Geol., Tulsa, Education Course Note No. 20.

PLAZIAT, J.C. (1981) Late Cretaceous to late Eocene palaeogeographic evolution of Southwest Europe. *Palaeogeog. Palaeoclimatol. Palaeoecol.* **36**, 263–320.

READING, H.G. (1980) Characteristics and recognition of strike slip fault systems. In: *Sedimentation in Oblique Mobile Zones* (Eds Ballance, P.F. & Reading, H.G.) pp. 7–26. Spec. Publs Int. Ass. Sediment. **4**.

RIBA, O. (1976) Syntectonic unconformities of the Alto Cardener, Spanish Pyrenees: a genetic interpretation. *Sedim. Geol.* **15**, 213–233.

SIGLEO, W. & REINHARDT, J. (1988) Paleosols from some Cretaceous environments in the southern United States. In: *Palaeosols and Weathering through Time: Principles and Applications* (Eds Reinhardt, J.A. & Sigleo, W.R.) pp. 123–142. Geol. Soc. Am., Boulder, Spec. Paper 216.

SIMO, A. & PUIGDEFABREGAS, C. (1985) Transition from shelf to basin on an active slope, Upper Cretaceous, tremp area, Southern Pyrenees. In: *Excursion Guidebook, 6th IAS European Regional Meeting, Lleida*, pp. 61–108.

SINGER, A. (1984) Pedogenic palygorskite in the arid environment. In: *Palygorskite–Sepiolite Occurrences, Genesis and Uses* (Eds Singer, A. & Galan, E.) pp. 169–176. Developments in Sedimentology 37. Elsevier, Amsterdam.

SITTLER, C. (1965) *Le Paléogène des Fossés Rhénan et Rhodanien. Etudes Sédimentologiques et Paléoclimatiques*. Mém. Serv. Carte Géol. Als. Lorr. 24. Strasbourg, 392 pp.

SITTLER, C. & MILLOT, G. (1964) Les climats du Paléogéne français reconstitués par les argiles néoformées et les microflores. *Geol. Rundsch.* **54**, 333–343.

SYLVESTER, A.G. (1988) Strike-slip faults. *Bull. Geol. Soc. Am.* **100**, 1666–1703.

WESTPHAL, M. & DURAND, J.P. (1990) Magnétostratigraphie des séries continentales fluvio-lacustres du Crétacé supérieur dans le synclinal de l'Arc (région d'Aix-en-Provence, France). *Bull. Soc. Géol. France* **8** (IV), 609–621.

WRIGHT, V.P. (1986). *Paleosols*. Blackwell Scientific Publications, Oxford, 313 pp.

Control of basin symmetry on fluvial lithofacies, Camp Rice and Palomas Formations (Plio-Pleistocene), southern Rio Grande rift, USA

G.H. MACK* and W.C. JAMES†

*Department of Geological Sciences, New Mexico State University, Las Cruces, NM 88003, USA; and
†Department of Geological Sciences, The Ohio State University, 125 South Oval Mall, Colombus, OH 43210, USA

ABSTRACT

Plio-Pleistocene fluvial strata of the Camp Rice and Palomas Formations were deposited in broad, low-sinuosity streams characterized by rapidly shifting channels in turn separated by episodically emergent bedforms and bars. Although the Ancestral Rio Grande was braided throughout the southern rift, important differences exist in channel characteristics, proportion of preserved floodplain strata and associated palaeosol deposits. These differences primarily reflect the influence of basin symmetry on fluvial sedimentation.

Compared to asymmetrical basins, fluvial strata in symmetrical basins: (i) were formed on a wider alluvial plain, (ii) have a much lower percentage of multistorey channel sandstones, (iii) have a significantly higher proportion of floodplain deposits, (iv) are characterized by palaeosols with greater maturity and abundance, and (v) contain channel sandstones with more planar crossbed sets and greater evidence of low-stage bar and bedform modification.

INTRODUCTION

Crustal extension associated with the Rio Grande rift in south-central New Mexico, USA began about 29 Ma ago (late Oligocene) and has continued intermittently to the present (Seager, 1975; Seager et al., 1984). Throughout much of the middle to late Cenozoic, sedimentation in the southern rift took place in closed basins floored by playa lakes. Approximately 4 Ma ago (early Pliocene), however, drainage in the rift was integrated by the Ancestral Rio Grande, which flowed southward 800 km from southern Colorado before emptying into large lakes in west Texas and northern Chihuahua, Mexico (Kottlowski, 1953, 1960; Ruhe, 1962; Reeves, 1965; Strain, 1966; Hawley et al., 1969; Repenning & May, 1986).

Mack & Seager (1990) documented the presence of interconnected symmetrical and asymmetrical basins within the southern Rio Grande rift (4.0–0.5 Ma) and the tectonic role exerted by each on the general distribution of non-marine facies. The purpose of our study is to build on this basic framework by delineating the details of sedimentary facies formed within the Pliocene–middle Pleistocene axial-fluvial system of the Ancestral Rio Grande (Fig. 1). Differences in axial-fluvial facies developed in symmetrical versus asymmetrical basins are assessed in terms of the: (i) specific character of channel deposits; (ii) ratio of channel to floodplain lithofacies; and (iii) abundance and distribution of palaeosols.

METHODS

Twelve stratigraphic sections, ranging in thickness from 20 to 130 m, were measured between Truth or Consequences and Las Cruces, New Mexico, USA (Fig. 1). Five sections have the complete thickness

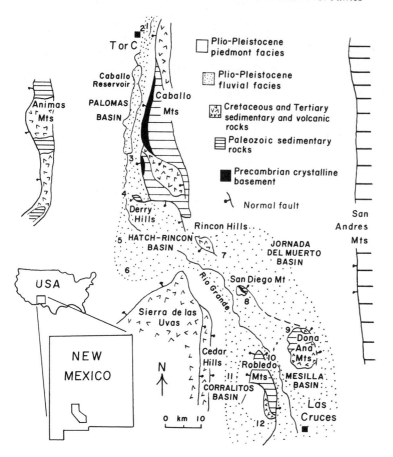

Fig. 1. Index map, southern Rio Grande rift, New Mexico, USA. Numbers refer to measured sections: 1, Truth or Consequences (T or C)–Highway 171; 2, Truth or Consequences–Cuchillo Negro Creek; 3, Apache Canyon; 4, Garfield; 5, Hatch Siphon; 6, Hatch Airport; 7, Rincon Arroyo; 8, San Diego Mountain; 9, Lucero Arroyo; 10, Northeast Robledo Mountains; 11, Corralitos graben; 12, Picacho Mountain.

of the Pliocene–middle Pleistocene Camp Rice and Palomas Formations (Hatch Siphon, Rincon Arroyo, Lucero Arroyo, Corralitos graben, Picacho Mountain); the other seven are partial sections (Figs 2 & 3). In addition to data collected during section measuring, lithofacies relationships within channel sandstone bodies were mapped on twelve two-dimensional photomosaics. Line drawings of

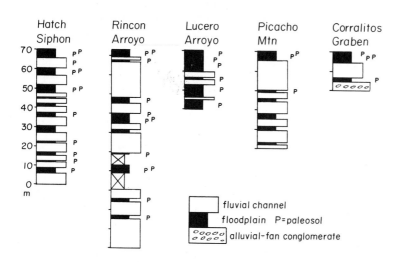

Fig. 2. Stratigraphic columns of fluvial strata in symmetrical basins. See Fig. 1 for location of columns. Each stratigraphic section contains the unconformable base and the constructional top of the Camp Rice Formation. P refers to stage II or III caliche palaeosols; the stage IV caliche palaeosol, which occupies the constructional top of each section, is not shown.

Basin symmetry control on fluvial lithofacies 441

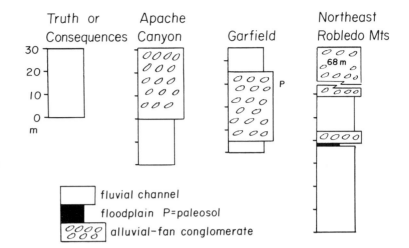

Fig. 3. Stratigraphic columns of fluvial and alluvial-fan strata in asymmetrical basins. See Fig. 1 for location of columns. None of the stratigraphic sections has the complete thickness of the Camp Rice or Palomas Formations. P refers to stage II or III caliche palaeosols.

four of the most instructive mosaics are illustrated in Fig. 4. The lithofacies code used in the line drawings is that of Miall (1978) and the hierarchy of bounding surfaces is from Miall (1988). The terminology for sandstone body geometry is that of Friend *et al.* (1979). Two hundred palaeo-

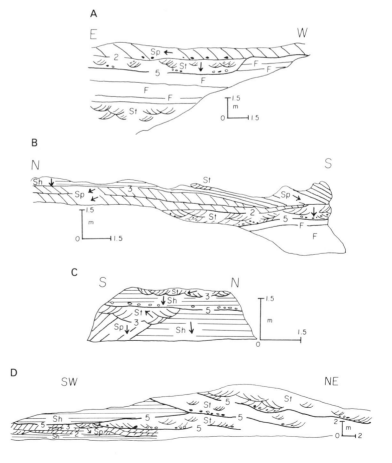

Fig. 4. Line drawings from photograph mosaics illustrating channel and floodplain lithofacies relationships. Lithofacies abbreviations follow Miall (1978); numbers 2, 3 and 5 refer to the classification of bounding surfaces of Miall (1988); arrows indicate palaeocurrent directions.
(A) Single-storey channel with stair-stepped type-5 lower bounding surface overlying floodplain deposits (F). Large-scale trough crossbeds occupy floor of deepest part of channel, and are overlain by a planar crossbed set that migrated normal to the direction of movement of the dune bedforms. Outcrop located at Picacho Peak section. (B) Lower part of single-storey channel floored by large-scale trough crossbeds and overlain by cosets of planar crossbeds. Uppermost planar crossbed overlies scour surface formed by low-stage flow. Picacho Peak section. (C) Parts of two storeys separated by a type-5 bounding surface. Upper part of each storey has a type-3 bounding surface overlain by small-scale trough crossbeds produced by low-stage flow. Hatch Siphon section. (D) Multistorey channel sandstone body internally defined by type-5 bounding surfaces; near Truth or Consequences.

current measurements were made on planar crossbeds ($n = 35$), trough crossbeds ($n = 157$), and parting lineations ($n = 8$). The method of DeCelles et al. (1983) was employed for most of the trough crossbed data, although a few measurements ($n = 5$) represent trough axes on bedding planes.

TECTONIC AND STRATIGRAPHIC SETTING

The southern Rio Grande rift developed in two distinct stages (Seager, 1975; Seager et al., 1984). The initial stage of extension began in late Oligocene and resulted in broad, northwest-trending uplifts and complementary closed basins. The second stage of rifting began between 9.6 and 7.1 Ma ago (late Miocene) and continues today (Seager et al., 1984). Unlike the first stage, the second resulted in more closely spaced, primarily north-trending uplifts and basins.

Five distinct stage-II rift basins, all of which contain fluvial strata, exist between Truth or Consequences and Las Cruces, New Mexico (Fig. 1). The Palomas basin is asymmetrical, as is the northern part of the Mesilla basin (Mack & Seager, 1990). The Hatch–Rincon, Corralitos, Jornada del Muerto and southern Mesilla basins are symmetrical.

The Palomas basin is an eastward-tilted half graben bounded by the Caballo Mountains (footwall block) on the east and the Animas Mountains (hanging wall block) on the west. Outcrops within the Palomas basin include two near Truth or Consequences, both of which are restricted to fluvial rocks, as well as outcrops in Apache Canyon and near Garfield, which consist of both fluvial and alluvial-fan strata (Fig. 1). The northern part of the Mesilla basin between the Robledo and Dona Ana Mountains is a westward-tilted half graben and is represented by an outcrop of mixed fluvial and alluvial-fan strata along the northeastern flank of the Robledo Mountains (Fig. 1). The Hatch–Rincon and Corralitos basins are symmetrical full grabens, whose margins are characterized by a series of down-to-the-basin normal faults (Fig. 1; Mack & Seager, 1990). Outcrops in the Hatch–Rincon basin include Hatch Siphon, Hatch Airport and Rincon Arroyo, whereas only one outcrop was examined in the Corralitos graben (Fig. 1). Deposition of fluvial sediment near Picacho Peak (section 12, Fig. 1) pre-dated major movement on the southern part of the East Robledo fault and thus took place in the symmetrical southern Mesilla basin (Seager et al., 1984). The southern part of the Jornada del Muerto basin between San Diego Mountain and the Doña Ana Mountains is also a full graben, and is represented by sections at Lucero Arroyo and San Diego Mountain (Fig. 1; Seager et al., 1987).

Basin-fill sedimentary rocks of the second stage are called the Palomas Formation in the Palomas basin and the Camp Rice Formation in basins to the south (Fig. 1; Strain, 1966; Seager et al., 1982; Lozinsky & Hawley, 1986a,b; Seager et al., 1987). Exposed thicknesses of the Palomas and Camp Rice Formations range from 20 m in the Corralitos basin to 130 m along the northeastern flank of the Robledo Mountains. Vertebrate faunas, radiometrically dated basalt, tephrochronology and magnetostratigraphy bracket the age of the Palomas and Camp Rice Formations between 4.5 and 0.5 Ma (Bachman & Mehnert, 1978; Hawley, 1981; Tedford, 1981; Seager et al., 1984; Lucas & Oakes, 1986; Repenning & May, 1986). Deposition of the Palomas and Camp Rice Formations ended approximately 0.5 Ma ago when the Rio Grande drainage system began to entrench its basins (Hawley & Kottlowski, 1969; Hawley, 1981). A stage IV caliche developed on the constructional top of the Palomas and Camp Rice Formations, forming a distinct resistant mesa throughout the study area (Ruhe, 1962, 1964; Gile et al., 1966, 1981).

FLUVIAL SEDIMENTOLOGY OF SYMMETRICAL BASINS

Channel sandstones

Channel sandstones in symmetrical basins consist of single-storey and multistorey sheets that range in thickness from 2 to 18 m (Fig. 2). The most common grain size is medium sand, although fine and coarse sands are also present. Granules and pebbles are scattered within the sandstones or make up distinct laminae or lenses a few centimetres thick. The basal contact of channel sandstones is erosional and either horizontal or gently concave upward. Locally, horizontal basal contacts are steeply stair-stepped, displaying up to 1 m of relief (Fig. 4A). The base of a channel is commonly overlain by up to 0.5 m of conglomerate consisting of variable amounts of granule-, pebble- and cobble-sized extraclasts, calcareous intraclasts, or intraclasts of brown or green mudstone. The upper contact of the channel sandstones is generally sharp and locally contains calcareous rhizoliths.

The width of channel sandstone bodies was determined along laterally continuous cliff faces oriented perpendicular to palaeoflow, such as at Rincon Arroyo, Lucero Arroyo, Hatch Siphon and Picacho Peak (Fig. 1). At these locations, channel bodies can be traced laterally for several kilometres, and it is uncommon for channels to extend less than hundreds of metres laterally. Approximately one-third of the channel sandstone bodies are multistorey, as indicted by the presence of one or more low-angle type-5 truncation surfaces floored by gravel and/or intraclasts.

The most abundant sedimentary structure in the channel sandstones is trough crossbeds, which have a range in set thickness from 10 to 150 cm. The coarsest, thickest sets occupy the basal parts of channels (Fig. 4A,B). In some cases, trough crossbed sets fine and thin upward within the basal 2 m of a channel or, less commonly, within the entire sand body. Palaeocurrents for the basal trough crossbeds have a low variance and are generally parallel to the trend of the alluvial plain (Fig. 5). These large-scale trough crossbeds are similar to those observed in modern low-sinuosity rivers and are interpreted to represent fields of dune bedforms that migrated down the deepest part of the channels (Harms et al., 1963; Cant & Walker, 1978; Miall, 1985; Collinson, 1986).

Smaller-scale, finer-grained cosets of trough crossbeds are located in the upper part of some channel bodies or storeys (Fig. 4C). In most cases the small-scale trough crossbeds are located above low-angle scour surfaces that extend only a few metres laterally and are not floored by a coarse lag. Consequently, these surfaces are not considered to be storey floors (type-5 surfaces of Miall, 1988), but rather represent type-3 surfaces produced by low-flow stage modification of bars and bedforms (Smith, 1971; Singh, 1977; Cant & Walker, 1978). Supporting this interpretation is the fact that palaeocurrents of trough crossbeds above the type-3 surfaces commonly deviate by as much as 90° from palaeocurrent directions below the type-3 surfaces (Fig. 4C). The abundance of these type-3 surfaces in channel bodies suggests relatively slow, probably seasonal discharge variations, which are the optimum conditions for modification and ultimately exposure of bars and bedforms (Jones, 1977).

Also common are beds of horizontally laminated sandstone that range in thickness from 0.5 to 2 m

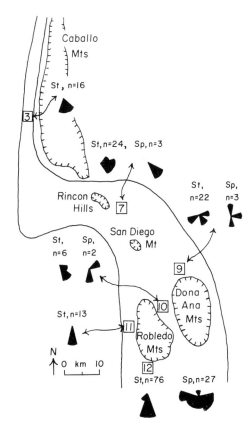

Fig. 5. Palaeocurrent rose diagrams for large-scale trough crossbeds (St) and planar crossbeds (Sp). Palaeocurrent roses in 30° intervals. Outline of alluvial plain, adapted from Fig. 1, is shown for reference.

and extend laterally for metres. Parting lineation is present on bedding planes and commonly parallels the palaeocurent direction of interbedded large-scale trough crossbeds. Coarse grain size, position near the base of channels and lack of evidence of shallowing or subaerial exposure suggest that most of the beds of horizontal laminae resulted from high-velocity flow near the peak of flood stage, rather than shallow flow during waning flood stage.

Also present in channel sandstones in symmetrical basins are individual sets or cosets (up to four stacked sets) of planar crossbeds. Laterally continuous for metres to tens of metres, planar crossbed sets range in thickness from 0.5 to 2 m, with an average of 1 m. Planar crossbeds are located in all positions within chananel bodies, but are most common in the upper part (Fig. 4A,B). Planar crossbeds are interbedded with trough crossbeds and horizontal laminae, and in some cases horizontal laminae change laterally into planar crossbeds (Fig. 4B). Palaeocurrent data from planar crossbeds show a wide variance and are parallel to or deviate by as much as 90° from palaeocurrents of basal trough crossbeds in the same channel body (Figs 4A,B & 5).

Single sets of planar crossbeds oriented at high angles to the downstream direction may be analogous to cross-channel bars, which in the South Saskatchewan River are mesoforms that develop where flow expands and velocity declines (Cant & Walker, 1978). In contrast, cosets of planar crossbeds may represent macroforms, such as sandflats in the South Saskatchewan River (Cant & Walker, 1978). Characteristics supporting a sandflat origin for the planar crossbed cosets include great lateral extent, crossbed sets with similar palaeocurrent orientations oblique to the downcurrent direction, and bimodal palaeocurrents symmetrically distributed with respect to the downstream direction (Fig. 5; Cant & Walker, 1978; Collinson, 1986).

However, convex-upward, type-4 bounding surfaces, considered by Miall (1988) to be diagnostic of sandflat macroforms, have not been recognized, perhaps due to inadequate outcrops or to modification of the macroform by low-stage flow.

The least common sedimentary structure within channel bodies is ripple cross-laminae. These asymmetrical ripples are restricted to fine sandstone and are located in the uppermost part of a channel body. These low-velocity structures formed either during low-stage modification of bar tops or just prior to channel abandonment.

Floodplain deposits

Fine-grained floodplain strata constitute approximately 35% of the total thickness of the fluvial deposits in the symmetrical basins (Figs 2 & 6). Individual units of floodplain deposits range in thickness from 0.1 to 4.5 m and can either extend laterally for hundreds of metres to kilometres or are abruptly truncated within a few metres by channel

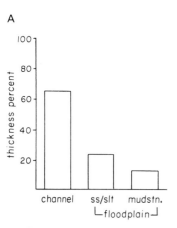
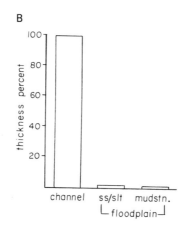
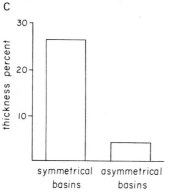

Fig. 6. Abundance of channel and floodplain lithofacies in symmetrical (A) and asymmetrical (B) basins. (C) Abundance of palaeosols in asymmetrical and symmetrical basins.

sandstones. Floodplain deposits consist of individual beds or interbeds of sandstone/siltstone and mudstone, both of which commonly display pedogenic features.

White to tan, coarse siltstone or very fine to fine sandstone beds, as much as 2 m thick, make up about 60% of the floodplain deposits and are commonly interbedded on a scale of 0.2–1.0 m with mudstone. This lithofacies is always distinctly finer grained than overlying or underlying channel sandstones. Sandstone/siltstone beds generally are internally structureless, but in a few places horizontal laminae or asymmetrical ripple cross-laminae are present. Interbedded mudstones are dark brown, red, or light green, rarely exceed 1 m in thickness and have a blocky fabric. Contacts between sandstone/siltstone and mudstone beds can be either gradational or sharp.

Calcareous palaeosols are common in the floodplain deposits (Figs 2 & 6). Palaeosols may be restricted to individual beds of sandstone/siltstone or mudstone or may cut across lithologic contacts. Palaeosols can be traced laterally for up to a kilometre, although thickness and degree of development can change along strike.

The palaeosols are characterized by a calcareous B horizon, which is 0.3–1.5 m thick and consists of scattered calcareous nodules and tubules (B_k horizon) or, less commonly, massive carbonate (K horizon) (Fig. 7; Gile *et al.*, 1966, 1981). Nodules range from 1 to 10 cm in diameter and are composed primarily of micrite and secondarily of veins of sparry calcite. Calcareous tubules range from 1 to 5 cm in diameter, taper downward and are oriented perpendicular to bedding, characteristics that suggest an origin as rhizoliths (Klappa, 1980; Cohen, 1982). Scattered nodules and tubules correspond to stage II caliche of Gile *et al.* (1966). Beds of massive carbonate, which correspond to stage III caliche, consist of grains of detrital sand, silt and clay floating in a matrix of white, micritic calcite. Irregular patches of host rock are present and pisoliths up to 1 cm in diameter are rare in the stage III caliches. The constructional top of the fluvial strata is characterized by a thick (3 m) caliche composed of massive to brecciated carbonate overlain by a thin (30 cm) zone of laminated carbonate. This stage IV caliche forms a resistant topographic surface throughout the study area (Ruhe, 1962, 1964; Gile *et al.*, 1966, 1981).

Palaeosols developed in sandstone/siltstone commonly have an argillic B horizon (B_t horizon) up to 1 m thick overlying the calcareous horizon (Fig. 7). The dark brown argillic B horizon has subangular blocky peds and vertical root traces, some of which are calcareous. In thin section, detrital sand grains display clay coats (argillans of Brewer, 1964), which formed as clay filtered downward from the A horizon into the Bt horizon (Buol *et al.*, 1980). Argillic B horizons are not as common in palaeosols developed in a mudstone host, although ped argillans are locally present.

'A' horizons are rarely preserved in the palaeosols. Where present, they are thin ($\leq$10 cm), light coloured, and contain fine root traces. The paucity of A horizons may be the result of erosion following soil development or overprinting by an overlying B_k or K horizon.

The presence of calcareous palaeosols in the Camp Rice and Palomas Formations suggests a relatively dry, probably semiarid palaeoclimate (Reeves, 1970; Cerling, 1984). A similar interpretation is provided by vertebrate fauna, which indicate that the palaeoclimate was frost free and that the floodplain was an open grassland (Lucas & Oakes, 1986).

FLUVIAL SEDIMENTOLOGY OF ASYMMETRICAL BASINS

Fluvial strata in the asymmetrical basins consist almost exclusively of multistorey channel sandstones (Fig. 3). Individual outcrops are crosscut by numerous low-angle erosional surfaces floored by gravel-sized extraclasts and/or mudstone intraclasts (Fig. 4D). These erosional surfaces are classified as type-5 bounding surfaces of Miall (1988) and constitute the base of individual storeys. Only rarely, because of the erosion surfaces, is it possible to see a complete channel in cross-section; where it is possible both ribbons and sheets are present.

The most common sedimentary structures in the channel sandstones are large-scale trough crossbeds and horizontal laminae, which are similar in grain size and scale to those in the symmetrical basins. Much less common than in the symmetrical basins are (i) planar crossbeds, which most commonly exist as single sets, and (ii) type-3 bounding surfaces overlain by fine-grained, small-scale trough crossbeds.

Floodplain deposits make up less than 1% of the total thickness of fluvial sediment in the asymmetrical basins (Figs 3 & 6). Floodplain deposits com-

Fig. 7. (A) Palaeosol at Rincon Arroyo showing upper argillic (B_t) and lower calcareous (B_k) horizons; Jacob's staff is 1.5 m. (B) Close-up of palaeosol shown in (A); Jacob's staff divided in decimetres.

posed of red mudstone and white siltstone to very fine sandstone generally are less than 1 m thick and rarely extend more than a few tens of metres laterally before being truncated by a channel sandstone. Locally, lenses of green mudstone up to 1 m thick and <10 m long occupy broad, shallow depressions within a channel sandstone. These green mudstones may have accumulated in small ponds during channel abandonment. Rip-up clasts of green mudstone are common at the base of storeys, suggesting that many lenses of mudstone were destroyed by subsequent channel erosion.

Palaeosols are uncommon and weakly developed in fluvial strata of the asymmetrical basins. The majority of palaeosols consist of zones of widely spaced (<10 cm) calcareous rhizoliths developed in the upper part of channel sandstones. These palaeosols lack the argillic and calcareous B horizons that are present in palaeosols of the symmetrical basins.

DEPOSITIONAL MODEL

Fluvial strata of the Camp Rice and Palomas

Formations were deposited by broad, low-sinuosity streams characterized by rapidly switching channels that were separated by periodically emergent bedforms and bars. This interpretation is supported by the presence of multistorey sandstone sheets, by complex lateral and vertical arrangement of channel lithofacies, by the paucity of floodplain deposits, and by the absence of lateral accretion sets (cf. Rust & Jones, 1987). The braided character of the Ancestral Rio Grande was probably a response to active uplift of adjacent highlands and semiarid palaeoclimate, both of which assured an abundant supply of sand and small gravel to the fluvial system. Data from experiments and from observations of modern streams suggest that abundant sandy bedload tends to produce braided channels with high width to depth ratios (Schumm, 1961; Schumm & Khan, 1972). Seasonal variability in discharge, indicated by low-stage modification of bars and bedforms, probably also contributed to the existence of braided streams (Schumm & Lichty, 1963; Baker & Penteado-Orellana, 1977).

Although the Ancestral Rio Grande was probably braided throughout the southern Rio Grande rift, basin symmetry exerted a strong influence on fluvial sedimentation, resulting in significant differences in the fluvial strata between the symmetrical and asymmetrical basins. Compared to the asymmetrical basins, fluvial strata in symmetrical basins: (i) were deposited on a broader alluvial plain, (ii) have a lower percentage of multistorey channel sandstones, (iii) have a smaller ratio of channel to floodplain deposits, (iv) have more palaeosols and palaeosols with a greater degree of maturity, (v) have more sets of planar crossbeds, and (vi) display more type-3 bounding surfaces overlain by small-scale trough crossbeds.

In the asymmetrical basins, the alluvial plain was narrow (≤ 5 km) and was positioned directly above the locus of maximum subsidence within a kilometer or less of the footwall block (Fig. 1; Mack and Seager, 1990). This relationship reflects the fact that channels preferentially avulse into the topographically lowest part of a basin, which in the case of half grabens is above the locus of maximum subsidence (Bridge & Leeder, 1979; Alexander & Leeder, 1987; Leeder & Alexander, 1987; Leeder & Gawthorpe, 1987). Because the alluvial plain was narrow, the braided channel moved relatively quickly across it by a combination of lateral migration and avulsion, producing multistorey channel sandstones. Floodplain sediment was not allowed to accumulate to a thickness capable of surviving subsequent channel erosion, nor were palaeosols allowed to develop to the stage of maturity common in the symmetrical basins.

In contrast, the alluvial plain in the symmetrical basins occupied a large part of the total width of the basin (Fig. 1; Mack and Seager, 1990). In the Hatch-Rincon basin, for example, fluvial strata are present within a few hundred meters of both the northern and southern margins of the basin. In a symmetrical basin, subsidence is determined by movement of faults on opposite sides of the basin. If the faults are active at the same time and have similar magnitudes of displacement, then the basin subsides uniformly across its width. If the faults are active at different times, or at the same time but with different magnitudes of displacement, then the zone of maximum subsidence will change position within the basin through time. Either way, the channel will be allowed to migrate across almost the entire basin. In the process of traversing the wide alluvial plain, the channel will be absent from portions of the floodplain for long periods of time. In the symmetrical basins of the southern Rio Grande rift, the channel was apparently absent from parts of the basin long enough to allow accumulation of up to 4.5 m of floodplain strata. Stage II and III caliche palaeosols in the floodplain strata further suggest that some areas of the floodplain experienced little or no sedimentation for thousands to tens of thousands of years (Gile et al., 1981). Such long periods of pedogenesis were not possible in the asymmetrical basins.

The fact that fluvial strata in the asymmetrical basins display fewer planar crossbeds sets and less evidence of low-stage modification of bars and bedforms compared to symmetrical basins is more difficult to explain. The paucity of these structures in the asymmetrical basins may reflect low survivability potential related to the greater degree of channel reworking. The paucity of these structures may also be the result of unique conditions of channel depth, velocity or geometry that cannot be determined or have not been recognized in outcrop.

SUMMARY AND CONCLUSIONS

Plio-Pleistocene fluvial strata deposited in symmetrical and asymmetrical basins were examined throughout the southern Rio Grande rift. The main conclusions are the following:

1 Channel sandstones in symmetrical basins are a mix of single-storey and multistorey sheets with the former dominant. Large-scale trough crossbeds, primarily near the base of channels, and horizontal laminae are common. Planar crossbeds are present throughout. Small-scale trough crossbeds overlying type-3 boundary surfaces are often present, especially in the upper part of channel sandstones.

2 Asymmetrical basins are characterized by multistorey channel sandstones. Large-scale trough crossbeds and horizontal laminae are very common with a paucity of planar crossbeds and type-3 boundary surfaces overlain by small-scale trough crossbed sets.

3 Floodplain lithofacies of fine sandstone to mudstone are ubiquitous in symmetrical basins and can extend laterally for kilometres. This facies is very rare, thin and discontinuous in fluvial strata deposited in asymmetrical basins.

4 Palaeosols are laterally continuous and about five times more abundant in symmetrical basins compared to asymmetrical basins. In symmetrical basins argillic B horizons and calcareous B horizons to K horizons are typical and associated with stages II and III caliches. Palaeosols in asymmetrical basins lack well-developed horizons and are mainly confined to intervals of widely spaced rhizoliths developed in the upper parts of channel sandstones.

5 Fluvial strata were deposited by broad, low-sinuosity streams characterized by rapidly switching channels separated by periodically emergent bedforms and bars. The braided character of the Ancestral Rio Grande in both symmetrical and asymmetrical basins was a response to high sand supply resulting from a combination of semiarid climate and active uplift of proximal highlands.

ACKNOWLEDGEMENTS

We are grateful to the New Mexico Bureau of Mines and Mineral Resources and the Ohio State University for support. We especially thank Helen Hayes for typing.

REFERENCES

ALEXANDER, J. & LEEDER, M.R. (1987) Active control on alluvial architecture. In: *Recent Developments in Fluvial Sedimentology* (Eds Ethridge, F.G., Flores, R.M. & Harvey, M.D.) pp. 243-252. Soc. Econ. Paleontol. Mineral., Tulsa, Spec. Publ. 39.

ALLEN, J.R.L. (1970) *Physical Processes of Sedimentation.* George Allen and Unwin, London, 248 pp.

BACHMAN, G.O. & MEHNERT, H.H. (1978) New K-Ar dates and late Pliocene to Holocene geomorphic history of the Rio Grande region, New Mexico. *Bull. Geol. Soc. Am.* **89**, 283-292.

BAKER, V.R. & PENTEADO-ORELLANA, M.M. (1977) Adjustments to Quaternary climatic change by the Colorado River in central Texas. *J. Geol.* **85**, 395-422.

BREWER, R. (1964) *Fabric and Mineral Analysis of Soils.* John Wiley & Sons, New York, 470 pp.

BRIDGE, J.S. & LEEDER, M.R. (1979) A simulation model of alluvial stratigraphy. *Sedimentology* **26**, 617-644.

BUOL, S.W., HOLE, F.D. & MCCRACKEN, R.J. (1980) *Soil Genesis and Classification.* Iowa State University Press, Ames, 406 pp.

CANT, D.J. & WALKER, R.G. (1978) Fluvial processes and facies sequences in the sandy braided South Saskatchewan River, Canada. *Sedimentology* **25**, 625-648.

CERLING, T.E. (1984) The stable isotopic composition of modern soil carbonate and its relationship to climate. *Earth Planet. Sci. Lett.* **71**, 229-240.

COHEN, A.S. (1982) Paleoenvironments of root casts from the Koobi Fora Formation, Kenya. *J. Sedim. Petrol.* **52**, 401-414.

COLLINSON, J.D. (1986) Alluvial sediments. In: *Sedimentary Environments and Facies* (Ed. Reading, H.G.) pp. 20-62. Blackwell Scientific Publications, Oxford.

DECELLES, P.G., LANGFORD, R.P. & SCHWARTZ, R.K. (1983) Two new methods of paleocurrent determination from trough cross-stratification. *J. Sedim. Petrol.* **53**, 629-642.

FRIEND, P.F., SLATER, M.J. & WILLIAMS, R.C. (1979) Vertical and lateral building of river sandstone bodies, Ebro basin, Spain. *J. Geol. Soc. Lond.* **136**, 39-46.

GILE, L., HAWLEY, J.W. & GROSSMAN, R.B. (1981) *Soils and Geomorphology in a Basin and Range Area of Southern New Mexico—Guidebook to the Desert Project.* New Mexico Bur. Mines Mineral Res., Socorro, Memoir 39.

GILE, L.H., PETERSON, F.F. & GROSSMAN, R.B. (1966) Morphological and genetic sequences of carbonate accumulation in desert soils. *Soil Sci.* **101**, 347-360.

HARMS, J.C., MACKENZIE, D.B. & MCCUBBIN, D.G. (1963) Stratification in modern sands of the Red River, Louisiana, *J. Geol.* **71**, 566-680.

HAWLEY, J.W. (1981) Pleistocene and Pliocene history of the international boundary area, southern New Mexico. In: *Geology of the Border—southern New Mexico-northern Chihuahua,* pp. 25-32. El Paso Geol. Soc., El Paso.

HAWLEY, J.W. & KOTTLOWSKI, F.E. (1969) Quaternary geology of the south-central New Mexico border region. *Circular New Mexico Bur. Mines Mineral. Res.* **104**, 89-115.

HAWLEY, J.W., KOTTLOWSKI, F.E., SEAGER, W.R., KING, W.E., STRAIN, W.S. & LEMONE, D.V. (1969) The Santa Fe Group in the south-central New Mexico Border. *Circular New Mexico Bur. Mines and Mineral Res.* **104**, 52-76.

JONES, C.M. (1977) Effects of varying discharge regimes on bed-form sedimentary structures in modern rivers. *Geology* **8**, 567-570.

KLAPPA, C.F. (1980) Rhizoliths in terrestrial carbonates: classification, recognition, genesis and significance. *Sedimentology* 27, 613–629.

KOTTLOWSKI, F.E. (1953) Tertiary–Quaternary sediments of the Rio Grande Valley in southern New Mexico. *Guidebook New Mexico Geol. Soc.* 4, 144–148.

KOTTLOWSKI, F.E. (1960) *Summary of Pennsylvanian Sections in Southwestern New Mexico and Southeastern Arizona.* New Mexico Bur. Mines Mineral Res. Bull. 66., Socorro.

LEEDER, M.R. & ALEXANDER, J. (1987) The origin and tectonic significance of asymmetrical meanderbelts. *Sedimentology* 34, 217–226.

LEEDER, M.R. & GAWTHORPE, R.L. (1987) Sedimentary models for extensional tilt-block/half graben basins. In: *Continental Extensional Tectonics* (Eds Coward, M.P., Dewey, J.F. & Hancock, P.L.) pp. 139–152. Geol. Soc., Lond., Spec. Publ. 28. Blackwell Scientific Publications, Oxford.

LOZINSKY, R.P. & HAWLEY, J.W. (1986a) Upper Cenozoic Palomas Formation of south-central New Mexico. *Guidebook New Mexico Geol. Soc.* 37, 239–247.

LOZINSKY, R.P. & HAWLEY, J.W. (1986b) The Palomas Formation of south-central New Mexico—a formal definition. *New Mexico Geol.* 8, 73–82.

LUCAS, S.G. & OAKES, W. (1986) Pliocene (Blancan) vertebrates from the Palomas Formation, south-central New Mexico. *Guidebook New Mexico Geol. Soc.* 37, 249–255.

MACK, G.H. & SEAGER, W.R. (1990) Tectonic control on facies distribution of the Camp Rice and Palomas Formations (Plio-Pleistocene) in the southern Rio Grande rift. *Bull. Geol. Soc. Am.* 102, 45–53.

MIALL, A.D. (1978) Lithofacies types and vertical profile models in braided river deposits: a summary. In: *Fluvial Sedimentology* (Ed. Miall, A.D.) pp. 597–604. Can. Soc. Petrol. Geol., Calgary, Memoir 5.

MIALL, A.D. (1985) Architectural-element analysis: a new method of facies analysis applied to fluvial deposits. *Earth Sci. Rev.* 22, 261–308.

MIALL, A.D. (1988) Facies architecture in clastic sedimentary basins. In: *New Perspectives in Basin Analysis.* (Eds Kleinspehn, K.L. & Paola, C.) pp. 67–89. Springer-Verlag, New York.

REEVES, C.C., JR (1965) Pluvial Lake Palomas, northwestern Chihuahua, Mexico and Pleistocene geologic history of south-central New Mexico. *Guidebook New Mexico Geol. Soc.* 16, 199–203.

REEVES, C.C. (1970) Origin, classification, and geologic history of caliche on the southern High Plains, Texas and eastern New Mexico. *J. Geol.* 78, 352–362.

REPENNING, C.A. & MAY, S.R. (1986) New evidence for the age of lower part of the Palomas Formation, Truth or Consequences, New Mexico. *Guidebook New Mexico Geol. Soc.* 37, 257–259.

RUHE, R.V. (1962) Age of the Rio Grande Valley in southern New Mexico. *J. Geol.* 70, 151–167.

RUHE, R.V. (1964) Landscape morphology and alluvial deposits in southern New Mexico. *Ann. Ass. Am. Geogr.* 54, 147–159.

RUST, B.R. & JONES, B.G. (1987) The Hawkesbury Sandstone south of Sydney, Australia: Triassic analogue for the deposit of a large, braided river. *J. Sedim. Petrol.* 57, 222–233.

SCHUMM, S.A. (1961) Effect of sediment characteristics on erosion and deposition of ephemeral-stream channels. US Geol. Surv. Prof. Paper 352-C.

SCHUMM, S.A. & LICHTY, R.W. (1963) Channel widening and flood-plain construction along Cimarron River in southwestern Kansas. US Geol. Surv. Prof. Paper 352-D.

SCHUMM, S.A. & KHAN, H.R. (1972) Experimental study of channel patterns. *Bull. Geol. Soc. Am.* 83, 1755–1770.

SEAGER, W.R. (1975) Cenozoic tectonic evolution of the Las Cruces area, New Mexico. *Guidebook New Mexico Geol. Soc.* 26, 241–250.

SEAGER, W.R., CLEMONS, R.E., HAWLEY, J.W. & KELLEY, R.E. (1982) Geology of northwest part of Las Cruces 1° × 2° sheet (scale 1 : 125 000), New Mexico. Geologic Map 53, New Mexico Bur. Mines Mineral. Res., Socorro.

SEAGER, W.R., HAWLEY, J.W., KOTTLOWSKI, F.E. & KELLEY, S.A. (1987) Geology of east half of Las Cruces and northeast El Paso 1° × 2° sheets, New Mexico (scale 1 : 125 000). Geologic Map 57, New Mexico Bur. Mines Mineral. Res., Socorro.

SEAGER, W.R. SHAFIQULLAH, M., HAWLEY, J.W. & MARVIN, R. (1984) New K-Ar dates from basalts and the evolution of the southern Rio Grande rift. *Bull. Geol. Soc. Am.* 95, 87–99.

SINGH, I.B. (1977) Bedding structures in a channel sand bar of the Ganga River near Allahabad, Uttar Pradesh, India. *J. Sedim. Petrol.* 47, 747–752.

SMITH, N.D. (1971) Transverse bars and braiding in the Lower Platte River, Nebraska. *Bull. Geol. Soc. Am.* 82, 3407–3420.

STRAIN, W.S. (1966) *Blancan Mammalian Fauna and Pleistocene Formations, Hudspeth County, Texas.* Bull. University of Texas at Austin, Texas Memorial Museum, 10.

TEDFORD, R.H. (1981) Mammalian biochronology of late Cenozoic basins of New Mexico. *Bull. Geol. Soc. Am.* 92, 1008–1022.

Simultaneous dispersal of volcaniclastic and non-volcanic sediment in fluvial basins: examples from the Lower Old Red Sandstone, east-central Scotland

P.D.W. HAUGHTON*

Department of Geology and Applied Geology, University of Glasgow, Glasgow G12 8QQ, UK

ABSTRACT

Fluvial basins flanking active volcanoes can preserve complex sequences in which intervals of volcaniclastic sediment alternate with mixed provenance or non-volcaniclastic sediments. Transport mechanisms, the pattern and rate of sediment supply, and depositional processes can be markedly different for these deposits. Deposition of polymict Lower Old Red Sandstone fluvial deposits in central Scotland was repeatedly interrupted by the dispersal of units of volcaniclastic sediment. These were emplaced by a combination of fluvial, debris flow and fluidal sediment flow processes. The nature of the transitions to and from these volcaniclastic intervals is documented and three recurring types of sequence are identified. In the first, contacts between volcaniclastic and mixed fluvial deposits are sharp and the structure, composition and palaeoflow of the fluvial sediments are unperturbed above and below the intervals of volcaniclastic sediment. In the second, the volcaniclastic intervals appear to determine a component of the structure of the enclosing non-volcanic deposits, with evidence for periods of aggradation and degradation associated with volcaniclastic and non-volcaniclastic sequences respectively. A third type of sequence, closely related to the second, shows that degradation on return to normal fluvial sedimentation was at first suppressed, with flat stratified units of mixed provenance intervening between the volcaniclastic and non-volcaniclastic deposits. The various sequences appear to be specific to one or other margin of a fault-controlled basin, and can be related to contrasting aggradation rates, the type of non-volcanic fluvial system and the nature of the volcaniclastic sources. The synchroneity or otherwise of eruption and volcaniclastic deposition was particularly important, with the latter recording both episodic, syn-eruption flushing of parts of the basin with volcanic sand, and normal degradation of volcanic sourcelands.

INTRODUCTION

Over the last decade, there has been considerable interest in the depositional response of fluvial systems subject to eruption-induced sediment loads (Walton, 1977; Davies *et al.*, 1978; Kuenzi *et al.*, 1979; Vessell & Davies, 1981; Harrison & Fritz, 1982; Mathisen & Vondra, 1983; Smith, 1986, 1987a,b, 1988a,b; Runkel, 1990). The deposits generated in this way differ from 'classical' fluvial sequences in that they often show evidence of unusually rapid aggradation. In addition, the distinctive texture of volcanic detritus and the high rates of sediment transport during and following explosive eruptions tend to produce departures from conventional fluvial facies. Thus Smith (1986, 1987a) has highlighted the importance of high sediment-concentration flows intermediate between normal stream flows and debris flows on volcaniclastic aprons, and the different implication of debris flow deposits in volcaniclastic and non-volcaniclastic successions. Volcaniclastic debris flows can have long run-out distances and are not necessarily diagnostic of an alluvial fan environment.

Rapidly subsiding, arc-adjacent basins which protect subaerial volcaniclastic sequences from ero-

*Present address: Badley Ashton and Associates Ltd, Winceby House, Winceby, Horncastle, Lincolnshire LN9 6PB, UK

sion are favourable sites for the preservation of older synvolcanic successions (Francis, 1983). As consequent and/or antecedent streams unconnected with the volcanic activity can also operate in these basins, particularly during periods of volcanic quiescence, the detrital record of ancient volcanic provinces is often one of volcaniclastic sediments repeatedly interbedded with mixed provenance and/or non-volcanic deposits. Transport mechanisms, the continuity of sediment supply and rates of aggradation can be markedly different for these various deposits.

Whilst previous studies have treated volcaniclastic and normal fluvial deposits separately, the purpose of this study is to explore the interface between the two, focusing in particular on the nature of the vertical transitions separating packages of volcaniclastic sediment from 'ambient' non-volcanic or mixed fluvial deposits. The Lower Old Red Sandstone succession of central Scotland contains many examples of interbedded volcaniclastic and polymict fluvial deposits, and it is the relationship between some of these intervals which will be addressed below. Specific aims include (i) to distinguish, in the absence of pyroclastic deposits, volcaniclastic intervals related to 'episodic' syneruption sediment supply from those formed by normal degradation of inactive volcanic relief; (ii) to examine the response (if any) of non-volcanic fluvial systems to the rapid aggradation often implied by the disparate structure of the interbedded volcaniclastic intervals. Might the rapid emplacement of syn-eruption volcaniclastic deposits influence deposition of non-volcanic conglomerates by, for instance, perturbing base levels?; and (iii) to consider how vertical cycles in non-volcanic conglomerates relate to intervals of volcaniclastic deposition.

LOWER ORS NON-VOLCANICLASTIC FLUVIAL DEPOSITS

Lower Old Red Sandstone (?Pridoli–Emsian) fluviatile rocks of the Strathmore region are found in a belt traversing central Scotland, flanked to the north for the most part by the Highland Boundary Fault. Their pattern of outcrop is largely determined by a pair of NE-SW striking folds, the Strathmore Syncline and Sidlaw Anticline (Fig. 1). The predominantly conglomeratic and sandy fluvial deposits are interstratified with a thick calc-alkaline lava sequence which is brought to the surface in the core of the Sidlaw Anticline. These lavas form part of a wider suite of late Caledonian volcanic rocks of this age (Thirlwall, 1981, 1988). The Old Red Sandstone succession is subdivided into five lithostratigraphical groups (Fig. 1), comprising in ascending order: the Dunnottar, Crawton, Arbuthnott, Garvock and Strathmore Groups (Armstrong & Paterson, 1970). Volcanism climaxed during deposition of the Arbuthnott Group, producing a lava sequence which thickens southeastwards, locally reaching 2400 m (Francis et al., 1970). However, thin lava flows are present throughout the lower part of the succession, attesting to the general synvolcanic nature of sedimentation at this level.

Volcaniclastic sequences of 1 m to >50 m in thickness are enclosed within non-volcaniclastic fluvial sediments in well exposed cliff and foreshore outcrops of the Dunnottar, Crawton and basal Arbuthnott Groups between Stonehaven and Gourdon on the east coast (Figs 1 & 2). Two contrasting 'non-volcaniclastic' fluvial sequences can be distinguished in this area. Both are dominantly conglomeratic and contain variable proportions of volcanic detritus, but this is generally subordinate to non-volcanic sediment. The first occurs throughout the stratigraphic slice considered here and has been described in some detail by Haughton (1989). It comprises thickly bedded sheets of generally imbricate boulder- and well-sorted pebble-conglomerate, together with rare cross-stratified conglomerate (foresets up to 4 m thick) and thin, laterally extensive sandstone beds. Textural and compositional maturity are high, and the conglomerates are characterized by a well-developed bimodal texture. Although some vertical megasequences (sensu Heward, 1978) have been recognized, these are subtle and there are thick vertical sequences over which there is little change in grain size and structure. A complex provenance is evident (Haughton et al., 1990), as the clasts include large volumes of clearly polycyclic, well-rounded quartzite gravel, together with granite, psammite, porphyry and volcanic detritus. Palaeoflow over the basal and most northerly part of the exposed section is characterized by major reversals of flow from NE to SW. In southerly outcrops, palaeoflows are consistently towards the south. Previously regarded as alluvial fan deposits draining a fault-scarp bounding a mountainous source region to the north of the basin, the structure of these conglomerates is more akin to that of braidplain deposits (Haughton, 1989). They are inter-

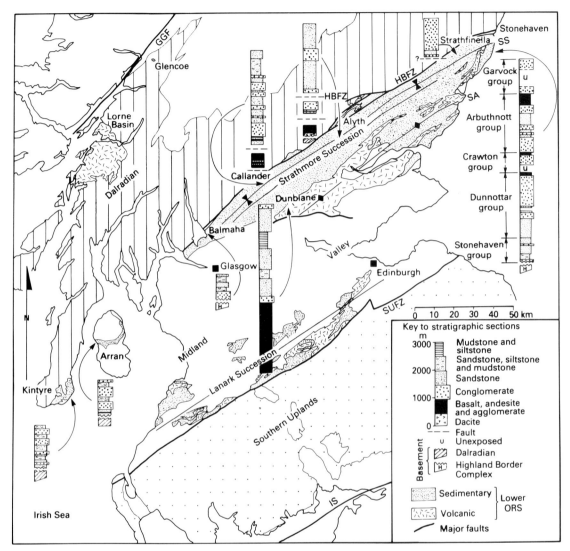

Fig. 1. Sketch map of central Scotland with distribution of Lower ORS sedimentary and volcanic rocks. Schematic sections for Lower ORS successions in the Strathmore region and its western extension have been compiled from Armstrong & Paterson (1970), Morton (1976) and Friend *et al.* (1963). GGF, Great Glen Fault; HBFZ, Highland Border Fault Zone; SUFZ, Southern Uplands Fault Zone; IS, Iapetus Suture; SS, Strathmore Syncline; SA, Sidlaw Anticline.

preted as a product of high-discharge, probably antecedent, braided streams entering the basin transversely from the north.

The second type of fluvial sequence occurs only in the upper part of the Crawton Group, cropping out between Whistleberry and Gourdon (Fig. 2). This is stratigraphically the oldest part of the succession to appear on the southern limb of the Strathmore syncline. Here, first-cycle, texturally immature conglomerates occur with clasts of grey-wacke, microconglomerate, lithic arenite, granite, andesite, hornfels and limestone. Clasts are generally subangular and the deposits polymodal. Instead of the dominant sheet geometry of the quartzite conglomerate sequences, channelling is widely observed and cross-stratified conglomerates are more abundant (Haughton & Bluck, 1988). In addition, textural segregation of coarse and fine gravel, and

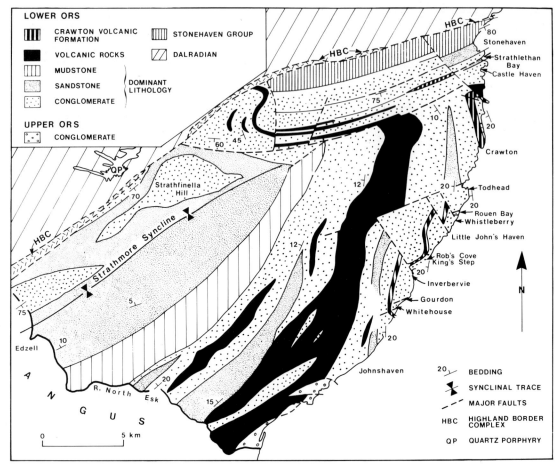

Fig. 2. Sketch map showing Old Red Sandstone geology of the NE Midland Valley, with localities discussed in text. (Modified from Campbell (1913) and Armstrong & Paterson (1970).)

imbricate clast fabrics are less well developed. Directional structures indicate dispersal towards the north and west, i.e. in the opposite direction to the braidplain deposits previously described. These conglomerates are interpreted as the deposits of streams issuing from the southeastern margin of a small, largely conglomerate filled sub-basin, hence their restriction to the earliest most southerly exposures. Their provenance was quite different to that tapped by the coeval transverse system supplying sediment across the northern margin of the sub-basin, implying that the basin lay on the boundary between quite different crustal blocks. The less mature textures suggest that deposition may have been on alluvial fans.

Braidplain conglomerates relating to the northern source rest with a sharp and locally unconformable contact on the northerly and westerly dispersed conglomerates and interbedded sandstones in faulted outcrops of the Crawton Group between Whistleberry and Gourdon (Fig. 2). The northerly and westerly dispersed succession does not appear on the northern limb of the Strathmore syncline, and where its base is found, at Little Johnshaven (Fig. 2), it is underlain by pebbly sandstones of northern provenance. It is seen as an interfinger, several hundred metres thick, passing northwards into sediments derived from the north. The prominent unconformable contact at the top of this wedge may represent the abandonment of the early sub-basin, with the southeasterly source overstepped by deposits of the northern rivers operating in a subsequently enlarged basin. The distinctive greywacke and granite bearing conglomerates do not appear

volcaniclastic intervals associated with the northerly and westerly dispersed fluvial succession (Fig. 4D). The former occur as either ungraded units up to 1.5 m thick which locally feather out laterally into tuffaceous sandstones, or thicker units (up to c. 5 m thick) in which coarse outsized clasts (including non-volcanic lithologies) are concentrated in the middle to upper part of the bed (Figs 4C & 6B). Clasts are angular to subrounded, with the more rounded examples (Fig. 7D) possibly reflecting bulking of existing volcaniclastic alluvium (cf. Scott, 1988).

The sheets of unsorted breccia form beds 0.03–1.5 m thick and have been traced for over 100 m laterally, the limit of outcrop. Some show evidence for lateral thinning by pinch-out along curviplanar bases. However, bases are generally non-erosive and the upper surfaces are sharp. The breccia sheets occur as isolated beds, or as packages of several beds. The angular fragments of andesite of which they are composed rarely exceed 2–3 cm in diameter, and there is no obvious matrix, grading, clast fabric or internal stratification. These features suggest *en masse* 'plug' flow deposition from extensive sheet-like flows emplaced over a relatively smooth depositional surface. The fine grain size, monomict composition and granular texture (i.e. lacking the mud which would be expected if they were derived from altered volcanic rock) are consistent with mass flow redeposition of primary pyroclastic sediments (e.g. airfall tuffs or flow deposits). Beds with similar features have elsewhere been interpreted as lahar runout facies on the basis of bed thickness– maximum clast size data (Walton & Palmer, 1988) but the deposits described here lack the faint stratification and gravel lenses of lahar runout units at Mount St Helens (Pierson & Scott, 1985). Some units contain small non-volcanic fragments of plutonic or sedimentary rock fragments. These unstructured breccia beds are closely associated with flat-stratified tuffaceous sandstone and pebbly sandstone units (Fig. 4D) resembling the 'hyperconcentrated flood flow' deposits described by Smith (1986; see below). This association is a feature of volcaniclastic aprons elsewhere.

Coarse clast-supported boulder conglomerates occupying deep channels (c.5 m) represent a third candidate for debris flow emplacement (Fig. 4B). Again these have only been observed in volcaniclastic sections enclosed in the northerly and westerly dispersed fluvial deposits. The channels, cutting tuffaceous sandstones, are filled with ungraded, disorganized conglomerates comprising subrounded to rounded clasts of a single volcanic lithology (the distinctive hornblende–biotite andesite). These are envisaged as non-cohesive flows which were diverted into or confined by channels which locally degraded the volcaniclastic surface. The degree of clast rounding may reflect bulking of pre-existing alluvium. On the whole, evidence for degradation and significant erosion within the volcaniclastic units is rare.

The evidence for debris flows during deposition of the volcaniclastic intervals is interesting, given that mass-flow deposits have not been recognized in the non-volcaniclastic fluvial deposits which enclose them. This may partly reflect the longer runout distances of volcanic debris flows in volcaniclastic aprons compared with those on conventional alluvial fan/braidplain surfaces (Smith, 1987a). The former can travel over 100 km from source, whilst the latter rarely extend off the fan surface (generally <10 km).

Fluidal sediment flow deposits

Nemec & Steel (1984) suggest that the term *fluidal sediment flow* be used to describe a continuum of poorly understood flows intermediate between true debris flows and normal fluvial flows. These include the intermediate-type, streamflood and hyperconcentrated flows of other workers. Poorly sorted, clast-supported, texturally immature cobble and boulder conglomerates alternating with coarse tuffaceous sandstone or breccia beds with horizontal stratification are collectively interpreted as a product of such flows (Fig. 4C). These deposits are often closely associated with debris flow units. The conglomerates form both extensive sheets, with rather diffuse contacts, and thinner lenticular units. Clast fabrics are not well developed. Some units show normal grading whilst others are ungraded and may be true debris flow units. A crude horizontal stratification is present in some cases. The associated beds of coarse sandstone are poorly sorted and commonly have outsized volcanic clasts. They may gradationally overlie conglomerate units, forming *graded-stratified* deposits (Fig. 4C, see Smith, 1986). These features, together with the geometry of the pervasive stratification (dominantly horizontal, thick laminae, with diffuse stratal contacts and a wide range of grain sizes) resemble hyperconcentrated flood flow deposits described by Smith (1986, 1987a). Horizontal stratification is accompanied locally by low-angle cross-stratication in sequences of thin (up to 50 cm thick) conglomerate sheets and lenses which alternate with pebbly sandstones. These may reflect deposition from unconfined flows towards the water-rich end of the water/sediment fluidal flow continuum.

Braided-stream deposits

Clast-supported pebble and cobble conglomerates alternating with planar and cross-stratified sandstones reflect normal fluvial transport and deposition of volcanic detritus. This facies is only well developed in volcaniclastic intervals associated with the northerly derived, non-volcanic braidplain deposits. As many of the volcanic clasts are equant, clast fabrics tend to be poorly developed. The conglomerates are generally flat-stratified with little evidence for local channelization. Clasts are mostly subrounded to well rounded. Cross-bedding in related sandstones is predominantly planar and directions show a large variance. Deposition from shallow, probably braided streams is indicated. The degree of rounding in some conglomerates suggests appreciable transport.

'Crystal-rich' volcaniclastic deposits

Whilst sandstones associated with volcaniclastic intervals are dominantly volcanic lithic, levels of thick bedded, moderately well-sorted sandstone characterized by a dominance of crystal fragments are locally developed in units interbedded with the southeastern fluvial system. Mechanisms of crystal concentration in volcaniclastic deposits are poorly understood, but may reflect either or both pyroclastic and epiclastic processes (cf. Cas & Wright, 1987). Lava clast textures in the associated monomict conglomerates show that many of the erupted magmas were relatively highly crystallized (20–30% phenocrysts). The 'crystal tuff' horizons were deposited by epiclastic flows, probably sheet floods, and thus they may owe part of their high crystal contents to epiclastic reworking. However, the restricted occurrence of these intervals in a succession where sandstones are dominantly lithic could reflect occasional reworking of primary 'crystal-rich' pyroclastics.

STRATIGRAPHY OF INTERBEDDED VOLCANICLASTIC AND FLUVIAL DEPOSITS

The relationship between the ambient fluvial and the volcaniclastic intervals have been examined in a series of well-exposed sections, with the aim of understanding the way in which source switching took place and the degree to which the stacking of disparate deposits has determined a component of their structure. If, for example, deposits with different aggradation rates were interleaved, evidence for episodic regrading and readjustment might be preserved. In the case of the Lower Old Red Sandstone outcrops examined, repeated stacking of 'background' fluvial and the volcaniclastic deposits (and vice versa) produced three distinct types of vertical sequence. These are described below and summarized in Fig. 5. Significantly, there is an important relationship between the nature of stacking and which of the two fluvial systems is involved. Polymict volcaniclastic intervals in fluvial sequences relating to the northern transverse source produced sequences which differ from those in the southern fluvial system with its monomict volcaniclastic intervals. This reflects a geographical variation in both the type of volcaniclastic sediment received by the basin, and the response of the ambient fluvial deposits to blanketing by this sediment.

Type I sequences

Sections illustrating the alternation of polymict volcaniclastic sediments and fluvial deposits of the northern transverse system are shown in Fig. 6. The fluvial deposits in both sections exhibit gradual vertical changes in grain size (as defined by maximum clast size) and structure (e.g. per cent sandstone in the sequence) over sections hundreds of metres thick, and represent megasequences (*sensu* Heward, 1978), one coarsening upward (CU), the other fining upward (FU). Volcaniclastic intervals punctuate the two megasequences. Whilst the Strathlethan South section (Fig. 6A) shows repetitive transitions to relatively thin volcaniclastic intervals (several metres to *c.*30 m thick) dominated by normal fluvial facies, the Todhead–Rouen Bay section (Fig. 6B) is interrupted by a single volcaniclastic interval (*c.* 55 m thick) composed of debris and fluidal flow deposits (Fig. 4C).

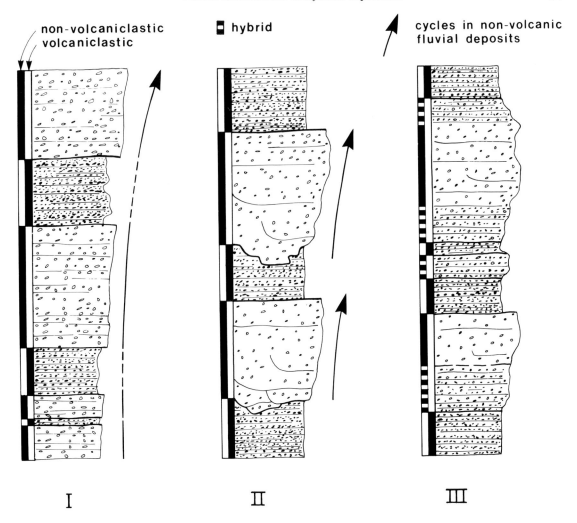

Fig. 5. Summary sections illustrating three different types of volcaniclastic/non-volcaniclastic sequence preserved in the Lower ORS of the NE Midland Valley.

Transitions between the mixed fluvial and volcaniclastic facies (and vice versa) are in all cases sharp and generally planar. In favourable exposures they have been traced laterally for up to several hundred metres (e.g. Fig. 7A), with no discordance of bedding. Gradual vertical grain size trends in the mixed fluvial deposits continue across the volcaniclastic intervals, irrespective of their thickness. This is well illustrated by the Todhead–Rouen Bay section (Fig. 6B), where the FU tendency evident beneath the volcaniclastic interval is continued in the deposits above it. In addition, there is no change in the structure or clast composition of the fluvial deposits immediately above or below the volcaniclastic intervals. Palaeoflow measurements based on well-developed imbrication in the fluvial deposits are generally consistent and show no divergence when followed up section towards the transition to volcaniclastics. The dispersal direction of the volcaniclastic sediment is less well constrained, but the small amount of non-volcanic material present in these deposits (dominantly quartzite and psammite) has an affinity with the northern transverse source tapped by the non-volcaniclastic

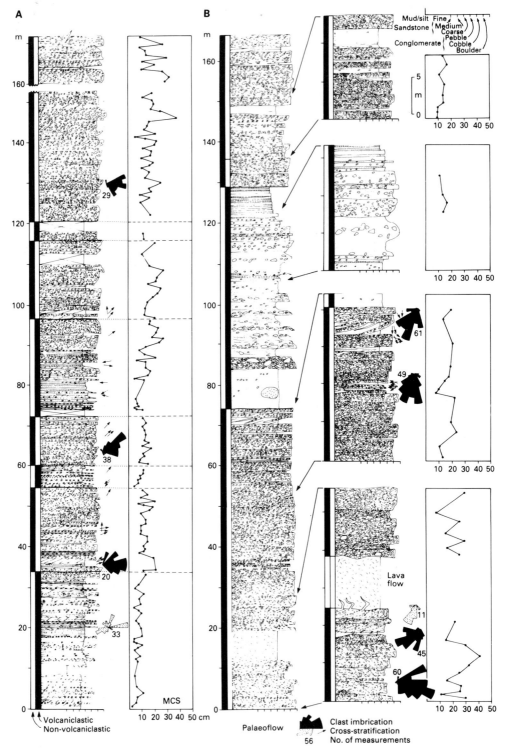

Fig. 6. Logged sections illustrating examples of Type I volcaniclastic/non-volcaniclastic transitions: (A) Strathlethan Bay; section from NO88058474 to NO88068464. (B) Todhead–Rouen Bay; section from NO86977660 to NO86357578. MCS is maximum clast size, the mean of the maximum diameters of the ten largest clasts over an area of 2 m^2. Roses summarize palaeoflow directions deduced from imbrication and cross-stratification.

Fig. 7. Examples illustrating volcaniclastic/non-volcaniclastic transitions. (A) Cliff-section at Rouen Bay showing Type I sequence. Volcaniclastic interval (2) is sandwiched within non-volcanic fluvial deposits (1 and 3). Note planar contact between volcaniclastic and fluvial deposits, highlighted by dashed line. Cliff is approx. 35 m high. Photograph taken looking north from NO86357575. (B) Detail of channel floor with evidence for rapid vertical degradation of the volcanic substrate on which the non-volcanic conglomerate-filled channel developed (NO84397291). Note large block of tuffaceous sandstone (1) entrained in the channel conglomerate and the undercut margin on the basal scour hollow (2). (C) View of sharp Type II volcaniclastic/non-volcaniclastic transition with channelled surface (highlighted). Note undercutting of channel margin and flat stratified and sand-prone nature of volcaniclastic unit (NO84697337). (D) Detail of undercut box-like channel margin (NO84477297) with non-volcanic fluvial conglomerates (1) incised into volcanic debris flow unit dominated by detritus of a single volcanic lithology (2). (E) Outsized volcanic clasts, all of a distinctive porphyritic andesite, concentrated along base of channel filled by non-volcanic fluvial conglomerates (NO85497429).

conglomerates in which they are enclosed. However, psammite blocks in the volcaniclastic intervals are of a lower metamorphic grade than those in the non-volcanic sediments, suggesting that the volcanic rocks may have occupied part of the source terrain not sampled by the drainage basin that supplied the transverse fluvial system.

The above observations suggest little mutual interaction between the volcaniclastic and non-volcaniclastic depositional systems. The fluvial system carrying mixed detritus to the basin did not respond in anticipation of the influx of volcaniclastic sediment, or as a consequence of the re-establishment of flow on a substrate of volcaniclastic materials. The lack of evidence for incision indicates that intrabasinal degradation of the volcaniclastic sediments was not important during the transition from volcaniclastic back to normal fluvial sedimentation. There may thus have been little scope for intrabasinal reworking of volcanic detritus and the subordinate volcanic component present in the fluvial deposits was probably achieved by mixing prior to entering the basin.

Type II sequences

A different style of volcaniclastic/fluvial alternation is seen in sections of the Crawton Group involving fluvial deposits of northerly and westerly dispersal. Key features of Type II sequences are shown in Figs 7B,C,D,8,9 & 10. These sequences are characterized by monomict volcaniclastic units up to 10 m thick enclosed in non-volcanic conglomerates up to 30 m thick. There is a strong compositional segregation, with only minor volcanic detritus incorporated in the non-volcanic conglomerates. The volcaniclastic units are strikingly more sand-prone than the enclosing conglomerates (not always the case in Type I sequences) and tuffaceous sandstones within them can be crystal-rich.

Two examples of coarsening-upward fluvial cycles, sandwiched between volcaniclastic units, are illustrated in Fig. 8. In these, the upward transition from volcaniclastic to non-volcanic sediment is achieved by the vertical coalescence of conglomerate-filled channels incised into the volcaniclastic deposits. In this instance, vertical trends in the structure and grain size of the non-volcanic deposits do not span volcaniclastic intervals.

The volcaniclastic intervals largely comprise debris and fluidal flow deposits, with a predominance of flat stratification and little evidence for erosional relief, apart from rare channels filled with monomict, clast-supported boulder conglomerate. Nested channels filled by non-volcanic conglomerate overlie the volcaniclastic intervals with erosional relief of several metres indicating important degradation of the volcaniclastic deposits. The density of non-volcanic channels, their depth and the grain size of their fill all show a tendency to increase upwards, with merging of the channel fills to form laterally extensive conglomeratic bodies. The channels at the base of these cycles show evidence of rapid incision into their volcaniclastic substrate, with irregular basal scours, sometimes with undercut margins, and with reworked blocks of volcaniclastic sediment floating within the channel fill (Figs 7B,C & 9). There is thus a marked contrast between the horizontal stratification of the volcaniclastic intervals and evidence for channelling in the non-volcanic deposits. Channels typically have box shapes with steep margins (also locally undercut, Fig. 7D) and large outsized volcanic clasts tend to line the floor to channels otherwise filled with non-volcanic clasts (Fig. 7E). Whilst both channel margins are in some cases preserved immediately above volcaniclastic intervals, those within the non-volcanic fluvial intervals tend to have only a single margin preserved, suggesting that the channels here may have been more mobile. The box-shaped channel geometry is typical of the rapid subaerial erosion of unlithified tephra and volcaniclastic sediment seen in other volcanic terrains (e.g. Segerstrom 1950; Branney, 1988). Similar channels have been figured by Smith (1987a, fig. 9a) from the Deschutes Formation flanking the Cascades, and by van Houten (1976, fig. 7) from the Gigante Formation flanking the Andean Cordillera Central in Columbia.

Transitions from non-volcanic to volcaniclastic sediment are generally sharp but, in at least one instance, there is evidence of a gradual compositional change in the bedload of the non-volcanic fluvial deposits lying beneath the transition. Figure 10 illustrates a gradual dilution of the greywacke–granite detritus by volcanic detritus (at constant grain size) within large ($c.6$ m thick) gravelly foresets sitting beneath a volcaniclastic interval. The compositional variation implies that the bedload, in what was probably an entrenched river (accounting for the thick foresets), was progressively modified during accretion of the gravel foresets. The bedload gradually became increasingly dominated by volcaniclastic detritus, and then

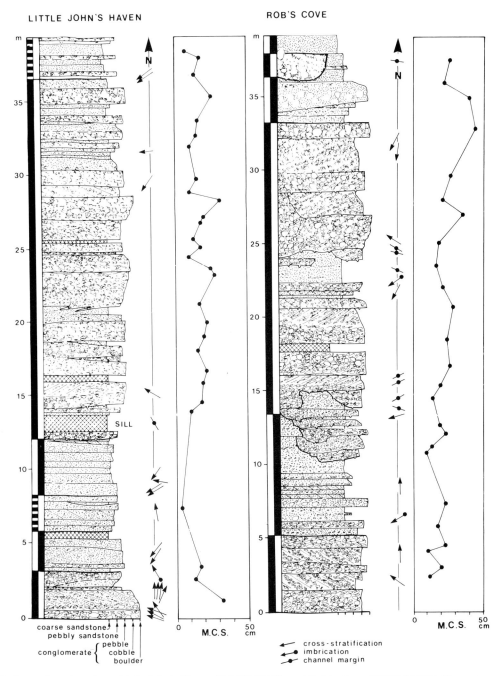

Fig. 8. Logged sections from Little John's Haven (NO85557430) and Rob's Cove (NO84457296) illustrating Type II alternation of sandy, flat-stratified volcaniclastics and gravelly, non-volcanic fluvial deposits.

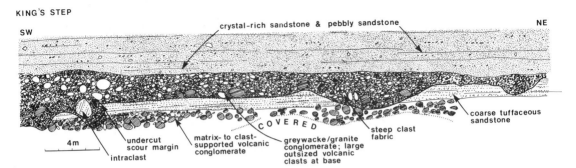

Fig. 9. Field sketch (NO84397291) illustrating oblique section through channel filled with non-volcanic conglomerate and incised into a volcaniclastic sequence comprising matrix-supported volcanic conglomerate overlain by coarse tuffaceous sandstone. Note the large outsized volcanic clasts along base of the channel (oblique shading). Channel is overlain by sequence of crystal-rich volcaniclastic sandstones.

switched suddenly into an aggradational phase with the transition from channelized to sheet volcaniclastic flows.

Type III sequences

These involve alternations of the same monomict volcaniclastic deposits and greywacke–granite bearing conglomerates found in Type II sequences but the non-volcanic deposits have a greater proportion of volcanic material (c.50%) and tend to be more sand prone. Whilst the contacts between volcaniclastic and these mixed provenance deposits remain well defined, they are planar with little evidence for the incision and channelling which characterize Type II sequences (Figs 11 & 12). The non-volcaniclastic intervals comprise conglomerates and coarse sandstones having a prominent flat or low angle stratification and the conglomerates often have a lenticular geometry. They either separate volcaniclastic intervals, or sit between a volcaniclastic interval and channelized non-volcanic conglomerates of the sort described for Type II sequences (Fig. 11). Type II and III sequences are closely related and may alternate in the same section.

INTERPRETATION

Factors determining the style of volcaniclastic/ fluvial alternations are going to include the difference in aggradation rate between volcaniclastic and fluvial deposits, the synchroneity or otherwise of the eruption and deposition of the volcaniclastic units, the overall subsidence rate, and the proximity to active volcanoes. If the volcaniclastic intervals were directly related to coeval eruption, they should show evidence for unusually rapid aggradation. Such aggradation is known to occur during an eruption, and for several decades later (Kuenzi *et al.*, 1979; Smith, 1987a, 1988a), and it can influence depositional systems far removed from the actual site of eruption (cf. Kuenzi *et al.*, 1979).

In the ancient record, evidence for unusually rapid volcanic-induced aggradation can come from examining the transition back to normal inter-eruption sedimentation. This change should involve a period of incision and regrading as streams adjust and re-establish their longitudinal profiles. Smith (1988a) has described transitions from broad sandy bedload systems to narrow, polymict gravel bedload streams in deep palaeovalleys from the volcaniclastic Neogene Ellensberg Formation flanking the Cascade Range. These contrasting fluvial styles are thought to reflect eruption-generated aggradation and inter-eruption degradation, respectively.

Smith & Vincent (1987) have distinguished sequences in which a low rate of subsidence prevents accumulation of an inter-eruption record (only erosion surfaces record inter-eruption deposition in their Type 1 sequences), from sequences in which long-term subsidence balances sedimentation rate, with rapid syn-eruption deposition followed initially by degradation, but then by accumulation of an inter-eruption record involving facies and compositions different from the eruption-related sediments (their type 2 sequences). Runkel (1990) contrasts two Eocene volcaniclastic aprons showing

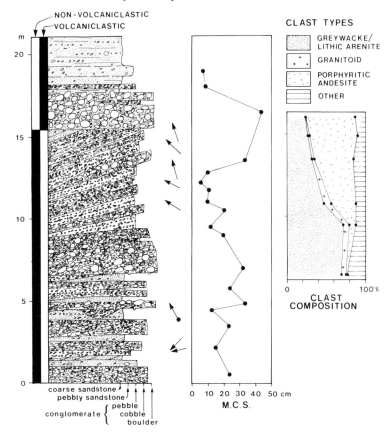

Fig. 10. Logged section showing gradual compositional change in conglomeratic foresets beneath transition to volcaniclastic facies, Robs Cove (NO84497303). Arrows represent tilt-corrected foreset dip azimuths; flighted arrow is flow direction inferred from clast imbrication. Compositions are based on counts of 100 clasts with maximum diameters between 5 and 15 cm.

this distinction, but prefers to see the preservation of inter-eruption deposits in terms of maintaining a high rate of sediment supply between major eruptions, so preventing incision. In addition, Runkel (1990) found that an episodic depositional record could not be identified in the distal reaches of these aprons as a result of a lower and less drastically fluctuating sediment supply there.

Turning now to the Lower Old Red Sandstone sequences. The most significant feature of the Type 1 sequences is the way in which the volcaniclastic intervals have had no influence on the structure of the non-volcanic fluvial deposits immediately above or beneath them. Thus, even where the facies and inferred depositional processes are markedly different, there is no evidence for regrading of the fluvial system immediately overlying the largely mass and fluidal flow emplaced volcaniclastic rocks. Instead, these sequences seem to record the interdigitation of separate volcaniclastic and fluvial dispersal systems and these must have aggraded at similar rates under conditions of rapid subsidence.

Parity of aggradation rates could be achieved by having inter-eruption deposition from a high discharge regional drainage system (thus matching high volcaniclastic aggradation rates) and/or suppressing the rate of supply of volcaniclastic materials to the basin, as when this is not directly connected with large explosive eruptions, or where an intervening basin acts as a buffer to sediment transfer. In this instance, dispersal of the volcaniclastic units was probably not from sites of coeval eruption. This could account for the wide range of volcanic lithologies, the dominance of normal fluvial facies and well-rounded clasts in some intervals, and the entrained non-volcanic material. High discharge antecedent streams (see Haughton, 1989) could have contributed to maintaining parity between aggradation of fluvial deposits and interbedded debris and fluidal flow dominated volcaniclastic intervals. The volcaniclastic units are thought to represent normal denudation of dormant volcanoes and/or epiclastic reworking of older synvolcanic detritus from hinterland volcaniclastic

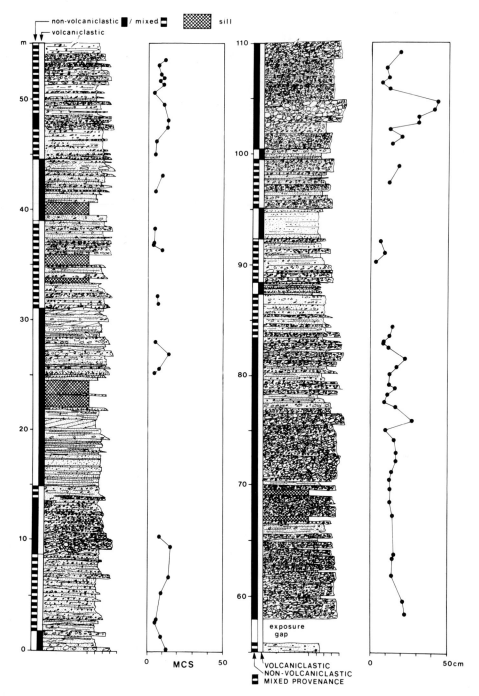

Fig. 11. Logged section (NO85207391 to NO85017390) illustrating Type III volcaniclastic/non-volcanic sequences.

Fig. 12. Example of Type III sequence. A thick-bedded volcaniclastic interval comprised of fine-grained debris and fluidal flow sheets is overlain by flat-stratified conglomerates and tuffaceous sandstones of mixed provenance. The contact is sharp and planar.

sequences on the northern margin of the basin (Fig. 13). Both volcaniclastic and non-volcaniclastic systems may have operated simultaneously, such that the alternating sequences preserved reflect the operation of both antecedent and consequent drainage systems tapping different parts of the northern source block.

In contrast, several features are consistent with a syn-eruption origin for the volcaniclastic intervals associated with the Type II and III sequences. Firstly, the monomictic nature of the volcanic detritus in these units is typical of syn-eruption deposits; secondly, hornblende–andesite and dacite producing eruptions of the type envisaged here are known to produce large volumes of syn-eruption sediment and extensive volcaniclastic aprons; thirdly, the crystal-rich sandstones, the fluidal and debris flow deposits, and the sandprone nature of the volcaniclastic intervals sandwiched between gravelly fluvial deposits (Fig. 8) are typical of syn-eruption sequences elsewhere. Smith (1987a) has shown that an important feature of volcanic-induced sedimentation is the deposition of large volumes of sand in areas normally occupied by gravel bedload rivers; and last, a syn-eruption origin might explain the structure of Type II sequences in terms of the evolution from rapidly aggrading volcaniclastic deposition to inter-eruption fluvial sedimentation, at least initially characterized by degradation.

Existing treatments of volcanism-induced sedimentation have concentrated on volcaniclastic fans and aprons where the inter-eruption record is also volcaniclastic, albeit polymictic and related to normal erosion. However, the radical change in composition (cf. Fig. 10) between inferred syn-eruption volcaniclastic deposition and the enclosing conglomerates in the Lower Old Red Sandstone indicates that this is not the case here. Instead, the intercalations of coarse non-volcanic conglomerates require repeated transfer to streams draining a hinterland of greywacke and granitoid lithologies lying to the east and south of the outcrop. The dispersal of the volcaniclastic units is less certain due to the poor development of directional structures. Rare cross-strata and channel orientations suggest a source which lay to the east or south-east. This is consistent with the lack of monomict hornblende–biotite andesitic detritus in the coeval succession along the inferred northern flank of the basin. Small greywacke fragments are scattered throughout many of the volcaniclastic intervals, suggesting upslope mixing (or country-rock entrainment during eruption) of the same lithologies cropping out in the hinterland to the non-volcanic river systems. Consequently, both the volcanogenic and non-volcanic provenances probably overlapped upslope. The coarse grain size and immature texture of the non-volcanic conglomerates suggest a nearby source, which, given the thickness of the succession, was probably the footwall to the fault system marking the southeastern boundary of the small sub-basin developed during deposition of the Crawton Group.

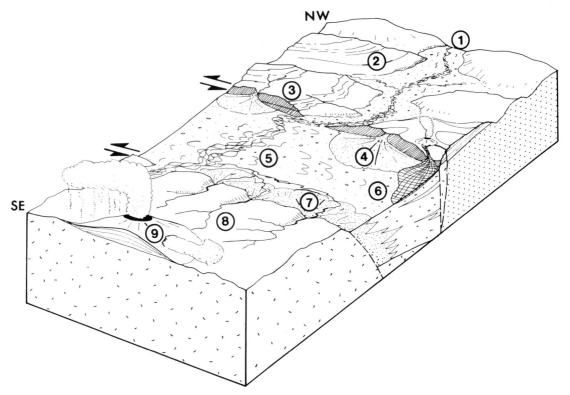

Fig. 13. Schematic Lower Old Red Sandstone palaeogeography for the NE Midland Valley of Scotland during the early Lower Old Red Sandstone (Crawton Group). Main elements are (1) a large antecedent stream with complex provenance (including pre-existing conglomerates/gravels) entering the basin from north, and reworking volcanic and/or volcaniclastic deposits (2) to provide the well-mixed and diverse volcanic detritus in the braidplain sequences. Local volcanic rocks (3) resting on Lower Greenschist facies basement (not sampled by the large braided streams) supplied volcaniclastic fans (4) with fluvial, fluidal and debris flow deposits. These interfingered with, but did not perturb the braidplain deposits (5). Occasional basaltic lava flows (6) blanketed the braidplain surface, and may locally have diverted streams (see base of Todhead–Rouen Bay section, Fig. 6B). Fans (7) deposited along the SE flank of the basin were supplied by streams draining a cryptic greywacke–granite source (8) exposed in the footwall to the faulted basin margin. These streams were episodically flushed with volcaniclastic detritus following eruption of a large andesitic volcano (9) located somewhere to the SE.

The following scenario is envisaged. Periodic and explosive eruption of a volcano located in the southeastern hinterland to the basin (Fig. 13) produced large volumes of fragmental volcanic material which subsequently overwhelmed the gravelly streams which operated during periods of volcanic quiescence. In some cases, the gravel bed-load streams may have accommodated the initial influx of volcaniclastic material, but these rapidly became choked as deposition switched to unchannelized flows leading to the emplacement of a smooth mantle of volcaniclastic breccias and tuffaceous sandstones. The absence of primary pyroclastic detritus suggests that the location of the eruption may have been some distance from the site of deposition. As the contribution of volcanic detritus waned, drainage of fault escarpments with non-volcanic lithologies in their footwalls gradually regained precedence. At first, narrow incised channels developed, degrading the volcaniclastic surface. Coarsening/deepening-upwards trends are consistent with progressive enlargement of channel networks. The subordinate and outsized volcanic material found in the non-volcanic conglomerates was probably largely lagged during initial degradation of the volcanic deposits. When fully developed, channel-filling conglomerate bodies merged to generate extensive non-volcanic conglomerate units. It is likely that these formed a series of footwall-sourced alluvial fans, but critical features indicative of fan deposition (radial palaeoflow, rapid downslope facies change, lithosome geometry) cannot be deduced from the available exposure. The debris flow deposits in the volcaniclastic intervals cannot be used to infer a fan origin for the enclosing conglomerates for the reasons outlined by Smith (1987a). Following initial degradation associated with attainment of channel grade and adjustments relating to the development of equilibrium channel densities, the non-volcaniclastic alluvium may have either prograded and/or retreated in the way that normal fault-controlled sequences do in response to pulsed subsidence. Dominant control of the overall sequence thus probably oscillated between volcanic and tectonic processes, with rapid subsidence ensuring that both eruption and inter-eruption sequences were preserved.

Figure 8 shows that eruption-related sequences 5–10 m thick alternate with packages of non-volcanic conglomerate $c.20$–25 m thick. Eruption frequencies for large-scale calc-alkaline eruptions (explosive energy $>10^{20}$ ergs with at least 2 m of pumice at 5 km or more from source) are 1.88×10^{-4} to 3.57×10^{-4} events per volcano per year for circum-Pacific arc-segments (Williams & McBirney, 1979), i.e. one to four large events per volcano in every 10 000 years. Single eruption events can raise river beds by 10–15 m over 40 km from the site of eruption (Kuenzi et al., 1979). If an eruption frequency of 10^4 years is assumed, and compaction ignored, accumulation rates of the order of 2 mm yr^{-1} are required to generate inter-eruption conglomerate sequences $c.20$ m thick. This falls within the range of positive accumulation rates (measured at timespans of 10^4 years) compiled by Sadler (1981) for fluviatile sediments, indicating that episodic syn-eruption deposition in an area normally occupied by a gravelly fluvial system can be reconciled with the scale of source switching observed in Type II sequences. Strike-slip basins of the sort envisaged for Old Red Sandstone deposition in the NE Midland Valley (Haughton & Bluck, 1988) can subside sufficiently rapidly to accumulate sediment deposited at rates of 3 mm yr^{-1}. Pitman & Andrews (1985) calculate subsidence rates of as high as 5 mm yr^{-1} for small pull-apart basins.

The related Type III sequences in which monomict volcaniclastic intervals are interbedded with flat-stratified, mixed provenance intervals showing no evidence for incision and readjustment imply that an aggradational mode was sometimes retained during the changeover from volcaniclastic to non-volcaniclastic deposition. This could reflect switching from syn- to inter-eruption deposition during periods of more rapid subsidence, so that streams transporting non-volcanic detritus never had to regrade. However, the mixed provenance suggests that perhaps parts of the hinterland drainage basin and marginal fault scarps were choked or blanketed with volcaniclastic detritus during eruption, and that initial inter-eruption erosion of this loosely consolidated material together with non-volcanic bedrock resulted in sediment yields which were still large enough to exceed the geomorphic threshold required for the generation of syn-eruption style sequences. Where the mixed provenance deposits are overlain by non-volcanic conglomerates (e.g. Fig. 11) incision is evident and regrading must have taken place as sediment yields fell.

DISCUSSION

The evidence outlined above suggests that early Lower Old Red Sandstone volcaniclastic intervals are polygenetic. Syn-eruption deposits have been identified in sections from the southern part of the Crawton sub-basin by considering the relationship between volcaniclastic and non-volcaniclastic intervals in the light of the rapid and episodic aggradation to be expected with syn-eruption deposition. Such an approach is particularly important in successions where primary pyroclastic deposits are absent, as in this case, although the syn-eruption origin here is also consistent with the monomict composition and nature of the volcaniclastic detritus, and the presence of crystal-rich sandstones.

In contrast, the nature of the polymict volcaniclastic intervals and the relationships they display with the enclosing northern fluvial deposits suggest that they were generated by normal erosion processes. If volcanism was coeval here, large explosive eruptions appear either not to have been important, or their influence screened in some way from the northern part of the basin. Thirlwall (1988) suggests that volcanism north of the Highland Boundary Fault (at Lorne and Glencoe, Fig. 1) took place slightly earlier ($c.420$ Ma) than that within the Midland Valley (415–411 Ma) and it is possible that the Old Red Sandstone sediments reflect the demise of the northern volcanic centres in favour of those in the central Midland Valley.

This Old Red Sandstone study emphasizes the importance of vertical cycles in evaluating eruption-induced perturbations of fluvial systems. Megasequences in fluvial sequences which span volcaniclastic intervals preclude such perturbations, whilst the stacking of syn-eruption volcaniclastic and non-volcaniclastic deposits can produce 'adjustment' cycles which relate to the re-establishment of non-volcanic drainage following inundation by rapidly emplaced volcaniclastic sediments. These cycles should not be confused with cycles produced by tectonic or other external factors.

SUMMARY AND CONCLUSIONS

Three types of volcaniclastic/non-volcaniclastic sequence are evident in the Lower Old Red Sandstone of the NE Midland Valley of Scotland. The first shows no interaction between the disparate systems delivering volcaniclastic and non-volcaniclastic sediment to the basin. The second can be explained in terms of episodic eruption-induced aggradation and inter-eruption degradation, with the aggradation–degradation cycles developed here differing from those previously described in that continuity with the volcanic source was not maintained during inter-eruption deposition. Instead, transfer to fault-controlled, non-volcanic fluvial deposits took place. The third reflects eruption to inter-eruption sedimentation in which overall aggradation was at least initially maintained during the change-over, perhaps by choking of hinterland drainage basins with syn-eruption sediment. The syn-eruption sequences appear to be restricted to the southern part of an early Old Red Sandstone sub-basin, the northern part of which preserves volcaniclastic intervals whose relationships to the enclosing fluviatile deposits suggest normal denudation processes. On a wider note, the sort of analysis explored here may have more general applications in looking at the relationship between disparate dispersal systems in fluvial basins. More attention needs to be paid to how these mutually interact to determine a component of the overall structure of basin fills.

ACKNOWLEDGEMENTS

I would like to thank Britoil, BP and the Royal Society of Edinburgh for support during this work, and J. Collinson for a helpful review.

REFERENCES

ARMSTRONG, M. & PATERSON, I.B. (1970) The Lower Old Red Sandstone of the Strathmore region. Rep. No. 70/12, Inst. Geol. Sci.

BRANNEY, M.J. (1988) The subaerial setting of the Ordovician Borrowdale Volcanic Group, English Lake District. *J. Geol. Soc. Lond.* **145**, 887–890.

CAMPBELL, R. (1913) The geology of south-eastern Kincardineshire. *Trans. R. Soc. Edinburgh Earth Sci.* **48**, 923–960.

CAS, R.A.F. & WRIGHT, J.V. (1987) *Volcanic Successions: Modern and Ancient. A Geological Approach to Processes, Products and Successions.* Allen & Unwin, London, 528 pp.

DAVIES, D.K. & VESSELL, R.K., MILES, R.C., FOLEY, M.G. & BONIS, S.B. (1978) Fluvial transport and downstream sediment modification in an active volcanic region. In: *Fluvial Sedimentology* (Ed. Miall, A.D.) pp. 61–83. Can. Soc. Petrol. Geol., Calgary, Memoir 5.

FRANCIS, E.H. (1983) Magma and sediment, II. Problems of interpreting palaeovolcanics buried in the stratigraphic column. *J. Geol. Soc. Lond.* **140**, 165–183

FRANCIS, E.H., FORSYTH, I.H., READ, W.A. & ARMSTRONG, M. (1970) *The Geology of the Stirling District.* Mem. Geol. Surv. UK (Sheet 39). HMSO, London.

FRIEND, P.F., HARLAND, W.B. & HUDSON, J.D. (1963) The Old Red Sandstone and the Highland Boundary in Arran, Scotland. *Trans. Edinburgh. Geol. Soc.* **19**, 363–425.

HARRISON, S. & FRITZ, W.J. (1982) Depositional features of March 1982 Mount St Helens sediment flows. *Nature* **299**, 720–722.

HAUGHTON, P.D.W. (1989) Structure of some Lower Old Red Sandstone conglomerates, Kincardineshire, Scotland: deposition from late-orogenic antecedent streams? *J. Geol. Soc. Lond.* **146**, 509–525.

HAUGHTON, P.D.W. & BLUCK, B.J. (1988) Contrasting alluvial sequences in the Lower Old Red Sandstone of the Strathmore Region, Scotland — implications for the relationship between late Caledonian tectonics and sedimentation. In *The Devonian of the World* (Eds McMillan, N.J., Embry, A.F. & Glass, D.J.) pp. 269–293. Can. Soc. Petrol. Geol., Calgary, Memoir 14.

HAUGHTON, P.D.W., ROGERS, G. & HALLIDAY, A.N. (1990) Provenance of Lower ORS conglomerates, SE Kincardineshire: evidence for the timing of terrane accretion in central Scotland. *J. Geol. Soc. Lond.* **147**, 105–120.

HEWARD, A.P. (1978) Alluvial fan sequence and megasequence models: with examples from Westphalian D–Stephanian B coalfields, northern Spain. In *Fluvial Sedimentology* (Ed. Miall, A.D.) pp. 669–702. Can. Soc. Petrol. Geol., Calgary, Memoir 5.

KUENZI, W.D., HORST, O.H. & MCGEHEE, R.V. (1979) Effect of volcanic activity on fluvial–deltaic sedimentation in a modern arc-trench gap, southwestern Guatemala. *Bull. Geol. Soc. Am.* **90**, 827–838.

MATHISEN, M.E. & VONDRA, C.F. (1983) The fluvial and

pyroclastic deposits of the Cagayan basin, northern Luzon, Philippines — an example of nonmarine volcaniclastic sedimentation in an interarc basin. *Sedimentology* 30, 369–392.

MORTON, D.J. (1976) *Lower Old Red Sandstone sedimentation in the north-west Midland Valley of Scotland*. PhD thesis, University of Glasgow.

NEMEC, W. & STEEL, R.J. (1984) Alluvial and coastal conglomerates: their significant features and some comments on gravelly mass-flow deposits: In: *Sedimentology of Gravels and Conglomerates* (Eds Koster, E.H. & Steel, R.J.) pp. 1–31. Can. Soc. Petrol. Geol., Calgary, Memoir 10.

PIERSON, T.T. & SCOTT, K.M. (1985) Downstream dilution of a lahar: transition from debris flow to hyperconcentrated stream flood. *Water Resources Res.* 21, 1511–1524.

PITMAN, W.C. & ANDREWS, J.A. (1985) Subsidence and thermal history of small pull-apart basins. In: *Strike-slip Deformation, Basin Formation and Subsidence* (Eds Biddle, K.T. & Christie Blick, N.) pp. 45–49. Soc. Econ. Paleont. Miner., Tulsa, Spec. Publ. 7.

RUNKEL, A.C. (1990) Lateral and temporal changes in volcanogenic sedimentation; analysis of two Eocene sedimentary aprons, Big Bend region, Texas. *J. Sedim. Petrol.* 60, 747–760.

SADLER, P.M. (1981) Sediment accumulation rates and the completeness of stratigraphic sections. *J. Geol.* 89, 569–584.

SCOTT, K.M. (1988) Origin, behaviour and sedimentology of prehistoric catastrophic lahars at Mount St Helens, Washington. In *Sedimentologic Consequences of Convulsive Geologic Events* (Ed. Clifton, H.E.) pp. 23–36. Geol. Soc. Am., Boulder, Spec. Publ. 229.

SEGERSTROM, K. (1950) Erosion studies at Paricutin, State of Michoacan, Mexico. *Bull. Geol. Surv. Am.* 965-A.

SMITH, G.A. (1986) Coarse-grained volcaniclastic sediment: terminology and depositional processes. *Bull. Geol. Soc. Am.* 78, 1385–1422.

SMITH, G.A. (1987a) Sedimentology of volcanism-induced aggradation in fluvial basins: examples from the Pacific Northwest, USA. In: *Recent Developments in Fluvial Sedimentology* (Eds Ethridge, F.G., Flores, R.M. & Harvey, M.D.) pp. 217–228. Soc. Econ. Paleont. Miner., Tulsa, Spec. Publ. 39.

SMITH, G.A. (1987b) The influence of explosive volcanism on fluvial sedimentation: the Deschutes Formation (Neogene) in central Oregon. *J. Sedim. Petrol.* 57, 613–629.

SMITH, G.A. (1988a) Sedimentology of proximal to distal volcaniclastics dispersed across an active foldbelt: Ellensberg Formation (late Miocene), central Washington. *Sedimentology* 35, 953–977.

Smith, G.A. (1988b) Neogene synvolcanic and syntectonic sedimentation in central Washington. *Bull. Geol. Soc. Am.* 100, 1479–1492.

SMITH, G.A. & VINCENT, K. (1987) Rates of sedimentation, subsidence, and volcanism as controls on facies architecture in terrestrial volcaniclastics. *Geol. Soc. Am. Abstr. Progs* 19, 849.

THIRLWALL, M.F. (1981) Implications for Caledonian plate tectonic models of chemical data from volcanic rocks of the British Old Red Sandstone. *J. Geol. Soc. Lond.* 138, 123–138.

THIRLWALL, M.F. (1988) Geochronology of Late Caledonian magmatism in northern Britain. *J. Geol. Soc. Lond.* 145, 951–967.

VAN HOUTEN, F.B. (1976). Late Cenozoic volcaniclastic deposits, Andean foredeep, Columbia. *Bull. Geol. Soc. Am.* 87, 481–495.

VESSELL, R.K. & DAVIES, D.K. (1981) Non-marine sedimentation in an active fore arc basin. In: *Recent and Ancient Nonmarine Depositional Environments: Models for Exploration* (Eds Ethridge, F.G. & Flores, R.M.) pp. 31–45. Soc. Econ. Paleont. Miner., Tulsa, Spec. Publ. 31.

WALTON, A.W. (1977) Petrology of volcanic sedimentary rocks, Vieja Group, Southern rim Rock Country, Trans-Pecos Texas. *J. Sedim. Petrol.* 47, 137–157.

WALTON. A.W. & PALMER, B.A. (1988) Lahar facies of the Mount Dutton Formation (Oligocene–Miocene) in the Marysville Volcanic Field, southwestern Utah. *Bull. Geol. Soc. Am.* 100, 1078–1091

WILLIAMS, H. & MCBIRNEY, A.R. (1979) *Volcanology*. Freeman, Cooper and Co., San Francisco.

Spec. Publs Int. Ass. Sediment. (1993) **17**, 473–488

Siliciclastic braided-alluvial sediments intercalated within continental flood basalts in the Early to Middle Proterozoic Mount Isa Inlier, Australia

K.A. ERIKSSON* and E.L. SIMPSON†

*Department of Geological Sciences, Virginia Polytechnic Institute and State University, Blacksburg, VA 24061, USA; and
†Department of Physical Sciences, Kutztown University, Kutztown, PA 19530, USA

ABSTRACT

The Eastern Creek Volcanics in the Mount Isa Inlier are up to 7 km thick and consist of two successions of subaerial basalts (5.5 km and 0.7 km thick) separated by a siliciclastic unit (up to 750 m thick). Predominantly tabular, siliciclastic units from 1 to 40 m thick are present in the upper half of the lowermost basaltic succession and throughout the uppermost basaltic succession. Siliciclastic debris was derived exclusively from the east from a provenance terrain consisting of quartz arenites, felsic volcanics and granites. With the exception of local aeolian facies, the siliciclastic units are exclusively of braided-alluvial origin. Facies analysis has identified two interacting alluvial systems: a relatively high-gradient and coarse-grained, transverse system that supplied siliciclastic debris from eastern highlands, and a lower-gradient and finer-grained, longitudinal or trunk system that reworked sediment down a south-to-north palaeoslope. Areal persistence of most siliciclastic units in the trunk system is attributed to continuous lateral switching of shallow, braided rivers. Lateral switching was promoted by slow subsidence attributed to cooling and thermal contraction of the volcanics. Siliciclastic units define prolonged hiatuses in volcanism; this interpretation is supported by the presence of a calcrete unit at the top of a basalt flow and below a siliciclastic unit.

INTRODUCTION

In the stratigraphic record, subaerial flood basalts and siliciclastic sedimentary rocks rarely are associated. Basalts typically consist of stacked flows of variable thickness that make up kilometres-thick successions (see papers in MacDougall, 1988, for examples). Similarly, braided-alluvial sediments, which dominate the Precambrian rock record at the expense of meandering river facies (Schumm, 1968), make up kilometres-thick successions with only subordinate shallow-marine or lacustrine facies. Examples include the Archean Moodies Group and Witwatersrand Supergroup and the Proterozoic Waterberg Group in South Africa (Tankard et al., 1982), and the Proterozoic Vallecito Conglomerate and Oronto Group in the USA (Daniels, 1982; Ojakangas & Morey, 1982; Ethridge et al., 1984). Less commonly, braided-alluvial as well as aeolian deposits are intercalated within basalts (see, for example, Merk & Jirsa, 1982; Clemmensen, 1988) and may reflect a different tectonic setting to that in which either thick basalts or thick siliciclastic sediments alone accumulated.

The Eastern Creek Volcanics in the Mount Isa Inlier, Queensland, Australia (Fig. 1) are up to 7 km thick and consist of the Cromwell Member, Lena Quartzite and Pickwick Member (Fig. 2). The Cromwell Member is up to 5.5 km thick (Derrick et al., 1977). Basalt dominates the lower half of the member, whereas the upper half consists of basalts with intercalated, mainly tabular, siliciclastic units between 1 and 40 m thick. The Lena Quartzite is made up entirely of siliciclastics and has an average thickness of 750 m. Basalts dominate the c. 700 m thick Pickwick Member that contains subordinate

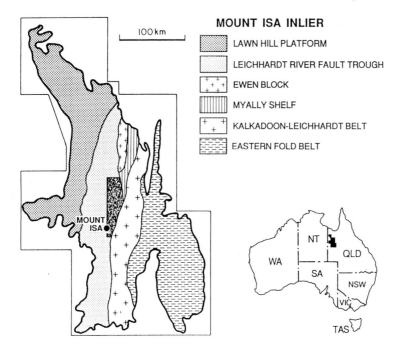

Fig. 1. Locality map of the Mount Isa Inlier showing tectonic framework. Cross-hatched rectangle is location of study area. (Modified from Blake (1987).)

5 to 10 m thick sandstone lenses (Derrick *et al.*, 1977). The Eastern Creek Volcanics thin to the east and west. Along the western flank of the Kalkadoon–Leichhardt Belt, thinner intervals of basalt are associated with coarse boulder-to-pebble conglomerates (Derrick, 1982).

This paper focuses on the Cromwell Member, with particular emphasis on the sedimentary units, and on the Lena Quartzite. Sediments in the Pickwick Member are thinner, but of comparable composition and depositional style to those in the Cromwell Member. We will show that the sedimentary units are almost entirely of braided-alluvial origin and were derived from extrabasinal sources including granites, felsic volcanics and an underlying interval of quartz arenites, as well as from the Eastern Creek Volcanics. Thick continental flood basalts, facies and petrographic data, and the stratigraphic distribution of siliciclastic units are incorporated into a genetic model for the Eastern Creek Volcanics that recognizes the interplay between volcanism, sedimentation and subsidence in an extensional setting.

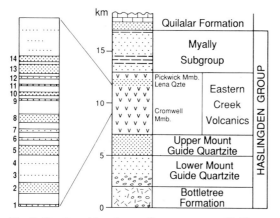

Fig. 2. Stratigraphic column of cover sequence II. The Haslingden Group represents a synrift succession and the Quilalar Formation a response to thermal subsidence. The upper Mount Guide Quartzite may represent an additional thermotectonic phase of sedimentation. Inset shows the stratigraphic distribution of siliciclastic units (1–14) in the upper Cromwell Member.

REGIONAL GEOLOGY

The Mount Isa Inlier is located on the North Australian Craton. The craton developed between 2000 and 1900 Ma (Page *et al.*, 1984); formation of the orogenic belt commenced around 1800 Ma and was terminated by deformation and regional metamorphism between 1610 and 1510 Ma (Page & Bel, 1986). The inlier consists of four north–south trending tectonic or structural elements, namely the Lawn Hill Platform, Leichhardt River Fault Trough including the Myally Shelf, Kalkadoon–Liechhardt Belt–Ewen Block and Eastern Fold Belt (Fig. 1).

Primary stratigraphic subdivision in the Mount Isa Inlier is between basement and cover. Basement is exposed mainly in the Kalkadoon–Liechhardt Belt but also in inliers in the Lawn Hill Platform and the Eastern Fold Belt. The basement consists of paragneiss, schist, quartzite and migmatite intruded by felsic plutonics (Blake, 1987). Cover rocks are divisible into three sequences. Cover sequence I is confined to the Kalkadoon–Liechhardt Belt and is composed mainly of felsic volcanic rocks dated at 1870 Ma (U–Pb zircon; Page, 1983a) that are intruded by cogenetic granite batholiths of $c.$1860 Ma age (Page, 1978). Cover sequences II and III consist of lower volcanic and sedimentary, and upper sedimentary intervals and are considered to represent two synrift to postrift phases of basin evolution (Blake, 1985; Wyborn et al., 1988).

The Eastern Creek Volcanics make up part of the synrift phase of cover sequence II that is represented by the $c.$16 km thick Haslingden Group (Fig. 2). The overlying Quilalar Formation represents a thermotectonic sag phase of cover sequence II (Eriksson et al., in press). Felsic volcanics of the Bottletree Formation at the base of the Haslingden Group are 1790–1800 Ma old (Page, 1983a) and provide a maximum age for the Eastern Creek Volcanics. Overlying stratigraphic units (Fig. 2) are the Mount Guide Quartzite ($c.$5 km), the Eastern Creek Volcanics ($c.$6 km) and the Myally Subgroup ($c.$5 km). The Mount Guide Quartzite is divided into a lower, 3 km thick interval of alluvial feldspathic sandstones and an upper, 2 km thick sequence of shallow-marine quartz arenites that are considered to represent an additional thermotectonic sag phase of sedimentation (Eriksson et al., in press). A minimum age for the Haslingden Group is 1740 Ma, the age of granites intruded into

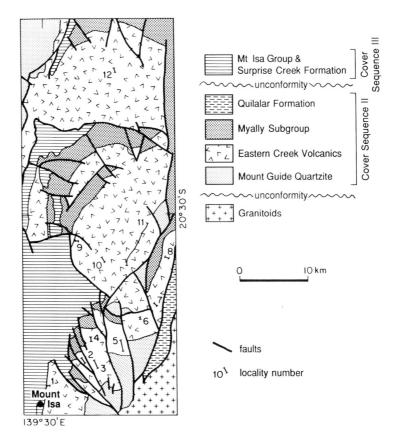

Fig. 3. Geological map of the study area showing study localities. (Modified from Blake (1987).)

correlatives of the Quilalar Formation (Page, 1983b; Fig. 2). Sedimentological and volcanological observations from the Eastern Creek Volcanics are from twelve locations east and northeast of Mount Isa (Fig. 3).

VOLCANOLOGY

Basalts in the Cromwell Member are mainly continental tholeiites (Glikson et al., 1976; Bultitude & Wyborn, 1985), although some samples with higher MgO contents resemble ocean floor basalts (Wilson, 1987). Individual flows range from 30 to 100 m thick and are thinner in the upper than the lower Cromwell Member (Derrick et al., 1977). Flows typically consist of a thin basal amygdaloidal zone, a thick massive middle zone, and a thin upper amygdaloidal zone. Tops of flows often are defined by volcanic breccias up to 10 m thick. Porphyritic lavas are rare except for a single flow near the top of the Cromwell Member that contains feldspar phenocrysts up to 1 cm long. Pillows and hyaloclastites are absent.

Rare, lenticular, mafic lapillistone to tuff units are present in the Cromwell Member. These make up 2–8 cm thick graded and stratified lapillistone layers with desiccated tuffaceous caps. Low-angle cross-stratification and water- or gas-escape structures are developed in some beds. The latter observations, together with evidence for desiccation, support a subaqueous, possibly lacustrine, as opposed to an airfall origin for these pyroclastic deposits (cf. Fisher & Schmincke, 1984). A 10–14 m thick breccia with lateral continuity of $c.2$ km is present near the top of the Cromwell Member. Clasts up to 50 cm in diameter are angular to subrounded and consist of massive, amygdaloidal and porphyritic basalt and rare quartz arenite. Matrix consists of mafic tuff. The breccia is poorly sorted and displays both matrix and clast support. The breccia resembles a lahar deposit with respect to its poor sorting, lateral continuity and mixture of pyroclastic and epiclastic debris (cf. Fisher & Schmincke, 1984).

The data presented above support a dominantly subaerial origin for the volcanics in the Cromwell Member. Subordinate, subaqueous, possibly lacustrine settings are inferred for the lapillistone and tuff units. The Cromwell Member thus represents a sequence of continental flood basalts; the stratigraphically higher Pickwick Member (Fig. 2) also consists of continental flood basalts (Bultitude & Wyborn, 1985). The geochemical characteristics of the volcanics imply a mantle source (Wilson, 1987).

SEDIMENTOLOGY

Stratigraphy and sediment body geometries

Basalts of the Cromwell Member contain fourteen (C1–C14) mappable siliciclastic units and an uppermost, $c.1$ km thick zone with numerous, thin, siliciclastic lenses (Fig. 2, inset; J.H.C. Bain & G.A.M. Henderson, pers. comm., 1988). No thickness trends to the siliciclastic units are apparent (Fig. 2, inset). Thicknesses of units vary; C1, C2, C5, C8, C13 and C14 range from 10 to 40 m, are rarely up to 100 m thick in the central part of the study area and thin to the east. The remaining units generally are less than 5 m thick and are most prominent in the central part of the study area. Thicker units are notably tabular and laterally persistent across the study area for distances of at least 25 by 20 km. In contrast, thinner units are lenticular over tens to hundreds of metres. At locality 5, C1 varies between 1 and 5 m thick over a strike length of 50 m. The Lena Quartzite varies in thickness between 280 and 1000 m across the study area. Conglomerates on the western flank of the Kalkadoon–Leichhardt Belt are developed at the same stratigraphic level as the Cromwell Member and Lena Quartzite to the west and are lenticular over distances of hundreds of metres. As a consequence of folding and faulting, siliciclastic units are exposed both parallel and perpendicular to the axis of the Leichhardt River Fault Trough.

Contacts between siliciclastic units and underlying volcanics are sharp and erosive. At locality 5, a $c.2$ m thick carbonate unit is present at the top of the basalt flow immediately below C5. The carbonate horizon contains spheroidal basalt 'corestones', carbonate peds, sheet cracks that are pytgmatically folded, and spar-filled voids (S.G. Driese, pers. comm., 1991). Quartz sand is present throughout the horizon but is most abundant in the upper few centimetres. The above characteristics are typical of calcretes (Retallack, 1981) and favour development of this horizon as a soil on top of the basalt flow prior to deposition of the overlying sandstone (S.G. Driese, pers. comm., 1991). Uper contacts of siliciclastic units with overlying basalts are less sharp, with sandstone often incorporated into the base of flows. This observation implies that at least the upper parts of siliciclastic units were unconsolidated at the time of volcanic outpouring.

Provenance

Within the study area, metre-thick conglomerate beds are confined to unit C2 at localities 2, 3 and 5 and to the base of the Lena Quartzite at location 7 (Fig. 3). In addition, conglomerates up to 15 m thick are developed along the western flank of the Kalkadoon–Leichhardt Belt to the east and northeast of the study area. Single-pebble to centimetre thick conglomerates are present in most siliciclastic units, especially in the southern half of the study area.

Clasts in unit C2 and in the thin conglomerates are dominated by mafic volcanics and siltstone-shale of local derivation. Also present are subordinate quartz arenite clasts with preserved, rounded grain boundaries and authigenic overgrowths. Similar quartz arenite clasts are the dominant component of conglomerates east and northeast of the study area. The Lena conglomerate contains mainly felsic volcanic, red granite and grey quartzite clasts. Compositionally and texturally mature sandstones of pre-Eastern Creek Volcanics are confined to the upper Mount Guide Quartzite (Fig. 2) that is the likely source of these quartz arenite clasts. The trace element geochemistry of felsic volcanic clasts matches volcanics of cover sequence I, whereas the granite clasts are akin petrographically to the cogenetic batholiths (D.H. Blake, pers. comm., 1989). Within grey quartzite clasts boundaries of grains are indistinct and display intense suturing. They closely resemble quartzites interbedded within cover sequence I.

Sandstones in the Cromwell Member range from volcaniclastic to epiclastic. Unit C2 contains up to 80% volcanic rock fragments dominated by mafic with subordinate felsic volcanic grains. Quartz is the dominant constituent of all siliciclastic units except C2 and comprises up to 95% of some samples. Both monocrystalline and polycrystalline quartz are present. The latter is represented by sedimentary rock fragments consisting of single or composite grains containing rounded cores with pre-depositional authigenic overgrowths. Subordinate constituents of units other than C2 in decreasing order of abundance are felsic and mafic volcanic rock fragments, plagioclase and albitized microcline, muscovite, angular intraformational sedimentary rock fragments, plutonic rock fragments, tourmaline and zircon. Magnetite generally is a minor component, but some samples contain distinctive, heavy mineral concentrations. Sandstones within the Cromwell Member become more feldspathic higher in the section, but lateral changes in framework mineralogy within individual siliciclastic units are not apparent. Most samples contain one or more of metamorphic calcite, biotite, actinolite, epidote and/or chlorite recording middle greenschist facies metamorphism.

The Lena Quartzite has a similar composition to the dominantly epiclastic units in the Cromwell Member with the exception that quartz arenite sedimentary and mafic volcanic rock fragments are not present. Monocrystalline quartz is the dominant constituent in the Lena Quartzite and includes unstrained and embayed forms. Subordinate components are plagioclase replaced by carbonate, albitized microcline, felsic volcanic rock fragments commonly silicified, muscovite, intraformational sedimentary rock fragments and magnetite. In general, the Lena Quartzite is mineralogically and texturally less mature in the east (locality 7) and more mature to the north (localities, 9, 11 and 12). Metamorphic minerals include calcite and chlorite.

Sandstone petrography is consistent with a similar provenance to that documented by the conglomerates. Mafic volcanic rock fragments and magnetite are exclusively of intrabasinal derivation. Quartz arenite sedimentary rock fragments represent recycled upper Mount Guide Quartzite, whereas felsic volcanic fragments, including unstrained and embayed quartz, and plutonic rock fragments reflect reworking of cover sequence I and related intrusive granites. These same granites also are a likely source for the monocrystalline quartz, feldspar and muscovite.

The absence of quartz arenite clasts and sedimentary rock fragments in the Lena Quartzite is the most notable compositional change upward within the stratigraphic column and most likely reflects earlier recycling of the upper Mount Guide Quartzite into siliciclastic units within the Cromwell Member. Consequent unroofing of granites intrusive into cover sequence I may be documented by the upward increase in feldspar in arenites of the underlying Cromwell Member. The presence of abundant quartz arenite clasts in conglomerates along the western flank of the Kalkadoon–Lichhardt Belt supports the suggestion by Derrick (1982) that these conglomerates are temporal equivalents of siliciclastic units within the Cromwell Member.

Facies analysis

Facies in siliciclastic units in the Cromwell Member and in the Lena Quartzite are discussed in terms of their associations.

Association 1

In the study area this association is confined to the Lena Quartzite at locality 7 (Fig. 3). It consists of polymictic conglomerates with thin, interbedded sandstones that are overlain by thick sandstone intervals (Fig. 4A). The conglomerate is gradational laterally over a distance of $c.1$ km into sandstone. Conglomerates are 5–6 m thick, massive and clast supported. Clasts are well rounded, up to 15 cm in diameter, and consist of, in order of decreasing abundance, felsic volcanics, red granite, grey quartzite, vein quartz and intraformational siltstone–shale. Interbedded, laterally impersistent arenites are 5–10 cm thick, medium to coarse grained, and horizontally stratified. Overlying coarse-grained sandstones contain scattered pebbles and are structured by medium-scale trough cross-beds. Conglomeratic intervals along the western flank of the Kalkadoon–Lichhardt Belt consist of similar facies with the exception that conglomerates are up to 20 m thick and clasts up to 50 cm in diameter.

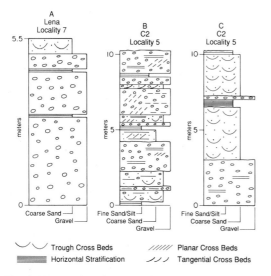

Fig. 4. Measured sections showing sequences of facies in associations 1 (A) and 2 (B,C).

The presence of clast-supported conglomerates and stratified sandstones in this association excludes a debris-flow origin. Instead, the association of facies resembles deposits of the proximal Scott outwash fan (Boothroyd & Ashley, 1978) and the proximal Donjek River (Rust, 1972, 1978) that are dominated by massive or horizontally stratified gravels and only minor, laterally impersistent, sandstone lenses. By comparison with these examples, the massive conglomerates are interpreted as longitudinal-bar deposits and the thin, horizontally stratified sandstone lenses as bar-top facies. Absence of bar-margin sandstone wedges probably is due to the low preservation potential of sandstone deposits in this setting. Vertical and lateral transition of conglomerates into sandstone reflects either cessation of gravel supply to the braided-alluvial system or waning of major floods. The latter control on the development of upward-fining sequences is favoured by Nemec & Steel (1984) in pre-Devonian settings.

Association 2

This association composes unit C2 at localities 2, 3 and 5. It consists of interstratified conglomerates, pebbly sandstones and sandstones containing multiple erosional surfaces (Figs 4B, C & 5). Conglomerates are 0.25–3.0 m thick and clast supported. Clasts consist mainly of well-rounded mafic volcanics up to 8 cm in diameter. Subordinate, intraformational siltstone–mudstone clasts are up to 20 cm long. Conglomerates are massive, horizontally stratified, cross-stratified and graded. Amalgamated, graded and stratified beds between 20 and 30 cm thick display an upward decrease in pebble size and transition into pebbly sandstones (Fig. 5). Pebble clusters are common on basal scour surfaces. Horizontally and wedge-planar, cross-stratified pebbly sandstone and sandstone lenses between 5 and 10 cm thick are present within some conglomerate beds (Fig. 6). Interstratified sandstone beds are 0.2–3.5 m thick (Fig. 4B, C) and consist dominantly of mafic volcanic rock fragments. Sedimentary structures include trough and tabular-tangential cross-bedding in medium- to coarse-grain sizes and horizontal stratification in fine-grained sandstone. Pebble stringers commonly define stratification. Horizontally stratified siltstones are subordinate facies in this association. The different facies in association 2 either are randomly interstratified (Fig. 4B) or define 5–7 m thick fining-upward sequences (Fig. 4C).

Fig. 5. Facies association 2, locality 5. Graded and stratified conglomerates and pebbly sandstones.

Limited thickness of individual conglomerate beds in this association, taken together with the presence of sandstone lenses intercalated within the conglomerates (Fig. 6), indicates relatively shallow, rapidly switching braided channels. The model developed by Hein & Walker (1977) for the Kicking Horse River in British Columbia may explain the diversity of stratification types exhibited by conglomerates in this association. They argue that under conditions of relatively high fluid and sediment discharge, massive and horizontally stratified conglomerates develop in response to downstream accretion of gravel sheets. In contrast, if fluid and sediment discharge decrease rapidly, bars aggrade and develop steep downstream faces favouring formation of cross-stratification. Additional evidence for rapidly fluctuating flow strengths during development of this association includes grading and interstratification of arenites within and between conglomerates. By comparison with the

Fig. 6. Facies association 2, locality 5. Massive conglomerates with lens of cross-stratified sandstone. Note scour surface at top of sandstone lens.

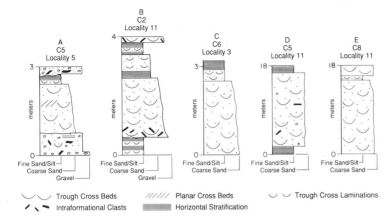

Fig. 7. Measured section showing sequences of facies in siliciclastic units in the Cromwell member: association 3.

proximal Scott and Donjek river systems discussed previously, the horizontally stratified and cross-stratified sandstone lenses are interpreted as bar-top and bar-edge deposits, respectively. The lenticularity of these arenites is in large part related to erosion. Pebble clusters at the base of conglomerate beds may be analogous to the initial deposits of lobes described by Southard et al. (1984). Lobes form by stalling and jamming of large clasts at downstream ends of chutes and grow by lateral and vertical addition of clasts. Occurrences of chutes in the system is favoured by the presence of multiple erosional surfaces within this association. Pebbly sandstones and sandstones that separate the conglomerates (Fig. 4B) similarly may represent the deposits of low-relief bars with slip faces. Horizontally stratified and massive conglomerates and cross-stratified arenites shown on Fig. 4C probably also represent bar deposits, but the thicknesses of the two homogeneous facies suggest somewhat greater water depths than inferred for the texturally diverse sequence on Fig. 4B. The upward-fining sequence (Fig. 4C) resembles sequences described by Rust (1978) and attributed to gradual abandonment of the braided system but equally could reflect waning floods (Nemec & Steel, 1984). The thin, fine-grained sandstone and siltstone interbeds (Fig. 4B,C) are intepreted as low-stage deposits.

Association 3

This is the dominant association in the study area and is recognized in all stratigraphic units. Facies typically are associated in 1–5 m thick, fining-upward sequences (Figs 7A,B,C & 8), but in some units no grain size trends are discernible within sandstone intervals greater than 15 m thick (Fig. 7D,E). Conglomerate lags are confined to the fining-upward sequences. Lags consist mostly of angular, intraformational siltstone–shale clasts up to 15 cm long. Subordinate clasts of well-rounded mafic volcanics, quartz amygdules and quartz arenite up to 5 cm in diameter are most common in unit C5 at locality 7 where internal horizontal stratification is apparent (Fig. 7A). Medium-scale trough cross-bedding in medium- to coarse-grained sandstone dominates this association. Cross-beds rarely decrease in scale upwards. Scattered intraformational and volcanic clasts rarely are present on foresets. In the Lena Quartzite in particular, the trough cross-beds display intense soft-sediment deformation

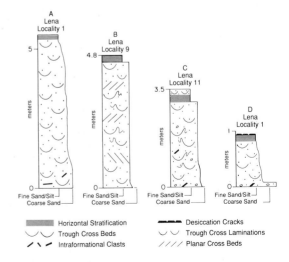

Fig. 8. Measured section showing sequences of facies in the Lena Quartzite: association 3.

(Fig. 8B,C). Subordinate tabular-planar to large-scale trough cross-beds are intercalated within the medium-scale trough cross-bedded intervals (Figs 7A & 8B). Internal stratification typically consists of tabular grainflows separated by magnetite concentrations (Fig. 9). Palaeocurrent azimuths from these cross-beds are oblique with respect to those for the dominant facies. Fining-upward intervals are capped by medium- to fine-grained sandstone containing either horizontal stratification with current lineations, or trough cross-laminae preserved on bedding planes as rib-and-furrow structure (Figs 7 & 8). Rare facies include wave ripples capping fining-upward sequences in unit C13 and desiccated mudstone layers capping thin sequences in the Lena Quartzite (Fig. 8D).

Fining-upward sequences in this association have Holocene counterparts in the Brahmaputra and South Saskatchewan Rivers (Coleman, 1969; Cant & Walker, 1978). Basal conglomerates, where present, represent channel lags derived by local reworking of underlying mafic volcanics and/or older sediments. Overlying intervals of trough cross-bedded sandstones reflect vertical aggradation and mid-channel bar formation in response to a decrease in stream competency. Cant & Walker (1978) noted that three-dimensional, subaqueous dunes are the most common bedform in the South Saskatchewan River at all stages of flow and that scale of bedforms relates to water depth. Soft-sediment deformation of the trough cross-beds is a result of dewatering and thus supports the notion that this facies formed by rapid aggradation. By analogy with the South Saskatchewan River, tabular-planar and large-scale trough cross-beds resulted from avalanching of grainflows off margins of mid-channel bars during periods of decreased discharge. Magnetite concentrations are a result of density segregation of particles within grainflows. Low-stage, bar-top deposits in this association are represented by upper flow-regime, horizontally stratified and lower flow-regime, trough cross-laminated, fine-grained arenites. Wave ripples capping fining-upward sequences reflect short-period waves related to wind shear in shallow, low-stage ponds. Desiccated mudstone layers are abandonment stage deposits; most of this facies is incorporated as lags into overlying sequences.

Thick intervals of trough cross-bedded sandstones lacking grain-size trends (Fig. 7D,E) may be a product of amalgamation of bar deposits due to erosion of low and abandonment stage facies. However, intraformational clasts of these facies generally are absent, suggesting that the trough cross-bedded intervals rather are a reflection of relatively constant, perennial discharge. Similar monotonous trough cross-bedded sequences are reported widely from the Precambrian rock record in particular (e.g. Eriksson, 1978; Eriksson & Vos, 1979; Krapez & Barley, 1987).

Association 4.

The Lena Quartzite at locality 12 contains the only

Fig. 9. Facies association 3, locality 5. Tabular-planar cross-bed sets consisting of grainflows defined by magnetite concentrations.

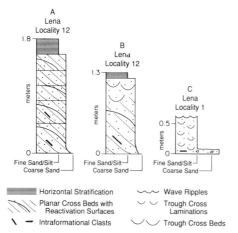

Fig. 10. Measured section showing sequences of facies in associations 4 (A,B) and 5 (C).

example of this association. The predominant facies above a scour base is tabular-planar to tangential cross-bedding in medium-grained sandstone. Sets are up to 90 cm and cosets up to 1.6 m thick (Fig. 10A,B). Within cosets, cross-bed sets decrease in scale upwards. Reactivation surfaces are common within cross-bed sets, and mudstone clasts up to 6 cm long are aligned along foresets. Medium- to small-scale trough cross-beds are erosional into tops of tabular cross-bed sets and cosets (Fig. 10B). Sequences between 1 and 2 m thick are capped by fine-grained sandstone containing horizontal stratification with current lineations (Fig. 11).

A Platte-type braided river system is inferred for this association of facies (cf. Smith, 1970; Miall, 1977). Tabular-planar to tangential cross-beds result from downstream migration of transverse and/or linguoid bars. Cosets develop when bars are superimposed. Trough cross-beds near the top of sequences and incised into tabular cross-bed sets and cosets reflect migration of dunes in channels incised into bars during falling-water stages. Horizontal stratification capping sequences records upper flow-regime conditions associated with a decrease in water depth during terminal stages of floods.

Association 5

The finest-grained facies in the study area consist of association 5, which is developed in units C13 and C14 at localities 4, 10 and 11, the lenticular units at the top of the Cromwell Member and at the base of the Lena Quartzite at locality 1. This association dominates the lenticular siliciclastic units in the Pickwick Member. Conglomerate lags rarely are present; these contain intraformational siltstone–mudstone and volcanic clasts. The predominant facies is fine-grained sandstone and siltstone containing trough cross-laminae (Fig. 10C) or horizontal stratification with current lineations. Isolated tabular-planar cross-beds sets are intercalated within the horizontally stratified intervals. Wave ripples rarely cap 0.5–1.0 m thick fining-upward sequences (Fig. 10C). Siltstone and mudstone inter-

Fig. 11. Facies association 4, locality 12. Tabular-planar cross-bed set capped by horizontal stratification.

vals between 2 and 5 m thick separate 0.5–2.0 m thick sandstone units, notably at locality 11. Mudstones typically are black and lack evidence of exposure.

The fine grain size of this association indicates significantly lower gradients than for associations 1–4. Thin, fining-upward sequences dominated by trough cross-laminae record similar depositional processes to those inferred for association 3, that is, erosion followed by aggradation and mid-channel bar formation. Sequences dominated by horizontal stratification, in contrast, reflect sheetflood processes characteristic of more ephemeral rivers (McKee *et al.*, 1967). Associated tabular-planar cross-beds are the product of waning flow. The dark colour of the intercalated mudstones and the absence of exposure features suggest a subaqueous depositional setting, possibly a lake. Wave ripples capping alluvial sequences probably record reworking during initial drowning of the alluvial plain.

Association 6

This association is present only in unit C1 at locality 5. Unit C1 is impersistent along strike and, where present, ranges from 1 to 3 m thick. Erosional relief up to 1 m is developed at the base of the unit that consists typically of locally derived, angular, mafic volcanic clasts. Overlying arenites contain low-angle stratification with distinctive 'pin-stripe' weathering. Individual laminae are 0.3–1.2 cm thick and display inverse grading from fine- to medium-grained sand.

Erosional relief and presence of conglomerates at the base of this association is indicative of fluvial reworking. Overlying arenites, in contrast, are of

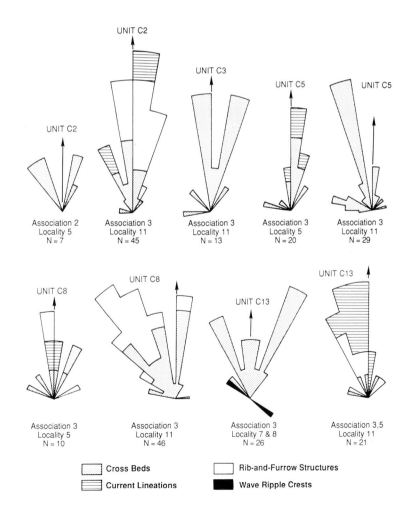

Fig. 12. Palaeocurrent data from siliciclastic units in the Cromwell Member. Current lineations are plotted as unidirectional structures.

484   K.A. Eriksson and E.L. Simpson

aeolian origin. Inversely graded laminae represent subcritically climbing translatent strata (cf. Hunter, 1977). Such strata are characteristic deposits of wind ripples in which ballistic processes predominate over avalanching (Hunter, 1977). Absence of internal cross-laminae is due to the low amplitude of wind ripples. The lack of cross-beds in this association may reflect insufficient time or insufficient sediment supply to develop aeolian bedforms before extrusion of basalt onto the wind-rippled surface.

Palaeocurrents

Palaeocurrent data are from units C2, C3, C5, C8 and C13 in the Cromwell Member and from the Lena Quartzite at different localities (Figs 3, 12 & 13). Most data are from association 3, with a limited number of measurements from associations 2, 4 and 5. The data are from inclined bedding plane exposures of cross-bed foresets, rib-and-furrow structures, current lineations and wave ripples. The first two structures show a consistent south to north flow; this is corroborated by data from current lineations that are plotted on Figs 12 & 13 as unidirectional structures. The rose diagram for the Lena Quartzite at locality 9 displays a large spread (Fig. 13). Cross-bed data are from medium- and large-scale troughs, with the medium-scale structures giving the northerly to northwesterly mode and the large-scale structures defining the northeasterly mode.

The depositional system

Coarsest facies (association 1) are confined to the eastern margin of the Lichhardt River Fault Trough. In the absence of cross-strata, palaeocurrent data are not available for this association. However, its areal distribution, lenticular geometry along a north–south outcrop belt at locality 7 and population of clasts are consistent with derivation from the east. Sedimentation took place on a relatively high-gradient alluvial plain akin to the proximal Scott Fan in Alaska. Associations 2–5, in contrast, record sedimentation on a lower-gradient braidplain with a south to north palaeoslope. This contention is supported by palaeocurrent data (Fig. 12) as well as by facies patterns. In particular, unit C2 is conglomeratic at localities 2, 3 and 5 in the south of the study area (Fig. 4B,C) and arenaceous at localities 9 and 11 to the north (Fig. 7B); unit C5 is coarser grained at localities 2, 3 and 5 (Fig. 7A) than at localities 9 and 11 (Fig. 7D); unit C13 is coarser grained in the south and southeast (localities 5 and 8) than in the north (localities 4, 10 and 11); and the Lena Quartzite consists mainly of South Saskatchewan-type sequences at localities

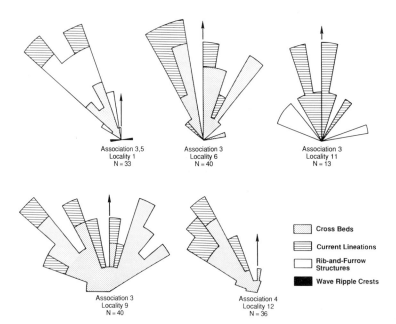

Fig. 13. Palaeocurrent data from the Lena Quartzite. Current lineations are plotted as unidirectional structures.

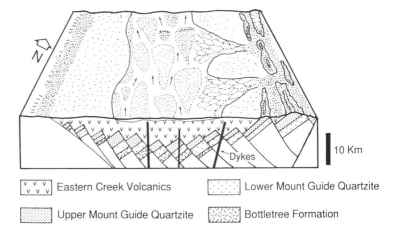

Fig. 14. Palaeogeographic model for siliciclastic units in the Cromwell Member illustrating interacting transverse and longitudinal alluvial systems. A similar scenario is envisaged for the Lena Quartzite and the Pickwick Member. Siliciclastic debris was derived exclusively from an eastern provenance consisting of the upper Mount Guide Quartzite, and cover sequence I consisting of felsic volcanics and related granites, and quartzite.

1, 9 and 11 (Fig. 8) and Platte-type sequences at locality 12 (Fig. 10A,B). In addition, the Lena Quartzite is mineralogically and texturally more mature in the north than in the south of the study area.

A palaeogeographic scenario thus envisages coarse-grained, higher-gradient, transverse braided rivers that supplied sediment from the east to a lower-graident, longitudinal trunk system; the depositional system was at least 30 km wide and 150 km long (Fig. 14). Clast populations and sandstone petrography support the contemporaneity of the transverse and longitudinal systems. A Holocene counterpart is the Kosi Fan and longitudinal Ganges River in India (Gole & Chitale, 1966; Wells & Dorr, 1987).

Facies analysis has shown that the siliciclastic units in the Cromwell Member and Lena Quartzite are almost entirely of braided-alluvial origin and consist of stacked, waning flood deposits with rare fine-grained, abandonment stage facies. The tabular geometry and lateral continuity of most siliciclastic units consisting of associations 2–5 imply that the high-bedload braided rivers migrated back and forth across the relatively low-gradient floodplain. Based on modelling, Paola (1988) argues that sheet gravels develop under conditions of slow, uniform subsidence. By analogy, the siliciclastic units in the Cromwell Member and the Lena Quartzite are considered to reflect slow subsidence. The braided rivers continuously filled to base level before switching laterally by avulsion. Local incision is recorded by intraformational lags and clasts; the clasts are a product of thalweg erosion of abandonment-stage deposits associated with the return of rivers to their earlier sites of deposition. Lenticular units in the upper Cromwell and Pickwick Members reflect either more rapid subsidence (cf. Marzo et al., 1988) or lower mean flow strength and shorter flood duration (Friend et al., 1979); the latter control is favoured by the finer grain size of the lenticular units.

VOLCANISM, SEDIMENTATION AND TECTONICS

Any model proposed for the origin of the Eastern Creek Volcanics, including the braided-alluvial sedimentary units, must account for the thick sequence of continental flood basalts, the presence of siliciclastic sedimentary units intercalated within the basalts, and the slow subsidence that favoured lateral switching and development of mostly tabular, siliciclastic units.

Continental flood basalts are related by many workers to thermal plumes in the asthenospheric mantle (e.g. White & McKenzie, 1989; Campbell & Griffiths, 1990). Associated with lithospheric thinning, decompression of the upwelling asthenospheric mantle produces partial melt that is either intruded into the lithosphere or extruded onto the crust. McKenzie & Bickle (1988) show that in the presence of mantle plumes, significant thicknesses (5–12 km) of tholeiitic basalts can be generated in this way with lithospheric extension (β) values of greater than 2.0. Thinning of the lithosphere is accomplished by either pure shear (McKenzie, 1978) or simple shear (Wernicke, 1981, 1985).

By analogy with the above, continental flood

basalts that make up most of the Eastern Creek Volcanics reflect an extensional setting. But what is the genetic relationship between volcanism and sedimentation? The absence of siliciclastic units in the lower half of the Cromwell Member has been attributed to a high frequency of volcanic eruptions (Derrick et al., 1977) and the total absence of any siliciclastic debris in this part of the stratigraphic column indicates no clastic supply to the volcanic environment. Sedimentary interbeds in the Cromwell and Pickwick Members, and the Lena Quartzite (Fig. 2), reflect episodic uplift and prolonged hiatuses in volcanism that permitted erosion of the volcanics and establishment of well-developed braided-alluvial systems. If, as seems likely, the calcrete horizon formed as a palaeosol on the top of the basalt flow, its presence implies a significant time lag between volcanism and sedimentation. The degree of carbonate development is similar to stage 4 to stage 6 calcretes (Gile et al., 1966; Machette, 1985) that require 10^4–10^5 years to form. In the absence of organic matter in Precambrian soils, these probably are minimum ages. In contrast to the long time lag at the base of siliciclastic units, incorporation of unconsolidated sediment into the base of overlying lava flows suggests a short hiatus between sedimentation and succeeding volcanism.

Uplift in the Kalkadoon–Liechhardt Belt to the east resulted in unroofing of the upper Mount Guide Quartzite and cover sequence I, including cogenetic granites, and their erosion and deposition in the upper half of the Eastern Creek Volcanics by high-gradient, transverse streams that flowed from east to west, and lower-gradient, longitudinal streams that reworked sediment down a south-to-north palaeoslope (Fig. 14). Slow subsidence that favoured deposition close to base level of individual braided-alluvial, siliciclastic units in the Cromwell and Pickwick Members is best attributed to cooling and thermal contraction of the volcanics. It is notable that thick siliciclastic units, reflecting greater amounts of thermal subsidence, overlie thick volcanic intervals whereas thin siliciclastic units overlie thin volcanic intervals. Cooling of the lithosphere is not a viable subsidence mechanism during development of the Eastern Creek Volcanics. Heat flow was high before and during volcanic eruptions, and volcanic hiatuses were considerably shorter than the millions to tens of millions of years required for lithospheric cooling (cf. McKenzie, 1978).

SUMMARY AND CONCLUSIONS

Most sequences of continental flood basalts are devoid of siliciclastic sediments suggesting a high frequency of eruption. Siliciclastic sedimentary units intercalated within the Eastern Creek Volcanics reflect prolonged hiatuses in volcanism that allowed erosion of the volcanics and episodic establishment of well-developed braided-alluvial systems. A calcrete horizon at the top of one basalt flow suggests that these hiatuses may have been as long as 10^4–10^5 years. With the exception of a thin aeolian deposit in unit C1 at two localities, the siliciclastic units are entirely of braided-alluvial origin. Facies analysis has identified a relatively high-gradient and conglomeratic, transverse braided-river system that supplied sediment from eastern highlands and a south-to-north flowing, lower-gradient and predominantly arenaceous, longitudinal braided-river system. Siliciclastic units that developed in the longitudinal system are mainly tabular. Cooling and thermal contraction of the subaerial volcanics created space for the siliciclastic units. The predominantly tabular shape of these units reflect slow subsidence that promoted lateral switching following aggradation to base level.

ACKNOWLEDGEMENTS

This study was supported jointly by the National Science Foundation through Grant EAR-87-07357 to K.A. Eriksson and by the Bureau of Mineral Resources, Geology and Geophysics, Australia. We benefited from discussion with J.H.C. Bain, D.H. Blake and G.A.M. Henderson and thank them for use of unpublished maps to guide us to critical localities. S.G. Driese made observations and provided comments on the calcrete horizon. We thank S. Chiang for drafting and M. Sentelle and L. Solowiej for word processing. Tim Cross and an anonymous reviewer provided valuable comments.

REFERENCES

BLAKE, D.H. (1985) Tectonic development of the Proterozoic Mount Isa Inlier, northwest Queensland. *Abstracts, Conference on Tectonics and Geochemistry of Early to Middle Proterozoic Fold Belts*, 7–14 August, Darwin. Bur. Min. Resources, Record 1985/28.

BLAKE, D.H. (1987) *Geology of the Mount Isa Inlier and*

Environs, Queensland and Northern Territory. Bur. Min. Resources, Bull. 225. Australian Government Publishing Service, Canberra.

BOOTHROYD, J.C. & ASHLEY, G.M. (1978) Processes, bar morphology and sedimentary structures on braided outwash fans, northeastern Gulf of Alaska. In: *Glaciofluvial and Glaciolacustrine Sedimentation* (Eds Jopling, A.V. & McDonald, B.C.) pp. 193–222. Soc. Econ. Paleont. Miner., Tulsa, Spec. Publ. 23.

BULTITUDE, R.J. & WYBORN, L.A.I. (1985) Distribution and geochemistry of volcanic rocks in the Duchess–Urandangi Region, Queensland. *Bur. Min. Resources, J. Austr. Geol. Geophys.* **7**, 99–112.

CAMPBELL, I.H. & GRIFFITHS, R.W. (1990) Implications of mantle plume structure for the evolution of flood basalts. *Earth Planet. Sci. Lett.* **99**, 79–93.

CANT, D.J. & WALKER, R.G. (1978) Fluvial processes and facies sequences in the sandy braided South Saskatchewan River, Canada. *Sedimentology* **25**, 625–648.

CLEMMENSEN, L.B. (1988) Aeolian morphology preserved by lava cover, the Precambrian Mussartût Member, Erikfjord Formation, South Greenland. *Bull. Geol. Soc. Denmark* **37**, 105–116.

COLEMAN, J.M. (1969) Brahmaputra River: channel processes and sedimentation. *Sedim. Geol.* **3**, 129–239.

DANIELS, P.A. (1982) Upper Precambrian sedimentary rocks: Oronto Group, Michigan–Wisconsin. In: *Geology and Tectonics of the Lake Superior Basin* (Eds Wold, R.J. & Hinze, W.J.) pp. 107–133. Geol. Soc. Am., Boulder, Memoir 156.

DERRICK, G.M. (1982) A Proterozoic rift zone at Mount Isa, Queensland, and implications for mineralisation. *Bur. Min. Resources, J. Austr. Geol. Geophys.* **7**, 81–92.

DERRICK, G.M., WILSON, I.H., HILL, R.M., GLIKSON, A.Y. & MITCHELL, J.E. (1977) *Geology of the Mary Kathleen 1 : 100 000 Sheet Area, Northwest Queensland.* Bur. Min. Resources, Bull. 193. Australian Government Publishing Service, Canberra.

ERIKSSON, K.A. (1978) Alluvial and destructive beach facies from the Archaean Moodies Group, Barberton Mountain Land, South Africa and Swaziland. In: *Fluvial Sedimentology* (Ed. Miall, A.D.) pp. 287–311. Can. Soc. Petrol. Geol., Calgary, Memoir 5.

ERIKSSON, K.A. & VOS, R.G. (1979) A fluvial fan depositional model for middle Proterozoic red beds from the Waterberg Group, South Africa. *Precambr. Res.* **9**, 169–188.

ERIKSSON, K.A. SIMPSON, E.L. & JACKSON, M.J. (in press) Stratigraphic evolution of a Proterozoic rift to thermal-relaxation basin, Mount Isa Inlier, Australia: constraints on nature of lithospheric extension. Spec. Publs Int. Assoc. Sedim.

ETHRIDGE, F.G., TYLER, N. & BURNS, L.K. (1984) Sedimentology of a Precambrian quartz-pebble conglomerate, southwest Colorado. In: *Sedimentology of Gravels and Conglomerates* (Eds Koster, E.H. & Steel, R.J.) pp. 165–174. Can. Soc. Petrol. Geol., Calgary, Memoir 10.

FISHER, R.V. & SCHMINCKE, H.-U. (1984) *Pyroclastic Rocks.* Springer-Verlag, New York, 472 pp.

FRIEND, P.F., SLATER, M.J. & WILLIAMS, R.C. (1979) Vertical and lateral building of river sandstone bodies, Ebro basin, Spain. *J. Geol. Soc. Lond.* **13b**, 39–14.

GILE, L.H., PETERSON, F.F. & GROSSMAN, R.B. (1966) Morphological and genetic sequences of carbonate accumulation in desert soils. *Soil Sci.* **101**, 347–360.

GLIKSON, A.Y., DERRICK, G.M., WILSON, I.H. & HILL, R.M. (1976) Tectonic evolution and crustal setting of the middle Proterozoic Leichhardt River fault trough, Mount Isa region, northwestern Queensland. *Bur. Min. Resources, J. Aust. Geol. Geophys.* **1**, 115–129.

GOLE, C.V. & CHITALE, S.V. (1966) Inland delta building activity of the Kosi River. *J. Hydrol. Div. ASCE* **92**, 111–126.

HEIN, F.J. & WALKER, R.G. (1977) Bar evolution and development of stratification in the gravelly, braided, Kicking Horse River, British Columbia. *Can. J. Earth Sci.*, **14**, 562–570.

HUNTER, R.E. (1977) Basic types of stratification in small eolian dunes. *Sedimentology* **24**, 361–388.

KRAPEZ, B. & BARLEY, M.E. (1987) Archaean strike-slip faulting and related ensialic basins: evidence from the Pilbara Block, Australia. *Geol. Mag.* **124**, 555–567.

MACDOUGALL, J.D. (Ed.) (1988) *Continental Flood Basalts.* Kluwer, Dordrecht, 341 pp.

MCKEE, E.D., CROSBY, E.J. & BERRYHILL, J.R. (1967) Flood deposits, Bijou Creek, Colorado, June, 1965. *J. Sedim. Petrol.* **37**, 329–351.

MCKENZIE, D.P. (1978) Some remarks on the development of sedimentary basins. *Earth Planet. Sci. Lett.* **40**, 108–125.

MCKENZIE, D. & BICKLE, M.J. (1988) The volume and composition of melt generated by extension of the lithosphere. *J. Petrol.* **29**, 625–679.

MACHETTE, M.N. (1985) Calcic soils of the southwestern United States. In: *Soils and Quaternary Geology of the Southwestern United States* (Ed. Weide, D.L.) pp. 1–21. Geol. Soc. Am., Boulder, Spec. Paper 203.

MARZO, M., NIJMAN, W. & PUIGDEFABREGAS, C. (1988) Architecture of the Castissent fluvial sheet sandstones, Eocene, South Pyrenees, Spain. *Sedimentology* **35**, 719–738.

MERK, G.P. & JIRSA, M.A. (1982) Provenance and tectonic significance of the Keweenawan interflow sedimentary rocks. In: *Geology and Tectonics of the Lake Superior Basin* (Eds Wold, R.J. & Hinze, W.J.) pp. 97–105. Geol. Soc. Am., Boulder, Memoir. 156.

MIALL, A.D. (1977) A review of the braided river depositional environment. *Earth Sci. Rev.* **13**, 1–62.

NEMEC, W. & STEEL, R.J. (1984) Alluvial and coastal conglomerates: their significant features and some comments on gravelly mass-flow deposits. In: *Sedimentology of Gravels and Conglomerates* (Eds Koster, E.H. & Steel, R.J.) pp. 1–31. Can. Soc. Petrol. Geol., Calgary, Memoir 10.

OJAKANGAS, R.W. & MOREY, G.B. (1982) Keweenawan sedimentary rocks of the Lake Superior region: a summary. In: *Geology and Tectonics of the Lake Superior Basin* (Eds Wold, R.J. & Hinze, W.J.) pp. 157–164. Geol. Soc. Am., Boulder, Memoir 156.

PAGE, R.W. (1978) Response of U–Pb and Rb–Sr total-rock and mineral systems to low-grade regional metamorphism in Proterozoic igneous rocks. Mount Isa, Australia. *J. Geol. Soc. Austr.* **25**, 141–164.

PAGE, R.W. (1983a) Timing of superposed volcanism in the Proterozoic Mount Isa Inlier, Australia. *Precambr. Res.*, **21**, 223–245.

PAGE, R.W. (1983b) Chronology of magmatism, skarn formation and uranium mineralization, Mary Kathleen, Queensland, Australia. *Econ. Geol.* **85**, 838–853.

PAGE, R.W. & BELL, T.H. (1986) Isotopic and structural responses of granite to successive deformation and metamorphism. *J. Geol.* **94**, 365–379.

PAGE, R.W., MCCULLOCH, M.T. & BLACK, L.P. (1984) Isotopic record of major Precambrian events in Australia. In: *Proc. 27th Geol. Congress, Vol. 5, Precambrian Geology*, pp. 25–72. VNU Science Press, Utrecht.

PAOLA, C. (1988) Subsidence and gravel transport in alluvial basins. In: *New Perspectives in Basin Analysis* (Eds Kleinspehn, K.L. & Paola, C.) pp. 231–243. Springer-Verlag, New York.

RETALLACK, G. (1981) Fossil soils: indicators of ancient terrestrial environments. In: *Paleobotany, Paleoecology and Evolution*, Vol. 1 (Ed. Niklas, K.J.) pp. 55–102. Praeger, New York.

RUST, B.R. (1972) Structure and process in a braided river. *Sedimentology* **18**, 221–245.

RUST, B.R. (1978) Depositional models for braided alluvium. In: *Fluvial Sedimentology* (Ed. Miall, A.D.) pp. 605–625. Can. Soc. Petrol. Geol., Calgary, Memoir 5.

SCHUMM, S.A. (1968) Speculations concerning the paleohydrologic controls of terrestrial sedimentation. *Bull. Geol. Soc. Am.* **79**, 1573–1578.

SMITH, N.D. (1970) The braided stream depositional environment: comparison of the Platte River with some Silurian clastic rocks, northcentral Appalachians. *Bull. Geol. Soc. Am.* **81**, 2993–3013.

SOUTHARD, J.B., SMITH, N.D. & KUHLE, R.A. (1984) Chutes and lobes: newly identified elements of braiding in shallow gravelly streams. In: *Sedimentology of Gravels and Conglomerates* (Eds Koster, E.H. & Steel, R.J.) pp. 51–59. Can. Soc. Petrol. Geol., Calgary, Memoir 10.

TANKARD, A.J., JACKSON, M.P.A., ERIKSSON, K.A., HOBDAY, D.K., HUNTER, D.R. & MINTER, W.E.L. (1982) *Crustal Evolution of Southern Africa.* Springer-Verlag, New York, 523 pp.

WELLS, N.A. & DORR, J.A. (1987) A reconnaissance of sedimentation on the Kosi alluvial fan of India. In: *Recent Developments in Fluvial Sedimentology* (Eds Ethridge, F.G., Flores, R.M. & Harvey, M.D.), pp. 51–61. Soc. Econ. Paleont. Miner., Tulsa, Spec. Publ. 39.

WERNICKE, B. (1981) Low-angle normal faults in the Basin and Range Province: nappe tectonics in an extending orogen. *Nature* **291**, 645–648.

WERNICKE, B. (1985) Uniform-sense normal simple shear of the continental lithosphere. *Can. J. Earth Sci.* **22**, 108–125.

WHITE, R.S. & MCKENZIE, D.P. (1989) Magmatism at rift zones: the generation of volcanic continental margins and flood basalts. *J. Geophys. Res.* **94**, 7685–7729.

WILSON, I.H. (1987) Geochemistry of Proterozoic volcanics, Mount Isa Inlier, Australia. In: *Geochemistry and Mineralization of Proterozoic Volcanic Suites* (Eds Pharaoh, T.C., Beckinsale, R.D. & Rickard, D.), pp. 409–423. Geol. Soc. Spec. Publ. 33. Blackwell Scientifc Publications, Oxford.

WYBORN, L.A.I., PAGE, R.W. & MCCULLOCH, M.T. (1988) Petrology, geochronology, and isotope geochemistry of the post-1820 Ma granites of the Mount Isa Inlier: mechanisms for the generation of Proterozoic anorogenic granites. *Precamb. Res.* **40/41**, 509–541.

Sedimentological response of an alluvial system to source area tectonism: the Seilao Member of the Late Cretaceous to Eocene Purilactis Formation of northern Chile

A.J. HARTLEY

Production Geoscience Unit, Department of Geology and Petroleum Geology, University of Aberdeen, Meston Building, King's College, Aberdeen AB9 2UE, UK

ABSTRACT

The Seilao Member of the Late Cretaceous to Eocene Purilactis Formation in northern Chile consists of alluvial fan deposits organized into a series of large-scale ($\geq$100 m), basin-wide, coarsening and fining upwards sequences (megasequences). Coarsening upwards sequences record a change from playa and sandflat sedimentation to poorly confined conglomeratic sheetflood and hyperconcentrated flood flow deposits and show a decrease in incision up section. Conversely, fining upwards megasequences are the reverse of this sequence. Fining upwards megasequences may be overlain by thick sequences dominated by small-scale (3–20 m) fining upwards cycles composed of pebbly braided stream deposits.

Megasequences developed in response to tectonism. Coarsening upwards cycles resulted from the easterly progradation of gravels during limited source area uplift/relative basin subsidence. Relative subsidence was sufficient to permit aggradation in proximal areas (with no incision) and gravel progradation in more distal areas of the basin. Fining upwards megasequences developed during periods of rapid source area uplift/relative basin subsidence. Small-scale fining upwards cycles result from periods when sediment supply and subsidence were close to equilibrium; consequently, allocyclic influence on alluvial architecture was minimal and autocyclic processes dominated.

Although difficult to constrain, the Late Cretaceous tectonic setting in the north Chilean Precordillera suggests that source area uplift resulting in relative basin subsidence was the most important control on sedimentation. Uplift resulted from inversion along pre-existing extensional faults (possibly associated with the development of an easterly propagating thrust front) following mid-Cretaceous compression (Peruvian Orogeny). Alternative allocyclic mechanisms had only a minimal influence on alluvial architecture. In particular, no change in macroclimate or source area composition took place during sedimentation and eustatic fluctuations were unimportant in this closed, intermontane basin.

INTRODUCTION

Allocyclic (extrabasinal) mechanisms such as climate, eustasy, source area relief and source area lithology determine features such as discharge, grain size of sediment supply, subsidence rates and base-level within a sedimentary basin. These factors in turn control autocyclic (intrabasinal) mechanisms such as avulsion, neck and chute cutoff and channel sinuosity and migration. The alluvial architecture of a basin is, therefore, the resultant of all the above mechanisms, but ultimately is controlled by allocyclic mechanisms. In order to assess the importance of any one allocyclic mechanism on the alluvial architecture of a basin-fill sequence the mechanism should be studied in isolation from others, a situation often difficult to achieve.

In this paper evidence is presented from the Seilao Member of the Purilactis Formation of northern Chile to show that the principal allocyclic control on alluvial architecture during sedimentation was source area tectonism (uplift) and relative basin subsidence. The detailed facies analysis of the Seilao Member provides an analogy for other basins where allocyclic controls on alluvial architecture may be more difficult to define or distinguish from other controls.

The Late Cretaceous to Eocene Purilactis Formation (Bruggen, 1950; Dingman, 1967; Flint et al., 1989) of northern Chile is located within the Precordillera of the Central Andean forearc (Fig. 1) where it comprises the upper part of the Purilactis Group (Charrier & Reutter, 1990). The entirely continental strata of the formation were deposited in alluvial fan, aeolian and lacustrine environments in a closed intermontane basin (Hartley et al., 1988), formed following the mid-Cretaceous (Cenomanian–Turonian) Peruvian Orogeny (Steinmann, 1929; Coira et al., 1982).

The alluvial, 1150 m thick Seilao Member is well exposed within the Purilactis Basin (Fig. 2), where it erosively overlies a 64 (± 10) Ma andesite (Flint et al., 1989). Throughout deposition, the Seilao Member was sourced from the west (Fig. 3), from a terrain composed of acidic to intermediate volcanic material and to a lesser extent Jurassic limestones (Felsch, 1933), red sandstones and acidic intrusives (Fig. 4). Clast types can be matched with Early Cretaceous acidic to intermediate volcanics and intrusives, and Late Jurassic (?) red beds which infilled an earlier carbonate-dominated Jurassic marginal basin located to the west of the Purilactis Basin in the Cordillera de Domeyko (Chong, 1977; Hartley et al., 1988).

FACIES ANALYSIS

Detailed facies analysis revealed the presence of six related facies associations, discussed below.

Description

Facies association A

Massive, poorly sorted, generally clast-supported, laterally extensive sheet conglomerates of pebble to boulder grade comprise this facies association (Fig. 5A). Basal bedding contacts are sharp and irregular with a relief of up to 0.75 m. Bed thicknesses vary between 0.2 and 2 m (average 0.6 m). The conglomerates have a loose to moderately tight packing and may be normally graded (cobble → pebble) in the top 20% of the beds. Matrix (up to granule grade) may be sufficiently abundant in some units to support clasts. Imbrication is occasionally developed at the top of beds, where pebble a-axes are parallel to flow and imbricated ($a(p)a(i)$). The con-

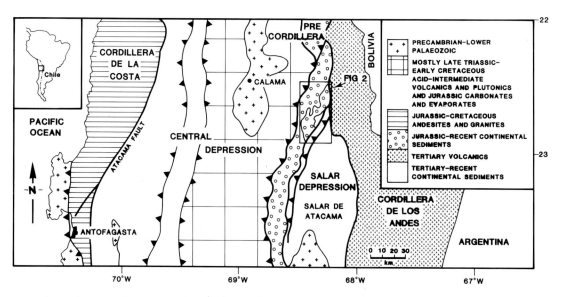

Fig. 1. Geological map of northern Chile showing the five morphotectonic units which comprise the Andean forearc. Box shows area of Fig. 2 with heavy line marking the position of the Purilactis Formation.

glomerates are commonly capped by medium to coarse grained sandstones up to 0.3 m thick, which drape the irregular topography of the underlying clasts. They are usually parallel laminated or display low angle (<6°) planar cross-stratification, often with a primary current lineation (PCL) parallel to the fabric in the underlying conglomerates.

Facies association B

This facies association is characterized by crudely horizontally stratified, normally graded, clast-supported, laterally discontinuous sheet conglomerates of pebble grade (Fig. 5B). Basal surfaces of the beds are sharp and irregular with relief of up to 0.35 m. Bed thicknesses are less than those of facies A varying between 0.05 and 1.2 m (average 0.2 m). The conglomerates are well sorted, moderate to tightly packed and normally graded (pebble → very coarse sandstone) and display moderate to well developed clast fabrics with a-axes transverse to flow and b-axes imbricated ($a(t)b(i)$). As with facies association A, the conglomerates are capped by medium to coarse grained, 0.1–0.25 m thick sandstones displaying parallel lamination and low angle planar cross-stratification together with rare trough cross-strata (preserved set size up to 0.2 m). The sandstones are laterally discontinuous due to erosion by the succeeding conglomerate.

Facies association C

Horizontal and cross-stratified conglomerates and pebbly sandstone bodies characterize this facies association (Figs 5C & 6). The bodies vary between 1 and 8 m in width and are up to 1.25 m thick. Conglomerates comprise the lower part of the bodies, commonly display $a(t)b(i)$ imbrication and contain cross-stratified sandstone lenses. Crude horizontal and/or trough cross-stratification is frequently developed. The conglomerates grade vertically and horizontally into planar and trough cross-stratified granular and pebbly sandstones with set sizes up to 0.65 m, but averaging 0.25 m. The pebbly sandstones may grade into well sorted, medium to coarse grained, horizontally stratified sandstones with a PCL. Planar and more rarely trough cross-stratification is developed in thicker sandstone beds (⩾ 35 cm). Internal erosion surfaces (up to 75 cm in relief and 10 m in width) are common within the facies association.

Facies association D

This facies association is composed of lenticular, coarse grained sandstone bodies with width : height ratios of ⩾ 30 : 1. The bodies are composed of individual storeys which are 0.5–4 m thick and comprise an erosive base overlain by a pebble lag with both intra- and extraformational clasts (Fig. 6A). The lag is succeeded by medium to very coarse grained sandstones dominated by low angle (⩽ 4°) planar cross-stratification, horizontal stratification with PCL, and, to a lesser extent, trough cross-stratification. Grading is restricted to the top of individual storeys where coarse sandstone grades to parallel laminated siltstone or mudstone containing disseminated anhydrite crystals, desiccation cracks, vertical burrows and raindrop pits.

Facies association E

Isolated fine to coarse grained sheet sandstone beds (traceable for over 100 m) characterize this facies association (Figs 5 & 6). Beds range between 0.05 and 0.6 m in thickness (average 0.15 m) and have sharp or mildly irregular basal bedding contacts. Horizontal to low angle (⩽ 6°) stratification with PCL is common and, more rarely, small-scale trough cross-stratification (0.05–0.1 m sets) and massive bedding is seen. Normal grading is occasionally developed. Sandstone tops are frequently ripple cross-laminated and draped by thin (0.02–0.1 m thick) mudstone laminae containing desiccation cracks and vertical burrows. Beds may be isolated or amalgamated to form sequences up to 5 m thick.

Facies association F

This facies association is characterized by red mudstones which are commonly massive and more rarely display millimetre-scale undulose or parallel laminae (Fig. 5). Desiccation cracks, disseminated anhydrite crystals and vertical and horizontal burrows are common. Thin (5 cm) discontinuous nodular anhydrite beds are occasionally developed.

Interpretation

Facies association A: hyperconcentrated flood flow deposits

The sheet-like, poorly sorted, sharply based, clast-supported (with sandy/granule matrix) nature of the

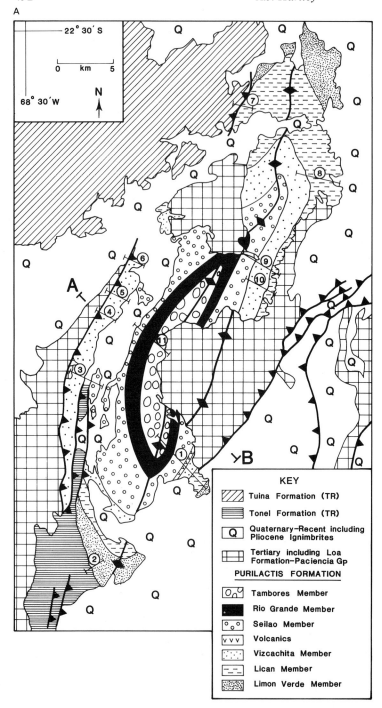

Fig. 2. (A) Map of the Purilactis Formation, showing the distribution of Members within the basin. Numbers refer to logged sections. (Modified from Marinovic & Lahsen (1984) and Hartley et al. (1988).)

conglomerate together with $a(p)a(i)$ imbrication suggests that they were deposited by unconfined high-density flood flows (hyperconcentrated flood flows of Smith, 1986). The above features and bipartite nature of facies A resemble conglomerates described by Todd (1989) from the Old Red Sandstone of SW Ireland, where the lower conglomeratic part represents a high-density gravelly traction carpet, decoupled from an overlying sand- and silt-laden turbulent flow. These high-density flood flows

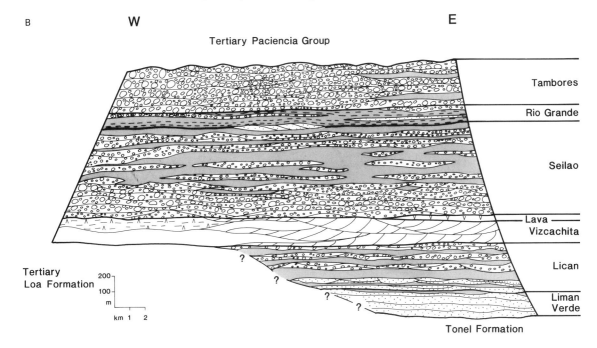

Fig. 2. (B) Schematic E–W cross-section across the central part of the Purilactis Basin Formation showing the general stratigraphy and relative position of the Seilao Member within the Purilactis Formation. Ellipses, conglomerate; dots, sandstone; dashes, shales; (∧), evaporites. The sides of the figure taper to account for tectonic dip of the measured sections.

are transitional between normal stream flows and debris flows (Smith, 1986, 1987; Flint & Turner, 1988) and are particularly characteristic of ephemeral (flashy) discharge in semiarid alluvial fan environments (cf. Allen, 1981; Ballance, 1984; Flint & Turner, 1988).

Facies association B: proximal sheetflood deposits

The conglomerates of facies association B are significantly better sorted, better graded and thinner than those of facies A. They are interpreted to represent deposition from poorly confined sheetflows where 'normal' grain-by-grain tractional bedload sedimentation predominated and a traction carpet was not developed. The sandstone tops are thought to represent the development of upper flow regime plane-bed conditions (parallel lamination), low relief transverse bars (low angle planar cross-stratification) and occasional sinuous-crested dunes (trough cross-stratification) during peak and waning flood stage. This facies association is thought to record ephemeral flood sedimentation in an alluvial fan environment, with features typical of sheetflow deposits in very shallow channels containing low relief longitudinal bars (represented by laterally restricted pebble sheets).

Facies association C: proximal channelized streamflow deposits

The presence of large-scale stratification and erosional relief together with the limited lateral extent of this facies association suggests deposition by streamflow processes in much more confined channels than either of facies associations A or B (pebbly braided stream deposits). The conglomerates located at the base of the incised bodies are thought to represent channel deposits, with the pebbly cross-stratified sandstones representing in-channel longitudinal bar forms with greater relief and lateral extent than those of facies association B. However, it should be noted that a complete continuum exists between facies associations B and C depending on the degree of channelization, and consequently the examples described here represent only end members of the spectrum.

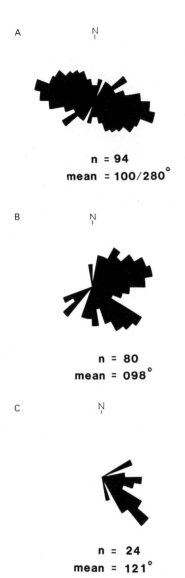

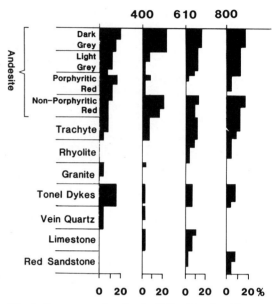

Fig. 3. Palaeocurrent data throughout the Seilao Member showing number of counts: (A) channel axes; (B) cross-strata; (C) imbrication $a(t)b(i)$ data.

Fig. 4. Provenance data for the Seilao Member. Note the dominance of acidic–intermediate lavas and little change in composition throughout (derived from analysis of 100 clasts at each point).

Facies association D: distal channelized streamflow deposits

The sandstone bodies represent alluvial channel deposits. The dominance of horizontal and low angle planar cross-stratification with a PCL indicates deposition of flat beds/low relief transverse bars at high flood stage, with trough cross-strata reflecting the development of sinuous-crested in-channel dunes formed at low flood stage. The above features, coupled with the presence of non-depositional indicators such as bioturbation, desiccation cracks and anhydrite crystals, indicates that deposition of facies association D was due to rapid, ephemeral flood events. Facies association D probably represents a distal equivalent to facies association C.

Facies association E: distal sheetflood deposits

The lack of channelization, large lateral extent, little thickness variation and dominance of horizontal laminae with PCL together with their mainly coarse grain size suggests that the sandstones of facies association E represent deposition by sheetflood events under upper flow regime conditions. The rippled tops suggest reworking by decelerating flood currents, possibly analogous to the sheetwash process described by Hardie et al. (1978). The presence of desiccation cracks and burrows testifies to the episodic nature of the flood events. The finer grain size, better sorting and limited thickness of the

Fig. 5. (A) Hyperconcentrated flood flow deposits of facies association A. Note poor sorting, absence of stratification, flat base and normally graded top. Height of flow = 2 m. (B) Facies association B forming well-cemented sheet-like bodies, interbedded with facies association F (dark areas). Height of cliff = 30 m. (C) Trough cross-stratified pebbly sandstones of facies association C. Set size = 50 cm.

sandstones of facies association E compared to the deposits of facies associations A, B and C suggests that facies association E sheetflows represent much more distal flood events (possibly similar to that described by McKee *et al.*, 1967) and that they may be laterally equivalent to the fluvial channels of facies association D. The close association of facies association E with the mudstones of facies association F and their deposition under upper flow regime conditions suggests that they represent distal sheetflows at the toes of alluvial fans in a sandflat environment (cf. Tunbridge, 1981; Hubert & Hyde, 1982).

Facies association F:
playa-flat mudstones

The presence of millimetre-scale laminae within the mudstones of facies association F indicate deposition from a standing body of water, probably following a flood event. The presence of desiccation cracks and anhydrite crystals suggests that any standing body of water was ephemeral. Facies association F is thought to represent a playa-mudflat subject to periodic flood events represented by facies association E and occasionally cut by the channels of facies association D.

FACIES ASSOCIATION ORGANIZATION

Figure 7 shows the position of the six facies associations and their envisaged lateral relationships in an alluvial fan, mountain front setting during deposition of the Seilao Member. It demonstrates the well-established relationship of decreasing grain size and change in depositional mechanism (from unconfined hyperconcentrated floodflows, through sheetflood to streamflow dominated sedimentation) with increasing distance from the alluvial fan source (e.g. Heward, 1978, and references therein).

Vertical sequence analysis (Fig. 8) reveals the presence of large-scale (≥100 m), basin-wide coarsening and fining upwards sequences (megasequences) within the Seilao Member. Coarsening upwards sequences show a transition from the more distal facies associations of D, E and F to the more proximal C and B and occasionally to the most proximal facies association A (it should be noted that facies association A forms a volumetrically minor part of the Seilao Member, but is particularly significant in determining the proximity to the

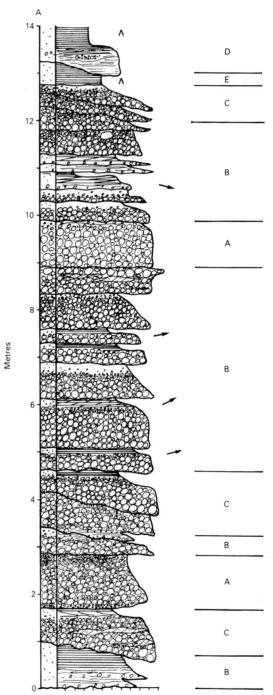

Fig. 6. (A) Lithological log beginning 287 m above the base of the Seilao Member (Quebrada Seilao, locality 1 of Fig. 2A). The section represents the top of a coarsening upwards megasequence, showing the development of proximal facies associations. Letters correspond to facies associations. Arrows, palaeocurrents; (∧) anhydrite.

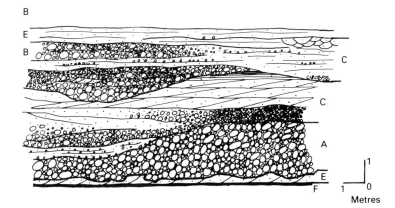

Fig. 6. (B) Sketch section from photographs and field observations. Letters correspond to facies associations; flow towards viewer; 410 m above the base of the Seilao Member (Quebrada Seilao, locality 1 of Fig. 2A).

alluvial fan source). Fining upwards sequences show the reverse of this facies association relationship. In the middle part of the Seilao Member a 300 m sequence is developed where only small-scale (3–20 m) fining upwards sequences are present, with facies associations C, D, E and F predominating.

The alluvial architecture of the Seilao Member is closely related to the development of the megasequences. At the top of coarsening upwards sequences laterally extensive coarse grained facies are developed with good interconnectedness, whereas at the top of fining upwards sequences and where only small-scale cyclicity is developed, laterally impersistent coarse grained facies predominate with a high percentage of mudstone.

CONTROLS ON SEDIMENTATION

The basin-wide occurrence of megasequences suggests that allocyclic mechanisms were responsible for controlling facies association distribution within the Seilao Member. A number of allocyclic mechanisms can be discounted as influencing megasequence development. In particular, macroclimatic fluctuations were unimportant as deposition by

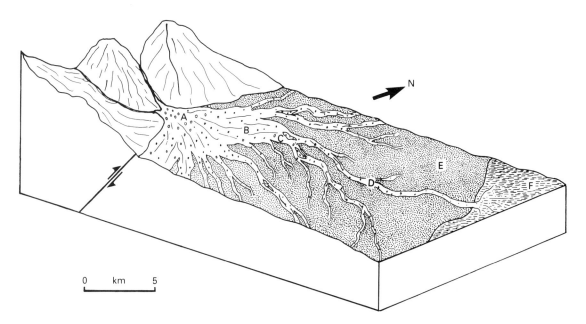

Fig. 7. General palaeogeographic reconstruction of Seilao Member showing envisaged position of facies associations (letters) relative to the source area.

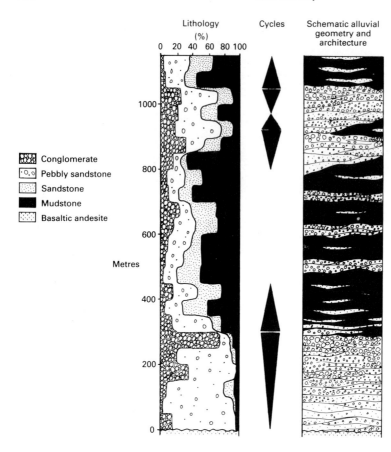

Fig. 8. Lithological and schematic architectural logs of the Seilao Member showing coarsening and fining upwards cycles. Percentage lithology was calculated by averaging lithologies over 50 m intervals from four logged sections. Conglomerate corresponds to facies associations A, B and C; pebbly sandstones to facies associations B, C and to a lesser extent D; sandstone to facies association E and to a lesser extent D; mudstone to facies association F.

ephemeral flash floods and the presence of evaporites and desiccation cracks throughout the Member suggests sedimentation under a constantly semiarid climate. Indeed, a semiarid climate predominated throughout central South America during the Cretaceous (Riccardi, 1987) and much of the Tertiary (Jolley et al., 1990). Changes in source area lithology did not significantly influence sedimentation as clast compositions remained constant throughout deposition of the Seilao Member (Fig. 4). Eustatic variations did not influence local base-level, as sedimentation took place within a closed, intermontane basin.

The above observations suggest that tectonic activity, whether in the form of source area uplift and/or basin subsidence, was responsible for megasequence development; a scenario favoured by the tectonic setting of the Purilactis Formation. As previously outlined, deposition of the Purilactis Formation commenced following the mid-Cretaceous Peruvian Orogeny. The orogeny resulted in the deformation and uplift of a marginal basin containing Early Cretaceous lavas and plutonics, Jurassic carbonates and redbeds. Although poorly constrained, regional considerations suggest that deformation took the form of compressional reactivation of pre-existing extensional faults and associated thrust faulting during the early part of the late Cretaceous (G.D. Williams, pers. comm., 1990). This resulted in the uplift of fault-bounded blocks to source the Purilactis Formation. Subsequently, further compressional deformation led to the development of an easterly propagating thrust front in the north Chilean Precordillera (Flint et al., 1989) which was periodically active throughout the Tertiary and has been up to the present day (Jolly et al., 1990).

Previous work (Allen, 1978; Blair & Bilodeau, 1988; Heller et al., 1988; Paola, 1988; Heller & Paola, 1989) suggests that coarsening upwards alluvial megasequences result from the reworking of proximal gravels during periods of decreased subsidence following tectonic activity; progradation of the gravel front is accompanied by increased incision in more proximal areas of the basin. Conversely, fining upwards megasequences occur during periods of increased subsidence related to renewed tectonic activity, resulting in the restriction of coarse grained detritus to the more proximal areas of the basin (Blair & Bilodeau, 1988; Heller et al., 1988).

Coarsening upwards megasequences in the Seilao Member almost certainly reflect periods of gravel progradation during reduced tectonic activity. However, the absence of extensive channelization at the top of coarsening upwards cycles suggests that gravel progradation was not accompanied by incision in the more proximal parts of the basin. It is possible that the observed megasequences in the Seilao Member were developed in more distal parts of the basin where only limited or no incision would be expected. However, the proximal nature of facies association A suggests that this is not the case. Alternatively, and more likely, it is possible that the more proximal areas of the basin continued to aggrade at a reduced rate during periods of decreased subsidence, without incision taking place, whilst the gravel front prograded eastwards (e.g. Paola, 1988).

The deposition of small-scale fining upwards cycles following fining upwards megasequences is thought to represent periods when relative subsidence and sediment supply were close to equilibrium. Consequently megasequences were not developed, more distal facies associations predominated and autocyclic processes such as avulsion and channel migration are thought to have controlled alluvial architecture.

SUMMARY AND CONCLUSIONS

1 The Seilao Member of the Purilactis Formation consists of six alluvial facies associations deposited on or adjacent to an ephemeral ('flashy') alluvial fan under a semiarid climate. They include facies associations: (A) hyperconcentrated flood flow deposits (unstratified conglomerates), (B) tractional sheetflow deposits (normally graded conglomerate → coarse sandstone), (C) channelized streamflow deposits (stratified conglomerates and pebbly sandstones), (D) channelized streamflow deposits (stratified coarse grained sandstones), (E) distal sheetflows deposited under upper flow regime conditions (fine → coarse sandstones), and (F) massive and laminated playa mudstones.

2 Facies association distribution reflects source area proximity, with the most proximal facies association A passing downslope into B and C, which in turn grade into facies associations D and E and ultimately F.

3 Facies associations are organized into a series of large-scale ($\geqslant 100$ m) coarsening and fining upwards megasequences. Fining upwards megasequences are overlain by sequences dominated by small-scale fining upwards cycles (3–20 m).

4 Coarsening upwards megasequences developed during periods of reduced source area uplift/subsidence following tectonic activity; continued aggradation in the proximal part of the basin prevented incision. Fining upwards megasequences developed during periods of increased source area uplift/relative subsidence, when gravels were restricted to the more proximal parts of the basin. Small-scale fining upwards sequences were produced by autocyclic processes during periods when sediment supply and relative subsidence were close to equilibrium. Alternative allocyclic mechanisms such as changes in macroclimate, eustasy and provenance had virtually no influence on sedimentation.

5 The limited structural data for the north Chilean Precordillera suggest that source area uplift/relative subsidence was related to inversion of fault blocks along pre-existing extensional faults as a result of the mid-Cretaceous Peruvian Orogeny.

ACKNOWLEDGEMENTS

This work was initiated through a NERC funded PhD. Thanks are due to Pete Turner, Liz Jolley, Guillermo Chong and Steven Flint for help in the field and for discussion. Thanks also to Graham Williams for discussion on structural aspects of the Chilean forearc, Jan Alexander who critically reviewed an earlier version of this manuscript and the constructive comments of the two referees P. Heller and M. John.

Note added in proof

For a more recent analysis of the tectonic setting of the Purilactis Basin, see Hartley et al. (1992).

REFERENCES

ALLEN, J.R.L. (1978) Studies in fluviatile sedimentation; an exploratory quantitative model for the architecture of avulsion-controlled alluvial suite. *Sedim. Geol.* **21**, 129–147.

ALLEN, P.A. (1981) Sediment and processes on a small stream-flow dominated Devonian alluvial fan, Shetland Islands. *Sedim. Geol.* **29**, 31–66.

BALLANCE, P.F. (1984) Sheetflow dominated gravel fans of the non-marine middle Cenozoic Simmler Formation, Central California. *Sedim. Geol.* **38**, 337–359.

BLAIR, T.C. & BILODEAU, W.C. (1988) Development of tectonic cylothems in rift, pull-apart and foreland basins: sedimentary response to episodic tectonism. *Geology* **16**, 517–520.

BRUGGEN, J. (1950) *Fundamentos de la Geologia de Chile.* Instit. Geograf. Militar., Santiago.

CHARRIER, R. & REUTTER, K.J. (1990) The Purilactis Group of northern Chile: link between arc and bak-arc during late Cretaceous and Palaeocene. In: *Abst. Int. Symp. Andean Geodynamics, Grenoble, France*, pp. 249–252. Orstom, Paris.

CHONG, G. (1977) Contributions to the knowledge of the Domeyko Range, in the Andes of northern Chile. *Geol. Rund.* **66**, 374–404.

COIRA, B., DAVIDSON, J., MPODOZIS, C. & RAMOS, V. (1982) Tectonic and magmatic evolution of the Andes of northern Argentina and Chile. *Earth Sci. Rev.* **18**, 303–332.

DINGMAN, R.J. (1967) *Geology and groundwater resources of the northern part of the Salar de Atacama, Antofagasta Province, Chile.* US Geol. Surv. Bull. 1219.

FELSCH, J. (1933) Informe preliminar sobre los reconocimientos geologicas de los yacimientos petroleros en la Cordillera en la Provincia de Antofagasta. *Minas y Petroleo Bol.* (Santiago, Chile) **3**, 411–422.

FLINT, S. & TURNER, P. (1988) Alluvial fan and fan-delta sedimentation in a forearc extensional setting: the Cretaceous Coloso Basin of northern Chile. In: *Fan Deltas: Sedimentology and Setting* (Eds Nemec, W. & Steel, R.J.) pp. 387–399. Blackie, London.

FLINT, S., HARTLEY, A.J., REX, D.C., GUISE, P. & TURNER, P. (1989) Geochronology of the Purilactis Formation, northern Chile. An insight into late Cretaceous/early Tertiary basin dynamics of the Central Andes. *Rev. Geol. Chile* **16**, 241–246.

HARDIE, L.A., SMOOT, J.P. & EUGSTER, H.P. (1978) Saline lakes and their deposits: a sedimentological approach. In: *Modern and Ancient Lake Sediments* (Eds Matter, A. & Tucker, M.E.) pp.7–41. Int. Assoc. Sediment. Spec. Publ. 2. Blackwell Scientific Publications, Oxford.

HARTLEY, A.J., FLINT, S. & TURNER, P. (1988) A proposed lithostratigraphy for the Cretaceous Purilactis Formation, Antofagasta Province, northern Chile. *Actas V Congresso Geol. Chileno, Santiago H83-99.*

HARTLEY, A.J., FLINT, S. TURNER, P. & JOLLEY, E.J. (1992) Tectonic controls on the development of a semiarid, alluvial basin as reflected in the stratigraphy of the Purilaetis group (Upper Cretaceous–Eocene), northern Chile. *J.S. Am. Earth Sci.* **5**, pp. 273–294.

HELLER, P.L. & PAOLA, C. (1989) The paradox of Lower Cretaceous gravels and the initiation of thrusting in the Sevier Orogenic Belt, Western Interior. *Bull. Geol. Soc. Am.* **101**, 864–875.

HELLER, P.L., ANGEVINE, C.L., WINSLOW, L.S. & PAOLA, C. (1988) Two-phase stratigraphic model of foreland-basin sequences. *Geology* **16**, 501–504.

HEWARD, A.P. (1978) Alluvial fan sequence and megasequence models: with examples from Westphalian D–Stephanian B coalfields, northern Spain. In: *Fluvial Sedimentology* (Ed. Miall, A.D.) pp. 669–702. Can. Soc. Petrol. Geol., Calgary, Memoir 5.

HUBERT, J.F. & HYDE, M.G. (1982) Sheet-flow deposits of graded beds and mudstones on an alluvial sandflat–playa system: Upper Triassic Blomidon redbeds, St Mary's Bay, Nova Scotia. *Sedimentology* **29**, 457–474.

JOLLEY, E.J., TURNER, P., WILLIAMS, G.D., HARTLEY, A.J. & FLINT, S. (1990) Sedimentological response of an alluvial fan system to Neogene thrust tectonics, Atacama Desert, northern Chile. *J. Geol. Soc. Lond.* **147**, 769–784.

MCKEE, E.D., CROSBY, E.J. & BERRYHILL, H.L. (1967) Flood deposits, Bijou Creek, Colorado, June 1965. *J. Sedim. Petrol.* **67**, 829–851.

MARINOVIC, N. & LAHSEN, A. (1984) Hoja Calama. Carta Geol. Chile 1 : 250 000. Serv. Nac. Geol. Min., Santiago.

PAOLA, C. (1988) Subsidence and gravel transport in alluvial basins. In: *New Perspectives in Basin Analysis* (Eds Kleinspehn, K. & Paola, C.). Springer Verlag, New York.

RICCARDI, A.C. (1987) Cretaceous palaeogeography of southern South America. *Palaeogeog. Palaeoclimatol. Palaeoecol.* **59**, 169–195.

SMITH, G.A. (1986) Coarse grained volcaniclastic sediment: terminology and depositional process. *Bull. Geol. Soc. Am.* **97**, 1–10.

SMITH, G.A. (1987) Sedimentology of volcanism-induced aggradation in fluvial basins: examples from the Pacific northwest, USA. In: *Recent Developments in Fluvial Sedimentology* (Eds Ethridge, F.G., Flores, R.M. & Harvey, M.D.) pp. 217–228. Soc. Econ. Palaeont. Miner., Tulsa, Spec. Publ. 39.

STEINMANN, G. (1929) *Geologie von Peru.* Winters, Heidelberg.

TODD, S.P. (1989) Stream driven, high-density gravelly traction carpets: possible deposits in the Traberg Conglomerate Formation, SW Ireland and some theoretical considerations on their origin. *Sedimentology* **36**, 513–530.

TUNBRIDGE, I. (1981) Sandy, high energy flood sedimentation—criteria for recognition, with an example from the Devonian of SW England. *Sedim. Geol.* **28**, 79–95.

Cyclicity in non-marine foreland-basin sedimentary fill: the Messinian conglomerate-bearing succession of the Venetian Alps (Italy)

F. MASSARI*, D. MELLERE* and C. DOGLIONI†

*Dipartimento di Geologia, Paleontologia e Geofisica,
Universita' di Padova, Via Giotto 1, Padova, Italy; and
†Dipartiments di Geologia e Geofisica, Campus Universitario, Bari, Italy

ABSTRACT

Messinian-age continental deposits in the sedimentary fill of the Venetian foreland basin show a cyclical organization. Three facies associations have been recognized: (i) alluvial fan conglomerate deposits, ranging from debris flows to shallow braided gravels and sands; (ii) channel sequences, dominated by thick sets or cosets of planar cross-bedded conglomerate, interpreted as the record of a master system of sinuous and relatively deep rivers, laterally confined by muddy banks and characterized by perennial flow and prolonged floodstages; (iii) fine-grained clastics deposited in a lacustrine basin that formed near the thrust front.

Two ranks of cyclicity have been recognized. Small-scale cycles 12–25 m thick are laterally persistent; each of them comprises a sharp-based lower member of alluvial-fan/fluvial conglomerate/sandstone and an upper fine-grained lacustrine member. These cycles have periods of less than 30 ka and are thought to record the alternating contraction and expansion of the lake system, as a response to fifth-order climatic fluctuations.

Large-scale cyclicity is expressed by unconformity-bounded megasequences hundreds of metres thick, with coarsening-upward and then fining-upward (progradational–retrogradational) trends; the younger of them are bounded by angular unconformities grading along-strike into paraconformities. The evolution of a thrust-cored anticline influenced the geometry of the unconformities: pure compression, generating a growth fold with horizontal axis, produced along-dip angular unconformities. Local transpressive motion resulted in both along-dip and along-strike angular unconformities. Eustatic modulation of large-scale cyclicity may be ruled out, due to the existence of a regional base level for continental sedimentation during the Messinian. The megasequences may record either the pulsatory growth of the thrust-cored anticline, or the interaction of relatively continuous deformation with some external factor, such as climatically modulated fourth-order cyclical changes in the regional base level.

INTRODUCTION

Jordan et al. (1988) observe that foreland basin strata may form the basis for analysing the deformational history of the mountain belt. However, they stress that, in order to examine foreland basin stratigraphy, the competing importance of thrust motion (first-order control), sea level changes, climate, bedrock lithology and autocyclic processes (second-order controls) must be considered. Due to the complexity of the interaction between different variables, it may be difficult to discern their specific roles accurately, although this task may be easier when the influence of some of the variables is known, or when some variables are known to have been virtually ineffective. Some of these conditions occur if: (i) the basin is structurally bounded and maintained at a regional base level, thus being insensitive to eustatic changes; (ii) the sedimentation pattern is known to have been primarily controlled by allocyclic factors; and (iii) source rocks are well known. The above conditions seem to be at least partly fulfilled in the studied foreland basin during the Messinian.

GEOLOGICAL AND STRATIGRAPHICAL SETTING

The Venetian Alps are part of the S-vergent thrust belt of the Southern Alps (Figs 1, 2 & 3). In a map view of the area, the overthrusts show an anastomosed pattern along strike, with a shortening of not less than 30 km (Doglioni, 1987). Local structural undulations in the general N 60–80° E structural grain of the chain are due to inherited features in the basement and Mesozoic sedimentary cover, reflecting an older (mostly Mesozoic) arrangement in almost N–S trending basins and swells (Doglioni, 1988). The influence of these inherited features on the tectonic evolution of the Venetian Alps is particularly clear at the margins of Mesozoic platform blocks (i.e. the Lessinian high and Friuli carbonate platform), where the boundary features were reactivated during the Neogene compression as transfer zones.

Fig. 1. Aerial view of eastern part of Southern Alps between Lake Garda and Tagliamento River. Main structural features may be seen in Fig. 2. Inset shows area of Fig. 3. (Photograph NASA Erts E-1218-09335-6 02, 26 February 1973.)

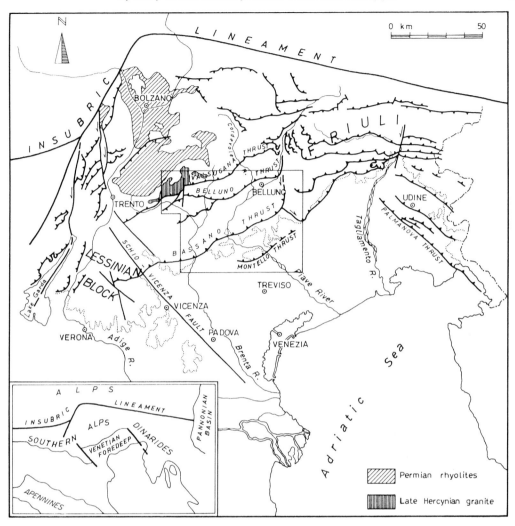

Fig. 2. Simplified structural map of eastern part of Southern Alps. Inset shows area of Fig. 3.

During Palaeogene and Early Miocene times the study area was located in the foredeep of the Dinaric thrust belt (Fig. 4) (Massari *et al.*, 1986; Doglioni & Bosellini, 1987) and underwent asymmetrical subsidence due to the load of the SW-vergent Dinaric thrust sheets. The regional NE-dipping ramp which developed at that time was inherited and involved in the SSE-vergent Neogene–Quaternary Southalpine deformation. The regional unconformity related to the main Dinaric diastrophic phase was folded and exposed at the surface by the later S-vergent Southalpine thrust belt (Fig. 4). The Venetian foredeep may consequently be defined as a composite foreland basin.

In the Venetian segment of the Alps, the geometry of the thrust belt is that of an imbricate fan of which the main elements are, in order from the internal parts to the foreland, the Valsugana, Belluno, Tezze and Bassano Thrusts (Figs 2, 3 & 5). The Neogene–Quaternary structural evolution of the thrust belt shows a general outward shift of deformation from the internal to the external overthrusts. However, the internal thrust sheets seem to have been reactivated in recent times (de Concini *et al.*, 1980; Zanferrari *et al.*, 1982; Slejko *et al.*, 1987).

The frontal part of the thrust belt is characterized by a triangle structure which generates a general S-dipping monocline (Fig. 5), characteristic of the

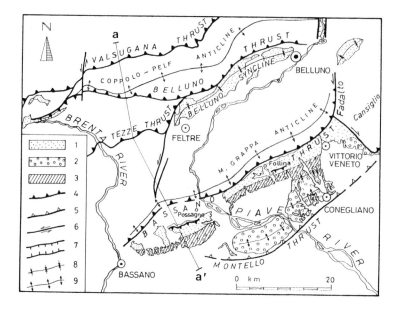

Fig. 3. Simplified geological map of Venetian Alps between Brenta River and Vittorio Veneto, with indication of main thrusts.
1, Pre-Messinian molasse deposits; 2, Montello Conglomerate; 3, study area; 4, thrust; 5, blind thrust; 6, transcurrent fault; 7, graben; 8, syncline; 9, anticline.

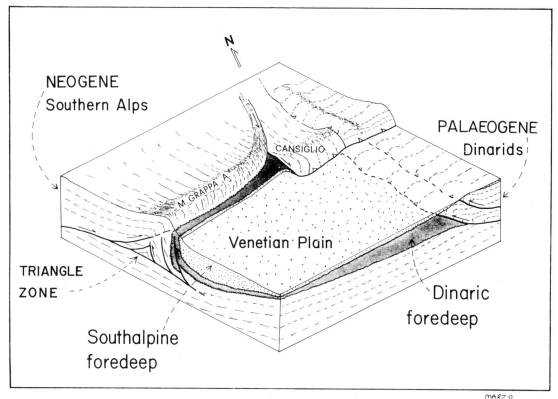

Fig. 4. Schematic block diagram (not to scale) showing how Venetian foreland basin underwent influences first of Dinaric thrusting during Palaeogene–Early Miocene and then of Southalpine thrusting during Neogene. Only the latter thrust system affected Messinian sedimentation.

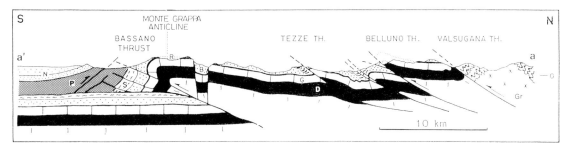

Fig. 5. Cross-section (location in Fig. 3) of Southalpine thrust belt showing general S-vergent imbricate fan shape. Note frontal triangle structure. F, Palaeozoic phyllites; Gr, Late Hercynian granite; D, Dolomia Principale (Norian); G, Calcari Grigi (Lias) and Vajont Limestone (Dogger); R, Rosso Ammonitico and Fonzaso Formation (Dogger and Malm); B, Biancone (Lower Cretaceous); S, Scaglia rossa (Upper Cretaceous); P, undifferentiated Palaeogene; N, undifferentiated Neogene and Quaternary.

Venetian foothills between Bassano and Vittorio Veneto. This triangle structure seems to be connected with a ramp-flat geometry of the deep-seated blind overthrust which generated a ramp growth fold (Monte Grappa Anticline) (Doglioni, 1990).

The beginning of thrust movements in the Venetian Alps is indicated by the rapid acceleration of subsidence in the foreland basin from the base of the Serravallian onwards (Massari et al., 1986). This event was immediately followed by the first massive appearance of extrabasinal carbonate grains in the sandstones (Stefani, 1987), which probably reflects initial deformation and uplift of the eastern Southalpine limestone thrust belt. Serravallian and younger rocks record gradual shallowing of the foreland basin (Stefanini, 1915) from upper bathyal, through prodelta and delta front, into fan delta and alluvial-fan/lacustrine depositional settings.

The Messinian deposits (Montello Conglomerate), which are the subject of this paper, show a maximum thickness of 1800 m, and an overall thickening and coarsening upward trend recording a regressive marine–non-marine transition. The parent lithologies of the conglomerates are well known in the eastern Southern Alps and mostly consist of Mesozoic limestones and dolostones. The vertical changes in composition reflect the progressive unroofing of the source areas: examples are the significant increase in the amount of Permian rhyolites in the upper part of the succession, and the late appearance of Hercynian granitic clasts (Figs 2 & 6).

The overall average rate of accumulation of Messinian sediments was at least 1.2 m/1000 years, which reflects a combination of high rates of subsidence and sediment supply. Messinian deposits show maximum thickness near the front of the mountain chain and are completely missing in the subsurface of the southern Venetian plain (ENI, 1969), suggesting strongly asymmetric subsidence. The activity of the thrust belt during the Messinian is indicated not only by this rapid subsidence, but also by the synsedimentary deformation of the inner part of the basin fill, as suggested by the presence of intraformational progressive angular unconformities (Fig. 6a,b). A typical feature of the conglomerates, believed to result from very early tectonic stresses, is the abundance of solution pits at the pebble contacts. This suggests that the Messinian clastic wedge was in an almost uncemented state during the first, synsedimentary stages of deformation. Lithification may have taken place during deformation, by precipitation in the strain-shadow areas of a calcite cement provided by pervasive pressure solution.

Deformation of the sedimentary succession continued into the Plio-Pleistocene and eventually led to the formation of a steeply dipping monocline. Due to this structural setting, the individual sedimentary bodies can be easily followed along strike, but exposure in the dip direction is very limited.

Although the non-marine deposits lack stratigraphically significant fossils, the Messinian age of the Montello Conglomerate is constrained by the unconformably overlying Upper Tabianian marine mudstones near Cornuda (Massari et al., 1976) and by the presence of a horizon of brackish-water mudstones in the lowermost part of the formation (star in Fig. 9, log CD), which contains an ostracod assemblage including forms indicative of the basal Messinian (*Neomonoceratina mouliana*, a displaced

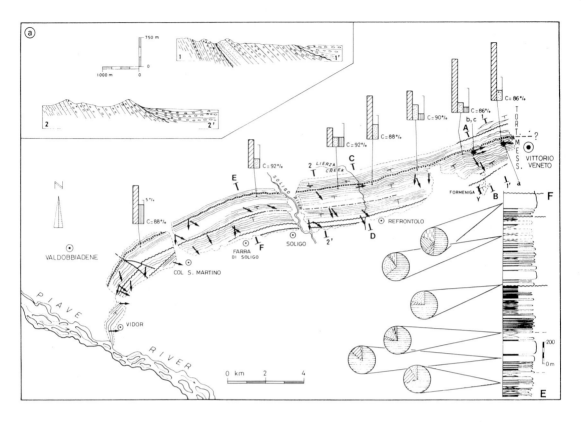
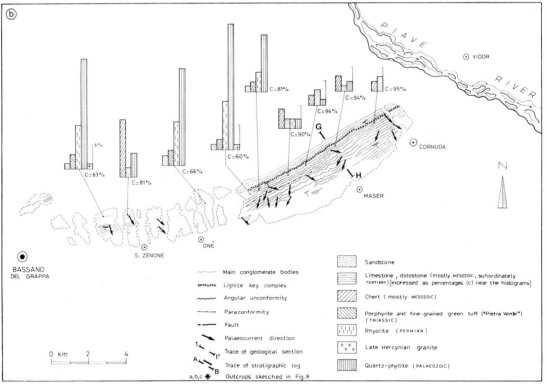

marine form, associated with autochthonous oligohaline and freshwater forms, including *Cyprideis* sp., *Paralimnocythere* sp. and rare *Hemicythere* sp.). Freshwater and terrestrial molluscs locally associated with freshwater to oligohaline ostracoda (*Candona* sp., *Iliocypris* sp.) occur in the overlying lacustrine mudstones throughout the continental succession (Wenz, 1942).

In contrast to other periadriatic basins, which were sites of evaporite deposition during the Messinian, the Venetian basin was kept free from evaporite precipitation due to the excess of freshwater runoff from the Southalpine uplands. It evolved into a freshwater basin as a result of the progressive cut-off of the connection with the Mediterranean Sea. The basin was bounded to the west by the Lessini high and to the east by the front of the Dinaric chain (Fig. 2). A connection with the sea on the southern side existed until the earliest Messinian, but the subsequent establishment of the Mediterranean salinity crisis, coupled with the northward tilting of the ramp of the Venetian foreland, led to the settling of a regional base level during most of the Messinian. As a consequence, the basin became virtually insensitive to eustatic base level variations. This may be a quite common evolutionary feature of composite foreland basins (Jordan et al., 1988).

Pliocene marine mudstones near Cornuda fill a deep ria cut into the Messinian conglomerates (Massari et al., 1976). These relationships suggest that, towards the end of the Messinian, the drastic drop in sea level known elsewhere in the Mediterranean (Ryan & Cita, 1978) caused the breakdown of the southern margin of the basin by headward fluvial erosion, with consequent incision of deep valleys into previous Messinian deposits.

Fig. 6. (See p. 506.) Schematic photogeological map of Montello Conglomerate between Vittorio Veneto and Piave River (a), and between the latter and Bassano del Grappa (b) (see Fig. 3 for location), displaying: lateral persistence of conglomerate bodies; angular unconformities (see also cross-sections) grading along strike into paraconformities; general pattern of palaeocurrents and compositional trend both laterally (histograms) and upsection (pie charts). Progressive unroofing of source areas is indicated by significant increase in amount of Permian rhyolites in upper part of succession and appearance of granitic clasts toward top. Local synsedimentary sinistral transpression, e.g. in the Piave River area between Valdobbiadene and Cornuda, is expressed by angular unconformities along both dip and strike.

DEPOSITIONAL ENVIRONMENTS AND FACIES

Facies association A: alluvial fan deposits

A1: mass-flow-dominated facies association

Facies terminology follows the code proposed by Miall (1977) and expanded by Miall (1978), Rust (1978) and Miall (1985). This facies association is typically developed in the uppermost part of the Messinian succession and is dominated by stacked sheets of clast-supported conglomerate 1.5–3 m thick, characterized by a wide range of clast sizes and a sandy to sandy–muddy matrix. This facies may be regarded as intermediate between Miall's (1985) facies Gms and Gm. The bed bases are usually planar, but locally show low-relief channelling. Outsize clasts up to 30–40 cm in diameter and large mudstone intraclasts up to 80 cm, sometimes armoured by small pebbles, are not uncommon. The fabric is disorganized, with no preferred clast orientation. Local clasts of conglomerate are evidence of cannibalism of older conglomerate deposits.

The above facies is interbedded with massive to crudely horizontally stratified beds with organized fabric (facies Gm), and rarely with lenses of fine to coarse sandstone displaying planar lamination locally marked by pebble stringers.

Interpretation

Framework-supported, poorly sorted and disorganized conglomerate is typical of low-viscosity, cohesionless debris flows. Interbedding of debris flow and stream-deposited conglomerate, as well as radial pattern of palaeocurrents (Massari et al., 1974) are consistent with deposition on an alluvial fan (Bull, 1972; Rust, 1978). Units characterized by channelled lower contacts may represent stream-flood deposits, whereas facies Gm is thought in most cases to represent longitudinal bars of shallow unconfined braided systems wandering on the fan surface during minor floods, or during the falling stage of major floods.

A2: Gm-dominated facies association

This association is dominated by massive to crudely horizontally stratified beds (facies Gm) which typically occur as stacked units (log x in Fig. 7). The Gm units are on average 70 cm thick (range: 45–

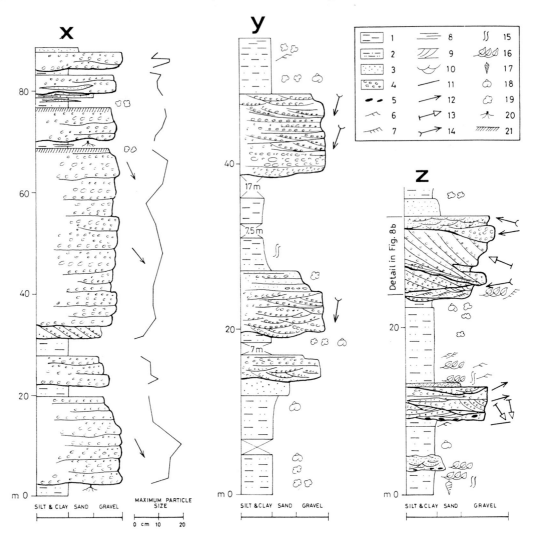

Fig. 7. Segments of stratigraphic logs showing examples of respectively Gm-dominated (x), conglomerate–sandstone (y) and Gp-dominated (z) facies associations. 1, Mudstone; 2, siltstone and sandy siltstone; 3, sandstone; 4, conglomerate; 5, mudstone clasts; 6, ripple drift cross-lamination; 7, climbing ripple lamination; 8, planar lamination; 9, planar cross-bedding; 10, trough cross-bedding; 11–14, palaeocurrent directions respectively inferred from ridges and furrows at the base of conglomerate bodies (11), pebble imbrication (12), planar cross-bedding (13) and trough cross-bedding (14); 15, bioturbation; 16, plant debris; 17, brackish-water gastropods; 18, freshwater and terrestrial gastropods; 19, small carbonate nodules; 20, root traces; 21, palaeosol. See Fig. 9 for location of x and z, Fig. 6 for y.

180 cm) and consist of clast-supported pebble or pebble–cobble conglomerate displaying moderate to good sorting and organized fabric. Horizontal stratification is distinguished by fine/coarse layers, sandstone lenses, or zones with well-developed imbrication. The lower contacts of beds range from planar to slightly irregular and channelled, and the top may be affected by local scours. Clast a-axes show a dominant mode transverse to flow (b-axes dipping upstream), locally accompanied by a secondary mode parallel to flow, mostly in small pebbles (Massari *et al.*, 1974). The palaeocurrent directions inferred from pebble imbrication are generally characterized by low variance on the outcrop scale and by radial pattern on the regional scale (Massari *et al.*, 1974) (Figs 6 & 7).

Planar- (Gp) and trough cross-bedded (Gt) gravel may occur in this facies association but is largely subordinate. Scour-fill conglomerate, sometimes with trough cross-bedding and sparse mudstone clasts, was locally observed, with the scour surface in places draped with a thin layer of sandstone. Sandstone occurs interbedded with conglomerate as minor, laterally impersistent lenses or wedges with planar lamination or trough cross-lamination, or as thin planar-laminated and/or rippled layers at the top of the Gm units.

Interpretation

Most of the Gm units probably reflect deposition by longitudinal bars in a relatively shallow flow. Isolated trough cross-sets may represent scour-fill features. Specifically, scours at the top of Gm units may reflect the incision of longitudinal bars during falling floodstages. Similarly, sandstone wedges were probably built out from bar lateral margins during falling stages or periods of low water, and horizontally bedded or rippled sandy deposits may have mantled the gravelly top of longitudinal bars. The rarity of sandstone lenses or layers reflects the poor preservation potential of sandy deposits (Kraus, 1984).

We suggest that this facies association represents the medial part of alluvial fans, characterized by a braided network of streams, with shallow unconfined or poorly confined flows activated during periodic floods; longitudinal bars were the dominant bedforms, stable at maximum floodstage. The low variance and radial pattern of palaeocurrents support this interpretation.

A3: conglomerate-sandstone facies association

Trough cross-stratified fine conglomerate (Gt), sandstone and pebbly sandstone (St) form the bulk of this facies association (y in Fig. 7). The association also includes rare sheet-like Gm units 50–100 cm thick, and abundant sandstone including horizontally laminated (Sh), planar cross-bedded (Sp) and rippled (Sr) facies, commonly rich in plant remains and mollusc fragments. Typical of this facies association is great textural and structural variety of facies, which complexly cross-cut one another. Individual sedimentation units are commonly lenticular, appearing as scour- or channel-fills. Horizontally stratified to low-angle cross-bedded pebbly sandstone, sometimes with stringers of clay chips, locally forms lenses interbedded between Gm conglomerate. Planar cross-stratified conglomerate (facies Gp) is much less common than in facies association B, and sets are never thicker than 1 m; this facies displays good sorting and thin regular foresets sometimes displaying normal grading. Lenses of mudstone locally alternating with fine sandstone may occur as the infill of shallow scours.

Interpretation

Palaeocurrent directions, finer average grain-size, and clast composition are consistent with this association being the distal equivalent of conglomerate association A2; in other words, the downstream reaches of alluvial fans. The nature of the depositional facies, lateral and vertical transitions and lithosome geometries are all indicative of braided stream deposition. Shallow flow depth may be deduced from the thinness of the inferred bar units, which appear partly eroded or with preserved bar tops, and of the planar cross-bed sets. Horizontally stratified conglomerate may reflect high stages of flow, whereas finely cross-bedded gravel and sand were probably laid down during falling floodstages (Bluck, 1979); shallow channelling followed by scour infill may have been produced under similar conditions. Discontinuous mud lenses may represent filling of abandoned channels.

**Facies association B:
Gp-dominated fluvial deposits**

This facies association (z in Fig. 7; Fig. 8) typically occurs in single-storey channelized sequences, and mainly consists of large single sets of planar cross-stratified conglomerate (Gp facies) up to 4.5 m thick (Fig. 8a) (range: 1.9–4.5 m) or cosets of compound cross-bedding (Fig. 8c). Individual sets within each coset are bounded by erosional surfaces and descend in a down-palaeoflow direction. Second-order erosional surfaces may occur within individual sets and separate bundles of foresets sometimes displaying different geometry (tangential and planar foresets) or different dip direction. Foreset dip angles may exceed 30°. Recognition of foreset bedding is made easier by the alternation of coarse- and fine-grained layers. Individual foresets range in thickness from 6 to 70 cm (average: 30 cm) and there is apparently a linear relationship between thickness and grain size. Thinner foresets are

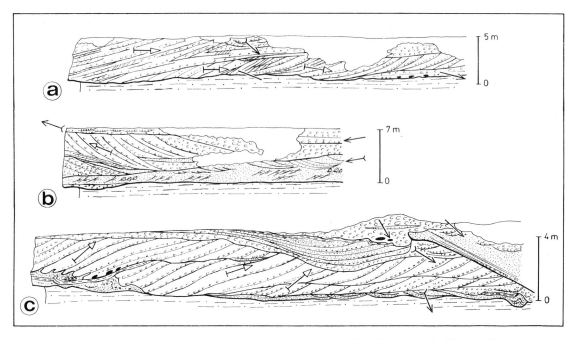

Fig. 8. Field sketches of Gp-dominated channel sequences. On left side of (c), planar cross-bedded conglomerates can be seen overlying a plastically folded lens of alternating mudstone, sandstone and discontinuous, thin, fine-grained gravel layers; a mudstone band, rich in plant remains, which drapes lower part of erosional between-set surface, appears to merge into this fine-grained lens; cut bank can be seen at extreme right. Further comments in text. For palaeocurrent symbols see Fig. 7. Location indicated in Figs 6 & 9.

very well sorted, commonly normally graded, and finer-grained (usually in the small-pebble size range), whereas thicker foresets are dominated by cobbles. In both cases, the conglomerate is generally somewhat finer-grained and texturally more mature than that forming horizontally stratified and trough cross-stratified units. The long axes of clasts are dominantly oriented parallel to the foreset dip, and may display high angles of imbrication dip. Convolutions generated either by water escape or by small-scale slumping down the slip-face are locally observed. Sandy or fine-gravelly lenticular intrasets 15–50 cm thick (Fig. 8a) with planar lamination or cross-laminae dipping in the same direction as the foreset beds rarely occur.

The poor quality of exposures only allows a limited number of observations on set orientation. However, it can be stated that the spread of foreset dip directions in the facies Gp is significantly higher than that of clast-imbrication dips in the facies Gm and sole markings at the base of the conglomerate bodies (log z in Fig. 7). The divergence may be as much as 70–80° from the average flow direction inferred from the above-mentioned low-variance indicators. In the lower part of the channel sequences, the facies Gp may directly overlie either the base of the body or a channel-floor facies represented by a discontinuous sandstone or gravel layer commonly bearing large mudstone clasts and locally displaying trough cross-bedding (Fig. 8c); mudstone bands are locally associated (Fig. 8c, lower left).

The basal contact of the channel sequence is sharply erosional and, in the case of a muddy substrate, large-scale loaded longitudinal sole markings occur; these appear as alternating ridges and furrows (Massari, 1983).

In some channel sequences (e.g. log z in Fig. 7) the palaeocurrent directions inferred from facies Gm and Gt (which usually are low-variance indicators) are subparallel to the basin axis, that is ENE–WSW; in others (e.g. Fig. 8a,c) they slightly diverge those that of alluvial-fan deposits.

In the upper part of the channel sequences the planar cross-stratified sets are commonly incised by broad, shallow channels infilled by trough cross-bedded or horizontally bedded conglomerate, pebbly sandstone and sandstone (log z in Fig. 7; Fig. 8). The top facies sometimes includes pebble or cobble sheets, which can be traced locally into the slip-face gravels of the facies Gp (Fig. 8a). The top facies is commonly coarser-grained than the underlying facies Gp; this may lead to an irregular coarsening-upward trend in the sequence.

The channel sequences are commonly characterized by multilateral organization and high lateral persistence, at least along the strike of the regional monocline.

Interpretation

The considerable scale of the cross-strata requires remarkable channel depth to produce sufficient bar-channel relief and non-ephemeral, although clearly variable, discharge (Steel & Thompson, 1983). Rust (1978) suggested that unusual magnitude of planar cross-beds requires floodstages that are deeper, more frequent and of longer duration than those of modern braid plains, to allow more stable conditions for bedform genesis. Unusual depth requires definite confinement of the stream system; in the case of Messinian channels, this is thought to have been provided by highly cohesive and probably vegetated muddy banks, as the channels are commonly incised into previously deposited lacustrine mudstone.

The relatively wide directional dispersion of planar cross-bedded gravel, when compared to the low variability of both clast imbrication in facies Gm and trough axes in facies Gt, suggests that it represents a form of in-channel variance related to differing bedform orientations (Bluck, 1976). Lateral transition from planar to tangential foresets may be due to a change in the geometry of the leeside separation flow from two-dimensional to three-dimensional, produced when bedform crestlines are skewed relative to the flow direction. We suggest that in most cases bars and riffles retained an oblique accretionary front, alternately facing opposite sides of the river. Local evidence of bar progradation into a stillwater pool (Fig. 8c, lower left) and the dissection of the upper bar surface by inferred chute channels suggest the activity of bank-attached bars developing on the inner side of channel bends and locally growing into a slough.

The major sets of planar cross-bedded gravels are interpreted either as bar tail sequences resulting from progradation of the bar-head avalanche face over the bar tail, or as bar platform structures built during the migration of a pool–riffle unit. The macroforms involved may have been of the linguoid or multiunit point bar types in the sense of Smith (1989). These point bar types were described in the River Tywi (South Wales), a flashy meandering gravel-bed river, displaying channel sinuosities of 1.5–2.2 in actively migrating reaches, and channel evolution by a combination of meander loop expansion, translation, chute cut-off and frequent avulsion (Smith, 1989). The internal structure observed in Messinian sequences seems to be particularly similar to that of multiunit point bars. Their riffles typically display lee-side high-relief avalanche faces that run obliquely across the channel. Consequently, the gravel foresets built during the migration of the pool–riffle unit are obliquely oriented with respect to the channel (Smith, 1989). Intrasets may record superimposition and downslope migration of smaller bedforms on the slip-face of the bars.

Erosional between-set surfaces dipping at low angles into the channel may represent reactivation surfaces resulting from washing out of the bars during falling stages, buried by slip-face accretion during subsequent high-stage flow.

Parallel-stratified gravel, and trough cross-bedded gravel infilling low-relief scours at the top of channel sequences may represent the bar supraplatform and reflect deposition from gravel sheets and the effect of local scouring of the bar top during major floods (chute channels?).

The fine-grained deposits occurring in places in the lower part of the sequences may represent slough-channel fills (e.g., that of Fig. 8c, lower left) or pool deposits overlain by the inclined gravel beds related to the riffle-face migration.

The assemblage of structures suggests that the channel sequences of facies association B record a major fluvial system with perennial regime and large, although variable, annual discharge. Palaeocurrent data suggest flow in axial directions (both directions, ENE and WSW, being apparently represented), as well as in the transverse direction away from the chain front. The initial downcutting stage leading to channel incision probably resulted from lowering of the base level. In addition to repeated avulsion around bars within the channel, reflected by chutes dissecting bar tops, the multilateral character of the bodies suggests frequent avulsions of the river.

Facies association C: lacustrine deposits

The fine-grained deposits of this facies association form intervals of variable thickness (from less than 1 m to several tens of metres) which alternate with conglomerate units; they show high lateral persistence and tend to wedge out gradually only in the body of major alluvial-fan complexes.

Four facies have been identified in this association.

1 Massive, either mottled or homogeneous, mudstone and muddy siltstone. Massiveness probably reflects thorough bioturbation. Colour mottling may be produced by the oxidation of original small pyritic concretions around plant remains, and stirring by burrowing. Fine shell hash and whole shells of terrestrial or freshwater molluscs are found either scattered or in local concentrations.

2 Varve-like finely laminated alternations of muddy siltstone and mudstone. Colour differences are due to various amounts of finely macerated plant debris, and to slight differences in grain size, the darker laminae being those with finer grain size and more carbonaceous material. Sparse, locally occurring fine tubes were probably produced by polychaete worms. Whole molluscan shells and fine shell hash are sparsely distributed.

3 Small-scale rippled and/or parallel-laminated, very fine-grained sandstone and siltstone, thinly interbedded with mudstone and silty mudstone. This heterolithic facies commonly shows irregularly wavy, convolute or contorted laminations and, locally, mudstone injections into sandstone. Small-scale cross-lamination in the very fine sandy or silty layers is unidirectional (current-generated), commonly of the climbing type, with set height ranging from less than 1 cm to 1.5 cm, and clay drapes locally occurring between the laminae. Variable angles of climb lead either to truncated ripple foreset cross-laminae or to complete rippleform laminae. In places, climbing ripple lamination is found on top of thicker (up to 30–40 cm) sharp-based layers of fine sand interbedded with mudstone and displaying occasional load-casting at the base, planar to convolute lamination, and rows of small clay chips. Bioturbation is sparse to common, but most of the original lamination is usually preserved. Plant fossils are well preserved on lamina surfaces. This facies displays many variations and is transitional into the other facies.

4 Bands of dark-grey to black mudstone, rich in allochthonous plant material, locally grading to true lignite seams.

All the facies within association C are blue-green to grey-green in colour and may contain ostracods and charophytic algal fragments, as well as small carbonate or pyritic concretions, commonly grown around small plant fragments. Mollusc remains are most abundant in the massive facies. Organic matter is mainly land-derived plant debris of variable size. Diffractometric analysis of the clay fraction in the Lierza section shows that trioctahedric smectites are largely dominant, whereas kaolinite sparsely occurs only in the upper part of the succession.

Interpretation

The widespread extent of this facies association, the lack of desiccation features and the rarity of rootlet beds suggest that it was deposited in a large lake complex. The alternation with alluvial-fan conglomerate suggests that the lake was relatively shallow, at least in the outcrop area of the Messinian succession. The abundance of mudstone indicates that the primary mode of sedimentation was from suspension. The varved and massive-bioturbated facies may record stages when the waters of the lake bottom were respectively anoxic and oxygenated. Silt and sand deposition may have characterized the prodelta and delta front environments. The sequence of structures of thicker sandy beds suggests waning flows and deposition from density underflows occurring during flood-related influxes of sediment.

The abrupt contacts of conglomerates on lacustrine deposits indicate that the lake retreated prior to fluvial or alluvial-fan deposition, probably as a result of significant lowering of its level ('forced regression').

The possibility that facies association C partly

Fig. 9. Simplified stratigraphic logs of Montello Conglomerate, with indication of palaeocurrents. Comparison of four sections shows important changes along strike, including major decrease in average grain size and increase in fine-member thickness from WSW to ENE. Cycles of different rank (small sequences, each consisting of a coarse/fine member couple, and unconformity-bounded megasequences) may be recognized. Thicker bodies result from amalgamation of small alluvial fan sequences. The poorly outcropping 'mudstone' intervals may actually include small-scale silty to finely sandy units as lateral (distal) equivalents of coarser units.

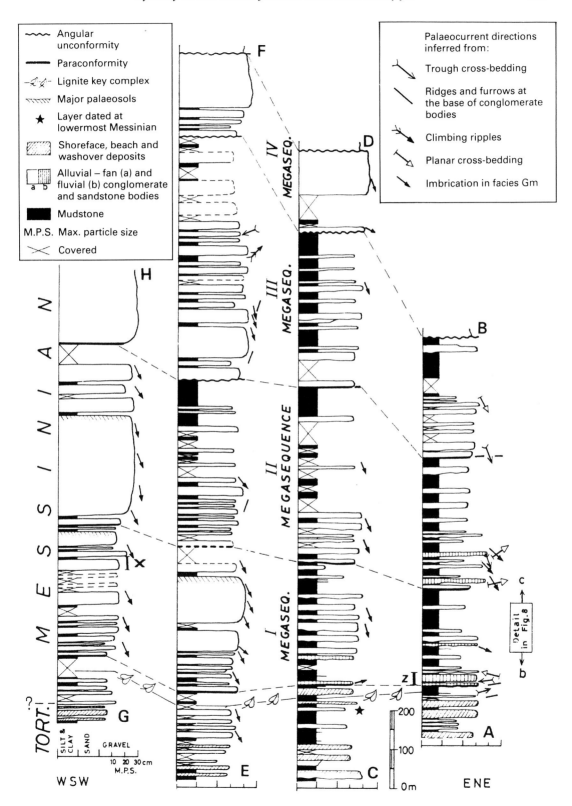

consists of overbank fines cannot be ruled out; this may be locally true where these deposits are associated with channel sequences. However, the monotony of the facies association and the thickness of the fine-grained intervals suggest that the bulk of facies association C was deposited in a large lacustrine basin.

STRATIGRAPHY OF THE MESSINIAN SUCCESSION

Four representative sections were measured, for recognition and analysis of vertical and lateral facies changes (Figs 6 & 9; locations shown in Fig. 6). The Messinian succession overlies a set of Upper Tortonian marine parasequences that show a progradational stacking pattern shallowing upwards from delta-front to fan-delta fringe settings. The lowermost Messinian deposits are represented by a number of cycles consisting of shallow-marine to brackish-water mudstone and sandstone, associated with wave-worked and/or fluvial/alluvial-fan conglomerate and sandstone; interbedded with brackish-water mudstone a number of closely spaced thin lignitic seams can be traced throughout the basin, forming a useful marker horizon (Figs 6 & 9). These lowermost deposits represent the last record of a marine influence in the Messinian succession.

The comparison of the four sections shows important along-strike changes. Channel sequences of facies association B are confined to the eastern area and occur mostly in the lower part of the continental succession. The remaining part consists of a more or less regular alternation of lacustrine deposits (facies association C) and alluvial-fan conglomerates (facies association A); the upward increasingly proximal character of the alluvial-fan deposits is particularly clear in western sections (Fig. 9). The latter (logs GH and EF in Fig. 9) are characterized by major alluvial-fan bodies, resulting from amalgamation of minor conglomerate sequences. Although the cores of these bodies are almost wholly conglomeratic, lacustrine mudstone units can be traced to fan-fringe and locally also mid-fan deposits as thin wedging-out tongues. Major fan bodies are apparently diachronous, suggesting a shifting through time of major entry points of coarse terrigenous influx into the basin (Fig. 9).

Towards the ENE, the muddy intervals increase significantly in importance, and the fan deposits themselves decrease in average thickness and grain-size, being represented by increasingly distal facies associations (logs CD and AB in Fig. 9).

The succession may be subdivided into sequences of two different ranks.

1 Coarsening and then fining upward megasequences some hundreds of metres thick, bounded either by paraconformities, or angular unconformities (mostly recognizable on aerial photographs), replaced along strike by paraconformities (Fig. 6). There are four such megasequences. The major and most mature palaeosols apparently formed near the close of coarsening-up tracts of the megasequences.
2 Small-scale sequences 12–25 m thick form the basic cyclic motif of the Messinian succession. These are mostly characterized by a sharp-based conglomerate member overlain by a mudstone member; in the lower part of the succession and rarely also in the middle-upper part, the fine-grained member may show a shallowing trend, with increasingly thicker layers of siltstone and silty to very fine sandstone interbedded with mudstone. However, in this case too, the transition to conglomerate is abrupt, i.e. marked by a sharp change in grain-size. In the case of the conglomerate member directly overlying the mudstone, its basal unit is locally represented by a planar cross-bedded channel fill.

The organization of the alluvial-fan members of the sequences is particularly significant, as a CU–FU trend in grain-sizes commonly occurs and traces of immature pedogenic modification are locally observed at the top. The topmost fine-grained gravel layers locally show thin, regular bedding and coexisting opposite imbrication dips.

INTERPRETATION OF SMALL-SCALE CYCLES

The lateral persistence of the small-scale cyclicity and the regular alternation of coarse and fine members in the Montello Conglomerate are apparent on aerial photographs. Although the muddy intercalations wedge out into the major fan bodies (Fig. 9), individual cycles can be traced along strike for several kilometres. The lateral persistence of the cycles and the abruptness of the basal contact of the conglomerate, even in the case of intervening shallowing-up prodelta–delta-front sequences, are considered good arguments for allocyclic control. Autocyclic events are not believed to be directly responsible for cyclicity, as their influence was probably limited to some internal features of the coarse members of cycles; for instance, the multilateral organization of channel conglomerates may result from repeated river avulsions; rapid or abrupt changes in grain-size within alluvial-fan bodies may reflect lateral shift of major distributaries, rapid abandonment or avulsive initiation of active fan segments, or fan-head entrenchment leading to downfan shifting of the locus of main deposition.

Due to the limited outcrop extension normal to the chain front, palaeocurrent evidence alone does not allow us to determine whether the lake was hydrologically open, with outflowing waters, or a hydrologically closed basin lacking outflow, or a basin subject to rhythmic fluctuations between open and closed conditions. Extreme water-level changes are characteristic of hydrologically closed lakes (Gore, 1989), and the abrupt contact of the conglomerate members of the cycles on the lacustrine mudstone does suggest that fluctuations in lake level were considerably severe. In addition, subsurface data (ENI, 1969) indicate that the Messinian deposits are confined to the internal part of the Venetian basin and are completely missing to the south, suggesting that sediments were trapped in the rapidly subsiding axial belt of the basin and that the southward rise of the foreland ramp (and forebulge?) formed a barrier preventing drainage into the Mediterranean. The lack of definite incised valleys infilled with Messinian sediments in the continental succession suggests that the regional base level was unaffected by the Messinian lowstands of the Mediterranean. Only at the end of the Messinian could stream equilibrium profiles become graded to a very low base level, external to the basin. In fact, near Cornuda, a valley at least 100 m deep, incised in the Messinian conglomerates, is filled with marine Pliocene mudstones (Massari et al., 1976) indicating that, prior to the Pliocene transgression, a definite lowering of the stream equilibrium profile took place. This also occurred in many other marginal areas of the Mediterranean (Rizzini & Dondi, 1978; Ryan & Cita, 1978).

The period of the short-term cyclicity cannot be determined with precision, as we do not know how much of the Messinian is represented by the sedimentary record in the study area; the contact between Messinian and Pliocene deposits is in fact erosional. The only constraining data are the earliest Messinian age of the ostracode-bearing brackish-water horizon underlying the continental succession and the Late Tabianian age of the valley-fill marine mudstones near Cornuda. The Messinian stage boundaries are not definitely dated; an age of about 6.44 Ma has been estimated for the Tortonian–Messinian boundary by Channell et al. (1990), and 4.83–4.84 and 4.93 Ma for the Messinian–Pliocene boundary by Zijderveld et al. (1986) and Channell et al. (1988) respectively. Taking into account that the Messinian succession in the study area consists of about sixty small-scale cycles (obviously recognizable only in the areas where alluvial-fan deposits are not amalgamated into major bodies), we may conclude that the average period of short-term cyclicity is less than 30 ka.

Tectonic control of small-scale cyclicity, such as variable rates of subsidence resulting from repeated motion on thrusts, seems to be unlikely, due to the high frequency and relative regularity of the cycles.

The pattern of the small-scale cycles in the Montello Conglomerate reflects numerous episodes of progradation of alluvial-fan clastics, punctuated by intervening episodes of fine-grained lacustrine deposition. Periods of lake contraction apparently alternated with periods of lake expansion. Although it is difficult to establish the role and expression of climatic fluctuations in the continental sedimentation because of the complexity of the factors which may have interacted, it is tempting to postulate that climatic changes were important in controlling the balance between chemical and physical weathering processes in the catchment area, and water and detrital influx into the basin. The predominance of Mesozoic carbonates in the catchment area of the Messinian rivers may have played an important role. A shift to a more humid climate would have favoured the development of vegetative cover (Schumm, 1977) and promoted chemical weathering of carbonates, while strongly limiting erosional and runoff processes. As a consequence, detrital influx into the basin would have been relatively low and finer-grained during such periods. Furthermore, a shift to a more humid climate would have produced an almost instantaneous response in the lake, leading to expansion of the lacustrine basin (see also Frostick & Reid, 1989), especially if the latter was hydrologically closed. On the other hand, a shift to more arid conditions would have promoted a rapid increase in drainage density, coarse sediment yield and flood magnitude (Schumm, 1977), and favoured fan progradation in concomitance with lake contraction. In reality, the response of the system to climatic changes may have been more complex than outlined above. Schumm (1977) pointed out that even in relatively small drainage basins, the presence of sufficient relief may produce a number of climatic zones within the watershed, ranging from arid or semiarid to humid or periglacial at the highest elevations. The role of periglacial processes in the production of rock fragments at the highest elevations in the catchment area cannot be entirely excluded, taking into account that, at the time of the salinity crisis, the Alpine chain could have been as high as it is today and that there are several pieces of evidence of Late Miocene cooling (Adams et al., 1977).

We conclude that we are possibly dealing with fifth-order cycles modulated by climatic fluctuations. In terms of sequence stratigraphy, the base of the conglomerate member marks a true basinward shift of the facies and may consequently be taken as representing the sequence boundary. The conglomerate member may be regarded as a small-scale lowstand wedge bounded at the base by an unconformity. The cross-bedded channel-fill locally occurring at the base of coarse members directly overlying lacustrine mudstone may reflect confined streams which cut into the exposed muddy lake margin during periods of falling lake level, preceding the unconfined flows of fan progradation. The CU–FU trend of coarse members seems to reflect fan progradation followed by abandonment, the latter locally marked by traces of pedogenic modification at the top. Thin, regular topmost beds with coexisting opposite imbrication dips suggest wave reworking during lake transgression. Overlying lacustrine mudstones with locally associated pro-delta and delta-front siltstones and fine sandstones are thought to represent transgressive and 'highstand' deposits.

The apparent lack of conglomerates obviously emplaced by subaqueous gravity flows and other coarse-grained fan-delta deposits may reflect very low gradients of the lake bottom, and rapid drops in lake level, leading to very large basinward displacement of the coastline.

INTERPRETATION OF MEGASEQUENCES

Jordan et al. (1988) point out the difficulties of attempting to define the dynamics of the interaction of thrust belts and foreland basins through interpretations of depositional facies. The overall distribution of sedimentary facies within a foreland basin reflects the interplay between tectonic subsidence, rate of sediment supply to the basin, type and relative abundance of rocks outcropping in the source area, source area relief, sea-level changes and climatic fluctuations. In the studied basin, the analysis is further complicated by the fact that three-dimensional data are not available and the geometry of major sedimentary bodies is unknown. However, the influence of some variables on the origin of the megasequences may be ruled out with sufficient confidence. As outlined in the previous section, the sedimentation pattern in the Venetian basin during most of the Messinian may have been insensitive to sea-level changes in the Mediterranean, as during this time the basin was structurally bounded and sedimentation was controlled by a regional base level. The apparent shifting in time of main entry points of coarse terrigenous influx into the basin (Fig. 9) suggests major changes in drainage organization. However, in spite of this clearly diachronous evolution, the general CU–FU trend of the megasequences can be equally recognized in different sections, suggesting that the controlling factors were external and allocyclic.

In any case, whatever the interpretation of the large-scale cyclicity, the fact that the megasequences are locally bounded by progressive and angular unconformities cannot be ignored (Riba, 1976; Anadon et al., 1986). The Monte Grappa Anticline, a ramp fold generated by a deep-seated blind overthrust, was clearly growing during the Late Miocene, because Tortonian and Messinian sediments onlap the southern limb of the anticline with a gradually smaller inclination (Massari et al., 1986). Megasequence boundaries in the southern fold limb are locally marked by angular unconformities, suggesting the coeval activity of the frontal fold. Angular along-dip unconformities in the fold forelimb occur where the fold axis is perpendicular to the regional maximum Neogene stress (sigma 1: N 20–30° W). Where there are structural undulations in the fold axis, as in the sinistral transpressive zones of Valdobbiadene–Cornuda and the greater Fadalto alignment, the unconformities are marked by angular relationships both along dip and strike (Figs 3 & 6). In other words, the influence of structural evolution on the geometry of the unconformities is evident: a growth fold with a constant horizontal axis, generated by pure compression, produces angular unconformities only along dip, while a growth fold generated by transpression produces angular unconformities both along dip and strike, as in this case strata rotate around a non-horizontal, slightly plunging axis. In addition, the latter structural setting would easily explain why syntectonic unconformities diminish laterally in magnitude and eventually die out.

Two interpretations are presented for the origin of the megasequences. The first attributes the cyclicity to pulsating tectonic episodes, the second to the interaction of some external cyclical factor with relatively continuous tectonic activity.

First hypothesis

Recent studies have concluded that flexural subsidence due to thrust-load emplacement would result in fine-grained sedimentation predominating in much of the proximal part of the basin, except in a narrow belt immediately adjacent to the thrust front (Beck & Vondra, 1985; Bilodeau & Blair, 1986; Blair & Bilodeau, 1988; Heller et al., 1988; Flemings & Jordan, 1990). Axial-fluvial systems and lacustrine or marine environments would occupy the basin margin depression, as they respond more quickly to tectonic subsidence than fans. Conglomeratic sheets prograding across basins could be primarily produced during post-tectonic lulls, when basin subsidence greatly decreases or ceases and is consequently exceeded by the rate of denudation in the source area (Blair & Bilodeau, 1988). However, the question remains controversial since, as Jordan et al. (1988) observe, depositional relief and the balance between vertical motions and denudation are dependent upon many factors (bedrock lithology, thrust motion, climate, base level). This increases the difficulty of predicting whether or not progradation will occur for any given increment of thrust thickening.

Younger megasequences are generally bounded by angular unconformities, whereas older ones are bounded by paraconformity surfaces (Fig. 9). The general trend of individual megasequences is at first coarsening and thickening upwards, and then fining and thinning upwards (the latter particularly developed in upper megasequences) (Fig. 9), pointing to progradation followed by retrogradation of the facies belt. Flemings & Jordan (1990) state that the stratigraphic record in non-marine foreland basins is characterized by a stairstepped facies package in which progradations are punctuated by retrogradations (toward the thrust), the latter occurring at the onset of thrust cycles. The upward change in character of the successive megasequences in the study area may reflect the basinward 'progradation' of the active margin, leading to increasing importance and duration of thrust-induced subsidence and increasingly angular character of the unconformities. The FU patterns may reflect accelerated subsidence due to the basinward tilting of the depositional surface produced by the growth of the thrust-cored anticline. The progressive angular unconformity should record the climax of tectonic activity accompanied by erosion of the actively rising structure (Riba, 1976). The following CU pattern may reflect the decrease and cessation of thrust-induced subsidence, allowing the deposition of a progradational clastic wedge onlapping onto the unconformity and extending into the basin. It should be noted that, in the hypothesis of pure tectonic control, unconformity-bounded CU–FU megasequences, as recognized in this study, would be out of phase with the thrust cycle, because retrogradational and progradational deposits are separated by the unconformity.

In the hypothesis that thrust movement is pulsatory rather than continuous, five tectonic pulses would have taken place in a time span of no more than 1.5 Ma.

Second hypothesis

The megasequences may result from the interaction between some external cyclical factor and a relatively continuous, rather than pulsating, tectonic activity. One major difficulty in the first hypothesis is that the average duration of tectonic cyclothems described in foreland basins ranges from 7 to 15 Ma (Blair & Bilodeau, 1988), i.e. much longer than the duration of the studied megasequences, which is of the order of a few hundred thousand years. It should be noted that the latter value may be in the range of orbital-related climatic cycles. It should also be noted in this connection that Bertolani Marchetti & Cita (1975) report remarkable variations in Messinian vegetational characteristics, suggesting climatic fluctuations. Müller (1985) presented a palaeotemperature curve for the Mediterranean, based on calcareous nannoplankton, which shows five distinct oscillations of surface water temperature in the Messinian. A sequential study of the carbonate platforms in the western Mediterranean domain led Saint-Martin & Rouchy (1990) to recognize four complexes, locally separated by unconformities, thought to be primarily controlled by eustatic fluctuations.

In this perspective, the CU–FU trends found in the studied megasequences may merely indicate the response of sedimentation to climatically modulated changes in the regional base level (fourth-order cycles?), the FU patterns reflecting long-term rising of lake level and CU patterns lowstand progradations. In the hypothesis of relatively continuous deformation, an angular unconformity would mark the discontinuity in sedimentation during the time of maximum rate of base-level fall. This interpretation is more in line with ideas concerning the interaction between tectonics and base-level changes (Vail *et al.*, 1977a,b).

Furthermore, the overall CU trend of the Messinian succession may reflect the interaction of tectonic activity (thrust 'progradation') with the persistence of a regional base level during most of the Messinian — an interpretation indirectly confirmed by the abrupt fining of grain sizes at the transition to the overlying marine Pliocene deposits.

ATTEMPT AT PALAEODRAINAGE RECONSTRUCTION

The spectrum of clast lithologies occurring in the alluvial fan/fluvial sediments indicates that most of

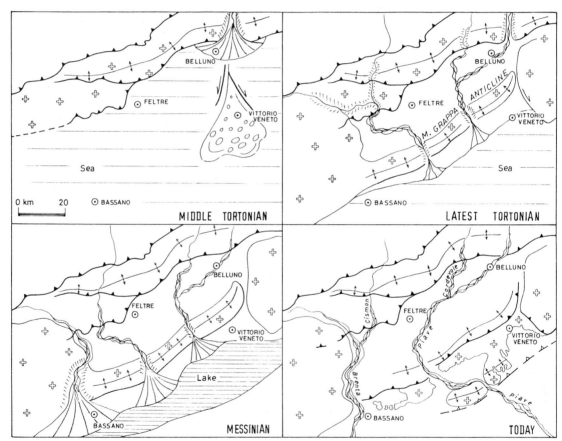

Fig. 10. Highly hypothetical reconstruction of palaeodrainage pattern from Middle Tortonian to present times in Venetian area, coevolving with active thrusting in Venetian Alps. During Middle Tortonian a fan-delta system fed by a palaeo-Piave River encroached on fault-bounded transtension-generated depression in eastern area, south of incipient Belluno thrust. Growth of thrust-cored M. Grappa Anticline coupled with transpression along former faults led to gradual narrowing and closure of this preferential pathway. Growth of M. Grappa Anticline controlled evolution of palaeodrainage pattern during Messinian times and generated intraformational unconformities. Note progressive westward activation of major entry points. Fan position has been reconstructed from palaeocurrent and compositional trends, and location of major conglomerate bodies. Crosses indicate rising areas.

the clastic sediments derived from the uplifted areas were transported by rivers having large drainage basins and flowing in transverse and longitudinal intramontane valleys (Fig. 10). Some were partly diverted parallel to the growing relief (Monte Grappa Anticline) until they could exit along a transverse structural feature. The growth of this relief caused abandonment of certain river courses, but in most cases rivers maintained their previous courses by deepening their valleys across the rising anticline (antecedent rivers). Virtually all the rivers draining the uplands and flowing out of the orogen built alluvial cones that coalesced along the active basin margin and prograded basinward. Parallel to the mountain front and the basin axis, a subsidence-induced elongated depression accommodated an axial system, represented by a large lacustrine basin and, at least in the early stages of the basin history, a master fluvial system with perennial regime. The position of most active entry points for alluvial fans may be reconstructed from palaeocurrent trends, clast composition and location of major conglomerate bodies. It is clear that the maxima of progradation are not coeval for different fans; this probably reflects drainage reorganization following the irregular growth of the structures. Particularly significant is the case of the Vittorio Veneto area: this was a differentially subsiding trough trending approximately N–S and bounded by inferred transtensional faults during the Middle Tortonian, allowing the accumulation of a gravity-emplaced complex of mass-flow conglomerates and turbidites, and the subsequent progradation of a system of Gilbert-type deltas fed by an ancestral Piave River; this structurally controlled preferred pathway was then affected by transpressive narrowing during the Messinian, which led to its almost complete abandonment (as indicated by the relatively fine-grained Messinian sedimentation in this area) and deflection of the palaeo-Piave River to the west.

ACKNOWLEDGEMENTS

This work was financially supported by the 'Centro C.N.R. per lo studio dell'orogeno delle Alpi Orientali', Padova, and by a 40% grant from the Italian Ministry of Education. We are grateful to Professor A. Bossio for determination of Messinian ostracod assemblages, to Professor D. Esu for examination of gastropod assemblages, and to Professor Gp. De Vecchi for identification of clay minerals. Critical revision of the manuscript by S.G. Crews and G. Nichols greatly improved its quality and is gratefully acknowledged. In addition, we thank F. Todesco for the careful drawings.

REFERENCES

ADAMS, C.G., BENSON, R.H., KIDD, R.B., RYAN, W.B.F. & WRIGHT, R.C. (1977) The Messinian salinity crisis and evidence of late Miocene eustatic changes in the world ocean. *Nature* **269**, 383–386.

ANADON, P., CABRERA, L., COLOMBO, F., MARZO, M. & RIBA, O. (1986) Syntectonic intraformational unconformities in alluvial fan deposits, eastern Ebro Basin margins (NE Spain). In: *Foreland Basins* (Eds Allen, P.A. & Homewood, P.) pp. 259–271. Int. Ass. Sediment. Spec. Publ. 8. Blackwell Scientific Publications, Oxford.

BECK, R.A. & VONDRA, C.F. (1985) Syntectonic sedimentation and Laramide basement thrusting. In: *Abstracts, International Symposium on Foreland Basins: Fribourg, Switzerland*, p. 36. Int. Ass. Sediment.

BERTOLANI MARCHETTI, D. & CITA, M.B., (1975) Studi sul Pliocene e sugli strati di passaggio dal Miocene al Pliocene. VII. Palynological investigations on Late Messinian sediments recorded at DSDP Site 132 (Tyrrhenian Basin) and their bearing on the deep basin desiccation model. *Riv. Ital. Paleont.* **81**, 281–308.

BILODEAU, W.L. & BLAIR, T.C. (1986) Tectonics and sedimentation: timing of tectonic events using sedimentary rocks and facies. *Geol. Soc. Am. Abstracts with Programs* **18**, 542.

BLAIR, T.C. & BILODEAU, W.L. (1988) Development of tectonic cyclothems in rift, pull-apart, and foreland basins: sedimentary response to episodic tectonism. *Geology* **16**, 517–520.

BLUCK, B.J. (1976) Sedimentation in some Scottish rivers of low sinuosity. *Trans. R. Soc. Edinburgh* **69**, 225–256.

BLUCK, B.J. (1979) Structure of coarse-grained braided stream alluvium. *Trans. R. Soc. Edinburgh* **70**, 181–221.

BULL, W.B. (1972) Recognition of alluvial-fan deposits in the stratigraphic record. In: *Recognition of Ancient Sedimentary Environments* (Eds Rigby, J.K. & Hamblin, W.K.) pp. 63–83. Soc. Econ. Paleont. Miner., Tulsa, Spec. Publ. 16.

CHANNELL, J.E.T., RIO, D. & THUNELL, R. (1988) Mio-Pliocene boundary magnetostratigraphy at Capo Spartivento (Calabria, Italy). *Geology* **16**, 1096–1099.

CHANNELL, J.E.T., RIO, D., SPROVIERI, R. & GLAÇON, G. (1990) Biomagnetostratigraphic correlations from Leg 107 in the Tyrrhenian Sea. In: *Proceedings of the Ocean Drilling Program*, Vol. 107 (Eds Kastens, K., Mascle, J. *et al.*) pp. 669–682.

CONCINI, C. DE, DE FLORENTIIS, N., GATTO, G., GATTO, G.O. & ILICETO, V. (1980) Movimenti attuali nelle Alpi Orientali rilevati mediante livellazioni ripetute. *Mem. Sci. Geol. (Padova)* **34**, 53–66.

DOGLIONI, C. (1987) Tectonics of the Dolomites (Southern Alps, northern Italy). *J. Struct. Geol.* **9**, 181–194.

DOGLIONI, C. (1988) Structure of the Venetian Southern Alps. In: *Tectonic Studies Group, 19th Annual Meeting. Abstracts Volume*. Cambridge.

DOGLIONI, C. (1990) The Venetian Alps thrust belt. In: *Thrust Tectonics* (Ed. McKlay, K.).

DOGLIONI, C. & BOSELLINI, A. (1987) Eoalpine and mesoalpine tectonics in the Southern Alps. *Geol. Rundschau* **76**, 735–754.

ENI (1969) *Enciclopedia del Petrolio e del Gas Naturale*, Vol. 6 (Ed. Colombo, C.), 1397 pp., ENI, Roma.

FLEMINGS, P.B. & JORDAN, T.E. (1990) Stratigraphic modelling of foreland basins: interpreting thrust deformation and lithosphere rheology. *Geology* **18**, 430–434.

FROSTICK, L.E. & REID, I. (1989) Climatic versus tectonic controls of fan sequences: lessons from the Dead Sea, Israel. *J. Geol. Soc. Lond.* **146**, 527–538.

GORE, P.J.W. (1989) Toward a model for open- and closed-basin deposition in ancient lacustrine sequences: the Newark Supergroup (Triassic–Jurassic), eastern North America. *Palaeogeog. Palaeoclimatol. Palaeoecol.* **70**, 29–51.

Heller, P.L., Angevine, C.L., Winslow, N.S. & Paola, C. (1988) Two-phase stratigraphic model of foreland-basin sequences. *Geology* **16**, 501–504.

JORDAN, T.E., FLEMINGS, P.B. & BEER, J.A. (1988) Dating thrust-fault activity by use of foreland-basin strata. In: *New Perspectives in Basin Analysis* (Eds Kleinspehn, K. & Paola, C.) pp. 307–330. Frontiers in Sedimentary Geology. Springer-Verlag, New York.

KRAUS, M. (1984) Sedimentology and tectonic setting of early Tertiary quartzite conglomerates, Northwest Wyoming. In: *Sedimentology of Gravels and Conglomerates* (Eds Koster, E.H. & Steel, R.J.) pp. 203–216. Can. Soc. Petrol Geol., Calgary, Memoir 10.

MASSARI, F. (1983) Tabular cross-bedding in Messinian fluvial channel conglomerates, Southern Alps, Italy. In: *Modern and Ancient Fluvial Systems* (Eds Collinson, J.D. & Lewin, J.) pp. 287–300. Int. Ass. Sediment. Spec. Publ. 6. Blackwell Scientific Publications, Oxford.

MASSARI, F., ROSSO, A. & RADICCHIO, E. (1974) Paleocorrenti e composizione dei conglomerati tortoniano-messiniani compresi tra Bassano e Vittorio Veneto. *Mem. Ist Geol. Min. Univ. Padova* **31**, 22 pp.

MASSARI, F., IACCARINO, S. & MEDIZZA, F. (1976) Depositional cycles in the Tortonian–Messinian of the Southern Alps (Italy): transition from fan-delta to alluvial fan sedimentation. In: *CNR Programma Geodinamica & UNESCO–IUGS Int. Geol. Correl. Progr., Project N.96—Messinian Correlation, Messinian Seminar No. 2, Gargnano, 1976, Field-trip Guidebook*, pp. 17–37.

MASSARI, F., GRANDESSO, P., STEFANI, C. & JOBSTRAIBIZER, P.G. (1986) A small polyhistory foreland basin evolving in a context of oblique convergence: the Venetian basin (Chattian to Recent, Southern Alps, Italy). In: *Foreland Basins* (Eds Allen, P.A. & Homewood, P.) pp. 141–168. Int. Ass. Sediment. Spec. Publ. 8. Blackwell Scientific Publications, Oxford.

MIALL, A.D. (1977) A review of braided-river depositional environment. *Earth Sci. Rev.* **13**, 1–62.

MIALL, A.D. (1978) Lithofacies and vertical profile models in braided river deposits: a summary. In: *Fluvial Sedimentology* (Ed. Miall, A.D.) pp. 597–604. Can. Soc. Petrol. Geol., Calgary, Memoir 5.

MIALL, A.D. (1985) Architectural-element analysis: a new method of facies analysis applied to fluvial deposits. *Earth Sci. Rev.* **22**, 261–308.

MÜLLER, C. (1985) Late Miocene to Recent Mediterranean biostratigraphy and paleoenvironments based on calcareous nannoplankton. In: *Geological Evolution of the Mediterranean Basin* (Eds Stanley, D.J. & Wezel, F.-C.) pp. 471–485. Springer-Verlag, New York.

RIBA, O. (1976) Syntectonic unconformities of the Alto Cardener, Spanish Pyrenees, a genetic interpretation. *Sedim. Geol.* **15**, 213–233.

RIZZINI, A. & DONDI, L. (1978) Erosional surface of Messinian age in the subsurface of the Lombardian Plain (Italy). *Mar. Geol.* **27**, 303–325.

RUST, B.R. (1978) Depositional models for braided alluvium. In: *Fluvial Sedimentology* (Ed. Miall, A.D.) pp. 605–625. Can. Soc. Petrol. Geol., Calgary, Memoir 5.

RYAN, W.B.F. & CITA, M.B. (1978) The nature and distribution of Messinian erosional surfaces — indicators of a several-kilometer-deep Mediterranean in the Miocene. *Mar. Geol.* **27**, 193–230.

SAINT-MARTIN, J.-P. & ROUCHY, J.-M. (1990) Les plates-formes carbonatées messiniennes en Méditerranée occidentale: leur importance pour la reconstitution des variations du niveau marin au Miocène terminal. *Bull. Soc. Géol. France* **8**, 83–94.

SCHUMM, S.A. (1977) *The Fluvial System*. John Wiley & Sons, New York, 338 pp.

SLEJKO, D., CARULLI, G.B., CARRARO, F., CASTALDINI, D., CAVALLIN, A., DOGLIONI, C., ILICETO, V., NICOLICH, R., REBEZ, A., SEMENZA, E., ZANFERRARI, A. & ZANOLLA, A. (1987) Modello Sismotettonico dell'Italia Nord-Orientale. Rend. n.1, Gruppo Naz. per la difesa dai terremoti. CNR., Trieste, 82 pp.

SMITH, S.A. (1989) Sedimentation in a meandering gravel-bed river: the River Tywi, South Wales. *Geol. J.* **24**, 193–204.

STEEL, R.J. & THOMPSON, D.B. (1983) Structures and textures in Triassic braided stream conglomerates ('Bunter' Pebble Beds) in the Sherwood Sandstone Group, North Staffordshire, England. *Sedimentology* **30**, 341–367.

STEFANI, C. (1987) Composition and provenance of arenites from the Chattian to Messinian clastic wedges of the Venetian foreland basin (Southern Alps, Italy). *Giornale di Geologia* (Bologna) **49**, 155–166.

STEFANINI, G. (1915) Il Neogene Veneto. Mem. Ist. Geol. Univ. Padova 3, pp. 340–662.

VAIL, P.R., MITCHUM, R.M., JR & THOMPSON III, S. (1977a) Seismic stratigraphy and global changes of seal level, part 3: relative changes of sea level from coastal onlap. In: *Seismic Stratigraphy — Applications to Hydrocarbon Exploration* (Ed. Payton, C.E.) pp. 63–82. Am. Ass. Petrol. Geol., Tulsa, Memoir 26.

VAIL, P.R., MITCHUM, R.M., JR & THOMPSON III, S. (1977b) Seismic stratigraphy and global changes of sea level. In: *Seismic Stratigraphy — Applications to Hydrocarbon Exploration* (Ed. Payton, C.E.) pp. 83–98. Am. Ass. Petrol. Geol., Tulsa, Memoir 26.

WENZ, W. (1942) Zur Kenntnis der fossilen Land- und Süsswassermollusken Venetiens. Mem. Ist. Geol. R. Univ. Padova 14, 51 pp.

ZANFERRARI, A., BOLLETTINARI, G., CAROBENE, L., CARULLI, G.B., CASTALDINI, D., CAVALLIN, A., PANIZZA, M., PELLEGRINI, G.B., PIANETTI, F. & SAURO, U. (1982) Evoluzione neotettonica dell'Italia nord-orientale. *Mem. Sci. Geol. (Padova)* **35**, 355–376.

ZIJDERVELD, J.D.A., ZACHARIASSE, J.W., VERHALLEN, P.J.J.M. & HILGEN, F.J. (1986) The age of the Miocene–Pliocene boundary. *Newsletter of Strat.* **16**, 169–181.

The impact of incipient uplift on patterns of fluvial deposition: an example from the Salt Range, Northwest Himalayan Foreland, Pakistan

T.J. MULDER* and D.W. BURBANK

Department of Geological Sciences, University of Southern California, Los Angeles, CA 90089-0740, USA

ABSTRACT

During much of the middle and late Miocene, the Northwest Himalayan foreland basin was dominated by a large, eastward flowing, axial fluvial system analogous to the modern Ganges drainage. As structural deformation encroached on the foreland during latest Miocene to Pleistocene, the foreland became increasingly partitioned. One of the earliest defined deformational events was the incipient uplift of the Salt Range along the southern margin of the Potwar Plateau. Initial deformation of the Salt Range at 5.7–5.8 Ma is manifest in the depositional record by the appearance of N–NE-flowing fan deposits that initially interfinger with and eventually replace large-scale sheet sandstone bodies of the east-flowing axial draining system. Detailed stratigraphic investigations serve to delineate the contrast between these two systems. Distinctive upward fining trends occur during this transition, and channel-belt width and depth dimensions decrease significantly as the local, smaller rivers begin to dominate the proximal depositional record. Within the temporal context defined by magnetostratigraphic data, a gradual northwards displacement of the ancestral, axial, Indus-like system in response to the initial stage of uplift can be delineated. Continuing deformation appears to have shunted the ancestral Indus system to the west and off the Potwar Plateau slightly after 5 Ma.

INTRODUCTION

Erosion of syntectonic strata, inadequate temporal control and incomplete exposures often preclude unambiguous analyses of the impact of structural events on an ancient depositional system. Recent studies in the Himalayan foreland basin of northern Pakistan have successfully dated episodes of movement along individual faults and well-defined fault zones (Burbank *et al.*, 1986; G. Johnson *et al.*, 1986; Burbank & Beck, 1989a). This success results from both the excellent exposures and the extensive suite of magnetostratigraphies (e.g. N. Johnson *et al.*, 1982; Raynolds & Johnson, 1985) that exist in the Potwar Plateau region (Fig. 1). The combination of stratigraphic indicators of deformation, such as tectonic rotations, unconformities and changes in provenance, palaeocurrents and subsidence rates (Burbank & Raynolds, 1988), with precise chronologies from previous palaeomagnetic studies, has served to delineate a detailed history of thrusting within the northwestern Himalayan foreland during the past 6 m.y. (Fig. 2).

Most of the structural events described in these previous studies have been defined through a synthesis of well-dated deformational indicators derived from numerous stratigraphic sections. Given this history of compressional deformation during the late Cenozoic in the Potwar Plateau region, it is now possible to examine in considerably more detail the sedimentological changes associated with individual thrusting events. This paper focuses on changes in patterns of fluvial deposition that were a result of initial deformation and uplift of the Salt Range. Sedimentological investigations have been carried out at four localities north and east of the present day Salt Range (Fig. 3). The results reveal distinct changes in the fluvial systems, unique to

*Present address: International Technology Corporation, Chester Towers, 11499 Chester Road, Cincinnati, OH 45246, USA.

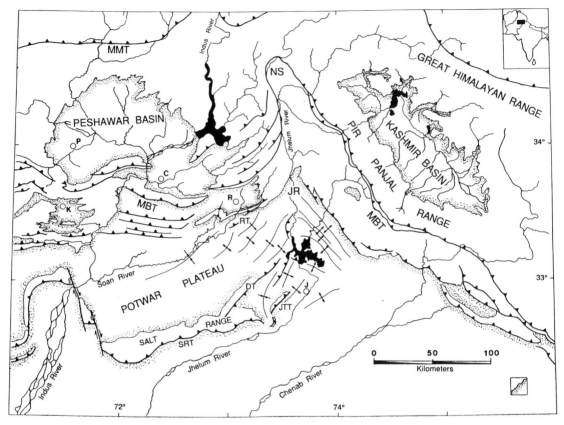

Fig. 1. Map of the northwestern portion of the Indo-Gangetic foredeep and the southern margin of the Himalaya and Hindu Kush in the vicinity of the Northwest Syntaxis (NS). The major anticlinal axes in the deformed molasse sediments, as well as major thrust faults (barbed lines) and strike-slip faults in the region surrounding the Jhelum Re-entrant (JR), are delineated. Major thrust faults: DT, Domeli Thrust; JTT, Jogi Tilla Thrust; MBT, Main Boundary Thrust; MMT, Main Mantle Thrust; SRT, Salt Range Thrust. Geographical locations: C, Campbellpore; J, Jhelum; K, Kohat; R, Rawalpindi; P, Peshawar.

each locality, that occurred during and immediately after uplift began ~5.8 Ma.

GEOLOGICAL SETTING

In northern Pakistan, the Northwest Syntaxis (Fig. 1) marks a change in the geometry of the Himalayan arc from a northwest–southeast structural trend to an east–west trend. Located at the southern end of the Syntaxis, the central and eastern Potwar Plateau lies at the distal edge of present-day Himalayan deformation and is an area containing gently folded molasse strata, ranging in age from 18 to 0 Ma (N. Johnson et al., 1982, 1985; Raynolds & Johnson, 1985).

Three aspects of the stratigraphy of the Potwar Plateau and the Salt Range are important to this study. First, a Precambrian/Cambrian evaporite overlies crystalline basement rocks and forms a weak horizon in which the décollement underlying the Potwar region rests. Second, the Palaeozoic succession includes the Permian Tobra Formation, which contains clasts of highly distinctive red–pink granites and rhyolites, known informally as Talchir clasts. The Tobra Formation crops out in the southernmost part of Domeli Ridge and in the central and western Salt Range, but not in the easternmost portion of the range (Gee, 1980; Yeats et al., 1984) or any place to the north of it (Fig. 3). Because the only known source for Talchir granitic clasts is in the Salt Range, their presence as exotic clasts within

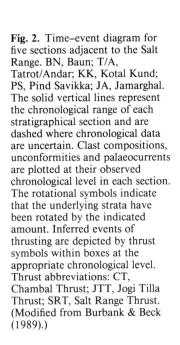

Fig. 2. Time–event diagram for five sections adjacent to the Salt Range. BN, Baun; T/A, Tatrot/Andar; KK, Kotal Kund; PS, Pind Savikka; JA, Jamarghal. The solid vertical lines represent the chronological range of each stratigraphical section and are dashed where chronological data are uncertain. Clast compositions, unconformities and palaeocurrents are plotted at their observed chronological level in each section. The rotational symbols indicate that the underlying strata have been rotated by the indicated amount. Inferred events of thrusting are depicted by thrust symbols within boxes at the appropriate chronological level. Thrust abbreviations: CT, Chambal Thrust; JTT, Jogi Tilla Thrust; SRT, Salt Range Thrust. (Modified from Burbank & Beck (1989).)

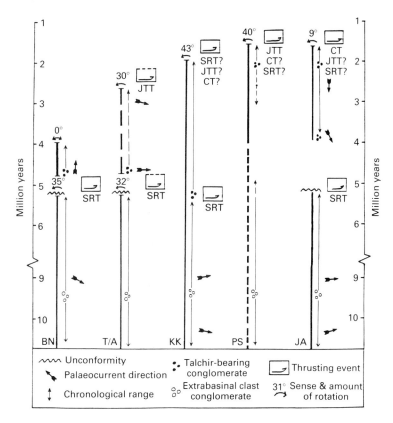

the late Cenozoic molasse strata helps to define a Salt Range source area for the enclosing beds. Third, following deposition of a varied Palaeozoic and Mesozoic sequence and an early Cenozoic carbonate sequence, the thick fluvial molasse strata of the Rawalpindi and Siwalik Groups (Fatmi, 1974) were deposited (Fig. 4). Strata of the Middle Siwalik Dhok Pathan Formation were deposited ~8–5 Ma, whereas the Upper Siwalik Soan Formation was deposited mostly during the Plio-Pleistocene. This study focuses on these two formations in the vicinity of the Salt Range. In general, strata of the Dhok Pathan Formation, like those of the underlying Chinji and Nagri Formations are thought to have been deposited by a large easterly flowing river analogous to the modern Indus River (Raynolds, 1980, 1981; Behrensmeyer & Tauxe, 1982; G. Johnson et al, 1986; Burbank & Beck, 1989a). The depositional setting of the Soan Formation is considerably more diverse. Physical disruption of the proximal foreland by progressive encroachment of structural deformation caused a partitioning of the previously monolithic basin (Raynolds & Johnson, 1985; Burbank & Raynolds, 1988). Consequently, smaller scale, locally sourced fluvial systems supplanted the previously axial system.

Structurally, the Potwar Plateau is an allochthonous sheet that has moved southward, with little internal deformation, on a low angle detachment (Seeber & Armbruster, 1979; Leathers, 1987; Baker et al., 1988; Lillie et al., 1987). The detachment ramps toward the surface near the northern margin of the Salt Range. The abnormally large width (60–100 km) of the detached, but little deformed, Potwar area is attributed to the regional extent of the underlying evaporite sequence. Total displacement (18–24 km) along the tip line of the Salt Range detachment decreases toward the east. Concomitantly, however, internal deformation of the thrust sheet increases in an eastwards direction (Fig. 3), being accommodated by closely spaced folds, thrust-cored anticlines, and both fore- and back-thrusts (Baker, 1987; Leathers, 1987; Pennock, 1988).

The temporal control for this study comes from

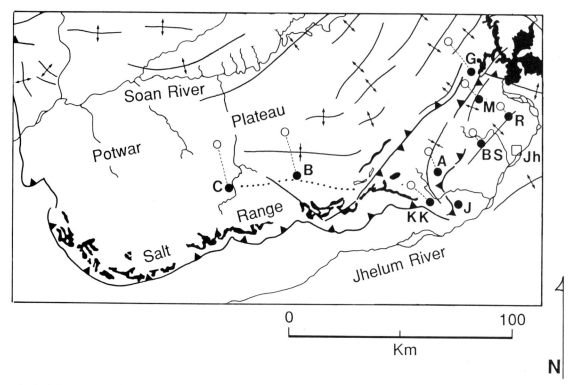

Fig. 3. Salt Range location map with relocated sight positions. (●) Present locations of magnetic polarity stratigraphies. (○) Relocated sights. Sights have been relocated with respect to a stationary datum in the Indian plate, a buried normal fault (Lillie *et al.*, 1987), represented by a dotted line. Estimates of structural shortening of the Potwar Plateau come from Baker (1987) and Pennock (1988). Shaded in black is the outcrop extent of the Permian Tobra Formation from which Talchir clasts originate. Note the lack of outcrop of Tobra Formation in the east where molasse rests unconformably on Cambrian strata. Locations of magnetic polarity stratigraphy sections studied here: B, Baun; KK, Kotal Kund; A, Andar; J, Jamarghal Kas. Locations of magnetic polarity sections studied elsewhere and referred to in text: BS, Basawa-Sangoi Kas; R, Rhotas; M, Mahesian; G, Ganda-Paik.

several (N. Johnson *et al.*, 1982) of the many published magnetic polarity stratigraphies (MPSs) that have been created in the central and eastern Potwar Plateau. These have been correlated to the magnetic polarity time scale (MPTS) through (i) recognition of distinctive reversal patterns (e.g. Opdyke *et al.*, 1979), (ii) consideration of index fossils (mammalian fauna) that have been elsewhere correlated to the MPTS (e.g. Barry *et al.*, 1982), and (iii) consideration of fission-track dates (G. Johnson *et al.*, 1982) determined for volcanic ashes present in the stratigraphic sections.

Two characteristics of the Late Cenozoic MPTS have assisted in the development of reliable correlations with the local magnetostratigraphies. First, a long, nearly unbroken interval of normal polarity spanning ~9.0–10.5 Ma and labelled variously as chron C5N, chron 11 (Berggren *et al.*, 1985) or chron 9 (LaBreque *et al.*, 1977) is a readily identifiable portion of the middle-to-late Miocene time scale. Similarly, the largely normal polarity of the Gauss chron (3.4–2.5 Ma) is also quite distinctive. Magnetic correlations in the Siwaliks have also been greatly assisted by the occasional presence of volcanic ashes. An apparently isolated ash appears near the middle of chron C5N (G. Johnson *et al.*, 1982), whereas as many as seventeen ashes are associated with the Gauss/Matuyama boundary (Frost, 1979; Raynolds, 1980; G. Johnson *et al.*, 1982). Two of the thickest and most prominent ashes in this sequence straddle this magnetic boundary.

Each of the magnetic sections used for time calibration in this paper includes near its base a long normal-polarity interval (Fig. 5) that appears correlative with magnetic chron C5N (e.g. N. Johnson *et al.*, 1982). Two of the studied sections also include

Age (Ma)	Group	Formation	Lithology	Thickness (m)
0 —	Upper Siwalik	Soan	Highly variable. Variably coloured sandstones and mudstones and conglomerates. A number of volcanic ashes.	Variable to 2000
4–5.5 —	Middle Siwalik	Dhok Pathan	Variable. White-grey to buff-brown sandstones with interbedded red-brown silts.	400 — Kotal Kund 1600 — Khaur
8 —		Nagri	White to blue-grey sandstone with subordinate red-brown silts.	400 — Kotal Kund 1300 — Khaur
10 — 14 —	Lower Siwalik	Chinji	Red-brown to bright red silts with subordinate white to grey sandstone.	500 — Tatrot 1300 — Khaur
17 —	Rawalpindi	Kamlial	Brown resistant sandstone with subordinate red-purple silts.	400+ — Kotal Kund 400 — Khaur
??? — N. Potwar Palaeocene — Murree area		Murree	Red to purple silts with brown and purple sandstone.	Absent — Kotal Kund 1300 — Khaur

Fig. 4. Siwalik stratigraphic nomenclature. Age and thickness estimates: Upper Siwaliks, Opdyke *et al.* (1979), Raynolds (1980); Middle Siwaliks and Chinji Formation, N. Johnson *et al.* (1982), Tauxe & Opdyke (1982); Kamlial Formation, Raza (1983), Fatmi (1974), our estimates; Murree Formation, Fatmi (1974), Bossart & Ottigar (1989). Age estimates for formational boundaries are approximate, and have been documented as being regionally time transgressive on the order of 10^5–10^6 years.

the Gauss chron, as evidenced by their magnetozonations and by the presence of radiometrically dated ashes (fission track method: G. Johnson *et al.*, 1982). The two sections at Tatrot-Andar Kas and Kotal Kund have been linked through tracing laterally extensive sheet sandstones between the two localities (N. Johnson *et al.*, 1982). Correlations of these MPSs with the MPTS are of high quality and allow for confidence in dating initial uplift of the Salt Range based on stratigraphic evidence. The correlations of the magnetostratigraphies used in this study with the MPTS (Fig. 5) are largely those of the original workers (N. Johnson *et al.*, 1982), with the exception of portions of the Baun and Jamarghal Kas sections. These latter correlations are tenuous in portions due to complex reversal patterns. However, the correlations presented here (Fig. 5) are most consistent with what is known of lithofacies ages in other parts of the Potwar Plateau. These chronological data have been used to constrain the timing of initiation of uplift of the Salt Range (Burbank & Raynolds, 1988; Burbank & Beck, 1989a, b).

Near Baun (Fig. 3), previous workers (N. Johnson *et al.*, 1982; Burbank & Beck, 1989a) have noted a 10° shallowing in dip and distinct sedimentological changes across the Dhok Pathan/Soan Formation boundary, which may be an unconformity. Sedimentological changes include the sudden appearance of coarse, immature sands, poorly rounded gravels with a Salt Range provenance, and a palaeocurrent change from easterly directed flow to more northerly directed flow. Magnetic data from Opdyke *et al.* (1982) define a differential tectonic rotation across this boundary, such that the stratigraphically higher Soan Formation is unrotated. The changes occurring at the Dhok Pathan/Soan boundary have been interpreted as a drainage reorganization resulting from incipient uplift of the Salt Range south of the study area (Burbank & Beck, 1989).

Although our suggested correlation of the Baun MPS with the MPTS would place this initial uplift event at ~5.4 Ma, the uncertainties in the correlation itself indicate that better constrained sections need to be used to date this event more unambiguously. The first appearance of distinctive sedimentological changes occurs at Kotal Kund at ~5.7 Ma. These distinctive changes are followed shortly by the first appearance of Talchir clasts at ~5.4 Ma.

526   T.J. Mulder and D.W. Burbank

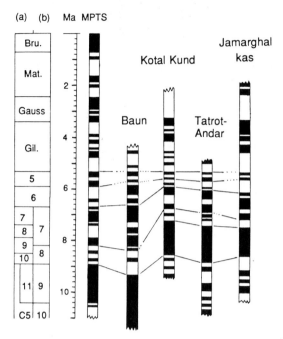

Fig. 5. Magnetic polarity stratigraphy correlation diagram for the localities relevant to this study along the flanks of the Salt Range. The correlations between the magnetostratigraphies and the magnetic polarity time scale (MPTS) are aided by (i) recognition of distinctive reversal patterns, particularly the recognition of long normal polarity zones correlative with magnetic chron 11 and the Gauss chron; (ii) faunal data; and (iii) consideration of fission-track dates on volcanic ashes associated with the Gauss/Matuyama boundary. Magnetic polarity stratigraphies are those of N. Johnson et al. (1982). Magnetic chron terminology: (a) Berggren et al. (1985); (b) Labreque et al. (1977). Dates on the MPTS are from Berggren et al. (1985). Lines of correlation between sections delineate chrons following divisions and terminology in Labreque et al. (1977).

Because the magnetic correlation for Kotal Kund (Fig. 5) is more straightforward than at Baun, it provides a more reliable minimum time constraint on the initiation of Salt Range uplift.

METHODOLOGY

At each locality initial attention was focused on examining conglomerate populations, palaeocurrents and internal and external sandbody geometries in order to identify and separate distinctly different fluvial systems. Further detailed examinations were directed at understanding the systems within the detailed temporal framework provided by correlative magnetostratigraphies.

The stratigraphic sections were plotted at a scale of 1:400 and were measured using a Jacob's staff and Abney level. Correlations of outcrop exposures to existing magnetostratigraphies were made through detailed comparisons to vertical logs and sample location maps from previously published studies. Additional use was made of original field notes and photographs provided by N. Johnson and G. Johnson.

Lateral-facies mapping (Allen, 1983; Miall, 1985) was undertaken at two different localities. Exposures are generally continuous, allowing most major sandbody margins to be examined. The weakly resistant nature of intervening fine facies makes precise lateral correlation of fine units difficult. Because many sandstones are also poorly cemented, small-scale sedimentary structures are often obscured.

Palaeocurrents were collected wherever possible during the course of measuring sections or creating lateral facies diagrams. Palaeocurrents were collected from large-scale (>0.5 m) trough crossbeds, being typically measured from axes of troughs exposed on dip surfaces. Palaeocurrents were also calculated from planar and trough crossbeds exposed in three-dimensional exposures within sandstone bodies. Where appropriate, magnetically defined tectonic rotations (Opdyke et al., 1982) were removed in order to determine palaeocurrents at the time of deposition. Through the course of this and companion studies, it was found useful to characterize depositional geometries at particular time-horizons. To do so, palaeocurrent data were collected from three to four sandstones adjacent to reversal boundaries defining the interval of interest.

Conglomerate compositions were examined using two techniques. First, point counts of ~150 clasts were made within the confines of a grid drawn on a suitable outcrop face. Second, concerted searches were made for distinctive clasts (e.g. Talchir clasts) whose presence or first appearance are important provenance indicators. We believe that this approach has successfully revealed the stratigraphically lowest occurrence of Talchir clasts within each section.

Finally, data were used to create three isochronous palaeogeographical/palaeotopographical reconstructions (7.0, 5.5 and 4.8 Ma) through consideration of chronological, palaeocurrent, petrological and sedimentological data. The data have been used to differentiate and compare the different fluvial systems present in the foredeep and to document the interactions of the fluvial systems within the basin, prior to and during disruption of the foreland.

RESULTS

Four sections are reported on here that provide insights into depositional changes resulting from initial uplift along the Salt Range décollement. The Baun, Kotal Kund and Andar Kas sections chronicle sedimentological changes north and east of the present day Salt Range. The Jamarghal Kas section is notable in that it records effects of uplift southeast of where incipient structures were growing. The Kotal Kund section provides the most detailed and well-correlated magnetostratigraphy, enabling a determination that initial Salt Range deformation began as early as 5.7 Ma (after MPTS of Berggren et al., 1985).

Lithofacies

Two groups of sandstone are readily differentiated in strata deposited between 7 and 4.5 Ma in the Salt Range area: white to light grey sheet sandstones and pale brown sandstones having comparatively reduced channel-belt dimensions of width and depth. White sandstones are distinctive from brown sandstones because of their greater bedform size, distinctive internal bedding structures, greater channel belt width, larger macroform size, higher degree of grain sorting, and distinctive pebble component.

White sandstones are a uniform colour throughout the study area, varying between light grey and white. White sandstones have relatively uniform grain size, dominantly medium grain in beds containing metre-scale cross-stratification and fine grain in planar stratified beds. Pebble conglomerates occur infrequently as thin (<20 cm) lags at the bases of erosional scours (particularly the bases of stories) and are of two types: those dominated by extraformational clasts and those dominated by intraformational clasts. Soil concretions and mudstone rip-ups constitute more than 95% of the intraformational conglomerates. Soil concretions are generally poorly rounded, and mudstone rip-ups are subangular with rounded edges. Extraformational conglomerates typically contain less than 10% intraformational clasts and are characterized by rounded to well-rounded intrusive, volcanic, metamorphic, quartzite and limestone clasts (Fig. 6). Heavy-mineral analyses of the white sandstones in the Potwar Plateau and Jhelum Re-entrant area (Raynolds, 1981; Cerveny, 1986) show these sandstones to be characterized by abundant blue-green hornblende (generally >30%). Two rivers draining the Kohistan arc terrain of the Himalaya, the modern Indus and Swat Rivers, are the only rivers flowing into the Himalayan foreland today that have similar high blue-green hornblende contents (Cerveny, 1986). This similarity, along with the scale of fluvial channels and channel belts and palaeocurrent indicators of easterly directed flow, has led to the conclusion that a large fluvial system draining the high Himalaya, the palaeo-Indus River, formerly flowed eastward across the Potwar Plateau region before entering the Ganges River drainage (Raynolds, 1981; G. Johnson et al., 1982).

White sandstones display well-developed bedding structures with large-scale (50–75 cm laminae sets) trough cross-stratification the dominant structure. Planar cross-stratification occurs less frequently and is also typically found in laminae sets of 50–75 cm height. Planar stratification is found frequently in the upper one-half of stories (after Friend, 1983).

Sandstone bodies always display sharp erosional lower contacts. Over lateral distances of 30 + m, channelling of sandstone bases more than 5 m is common. The erosional channels of some sandstone bodies are infrequently as deep as 15 m. Sandstone bodies are generally multistoreyed, with individual storeys of the order of 2–3 m thick and lateral dimensions often in excess of 50 m. Storeys are generally distinguished by their sharp erosive contacts with underlying units. Some storeys viewed in outcrops oriented perpendicular to palaeoflow can be clearly identified as channel-scour fill units. In exceptional outcrops, lateral accretion bed-sets are distinguishable. Within some storeys, sandstone- and siltstone-filled channels exhibit dimensions of the order of 3–5 m depth and 30–75 m width. Storeys observed in exceptional outcrops oriented parallel to palaeoflow directions can often be traced for distances of 50–100 m. Upper contacts of storeys or sandstone bodies with overlying silt and clay units are typically gradational. In most cases, bedding structures within the transition from sandstone to silt and clay floodplain deposits is obscured by disruption, particularly bioturbation of bedding.

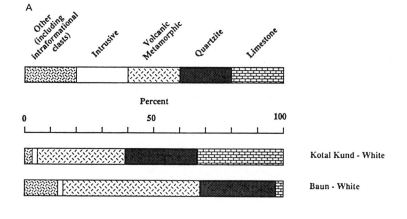

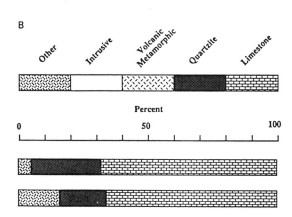

Fig. 6. Conglomerate clast compositions for pre-thrust and post-thrust molasse strata at Baun and Kotal Kund. (A) Clast count data from 'white' sandstone complexes. The category labelled as 'other' includes intraformational clasts such as carbonate soil nodules, sandstone clasts and mudstone clasts. (B) Clast count data from 'brown' sandstone complexes. Conglomerates in brown sandstone complexes are dominated by intraformational clasts (forming 90–95% of the total population). For clarity, we present here only the exotic clast compositions for 'brown' sandstones.

White sandstones display sheet morphologies and can in places be traced along outcrop for distances greater than 10 km. The average sandstone thickness varies within the study area, but averages 5–8 m. At each locality, considerable variability in sandstone thickness exists as sandstones vary from strongly multistoreyed to single-storeyed. Sandstone bodies alternate with clay and silt floodplain facies. Silt and clay units generally lack depositional structures, but do contain extensive bioturbation (e.g. root casts) and soil structure.

In contrast with white sandstones, brown sandstones display significantly greater variability of bedding characteristics within stratigraphic sections and between sections. Two distinct facies are recognized: proximal fan facies and floodplain (distal fan) facies. Sandstones of proximal fan facies show closely spaced and sharp, lateral and vertical facies transitions from sand-dominated facies to gravel-dominated facies and back. Such changes occur within individual storeys and represent discrete macroforms within an evolving channel. Sandstones within floodplain facies show comparatively less lateral and vertical variability in grain size and structure of macroforms. Characteristics of brown sandstones are discussed individually for each section studied.

Baun/Sauj Kas

Observations

A 1700 m thick section in Sauj Kas (*kas*: local term for ravine) near Baun was first described and dated by N. Johnson *et al.* (1982). The study site lies ~25 km north of the tipline of the main Salt Range thrust. The Dhok Pathan (white sandstone) portion of the section is characterized by alternating mudstones and sandstones and is typical of Dhok Pathan sections elsewhere on the southern Potwar

Plateau (Tatrot-Andar Kas, Kotal Kund and Jamarghal Kas). The magnetic polarity stratigraphy has been interpreted as spanning ~10–4 Ma.

The upper 10 m of Dhok Pathan section (Fig. 7A) at Baun consists of massive and stratified red-brown mudstone horizons interspersed with two 2 m thick palaeosols, all typical of the underlying section. The stratigraphically highest 'white sandstone' complex occurs immediately below this palaeosol complex. Directly overlying the 'white sandstone' and palae-

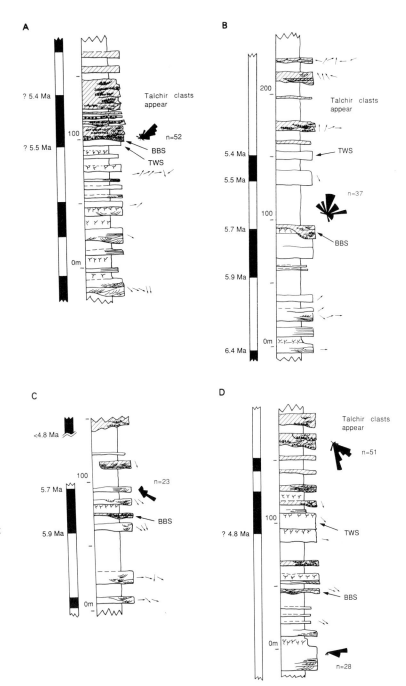

Fig. 7. Vertical stratigraphic sections from (A) Baun, (B) Kotal Kund, (C) Andar Kas and (D) Jamarghal Kas. Brown and beige coloured sands having a Salt Range source are stippled with grey pattern. First occurrences of sands having a Salt Range source (basal brown sandstone, BBS) and last occurrences of white, 'palaeo-Indus' sands (top white sandstone, TWS) are indicated. Palaeocurrents collected during measurement of the sections are presented as both individual measurements (small arrows) and summary rose diagrams.

osol complex, the basal 30 m of what have been mapped as Soan Formation strata consist of brown sandstone: coarse sand, gravelly sand and pebbly conglomerates. Palaeocurrent indicators document an eastward palaeoflow for white sandstones of the Dhok Pathan section. Palaeocurrent indicators collected from brown sandstone strata indicate a clear NNE flow direction (Figs 7 & 8A). The basal brown sand stratum rests on an erosional base which has been previously interpreted as an unconformity (Opdyke et al., 1979, 1982). This contact was traced laterally (E–W) over 2 km, and it showed no angular discordance with the underlying strata, although it did show channelling on a scale of 1–5 m depth.

Lithofacies within macroforms in the first 30 m of brown sandstones show closely spaced and sharp, lateral and vertical facies transitions from sand-dominated facies to gravel-dominated facies and back. Trough-crossbedded sets attain heights of 2 m, but heights of 50 cm are the norm. Channel and storey boundaries are difficult to observe in outcrop. The dominant macroforms observable in outcrop are cut-and-fill structures 50 cm to 2 m high and 5–20 m wide. The 30 m thick sandstone extends more than 4 km along strike (E–W) and contains very few intervals of silt and clay facies over that distance. Conglomerate compositions consist of predominantly intraformational soil nodules, mudstone clasts and sandstone clasts. A subordinate extraformational population includes Palaeocene Nummulitic limestone clasts and a trace amount of Talchir clasts; the latter's only known source lies within the Permian Tobra Formation exposed in the Salt Range (Gee, 1980). Clasts are angular to subangular and, in comparison to the well-rounded clasts of the underlying Dhok Pathan, have been transported a significantly shorter distance.

Above the basal 30 m of sand- and gravel-dominated strata is a >200 m section characterized by extensive thicknesses of massive mudstones, soil

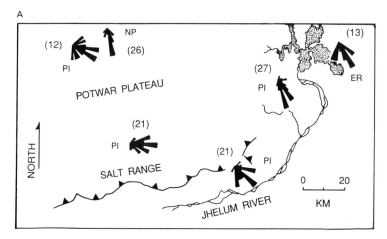

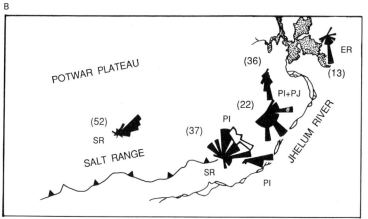

Fig. 8. Palaeocurrent data for isochrons at (A) 7.5 and (B) 5.5 Ma. Palaeocurrent measurements at each locality were collected throughout a 50–75 m interval, containing multiple sandbodies, in order to obtain representative samplings of current directions. Palaeocurrent data presented here have been corrected for later tectonic rotations (after Raynolds, 1980; Opdyke et al., 1982). Numbers in parentheses adjacent to palaeocurrent roses indicate the number of measurements for the respective rose. Palaeocurrent measurements at 7.5 Ma were taken from sandstones that represent the deposits of several fluvial systems: PI, palaeo-Indus River (master axial system); NP, northern Potwar Plateau system (transverse drainage); ER, eastern re-entrant system (transverse drainage). Palaeocurrent data at 5.5 Ma include data collected from additional fluvial systems not present in the study area at 7.5 Ma: PJ, palaeo-Jhelum River (transverse drainage with a source in the Jhelum Re-entrant); SR, Salt Range drainage.

horizons and well-stratified silt, interbedded with thin (3–5 m) fluvial bodies. The thin brown sandstones have relatively uniform grain size, ranging from medium to fine. Internal structures are difficult to distinguish due to poor exposures, but it can be discerned that grain size is relatively constant throughout outcrops and displays none of the rapid lateral and vertical changes described above. Poor exposures of this sequence have prevented the collection of palaeocurrent data from this interval.

Interpretation

The changes occurring at the Dhok Pathan/Soan boundary have previously been interpreted as an angular unconformity and drainage reorganization resulting from incipient uplift of the Salt Range south of the study area (N. Johnson *et al.*, 1982; Burbank & Beck, 1989a, b). Field observations of the boundary imply that the observed change in dip is an artefact of a shallowing in dip that occurs through that portion of the section and not an angular unconformity. The boundary is considered to be a disconformity that represents a short, but indeterminate, period. Changes (provenance, sandbody characteristics and palaeocurrents) that occur across the boundary represent a geologically instantaneous change from sedimentation by the axially oriented palaeo-Indus River to proximal fan sedimentation from fluvial systems that flowed transverse to the juvenile Salt Range. Strata that lie greater than 30 m above the boundary display higher proportions of silt relative to sand, as well as sandstones that are relatively less coarse. Strata above the proximal fan deposits appear to represent deposition within low-energy, low-gradient systems characterized by small, sluggish channels. These strata appear to represent ponded or relatively more distal facies.

Kotal Kund

Observations

The Kotal Kund section lies at the eastern termination of the Salt Range within the western limb of the Kotal Kund syncline and ~5 km from the tipline of the Salt Range thrust (Fig. 3). At Kotal Kund, a continuous section exists from the Chinji Formation through the Nagri, Dhok Pathan and Soan Formations. The MPS at Kotal Kund has been interpreted as spanning an interval from ~11 to 2 Ma. (Fig. 5). Unlike the section at Baun, the Dhok Pathan/Soan Formation boundary is indistinct and occurs over a 150 m thick interval. This zone is marked by interfingering of white sandbodies with brown-sand channel-belts of smaller dimensions that contain conglomeratic clast populations indicating a Salt Range source area. Palaeocurrent data document a change from southeasterly directed flow below the boundary to northeasterly directed flow above the boundary (Figs 7B & 8).

A panel diagram (Fig. 9) showing lateral and vertical facies variability and encompassing the interval marking the Dhok Pathan/Soan Formation boundary serves to illustrate the large-scale character of changes in fluvial style that occurred at this locality. The lowermost 120 m are marked by extensive sheets of poorly cemented white-grey sandbodies alternating with overbank silt and soil units. Channel belts, 5–15 m thick, typically consist of stacked units of individual channels with storey thicknesses of 3–5 m. Though exposure is greater than 80%, the poorly cemented sandstones create inadequate outcrops, preventing determination of lateral dimensions for individual channels. The widths of the channel belts are also poorly constrained, but are usually larger than the lateral scale of observation, i.e. >4 km.

The basal brown sandstone described here (Figs 7B & 9) lies above a well-developed soil, is laterally discontinuous, and is erosionally scoured as deeply as 10 m into the underlying strata. The well-developed soil horizon is only present where it has not been removed by channelling of the overlying sand, i.e. erosion followed soil development. Within this predominantly medium grain sandbody, there are numerous conglomeratic lags of pebble size material that are exclusively intraformational, comprising soil nodules, mudstone rip-ups and angular Siwalik sandstone clasts. Within the overlying brown sandstone, there are also extensive pebbly conglomeratic lags dominated by intraformational material, but they also contain a subordinate population (<5% of the total) of extraformational material. This extraformational material primarily contains assorted limestone, Eocene Nummulitic limestone, quartzite and pink Talchir clasts. As seen at Baun, the Talchir clasts indicate a clear Salt Range provenance.

Above the top white sandstone (Figs 7B & 9), macroforms in brown sandstones display rapid lateral and vertical facies changes, from gravels to

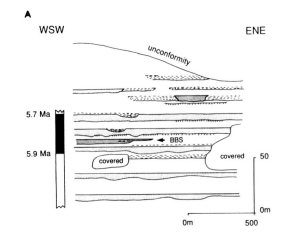
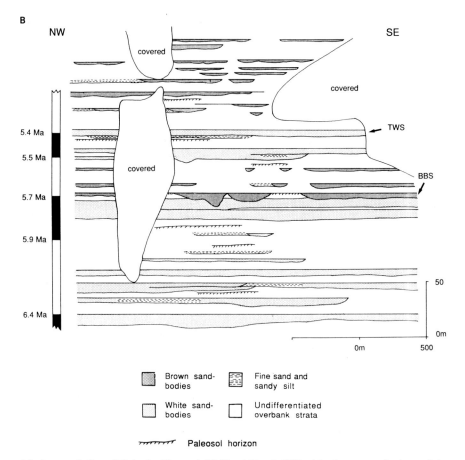

Fig. 9. Lateral facies panels from (A) Andar Kas and (B) Kotal Kund. Difficulties in accurately determining stratigraphic thicknesses along outcrop faces necessitated drawing sandbodies with horizontal tops, whereas in fact the upper surfaces undoubtedly had some topography. However, the topography of upper surfaces is comparatively less than that of sandbody bases. Difficulties were encountered in correlating the magnetostratigraphic log of N. Johnson *et al.* (1982) with exposures in the upper portion of the lateral panel constructed at Kotal Kund, and therefore the MPS for this locality is only well correlated for the lower portion. Similarly, the lateral panel at Andar Kas has been positively correlated with the magnetostratigraphy of N. Johnson *et al.* (1982) only in the lower portion. The magnetostratigraphy at Andar Kas ends at the level shown and is continued 5 km away, near the village of Tatrot. BBS, Basal brown sandstone; TWS, top white sandstone.

pebbly sands, to exclusively sands. Sandbodies are laterally discontinuous, occurring as thin channelized sheets, indicating comparatively small channel-belt proportions. The basal brown sandstone does not show closely spaced and rapid facies changes as the overlying brown sandstones do.

Interpretation

The bottom brown sandstone, though lacking extraformational clasts indicating a Salt Range provenance, is best explained as being the first expression of a new fluvial regime within the area, because (i) all other occurrences of anomalous coloured sandstones within Dhok Pathan strata on the Potwar Plateau can be attributed to fluvial systems different from the palaeo-Indus, (ii) sandstones with a definite Salt Range source make their first appearance only slightly above the bottom brown sandstone and (iii) the bottom brown sandstone and later brown sandstones have northeast-directed palaeocurrents, whereas earlier deposited white sandstones have clear east-southeast palaeocurrents. The displacement of the white sand system and the appearance of the new system is best explained as a manifestation of topographic expression of the Salt Range to the southwest. The sandstone's location straddling the reversal boundary between what is identified as magnetic chrons 5.1r and 5.2 (Berggren *et al.*, 1985) suggests that initiation of uplift within the Salt Range had occurred by ~5.7 Ma. Talchir clasts first appear 150 m above the basal brown sand. The delayed appearance of Talchir clasts corresponds well with the expected unroofing sequence of the Salt Range. The unroofing of Permian Tobra Formation strata (the source of Talchir clasts) would have required the stripping of 2–3 km of overlying strata (primarily Murree and Siwalik strata) (Leathers, 1987).

The basal brown sand is of significantly larger dimensions than later brown sands. The basal brown sand is apparently an amalgamated channel-belt, while the later brown sandstones are reduced in size and represent small channel belts or individual channelling events. Several explanations may be proposed for the difference. One explanation is that the basal brown sand is a contaminated palaeo-Indus deposit. This suggests that the white sands of the palaeo-Indus were diluted by locally high input of detritus from the juvenile Salt Range. Alternatively, the anomalously large dimensions of the basal brown sand may simply reflect a reduced subsidence rate acting during deposition of sands that are entirely derived from the juvenile Salt Range. A third explanation would be to consider the basal brown sand as representative of a depositional environment intermediate between an alluvial fan environment (e.g. the latter, smaller scale brown sands) and the large axial system north of the Salt Range.

Tatrot-Andar Kas

Observations

The Tatrot-Andar Kas section is situated 1–3 km north of the Jogi Tilla thrusts in the southeastern limb of the Kotal Kund syncline (Fig. 3). The section consists of two parts, separated by a distance of 5 km along strike. The two sections have been connected by means of tracing between the two sections a continuous sandstone bed and a specific normal polarity zone (McMurtry, 1980; N. Johnson *et al.*, 1982). Only the Andar Kas sequence was examined in detail during this study.

The Dhok Pathan Formation at Kotal Kund and Andar Kas are similar in many respects. Some channel sandstone units are traceable between the two sections (N. Johnson *et al.*, 1982). Within exposures of the Kotal Kund syncline, individual sandbodies of the Dhok Pathan lithofacies can in exceptional cases be traced laterally for 10 km.

Figure 9A is a panel diagram constructed on a sequence of outcrops oriented approximately perpendicular to palaeoflow directions for the white sand system of the Dhok Pathan. Notable in the diagram is an upward fining trend. Below the strata shown in the panel there is a thick Dhok Pathan section containing laterally extensive white sandstone channel belts. Throughout the Dhok Pathan interval, the white sandstones are characterized by channel belts ranging from 5 to 15 m in thickness and having a lateral extent of 2 to >10 km. The top of the Dhok Pathan Formation is marked by a slight unconformity at Tatrot village (Opdyke *et al.*, 1979; N. Johnson *et al.*, 1982). This unconformity appears considerably more pronounced at Andar Kas, where it cuts stratigraphically downwards across the top of the panel (Fig. 9). The strata overlying the unconformity have an apparent Salt Range source as evidenced by the presence of minor quantities of Talchir clasts near Tatrot village (Burbank & Beck, 1989a). The strata above the unconformity are normally magnetized (N. Johnson *et al.*, 1982) and likely are younger than the middle Gilbert (<4.5 Ma). Although the older strata are rotated >30° counter-clockwise (Opdyke et al., 1982), the rotation at Andar did not occur during the initial rotation of the Salt Range thrust (as it apparently did at Baun), but during the past 2 m.y. (Burbank & Raynolds, 1988; Burbank & Beck, 1989a).

Two distinctive sandbodies are present in the upper half of the panel (Fig. 9). The first (stippled sand at 65 m level) is a brown-beige sandbody which contains intraformational soil nodules and concretions within numerous channel lags and lenses, as well as scattered clasts within trough cross-bedded and parallel laminated sandstone facies. The second is a 200 m wide channel in the upper right portion of the diagram. The latter sandbody is erosionally scoured over 10 m into the underlying strata and has lacustrine clays and silts within the lowermost 2 m of channel-fill. The upper portion of the channelized unit consists of well-stratified fine–medium grain sand.

Interpretation

As at Kotal Kund, the bottom brown sandstone (Figs 7C & 9) does not contain Talchir clasts. The paucity of extraformational lithologies and abundance of soil concretions as clasts within it suggests derivation through denudation of well-developed soils in localities not far removed from the sight of deposition. The abrupt introduction of brown sandstones, their striking contrast with the white sandstones, and their similarity in appearance to the brown sandstones found at Kotal Kund and Baun suggests that these, too, resulted from initial uplift of the Salt Range and displacement of the white sandstone system to the north. Based on the magnetic correlation used here (Fig. 5), the age of the basal brown sandstone is ~5.8 Ma (Fig. 7C), an age in agreement with that determined at Kotal Kund (Fig. 7B). The pre-deformation position of Andar Kas (Fig. 3) is ~25 km northwest of the Kotal Kund site. Consequently, brown sandstones at Andar Kas are likely distal equivalents of coarse sands and gravels at the more proximal Kotal Kund location.

Jamarghal Kas

Observations

N. Johnson and others (1982) measured a 2100 m thick section along Jamarghal Kas near Jalalpur (Fig. 3), which spans a moderately dipping section containing Chinji, Nagri, Dhok Pathan and Soan Formation strata. The magnetostratigraphy (Fig. 5) contains a long normal magnetozone at the base of the section that has been correlated to magnetic chron C5N (8.9–10.4 Ma) on the basis of its distinctive reversal pattern (N. Johnson et al., 1982), and the Gauss chron (2.5–3.4) has been confidently identified in the upper part of the section (G. Johnson et al., 1982). Like the Baun section, the intervening interval contains a complex reversal pattern which yields an ambiguous correlation to the MPTS.

The Dhok Pathan portion of the stratigraphy contains channel sandstones that are similar in both internal characteristics and scale to the white sandstones at Baun, Kotal Kund and Tatrot-Andar Kas. In comparison to these other sites, however, a marked reduction in both channel sandstone frequency and the degree of sandstone amalgamation is apparent.

Soan Formation strata at Jamarghal Kas are distinctly different from those observed at any of the previously described sections. The upper 60 m of Dhok Pathan strata are marked by alternations of white sandstone bodies and smaller scale beige coloured sandstone bodies (Fig. 7D). The brown sandstones are smaller, being typically of the order of <1 km in width, as opposed to generally >4 km widths for the white sandstone system. The channelized brown sandstones are, on average, thinner than coeval white sandbodies, with typical thicknesses of 4–10 m and 8–20 m, respectively.

Above the last white sandstone, channel belts are multistoreyed, having typical thicknesses of 7–10 m (Fig. 7D). These sandstones are significantly more resistant than the underlying white and beige sandbodies. They have numerous conglomerate lags at their channel bases, and are laterally extensive, with individual sandbodies being sufficiently large to be traced obliquely to mean southeast flow directions for more than 5 km. Conglomerate clast populations have maximum clast dimensions of 5 cm, with average clasts being of the order of 1–2 cm. Intraformational material consists of soil concretions and nodules, and sandstone clasts make up from 80 to 95% of the conglomeratic material. The remaining material is dominated by limestone clasts, including Palaeocene Nummulitic limestone clasts. Trace quantities of pink Talchir clasts are present, but in such reduced quantities that extensive searching is required to find them.

Palaeocurrents collected from white sandbodies

at 5.5 Ma show an east flow direction (Figs 7D & 8). Palaeocurrents collected from brown sandbodies above the last white sandstone document a southeast flow direction.

Interpretation

Two reasonable correlations of the magnetic reversal pattern may be made for the age of the white/brown sandstone correlation. First, the two normal magnetozones (100–150 m level: Fig. 7D) may be correlated with magnetic chron 5. This correlation is not preferable, because it places the first occurrence of brown sands at ~6.0 Ma. If the displacement of the white sands and appearance of the brown sands is considered to be linked with changes at either Kotal Kund or Rhotas, a date of ~6.0 Ma is inconsistent with what is known at these other areas. Alternatively, the two normal magnetozones within the transition may be correlated with the two lowest normal polarity magnetozones in the lower half of the Gilbert (Fig. 5). This correlation is preferred, because it places the first brown sand appearance at ~5.2 Ma, a date in concordance with sedimentological changes observed at Kotal Kund, Andar Kas and Rhotas (discussed below), and because the dimensions of the brown sandbodies above the last white sandstone are considerably larger than in the other studied sections. This suggests that the brown sandstones at Jamarghal Kas may represent a younger fluvial system that has integrated flow from the Salt Range with the ancestral Jhelum River (Raynolds, 1980).

Relative to other studied sites, the reduced channel-sandstone frequency in the Jamarghal Kas section may result from its location farther to the south and, consequently, farther from the flexure-inducing load of the Himalaya, such that subsidence rates were lower (Burbank & Beck, 1989c). As a result, there may have been a lower frequency of palaeo-Indus channel migration across the Jamarghal Kas region. Such a situation has been predicted by depositional models for asymmetrically subsiding basins (Leeder & Gawthorpe, 1987). Additionally, the difficulty in matching the Jamarghal MPS with the MPTS may reflect lower subsidence rates. With slower subsidence, strata remain closer to the depositional surface for longer intervals, and as a result they are more likely to be removed by subsequent erosional events or to be magnetically overprinted due to complex pedogenic processes.

Beige sandbodies that are interfingered with white palaeo-Indus River sandbodies are attributed to fluvial systems that drained from the juvenile Salt Range, and larger brown sandbodies that occur above the last white sand are attributed to a larger fluvial system, probably the palaeo-Jhelum River. The last appearance of the white palaeo-Indus sands on the Potwar Plateau occurs in the lower Gilbert chron (Rhotas, Ganda Paik) (Raynolds, 1981). Large-scale brown channel sands, equivalent in scale to the preceding white sandstones, are present throughout the Gilbert at Jamarghal Kas and on the Rhotas and Mahesian structures (Raynolds, 1980, 1981; Burbank et al., 1986). In consideration of this last observation, thick, laterally extensive brown sands at Jamarghal Kas are most likely the deposits of a palaeo-Jhelum system, such as that seen on the Rhotas and Mahesian structures. After the palaeo-Indus was displaced to its present course west of the Potwar Plateau, the palaeo-Jhelum system of N–S drainages flowed freely across the as yet relatively undeformed eastern Potwar Plateau.

SUMMARY AND CONCLUSIONS

Along the northern flank of the Salt Range, distinct sedimentological changes have been interpreted in the past as representing incipient uplift of the Salt Range at ~5 Ma (Opdyke et al., 1979; N. Johnson et al., 1982; G. Johnson et al., 1986; Burbank & Beck, 1989). Burbank & Beck (1989a) combined data on conglomerate compositions, palaeocurrents and palaeomagnetically defined tectonic rotations in order to date and describe initial uplift of the Salt Range. Results presented here more precisely date the initiation of uplift and provide a more detailed picture of fluvial system responses to uplift.

Two distinct phases of sedimentation are observed at each of the studied localities. The older phase is characterized by white sandstones deposited by a very large, easterly flowing, axial fluvial system. Channel-belt sandstones are often >10 m thick and have lateral dimensions exceeding several kilometres (Fig. 9). Some channel-belt sandbodies can be traced laterally for 10 km or more. The predominance of blue-green hornblende in the heavy-mineral suite and the 'Himalayan' lithologies found in conglomerates suggests deposition from an ancestral Indus River system, because this is the only *major* river system in the Himalaya today that carries similar abundances of blue-green hornblende (Cerveny, 1986). The dimensions of the white sandstone system in the southern Potwar region during Dhok Pathan deposition are comparable to those defined for slightly older strata (~8 Ma) in the north-central Potwar area near Khaur (Behrensmeyer & Tauxe, 1982), where a palaeo-Indus River is also suggested as the depositional agent for extensive sheet sandstones. Syntheses of palaeocurrent data from isochronous horizons across the southern and central Potwar Plateau (Fig. 8) indicate that the white sandstones deposited ~7.5 Ma were part of an ESE-flowing fluvial system. In conjunction with other palaeocurrent data from the Jhelum Re-entrant (Raynolds, 1980, 1981), this finding supports the contention that prior to ~5 Ma the ancestral Indus River flowed eastwards across the Potwar Plateau region and joined the Ganges River system farther east.

Results at Baun and Kotal Kund suggest that, following incipient motion on the Salt Range thrust, a north-northeast oriented drainage network developed on the northern side of an eroding hangingwall block in the vicinity of the modern day Salt Range. The magnitude of the slip needed only to be large enough to bring Permian Talchir clasts to the erosional surface, i.e. 4–5 km. Data from the Soan syncline to the northeast (Raynolds & Johnson, 1985) indicate that fluvial molasse sediments can be stripped very rapidly (mean rates >15 mm yr^{-1}) from uplifting surfaces. Consequently, if it is assumed that erosion nearly kept pace with the rate of uplift, the amount of relief developed within the Salt Range area due to 4–5 km of shortening was probably very subdued. It was sufficient, however, to cause a reorganization of the pre-existing drainage network (Fig. 8), to deflect the ancestral Indus River into a more northerly position, and to create a new source area with distinctive lithologies to the south (Fig. 10).

The younger brown sandstones with their abundant soil concretion clasts and variable amounts of Talchir clasts represent the drainage system that formed along the southern edge of the newly created piggy-back basin (Ori & Friend, 1984) during deposition of the Soan Formation. Through extensive field observations, a stratigraphically lower first appearance has been determined for Talchir clasts at Kotal Kund. Additionally, previously unrecognized differences in sandbody geometries have been used to delineate changes in depositional geometries whose patterns can only be explained as reflecting the replacement of a large fluvial system with smaller fluvial systems. Whereas a younger date of 5.0–5.4 Ma (first appearance of Talchir clasts) had previously been taken to mark the inception of Salt Range uplift (Burbank & Raynolds, 1988; Burbank & Beck, 1989), the earliest brown sandstones also reflect the same tectonic event, but at an earlier stage, probably before Talchir source beds were brought to the surface. Based on the age of the brown sandstones, the initial Salt Range thrusting should, therefore, be redefined as beginning ~5.7–5.8 Ma. Talchir clasts appear in strata at Baun coincident with the first appearance of brown sandstones. No lag in the appearance of Talchir clasts is seen, because either the tectonism at the inception of Salt Range thrusting was of a configuration such that fluvial systems draining the Salt Range did not initially extend far north or the unconformity beneath the basal brown sandstone removed a significant sequence of strata.

The geometries and distributions of the brown sandbodies show dramatic contrasts with the white sandstone system. The average channel-belt depth decreased ~50% in the brown sandstones, and the mean channel-belt width decreased three- to tenfold. Additionally, dimensions of individual channels decreased dramatically in brown sandbodies. In comparison to the white sandstone system, the abundance of sand lithofacies (relative to overbank lithofacies) also decreased significantly during deposition of the brown sandstones (Fig. 9). Generally slower rates of deposition, higher percentages of unchannelized and overbank fine-grained sediments and increased pedogenesis characterize the brown sand system.

Although there are a few strong differences visible in the white sandstone facies at the various localities, the brown sandstone system shows some distinctive variability that appears to be related to proximity to the uplifted source area. Baun and Kotal Kund are in the most proximal positions relative to the Salt Range thrust, whereas Andar Kas is intermediate and Jamarghal Kas is in a more distal position. The angularity of the clasts, the percentage and size of the conglomerates and the complexity of the channel geometries are all greater in the more proximal settings. The first appearance of brown sandstones at Andar Kas and the first small channels at Jamarghal Kas likely record a depositional history of small, axially oriented fluvial systems flanking the Salt Range. These systems would presumably have been distal counterparts of gravel-dominated sheet-flood systems, such as those present at Baun and Kotal Kund. The inception of brown sandstone deposition at Jamarghal Kas apparently occurs ~0.5 m.y. later than it does at the

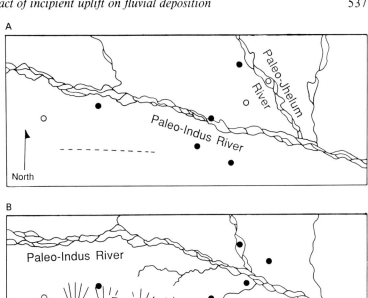

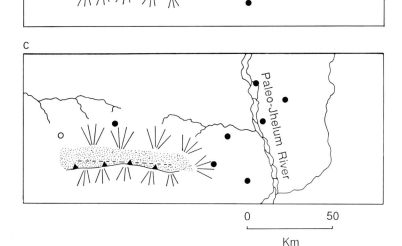

Fig. 10. Palaeogeographical reconstructions for (A) 6.0, (B) 5.2 and (C) 4.8 Ma. (---) Position of the basement normal fault used as a reference frame. (A) Prior to initiation of deformation in the Salt Range, the palaeo-Indus system migrated (north–south) freely across the Potwar Plateau region while flowing ESE across the region to the Ganges River system. (B) Subsequent to initial uplift in the vicinity of the normal fault that apparently localized ramping of the Salt Range décollement, small drainages developed and shed sediments onto fan systems which developed along the flanks of the range. The palaeo-Indus River was confined between transverse drainages to the north and the Salt Range to the south. Data from the MPS sections on the Rhotas anticline indicate that white sands of the 'palaeo-Indus' were deposited throughout the period of fluvial system interfingering along the flanks of the Salt Range. Deposits at Andar Kas and Jamarghal Kas are interpreted as being distal equivalents to fan systems observed at Kotal Kund and Baun. (C) Continued deformation of the Potwar allochthon shunted the palaeo-Indus River off the Potwar Plateau to its present southerly flowing course west of the study area. The palaeo-Jhelum River subsequently flowed freely across the eastern Potwar Plateau.

three more proximal sites. This suggests that the displacement of the ancestral Indus system to the north was accomplished through a gradual shifting of the drainage axis to the north prior to the eventual permanent shunting of the palaeo-Indus system off the Potwar Plateau. This interpretation is reinforced by the stratigraphy at Kotal Kund, where white and brown sandbodies interfinger for ~0.3 m.y. (Figs 7B & 9) before the white sand system is finally displaced.

Lithofacies relationships in sections (Raynolds, 1980) on the Rhotas and Mahesian structures suggest that the palaeo-Indus system (as represented by the white sand system) persisted in the Potwar Plateau region, north of the present day Salt Range, as late as the early (4.8–5.3 Ma) Gilbert magnetic chron (Fig. 5). If the palaeo-Indus debouched into the foredeep in the vicinity of the Kohat Plateau to the west, as suggested by Johnson et al. (1985), then the implication is that deformation and uplift within the Potwar allochthon, during its initial existence from ~5.8–5.0 Ma, was insufficient to redirect the palaeo-Indus to its present course west and south of the Potwar Plateau. Data described from Kotal Kund, Andar Kas and Jamarghal Kas reinforce this conclusion, documenting interfingering between facies having a Salt Range provenance and white sandstones having almost identical and sedimentological and petrological characteristics (Raynolds, 1980; Cerveny, 1986) to those observed at Rhotas and Mahesian. Large-scale brown sandbodies at Jamarghal appear after a period of interfingering between white sandbodies and small-scale beige bodies. These large sandbodies likely are deposits of the palaeo-Jhelum River system which was confined to the easternmost portion of the Potwar Plateau, in contrast to the white sand system which traversed the Potwar Plateau (Raynolds, 1980). Following the tectonic shunting of the palaeo-Indus River off the Potwar Plateau, the palaeo-Jhelum River would have been free to migrate across the undeformed eastern Potwar region, giving rise to the large sandbodies observed at Jamarghal Kas.

Because of the efficient and very gently sloping detachment underlying the Potwar Plateau, there was little deformation in the allochthon during southward transport. This permitted large-scale fluvial systems to persist within the newly formed piggy-back basin during its development. Also because there was little structural closure to the east and the west, largely unconfined longitudinal systems could still traverse the basin. This stands in contrast to the ponded drainage pattern predicted for structurally closed basins (Burbank & Raynolds, 1988) and observed in thrust-bounded intermontane basins like the Kashmir basin (Burbank & Johnson, 1983). Eventually, however, continued structural disruption of the foreland shunted the Indus River to the west and focused the Jhelum River along the structural re-entrant of the Northwest Syntaxis.

ACKNOWLEDGEMENTS

Financial support for this research was provided by National Geographic Field Research Grant 3677-87 and NSF Grant EAR-8720970 to D.W. Burbank. Additional financial support was provided in the form of a Chevron Fellowship, USC Graduate Student Research Award and USC Graduate Student Travel Fund Award to T.J. Mulder. Richard Beck is gratefully acknowledged for his assistance in providing moral and logistical support in Pakistan and at USC. Additional logistical support was provided by the National Center for Excellence in Geology at Peshawar University, particularly Drs Javed Khan, Rashid A.K. Tahirkheli and Qasim Jan.

REFERENCES

ALLEN, J.R.L. (1983) Studies in fluvial sedimentation: bars, bar complexes, and sandstone sheets (low sinuosity braided streams) in the Brownstones (L. Devonian), Welsh Borders. *Sedim. Geol.* **33**, 237–242.

BAKER, D.M. (1987) *Balanced structural cross-sections of the central Salt Range and Potwar Plateau of Pakistan: shortening and overthrust deformation.* MS thesis, Oregon State University.

BAKER, D.M., LILLIE, R., YEATS, R., JOHNSON, G.D., YOUSUF, M. & ZAMIN, A. (1988) Development of the Himalayan frontal thrust zone; Salt Range, Pakistan. *Geology* **16**, 3–7.

BARRY, J.C., LINDSAY, E.H. & JACOBS, L.L. (1982) A biostratigraphic zonation of the middle and upper Siwaliks of the Potwar Plateau of northern Pakistan. *Palaeogeog. Palaeoclimatol. Palaeoecol.* **37**, 95–130.

BEHRENSMEYER, A.K. & TAUXE, L. (1982) Isochronous fluvial systems in Miocene deposits of northern Pakistan. *Sedimentology* **29**, 331–352.

BERGGREN, W.A., KENT, D.V., FLYNN, J.J. & VAN COUVERING, J.A. (1985) Cenozoic geochronology. *Bull. Geol. Soc. Am.* **96**, 1407–1418.

BOSSART, P. & OTTIGER, R. (1989) Rocks of the Murree Formation in northern Pakistan: indicators of a descending foreland basin of late Paleocene to middle Eocene age: *Ecolgae Geologicae Helvetica* **82**, 133–165.

BURBANK, D.W. & BECK, R.A. (1989a) Early Pliocene uplift of the Salt Range; temporal constraints on thrust wedge development, northwest Himalaya, Pakistan. In: *Tectonics and Geophysics of the Western Himalaya* (Eds Malinconico, L.L. & Lillie, R.J.). Geological Society of America, Special Paper 232.

BURBANK, D.W. & BECK, R.A. (1989b) Comment to Baker *et al.*, Development of the Himalayan frontal thrust zone, Salt Range, Pakistan. *Geology* **17**, 378–380.

BURBANK, D.W. & BECK, R.A. (1989c) Synchronous sediment accumulation, decompaction, and subsidence in the Miocene foreland basin of northern Pakistan. *Geol. Bull. University of Peshawar* **22**, 11–24.

BURBANK, D.W. & JOHNSON, G.D. (1983) The Late Cenozoic chronologic and stratigraphic development of the Kashmir intermontane basin, northwestern Himalaya. *Palaeogeog. Palaeoclimatol. Palaeoecol.* **43**, 205–235.

BURBANK, D.W. & RAYNOLDS, R.G.H. (1988) Stratigraphic keys to the timing of deformation: an example from the northwest Himalayan foredeep. In: *New Perspectives in Basin Analysis* (Eds Kleinspehn, K. & Paola, C.) pp. 331–351. Springer-Verlag, New York.

BURBANK, D.W., RAYNOLDS, R.G.H. & JOHNSON, G.D. (1986) Late Cenozoic tectonics and sedimentation in the northwestern Himalayan foredeep, II. Eastern limb of the northwest syntaxis and regional synthesis. In: *Foreland Basins* (Eds Allen, P. & Homewood, P.) pp. 293–306. Int. Ass. Sediment. Spec. Publ. 8.

CERVENY, P.F. (1986) *Uplift and erosion of the Himalaya over the past 18 million years; evidence from fission-track dating of detrital zircons and heavy mineral analysis.* MS thesis, Dartmouth College.

FATMI, A.N. (1974) *Lithostratigraphic units of the Kohat-Potwar province, Indus basin, Pakistan.* Geological Survey of Pakistan Memoir 10.

FRIEND, P.F. (1983) Toward a field classification of alluvial architecture or sequence. In: *Modern and Ancient Fluvial Systems* (Eds Collinson, J.D. & Lewin, A.J.) pp. 345–354. Spec. Publs Int. Ass. Sediment. 6.

FROST, C.D. (1979) *Geochronology and depositional environment of a Late Pliocene age Siwalik sequence enclosing several volcanic tuff horizons, Pind Savikka area eastern Salt Range, Pakistan.* AB thesis, Dartmouth College.

GEE, E.R. (1980) Pakistan geological Salt Range series, 1:50 000, 6 sheets. Geological Survey of Pakistan, Quetta.

JOHNSON, G.D., RAYNOLDS, R.G. & BURBANK, D.W. (1986) Late Cenozoic tectonics and sedimentation in the northwestern Himalayan foredeep, I. Thrust ramping and associated deformation in the Potwar region. In: *Foreland Basins* (Eds Allen, P. & Homewood, P.) pp. 237–291. Spec. Publs Int. Ass. Sediment. 8.

JOHNSON, N.M., OPDYKE, N.D., Johnson, G.D., LINDSAY, E. & TAHIRKHELI, R.A.K. (1982) Magnetic polarity stratigraphy and ages of the Siwalik group rocks of the Potwar Plateau, Pakistan. *Palaeogeog. Palaeoecol. Palaeoclimatol.* 37, 17–42.

JOHNSON, N.M., STIX, J., TAUXE, L., CERVENY, P.F. & TAHIRKHELI, R.A.K. (1985) Paleomagnetic chronology, fluvial processes and tectonic implications of the Siwalik deposits near Chinji village, Pakistan. *J. Geol.* 93, 27–40.

JOHNSON, G.D., ZEITLER, P., NAESER, C.W., JOHNSON, N.M., SUMMERS, D.M., FROST, C.D., OPDYKE, N.D. & TAHIRKHELI, R.A.K. (1982) The occurrence and fission-track ages of Late Neogene and Quaternary volcanic sediments, Siwalik Group, Northern Pakistan. *Palaeogeog. Palaeoclimatol. Palaeoecol.* 37, 63–93.

LABREQUE, J.L., KENT, D.V. & CANDE, S.C. (1977) Revised magnetic-polarity time scale for Late Cretaceous and Cenozoic time. *Geology* 5, 330–335.

LEATHERS, M. (1987) *Balanced structural cross section of the western Salt Range and Potwar plateau: deformation near the strike-slip terminus of an overthrust sheet.* MS thesis, Oregon State University.

LEEDER, M.R. & GAWTHORPE, R.L. (1987) Sedimentary models for extensional tilt block/half graben basins. In: *Continental Extensional Tectonics* (Eds Coward, M.P., Dewey, J.F. & Hancock, P.L.) pp. 139–152. Geol. Soc., Lond., Spec. Publ. 28.

LILLIE, R.J., JOHNSON, G.D., YOUSUF, M., ZAMIN, A.S.H. & YEATS, R.S. (1987) Structural development within the Himalayan foreland fold and thrust belt of Pakistan. In: *Sedimentary Basins and Basin Forming Mechanisms* (Eds Beaumont, C. & Tankard, A.J.) pp. 379–392. Can. Soc. Petrol. Geol., Calgary, Memoir 12.

MCMURTRY, M.G. (1980) *Facies changes and time relationships along a sandstone stratum, Middle Siwalik Group, Potwar Plateau, Pakistan.* AB thesis, Dartmouth College.

MIALL, A.D. (1985) Architectural element analysis: a new method of facies analysis applied to fluvial deposits. *Earth Planet. Sci. Rev.* 22, 261–308.

OPDYKE, N.D., JOHNSON, N.M., JOHNSON, G.D., LINDSAY, E.H. & TAHIRKHELI, R.A.K. (1982) Paleomagnetism of the Middle Siwalik Formations of Northern Pakistan and rotation of the Salt Range Decollement. *Palaeogeog. Palaeoclimatol. Palaeoecol.* 37, 1–15.

OPDYKE, N.D., LINDSAY, E.H., JOHNSON, G.D., JOHNSON, N.M., TAHIRKHELI, R.A.K. & MIZRA, M.A. (1979) Magnetic polarity stratigraphy and vertebrate paleontology of the Upper Siwalik Subgroup of northern Pakistan. *Palaeogeog. Palaeoclimatol. Palaeoecol.* 27, 1–34.

ORI, G. & FRIEND, P.F. (1984) Sedimentary basins formed and carried piggy-back on active thrust sheets. *Geology* 12, 475–478.

PENNOCK, N. (1988) *Structural interpretation of seismic reflection data from the eastern Salt Range and Potwar Plateau, Pakistan.* MS thesis, Oregon State University.

RAYNOLDS, R.G.H. (1980) *Plio-Pleistocene structural and stratigraphic evolution of the eastern Potwar Plateau, Pakistan.* PhD thesis, Dartmouth College.

RAYNOLDS, R.G.H. (1981) Did the ancestral Indus flow into the Ganges drainage? *Geol. Bull. Univ. Peshawar* 14, 141–150.

RAYNOLDS, R.G.H. & JOHNSON, G.D. (1985) Rates of Neogene depositional processes, northwest Himalayan foredeep margin, Pakistan. In: *The Chronology of the Geologic Record* (Ed. Snelling, N.J.) pp. 297–311. Geol. Soc., Lond., Memoir 10.

RAZA, S.M. (1983) *Taphonomy and paleoecology of Middle Miocene vertebrate assemblages, southern Potwar Plateau, Pakistan.* PhD thesis, Yale University.

SEEBER, L. & ARMBRUSTER, J. (1979) Seismicity of the Hazara arc in Northern Pakistan: decollement vs basement faulting. In: *Geodynamics of Pakistan* (Eds Farah, A. & DeJong, K.A.) pp. 131–142. Geological Survey of Pakistan, Quetta.

TAUXE, L. & OPDYKE, N.D. (1982) A time framework based on magnetostratigraphy for the Siwalik sediments of the Khaur area, northern Pakistan. *Palaeogeog. Palaeoclimatol. Palaeoecol.* 37, 43–61.

YEATS, R.S. (1984) Tectonics of the Himalayan thrust belt in northern Pakistan. In: *Marine Geology and Oceanography of the Arabian Sea and Coastal Pakistan* (Eds Haq, B.U. & Milliman, J.D.) pp. 177–198. Van Nostrand Reinhold, New York.

YEATS, R.S., KHAN, S.H. & AKHTAR, M. (1984) Late Quaternary deformation of the Salt Range of Pakistan. *Bull. Geol. Soc. Am.* 95, 958–966.

Ores

Principles of a sediment sorting model and its application for predicting economic values in placer deposits

M. NAMI *and* S.G.E. ASHWORTH

Exploitation Technology, COMRO, PO Box 91230, Auckland Park, Johannesburg, Republic of South Africa

ABSTRACT

Mining geologists currently have to base their deductions regarding the distribution of economic minerals in sedimentary deposits on descriptive science, experience and empirical rules. A need for improved quantitative prediction of gold grade distribution in Witwatersrand gold-bearing deposits based on geological information has led to the development of a numerical simulation model of sediment transport and sorting in natural systems.

In this paper the fundamental principles of this model are described. The main processes considered are entrainment and deposition of sediment, and its transport rate as bedload and suspended load. The sorting of the sediment, both in size and density, can be predicted as a function of selective entrainment, differential transport and differential settling with the results being dependent on the flow conditions imposed. The model has been verified by comparison with published data and by conducting experiments in a large-scale flume.

This paper also considers the application of the model for predicting the location of economic values in the gold-bearing placer deposits of the Witwatersrand Basin. Geological observations of a placer, coupled with palaeohydraulic reconstruction techniques or back calculation from model runs allowed deduction of the hydraulic conditions and geomorphology that prevailed during deposition of the placer. The sediment transport model was then used to predict quantitatively the mineral composition of a placer deposit to assist in mine planning functions on various scales. Gold grade distribution and sedimentological variations have been modelled at selected sites on the gold mines of the Witwatersrand Basin and the results of these studies are presented.

INTRODUCTION

A reliable knowledge of the distribution of economic minerals in a sedimentary deposit is of greatest importance for planners of mining operations. Such information is crucial for determining the economic viability of opening a new mine, for siting shafts so as to obtain rapid access to richer areas and for planning which areas will be extracted in a particular sequence once mining operations have commenced.

Statistical treatment of mineral content analyses are normally used to estimate and predict values, but are subject to wide uncertainty. Accordingly, quantitative techniques are required to enable the use of geological information to maximum effect in predicting patterns of mineralization within a deposit.

In the case of gold distributions within the placer deposits of the Witwatersrand Basin, the difficulties are especially severe. Gold distributions are, by nature, extremely sporadic on a centimetre scale. As a result samples taken from a deposit normally have gold values that are not representative of the area being sampled. This phenomenon is frequently referred to as the nugget effect by geostatisticians.

Although geologists have obtained much experience of the Witwatersrand deposits through many years of mining activity, it is normally discovered that the experience gained on one deposit is not applicable to another. Thus, predictions of gold distributions are liable to considerable uncertainty.

Sedimentological facies can be identified within mined out portions of Witwatersrand placers, either on a qualitative or a quantitative basis. However, the sedimentology and the gold distribution in areas to be mined cannot be reliably extrapolated into unmined areas. This is mainly because there is insufficient physically based understanding of the fluvial processes responsible for forming these deposits.

COMRO (Chamber of Mines Research Organization) has been involved in collaboration with various universities for several years in studies aimed at understanding mechanistically the transport of coarse-grained sediments containing small fractions of dense material. Previous researchers' findings, as well as the results from flume experiments conducted to investigate specific processes, have been incorporated into a mechanistic model, which describes how a bed of mobile sediment responds to imposed hydraulic conditions.

The purpose of this paper is to first describe the operation of the model; secondly to show how it has been verified by comparing its results with data from flume experiments; and thirdly, to demonstrate how sedimentological characteristics and gold distribution in a Witwatersrand placer could be reproduced.

PRINCIPLES OF MIDAS

The sediment transport model developed by COMRO has been given the acronym MIDAS (Model Investigating Deposition and Sorting of Sediments) after the mythological character whose touch turned everything to gold. MIDAS has been developed to model how a sediment of an initial composition will respond to a hydraulic flow and create a sedimentary deposit. Furthermore, MIDAS models how sediments of different densities are sorted according to flow conditions.

MIDAS has been developed specifically for conglomerate placer deposits in the Witwatersrand Basin. Of specific importance in its formulation are the motion of coarse-grained sediments with a wide range of size distribution; and the effects of density on the erosion, transport and deposition of sediment. This mechanistic model allows the quantitative prediction of aspects of sedimentary deposition which will result from a given sequence of hydraulic conditions. In addition, the improved understanding of the sedimentary sorting processes assists geologists in interpreting the significance of certain observations in sedimentary deposits.

Calculation of flow conditions

The depth and velocity of flow are calculated over the length of the channel being considered using a standard backwaters calculation based on conservation of the flow energy and energy loss due to friction on the bed. A logarithmic velocity profile is assumed, as this is appropriate for quasi-uniform flow conditions. The energy loss in the flow is divided into two components: that due to grain roughness, and that due to form roughness after Einstein & Barbarossa (1952). This allows the force or shear stress exerted on particles in the bed to be determined by subtracting the shear stress due to the form roughness from the total shear stress.

Initiation of sediment motion

To initiate the motion of particles, the shear stress exerted on them by the flow must exceed their critical shear stress. The critical shear stress depends not only on the particle diameter, density and shape, but also on the size distribution of material in the bed. In general, large and dense particles are more difficult to entrain, but grain hiding causes higher shear stresses to be required for small particles than that predicted by Shields' criterion. The functions proposed by Komar (1987) and Egiazaroff (1965), which both account for grain hiding, have been incorporated into MIDAS.

If the shear stress exceeds a value calculated according to the criterion that shear velocity must be greater than fall velocity (after Engelund, 1965), then the particles entrained will move as suspended load; otherwise they will move as bedload.

As shown in Fig. 1, the shear stress exerted by the fluid on the bed is not assumed to be constant, but rather varies in response to turbulence in the flow, or to short-term fluctuations in the flow conditions. Thus, for a particular average flow condition, the critical shear stress will not be exceeded for a proportion of the time as represented by area 1; for the proportion of the time when the critical shear stress is exceeded, motion will take place as bedload (area 2a) or as suspended load (area 2b), depending on the shear stress exerted.

Sediment transport

For the portion of time when sediment transport takes place as bedload, the rate of sediment transport is calculated for each size and density of sediment using the equation proved by Bridge & Dominic (1984). In order to make this equation suitable for application to a heterogeneous sediment, the critical shear stress calculated for each size and density of sediment is used rather than the average for the bed. Furthermore, the mass flux calculated is multiplied by the volumetric proportion of each sediment size and density present in the bed.

For the proportion of time when sediment is transported in suspension a concentration profile is calculated by solving a set of partial differential equations referred to as the convection diffusion equations. By convolving this concentration profile with the logarithmic velocity profile of the flow, the mass flux of each size, and density of material in suspension is calculated, as illustrated in Fig. 2.

Deposition and erosion

Having calculated the mass flux for each size and density of sediment at each point along the length of channel considered, the amount of deposition or erosion can be calculated for each size and density of sediment at each point, by performing a mass balance as illustrated in Fig. 3. MIDAS keeps track of the resulting bed elevation, and uses this in the calculation of future flow conditions. Furthermore, the new bed composition is determined by adding deposited material to or subtracting eroded material from the material which was present in the active layer at the surface of the bed. By these means, MIDAS is able to model a stratigraphic record of preserved sediments.

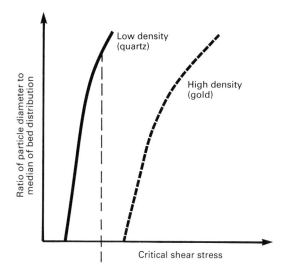

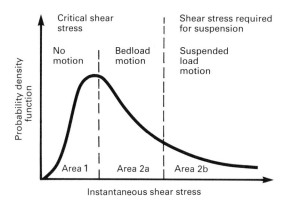

Fig. 1. Shear stress required for entrainment.

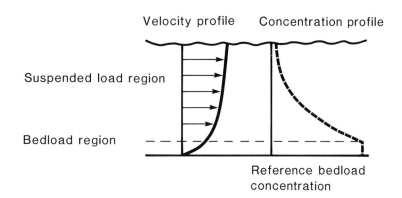

Fig. 2. Profiles for suspended motion.

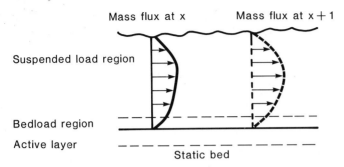

Fig. 3. Mass flux components.

Sorting modelled by MIDAS

From the description of the operation of MIDAS, it is apparent that various types of sorting behaviour can be modelled. Sorting due to differential settling is modelled as a result of solving the convection diffusion equations. Sorting due to differential entrainment is modelled as a result of applying either the Komar or Egiazaroff functions. Sorting due to differential transport is modelled by the methods used for calculating the mass flux of each different size and density of sediment. Indeed, of the four main sorting mechanisms identified by Slingerland (1984), only shear sorting, or sorting of material in a mobile mat, has not been incorporated at this stage of development.

Furthermore, MIDAS operates as a two-dimensional model analysing the changes in sorting in the flow direction. Channel geometries are only taken into account by changing the width of the channel along its length. This gives rise to altered flow conditions and therefore to changes in the pattern of sorted sediments. However, no attempt is made to analyse any changes in the nature of a sedimentary deposit across the width of a channel. This could be achieved by performing a succession of longitudinal and transverse model runs to simulate the formation of a three-dimensional deposit, as described by Nami & James (1987).

VERIFICATION OF MIDAS' PERFORMANCE

The physical realism of the principles incorporated in MIDAS have been tested by comparing its results with those obtained in a variety of flume experiments. The hydraulic conditions, flume geometries and initial sediment conditions were used as input to MIDAS, and various parameters produced were compared with the observations.

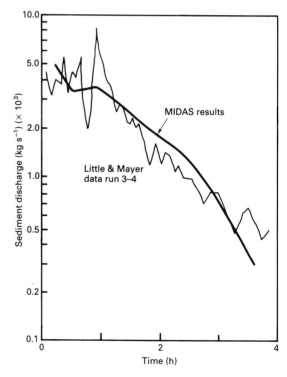

Fig. 4. Sediment discharge rate modelled for data from Little & Mayer (1976).

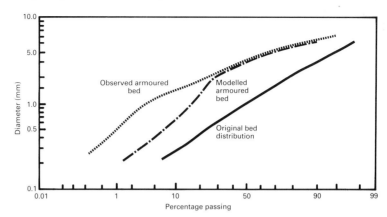

Fig. 5. Little & Mayer (1976) armoured bed composition.

Little & Mayer flume experiments

Little & Mayer (1976) conducted a suite of experiments in a straight flume with an equilibrium slope using a heterogeneous mixture of different sediment sizes. Their main objective was to determine the composition of the armoured bed produced when sediment motion had almost ceased. In addition to measuring the size distribution of the armour layer developed, Little & Mayer (1976) provided data on the mass flux of sediment transport at the downstream end of the flume during the course of the experiments.

Figure 4 shows how MIDAS was able to reproduce the sediment discharge rate quite accurately. The initial surge of sediment observed followed by a subsequent spike were both present in MIDAS' results. Furthermore, the decay in sediment discharge was adequately modelled.

The armoured bed composition was also reproduced by MIDAS quite accurately, as shown in Fig. 5, with the exception of the fine material, where MIDAS overpredicted its concentration. This is probably due to a minor incompatibility between MIDAS and Little & Mayer's sampling method. MIDAS' results are appropriate for a constant thickness of material, whereas Little & Mayer's sampling technique produced information relevant only for exposed particles at the surface. Thus, an overestimation by MIDAS would be in line with expectations.

COMRO flume experiments

COMRO has conducted a number of experiments in a 70 m long, 3 m wide flume using a sediment with a distribution of sizes ranging from 0.3 mm to over 20 mm. In order to test the depositional features of MIDAS in addition to the erosional aspects, a diverging channel was established as illustrated in Fig. 6. An experiment during which no sediment was fed into the flume was simulated using MIDAS.

MIDAS predicted that sediment would be eroded in the high flow energies at the narrowest portion of the flow, causing a coarsening of the bed, as shown in Fig. 7. As the flow energy decreased in response to the widening of the channel, the stream would become overloaded, and deposition would take place. MIDAS predicted a downstream fining of the sediment being deposited. As the bed at the constriction became armoured, and sediment was no longer available for transport, the sediment which had previously been deposited could now be transported in an unsaturated flow. Thus, a progradation of the sediment wave down the flume was predicted

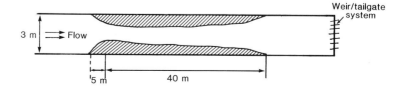

Fig. 6. Plan of flume geometry for experiments.

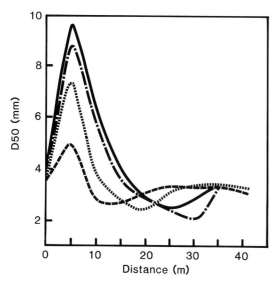

Fig. 7. Variation in median particle size.

by MIDAS. This behaviour reproduced qualitatively what was observed during the course of the flume expriments.

USE OF MIDAS ON A WITWATERSRAND PLACER

A study was conducted on a portion of a Witwatersrand placer deposit measuring 8 × 4 km which was interpreted to have been deposited within a braided river environment on an alluvial fan. This placer, like most of the other gold placers in the Witwatersrand Basin, is deposited on an angular unconformity formed by erosion of previously deposited sediments. The different footwall lithologies have a significant impact on the nature of the placer; where a shale footwall is found, more deeply incised palaeochannels are located, but on a sandstone footwall, topographic relief on the footwall contact is more limited.

Regionally, this placer is composed of two distinct events, which were found to be derived from different source materials by geochemical studies. The first event generally comprises a well-sorted conglomerate, ideally comprised of a fining upward sequence grading into sandstone. However, this sequence is normally truncated due to the erosional contact between the two events, which is interpreted to have been caused by a regional tectonic uplift of the hinterland. The second event is composed of more poorly sorted conglomerates, which may have been deposited in a flood event. The first event carries the higher gold concentrations, due to its better degree of sorting. Accordingly, this event was selected for modelling using MIDAS.

Measurements of directionally significant sedimentary structures, such as planar and trough cross-bedding, were used to determine palaeocurrent directions. From these, the angle of divergence of the fan, as well as the location of the apex of the fan was deduced. These findings were supported by identification of the lateral extent of the fan. Furthermore, this agrees with the location of debris flows which were interpreted to have been deposited close to the apex.

Because of the age of the sediments, about 2700 Ma, many of the primary features of the placer have been obliterated by subsequent tectonic and sedimentary events, which have eroded the most proximal parts of the placer. Only the upper to lower mid-fan region is available for study. The distal portions of the deposit are too deep below surface, greater than 3000 m, to allow access. The placer also underwent lower greenschist facies metamorphism, causing at least partial recrystallization of various minerals, including gold grains as well as the quartz matrix. Furthermore, several generations of faults, especially bedding-plane faults as well as normal faults, have been observed, making it extremely difficult to reconstruct the location of exposures at the time of deposition.

Despite these difficulties in studying such an ancient deposit, geological studies of this portion of the placer allowed the collection of much data such as pebble and sand measurements, sedimentary textures, mineralogy including heavy-mineral assemblages, geochemistry and, most importantly, gold concentrations. The location of the observations within the alluvial fan at the time of deposition was estimated.

The sedimentological data gathered from the conglomerate units hosting the gold were related to the distance from the apex of the fan, and a distinct downstream fining trend could be observed, as shown in Fig. 8. Variations transverse to the flow direction were also observed, but were ignored for the purpose of this exercise. Transverse changes are interpreted to be a result of varying topography due to changes in the footwall lithology, as well as due to lateral channel migrations.

Computer runs were conducted using MIDAS to simulate the downstream changes in the nature of this deposit. The divergence of the fan was modelled by increasing the width of the active

Principles of a sediment sorting model   549

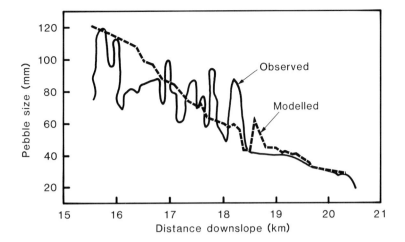

Fig. 8. Maximum pebble sizes.

channel. The rationale for modelling the deposition in this manner was that the same proportion of the total width of the fan would be occupied by channels at any downstream location. The unit discharge was also specified according to the results of a variety of palaeohydraulic reconstruction techniques. The system was allowed to develop its own slope which tended to about 0.005. This agrees with the slope deduced from palaeohydraulic reconstructions employing various features of deposit. An initially unsorted sediment was allowed to enter the system and the deposit created by MIDAS was compared with the observations. Figure 8 also shows the results of the MIDAS run where a good fit between the modelled and observed trends can be observed.

Detailed studies of the size distribution of the sediment in the placer were also conducted at a limited number of localities. Figure 9 illustrates the comparison between the observed and modelled

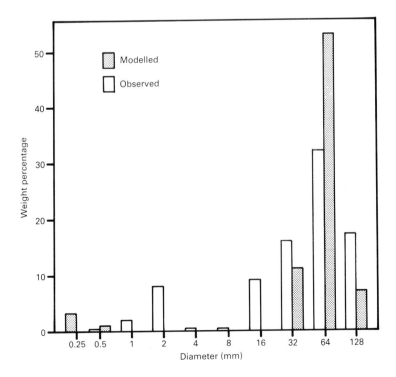

Fig. 9. Quartz grain size distributions.

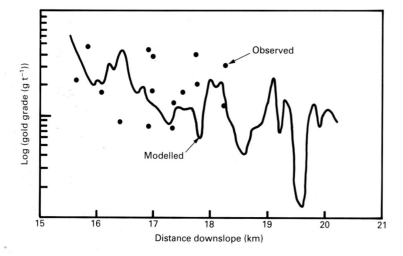

Fig. 10. Gold grade distribution down palaeoslope.

size distributions. A bimodal sediment size distribution was both observed during underground studies and predicted by MIDAS. The peak in the pebble range was extremely well reproduced, but a minor discrepancy in the location of the peak between the observations and the model was noted in the sand-sized fraction. This is probably due to a deficiency in modelling the effects of grain hiding by MIDAS, but may also be due to difficulties in measuring sand sizes as a result of quartz recrystallization of the sand size fraction of a 2700 Ma sediment.

A small amount of gold was also introduced in the sediment fed into the system during computer runs. Figure 10 shows a comparison between the observed trend in the gold concentrations and that predicted by MIDAS. The observations are very scattered, as is typical for the gold-bearing deposits of the Witwatersrand Basin, but they cluster around the modelled trend predicted by MIDAS. It is particularly interesting to note that MIDAS does not itself predict a smooth curve for the distribution of gold. Small variations in the flow conditions and the bed composition cause MIDAS to allow increased gold deposition at specific localities, and surges of sediment highly concentrated with gold were simulated to move down the system.

SUMMARY AND CONCLUSIONS

MIDAS, a model of sediment sorting by size and density, based on fundamental physical principles, has been shown to be capable of reproducing the observed behaviour of sediments in response to a hydraulic flow. Furthermore, it has been shown to be capable of reproducing trends in the sedimentology of an ancient rock record, as well as the distribution of an economically important mineral, gold. The authors consider that this technique will not only prove to be of great importance for predicting the location of economic values in unmined areas of a placer deposit, but that the traditional geological techniques, such as facies analysis, will also benefit greatly from the improved physically based understanding which can be achieved from the use of the MIDAS model.

ACKNOWLEDGEMENTS

The work described in this paper forms part of the COMRO's research programme. Their permission for its publication is gratefully acknowledged.

The authors would like to acknowledge the contribution of a large number of people who have been involved in the development and verification of MIDAS and in its use on Witwatersrand placers. On the development and verification Dr C. James and Mr A. van Niekerk, both of the University of the Witwatersrand, Johannesburg, South Africa, Professor R. Slingerland and Mr K. Vogel, both of Pennsylvania State University, USA, and Professor J. Bridge of the State University of New York, USA, have been responsible for the majority of the technical development. On the interpretation and modelling of the Witwatersrand placer described in this study Mr J. Schweitzer of COMRO, Johannesburg,

South Africa, and Mr K. Vogel of Pennsylvania State University, USA, are thanked for their major input into the work. Dr J. Petschnigg is also thanked for his assistance and ideas regarding modelling the Witwatersrand placer.

REFERENCES

BRIDGE, J.S. & DOMINIC, D.F. (1984) Bed load grain velocities and sediment transport rates. *Water Resources Res.* **20** (4), 476–489.

EGIAZAROFF, I.V. (1965) Calculations of non-uniform sediment concentration. *J. Hydr. Div., ASCE* **91** (HY4), 225–247.

EINSTEIN, H.A. & BARBAROSSA, N. (1952) River channel roughness. *Trans., ASCE* **117**, 1121–1146.

ENGELUND, F. (1965) A criterion for the occurrence of suspended load. *La Honelle Blanche* **8**, pp. 802 ff.

KOMAR, P.D. (1987) Selective grain entrainment by a current from a bed of mixed sizes: a re-analysis. *J. Sedim. Petrol.* **57** (2), 203–211.

LITTLE, W.C. & MAYER, P.G. (1976) Stability of channel beds by armouring. *J. Hydr. Div., ASCE* **112** (HY11), 1647–1662.

NAMI, M. & JAMES, C.S. (1987) Numerical simulation of gold distribution in the Witwatersrand placer. In: *Recent Developments in Fluvial Sedimentology* (Eds Ethridge, F.G., Flores, M. & Harvey, M.D.) pp. 353–357. Soc. Econ. Paleont. Miner., Tulsa, Spec. Publ. 39.

SLINGERLAND, R. (1984) Role of hydraulic sorting in the origin of fluvial plains. *J. Sedim. Petrol.* **54** (1), 137–150.

Alluvial basin-fill dynamics and gold-bearing aspect of Early Proterozoic strike-slip basins in French Guiana

E. MANIER*, D. MERCIER* and P. LEDRU†

*Ecole des Mines de Paris, CGGM, Laboratoire de Sédimentologie,
35 rue Saint-Honoré, 77305 Fontainebleau Cédex, France; and
†Bureau de Recherches Géologiques et Minières, Département GEO,
BP 6009, 45060 Orléans Cédex, France

ABSTRACT

As a consequence of the Trans-Amazonian collisional orogeny (2.2–2.0 Ga BP), the North Guyana Fault Zone developed as a major sinistral wrench-fault lineament with associated en échelon continental sedimentary basins. The pattern of infilling and the alluvial facies of these strike-slip basins show significant differences. The Mana Basin was formed as a graben-like feature along a relatively straight–linear segment of the strike-slip fault zone. There are no major escarpments at the basin margins, and the fluvial sedimentation apparently kept pace with semicontinuous basin-floor subsidence. The basin-fill alluvium (Arouany Formation), comprised of sandy braided stream deposits, is more than 5000 m thick. It shows no obvious cyclic organization due to tectonic pulses and contains no 'proximal' (fault scarp-derived) conglomeratic facies. The Régina Basin was formed along a curvilinear, southeastern segment of the strike-slip fault zone. The transpressional tectonic regime here created a fault escarpment along one of the basin margins. From there coarse debris-flow dominated alluvial fans repeatedly prograded into the fluvial basin's interior. The basin-fill alluvium (Tortue Formation) is 2000–3000 m thick, markedly conglomeratic along the faulted margin and shows cyclic organization (CU–TU sequences; 15–50 m thick) attributable to spasmodic rejuvenation of the fault escarpment due to transpressional stress build-up and release. Backward thrusting along the faulted margin supports this interpretation.

The 'proximal' (fault scarp-derived) conglomeratic alluvium in the Régina Basin contains gold, whose particles appear to be associated with the unsorted sandy matrix of clast-supported debris-flow conglomerates. The chemical composition and morphoscopic characteristics of gold particles are not consistent with a simple detrital origin. It is suggested that the gold, if originally of detrital provenance, has been remobilized and redistributed by hydrothermal processes, or was hydrothermally derived from external sources altogether. This notion is supported by the gold-bearing mineral paragenesis (quartz veins with pyrite, chlorite and tourmaline), which suggests hydrothermal derivation of both Fe–Ti oxides and the gold.

The study calls for a reassessment of the origin of gold occurrences in other, analogous Proterozoic deposits, such as those in West Africa and in other parts of South America, where a detrital ('placer') genesis has so far been adopted rather uncritically.

INTRODUCTION

The development and tectonic evolution of Early Proterozoic (2.6–1.6 Ga) sedimentary successions typically involved several phases of continental crustal building (Gruau *et al.*, 1985) followed by collisional events as documented from the South American, East African and Congolese cratons (Shackleton, 1986; Ledru *et al.*, 1989). In the Guyana shield, the Early Proterozoic tectonics and sedimentation were related to the Trans-Amazonian orogeny (Hurley *et al.*, 1968) which took place between 2.2 and 1.9 Ga ago (Gibbs & Olszewski, 1982; Swapp & Onstott, 1989). Our

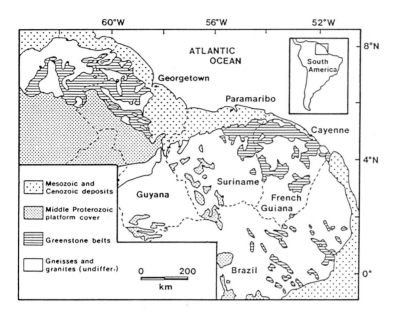

Fig. 1. Early Proterozoic greenstone belts of the Guyana Shield. (Modified from Gibbs & Olszewski (1982).)

observations from French Guiana indicate that the orogeny comprised two phases: an early collisional stage, and a later stage of sinistral strike-slip faulting probably due to lateral movements of crustal blocks.

It is generally agreed that the Lower Proterozoic of the Guyana Shield (Fig. 1) comprises two broad lithostratigraphic units (Kalliokoski, 1965; Choubert, 1974; Bosma et al., 1984), a lower formation of schists and an upper formation of coarse clastic metasediments, although their actual boundary somewhat varies according to different authors. The clastic sedimentary unit is of economic importance because of its gold-bearing aspect.

We describe here the upper clastic group, dominated by fluvial/alluvial deposits, and discuss the relationship between the sedimentation dynamics and basinal tectonic evolution. The occurrence of gold in these deposits is then reviewed and its origin is discussed in terms of a hydrothermal (non-detrital) hypothesis.

STRUCTURAL SETTING

The results of our field work in French Guiana (Ledru et al., 1987) have led us to modify the Lower Proterozoic succession (Barruol, 1961; Choubert, 1974). The lithostratigraphic scheme we propose (Lasserre et al., 1989; Ledru et al., 1991) takes into account the tectonic deformation and structural/metamorphic evolution of the rock succession.

The Paramaca Formation

This formation, together with associated granites, crops out over most of the French Guiana territory and is comprised of basic to intermediate metavolcanics overlain by metasediments. The metavolcanics are basalts and tholeiites, intercalated with amphibolites of komatiitic composition. Their radiometric age is 2.11 ± 0.03 Ga (Gruau et al., 1985). The metasediments are marine shales, typically black with graded arenitic sandstone beds a few centimetres thick and locally with tuffitic intercalations.

The entire rock assemblage experienced two major phases of deformation (referred to as D_1 and D_2). The earliest regional schistosity (S_1) is marked in the volcanics by the crystallization of amphibole, mica and chlorite in greenschist facies conditions; stratification in the sediments has been transposed in schistosity (S_1). This regional evolution is attributed to collisional tectonics and related to the crustal plate suture in southern Guyana, where granulite terrains overthrust Paramaca volcanics. A younger schistosity (S_2) developed due to greenschist facies metamorphism in the axial zones folds (F_2), or to strike-slip faulting (E–W trend). Asymmetric microfolds indicate that the second phase of deformation (D_2) was related to sinistral strike-slip movements.

Continental clastic deposits: Arouany Formation and Tortue Formation

The Paramaca Formation is overlain by conglomerates and sandstones that have experienced only one major phase of deformation (D_2). These deposits comprise detrital zircons with radiometric ages of 2.14–2.11 Ga BP (Manier, 1990) and are cut by granites dated at 1.9 ± 0.8 Ga BP (Choubert, 1974). The clastic deposits occur as an infill of separate basins bounded by sinistral strike-slip faults, along the North Guyana Fault Zone (Fig. 2). Because there are marked facies differences between these basins we refer to the sandstone sequence in the Mana Basin as the Arouany Formation, and to the conglomerate and sandstone sequence in the Régina Basin as the Tortue Formation.

Our geometrical reconstruction, based on structural analysis and sediment polarity measurements indicate that all basins are now in synclinal positions (Fig. 2). The synclinal structure is apparently related to strike-slip faults, and the actual deformation style will thus be different if the basins experienced transtension or transpression. The Mana Basin was deformed in transtensional regime; its southern boundary fault (N 100° E) is a sinistral shear feature with a normal strain component. Schistosity (S_2) is heterogeneous and related either directly to the sinistral strike-slip shear or to extensional microshear parallel to bedding planes. In the Régina Basin, the southern boundary fault (N 160°–N 140° E) had an inverse strain component that led to northeasterly overthrusting of the Paramaca Formation upon the Tortue Formation (Fig. 2). This caused crystallization of biotite and kyanite (sometimes growing on andalusite) in the conglomerates. The northern part of Régina Basin is bounded by a strike-slip fault similar to those in the Mana Basin. The second stage of deformation (D_2), which affected the clastic sedimentary formations, appears to have been controlled by major strike-slip shears.

The Arouany Formation in Mana Basin

The Mana Basin, bounded by sinistral strike-slip faults, has a rhomboidal shape with a length of

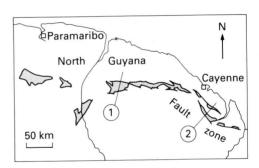

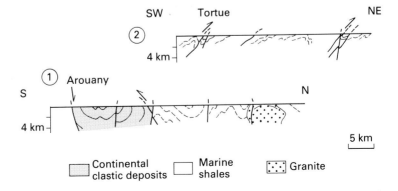

Fig. 2. Distribution of Early Proterozoic non-marine basins along the North Guyana Fault Zone of the Trans-Amazonian orogeny. Note the synclinal position of the deformed basins; (1) section through the Mana Basin; (2) section through the Régina Basin.

50 km and a width of 20 km. The Arouany Formation is a thick (more than 5000 m) basin-fill succession of sandstone facies with very little lateral or vertical variation.

Lithofacies

For simplicity and ease of comparison, Miall's notation (1977, 1978) is used as lithofacies labels in the following description. A representative example profile of the Mana basin-fill deposits is shown in Fig. 3. Conglomerates (facies Gm, rarely Gt) contain rounded pebbles of vein quartz (90%) and brown schists (10%) set in a sandy matrix: clast sizes never exceed 5 cm. The conglomerates occur as thin, lens-shaped units. Most of the basin-fill deposits (95 vol.%) are represented by silicified whitish sandstones with dark heavy-mineral laminae that mark primary internal stratification. Sandstone sequences a few metres thick comprise trough cross-stratification (facies St), planar cross-stratification (Sp) and less frequently horizontal stratification (Sh) and ripple cross-lamination (Sr) at the top. First-order sequences are fining upwards and have weakly scoured erosive bases (Fig. 4).

Palaeoenvironmental interpretation

The volumetric importance of facies Sp indicates that the sand was transported and deposited mainly as transverse bars, or as relatively straight-crested (two-dimensional) dunes or 'sandwaves'. Such features occur in broad shallow rivers that often lack well-defined physiographic differentiation of 'channel' and 'interchannel' areas. The depositional environment of the Arouany Formation was similar to the Platte River (Smith, 1970). The lack of any major vertical facies variations, in spite of the great thickness of the basin-fill (5000 m), implies a remarkable long-term balance between the rate of basin-floor subsidence and the rate of sediment supply and accumulation. Palaeocurrent directions near the southern boundary fault are perpendicular to the basin margin, but are more westerly in the basin's central and northern part (Fig. 5). The characteristics mentioned above are accounted for by the tectono-sedimentary model shown in Fig. 6.

The Tortue Formation in Régina Basin

The original size of the Régina Basin, now *c.* 100 km long and *c.*30 km wide, was reduced by

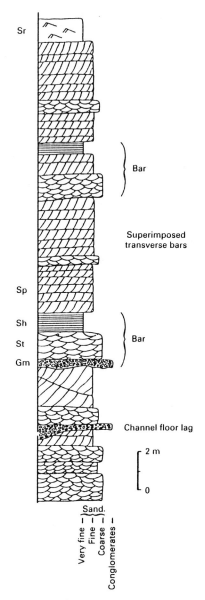

Fig. 3. A representative log through the braided-stream facies assemblage of the Arouany Formation. Lithofacies notation according to Miall (1977, 1978); Gm, massive or crudely stratified conglomerate; St, trough cross-stratified medium to coarse sandstone; Sp, planar cross-stratified fine to coarse sandstone; Sr, ripple cross-laminated very fine to medium sandstone; Sh, very fine to medium sandstone with horizontal stratification.

folding and thrusting. The entire western part of the basin has been concealed by the Paramaca Formation over the Tortue Formation (Fig. 7). This basin

Fig. 4. Channel-fill sequences in the Arouany braided-stream sandstones (hammer, 30 cm, for scale). The sequence comprises major and minor erosive surfaces (E and e, respectively), planar cross-stratified sandstone (Sp) and horizontally stratified sandstone (Sh).

is characterized by the presence of coarse conglomeratic alluvium along its southern (S and SW) margin, in addition to a sandy alluvium (Fig. 7). The total thickness of the basin-fill preserved here is c. 2500–3000 m.

Lithofacies

A typical profile through the coarse, basin-margin alluvium is shown in Fig. 8 (for locality see 'Montagne Tortue' area, Fig. 7). The succession is dominated by composite, amalgamated units of matrix-supported, unstratified conglomerates (facies Gms). Mean clast sizes vary from pebbles to cobbles, with

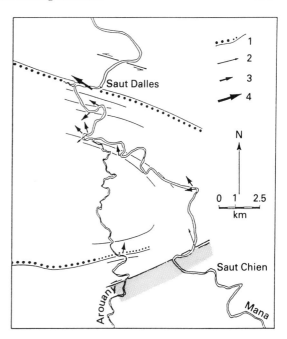

Fig. 5. Palaeocurrents in the fluvial Arouany sandstones in Mana Basin (as defined by its present day margins). 1, Major conglomerate bodies; 2, measurements from trough cross-sets; 3 & 4, measurements from planar cross-sets on a few tens of and more than a hundred data respectively.

maximum sizes reaching 40 cm. Conglomerates are very poorly sorted, and have a coarse matrix of ill sorted finer gravel and muddy sand (Fig. 9A). The volumetric ratio of vein-quartz clasts to schist and tuff clasts is lower than 1, commonly below 1/4. Conglomerate units are 2–6 m thick, show no obvious scour or channelling and lack internal stratifications. They are interpreted as debris-flow deposits (cf. Larsen & Steel, 1978; Nielsen, 1982). Beds typically have flat bases, except where they occur as a secondary infill of stream channels (see Fig. 8). They show inverse coarse tail grading in the lower to middle parts (Fig. 9B,C), and normal grading in the upper parts (Fig. 9C). The inverse grading is attributed to differential laminar shear in the flowing sediment mass, with the strength and clast-support competence of matrix material most affected in the flow's lower part (see Naylor, 1980; Nemec & Steel, 1984), while the normal grading at the top is ascribed to deposition from a more watery, turbulent upper part of the debris-flow or from a closely-following, shallow stream-flood

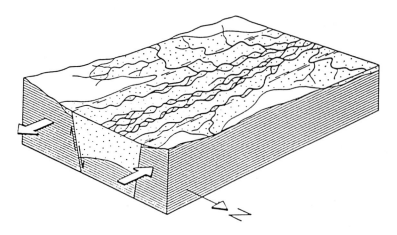

Fig. 6. Tectono-sedimentary model for the Mana Basin. For discussion see text.

current (see Heward, 1978a,b; Nemec & Steel, 1984).

Facies Gm is pebble to cobble sized conglomerate with angular to subrounded clasts of mainly quartz. Matrix is medium-grained quartz arenite. Beds are 1–3 m thick, massive or with crude horizontal stratification. Clast fabric appears to have been extensively modified by dissolution–recrystallization processes (schistosity S_2) that caused stretching of the clasts and altered their primary spatial relationships. Clasts of schists and metatuff are angular to subangular and their sizes do not exceed 15 cm. The volumetric ratio of quartz clasts to schist and volcanic clasts is high, usually well above 1, probably a result of material abrasion during transport. The conglomerate facies Gm is interpreted as fluvial bedload deposits. These massive, relatively thick, crudely stratified sheets of gravel can be interpreted as channel-floor pavement and longitudinal gravel bars (Hein & Walker, 1977). Sheets with scoured bases probably represent rapid filling of shallow channels during 'flashy' flood events. Conglomerate facies Gt can be attributed to linguoid bars or gravelly three-dimensional dunes, formed by flow less heavily loaded with sediment or during waning stages of stream-flood events. The associated sandstone facies (St, minor Sp) probably represent sand deposited as lateral wedges across the gravel bars during shoaling-water conditions of a waning flood flow or a low stage flow. Rare massive units of thin black mud-shales (Fm) at the top of some debris-flow beds represent suspension settling from ponded flood water in shallow topographic depressions.

Palaeoenvironment and vertical facies sequences

The sedimentary facies appear to grade laterally, along the palaeoflow direction (Fig. 7) from what we interpret as 'proximal' alluvial-fan deposits (facies Gms and Gm) into the more mature fluvial conglomerates (facies Gm and Gt) and sandstones (facies St and Sp) toward the north and north-west. The northwestern end of the outcrop belt (Fig. 7) consists of sandstones representing broad and shallow streams (mainly facies Sp; Fig. 10), very similar to those in the Mana Basin. Alluvial fans, with radii

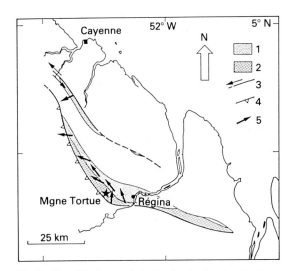

Fig. 7. Simplified map of the Régina Basin, showing its present day margins and the distribution of the main types of alluvium. 1, Mainly sandstones; 2, mainly conglomerates; 3, sinistral strike-slip faults; 4, northeastward thrust; 5, generalized sediment transport direction. The Montagne Tortue area (star) is referred to in the text and other figures (Figs 8 & 11).

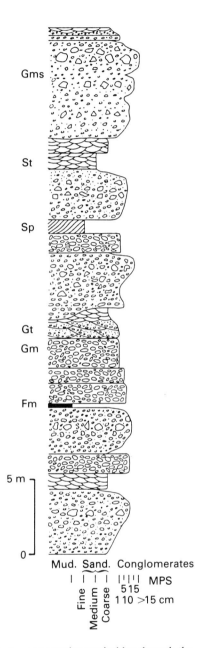

Fig. 8. A representative vertical log through the 'proximal' basin-margin alluvium of the Tortue Formation in Montagne Tortue area (see Fig. 7). Lithofacies are: Gms, massive matrix-supported conglomerate; Gm, massive or crudely stratified conglomerate; Gt, trough cross-stratified conglomerate; St, trough cross-stratified, medium to coarse or pebbly sandstone; Sp, planar cross-stratified, fine to coarse sandstone; Fm, massive mudshale.

of the order of a few kilometres, seem to have developed in the southeastern part of the Régina Basin. The debris-flow dominated fan alluvium is rapidly overlain by braided-stream deposits.

Vertical logs from the Montagne Tortue area (Fig. 7) show that the fan alluvium along the basin margin is organized into asymmetric, coarsening- to fining-upward (and thickening- to thinning-upward) sequences 15–50 m thick (Fig. 11). Their coarsening-upward lower parts represent progradation of debris-flow dominated alluvial-fan lobes probably in response to tectonic movements, whereas the fining-upward upper parts are mainly stream deposits that reflect the declining sediment influx associated with fan lobe abandonment (see Steel et al., 1977; Heward, 1978a). This type of facies organization may reflect the depositional effects of strike-slip basin-margin tectonics (Steel, 1988). The overall basin-fill trend is upward fining, and this may well be attributed to a horizontal migration of the basinal depocentre relative to its source terrain (Steel & Gloppen, 1980). Following these inferences, a tectono-sedimentary model for the Tortue Formation is proposed in Fig. 12.

Tectonics and basin formation

The North Guyana Fault Zone (Fig. 2) is a major tectonic lineament related to the Trans-Amazonian orogeny. Its activity had led to the development of a series of en échelon sedimentary basins controlled by strike-slip faulting. This transcurrent stage of wrench faulting followed collision manifested in South Guyana by the thrusting of a high-grade metamorphic complex (granulitic facies) on to the Paramaca volcano-sedimentary rock complex. Such an evolution, from thrust tectonics involving relatively deep crustal complexes to wrench tectonics (transcurrent faulting), is quite common in plate collision settings (Shackleton, 1986). In the present case, the transcurrent displacement of crustal blocks involved sinistral faults trending E–W, and was related to regional shortening. The strike-slip basins thus developed under the control of pre-existing crustal heterogeneities and tectonic features.

The structural boundaries of the basins appear to have been defined quite early as indicated by the development of schistosity in extensional shear zones. The varied pattern of alluvial sedimentation in the basins as shown by the two examples described above was probably related to differenttectonic styles of deformation and subsidence. The

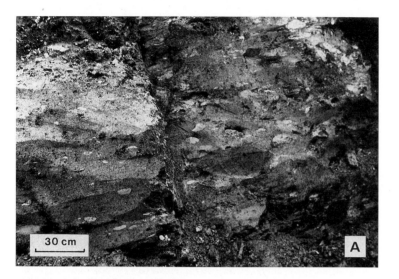

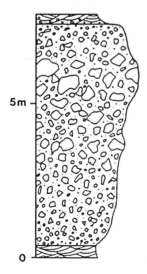 

Fig. 9. (A) Facies Gms conglomerates of the Tortue Formation, within a coarsening-upward fan-lobe sequence; note the massive matrix supported character of the conglomerates. (B) Basal part of a debris-flow conglomerate unit (Gms), showing coarse-tail inverse grading; stratigraphic top indicated by the arrow (15 cm). (C) Clast-size grading in a thick debris-flow conglomerate unit; maximum particle size (MPS) is the mean size of ten largest clasts per given level.

Early Proterozoic gold-bearing alluvium   561

Fig. 10. Fluvial sandstones of the Tortue Formation in the northwestern part of the Régina Basin. Note the predominance of facies Sp. The arrow (15 cm) indicates the palaeoflow direction.

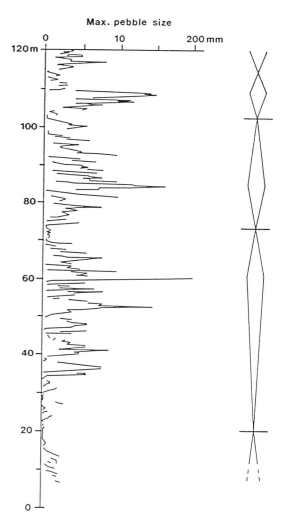

Fig. 11. Cyclic organization of the Tortue Formation in Montagne Tortue area showing CU–FU sequences defined by maximum particle sizes of the alluvial conglomerates. Note the upward thinning of the successive sequences.

Mana Basin developed along a relatively straight segment of the strike-slip fault zone (Fig. 2), whereby the bounding 'normal' faults of the graben promoted continuous gradual subsidence of its floor (Fig. 6), without any major tectonic escarpment to shed coarse debris; the thick basin-fill here comprises sandy braided-stream alluvium. The Régina Basin, in contrast, was formed along a strongly curvilinear segment of the strike-slip zone (Figs 2 & 7), whereby the resultant tectonic escarpment shed abundant coarse alluvium (Fig. 12) and the pulsatory build-up and release of transpressional stresses caused repetitive progradation of fan lobes (Fig. 11). The sinistral strike-slip movements eventually led to the closure and deformation of basins, locally with an important crustal thickening due to overthrusting.

Other early post-collisional strike-slip fault systems of the same age (2.0–1.9 Ga BP) are known from Venezuela (the Guri and Pisco-Jurua fault zones; Swapp & Onstott, 1989) and West Africa (Vidal, 1986; Ledru et al., 1989). The origin of numerous Early Proterozoic sedimentary basins in Africa and in South America could also be related to analogous fault systems. In particular, the lithological, structural and metamorphic analogies to the Tarkwaian series in Ghana are worth emphasizing (Milési et al., 1989). Another common characteris-

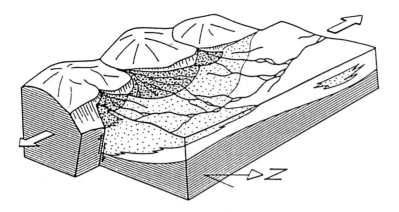

Fig. 12. Tectono-sedimentary model for the Régina Basin. Coarse 'proximal' (faulted-margin) alluvium grades rapidly into sandy alluvium; the Paramaca Formation is shown in white. For discussion see text.

tic of these Early Proterozoic basin-fills is their gold-bearing aspect.

GOLD MINERALIZATION IN CONGLOMERATES

The northern region of French Guiana is known for the occurrences of gold in Early Proterozoic sandstones and conglomerates. All these numerous occurrences, as far as we know, are located in proximity to the strike-slip faults of the North Guyana Fault Zone (Fig. 13). At least some of them (see Fig. 13, relevant localities are shown there as ●) are clearly related to the hydrothermal processes associated with stage D_2 of tectonic deformation, and the distribution of gold appears to have been controlled by sinistral shear. Gold occurs in concor-dant thin lodes of hydrothermal quartz and the characteristic mineral paragenesis is:

(1) Quartz + Calcite + Chlorite + Tourmaline + Pyrite + Gold, or
(2) Quartz + Galena + Tellurides (Hessite, Altaite, Petzite) + argentiferous Gold.

In other occurrences (localities shown as ▲ on Fig. 13), mineral paragenesis associated with gold is not easy to determine. We discuss below the mineralization in one of these occurrences, Montagne Tortue section, whose coarse alluvium has been described earlier in the text (see Figs 7–12).

The gold-bearing alluvium in the Montagne Tortue area

The Montagne Tortue section comprises 'proximal'

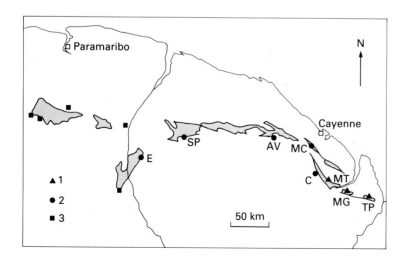

Fig. 13. Distribution of gold occurrences within the North Guyana Fault Zone of northern Guyana and Suriname. 1, Related to late deformation (stage D_2) hydrothermal processes; 2, gold present as isolated particles in conglomerates; 3, gold of unspecified nature. The localities, referred to in the text are: E, Espérance; SP, Saint-Pierre; AV, Adieu-Vat; C, Changement; MC, Montagne des Chevaux; MT, Montagne Tortue; MG, Montagne Geneviève; TP, Montagne des Trois Pitons. (From Bosma et al. (1984) and the present authors' studies.)

(basin-margin) alluvium of the Régina Basin, characterized by vertically stacked fan lobes built of debris-flow deposits interbedded with braided-stream deposits. Gold soil anomaly (> 200 p.p.b. Au) has an extent of $c.6$ km in length and $c.1.5$ km in width, with maxima of 1.5 g t^{-1}. By comparing the surficial extent of the clastic rock bodies and the Au anomaly in the soil (Fig. 14), one can observe strong areal correlation between the conglomeratic debris-flow and the highest Au mineralization. Bulk-rock assays along trenches further confirm that the gold is preferentially hosted by debris-flow conglomerates and located in the matrix. A parallel recent study in the Montagne des Trois Pitons area (TP in Fig. 13), where the gold is hosted by debris-flow deposits as well, has shown that the precious metal occurs in small lodes or as impregnations in the conglomerate matrix. In order to understand the actual origin of gold in these deposits both gold particles and associated minerals have been analysed and the results are presented below.

Minerals associated with gold

Most of the gold-bearing rocks are now unconsolidated and hence it was possible to pan concentrates and to extract and study both gold particles and the associated heavy minerals.

The most abundant heavy minerals are haematite and rutile, most of them (80–90%) recrystallized. Thin lodes of haematite are commonly seen cutting stratification and schistosity as well as broken

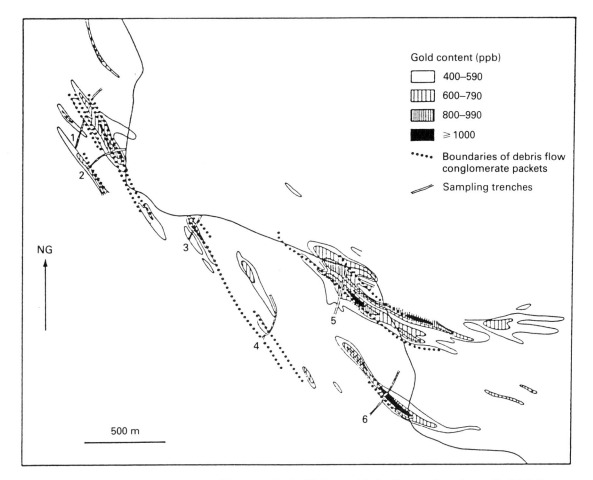

Fig. 14. Comparison of gold contents in soil (p.p.b. values) with the spatial distribution of conglomeratic debris-flow bodies.

quartz clasts. Systematic measurements of haematite content and gold content show that there is no recognizable correlation between the occurrence of the two minerals. The statistical analyses on bulk-rock assays show that Fe is mainly correlated with Ti ($r = 0.81$), Va ($r = 0.73$) and Zr ($r = 0.69$), whereas Au is associated with La and Ce ($r = 0.61$), Zn and Pb ($r = 0.55$) and more weakly with Cr ($r = 0.40$) and B ($r = 0.38$).

In concentrates of rich Au-bearing rock-units (2–12.5 g t^{-1} Au content) one can observe under a microscope, gahnite grains with authigenic crystal outlines. The mean composition of gahnite, based on 110 analyses, is (Fe0.8, Mg1.0, Mn0.1, Zn6.1) O$_8$, Al$_{16}$, O$_{24}$. There are two possibilities to account for the formation of gahnite: (i) development from a primary Zn oxide phase during metamorphism (Segnit, 1961), or (ii) precipitation of spinel from metamorphic or hydrothermal solutions (Wall, 1977). In each of these possibilities, the spatial association of gold and gahnite would be explained by their at least partially common history of crystallization or chemical remobilization through the metamorphic or hydrothermal processes.

Gold particle characteristics

Panned mineral samples were not milled, so as to avoid abrasion and deformation of gold particles. About 350 grains of gold from twelve samples, collected in the Montagne Tortue area a long the palaeoflow direction, have been observed. Particle sizes range from 10 to 250 µm, with a median at 70 µm. They all show similar morphology (Manier & Mercier, 1989), with irregular, jagged morphometry and authigenic crystal outlines (Fig. 15). Electron microprobe analyses of gold particles were performed to recognize features such as zoning, overgrowths or compositional heterogeneities.

Analytical techniques

Analytical data were obtained on a Cameca Camebax electron microprobe operated at an accelerating potential of 25 kV and using an LIF crystal for Au, Cu, Fe, Ni and Hg, a PET crystal for Ag, Pb and S, and a TAP crystal for As. Standards used were pure gold, silver, copper, arsenic and iron, for Hg, Pb, cinnabar and galena, and pyrite for S. To avoid the problem of high accuracy usually required in spectrometer motions we used the method of 'double'

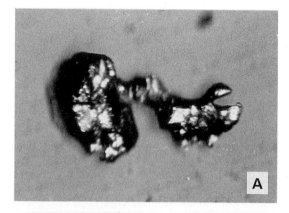

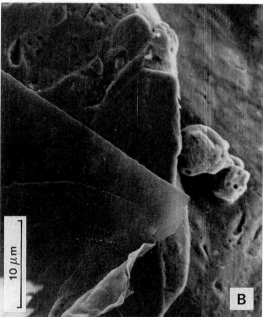

Fig. 15. Morphoscopic appearance of gold particles separated from conglomerate matrix; Montagne Tortue section, Régina Basin. (A) Typical shape of gold particles (150 µm in size) seen under a microscope. (B) Angular form of a gold particle observed under a Sweap Electronic Microscope.

measurement that provides sufficiently good quantitative results. This method, based on a double measurement of X-ray intensity on each side of the peal instead of a single measurement at its top (Marion & Vannier, 1983), allows better analytical reproducibility and a permanent control of the performance of the spectrometer. Analytical totals (Fig. 16) are somewhat lower than expected, and are

Early Proterozoic gold-bearing alluvium 565

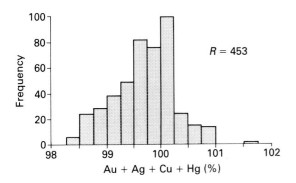

Fig. 16. Electron microprobe analyses of gold particle totals (Au + Ag + Cu + Hg) determined for the occurrences in the Montagne Tortue area, Régina Basin.

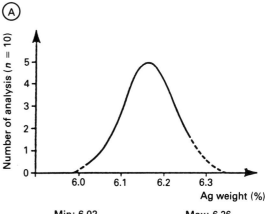

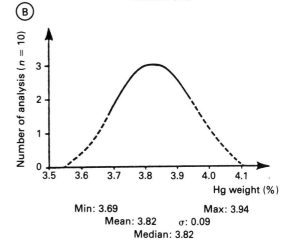

Fig. 17. Composition of gold particles. (A) Ag content in a particle from sample no. 2/6. (B) Hg content in a particle from sample no. 1/93. The normal distribution of the analyses through the particle is related to analytical dispersion.

thought to be underestimated. This may be due to the absorption coefficient for Ag $L\alpha$ by Au used in the absorption formula. We used the value of 1961 (from Heinrich, 1966), but the higher value of 2400 (Reid et al., 1988) might be more adequate. However, this sort of systematic error in analytical totals is not critical here, especially since the values are used mainly for comparative purposes, where relative rather than absolute values are crucial.

The analytical results have shown that the gold particles are four component alloys of Au–Ag–Hg–Cu, with a wide range of composition: 0–23 wt% Ag, 0–1.8 wt% Cu, and 0–4.4. wt% Hg. No relationship has been found between the contents of silver, mercury and copper. Each particle appears to be homogeneous, lacking significant compositional variations from the centre outwards (Fig. 17), although the particles differ both within and between samples. Particles with 15 wt% Ag and 0.1 wt% Cu, for instance, occur together with those having 0.5 wt% Ag and 1.3. wt% Cu.

The origin of gold: a discussion

In the Early Proterozoic alluvium of the Montagne Tortue area, mineralization appears to be related to the sedimentary characteristics of the rocks; gold occurs preferentially in the matrix-supported, debris-flow conglomerates. A sedimentary control on gold distribution in Precambrian conglomerates has been invoked by many authors for the Witwatersrand placers (Minter, 1976; Nami, 1983) and for the occurrences of this metal in conglomerates of comparable age in West Africa (Sestini, 1973; Kesse, 1982) and Brazil (Sims, 1977). In most of these other cases, the 'typical auriferous conglomerates' are far more mature, oligomictic (dominated by clasts of vein quartz and cherts) and clast-supported, with clasts generally well rounded, relatively well sorted and well packed (Pretorius, 1981). They typically represent braided-stream deposits of channel-floor or channel-bar origin. Very few occurrences of gold are known from debris-flow fanglomerates. Krapez (1985), for example, reported mean gold contents reaching 17 p.p.m. in the debris-flow deposits of the Ventersdorp Contact Placer (South Africa).

There are several possibilities to explain the origin of gold in conglomerates.

1 A detrital origin of gold, with concentrations related to hydraulic processes, as in modern and ancient placers (Smith & Minter, 1980; Nami, 1983).
2 A metamorphic remobilization and recrystallization of earlier detrital gold. In this model, the associated metals like silver, mercury and copper would not be inherited simply from the primary clastic sources. For example, Oberthür & Saager (1986) have suggested that mercury in the Carbon Leader Reef Placer (South Africa) was derived in solution from the surrounding sediments and subsequently amalgamated with gold due to metamorphism.
3 A similar derivation of gold, in solution from external sources, due to metamorphism or hydrothermal processes.

In spite of the apparent sedimentary control of gold occurrence in the present case (Fig. 14), both the morphoscopic characteristics and chemical homogeneity of the gold particles are not in favour of the first hypothesis. Detrital gold during transport suffers from mechanical abrasion, losing authigenic crystal shapes and gaining rounded edges, percussion marks and related features (Herail, 1984). Detrital gold particles also experience chemical weathering with silver leached towards the particle surface and possibly with neoformation of pure-gold crystal shape outlines.

We know, in the present case, that important mobilization of metals has taken place and led to the recrystallization of most (80–9%) of the Fe–Ti oxides, and this process would almost certainly affect detrital gold grains. The metamorphic mineral parageneses indicate that the metamorphism temperatures reached 350°C (for muscovite, brown biotite, garnet assemblage) or even more (for andalusite, kyanite), sufficient to cause remobilization of detrital gold (hypothesis 2 above). This would explain, according to the experiments by Czamanske et al. (1973), the compositional homogeneity of individual gold particles observed.

The third hypothesis refers to gold being brought into conglomerates by hydrothermal fluids from external sources. Notably, there is a close spatial relationship between the strike-slip faults and the occurrence of gold in the present case (Fig. 13). At least some of the occurrences of gold in alluvial sandstones and conglomerates (localities shown as ● in Fig. 13) are related to hydrothermal processes (association with chlorite, tourmaline and pyrite). In some of the gold mines in the Paramaca Formation (e.g. Changement in Fig. 13) gold is associated with calcite and tourmaline in small quartz lodes formed during tectonic deformation (stage D_2). Similar observations have been made by Ledru et al. (1988) in other gold-bearing Early Proterozoic conglomerates (such as the Tarkwaian in Ghana), where gold is either present in small calcite lodes or associated with recrystallized magnetite.

In the Montagne Tortue alluvium, a major role of hydrothermal processes is indicated by the presence of gahnite and by a tourmalization observed close to the southern basin margin. Gold seems to have been remobilized or even brought by the hydrothermal fluids. The likely influence of the conglomerate matrix on the physical behaviour of hydrothermal fluids (chemical reduction, limited permeability) might then explain why the mineralization is associated with debris-flow conglomeratic units.

SUMMARY AND CONCLUSIONS

During the Early Proterozoic Trans-Amazonian orogeny, en échelon sedimentary basins were formed due to sinistral strike-slip faulting in the North Guyana Fault Zone. The basins appeared as early post-collisional features, and the varied geometry of the strike-slip lineament resulted in a variety of basin types. In the Mana Basin, the straight–linear fault geometry resulted in tectonic subsidence controlled by a normal, dip-slip stress component, with no major fault escarpments and with the fluvial basin-fill sedimentation keeping pace with the subsidence rate; the basin fill lacks coarse margin alluvium, is sandy and shows no obvious cyclicity. In the Régina Basin, the curvilinear fault geometry resulted in a transpressional tectonic regime and the development of a repeatedly rejuvenated escarpment along the basin margin; coarse, immature, polymict debris was shed from the fault scarp and repeatedly prograded into the fluvial part of the basin as alluvial-fan lobes (basin-fill cyclicity). It is not unlikely that the two basins were initially connected, sharing an extensive axial drainage system in the form of a sand-bed braided stream belt. Future study of outcrops along the Maroni River, in the western part of French Guiana (see Fig. 2), will help to evaluate this latter possibility. Since similar strike-slip fault systems of comparable age (2.0–1.9 Ga BP) are known to have been active in Africa and South America, it is possible that important Early Proterozoic sedimentary basins were associated with these lineaments; they may be buried today, but would be worth exploring — also for economic purposes.

The alluvial-fan conglomerates of the Tortue Formation in the Régina Basin have an important gold-bearing aspect. However, the nature of the occurrence of this precious metal here does not match the classical, 'typical' occurrences in the Early Proterozoic auriferous conglomerates (cf. Pretorius, 1981). The gold particles in the present case occur in the matrix of debris-flow conglomerates and the characteristics of the particles indicate a non-detrital rather than purely detrital origin. The gold seems to have been remobilized in hydrothermal solution and recrystallized during metamorphism, along with Fe–Ti oxides, or could have been derived in solution from external sources to precipitate in the conglomerates as a purely post-depositional component. Post-schistosity tourmalization in the Montagne Tortue conglomerates, analogous to that in other localities along the North Guyana Fault Zone where gold is clearly of a hydrothermal origin, and correlation of Au with Zn, Pb and B points to the possibility that the gold in the present case was brought 'externally' by hydrothermal fluids. The considerable variation in the chemical composition of gold particles in such cases could be explained by several hydrothermal fluid pulses. Further bulk-rock assays and analyses of gold, tourmaline and gahnite grains are in progress to evaluate this latter model.

The present results from the Early Proterozoic conglomerates in French Guiana call for a serious reassessment of the origin of gold 'placers' of similar age in many other regions, particularly in the Tarkwaian strata of Ghana, which are the African equivalent of the deposits described here.

ACKNOWLEDGEMENTS

We wish to thank the Inventaire Minier de la Guyane Française, Bureau de Recherches Géologiques et Minières (BRGM) and Ecole Nationale Supérieure des Mines de Paris (ENSMP) for their financial support. Field work was done in collaboration with J.L. Lasserre and J.P. Milési (BRGM). We are grateful to Professor B. Beaudoin (ENSMP) for constructive discussions during the course of this research project. Reviews by W. Nemec and W.E. Minter were very helpful in clarifying points and correcting the English within the manuscript.

REFERENCES

BARRUOL, J. (1961) Le Bonidoro en Guyane française (Zone NW). In: *Proc. 5th Inter-Guiana Geol. Conf., Georgetown 1959*, pp. 57–67. Geol. Mijnbouw Dienst Sur. Jaarboek 1956–58.

BOSMA, W., KROONENBERG, S.B., VAN LISSA, R.V. & DE ROEVER, E.W.F. (1984) An explanation of the geology of Suriname. *Geol. Mijnbouw* 27, 33–82.

CHOUBERT, B. (1974) *Le Précambrien des Guyanes.* Mém. Bur. Rech. Geol. Minières 81. Orléans, 204 pp.

CZAMANSKE, G.K., DESBOROUGH, G.A. & GOFF, F.E. (1973) Annealing history limits for inhomogeneous, native gold grains as determined from Au–Ag diffusion rates. *Econ. Geol.* 68, 1275–1288.

GIBBS, A.K. & OLSZEWSKI, W.J., JR (1982) Zircons U/Pb ages of Guyana greenstone–gneiss terrane. *Precambrian Res.* 17, 199–214.

GRUAU, G., MARTIN, H., LÉVÊQUE, B., CAPDEVILA, R. & MAROT, A. (1985) Rb–Sr and Sm–Nd geochronology of Lower Proterozoic granite–greenstone terrains in French Guiana, South America. *Precambrian Res.* 30, 63–80.

HEIN, F.J. & WALKER, R.G. (1977) Bar evolution and development of stratification in the gravelly, braided, Kicking Horse River, British Columbia. *Can. J. Earth Sci.* 14, 562–570.

HEINRICH, K.F.J. (1966) X-ray absorption uncertainty. In: *The Electron Microprobe* (Eds McKinley, T.D., Heinrich, K.F.J. & Wittry, D.B.) pp. 296–377. John Wiley and Sons, New York.

HERAIL, G. (1984) *Géomorphologie et gîtologie de l'or détritique (piedmont et bassin intramontagneux du NO de l'Espagne).* CNRS Ed., Paris. 456 pp.

HEWARD, A.P. (1978a) Alluvial fan and lacustrine sediments from the Stephanian A and B (La Magdelena, Cinera–Matallana and Sabero) coalfields, northern Spain. *Sedimentology* 25, 451–488.

HEWARD, A.P. (1978b) Alluvial fan sequences and megasequences models, with examples from Westphalian D–Stephanian B coalfields, Northern Spain. In: *Fluvial Sedimentology* (Ed. Miall, A.D.) pp. 669–702. Can. Soc. Petrol. Geol., Calgary, Memoir 5.

HURLEY, P.M., KALLIOKOSKI, J., FAIRBAIRN, N.W. & PINSON, W.H. (1968) Some orogenic episodes in South America by K/Ar and whole rock Rb/Sr dating. *Can. J. Earth Sci.* 5, 633–638.

KALLIOKOSKI, J. (1965) Geology of North Central Guiana Shield. *Bull. Geol. Soc. Am* 7, 1027–1049.

KESSE, G.O. (1982) The occurrence of gold in Ghana. In: *Gold '82. The Geology, Geochemistry and Genesis of Gold Deposits; University of Zimbabwe, Harare* (Ed. Foster, R.P.) pp. 645–659. Spec. Publ. 1. A.A. Balkema, Rotterdam.

KRAPEZ, B. (1985) The Ventersdorp Contact placer: a gold-pyrite placer of stream and debris-flow origins from the Archean Witwatersrand Basin of South Africa. *Sedimentology* 32, 223–234.

Larsen, V. & Steel, R.J. (1978) The sedimentary history of a debris flow-dominated Devonian alluvial fan: a study of textural inversion. *Sedimentology* **25**, 37–59.

Lasserre, J.L., Ledru, P., Manier, E. & Mercier, D. (1989) Le Protérozoïque inférieur de Guyane: révision lithostructurale. Rapport 89GUF023. Bur. Rech. Geol. Minières, Orléans.

Ledru, P., Milesi, J.P., Vinchon, C., Lasserre, J.L. & Manier E. (1987) Etude structurale et gitologie de l'or dans les séries conglomératiques de l'Orapu (Guyane française). Rapport 87GUF248. Bur. Rech. Geol. Minières, Orléans.

Ledru, P., Milési, J.P., Vinchon, C., Johan, V., Marcoux, E. & Ankrah, P. (1988) Géologie et gîtologie de l'or des séries birrimiennes du Ghana. Rapport 88AFO122. Bur. Rech. Geol. Minières, Orléans.

Ledru, P., Eko N'Dong, J.E., Johan, V., Prian, J.-P., Coste, B. & Haccard, D. (1989) Structural and metamorphic evolution of the Gabon Orogenic Belt: collision tectonics in the Lower Proterozoic. *Precambrian Res.* **44**, 227–241.

Ledru, P., Lasserre, J.L., Manier, E. & Mercier, D. (1991) Le Protérozoïque nord guyanais: révision de la lithologie, tectonique transcurrente et dynamique des bassins sédimentaires. *Bull. Soc. géol. France* **162**, 627–636.

Manier, E. (1992) *Les conglomérats aurifères de Guyane Française (Protérozoïque inférieur): dynamique des bassins sédimentaires et contrôles des minéralisations.* Thèse Doct. Ecole Nationale Supérieure des Mines de Paris 1990. Mém. Sci. Terre 17.

Manier, E. & Mercier, D. (1989) Origin of gold in the Orapu Precambrian conglomerates, French Guyana. In: *Abstracts, Int. Symp. 'Gold 89 in Europe', Toulouse*, p. 43. Blackwell Scientific Publications, Oxford.

Marion, G. & Vannier, M. (1983) Dosage par double mesure au microanalyseur à sonde électronique automatisé. *J. Microsc. Spectrosc. Electron.* **8**, 31–46.

Miall, A.D. (1977) A review of the braided river depositional environment. *Earth Sci. Rev.* **13**, 1–62.

Miall, A.D. (1978) Lithofacies types and vertical profile models in braided river deposits: a summary. In: *Fluvial Sedimentology* (Ed. Miall, A.D.) pp. 597–604. Can. Soc. Petrol. Geol., Calgary, Memoir 5.

Milési, J.P., Feybesse, J.L., Ledru, P., Dommanget, A., Ouedraogo, M.F., Marcoux, E., Prost, A., Vinchon, C., Sylvain, J.P., Johan, V., Tegyey, M., Clavez, J.Y. & Lagny, P. (1989) Les minéralisations aurifères de l'Afrique de l'Ouest. *Chron. Rech. Min.* **497**, 3–98.

Minter, W.E.L. (1976) Detrital gold, uranium and pyrite concentrations related to sedimentology in the Precambrian Vaal Reef placer, Witwatersrand, South Africa. *Econ. Geol.* **71**, 157–176.

Nami, M. (1983) Gold distribution in relation to depositional processes in the Proterozoic Carbon Leader placer, Witwatersrand, South Africa. In: *Modern and Ancient Fluvial Systems* (Eds Collinson, J.D. & Lewin, A.J.) pp. 563–575. Int. Assoc. Sediment. Spec. Publ. 6. Blackwell Scientific Publications, Oxford.

Naylor, M.A. (1980) The origin of inverse grading in muddy debris flow deposits — a review. *J. Sedim. Petrol.* **50**, 1111–1116.

Nemec, W. & Steel, R.J. (1984) Alluvial and coastal conglomerates: their significant features and some comments on gravelly mass-flow deposits. In: *Sedimentology of Gravels and Conglomerates* (Eds Koster, E.H. & Steel, R.J.) pp. 1–31. Can. Soc. Petrol. Geol., Calgary, Memoir 10.

Nielsen, T.H. (1982) Alluvial fan deposits. In: *Sandstones' Depositional Environments* (Eds Scholl, P.A. & Spearing, D.) pp. 49–86. Am. Ass. Petrol. Geol., Tulsa, Memoir 31.

Oberthür, T. & Saager, R. (1986) Silver and mercury in gold particles from the Proterozoic Witwatersrand placer deposits of South Africa: metallogenic and geochemical implications. *Econ. Geol.* **81**, 20–31.

Pretorius, D.A. (1981) Gold and uranium in quartz-pebble conglomerates. *Econ. Geol.*, 75th anniv. vol., 117–138.

Reid, A.M., Le Roex, A.P. & Minter, W.E.L. (1988) Composition of gold grains in the Vaal placer, Klerksdorp, South Africa. *Miner. Deposita* **23**, 211–217.

Segnit, E.R. (1961) Petrology of the zinc lode, New Broken Hill Consolidated Ltd, Broken Hill, New South Wales. *Proc. Aust. Inst. Mining Metall.*, **199**, 87–112.

Sestini, G. (1973) Sedimentology of a paleoplacer: the gold-bearing Tarkwaian of Ghana. In: *Ores in Sediments* (Eds Amstutz, G.C. & Bernard, A.J.) pp. 275–305. Springer-Verlag, Heidelberg.

Shackleton, R.M. (1986) Precambrian collision tectonics in Africa. In: *Collision Tectonics* (Eds Coward, M.P. & Ries, A.C.) pp. 329–349. Geol. Soc. Lond. Spec. Publ. 19. Blackwell Scientific Publications, Oxford.

Sims, J.F.M. (1977) A geologia da Série Jacobina aurífera nas vizinhanças de Jacobina; Bahia, Brasil [The geology of the auriferous Jacobina serie in Jacobina area, Bahia, Brazil] *Bol. de Estudos Sociedade de Intercâmbia Cultural e Estudos Geológicos* **17**, 223–282 (in Portuguese).

Smith, N.D. (1970) The braided stream depositional environments: comparison of the Plate River with some Silurian clastic rocks, north central Appalachians. *Bull. Geol. Soc. Am.* **81**, 2993–3014.

Smith, N.D. & Minter, W.E.L. (1980) Sedimentological controls of gold and uranium in two Witwatersrand paleoplacers. *Econ. Geol.* **75**, 1–14.

Steel, R.J. (1988) Coarsening-upward and shewed fan bodies: symptoms of strike-slip and transfer fault movement in sedimentary basins. In: *Fans Deltas — Sedimentology and Tectonic Settings* (Eds Nemec, W. & Steel, R.J.) pp. 75–83. Blackie, London.

Steel, R.J. & Gloppen, T.G. (1980) Late Caledonian basin formation, western Norway: evidence for strike-slip tectonics during infilling. In: *Sedimentation in Oblique-slip Mobile Zones* (Eds Balance, P.F. & Readings, H.G.) pp. 79–103. Int. Ass. Sediment. Spec. Publ. 4. Blackwell Scientific Publications, Oxford.

Steel, R.J., Maehle, S., Nilsen, H., Roe, S.L. & Spinnangr, A. (1977) Coarsening-upward cycles in the alluvium of Hornelen Basin (Devonian), Norway: sedimentary response to tectonic events. *Bull. Geol. Soc. Am.* **88**, 1224–1134.

Swapp, S.M. & Onstott, T.C. (1989) P-T-time characterization of the Transamazonian orogeny in the Imataca complex, Venezuela. *Precambrian Res.* **42**, 293–314.

Vidal, M. (1986) Les déformations éburnéennes de l'unité Birrimienne de la Comoé (Côte d'Ivoire). *J. Afr. Earth Sci.* **6**, 141–152.

Wall, V.J. (1977) Non-sulphidic zinc-bearing phases and the behavior of zinc during metamorphism. In: *Abstracts, Second Aust. Geol. Con.*, p. 70. Geol. Soc. Aust.

Index

References to figures appear in *italic type*.
References to tables appear in **bold type**.

abrasion 23, 26, 28, 30, 31
accretion
   vertical 360, 372, 375, 406, 428
   see also lateral accretion; stacking, vertical
adhesion lamination 405, 406
adjustment cycles 470
aeolian deposits, interchannel 410
Afon Dyfi 24–34
Agda Dal Formation 413, 420
aggradation 451, 465, 469
   channel 161, 163, 375
   floodplain 321–5
   Raba River 307
   sudden switch to 462, 464
   vertical 379, 481
Agios Nektarios fan complex *245*
Aix en Provence structure 426
Albany River, Ontario
   cyclicity in deposits 73–4
   postglacial subarctic systems features 74
   sediment ice rafting 63–76
   suspended and solution load 64
   vertical bank sections 71, *72*, *73*
Alcotas Mudstones and Sandstones Formation 363, 377
algal mats 428
allocyclic mechanisms 489, 490, 516
   controlling facies
      association distribution, Seilao Member 497–9
   and small-scale cyclicity 514–15
alluvial aprons 289
alluvial architecture 319
   model described 320–1, **321**, **322**
   Seilao Member 497
alluvial deposits
   facies distributions in 211
   reworking of 82, 86
alluvial fans 47, 48, 411, 469, 509, 514
   aggradation by fan-lobe stacking 266
   cascading 264, *265*, 272
   in dry regions 290
   entrenched 246
   ephemeral flood sedimentation 493
   fan-fringe deposits 246
   gold deposits in 548–50
   lateral relationships in, Seilao Member 496, *497*
   minor/small 224, 289, 290
   outer fan sediments 433
   point-source, Bermejo basin 225–6, 229–30, 232
   Pre-Kaczawa alluvial fan 293–303
   proximal 433, 435
   deposits 432–3, 558
   Quaternary, Crete 235–73
      alluvial-fan complexes 241–71
      depositional setting 236–41
      as periglacial features 262–3
      Recent stage of destruction 267
      Régina Basin 558–9
      sieve deposits on 270
      size of 272
      stacked 241, 246, 265, 267, 271, 272
         telescoping mode 266
         see also fan lobes, stacked
      steepness of 247, 263–4, 268
      variation in deposits 224–5
   see also fan
alluvial plains 290, 406, 420, 484
   facies assemblages *281*, 282–3
alluvial sediment, in-transport modification of 23–34
alluvial systems
   with aeolian interfluves 410
   NE Madrid Basin 278–91
      Baides fluvial distributary system 285–6, 290
      Jadraque system 283–5, 290
      La Alarilla fluvial distributary system 279–83
      Las Inviernas fan 289
      minor fans and slope deposits 289
      Tajuña fluvial distributary system 286–8
alluvial-fan complexes, Crete
   debris flow-dominated fans 242–4
   Holocene fans and trench deposits 267–71
   marine incursion record 244–6
   stream flow-dominated fans 246–67
alluvium, gold-bearing, Early Proterozoic 553–67
Alveoline Limestone Formation 338
Alveoline Limestone Group 347
Amazon River mouth bar 207
amphibolite 554
anastomosing channels/rivers *65*, 74, 168
anchor ice 68
   pebbles from 69
Andersson Land Formation 411–13, *412*, 414, 419
Andes 222–3
andesite, in breccia 457
anhydrite crystals 491, 494
anoxic waters 512
antecedent rivers/streams 226, 228, 229, 451–2, 518
   high-discharge 465
Arbuthnott Group 452
Arenicolites 199
arenites 480, 483–4
Aridisols 257, *262*, 263, 266, 271, 356, 360

armoured bed composition 547
armouring 37–8, 40
   Nahal Hebron 45, 47, 48
   of newly scoured channel-floor surfaces 250–1
   poor development of 43
armouring ratio 47
   changes in 40–1
Arouany Formation 555–6, *557*
arroyos 338, 349, *350*, 351–2, 357, 359, 360
asymmetrical basins
   fluvial sedimentology 445–6
   narrow alluvial plains 447
Au anomaly *see* gold soil anomaly autocyclic
      mechanisms 489, 499
   not responsible for cyclicity 514
avalanching 387
avulsion 320–1, 327, 331–2
   channel 189
   downtilt 414
axial channel systems 216

back-filling, of fan trench 266, 267
Baides anticline 285
Baides fluvial distributary system 285–6, 290
bajada apron 224, *226*
bajada facies, Bermejo basin 224–5, 229, 230
bajada fans 224, 231
bank collapse 159
bankfull width, for Mill Pond channel deposit 136–7
bar complexes 163, 164, 165, 282
   sequences 161, *162*
bar-edge deposits 480
bar-head zone 252, 253, *258*, 371
bar migration 289, 371, 372, 373, 386
bar platforms
   deposits 308
   structures 511
bar supraplatforms 511
bar surface paving 314–15
bar-tail zone 252–3, 253, *258*
bar tails 312, 511
bar-top deposits 478, 480, 481
Barents Sea region 151–2
barrier-spit deposits 203–4, 208
bars 25, **29**, 153, 164, 286, 371, 511, 558
   alternate 39
   bank-attached 92, 96, 511
   braid bars 112
   channelized *429*
   combined cross- and planar-stratified 155, 155, *156*, 157
   composite *158*, 158–9, 159, 165, 379
   composite-compound 156
   compound 156, 158, 159, 165
   counterpoint 82
   cross-channel 444
   cross-stratified *154*, 154–5, 157
   in-channel 296
   lateral 363
   mid-channel *see* mid-channel bars
   muddy, stacked 390
   multiplication of 253
   planar-stratified *154*, 155, 156
   sheet 253
   simple *155*, 156, 159, 161, *162*, 165
   see also channel bars; gravel bars; longitudinal bars; mouth bars; point bars; side bars; transverse bars
   basal 554
Cromwell Member 476
   see also flood basalt
basin-fill trend, upward-fining 559
basin management, and sediment supply 315
basin symmetry, control on fluvial lithofacies 439–48
basins
   arc-adjacent, rapidly subsiding 452
   asymmetrical *see* asymmetrical basins
   controls on alluvial architecture of 489–90
   en échelon 559, 566
   subsidence of 490, 517
      see also drainage basins; foreland basins; subsidence
   symmetrical *see* symmetrical basins
Båsnæring Formation 151, *152*, 152
Bassano Thrust *503*, 503, *504*, *505*
batholiths, cogenetic 477
   unroofing of 477
Beatton River
   deposits 129, 130
   meander growth-patterns 127
   point bars 148
   scroll bars 128, 145
      initiation of 133–4
bed configuration, related to particle mobility and grain size 16–18
bed load 544–5
   deposits 558
   increase in, Raba River 310
   materials of 77
   and mean velocity of water discharge 84, *85*
   sensitivity of 77
   textural features, Parsęta River 79–81
   transport forms of 82–3
bed load streams 310, 315
bed load transport 16, 86, 379
   effective width of 83
   efficiency index 83
   Parsęta River 77–87
      morphodynamic transport zone 81–2
      textural features of bed load 79–81
   and secondary circulation 51
bed load transport rate 83–4, 86
   universal formula for, Parsęta River 84–6
bed material
   changes, ephemeral streams 37–48
      vertical exchange 43–4
   Nahal Hebron 39–41, *42*
   Nahal Og 41, *42*
bed roughness 4
   lower- and upper-regime 308, 309
bed topography 102, 109
bedding planes, master 203
bedding sequences/cycles
   relationships between 127, *128*
   thickening 120, *121*, *122*, 124, 130

bedding sequences/cycles (*cont'd*)
  thickening–thinning 120, *121*, *122*, 124–5, 125, 127, 128, 129, 130
  thinning *122*, 124, 130
bedding trends 128
bedform migration 16, 93, 98, 134, 371–2, 374
  in a meander bend 51–60
  see also bar migration; dune migration; ripples, migration of
bedform stability diagram, new 11–20
  flume experiment 13–16
  stability fields of bedforms under unidirectional currents 16–18
bedform stability diagrams, disadvantages of existing diagrams 16
bedforms
  affecting sediment movement 43
  classical concept of transport *52*, 52
  orientation of, and lateral sediment transport 57
  rotation of 57, 60
  stability fields under unidirectional current 16–18
  upstream migration of 97
  and water depth 481
Begudian Stage 426, 428
Belluno Syncline *504*
Belluno Thrust *503*, 503, *504*, *505*
Bermejo basin 221
  a closed basin 226
  controls on facies patterns and sediment composition 229–31
    alluvial-fan dimensions 230–1
    drainage basin controls 230
    implications of drainage patterns and sedimentary facies 231–2
Bermejo River, headwaters of 226, 226–7
Bermejo watershed 222, 224
  sub-basins of 226–9
Bighorn Basin 383–97
biotite 555
bioturbation 145, 198–9, 199, 202, 203, 204, 205, 284, 285, 390, 392, 494, 512
black ice 68
Boniches Conglomerates Formation 363, 377
Bottletree Formation 475
boundary shear stress 4
bounding surfaces 98
  as channel erosion surfaces 161
  major (MBS) 365, 366, *369*, 374, 378
  minor *367*, *368*
  significance of 159–63
  type-3 445, 448
  type-4 444
  type-5 445
Bowen Basin 195, 196, *197*, 207
Brahmaputra River 151, 163, 164
  confluence mouth bar *93*, 93
  confluence scour 95–6
  dune-generated cross-sets 158
  reverse currents 97
  stage fluctuations 164, 165
braid belts 355
braided channels 411
  ephemeral 289, 290

rapid switching 479
braided river/stream deposits 99, 349, 410, 413, 458, 473, 565
  Arounay Formation 556
  ephemeral streams 407
  Esplugafreda Formation 354–5, 360
  from extrabasinal sources 474
  Régina Basin 559, 565
braided river/stream systems 283–4, 288, 429–30, 486, 507, 509, 548
  Cañizar Sandstones 371, 373, 379
  Platte-type 482, 485, 556
braided rivers/streams 91, 96, 116, 163–4, 164, 253, 285, 286, 306–7, 447, 485
braiding 253
braidplains 290, 403, 413, 417, 452, 454–5, *468*, 468, 484
Brandywine Creek, Pennsylvania 322
breakups 66–7
breccia 285, 289, 435
  carbonate 43
  ferruginous 430
  horizontal stratification 457
  unsorted 456–7
  volcanic 476
  volcaniclastic 468
breccia flows 432
Brent Group, mouth bar sandbodies 207
Brownstone Beds 156
Buffalo River 186
Buntsandstein 363, 373–4
Burdekin River delta 207
burial depth 327, 329
burrows 491

Cailleux Roundness Index 25, **29**
Calatayud-Teruel basin 363
Calcaire de Rognac 430
Calcaire de Rognac Formation 434, 436
Calcaire de Vitrolles Formation 430, 430–1, 434
calcrete 406, 476, 486
caliche 347, 356, 442, 445, 448
Camp Rice Formation 440, 442, 445
  depositional model 446–7
Cañizar Sandstones Formation 363–4
  climatic control 377
  composition and age 364
  facies associations 373–4
  fluvial facies architecture 365–72
    sandy bedforms and elementary facies interpretation 371–2
  fluvial styles 374–6
  tectonic control 377–9
carbonate concretions 432
carbonate encrustations *see* pedogenic encrustations
catchment basins 231, 232
Catskill Magnafacies, interpretation of bedding geometry 104–9, 111–12
Celsius Bjerg Group 401, *402*, 403, 404, 420
cement
  calcite 246, 505
  carbonate 246

cement (cont'd)
   goethite 388
channel abandonment 106
channel-bar sheets 250, 251–3, 255, *256*, *258*, *259*, *261*, 272
channel bars 101, 308, 309
channel bed, changes in position of 82
channel-belt depth and width 536
channel belts 286, 288, 290, 320–1, 527, 533
   multistoreyed 534
   width of 201
channel-bend migration 127
channel-bend wavelength, variations in 103–4
channel bends
   asymmetry within 104
   downstream translation of 104
channel complexes 286, 393
   multistorey, multilateral 390–1
channel confluences, morphology and facies models of 91–9
channel cutoffs 392
channel deposit proportion 325–6, 330, 332
channel deposits 73, 217, 320–1, 323, 325, 366, 493
   gravel-bed 295, *297*
   low-sinuosity 116
   multistorey character 329–31, 334
   see also channel sandstones
channel depth 201
channel discharge
   fluctuations in 111
   reduction in 109
channel entrenchment/incision 350, 352–3, 357, 511
channel-fill sequences 164, 375, 427–8
   Arouany Formation *557*
   Greybull interval 386–92, 394–6
   a transitional deposit 428
channel fills 284, 363, 385, 407, 509, 516
   German Creek Formation 208
   gravel 287
   laterally accreted 200
   underwater 260
channel–floodplain lithofacies relationships *441*
channel-floor pavement 558
channel flow 201
channel form and fill 159
   Fugleberget Formation *160*, *161*
   Hestman Member *161*
channel geometry 536–7
   box-shaped 462
   changes in, Raba River 206–8, 310–11
channel location
   for overbank deposits 189–90
   unknown, use of grain size–distance plots 192–3, *193*
channel migration 101, 320, 331–2, 447
   by expansion 109
   lateral 414
   lower frequency of 535
   by translation and expansion 108–9
channel-mouth sands 204
channel networks, enlargement of 468
channel plugs 199
channel sandstone wing *180*, *181*, 181

channel sandstones 168, *180*, *181*, 181, 387–9, 427–8, 446, 534
   asymmetrical basins 448
   constraints on location of 182
   multistorey 443, 445, 448
   symmetrical basins 442–4, 448
channel sequences *162*, 165, *510*, 510, 514
   incised by shallow channels 510–11
channel stability, Albany River 74
channel switching 447, 448, 479
channel width 83, 201
channel width/depth ratios 103, 164, 201
   Raba River 307, 311
channelcuts, ribbon-like sandbodies 355
channelization
   decrease in 350
   Raba River 310, 311, 315
channelized deposits, probability of 215
channelized sequences, single-storey 509
channels
   braided 282, 283, 363, 376
   conglomerate-filled 457
   cyclic variation in size 412
   degradation of 315
   gravel, imbricated 288
   intrabar 39
   large, sandstone-dominated 386, 387–9
   nested 462
   stacked 386, 531
   transverse fill cross-stratification 282
Chinji Formation 523, *525*, 531, 534
Chinle Formation 358
Chondrites 199, 202
chronosomes 416–21
   Gunnar Anderson Land 416–21
chute channels 105, 145, 146, 373, 375, 511
chute cutoff 111
chute infills 312
chute levee deposits 138
chutes 146, 480
Claret Formation 338, 339, 340, *342*, 347
clast deposits, continental 555–9
clast fabric, disorganized 242, 243
clast volume 31
clasts
   abrasion by granular removal or breakage 30
   anomalously large 71
   bladed 30–1, 34
   breakage of 30, 33, 34
   carbonate 432
   chert 387
   coarse 75, 252–3
   extraformational 410
   graded 248, 255
   intraformational 410
   La Alarilla system 283
   Lena conglomerate 477
   natural weight loss 33
   Patsianos fan 245
   plates 30–1
   quartz arenite 477
   roundness of 28–9
   volcanic breccias 476

clasts (cont'd)
  see also Talchir clasts
clay plugs 392
climate
  affecting Provence Basin 436–7
  Quaternary, Crete 238
  and soil formation 357
climatic change
  affecting Gunnar Andersson Land 417–18, 419, 420, 421
  and Messinian deposits 515
climatic cycles, orbital-related 517
climbing strata 484
Cloughton Formation 169
Cloverly Formation 383, 385
clusters, effects of 8–9
coal seams, German Creek Formation 198, 205
coarse member deposits, Esplugafreda Formation 341, *342*, 343, *344*, 347, 348, 349, 355–6
coarsening-upwards facies 371
coarsening-upwards megasequences 498–9, 514
coarsening-upwards sequences 207, 331, 332, 458, 462, 496, *498*
coarsening-upwards trends 468
coastal retreat, Holocene 260
collisional tectonics 554, 559
compaction
  differential 432
  effects of 327, 329, 334–5
  Gunnar Andersson Land 419
'compensation beds' 129
'compensation cycles' 123
compression 239, 357, 359, 378, 498
  glacitectonic 303
confluence separation zone bars 96, 99
Conglomerat des Touars 433
conglomerate 279, 338, 341, 354, *429*, 435, 442, 452, 556
  basal 477, 481
  boulder 456, 457
  boulder–pebble 474
  Crawton Group 453–4
  fine-grained *297*
  French Guiana, gold mineralization in 562–6, 567
  intraformational 157, 159, 161
    and extraformational 527, 530, 531
  limestone 246
  massive 478
  non-volcanic 467
  pebble 527
  pebble–cobble 458, 508, 511, 557, 558
  planar- and cross-stratification 366, 370, 371
  polymictic 478
  quartzite 364
  Rødsten Formation 409
  Seilao Member 490–1, 492–4, 499
  stacked beds 284, 507
  tabular beds 287
  trough cross-stratified 509
  Witwatersrand placer 548, 549–50
conglomerate bodies, multistorey 279
connectedness 326–7, *328*, 330, 332, 334
  see also interconnectedness

consequent streams 451–2
convection structures 297–9, 301
Copper River delta 207
Corralitos basin 442
Crawton Group 452, 453, 454, 462, 467
Cretan block, tilting of 241, 242
crevasse beds 356
crevasse channel sandstone 182
crevasse channel/splay lobe width ratio 181
crevasse channels 168, 178
crevasse splay lobes 168, 174, 178, 179, 182, 183
  coalescence of 177
crevasse splay sandstones
  composite 172–4, 182
  estimated volumes 179–80
  geometries of 167–83
  length/thickness ratios 172, 174
  shape of 168, 174, 178
  single splay 182
  stacked 172, 174, 176
  width/thickness ratios 171, 174, 176
crevasse splays 332
  deposition 168, 202
    types of 179
  evolution of *177*, 178
  in oil wells 182
  relations with fluvial channels 180–2
crevasse subdelta development 177–8, 179, 182
Crocker Formation 196
Cromwell Member 473, 474, 486
  facies analysis 478–84
  siliciclastic units, palaeogeographic scenario for *485*, 485
cross-beds/bedding 146, 172, 201, 352, 355, 458, 482
  diffuse 284
  planar 199, 443–4, 448
    cosets of 444
    sets 375, 482
    stacked 145
  tabular 370
  trough 199, 295, 404, 413, 429, *430*, 443, 445, 448
    small-scale 201
cross-lamination 201, 405, 411, 413
  flat 201
  ripple drift 428
  see also ripple cross-lamination
cross-sets 152
  deformed *154*, 155, 157
  descending 158
  dune-generated *154*, 161, *162*
  migration up IHS surfaces 144
  planar 135, 139, 144, 145, 146, 148
  planar-stratified 159
  trough 140
    small-scale 159
cross-strata
  interdigitating *154*, 157, 158
  large-scale 154, 510–11
  overturned *154*, 155, 158
  planar, sets of 142
cross-stratification 93, 96, 98, 99, 287, 288
  bipolar (herringbone) 393
  climbing ripple 122, 123, 124, 125

cross-stratification (*cont'd*)
  cosets of *158*, 158
  foreset 119
  large-scale 105, 109
  low-angle 253, 458
  planar 282, 287, 288, 308, 527, 556
    climbing *141*
    large-scale 119, 122
    low-angle 491, 494
    ripple 123
    small-scale 105, 106, 108, 109, 111, 119, 122
  trough 140, 282, 287, 288, 308, 491, 527
    large-scale 108, 109, 111, 119, 122, 308, 393
    small-scale 373
  units, in ribbons 283
  see also hummocky cross-stratification
crustal extension 439
crustal shortening 559
crustal thickening 561
crusts
  calcareous 287
  calcrete 289
  carbonate, pisolithic and laminated 285
  pisolithic 285, 289
current ripples 371, 375
cut-and-fill pebble gravel 308
cut-and-fill structures 530
cycles
  fifth-order 51–6
  in Messinian deposits 514–16
  small-scale 497, 514
    fining-upwards 499
    interpretation of 514–15
cyclicity, in non-marine foreland-basin sedimentary fill 501–19

dacite, hornblende-biotite 457, 467
debris flow deposits 428, 469, 557–8
  Cretan alluvial fans 242–4, 255–7, 259, 271
  Lower Old Red Sandstone 451, *455*, 456–7
debris flows 310, 507, 548
  clast-supported 285
  volcaniclastic 451
deformation 498, 523
  Caledonian 403
  Gunnar Andersson Land 419
  schistosity 554–5, 559
  soft sediment 152, 202, 203, 295, 480, 481
  synsedimentary 505
  in transtensional regime 555
deformational structures 293, 294, 297–301
degradation, inter-eruption 464
deltas, modern 207
depocentre migration 435
deposition 53
  in 8 November 1986, Nahal Hebron flow event 44–7, 48
  Bermejo basin 224
  rapid, from hyperconcentrated flow 159
  selective 24
  from suspension 159, 185, 202
  tractional 255
  see also fluvial deposition
deposition rates, determination of 334
depositional processes, and fluvial geometry 348–56
deposits, ice rafted 68–9, 71
  deposition in preferred environments 67, 68, 71
  reworking of 71, *72*
  surface occurrences 68–9, 71
  vertical bank sections 71, *72*, *73*
Deschutes Formation 462
desiccation 476, 481
desiccation cracks 406, 409, 411, 413, 491, 494
detachments 523
Dhok Pathan Formation 523, *525*, 528–9, 533, 534
Dhok Pathan/Soan boundary, changes at 525, 531
diapirism 433, 435
dilated framework gravels *313*
Dinaric thrust belt 503
discharge
  alterations in, Raba River 310
  and bed load 84, *85*
  channel 109, 111
  lowered 158
  surge-type 270, 272
  variability of 164
disconformities 111, 531
dissolution–recrystallization processes 558
dolostones 505
Donjek glacio-fluvial model 73, 74
draas *404*, 406
drainage basins, control on facies and sediment composition 230
drainage geometries, and sediments deposited, Bermejo basin 224–6, 232
drainage networks 222, 231–2
  large 224, 226
drainage reversal, local 433
drapes 239
  clay 387, 512
  mud 170, 174, 371, 372, 376
  mudstone 491
  sandstone 509
  silt 69, *70*, 71, 119–20
  siltstone 159, *160*
dune height
  change during migration *52*
  secondary dunes 56
dune height–channel depth ratio 82–3
dune length, correlation with depth 82
dune migration 52, 96, 308, 406, 429, 443, 482
  direction of 55, 56, 57
  rate of 54, 55
  secondary dunes *56*
  upstream 99
dune orthogonals 52–4
  in a survey field 55–6
dunes 19, 82, 144, 145, 387–8, 390, 406, 410, 493, *556*, 558
  aeolian 410
  barchanoid 406
  coalescing 158
  in-channel 494
  inward rotation of crestlines 109

dunes (cont'd)
  secondary
    and lateral transport   55-6
    and transverse transport   54
Dunnottar Group   452

East Robledo fault   442
Eastern Creek Volcanics   473, 474, 475-6, 485-6
Eastern Fold Belt   475
Ellensberg Formation   464
Eller Beck Formation   169
Elsa Dal Formation   413-14, 420
entrainment
  of clasts   23-4
  selective   27
  shear stress for   544, *545*
  of spheres   3-10
entrenchment   352-3, 357
ephemeral stream/channel deposits   338, 360, 406, 410, 413
  distinguishing between ribbon- and sheet-like deposits   355
  facies types, Esplugafreda Formation   **345-6**
ephemeral stream-clay playa complex   339
ephemeral stream interpretation, criteria for   346-8
ephemeral streams   285, 338, 411
  bed material changes in   37-48
  non-coincident flood waves   98
  surge-type discharge   270, 272
  thin-bedded facies of   410, 411
epsilon cross-bedding   414-16
erg, Udkiggen Formation as   406
erosion, stoss-side   13, 14
erosion surfaces   105, 106, 107, 109, 355
  channel   165
  inclined   411
  second-order   509
erratics   73
Eslida Mudstones and Sandstones Formation   364
Esplugafreda Formation, fluvial architecture   338-60
  depositional model   357-60
    four-stage tectonosedimentary model   357-60
    tectonic influence on megacycle development   357
  facies-related building blocks   348-57
  geological setting and facies types   338-48
  fluvial architecture outline   341-7
estuarine tidal channel complexes, tripartite   394-5, 397
eustasy, and non-marine unconformities   383
evaporites   507, 522, 523
Ewen Block   475
extension   239
  lithospheric   485

Fadalto alignment   516
Fall River Sandstone-Lakota Formation contact   385
'false ponding'   121
fan apices   271
  connection to palaeovalley outlets   241-2
fan channels   267
fan core facies   288

fan deltas   246, 260, 265
fan lobes   246, 433, 559, 566
  carbonate-encrusted   246
  deltaic deposits   295
  shingling arrangement of   303
  stacked   246-7, 263, 266, 271
fan-toe deposits
  in a sandflat environment   496
  subaqueous   260, 262, *264*
fan trenching/trenches   266, 267, 270, 271, 272, 286-7
  and piedmont uplift   266
fanhead deposits, coarse   287
fault dip, influence of   213-14
fault orientation and palaeoslope   215
fault systems, Provence Basin   435-6
faults   228, 334, 548
  active, alluvial suite on hangingwall   216
  extensional   299-300, 301, 302
  graben boundary   378
  listric   239
    normal   300, *301*, 301
  movement on   211-12
  normal   239, 442
  planar normal   299-300, *301*
  properties controlling flow line concentrations   213-15
  reverse   285, 286, 290
    movement on   223
  sinistral   559
  strike-slip   554-5, 559, 566
    active — sedimentation along   293-303
    sinistral   555-6
  thrust   360
  see also thrusting
  transcurrent   559
ferruginous weathering horizon   430
filtration flow, intrastratal   259
fine member deposits, Esplugafreda Formation   341, 343, 348
fining
  lateral   109, 252, 351
  vertical   108, 109
fining-thinning upward sequences   279, 287, 352
fining upwards   134
  and accelerated subsidence   517
  Agda Dal Formation   413
  Andersson Land Formation   411
  channel complexes   386, 391-2
  channel margin/overbank deposits   388-9
  sandstone beds   406
  Sykes Mountain Formation   393
fining upwards megasequences   499, 514
fining upwards sequences   119, 295, 332, 356, 375, 389, 458, 496, *498*, 548
  Lena Quartzite and Cromwell Member   478, 480, 482, 483
fining upwards tendencies/trends   105, 109, 111, 279, 287, 343, 354, 370, 374, 378-9, 533
first-order sequences   556
fission-track dating   524
flaser bedding   205, 389
flash floods   310, 315, 347, 352
flexural subsidence   517

flood basalts, continental 473, 485
  intercalated siliciclastic units 473
'flood cyclothems' 145
flood discharge, growth in 310
flood events
  and disconformities 111–12
  ephemeral 494
flood waves, non-coincident 98
floodbasin deposition 134
floodplain deposits 282–3, 363, 435, 447, 448
  asymmetrical basins 445–6
  distal 355–6
  fluvial, Bermejo basin 226
  overbank sediment 188–9
  proximal 355–6
  symmetrical basins 444–5
floodplain lithofacies, Provence Basin 427–30
Floodplain Simulation Model 320–1
floodplains 185, 187
  aggradation, modelled 321–5
floods 168, 315
flow
  areas of high probability 216
  effect of changes in stage 97–8
  effects of fluctuation in strength 479
  gravity induced 290
  stream-dominated 288
flow conditions, calculation of 544
flow depth, decrease in 109
flow events 347
  8 November 1986, Nahal Hebron 44–7, 48
  and burial depth 43
  changing particle positions 48
  and distance of movement 41, 43
  and frontal splay sheetflood deposits 350, 352
  and scroll-bar deposits 148
flow lines
  computer modelling of 211–17
  'ponding' of 216
flow separation, and confluence scour 94
flow separation zones 96
  asymmetrical junctions 98–9
  confluence mouth bars 93
fluidal sediment flow deposits 455, 457–8
fluvial architecture
  Cañzar Sandstones Formation 363–79
  Esplugafreda Formation 338–60
fluvial belts, migration of 378
fluvial channels 527
  Hayburn Wyke 179
  relations with crevasse splays 180–2
  stacked 410
fluvial deposition
  impact of incipient uplift on 521–38
  structural and climatic controls on fluvial systems 401–21
fluvial deposits
  cold climate, Albany River, Ontario 63–75
  dominant geometries of sedimentary bodies 372–3
  interbedded with volcaniclastic deposits 541–70
fluvial discharge, increased 395–6
fluvial fans see alluvial fans; Tórtola fan; Villalba de la Sierra fan

fluvial lithofacies, control of basin symmetry on 439–48
fluvial sequences, non-volcaniclastic (ORS)
  differences between associated volcaniclastic sediments 456
  type I 452–3, 454–5
  type II 453–4
fluvial systems
  Provence Basin 433–4
  terminal 407, 413, 417, *418*, 419
fluvioglacial outwash 296, 302
fold–thrust belts 385
foreland basins 196, 521
  modern, controls on facies patterns and sediment composition 221–32
  Venetian, cyclicity in non-marine sedimentary fill 501–19
foreset bedding, recognition of 509–10
foresets
  planar 429
    large-scale 387
foreshore deposits 203
fossils, Greybull interval 389
framework gravels 309
frazil ice 68
French Guiana
  gold mineralization in the conglomerates 562–6
  Lower Proterozoic, structural setting of 554–62
  Trans-Amazonian orogeny 553–4, 559, 566
Friuli carbonate platform 502, *503*
Frontal Cordillera 222–3, 227
fugichnia *389*, 389, 393
Fugleberget Formation 151
  sheet sandstone bodies 153–9
Fuvelian Stage 426, 427

gahnite 564, 566
Garvock Group 452
Gauss chron 524–5, 534
German Creek delta system *206*, 207
German Creek Formation
  as a coarse clastic wedge 198, 208
  depositional setting 196
  divisions of 196
  economic seams 198
  facies analysis 198–208
    distal mouth bar facies 204–5
    distributary channel facies 199–201
    foreshore facies 202–4
    interdistributary bay facies 199, 201–2
    peat mires 205
    proximal mouth bar facies 204
    repeated progradation and retreat of deltas 207
Gigante Formation 462
glacial–postglacial transition 71
glaciation, Lefka Ori massif 237–8, 262, 266–7, 272
glacigenic deposits 295
glaciokarstic features 238
goethite 387, 388, 389
gold
  brought in by hydrothermal fluids 566, 567

gold (cont'd)
  detrital 566
    remobilization of 566
  minerals associated with 563–4
  origin of 565–6
gold concentrations, Witwatersrand placer 548
gold mineralization, French Guaiana
    conglomerates 562–6, 567
gold particle characteristics 564–5
gold soil anomaly 563
Gozdnica Formation 295
  relation to Pre-Kaczawa fan 301–3
grabens 442
  extensional 299–300
  synsedimentary 295
graded bedding 204
graded-stratified deposits 457–8
grading, inverse 285, 557, *560*
grain diameter, change in 81
grain-flow processes 374, 481
grain hiding 544, 550
grain roughness 16, 253, 544
grain size 163, 199, 354
  and bed response 164
  channel-bar sheets 251
  CU–FU trend 514
  decreasing 496
  distribution of in overbank sediments 185–93
  point bars 102
  variability of 429
  vertical distribution 412
  vertical variation in 111, 458, 459
grain size–distance relationship 186, 189, 190–2
grainfall strata 409, 410
granular removal 30, 33
granulite terrains 554
gravel 242–3, 262, 289, 312, 428
  braid-bar 272
  channel-lag *262, 263, 297*
  coarsening upwards 252, 253
  cobble 308
  cross-stratified 312
  dilated framework *313*
  framework 309
  Holocene 267–8
  horizontally bedded 285
  from intramontane valleys 241
  limestone 236, 247–8, 268, 271
  matrix-supported 285
  openwork 312, 314
  pebble 308–9
  planar and trough cross-bedded 509
  tabular beds 279, 285
gravel bar/sheet deposits 295
gravel bars 410, 558
  deposits 125, 296
  migration of 119
gravel bodies, Baides system 285
gravel infill, multistorey 285
gravel lobes 241, 267
  in-trench 268, 270, 272
  openwork texture 268, *269*
  see also sieve deposits

gravel-plug deposits 308
gravel sheets 248, *250*, 250, *251, 252, 254–5, 256*, 288, 428, 558
  coarsening upwards, misinterpretation of 253, 255
  diffuse 282
  downstream accumulation of 479
  multistorey 285
  ribbon-type geometry 250
greenstone belts, Guyana Shield *554*
Grenville orogeny 403
Greybull interval 383, 385
  channel-fill sequences 386–92
  effects of relative sea-level change 394–6, 397
  tectonic effects 396–7
Gristhorpe Member 169, 178
  composite crevasse splay sandstone 172–4, *175*, 176
  saline influences 176–7
gullies, on fans 267, 270, 272
Gunnar Andersson Land, NE Greenland
  lithostratigraphic units 403–13
  stratigraphic architecture *403*, 416–21
  structural evolution of 417–20
gutter casts 170, 172
gypsum 115, 347, 356

haematite 563–4
half grabens 442
hardground, iron encrusted 430
Haslingden Group 475
Hatch–Rincon basin 442, 447
HCS see hummocky cross-stratification (HCS)
heavy mineral concentrations 477
heavy mineral laminae 556
helical flow cells 93, 94
helicoidal flow, in meander loops 412
Hellenic Arc 239
Hestman Member 151
  sheet-sandstone bodies 153–9
hiatuses 364, 486
high sediment-concentration flows 451
Highland Boundary Fault 452
Himalayan foreland basin 521
Himes Member 383, 385
hornblende, blue-green 527, 535–6
hornblende-andesite 467
Huanghe (Yellow River) 151, 163, 164
Hudson Land phases 416–17
Huermeces reverse fault 285, 286, 290
human activity, diminishing basin retention
    potential 310
hummocky cross-stratification (HCS) 204, 205, 392, *394*
hydraulic conditions, temporal variation in 104
hydraulic sorting 23, 26, 29–30
hydrocarbon accumulations, North Sea 338
hydrothermal processes, and gold 566, 567
hyperconcentrated flood flow deposits 159, 457, 458, 491–2, 499

Iberian Basin 377

Iberian Ranges 363, 377
ice floes
　　dispersal of 66
　　partial removal of rafted load 67
　　preferred grounding areas 67, 68, 71, 75
ice jams 66, 67, 69, *70*
ice lensing 74
ice types, and rafted sediments 68
Iglesia piggyback basin 223, 228
Iglesia valley 227–8
IHS *see* inclined heterolithic stratification
imbrication 96, 279, *281*, 285, 287, 288, 295, *313*,
　　428, 490, 491, 492, 508
inclined heterolithic stratification 134, 135, 136, 142,
　　146, **147**, 201, 390–1, 395
　　surfaces 138–9, 143
　　　　as bounding surfaces 143
inclined stratification 134, 135, 140, 145, 146
palaeo-Indus River 527, 531, 535, 536, *537*, 537
　　shunted off the Potwar Plateau 538
infiltration rates, and sieve deposits 269, 270
inter-eruption deposits, preservation of 464–5
interchannel areas, Rødsten Formation 411
interconnectedness *343*, 354
interdigitation
　　of volcaniclastic and fluvial dispersal systems 465,
　　　　467
　　of white and brown sandstones 531, 535, 537, 538
interdune deposits *404*, 406
interflow areas, between Greybull channels 390–1
interlamination 200
intraclasts
　　claystone 135, 148
　　mudstone 140
IS *see* inclined stratification
isostatic rebound 359

Jadraque system 283–5, 290
　　Jadraque fluvial distributary system 283–4
　　Santiuste and Negredo–Cendejas minor fans 284–5
palaeo-Jhelum River 535, 538
Jogi Tilla thrusts 533

Kalkadoon-Leichhardt Belt 474, 475
　　conglomerate on W flank of 476, 477, 478
　　cover sequence I 475
　　uplift in 486
Kamlial Formation *525*
Kap Graah Group 401, *402*, 403, 404, 420
Kap Graah syncline 417, 420
Kap Kolthoff Group 403, 417
Kapsodasos fans 247
Keuper facies 363
key marker horizons 431, 433, 437
Kicking Horse River model 479
Klaralven River, scroll bars 148
Komitades fan complex 242, *245*
Kotal Kund syncline 531, 533
kyanite 555

La Alarilla fluvial distributary system 279–83, 290
lacustrine deposits 295, 512
lacustrine facies, Provence Basin 430–1
lag times 358
lags 125, *429*
　　basal 244–5
　　channel 295, 481, 534
　　　　sheets 250–1, *256*, *257*, *261*, 272
　　conglomerate 480, 531, 534
　　deflation 257
　　gravel 428
　　pebble 428, 491
　　top surface 202
lakes 216
　　expansion and contraction 515
　　hydrologically close 514–15
laminar shear 243
lamination
　　adhesion 405, 406
　　climbing-ripple 308, 512
　　flat 168, 202–3
　　flat-bedded 406
　　　　parallel 408, 411, 413
　　horizontal 170, 205, 405, 406, 443, 444, 445
　　low-angle 202–3, 405, 406
　　parallel 371, 406, 408, 491, 493
　　plane parallel 172
　　ripple 168, 386, 389, 390, 392
　　trough 168
　　wave ripple 174
Langmuir circulation 128
lapillistone 476
Laramide block uplifts 396
Las Inviernas fan 289
　　change of depositional style 290
lateral accretion 200, 373, 376, 379
　　deposits 101, 102, 391, 412, 413
　　point bars 127–9, 130, 134, 308
　　sequences, bedding disconformities 111
　　surfaces 119, 122, 123, 124, 146, 199, 386, 390
　　units 117, 120, 122–3, 129–30, 392
lava, calc-alkaline 452
Lawn Hill Platform 475
Lefka Ori massif 271
　　piedmont geology and geomorphology 236–41
　　Pleistocene glaciation 237–8, 262
Leichhardt River Fault Trough 475, 476, 484
Lena Quartzite 473, 474, 476, 486
　　basal conglomerate 477
　　composition of 477
　　facies analysis 478–84
lenticular bedding 389
Lessinian high 502, 507
levees 74, 75, 331, 332
　　accretion of 181
lignite seams, as marker horizons 514
limestone 505
liquefaction, intermittent 157
lithification, of Messinian clastic wedge 505
lithology, and abrasion 28
Little Ice Age 310, 315
Little Sheep Mudstone Member 385
loess 185

logs, coalified 203
longitudinal bars 253, 262, *281*, 282, 284, 286, 288–9, 363, 478, 493, 507, 509
   see also scroll bars
Loranca Basin, geometry and lateral accretion patterns in meander loops 115–30
Lower Old Red Sandstone (ORS) (Scotland)
   interbedded volcaniclastic/fluvial deposits 452
   non-volcaniclastic fluvial deposits 452–5
   proposed palaeogeographical scenario *468*, 468–9
   stratigraphy of interbedded deposits 458–69
   volcaniclastic facies 455–8
lutites 287, 289

macroturbulent bursts 157
Madrid Basin 277
   Miocene deposits of 277–8
   North East, alluvial systems of 278–91
magnetic polarity stratigraphies (MPSs)
   correlation with magnetic polarity time scale 525–6
   at Kotal Kund 531
   Potwar Plateau 524
magnetic polarity time scale (MPTS) 524
magnetic reversals 524, 533, 534
magnetite 477, 481
magnetozones 534, 535
Mana Basin 555–6, 561, 566
mantle plumes, and continental flood basalts 485
marine fauna, lower German Creek Formation 198–9
marine incursion, record of, Crete 244–6
marine–non-marine transition 505
marine regression and transgression 239, 241, 271
mass flows 291, 349–50
Maure Massif 433
MBS *see* bounding surfaces, major
mean grain/particle size 25, 26–8, 312
   alluvial-fan deposits 248
   downstream decline in *26*, 26–7, *28*
   Tremp Group 341, *344*
meander belt deposits 414–16
   exhumed **147**
meander bends
   bedform migration in 51–60
   lateral transport in 51
meander loop complexes 117, 123
meander loops, ancient, accretion pattern in 115–30
meander wavelength, Mill Pond channel deposit 137
meanders 186–7
megaripples 376
   migration of 372, 373, 374, 475
   progradation of 371
megasequences 458, 470, *496*, 496, 498–9, 514, 559
   interpretation of 516–18
Mesilla basin 442
Messinian deposits, Venetian foreland basin 505, *506*, 507
   depositional environments and facies 507–12
   interpretation of megasequences 516–18
   small-scale cycles 514–15
   stratigraphy of the succession 513–14
metamorphism, greenschist facies 548, 554–5
metasediments 554

microalluvial fans 69
Microcodium rosettes 347–8
microfolds, asymmetric 555
mid-channel bars 92, 164, 253, 309, 481
   postconfluence 92, 96, 97, 99
MIDAS (Model Investigating Deposition and Sorting of Sediments)
   principles of 544–6
   use of on a Witwatersrand placer 548–50
   verification of performance 546–7
Mill Pond channel deposit 135–48
   ridges 138–44
Mississippi delta, crevasse subdelta deposition 169, 177–8, 179
Mississippi River 186, 322
   channel position for overbank sediment 190
   flood deposition 145–6
   overbank sediments 190–1
Missouri River, flood deposition 145–6
mixed load streams 310, 315
molasse 522, 523
monoclines 503, 505
Montagne Tortue area, gold-bearing alluvium in 562–6
Monte Grappa Anticline *504*, *505*, 505, 516, 518
Montello Conglomerate *see* Messinian deposits, Venetian foreland basin
Montello Thrust *504*
Moodies Group 473
Moranbah Coal Measures 198
Morien Group, ridge-and-swale topography 133–48
morphodynamic transport zone, nature of 81–2
Mount Guide Quartzite 475
   upper 477, 486
Mount Isa Inlier 473
   regional geology 474–6
   sedimentology 476–85
   volcanism, sedimentation and tectonics 485–6
   volcanology 476
mouth bars 169, 260
   avalanche slopes on 92, 97, 98
   bedform orientation on top of 96
   confluence 92–3
   distal 204–5
   migration of 99
   progradation of 98
   proximal 204, 207
   tributary
      avalanche-bounded 93, 96
      sedimentary structures 97–8, *98*
   wave-dominated 207
mud cracks, sand-filled 389
muddy fill deposits 386, 390–1
mudstone 290, 371, 372, 491, 529, 531
   carbonate 430, 431
   lacustrine 512, 514, 516
   marine 505
   playa-flat 496, 499
   silty 279
Murillo point bar, ridge-and-swale topography *134*
Murree Formation *525*
Muschelkalk 363, 364
Myally Shelf 475
Myally Subgroup 475

Nagri Formation 523, *525*, 531, 534
  interpretation of bedding geometry 109, *110*, 111
Negredo–Cendejas system 284–5
Neomonoceratina mouliana 505, 507
Ness Formation 168
  reservoirs 182
nested channels 462
nodules
  calcareous 445
  carbonate 279, 285, 288, 427
Nomikiana fan complex *239*, 242, *245*, *268*, *269*
non-volcanic/volcaniclastic transitions, sharp *461*, 462
normal polarity intervals 524–5
North Australian craton 474–5
North-East Greenland, Structural and climatic controls on fluvial depositional systems 401–21
North Guyana Fault Zone 555, 559, 562, 566
  distribution of gold occurrences *562*
North Pyrenean trough 426
Northwest Syntaxis 522
nugget effect 543

offset across faults, influence of 214–15
oncolites 428
onlap 119
organic material 199, 512
outwash fans 262, 263, 272
  Scott outwash fan 478, 484
overbank deposits 73, 74, 135, 181, 189–90, 202, 308, 355, 372, 407
  in alluvial architecture 319, 320–1, 325, 331–2, 334
  Greybull interval 386, 388–9, 390
  see also crevasse splay sandstones; floodplain deposits; sheetflood deposits
overbank flow, Yellow River 14
overbank sediment, grain-size distribution of 185–93
overpassing mechanism 313–14
overstepping, of sandstone units 174
overthrusting 561

palaeobeaches, raised 241
palaeocliffs 241, 244
palaeocurrents (directions) 331, 414
  Agda Dal Formation 413
  Andersson Land Formation 411, *416*
  Arouany Formation 556
  Cañizar Sandstones Formation 364, 366, 374, 379
  Catskill Magnafacies 105, 107–8, 111
  change in 161
  Cromwell Member and Lena Quartzite 484
  Elsa Dal Formation 413
  Fugleberget and Hestman deposits 163
  German Creek Formation 200, 201, 202
  Greybull interval 387, 389, 396
  Gristhorpe Member 174
  Loranca Basin 115, 123
  Messinian deposits 508, 509, 519
  Nagri Formation 109
  Potwar Plateau 526
  Rødsten Formation 409, 410
  Salt Range 527, *529*, *530*, 530, 535
  Tajuña system 287
  Tremp–Graus Basin 339
  unidirectional 152
  variation at point bars 103
  Witwatersrand placer 548
  Zoologdalen Formation 406
palaeoflow 459
  Mill Pond channel deposit 135–6, 137, 140
  reversals of 452
palaeoslope reversal 420, 421
palaeosols 135, 246, 248, 251, 267, 290, 357, 486, 514, 529
  Camp Rice and Palomas Formations 445, 447, 448
  Esplugafreda Formations 355–6, 360
  mature 338, 343, 356–7, 359
  Sfakia fan complex 256–7, 259, *261*, *262*, *263*
  Tremp Group 339
palaeovalleys
  hanging 237, *239*, 241, 264
  and Jadraque fluvial distributary system 283–4
palaeowinds 410, 411
Palomas basin 442
Palomas Formation 440, 442, 445
  depositional model 446–7
palygorskite 431
paraconformities 514, 517
'paraglacial' effects 271
parallel bedding 392
Paramaca Formation 554–5
parasequences, marine 514
particle clustering, inhibiting entrainment 3
particle–particle impact 33
particle size 3
  and fan slope steepness 263
particles, vertical exchange in ephemeral streams 43–4
Patsianos fan
  fan-toe alluvium 242, 243, 260
    sedimentary sequence 244–6
Patsianos fan system 247
pause planes *347*, 347, 352, 355, 360
pavement horizons 314
PCL *see* primary current lineation
peat accumulation 212
peat mires 205
pebble clusters
  basal scour surfaces 478
  at base of conglomerates 480
pebble deposits, from anchor ice 69
pebble stringers 478, 507
pedogenesis 447
  calcimorphic 431
pedogenic carbonate 271, 347, 356, 363
pedogenic encrustations 259, 263, 265, 267
pelecypods 392–3
periglacial processes 515
Peruvian Orogeny 498
petrocalcic horizon 257, *262*, *263*
phenocrysts, feldspar 476
Pickwick Member 473, 473–4, 476, 482, 486
pisoliths 445
placer deposits 95
planar-strata
  large-scale 154

planar-strata (cont'd)
  low-angle inclined *154*, 155, 157
Planolites 199, 202
Platte River bars 374
playa-flat mudstones 496, 499
playa lakes 439
playa mudflat deposits 385
plug flow deposition 457
point bar lenses 136
point bar platform 55
  bedform migration on 56–7
point bars 24, 27, 39, 55, 82, 144, 390–1, 511
  defined 116–17
  deposits 312, 315, 412
    ancient, bedding geometry of 101–13
    preservation of 148
  geometries of 116–17
  in a meandering river system 4–12
  migration of 102, 104
    by translation and expansion 105–6
  in Mill Pond channel deposit 146, 148
  multiunit 511
  orientation of 414–15
  origination of 311, *312*
  Tórtola fan 119–20, 130
  Villalba de la Sierra fan 120–5, 130
pool–riffle units 511
porosity–depth expressions 327
'potholes' 13
Potwar Plateau 521, 522
  an allochthonous sheet 523
  magnetic polarity stratigraphies 524
Poudingue de la Galante Formation 428, *429*, *430*, 433, 437
Poudingue des Touars Formation 428, 433, 437
Poznań Formation 295
Poznań Lake 295
Pre-Kaczawa alluvial fan 293–303
  deformational structures 297–301
  sedimentary facies 295–7
    alluvial-fan facies 295
    extraglacial aeolian facies 297
    proglacial fluvial facies 296
    proglacial lacustrine–deltaic facies 295–6
precipitation, chemical 430
Precordillera 223
primary current lineation (PCL) 392, 491, 494
progradation 375, 515
  gravel 499
  punctuated by regression 517
progradational sequences, Varanger Peninsula 152
Provence, continental deposits 425, 426–7
Provence Basin
  lithofacies and depositional environments 427–33
    floodplain lithofacies 427–30
    lacustrine facies 430–1
    proximal alluvial-fan deposits 432–3
  palaeogeographic reconstruction 433–5
    fluvial system anatomy 433–4
    lakes extension 434–5, 437
  sedimentation controls 435–7
Pryor Conglomerate 385
Pryor uplift 396

psammite blocks 462
Purilactis Formation *see* Seilao Member, Purilactis Formation
pyroclastic deposits
  Cromwell Member 476
  mass flow redeposition of 457

quartz
  hydrothermal 562
  monocrystalline 477
Quilalar Formation 475

Raba River, hydrologic changes 305–16
  nineteenth-century evolution 306–10
  twentieth-century evolution 310–15
radiometric dating 555
ramps, lateral 228, 360
Ravenscar Group 168, 169
  crevasse splay sandstones 167–79
    composite 172–4
    length, width and thickness data 171, *172*
Rawalpindi Group 523
reactivation surfaces 120, 127–8, 155, 157, 482, 511
  Cañizar Sandstones Formation 370, 372, 375–6, 377
  interpretation of 126–7
  meander loop 117–19, 130
  mouth bars 93
reattachment zone 51–2
recirculating flow 97
recrystallization 548
Régina Basin 555, 556–9, 561, 563, 566
  reduction in size of 556–7
  tectonosedimentary model for *562*
regression, forced 512
reverse flow 96–7, 99
  flow separation zone 95
reworking 498–9
  aeolian 413
  erosional 251
  Patsianos fan 245–6
  of sandflats and bars 375–6
  wave 203
rhizocretions 356
rhizoliths 135, 443, 445
Rians anticline 433
Rians–Salernes structure 426
rib-and-furrow structures 481
ribbon-bodies
  multistorey ephemeral *352*, 353–4, 358, 360
  single-storey ephemeral 352–3
ribbon complexes 283, 290
ribbon rivers 64, 75
ribbons 349, 445
  gravel 283, *284*
  sand 283
ridge-and-swale topography
  on modern alluvial floodplains 14
  Morien Group 133–48
  multiple origin of 148
  origin of 144–6
  types and inferred origin *139*

ridges, Mill Pond channel deposit 138–43
  asymmetric 138–40
    origins of 144–5
  symmetric 140–3
riffles 40, 511
  formation of 45
Rillo de Gallo Sandstones 376
Ancestral Rio Grande 439, 447, 448
Rio Grande rift 439
  rift basin development 442
Río Huaco drainage basin 228–9
Río Huaco valley, antecedence or superposition 229
Río Jáchal
  Quaternary tilting of valley 229
  structural control of 228
Río Jáchal drainage basin 227–8
Río La Troya sub-basin 226
Río Vinchina sub-basin 226
rip-up clasts 446
  goethite-coated 387
ripple cross-lamination 135, 139, 140, 142, 199, 201, 205, 296, 556
  asymmetrical 445
ripple marks 82
ripple movement 13
ripples 99, 154
  asymmetrical 392
  climbing 15, 130, 199, 389
  counter-current *387*, 387
  current 169, 172, 174
  interference 392
  migration of 14, *15*, 93, 372
  oscillation 392
  small-scale 13
  starved 405, 406
  symmetrical 202, 203, *204*
river ice, effect of on sediment dispersal 63–75
river systems, Late Precambrian 163–4
rivers
  anastomosing 91
  braided *see* braided rivers/streams
  lateral migration 315–16
  shortening of, Raba River 308
Rødsten Formation 407–11, *409*, 417, 419
Rognacian Stage 426, 427
roof rolls 203
rotation
  within fans 435, 436
  in the Salt Range 534
  tectonic 526
rotational slides 300, 301
RST *see* ridge-and-swale topography
run-off, rapid 310
rutile 563

Salernes zone 433
Salt Range
  geological setting 522–6
  lithofacies 527–35
    Baun/Sauj Kas section 528–31
    Jamarghal section *529*, 534–5, 537
    Kotal Kund section *529*, 531–3

    Tatrot–Andar Kas section *529*, *532*, 533–4
  uplift causing changes in pattern of fluvial deposition 521–38
Saltwick Formation 169, 171
  composite sandstone sequence 174, 178–9
  overbank deposits 181
  saline influences 178
  width/thickness ratios 171–2
sand bars/sheetflood deposits 295
sand-flow structures, tongue-shaped 387–8
sand/shale ratios 358, 359
sand-sheets, aeolian 406
sandbodies
  channel-to-sheet transitional 350–1
  channel-within-channel 354
  density of, Andersson Land Formation 414
  of ephemeral streams 338, 360
  interconnectedness of 182
  mouth bar 208
  sheetflood 360
    stacked 351–2
sandflats 296, 371, 375, 379, 428
sands
  cross-stratified 285
  rippled 287
sandstone 341, 509
  brown, Salt Range 527, 530–1, 533, 534, 535, 536–8
    basal 533, 534
  channel-belt 319
  coarse-grained 478, 491
  crevasse splay 167–83
  in Cromwell Member 477
  cross-bedded 403, *404*, 406
  cross-stratified 458, 493, *495*
  crystal-rich 458, 469
  current-rippled 371, 372, 373
  fine-grained 152, 159
  horizontally laminated 443
  large-scale planar cross–bedded 370, 372, 373
  multistorey 363
  parallel-bedded, stacked sets 392, *393*
  parallel laminated 371, 372
  pebbly 491
  planar and ripple cross–laminated, interbedded sets 144
  planar-bedded 135
  planar stratified 458
  planar tabular cross-bedded 371, 372, 373
  ripple cross-laminated 141, 142
  sites of probable deposition 212
  tabular 410, 411
  tabular cross-bedded 370, 372
  trough cross-bedded 93, 96, 98, 99, *140*, 141, 370, 371, 373, 480, 481
    small-scale 370, 371–2
  trough cross-stratified 143, 145, 159
  tuffaceous 468
  white, Salt Range 527–8, 530, 531, 533, 535, 536
  see also channel sandstones; crevasse splay sandstones; sheet sandstones
sandstone bodies 199, 393, 396
  geometry, and channel system position 201
  multistorey 107, 109, 135–46, 279, 527

sandstone bodies (*cont'd*)
  pebbly 491
  predictions for 211
  sheet-like 202–3
  single- and multistorey 406, 411
  see also Cañizar Sandstones Formation; sheet-sandstone bodies
sandstone/siltstone ratio 406
Santiuste fan 284, 285
Scalby Formation 169, 174
Scarborough Formation 169
schistosity 554–5, 559
scour-and-fill cycles *15*, 15–16
scour-and-fill sequences/structures 13, 19, 98
scour channels 428
scour depth, confluence scour 95–6
scour fill 141, 352, 392, 428, 509
scours/scouring 13, 15, 80, 92, 159, 347, 350, 371, 432, 509, 534
  of bar tops 511
  basal 462
  confluence 93–6, 97, 98, 99
  conglomerate-filled 154
  ephemeral streams 43, 44, 45, 47
  in-channel 428–9
scour surfaces
  basal 478
  draped 509
  low-angle 443
  minor 355
scroll-bar deposits 138
scroll bars 308, 391–2, *392*
  in alluvial ridge-and-swale topography 133–4, 145, 146, 148
  in ancient meander loops 117, 120, 123, 126, 130
  formation of 128, 129
sea-level changes
  glacioeustatic 267
  relative 239
    effects of on a fluvial sequence 383–97
secondary channels, and grounded ice 68, 69
secondary circulation 57
  and flow in meanders 51
secondary flow 120
sediment
  accumulation rates in the Tremp Group 339–41
  acquisition of by ice 68
  changes noted in the Raba River 308–9, 312–14
  coarse 27–8, 34
  fine 96
    pelletized 69
    and tectonic activity 231
  ice rafting of, Albany River, Ontario 63–75
  spatial variability in size 40, 41, *42*
  storage of Crete piedmont area 239
    headwater regions, point-source fans 231
  supply/accommodation ratio 354, 355
sediment bypass processes 260, 262
sediment discharge, low 216
sediment flow
  downslope 157
  and point-source fans 231
  spatial and temporal variability of 221–2

sediment influx 340–1
  Provence Basin 437
sediment liquefaction 298
sediment load
  eruption-induced 451
  influence of 433–4
  reduced, Raba River 314, 315
  White, Buffalo and Mississippi Rivers 185, 187
sediment mobilizing events 37
  see also flow events
sediment size/character jumps 27, 34
sediment sorting model, principles of 543–50
sediment transport 544–5
  and confluence mouth bars 93
  lateral 52, 55–6
    in the meander bend 57–60
  longitudinal 56, 59
  net transverse 53
  up point bars 119–20
sediment transport rates 43, 52
sedimentary facies, distribution within a foreland basin 516
sedimentation
  along an active strike-slip fault 293–303
  controls on, Seilao Member, Purilactis Formation 497–9
  modern, foreland-basin 221–32
  related to slow subsidence 379
  response to climatically modulated change in regional base level 517
  and river confluences 91–9
  tractional bed load 493
Seilao Member, Purilactis Formation 489, 490
  controls on sedimentation 497–9
  facies analysis 490–6
  facies association organization 496–7
semi grabens 377, 378
'set boundary' 117
Severn, R. 24–34
Sevier fold–thrust belt 385
Sfakia fan complex 242, 246, *248*, *249*, *250–1*, *252*, *254–5*, *256*, *257–63*, 266, 270
  depositional history reconstructed *247*
  lacking grain-size trends 263
shape, sorting by 29–31
shear stress 4, 544
  critical dimensionless (Shields parameter) 6–7
shear velocities 4–5, **6**, 544
shear zones, vertical 300, 301
sheet-sandstone bodies
  Fugleberget and Hestman seqeunces 153–9
    type A sandstone bodies 154–7
    type B and C sandstone bodies 157–9
    wedge- and lenticular-shaped bodies 159
  scales of 165
sheet-sandstone complexes, multilateral, multistorey *see* Cañizar Sandstones Formation
sheet sandstones 409, 445, 491, 536
  multistorey 447
  Sykes Mountain Formation 392–4
sheetflood deposits 338
  distal 494, 496, 499
  frontal splay *350*, 351–2

sheetflood deposits (cont'd)
　proximal　493, 499
sheetflood processes　483
sheetfloods　410, 411
shooting flow　310
side bars　82, 112, 311
　laterally accreting　159
　migrating　159
sidebar-complexes　164
Sidlaw Anticline　452
Sierras Pampeanas　223–4
sieve deposits　241, 268–70, 271, 272
siliciclastic sediments, within continental flood basalts　473–86
silt (fine), and grain size–distance relationship　191–2
siltstone　201, 389, 392, 406, 413, 427, 445, 512
　ripple cross-laminated　142, 143
siltstone laminae　200
Siwalik Group　523
　magnetic correlations in　524
Skaloti fan system　247, *264*, 265
Skolithos　203
Slave River delta　207
Slippen anticline　417, 419, 420
slope-scree deposits　284, 289, 290, 291
slough-channel fills　511
slump features　390
snow slush　68
Soan Formation　525, 530, 531, 534
　depositional setting　523
Soan syncline　536
Sofia Sund Formation　403, 417
soft sediment deformation　152, 202, 203, 295, 480, 481
soil nodules　530, 534
soils　531
　calcimorphic　431
　and overbank deposits　331
sole marks　510
solifluction　69
sorting
　hydraulic　23, 26, 29–30
　modelled by MIDAS　546
source areas, unroofing of　505, 533
South Saskatchewan River　373, 376, 481
Southalpine thrust belt　503, *505*
sphere position, effect of　8–9
spheres, entrainment of　3–10
stacking
　of channels　386, 531
　of conglomerates　284, 507
　of crevasse splay sandstones　172, 174, 176
　and ephemeral streams　347–8
　of fan lobes　266
　of fluvial and volcaniclastic deposits　458
　of fluvial channel deposits　410
　frontal splay sheetflood sandbodies　351–2
　of planar foresets　387
　progradational pattern　514
　of syn-eruption deposits　470
　vertical　352, 353
　　of channel units　428
　　into cosets　370
　　of fan lobes　563

of waning flow deposits　485
starved ripple strata　409, 410
Strathmore Group　452
Strathmore Syncline　452
stratification
　horizontal　457, 478, 482, 556
　planar　105, 108, 109, 164–5
　plane parallel　252–3
streamflood deposits　507
streamflow deposits, channelized, proximal and distal　494, 499
sub-Dakota unconformity　383
sub-Greybull unconformity　385–6
subsidence
　asymmetric　503, 505
　continual　561
　flexural　517
　quick response to　231
　rate of　464–5, 535
　slow　485, 486
　tectonic　516, 566
Sudetic Marginal Fault (SMF)　293, 301
　formation of　294–5
supraplatform bar deposits　308, 312
surface water temperature oscillations, Messinian　517
surficial sediments, fine　68–9
suspended load streams　310
suspension　53
swash–backwash processes　204
Sycarham Member　169
Sydney Mines Formation, geological setting　134–5
Sykes Mountain formation　383, 385, 392–4
symmetrical basins
　fluvial sedimentology of　442–5
　subsidence in　447
syn-eruption sequences　467, 469, 470
synaeresis cracks　202
synsedimentary deformation　505

Tajuña fluvial distributary system　286–8
Talarn Formation　343
Talchir clasts　522–3, 525, 530, 531, 533, 534, 536
tectonic activity
　breccias a response to　432–3
　and megasequence development　498
tectonic cyclothems　517
tectonic movements
　Cretan blocks　239
　Gunnar Andersson Land chronosomes　419, 420
tectonic pulses　433, 434
tectonic tilting　320, 377, 378
　Crete　264, 266
　and style of fan progradation and stacking　264–5, 267, 272
　effects of　332–4
tectonic uplift　353
　pulses of　271
tectonics
　and basin formation　559, 561–2
　collisional　554, 559
　compressional　357, 369
　and fan stratigraphies　290

tectonics (cont'd)
  and the Greybull interval 396–7
  and small-scale cyclicity 515
  source area, response of an alluvial system to 489–99
tectonosedimentary cycles 357
terminal fan deposits 406
terra rossa soil material 256–7
terraces 187
  palaeochannel site 190
Tethys transgression 364
Tezze Thrust *503*, 503, *504*, *505*
Thematic Mapper (TM) images 224
Thermopolis Shale 385, 394
thinning-upward sequences 161
tholeiite 476, 485, 554
thrust belts 224
thrusting 433, 436
  active 357, *359*
  pulsatory 517
  and sediment dispersal 231
Tobra Formation 522, 530
  unroofing of 533
toesets, of counter-current ripples 387
Tongue River lineament 396
Tórtola fan 115–16, 119–20
Tortue Formation 555, 556–9, *560*, *561*
tourmalinization 566, 567
trace fossils 389
traction carpet 492–3
tractional fabric 250
tranquil flow 310
Trans-Amazonian orogeny 553–5, 559, 566
transition bedform state 19
translatent strata 409, 410, 412
transport, selective 24, 27
transpression 561, 566
transpressive zones 19
transverse bars 82, 83, 282, 284, 311, 379, 428, 437, 493
  downstream migration of 482
  inward migration of 133
  ridge-and-swale topography 144, 146, 148
  sorting of bed material 312–14
Tremp–Graus Basin 338–9, *340*, 358, *359*
Tremp Group 338–41
  facies associations 341–7
trenching 242–4
  see also entrenchment; fan trenching/trenches
tributary gradients, steepness of 27–8
Trollfjord-Komagelv Fault Zone, postsedimentation movement 152
truncation surfaces, type-5 443

Udkiggen Formation 404–6, 417, 419
  as an erg 406
  flat-bedded facies 404–6
unconformities 385, 533
  angular 119, 435, 514, 548
    intraformational progressive 505, 516
  non-marine 383
  and structural evolution 516
  sub-Greybull 385–6

Upper Lunde Formation 338
upwarping *359*

Vadsø Group 151, *152*, 152
Valdonnian Stage 426, 427
valleys, intramontane 237, 244, 266, 272
  as drainage and sediment-yield areas 262, 263
  and Holocene deposits 170
  occurrence of fans at outlets of 247
  and Pliocence transgression 239, 241
Valsugana Thrust *503*, 503, *504*, *505*
Varanger Peninsula, Norway 151–3
varves 512
vegetation, and scroll-bar growth 145
Venetian Alps 502, 505
Venetian foreland basin
  attempted palaeodrainage reconstruction 518–19
  geological and stratigraphical setting 502–6
  insensitive to sea-level changes 516
  see also Messinian deposits
vertical sequences, significance of 159–63
Vertisols 356
Villalba de la Sierra fan 115–16, 120–5
Vitrollian Stage 426
Vittorio Veneto area 519
volcaniclastic aprons 457, 464–5
volcaniclastic deposits
  crystal-rich 458
  fans *468*
volcaniclastic facies, and depositional mechanisms *455*, 455
volcaniclastic/fluvial deposits, interbedding of 452
  interpretation 464–9
  Type I sequences 458–62
  Type II sequences 462–4, 469
  Type III sequences 464, *466*, *467*, 469
volcanism
  calc-alkaline eruptions, frequency of 469
  hiatuses in 486
  north of Highland Boundary Fault 469–70
  Old Red Sandstone 452
Vraskas fan complex 245, 247

Wabash River, scroll bars 145, 146
wadis *see* arroyos
waning flow deposits 483, 558
  stacked 485
water, in debris flows 243
water stage and discharge, variability of, Raba River 305–6
Waterberg Group 473
wave-cut platform 246
wave ripples 481, 482, 483
weathering, and exhumed meander belts 146
White River 186
  overbank deposit 189–90
width/thickness ratios
  crevasse splay sandstones 171, 174, 176
  sandbodies, Esplugafreda Formation 348, *349*, 350, 351, 352, 353, 354–5, 359, 360

Witwatersrand Basin
  gold distribution in placer deposits 543
  use of MIDAS 544
Witwatersrand Supergroup 473

Ymer Ø deformation phase 417, 420

Zingg shape changes 31–3, *33*
Zingg Shape Classification 25, *30*, 30
zircon 555
Zoologdalen Formation 406–7, *408*, 417, 419
Zoologdalen syncline 417